Handbook of Research on Literacy in Technology at the K-12 Level

Leo Tan Wee Hin
National Institute of Education, Singapore

R. Subramaniam
National Institute of Education, Singapore

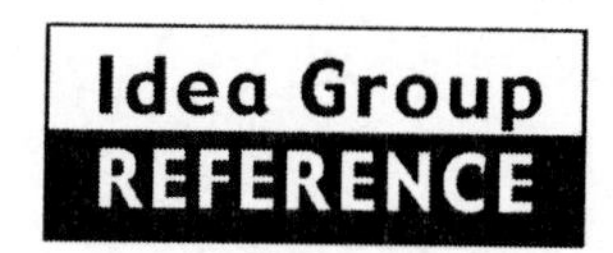

IDEA GROUP REFERENCE
Hershey · London · Melbourne · Singapore

KH

Acquisitions Editor: Michelle Potter
Development Editor: Kristin Roth
Senior Managing Editor: Amanda Appicello
Managing Editor: Jennifer Neidig
Copy Editor: Maria Boyer
Typesetter: Diane Huskinson
Support Staff: Wendy Catalano
Cover Design: Lisa Tosheff
Printed at: Yurchak Printing Inc.

Published in the United States of America by
Idea Group Reference (an imprint of Idea Group Inc.)
701 E. Chocolate Avenue, Suite 200
Hershey PA 17033
Tel: 717-533-8845
Fax: 717-533-8661
E-mail: cust@idea-group.com
Web site: http://www.idea-group-ref.com

and in the United Kingdom by
Idea Group Reference (an imprint of Idea Group Inc.)
3 Henrietta Street
Covent Garden
London WC2E 8LU
Tel: 44 20 7240 0856
Fax: 44 20 7379 3313
Web site: http://www.eurospan.co.uk

Library of Congress Cataloging-in-Publication Data

Handbook of research on literacy in technology at the K-12 level / Leo Tan Wee Hin and R. Subramaniam, editors.
p. cm.
Summary: "This book focuses on issues in literacy and technology at the K-12 level in a holistic manner so that the needs of teachers and researchers can be addressed through the use of state-of-the-art perspectives"--Provided by publisher.
Includes bibliographical references and index.
ISBN 1-59140-494-0 (hardcover) -- ISBN 1-59140-496-7 (ebook)
1. Education--Data processing. 2. Computers and literacy. 3. Computer-assisted instruction. 4. Computer managed instruction. I. Leo Tan, 1944- . II. Subramaniam, R. (Ramanthan), 1952- .
LB1028.43.H36 2006
607.1--dc22
2005022461

British Cataloguing in Publication Data
A Cataloguing in Publication record for this book is available from the British Library.

All work contributed to this book is new, previously-unpublished material. The views expressed in this book are those of the authors, but not necessarily of the publisher.

8/16/06

Editorial Advisory Board

List of Contributors

Contents

Section II
Teaching and Learning with Technology

Foreword

It is timely in the increasingly technology-driven 21st century that the issues and challenges of literacy in technology should be raised in this book. Today's world is moving towards a more open and global society. In order to deal with its changing demands, people need to learn how to cope with change and at the same time to interact constructively with it and retain control of the processes involved. Already information technologies are changing the ways we live, work, and learn. Technological development has also created a demand for new skills and at the same time is providing powerful new learning tools and opportunities. Not least, technology is introducing radical changes, with non-formal and informal education assuming an important place in addition to formal education.

Non-formal education is already improving because of the advantages of technology. It has proved to be effective in reaching out to vast and potentially otherwise inaccessible school-age populations. It stimulates learners to be more creative and innovative. In fact, it revolutionizes the way we handle information, from teaching to self-directed learning and from learning as a one-time event to a lifelong learning process. Thus, technological literacy more than ever needs to take on a prominent role in the K-12 classroom. Its potential impact is wide ranging and can be consolidated into pervasive benefits, which themselves can be challenges that the global classroom can and should be taking on board.

Firstly, technological literacy contributes toward narrowing the digital divide. All technology, not just computers and the Internet, empowers those who own it and understand it and can place others at a distinct disadvantage. If overall technological literacy is not introduced into the classroom, particularly among the technological have-nots, it is inevitable that the "technological divide" will increase. It is important to think about this possible marginalization, and the creation of new zones of power through the advantages being created for some using technology. For this technology to play its full role, it should be accessible to those who have been deprived of it. However, simply giving teachers and students broader access to hardware and software will not improve learning. Creating effective and economical strategies for using technology to support education in developing countries is a major challenge. In tackling this task, much will be gained by comparing and sharing experiences among both developed and developing nations.

Secondly, one of the obvious benefits of technological literacy is tied to the marketplace. Technology, particularly in the high-tech sector, has been driving much of the economic growth in nations across the world, and consequently has created an increasing percentage of occupations that require technological skills. Technology is everywhere in the world of work, from the home office to the medical examination room, from transport systems to mobile networks. The rapid introduction of new technologies is further changing the ways people interact, and the modern economy will continue to shift towards knowledge services and e-business. There is the inevitable need for a highly educated workforce, skilled in the use and application of technology, to stay innovative and competitive.

Thirdly, technological literacy can enhance well-being within a social and wider community. Technologically literate people with a broad comprehension of technology are able to adapt and make decisions effectively. This is demonstrated by the thinking skills and thought processes that the use of technology can provoke. Critical thinking, reflections on social values, and abstract reasoning can all reach new levels when technology is properly applied. It also empowers people by giving them the tools to make sense of their world, even if it is constantly changing around them.

At a pedagogical level, technological literacy has the potential to impact on and transform teaching, learning, and child development. This helps in enabling K-12 children to meet their highest expectations, connecting with hard-to-reach groups in new ways, opening up education to partnerships with other organizations, and moving to a new level of efficiency and effectiveness in classroom delivery.

Technological literacy can transform teaching, learning, and child development by enabling exciting lessons. Both presentations and productions (e.g., videos, multimedia projects, 3D animations, etc.) are expressions of a student's thought or interpretation on a particular topic. With the use of technology, these views can be more clearly and creatively illustrated; this use of imagination and illustrative structure potentially allows ideas to flow in a more understandable format. Such use of technological literacy can further personalize children's experiences of learning, whereby online exercises can adapt each new task to a pace appropriate to the learner. It is possible to provide live access online to learners in other places, including other countries, in order to make learning more relevant to their lives.

Importantly, the introduction of technology in the classroom can offer extra support to those with special needs; it can re-motivate those who have dropped out of formal education; it can give learners more choice about where and when they learn; and it can support children and young people in the transition to the workplace, within the community at large, and in their progression to the next stage of learning. All of this forms the building blocks toward an open access that can provide children and learners with a much wider range of options, including greater opportunities for different schools and school programs to work more closely together.

The mastery of any new knowledge is strengthened when there is active collaboration with others to communicate an understanding of what has been learned. With technological literacy, the extent of learning and the effectiveness of teaching need no longer be limited by the amount of time spent in the classroom or by the resources of a particular school. Teachers and students can tap into vast digital libraries and a wealth of texts, images, video, music, arts, and languages. They can work with scientists and scholars around the globe who can help them use experimental research, primary historical documents, and authentic learning in real-life settings to improve their understanding of physical phenomena and world events. The growth in access by schools to broadband and ubiquitous computing supports these various possibilities for extending the time, the places, and the resources available for learning. For instance, m-learning (mobile learning) is taking education into the community and into wider learning spaces with the use of personal digital assistants (PDAs) and other mobile devices for interactive class-work scenarios.

In this way, technological literacy not only supports the rise in options available to teachers and students, but additionally offers a means of achieving greater efficiency and effectiveness in learning both in and outside of the classroom, and the opportunity to work more productively through these available resources, tools, and shared good practice.

A critical proviso is, however, that adopting technology demands a process of selection and decision making about which technology is appropriate, by and for whom it is to be used, and what kind of communication and content is most relevant. For example, the e-learning market is rapidly moving into schools by providing portals, community sites, content repositories, and a broad array of products and services for classroom use. Teachers must themselves be technologically and pedagogically aware of developments since they represent the key integrators of the process and a major driving force in the acquisition and deployment of technological literacy. When the process is at its most successful, it should crystallize ideas, create visions, and motivate greater numbers to pursue literacy through technology, regardless of the specific delivery platform or format.

The material presented in this book provides a synthesis of these issues, underlining the main hypothesis that: using technological literacy effectively, K-12 students can be given the opportunity to process information in critical ways, causing new ideas and avenues of thought to develop, and the linking of new learning communities. We hope it will prompt discussion about the field and improvements for the future.

Dr. Ann Borda and Professor Jonathan P. Bowen
Institute for Computing Research
London South Bank University, UK

Preface

Advances in information, communication, and computer technologies are impacting on the educational system in multidimensional ways. They have started to redefine the way in which teaching and learning are being dispensed, with the result that curriculum and technology have started to merge so as to stretch the definition of literacy. Traditional metrics of literacy—numeracy (simple arithmetic) and language (reading, writing, listening and speaking)—are now being endowed with a technological dimension that cannot be overlooked.

The new age metrics for literacy leverages on a range of computer-related skills. These include, but are not limited to, word processing skills (in place of writing); e-mailing (in place of composing letters); information retrieval skills using the Internet (in place of going to the library to source for information); online collaboration skills (on top of working as part of a team in the real word); familiarity in using digital media such as computers, digital cameras, video cameras, mobile phones, and e-books; data analysis skills using a suite of software programs; Web site construction skills, including ability to creatively combine audio, video, data, multimedia, and text; creative PowerPoint presentation skills; ability to use spreadsheets and databases; and so on. New forms of literacy are also emerging as technologies evolve.

Other technologies have also started to impact on the educational system—for example, broadband technologies such as asymmetric digital subscriber line and hybrid fiber coaxial cable modem, and wireless technologies. Their nexus with literacy opens up more possibilities to maximize learning.

Reformatting the curricula along lines that promote acquisition of the aforementioned skills will be very helpful in endowing students with the necessary suite of literacy skill sets required to support the needs of the emerging knowledge-based economy, which some say is already here. There is evidence that learning in a wide range of technological settings provides rich conceptual experiences for students.

The ubiquity of the personal computer, the nature of the client-server architecture on the Internet, the low cost of logging onto the Internet, and the scope for simultaneous access are all factors that have helped to fuel the evolution of various genres of learning on the technological platform, and thus help to contribute towards the acquisition of literacy skills in technology.

Recognizing that the K-12 classroom is a strategic platform to entrench more firmly the various suites of literacy skill sets in technology among students, this handbook aims to provide an overview of the state-of-the-art developments in the field of literacy in technology at the K-12 setting and to address the needs of practitioners in this fast-developing field. The various chapters draw upon best practices from practitioners and researchers in the field, and provide a holistic perspective of issues of literacy in technology at the K-12 classroom.

The international flavor of this handbook can be seen from the fact that the 35 chapters accepted for publication represent the contributions of 51 authors from 37 institutions in nine countries: Australia, Canada, Cyprus, Hong Kong (China), Ireland, Israel, Italy, the UK, and the United States

The 35 chapters in the handbook span a useful spectrum of topics in the field. For convenience, they are grouped under three broad sections. Though all chapters are distinctly different, inevitably there will be some mirroring of content in the various sections and even between chapters. Herein, our approach has been that it is good for readers to get a diversity of perspectives from different experts as they are likely to approach issues from different angles and thus offer valuable insights from the lens of their experience.

The chapters in Section I present Perspectives on Technological Literacy from various experts. Such perspectives are useful in defining the directions that technological literacy is moving towards and how experts perceive its multifaceted dimensions and conceptual underpinnings. In the chapter "New Paradigm of Learning and Teaching in a Networked Environment: Implications for ICT Literacy," Yin Cheong Cheng presents a model that aims to develop students' contextualized multiple intelligence and their lifelong independent learning through a triplization process leveraging on individualization, localization, and globalization in both teaching and learning. The chapter "Technologies Challenging Literacy: Hypertext, Community Building, Reflection, and Critical Literacy," by Agni Stylianou-Georgiou, Charalambos Vrasidas, Niki Christodoulou, Michalinos Zembylas, and Elena Landone, discusses how new genres of texts transform conceptualization of literacy development and present new challenges for reading and writing.

Jared V. Berrett presents a framework for understanding perspectives on technological literacy in his chapter "Technological Literacy, Perspectives, Standards, and Skills in the USA" and argues that technology educators have a role to play in preparing K-12 students to be technologically literate citizens. In the process of meeting the necessary standards of technological literacy, Leonard J. Waks, in his chapter "Globalization, At-Risk Students, and the Reconceptualization of Technological Literacy," offers a revision of the concept of technological education that assigns important new roles for non-governmental organizations serving low-tier workers and that places hands-on use of networked computers at its core. Aidan Mulkeen, in her chapter "ICT in Schools: What is of Educational Value?", argues that many uses of ICT have little educational value. The real value of ICT in schools lies in enabling more challenging learning activities that develop higher-order thinking skills among students, she says. In the chapter "Web-Based Technologies, Technology Literacy, and Learning," Wan Ng illustrates how some Web-based technologies can be used to foster not only the desired suite of technology literacy skill sets, but also promote constructivist learning in the K-12 classroom.

The chapter "New Media Pathways: Navigating the Links between Home, School, and the Workplace," by Helen Nixon, Stephen Atkinson, and Catherine Beavis, presents the viewpoint that school-based courses in new media are important because they increase student retention and the chance of success in post-school employment; in this context, the authors stress that school curriculum and pedagogy have much to learn from young people's informal and leisure-based learning. Elizabeth A. Buchanan and Tomas A. Lipinski, in their chapter "Responsible Technologies and Literacy: Ethical and Legal Issues," stress the proper use of technology in the classroom and provide recommendations for creating greater awareness of the ethical and legal implications surrounding technology use and copyright in particular. Cushla Kapitzke professes the viewpoint that unsympathetic social policies and the increased surveillance of physical environments have contributed to the uptake of virtual space and online chatrooms as a means of social contact and engagement by youths in her chapter "Internet Chatrooms: E-Space for Youth of the Risk Society." It is clear that the potential of such avenues for promoting technological literacy cannot be overlooked.

Colin Baskin, in his chapter "Transforming the K-12 Classroom with ICTs: Recognizing and Engaging New Configurations of Student Learning," raises questions as to where and how the discourses of literacy, education, and technology are converging in the ICT-enabled classroom, and urges that before the transformative values of ICTs in teaching and learning are discounted, a case can be made for a new definition of student learning that focuses on the demands of the new world environment.

With the pervasiveness of technology in the classroom, evaluating its impact is often fraught with challenges — in this context, the chapter "The Complexities of Measuring Technological Literacy" by Marcie J. Bober draws in part on the best of many standards systems and offers strategies for improving the operationalizing of technological literacy as a construct. In the chapter "Systemic Innovations and the Role of Change-Technology: Issues of Sustainability and Generalizability," Chee-Kit Looi, Wei-Ying Lim, Thiam-Seng Koh, and Wei-Loong David Hung discuss the challenges that lie in the effective implementation of IT into the curriculum for meaningful student learning.

The series of chapters in Section II focus on Teaching and Learning with Technology; technology tools which afford immense scope for classroom use are explored here. In the chapter "Digital Video in the K-12 Classroom: A New Tool for Learning," Christopher Essex describes how digital video can provide tangible, real-world benefits in learning, as it requires that students work actively and collaboratively in authentic real-world tasks. Ann E. Barron, J. Christine Harmes, and Katherine J. Kemker, in their chapter "Technology as a Classroom Tool: Learning with Laptop Computers," call for the integration of laptop computers into the curriculum as it can create collaborative, student-centered learning environments that increase student and teacher technology literacy. Handheld computers can have a tremendous impact on teaching and learning given the right context — this is the subject of the chapter "Tapping into Digital Literacy: Handheld Computers in the K-12 Classroom" by Mark van 't Hooft, who discusses how they can be integrated into the classroom. In the chapter "Digital Literacy and the Use of Wireless Portable Computers, Planners, and Cell Phones for K-12 Education," Virginia E. Garland addresses the use of wireless technologies as instructional and managerial tools in the classroom, and suggests that they have the potential to shape the learning environment tremendously.

With the ubiquity of the personal computer and increasing access to a networked environment, Susan E. Gibson, in her chapter "Using WebQuests to Support the Development of Digital Literacy and Other Essential Skills at the K-12 Level," suggests that WebQuests can promote constructivist learning when they are carefully designed to encourage student-directed learning, problem solving, higher-order thinking, perspective taking, and collaborative learning on authentic real-world tasks. David A. Huffaker introduces the use of blogs as an educational technology in the K-12 classroom in his chapter "Let Them Blog: Using Weblogs to Advance Literacy in the K-12 Classroom." He argues that blogs can promote verbal, visual, and digital literacy skills through storytelling and collaboration. The potential of the Internet to foster technological literacy skills needs to go into higher gear — this is the thrust of the chapter "Information Problem-Solving Using the Internet" by Steven C. Mills, who describes how the vast resources of the Internet can supply communication tools and information resources that facilitate the application of a robust set of instructional methodologies in the classroom. In his chapter "How to Teach Using Today's Technology: Matching the Teaching Strategy to the E-Learning Approach," Moti Frank reviews the benefits and challenges of a number of approaches for integrating technology and teaching. A case study of how teachers integrate computer literacy into subject teaching is explored in the chapter by Allan Yuen and Patrick Wong entitled "Integrating Computer Literacy into Mathematics Instruction," where the focus is on the Hong Kong experience.

One of the new tools that have impacted the educational space in very recent times is the Tablet PC — this is the subject of the chapter "Teaching and Learning with Tablet PCs", drawing upon the experience of a school which introduced this tool to an entire cohort of students, the editors suggest that Tablet PCs are a utility tool for further enhancing the effectiveness of teaching and learning with technology. Lyn C. Howell, in her chapter "Using Technology to Create Children's Books for Students by Students," describes a project in which high school students used technology to create e-books for younger students. In her chapter "Electronic Portfolios and Education: A Different Way to Assess Academic Success," Stephenie M. Hewett argues on the need to use e-portfolios as another tool for assessment, especially as technology is rather pervasive in teaching and learning.

Section III focuses on Issues Related to Teacher Education and School-Based Matters. The success of technology literacy initiatives at the K-12 level depends very much on the extent to which teachers are technology enabled to infuse the desired suite of attributes among their students, as well as how the school administration prioritizes these concerns in their scheme of things. Karen Cadiero-Kaplan focuses on the pedagogy necessary in critically considering technology development for K-12 teachers and their students in her chapter "Teachers and Technology: Engaging Pedagogy and Practice," and outlines techniques and strategies that have been implemented successfully in building capacity among new and experienced teachers when using technology for lesson planning, teaching enhancement, and portfolio development. Kristina Love, in her chapter "Literacy in K-12 Teacher Education: The Case Study of a Multimedia Resource," identifies three areas faced by teachers in dealing with the complex forms of literacy that are increasingly required for success across the K-12 curriculum in Australia, and describes a video-based interactive CD-ROM that has been developed to address teacher concerns in these areas.

Paul Adams introduces constructivism as a pedagogical context from which educational professionals can analyze new technology exploiting learning-teaching interactions in his chapter "Demystifying Constructivism: The Role for the Teacher in New Technology Exploiting Learning Situations." Kendall Hartley describes the reality of K-12 classroom practice and of how this compares to common tenets in the field of instructional design in her chapter "K-12 Educators as Instructional Designers." She suggests how changes in teacher preparation and access to the appropriate tools could facilitate increase in student achievement. Tamara L. Jetton and Cathy Soenksen, in their chapter "Creating a Virtual Literacy Community between High School and University Students," describe a project in which a university professor and a high school English language teacher redesigned the curricula of a class so that their students could participate in a literacy project that focused on computer-mediated discussions of literature, in the process creating a virtual literacy community in which high school and university students incorporated traditional literacies of reading and writing within a virtual environment that facilitated communication, collaboration, and learning with text.

In the chapter "Online Learning Communities: Enhancing Learning in the K-12 Setting" by Chris Brook and Ron Oliver, a design framework intended to support and guide teachers in advancing K-12 literacy through collaborative learning and development of online learning communities is discussed. In the chapter "Knowledge Management, Communities of Practice, and the Role of Technology: Lessons Learned from the Past and Implications for the Future," Leo Tan Wee Hin, Thaim-Seng Koh, and Wei-Loong David Hung discuss the role of technologies and the issues of literacy in technology from the context of communities of practice and knowledge management. They draw implications on how teachers and students can be a community of learners-practitioners through technologies which support their work and learning processes. Christopher O'Mahony uses an Australian example to look at the decisions and dilemmas facing schools in developing their own virtual learning and managed learning environments in his chapter "The Emerging Use of E-Learning Environments in K-12 Education: Implications for School Decision Makers."

An exploratory study undertaken in Cyprus schools that examines the status of using ICT from the perspective of socio-technical systems is afforded in the chapter "A Socio-Technical Analysis of Factors Affecting the Integration of ICT in Primary and Secondary Education" by Charoula Angeli and Nicos Valanides. Kate Mastruserio Reynolds, Ingrid Schaller, and Dale O. Gable conclude the handbook with their chapter "Teaching English as a Second Language with Technology: Making Appropriate Pedagogical Choices," in which they outline the various constraints and challenges faced by K 12 teachers in attempting to include technology in the classroom and suggest a variety of ways to integrate technology into the K-12 classroom.

With the 35 chapters supporting this handbook, we trust that it would be a useful resource book and reference text for teachers, teacher educators, educational policy makers, and other practitioners.

Leo Tan Wee Hin and R. Subramaniam
National Institute of Education, Singapore

Acknowledgments

We thank all the authors for their assistance and cooperation in bringing out this book. Most of the authors also served as referees for the chapters. Their valuable feedback have helped to improve all chapters, and this is gratefully appreciated.

We thank the National Institute of Education, Nanyang Technological University for their assistance and support in the course of working on this book project for a year.

The staff at Idea Group Inc. has been extremely supportive in the process of bringing out this book. Our special thanks go to Ms. Michele Rossi, Ms. Jan Travers, Ms. Amanda Appicello, and Ms. Kristin Roth.

We are grateful to Dr. Ann Borda and Professor Jonathan Bowen of London South Bank University for writing the Foreword for this book.

We would also like to place on record our special thanks as well as gratitude and appreciation to Dr. Mehdi Khosrow-Pour of Idea Group Inc. for giving us this great opportunity to edit this book.

Leo Tan Wee Hin and R. Subramaniam
National Institute of Education, Singapore

Section I

Perspectives on Technological Literacy

Chapter I
New Paradigm of Learning and Teaching in a Networked Environment: Implications for ICT Literacy

Yin Cheong Cheng
Hong Kong Institute of Education, Hong Kong

ABSTRACT

This chapter introduces a new paradigm of learning and teaching that aims to develop students' contextualized multiple intelligence (CMI) and create unlimited opportunity for students' lifelong independent learning through a triplization process including individualization, localization, and globalization in teaching and learning. In particular, the chapter illustrates how students' self-learning can be motivated, sustained, and highly enhanced in an individually, locally, and globally networked human and ICT environment. Different from the traditional emphasis on delivery of knowledge and skills in planned curriculum, the new paradigm pursues the extensive application of ICT and enhancement of teachers and students' ICT literacy in building up a networked environment for students' individualized, localized, and globalized learning and CMI development. It is hoped that students equipped with the necessary ICT literacy can become borderless learners with unlimited opportunities for learning and development in a networked environment.

INTRODUCTION

In the new millennium, challenges such as rapid globalization, the tremendous impacts of information technology, international transformation towards knowledge-driven economy, strong demands for societal developments, and international and regional competitions have driven numerous educational changes in different parts of the world (Cheng & Townsend, 2000). Policymakers and educators in each country have to think how to reform education for preparing their young leaders to more effectively cope with the challenges in the new era.

In such a challenging context, paradigm shift in education becomes necessary in the new millennium. Adapted from the key theories in my previous work (Cheng, 2000), this chapter aims to illustrate how education at the K-12 level should be transformed from a traditional site-bounded paradigm towards a new paradigm including globalization, localization, and individualization in education with the support of information and communication technology (ICT) and international networking. In particular, the chapter also elaborates how unlimited opportunities for teaching and learning can be created in an individually, locally, and globally networked environment, what paradigm shift should be necessary in applying ICT in education, and what implications for literary in technology will be at the K-12 level. It is hoped that the proposed new paradigm of education in a networked environment will provide innovative ideas and possibilities for enhancing the effectiveness of K-12 education in different parts of the world to meet the challenges of the future.

TRIPLIZATION IN EDUCATION

Rapid globalization is one of the most salient aspects of the new millennium, particularly since the fast development of information technology in the last two decades (Brown, 1999). Inevitably, how education should be responsive to the trends and challenges of globalization has become a major concern in policy making in these years (Ayyar, 1996; Brown & Lauder, 1996; Green, 1999; Henry, Lingard, Rizvi, & Taylor, 1999; Jones, 1999; Pratt & Poole, 2000; Curriculum Development Council, 1999). Cheng (2000) argued that not only globalization but also localization and individualization are necessary in ongoing educational reforms. All of these processes as a whole can be taken as a *triplization process* (i.e., triple + izations) that can be used to consider educational reforms and formulate the new pedagogic methods and environment necessary to implement education at the K-12 level.

PARADIGM SHIFT IN LEARNING

With the concepts of triplization, a paradigm shift in K-12 education can be initiated from *the traditional site-bounded paradigm* to *the new triplization paradigm of education* (see Table 1).

Traditional Paradigm of Site-Bounded Learning

In traditional thinking, students' learning is part of the reproduction and perpetuation process of the existing knowledge and manpower structure to sustain developments of the society, particularly in the social and economic aspects (Cheng, Ng, & Mok, 2002; Blackledge & Hunt, 1985; Hinchliffe, 1987; McMahon, 1987). Education is perceived as a process for students, and their learning is being "reproduced" to meet the needs of manpower structure in the society.

Reproduced Learning

In traditional K-12 education, students are the followers of their teachers. They go through standard programs of education, in which students are taught in the same way and same pace even though their ability may be different. Individualized programs seem to be unfeasible. The learning process is characterized by absorbing certain types of knowledge: students are followers of their teachers, and they absorb knowledge from their teachers. Learning is a disciplinary, receiving, and socializing process such that close supervision and control on the

Table 1. Paradigm shift in learning

New Paradigm of Triplized Learning	Traditional Paradigm of Site-Bounded Learning
Individualized Learning: • Student is the Center of Education • Individualized Programs • Self-Learning • Self-Actualizing Process • Focus on How to Learn • Self-Rewarding	**Reproduced Learning:** • Student is the Follower of Teacher • Standard Programs • Absorbing Knowledge • Receiving Process • Focus on How to Gain • External Rewarding
Localized and Globalized Learning: • Multiple Sources of Learning • Networked Learning • Lifelong and Everywhere • Unlimited Opportunities • World-Class Learning • Local and International Outlook	**School Site-Bounded Learning:** • Teacher-Based Learning • Separated Learning • Fixed Period and Within Institution • Limited Opportunities • Site-Bounded Learning • Mainly Institution-Based Experiences

learning process is necessary. The focus of learning is on how to gain some professional or academic knowledge and skills. Learning is often perceived as hard working to achieve external rewards and avoid punishment.

Site-Bounded Learning

In the traditional paradigm, all learning activities at K-12 levels are school site bounded and teacher based. Students learn from a limited number of school teachers and their prepared materials. Therefore, teachers are the major sources of knowledge and learning. Students learn the standard curriculum from their textbooks and related materials assigned by their teachers. Students are often arranged to learn in a separated way and are kept responsible for their individual learning outcomes. They have few opportunities to mutually support and learn. Their learning experiences are mainly school experiences alienated from the fast-changing local and global communities. Learning happens only in school within a given timeframe. Graduation tends to be the end of students' learning.

New Paradigm of Triplized Learning

In the new paradigm, learning at the K-12 level should be borderless and characterized by individualization, localization, and globalization with the support of ICT and a networked environment. It is a triplized (i.e., globalized, localized, and individualized) learning.

Individualized Learning

The student is the center of education. Students' learning should be facilitated to meet their needs and personal characteristics as well as develop their potentials, particularly contextualized multiple intelligence (CMI)[1] in an optimal way. Individualized and tailor-made programs (including targets, content, methods, and schedules) with the support of ICT for different students are necessary and feasible. Students can be self-motivated and self-learning with appropriate guidance, ICT literacy, and facilitation. Learning is a self-actualizing, discovering, experiencing, and reflecting process. Since the information and knowledge are accumulated in an unbelievable speed but outdated very quickly, it is nearly impossible to make any sense if education is mainly to deliver skills and knowledge, particularly when students can find the knowledge and information easily with the help of information technology and the Internet. Therefore, the focus of learning is on learning how to learn, think, and create. In order to sustain lifelong learning, it should be facilitated in an enjoyable and self-rewarding manner (Mok & Cheng, 2001).

Localized and Globalized Learning

K-12 students' learning should be facilitated in such a way that local and global resource, support, expertise, and network can be brought in to maximize the opportunities for their developments during the learning process. Through localization and globalization, there are multiple sources of learning. Students can learn from multiple sources inside and outside their schools, locally and globally, and not limited to a small number of teachers in their own schools. Participation in local and international learning programs (e.g., learning activities conducted in the local community; overseas study visits or language immersion) can help them achieve the related local and global outlook and experience beyond schools. Now, more and more examples of such kind of programs can be found in France, Hong Kong, Japan, Singapore, and the United States. In addition, their learning is a type of networked learning. They are grouped and networked locally and internationally with the support of various types of ICT networks. Tan, So, and Hung (2003) of Singapore and Yuen (2003) of Hong Kong provided two typical examples of using ICT to network learners and create collaborative learning communities among learners. Learning groups and networks will become a major driving force to sustain the learning climate and multiply the learning effects through mutual sharing and inspiring. We can expect that each student can have a group of lifelong partner students in different corners of the world to share their learning experiences.

It is expected that learning happens everywhere and is lifelong. Education at the K-12 level is just the preparation for high-level lifelong learning and discovery (Liu, 1997; Mok & Cheng, 2001). Learning opportunities are unlimited. Students can maximize the opportunities for their learning from local and global exposures through the Internet, Web-based learning, videoconferencing, cross-cultural sharing, and different types of interactive and multimedia materials (Ryan, Scott, Freeman, & Patel, 2000; Education and Manpower Bureau, 1998). Numerous examples of such initiatives with the support of ICT can be found in Lee and Mitchell (2003). With the support of ICT and networking, students can learn from world-class teachers, experts, peers, and use learning materials from different parts of the world. In other words, their learning can be a world-class learning.

PARADIGM SHIFT IN TEACHING

The paradigm shift in learning inevitably requires a corresponding paradigm shift in teaching. The major changes are summarized in Table 2.

Table 2. Paradigm shift in teaching

New Triplization Paradigm	Traditional Site-Bounded Paradigm
Individualized Teaching:	**Reproduced Teaching:**
• Teacher is the Facilitator or Mentor to Support Students' Learning	• Teacher is the Center of Education
• Contextualized Multiple Intelligence Teacher	• Partially Competent Teacher
• Individualized Teaching Style	• Standard Teaching Style
• Arousing Curiosity	• Transferring Knowledge
• Facilitating Process	• Delivering Process
• Sharing Joy	• Achieving Standard
• As Lifelong Learning	• As a Practice of Previous Knowledge
Localized and Globalized Teaching:	**Site-Bounded Teaching:**
• Multiple Sources of Teaching	• Site-Bounded in Teaching
• Networked Teaching	• Separated Teaching
• World-Class Teaching	• Bounded Teaching
• Unlimited Opportunities	• Limited Opportunities
• Local and International Outlook	• Mainly Institutional Experiences
• As World-Class and Networked Teacher	• As Site-bounded and Separated Teacher

New Paradigm of Triplized Teaching

In the new triplization paradigm, teaching at the K-12 level should be triplized: individualized, localized, and globalized.

Individualized Teaching

Teaching is considered as a process to initiate, facilitate, and sustain students' self-learning, self-exploration, and self-actualization; therefore, the teacher should play a role as a facilitator or mentor who supports students' learning. The focus of teaching of the individual teacher is to arouse students' curiosity and motivation to think, act, and learn. In addition, teaching is to share with students the joy of the learning process and outcomes. To each teacher himself/herself, teaching is a lifelong learning process involving continuous discovery, experimenting, self-actualization, reflection, and professional development. Each teacher is a CMI teacher who can set a model for students in developing their multiple intelligences. Each teacher has his/her own potential and characteristics, and different teachers can teach in different styles to maximize their own contributions.

Localized and Globalized Teaching

Local and global resources, supports, and networks can be brought in to maximize the opportunities for teachers' contribution to students' learning. Through localization and globalization, there are multiple sources of teaching, for example, self-learning programs and packages, Web-based learning, outside experts, and community experiential programs, inside and outside their schools, locally and globally. Teachers can maximize the opportunities to enhance effectiveness of their teaching from local and global networking and exposure through the Internet, Web-based teaching, videoconferencing, cross-cultural sharing, and different types of interactive and multimedia materials (Holmes, 1999; Ryan et al., 2000; Education and Manpower Bureau, 1998). With teachers' help, students can learn from world-class materials, experts, peers, and teachers in different parts of the world such that teaching can become world-class teaching. Through participation in local and international development programs, teachers can achieve global and regional outlook and experiences beyond their schools. Two examples of initiatives for promoting professional development and teaching effectiveness of teachers through a networked environment are the Inquiry Learning Forum (http://ilf.crlt.indiana.edu) and the Web-based Peer Learning and Sharing (Yeung, Cheng, & So, 2003).

Furthermore, their teaching is a type of networked teaching. Teachers are grouped and networked locally and globally to develop and sustain a new professional culture and multiply their teaching effects through mutual sharing and inspiring. They become world-class and networked teachers through localization and globalization. It is not a surprise that each teacher can have a group of lifelong partner teachers in other parts of the world to continuously share and discuss their experiences and ideas of professional practice.

New Vision of Teaching

With the above paradigm shifts in teaching and learning, there should be a new vision of teaching that includes at least three major key components: (1) to facilitate students to experience a paradigm shift towards the triplized learning; (2) to provide a triplized environment (including a networked human and IT environment) for students' self-learning; and (3) to develop students' triplized self-learning ability and contextualized multiple intelligence. In other words, teachers are effective if they can achieve this new vision of teaching.

Role of Teacher in the New Paradigm

According to the new paradigm of teaching, teachers' roles may be different from the traditional paradigm. The different roles teachers play in the teaching process may shape differently the roles and qualities of students in the learning process that can vary from the very passive way to the active self-learning and self-actualization mode as shown in Table 3 (Weaver, 1970; Cheng, 2001a).

When the teacher plays a role as appreciator, partner, patron, guide, questioner, tutor, counselor, moulder, instructor, or exemplar, students will correspondingly play a role as searcher, partner, designer, explorer, investigator, thinker, client, subject, memorizer, or trainee. As teachers' roles are different, the nature of teaching and learning processes will be different, including those as determined by students, participation, making, searching, experimentation, reflection, expression of feeling, conditioning, transfer of information, or imitation. Correspondingly, the likely students' qualities' from the learning processes will also be different, such as self-determination, responsibility, creativity, adventurousness, investigation skill, understanding, insight, habits, possession of information, and skills.

There is an ecological relationship between roles of teachers and students. As teachers tend to be more teacher centered in teaching (towards roles 8, 9, and 10 as in Table 3), students become more passive in their learning and the qualities tend to be habits, possession of information and skills. It is often referred as *low-order learning*. As teachers tend to use student-centered approaches and play roles 1,

Table 3. Teachers' roles and corresponding students' roles and outcomes (adapted from Weaver, 1970 and Cheng, 2001a)

Teacher's Role		Student's Role		Teaching/ Learning Process	Likely Student Quality as Outcome
1. Appreciator	Student- Centred	1. Searcher	High-Order Learning	1. As determined by students	1. Self-Determination
2. Partner		2. Partner		2. Participation	2. Responsibility
3. Patron		3. Designer		3. Making	3. Creativeness
4. Guide		4. Explorer		4. Searching	4. Adventurousness
5. Questioner		5. Investigator		5. Experimentation	5. Investigation Skill
6. Tutor		6. Thinker		6. Reflection	6. Understanding
7. Counsellor	Teacher-Centred	7. Client	Low-Order Learning	7. Expression of feeling	7. Insight
8. Moulder		8. Subject		8. Conditioning	8. Habits
9. Instructor		9. Memorizer		9. Transfer of information	9. Possession of information
10. Exemplar		10. Trainee		10. Imitation	10. Skills

2, 3, 4, 5, and 6 in the teaching process, students have more opportunities to be active in self-learning and achieve the higher qualities of learning outcomes such as self-determination, responsibility, creativeness, adventurousness, investigation skill, and understanding, all of which are important in the new paradigm of triplized education and crucial to the future of students in the new century. This is a *high-order learning*.

We understand the educational aims and processes are complex and the role of teachers should be dynamic and complicated, including multiple roles ranging from roles 1 to 10—from total direct instruction to total student self-determination in the daily educational practices. A mix of multiple roles played by teachers in daily educational practices is often a fact of school life. What is important for teachers and educators is to keep in mind what educational aims we want to pursue. If we want to achieve a new paradigm of education for the future of our students, we should encourage the mix of multiple teacher roles to be more student centered and less teacher centered in the whole teaching and learning process at the K-12 level.

LEARNING IN A NETWORKED ENVIRONMENT

According to the above new paradigm of triplized education, we should emphasize K-12 students' continuous self-learning and development of CMI with the support of localization and globalization through information technology and various types of international and local networking. Mok and Cheng (2001) have proposed a theory of self-learning in a networked human and technology environment to show how students' individualized self-learning can be motivated, sustained, and optimized through the wide local and international support from the borderless and networked human and technological environment. The key concepts are summarized from Mok and Cheng (2001) as follows.

Self-Learning Cycle

The understanding of the nature of self-learning is important in implementing a new paradigm of triplized learning. Based on the concepts of action learning (Yuen & Cheng, 1997, 2000; Argyris & Schön, 1974; Argyris, Putnam, & Smith, 1985), Mok and Cheng (2001) conceptualized the process of self-learning as a cyclic process in a networked human and IT environment as shown in Figure 1. It subdivides a learning episode into a sequence of three components such as mental condition (mindset), action, and outcome, linked by four processes including planning, monitoring, feedback to mental condition, and feedback to action. There are two types of feedback from the monitoring process and outcomes to the learner: one to the mindset and the other one directly to the action. The feedback to mindset will help the learner to reflect on and change his/her own mental models including meta-cognition, thinking methods, meta-volition, and knowledge, and then to change

Figure 1. Self-learning cycle in a networked human and ICT environment (adapted from Mok & Cheng, 2000)

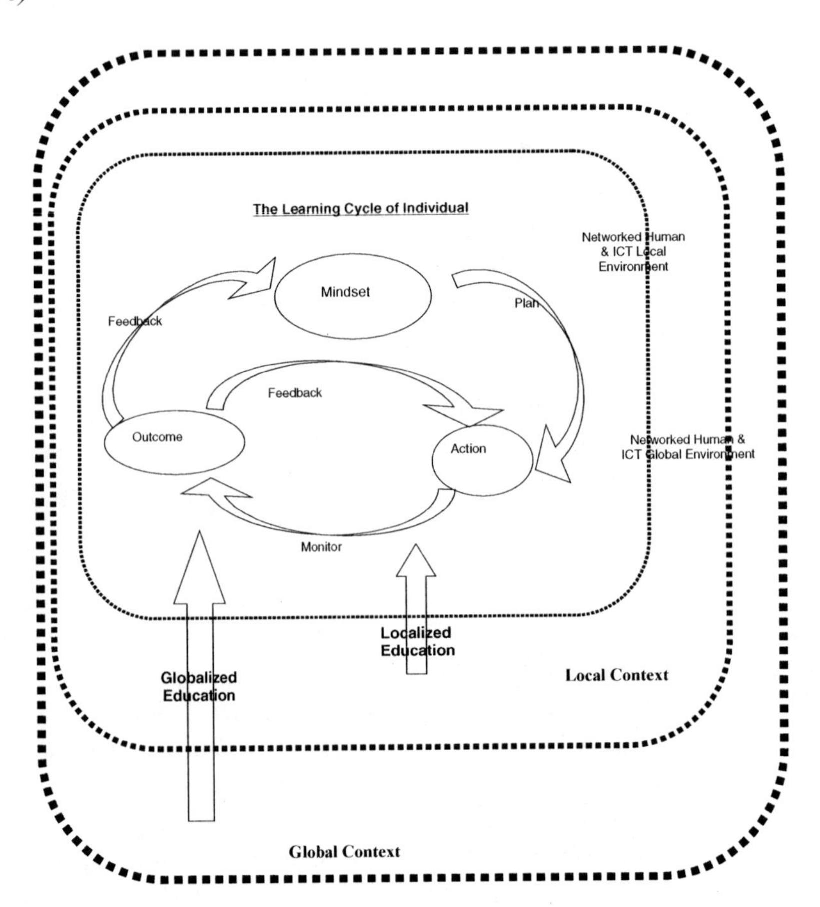

the planning process as well as the action of learning. The learning associated with change in mental-set or mental models is often referred as *high-order learning*.

The feedback directly to the action of learning will help the learner to adapt his/her learning behaviors. The learning associated with change in behaviors or actions is often referred to as *low-order learning*. Since this type of learning has not changed the mental conditions of the learner, it may not produce long-lasting learning effects at a higher level.

How to sustain the cyclic process of self-learning by the learners themselves continuously and throughout their life-span is really the core issue of current education reforms. According to the literature of learning environment, both human environment and technological environment are important in facilitating and sustaining self-learning (Garrison, 1997; Henderson & Cunningham, 1994). Particularly, how the human and IT environment can be designed, developed, and used to facilitate such a continuous lifelong self-learning inevitably becomes an important question to guide the development of self-learning theory in the context of the networked human and IT environment. Mok and Cheng (2001) have explained a theory of self-learning in a networked human and IT environment that can be used to support the new paradigm of triplized education at the K-12 level, as follows.

IT Environment

Due to the tremendous developments in IT, the Internet, and global networking, recently there has been a great demand for developing an ICT environment in order to support the paradigm shift in learning and teaching. Computer technology makes it possible for multiple learners to be networked and participate in the learning task, thus greatly enhancing the social interactions, sharing of learning experiences, and resources in a very convenient way. Information technology can also facilitate and accelerate the monitoring, assessment, and feedback processes in a very fast and efficient way (Embretson & Hershberger, 1999).

There may be four important aspects in which new technology can contribute to the development of a powerful ICT environment that can facilitate the self-learning cycle:

1. Computer technology revolutionized both the speed and access to information (Hallinger, 1998). Information is interpreted in its broadest sense, including resource materials for the learner as well as feedback concerning how well the learner has learned. With the help of the Internet, learners can access the best quality of Web-based learning materials in different parts of the world. Further, because of the high speed of information technology, feedback can be immediately generated for each step of learning tasks and activities as well as for the overall proficiency of learning. The fast feedback to the learner's mental conditions and learning behaviors, in fact, accelerates the speed of learning, including cognitive changes and behavioral changes in the learner.
2. Developments in IT make it possible for the application of measurement theory to assessment tasks during the self-learning process. Technology is now available for real-time scoring (Herl, Baker, & Niemi, 1996), computer adaptive testing (CAT), automated data logging (Chung & Baker, 1997), and computer item construction (Bennett, 1999). The advanced assessment methods can greatly improve the quality and accuracy of monitoring and feedback such that the quality and opportunity of learning can be ensured.

3. Developments in IT enable assessment to move away from the paper-pencil format to rich imagery laden multimedia task presentation and submission (Bennett, 1999; Chung & Baker, 1997) that can capture richly contextualized performances in the learning process (Bennett, 1999). For example, Chung and Baker (1997) described the scoring of complex concept maps constructed by students, based on information stored in Web pages. They were able not only to measure the quality of the finished product, but also capture, unobtrusively, the process of how students learned. Students' process of learning was monitored, using Web page access log, including information students considered important to the task, the amount of time searching the Web for relevant information, time students spent on each Web page, modification to the concept map under construction, and so forth. All this information would be useful for understanding the complex nature of the learning process and in turn improve learning strategies, activities, and outcomes.
4. The IT environment breaks down distance barriers of access to education and creates connectivity among learners (Mok & Cheng, 2000a). When learners, teachers, parents, resource people, and other related experts can be networked through IT, more opportunities will be available for social interactions, experience sharing, and information flow. With this, a networked human environment can be created to sustain and support self-learning of individual learners.

Networked Human Environment

The meaningfulness of learning is often constructed within a human environment that comprises the teacher, peers, parents, and other adults, and reflects to a certain extent the education values espoused by the social actors (Garrison, 1997). The human environment plays a significant role in all aspects of self learning: pedagogical, psychological, and behavioral (Schunk, 1998). In particular, Zimmerman (2000) highlighted the interdependent role of social, environmental, and self, and their bidirectional influences in self-learning.

In K-12 education, the human environment itself can be designed to become an important source of pedagogical information. The teacher, as a key actor in the human environment of learning, helps the learner to develop attitudes and skills for goal-setting, self-management, self-monitoring, and self-evaluation which are essential to the success of self-learning. For example, in this IT age, there is no shortage of information, but the learner needs to make judgment about the information. Consequently, the learner has to develop critical thinking skills to validate and authenticate the quality of instructional materials, such as those downloadable from the Web. Further, the teacher as a proficient adult provides appropriate learning references or guides the learner to these materials. Winne and Perry (2000) identified the unique position held by teachers in judging the quality of the student's self-learning and providing guidance where appropriate. The learner also learns from peers, parents, and other adults by observation and emulation (Schunk, 1987, as cited in Schunk, 1998).

Self-learning is a complex process and the endeavor can result in non-accomplishment, frustration, or even failure. In such instances, the empathy and social support from the teacher, parents, and peers acts as an emotional safety net for the learner. A strong social climate gives strength to the learner to continue engagement in the task, analyze strategies, and manage the failure and frustration in a positive way.

It is now possible, with developments in IT, to network the learner with the teacher, parents, peers, and other adults or professionals in the community such that influence of the human environment on self-learning can be maximized (Mok & Cheng, 2000a).

When individual learners are networked with the support of IT, as shown in Figures 2 and 3, there may be a multiplying effect on the amount of available information as well as human touches and interactions that will become fruitful stimulus to students' self-learning. The networked individual learners, teachers, parents, and other professionals may form a learning system to support students' continuous self-learning. In a learning society, each learner is self-motivated and generates a learning cycle of self-learning and self-evaluation. Learners, teachers, and parents are networked to form a learning classroom; classrooms are networked to form a learning school; schools and the community are networked to form a learning society; learning societies are networked across nations (Mok & Cheng, 2000b).

IT speeds up the process of providing social messages and informative feedback to the learners and members in the learning system. This speed, coupled with the massive amount of information available via the informative network, not only means that this will be the information-rich era, but also it implies that a closely networked social environment needs to be in place for promoting and supporting self-learning of individual learners. Self-learning is no longer the acquisition of information of individual learners in an isolated context. Instead, effective self-learning occurs in the human environment that can facilitate a higher level of intelligence and motivation of learners as well as other members in the human network in the selection, management, transfer, creation, and extension of knowledge (Mok & Cheng, 2000a).

Figure 2. Individually networked within school

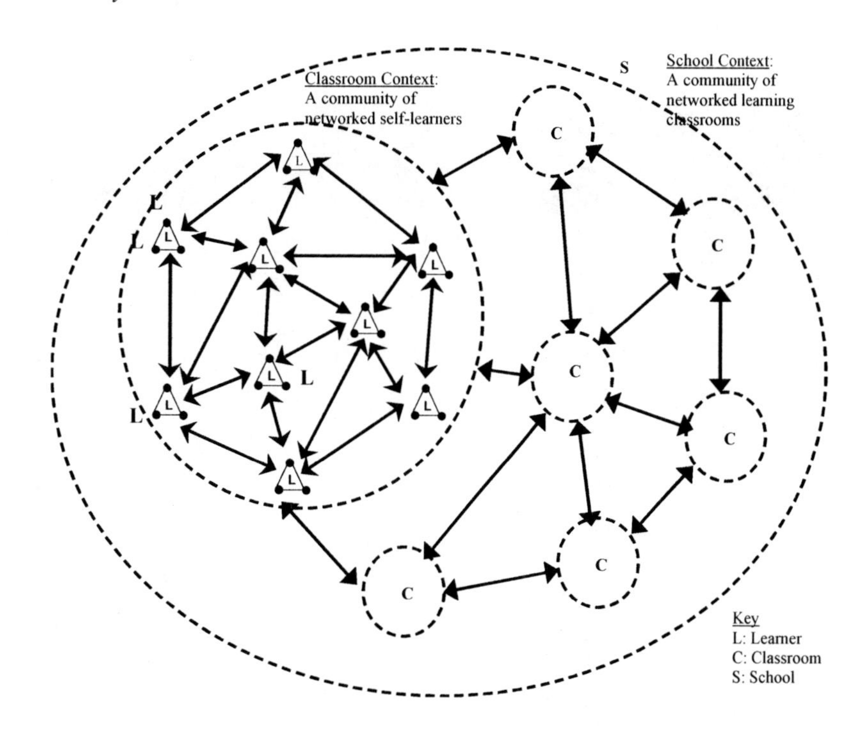

Teacher Facilitates Students' Self-Learning Cycle

Teachers can play a key role in building up a strong and direct linkage between each stage of students' self-learning cycle and the networked learning environment in daily educational practice. From the aforementioned nature of learning cycle and networked human and ICT environment, teachers should prepare themselves to have a new kind of professional competence that can facilitate students to initiate and sustain each stage of the self-learning cycle continuously to achieve effective learning with the support of a networked human and ICT environment. If a teacher can facilitate students' self-learning cycle with the support of such a networked human and ICT environment, he or she may be considered as an effective teacher.

School-Based and Central Platforms for Borderless Education

How to build up such a networked human and technological environment for borderless education at the K-12 level is very challenging to both educators and reformers. According to Cheng (2002, 2001a, 2001b), the development of a networked human and technology environment can be supported by the school-based platform and central platform (see Figure 4).

Platform is a new powerful concept in conceptualizing and organizing various types of existing resources, technology, knowledge, and even social and cultural capital from local and

Figure 3. Locally and globally networked with countries and communities

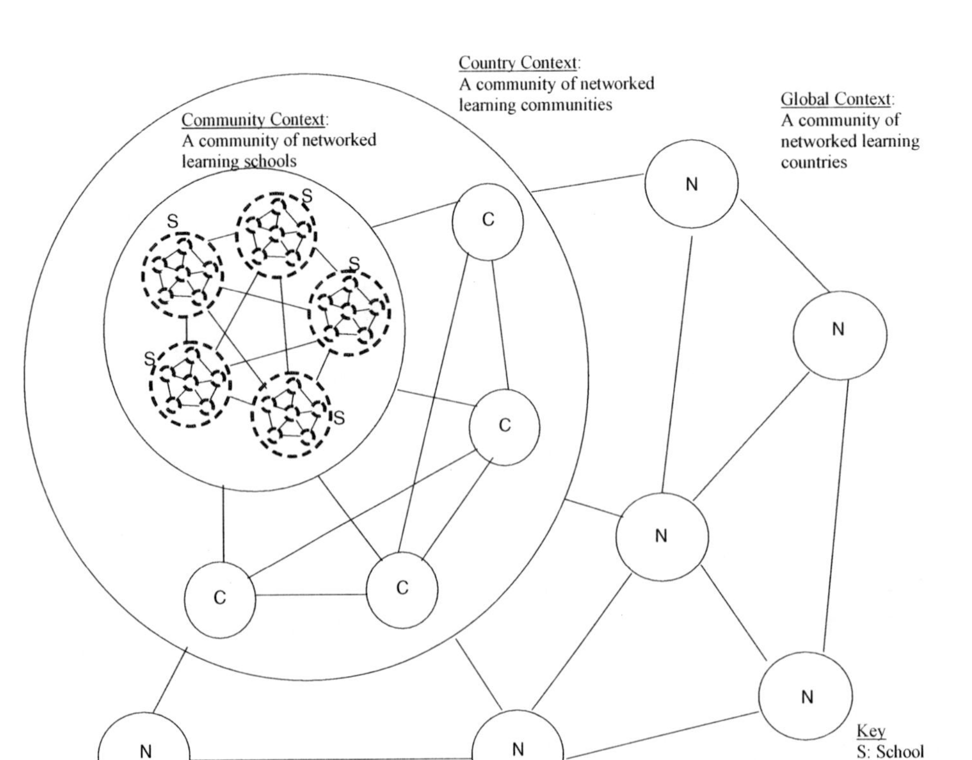

Figure 4. Platform theory for effective learning, teaching, and schooling

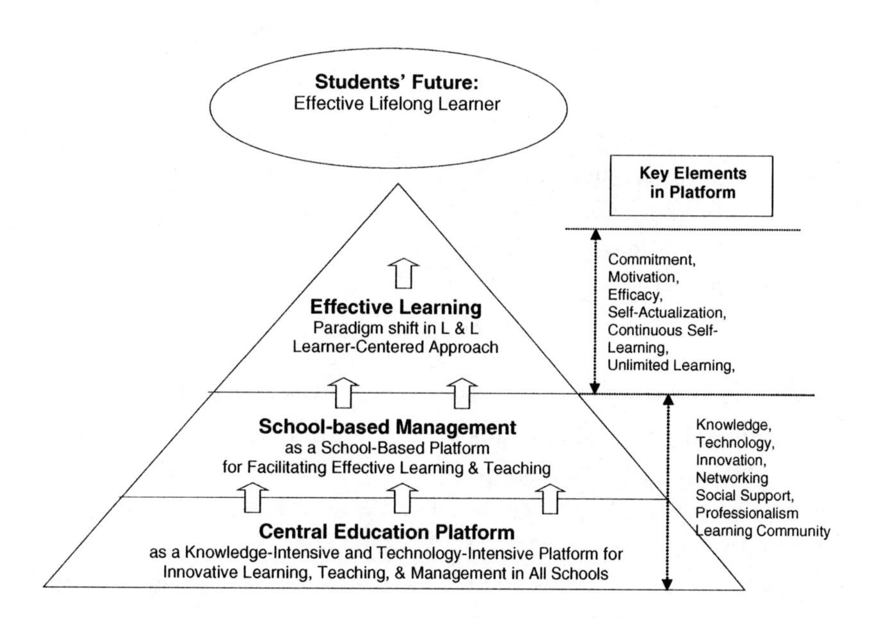

global sources to form an intelligence-intensive and technology-intensive platform or supporting environment, that can enable and facilitate people to work and perform in a smart and optimal way. Unlike the traditional concept of organizational structure with focus on control and coordination, platform is mainly for supporting people with the necessary knowledge, technology, and social environment such that they can have the maximum opportunity to develop themselves and perform at their highest potential in their work.

School-based management or educational decentralization should aim at developing as a *school-based platform* that can accumulate, organize, and apply the necessary knowledge and technology, useful experiences, networks, various types of internal and external resources, and social support to foster educational practice and innovation for effective teaching and learning, facilitate organizational learning, and develop a culture of professionalism within a school. To a great extent, a school-based platform is a powerful way to pool resources and network people for effective learning through localization and globalization. A detailed account of how school-based management can develop such a platform or mechanism for continuous development and effectiveness has been given in Cheng (1996).

At the system or regional level, a *central education platform* should be formed with the support of information technology and various types of local and global networking. This central platform aims to pool the most powerful and relevant knowledge, expertise, and resources from local and global sources to create a more knowledge-intensive, technology-intensive, and intelligence-intensive central base for supporting the development of all types of school-based platform and related initiatives. On this central platform, schools, teachers, and students can work on a higher level of knowledge to develop their school-based initiatives and avoid unnecessary wastage of time, resources, and efforts due to repeated "re-inventing of the wheel" or "start from scratch" initiatives. This central platform is also a huge network or learning community for sharing advanced knowledge, best practices, and experiences of success and failure among schools, educators, and experts (Mok & Cheng, 2001).

The key elements of the school-based platform and central education platform are accumulation, dissemination, and application of knowledge and technology to promote various types of innovation, networking, and social support, and develop a culture of professionalism and learning community in education. All this can support a paradigm shift in education and in effective learning, teaching, and schooling at the K-12 level.

As shown in Figure 4, with the support of these platforms, the key elements in effective learning and teaching are students' and teachers' commitment, motivation, and efficacy to achieve learning as continuous self-actualization and create unlimited opportunities for learning, developing learning groups, and evolving a learning culture among students and teachers (Cheng, 2001a, 2001b; Mok & Cheng, 2001).

PARADIGM SHIFT IN ICT AND ICT LITERACY AT THE K-12 LEVEL

Given the paradigm shift in learning and teaching, there is also a corresponding paradigm shift in ICT and ICT literacy at the K-12 level of education (see Table 4).

Since the traditional paradigm emphasizes the delivery of knowledge and skills in learning and teaching within the school-site boundary, the use of ICT and ICT literacy in education at the K-12 level is limited mainly to improving the efficiency of delivery of the necessary knowledge and skills in a planned curriculum. Whether ICT and ICT literacy can facilitate paradigm shift in K-12 education is not a concern. The discussion of literacy in ICT at the K-12 level is often focused on the following questions:

1. **How well is teachers' literacy in ICT to deliver the knowledge and skills required in a curriculum to students through ICT?** It assumes that teachers should have the necessary and relevant literacy in ICT to deliver a curriculum to students, particularly through ICT.
2. **How well is students' literacy in ICT to learn the knowledge and skills required in the curriculum from their teachers particularly through ICT?** It also assumes that students should be prepared with the necessary literacy in ICT to learn the knowledge and skills, no matter whether in a technology curriculum or not.
3. **How well can the delivery of knowledge and skills to students be ensured through the improvement of teaching and learning with ICT and the literacy in ICT of teachers and students?** It assumes that the application of ICT and the enhanced ICT literacy of teachers and students can help to improve the process of teaching and learning and ensure the effectiveness of delivering a curriculum.
4. **How well can students' performances meet given standards in the internal and external examinations with the support of ICT and their ICT literacy?** It expects that the ICT literacy and use of ICT can help students to enhance performance and outcomes of learning and meet

Table 4. Paradigm shift in applying ICT and enhancing ICT literacy

New Paradigm of Applying ICT & Enhancing ICT Literacy at the K-12 Level	Traditional Paradigm of Applying ICT and Enhancing ICT Literacy at the K-12 Level
• The extensive application of ICT and enhancement of ICT literacy of teachers and students in building up a networked environment for teachers' individualized, localized, and globalized professional learning, and CMI development is crucial and necessary. • ICT and related ICT literacy play a key role to facilitate paradigm shift in education at the K-12 level. • The effectiveness of applying ICT and enhancing ICT literacy in education at the K-12 level depends on: 1. How well can the application of ICT and enhancement of ICT literacy help globalize, localize, and individualize teachers' teaching and students' learning? 2. How well can the use of ICT and enhancement of ICT literacy maximize K-12 students' learning opportunities through establishing the borderless ICT environment, local and international networking, and various types of innovative learning programs? 3. How well can the use of ICT and enhancement of ICT literacy facilitate and ensure students' learning to be sustainable as potentially lifelong? 4. How well can the use of ICT and enhancement of ICT literacy ensure and facilitate the development of students' ability to triplize their learning and development? 5. How well can the application of ICT and enhancement of ICT literacy facilitate the development of a CMI pedagogical environment, in which students are immersed and inspired to be self-actualizing and developing CMI themselves?	• The use of ICT and ICT literacy in education at K-12 is limited, mainly on improving the efficiency of delivery of the necessary knowledge and skills in a planned curriculum. • Whether ICT and ICT literacy can facilitate paradigm shift in K-12 education is not a concern. • The effectiveness of applying ICT and enhancing ICT literacy in education at the K-12 level depends on: 1. How proficient is teachers' literacy in ICT in order to be able to deliver the knowledge and skills required in a curriculum to students through ICT? 2. How proficient is students' literacy in ICT in order to be able to learn the knowledge and skills required in the curriculum from their teachers, particularly through ICT? 3. How well can the delivery of knowledge and skills to students be ensured through the improvement of teaching and learning with the application of ICT and the enhancement of literacy in ICT of teachers and students? 4. How well can students' performance meet given standards in the internal and external examinations with the application of ICT and enhancement of their ICT literacy?

the explicitly planned goals and standards in examinations.

Clearly, these four questions are concerned with how the ICT literacy of students and teachers and the application of ICT in learning and teaching are related to the effectiveness of delivery of knowledge and skills in education at the K-12 level.

But the paradigm shift towards triplization in K-12 education induces a new thinking of ICT and ICT literacy in education because the aims, content, and process of education at K-12 are completely different from the traditional thinking. The extensive application of ICT and enhancement of ICT literacy of teachers and students in building up a networked environment for students' individualized, localized, and globalized learning and CMI development is crucial and necessary. ICT and related ICT literacy play a key role to facilitate the paradigm shift in education at the K-12 level.

The application of ICT and enhancement of ICT literacy in education can be based on the following major questions:

1. **How well can the application of ICT and enhancement of ICT literacy help globalize, localize, and individualize teachers' teaching and students' learning?** This question is proposed to ensure how the use of ICT and enhancement of ICT literacy of teachers and students effectively places teachers' teaching and students' learning in a globalized, localized, and individualized context. It assumes that delivery of knowledge and skills in K-12 education at the site level is not sufficient to ensure educational relevance to globalization, localization, and individualization for the future development of students.
2. **How well can the use of ICT and enhancement of ICT literacy maximize K-12 students' learning opportunities through establishing the borderless ICT environment, local and international networking, and various types of innovative learning programs?** This question is proposed to ensure how effective ICT and ICT literacy are in maximizing opportunities for students' learning and development in a triplized learning environment. The concern is not only on how much the internal processes of teaching and learning can be improved and how much the strategic stakeholders are satisfied, but also on how large and how many opportunities can be created for students' continuous learning.
3. **How well can the use of ICT and related ICT literacy facilitate and ensure students' learning to be sustainable and potentially lifelong?** This question focuses on ensuring that the use of ICT and enhancement of ICT literacy can effectively facilitate and ensure sustainable lifelong learning that is a core part of the new paradigm of education at the K-12 level. It assumes that short-term internal improvement in practice and short-term stakeholders' satisfaction with standards or performance may not be so important and relevant to the future development of students if students themselves cannot sustain their learning as a lifelong process with the support of ICT.
4. **How well can the use of ICT and enhancement of ICT literacy ensure and facilitate the development of students' ability to triplize their learning and development?** This question is proposed to ensure the influence of ICT and enhancement of ICT literacy in K-12 education relevant to the development of students' ability of triplizing their own learning. It is very important and necessary for students to develop their own ability for maximizing learning opportunities and sustaining their learning through globalization, localization, and individualization.
5. **How well can the application of ICT and enhancement of ICT literacy facilitate the development of a CMI pedagogical environment, in which students are immersed and inspired to be self-actualizing and developing CMI themselves?** The question focuses on how the use of ICT and enhancement of ICT literacy can ensure the outcomes of learning relevant to the development of students' CMI, including technological, economic, social, political, cultural, and learning intelligences that are crucial for students to meet the challenges in the future. This is one of the main concerns of

the new paradigm in education at the K-12 level.

From the above discussion, the implications for a paradigm shift in applying ICT and enhancing ICT literacy in K-12 education are substantial. The effectiveness of applying ICT and enhancing ICT literacy depends heavily on whether they can facilitate students to develop themselves successfully in terms of contextualized multiple intelligences and ability for triplization in learning.

CONCLUSION

The new paradigm of learning and teaching in a networked environment, that is contrastingly different from the traditional thinking, can be used to rethink education, enhancement of ICT literacy, and application of ICT at the K-12 level.

We expect our education to be triplized in the new century. In fact, the ongoing educational reforms in different parts of the world have already provided evidence that many countries are making an effort in this direction through various types of initiatives in globalization, localization, and individualization in education. We believe our learning and teaching will be finally borderless and characterized with globalization, localization, and individualization with the help of the information technology and boundless multiple networking.

We should use a new theory to promote self-learning in a networked borderless human and ICT environment. Particularly through localization and globalization, we should build up school-based and centralized platforms to pool local and global resources and intellectual assets, and form a networked borderless human and ICT environment to support learning and teaching. Through these platforms and the new paradigm of learning, we will create unlimited opportunities and multiple global and local sources for lifelong learning and development of both students and teachers. We believe new reforms in education should facilitate the triplized learning and make students' learning process interactive, self-actualizing, discovery-based, enjoyable, and self-rewarding.

With the paradigm shift in education, correspondingly there should be a paradigm shift in the application of ICT and the enhancement of ICT literacy in teaching and learning at the K-12 level. Different from the traditional emphasis on delivery of knowledge and skills in planned curriculum, the new paradigm pursues the extensive application of ICT and enhancement of ICT literacy of teachers and students in building up a networked environment for students' individualized, localized, and globalized learning and CMI development. It is hoped that our students equipped with the necessary ICT literacy can become borderless learners with unlimited opportunities for learning and development in an individually, locally, and globally networked environment. They can fully enjoy lifelong learning and self-actualization as CMI citizens for the new world.

REFERENCES

Argyris, C., & Schön, D. A. (1974). *Theory in practice: Increasing professional effectiveness*. San Francisco: Jossey-Bass Classics.

Argyris, C., Putnam, R., & Smith, D. M. (1985). *Action science*. San Francisco: Jossey-Bass

Ayyar, R. V. V. (1996). Educational policy planning and globalisation. *International Journal of Educational Development, 16*(4), 347-354.

Bennett, R. E. (1999). Using new technology to improve assessment. *Educational Measurement, 18*(3), 5-12.

Blackledge, D., & Hunt, B. (1985). *Sociological interpretations of education.* Sydney: Croom Helm.

Brown, P., & Lauder, H. (1996). Education, globalization and economic development. *Journal of Education Policy, 11*(1), 1-25.

Brown, T. (1999). Challenging globalization as discourse and phenomenon. *International Journal of Lifelong Education, 18*(1), 3-17.

Cheng, Y. C. (1996). *School effectiveness and school-based management: A mechanism for development.* London: Falmer.

Cheng, Y. C. (2000). A CMI-triplization paradigm for reforming education in the new millennium. *International Journal of Educational Management, 14*(4), 156-174.

Cheng, Y. C. (2001a, October 20). The strengths and future of Hong Kong education. *Proceedings of the Conference of the Leadership Development Network on "Facing up Educational Reforms: Leading Schools for Quality Education."* Organized by the Asia-Pacific Centre for Education Leadership and School Quality.

Cheng, Y. C. (2001b, July 2-6). Three waves of education reform: Paradigm shifts for the future in Hong Kong. *Proceedings of the 11th World Congress of Comparative Education,* Chungbuk, Korea.

Cheng, Y. C. (2002, September 2-5). Linkage between innovative management and student-centered approach: Platform theory for effective learning. *Proceedings of the 2nd International Forum on Education Reform: Key Factors in Effective Implementation.* Organized by Office of National Education Commission in collaboration with UNESCO and SEAMEO, Bangkok, Thailand.

Cheng, Y. C., Ng, K. H., & Mok, M. M. C. (2002). Economic considerations in education policy making: A simplified framework. *International Journal of Educational Management, 16*(1), 18-39.

Cheng, Y. C., & Townsend, T. (2000). Educational change and development in the Asia-Pacific region: Trends and issues. In T. Townsend & Y. C. Cheng (Eds.), *Educational change and development in the Asia-Pacific region: Challenges for the future* (pp. 317-344). Lisse, The Netherlands: Swets and Zeitlinger.

Chung, G. K. W. K., & Baker, E. L. (1997). *Year 1 technology studies: Implications for technology in assessment.* ERIC Document Reproduction Services ED 418 099.

Curriculum Development Council. (1999, October). *A holistic review of the Hong Kong school curriculum proposed reforms (consultative document).* Hong Kong: Government Printer.

Education and Manpower Bureau. (1998, November). *Information technology for learning in a new era: Five-year strategy 1998/99 to 2002/03.* Hong Kong: Government Printer.

Embertson, S. E., & Hershberger, S. L. (1999). *The new rules of measurement: What every psychologist and educator should know.* Mahwah, NJ: Lawrence Erlbaum.

Garrison, D. R. (1997). Self-directed learning: Toward a comprehensive model. *Adult Education Quarterly, 48*(1), 18-30.

Green, A. (1999). Education and globalization in Europe and East Asia: Convergent and divergent trends. *Journal of Education Policy, 14*(1), 55-71.

Hallinger, P. (1998). Educational change in Southeast Asia: The challenge of creating learning systems. *Journal of Educational Administration, 36*(5), 492-509.

Henderson, R. W., & Cunningham, L. (1994). Creating interactive sociocultural environments for self-regulated learning. In D. H. Schunk & B. J. Zimmerman (Eds.), *Self-regulation of learning and performance*. Hillsdale, NJ: Lawrence Erlbaum.

Henry, M., Lingard, B., Rizvi, F., & Taylor, S. (1999). Working with/against globalization in education. *Journal of Education Policy, 14*(1), 85-97.

Herl, H. E., Baker, E. L., & Niemi, D. (1996). Construct validation of an approach to modeling cognitive structure of U.S. history knowledge. *Journal of Educational Research, 89*, 206-218.

Hinchliffe, K. (1987). Education and the labor market. In G. Psacharopoulos (Ed.), *Economics of education: Research and studies* (pp. 315-323). Kidlington, Oxford: Pergamon Press.

Holmes, W. (1999). The transforming power of information technology. *Community College Journal, 70*(2), 10-15.

Jones, P. W. (1999). Globalisation and the UNESCO mandate: Multilateral prospects for educational development. *International Journal of Educational Development, 19*(1), 17-25.

Lee, K. T., & Mitchell, K. (Eds.). (2003). *Proceedings of the International Conference on Computers in Education,* Hong Kong.

Little, A. W. (1996). Globalization and educational research: Whose context counts? *International Journal of Educational Development, 16*(4), 427-438.

Liu, S. S. (1997). *Trends in Hong Kong University management: Towards a lifelong learning paradigm.* Hong Kong: Hong Kong Baptist University.

McMahon, W. W. (1987). Consumption and other benefits of education. In G. Psacharopoulos (Ed.), *Economics of education: Research and studies* (pp. 129-133). Kidlington, Oxford: Pergamon Press.

Mok, M., & Cheng, Y. C. (2000a, January 4-8). Self-learning driven assessment: A new framework for assessment and evaluation. *Proceedings of the 13th International Congress for School Effectiveness and Improvement (ICSEI 2000),* Hong Kong.

Mok, M., & Cheng, Y. C. (2000b, December 12-15). Global knowledge, intelligence and education for a learning society. *Proceedings of the 6th UNESCO-ACEID International Conference, "Information Technologies in Educational Innovation for Development: Interfacing Global and Indigenous Knowledge,"* Bangkok, Thailand.

Mok, M., & Cheng, Y. C. (2001). A theory of self-learning in a human and technological environment: Implications for education reforms. *International Journal of Education Management, 15*(4), 172-186.

Pratt, G., & Poole, D. (2000). Global corporations "R" us? The impacts of globalisation on Australian universities. *Australian Universities' Review, 43*(1)/*42*(2), 16-23.

Ryan, S., Scott, B., Freeman, H., & Patel, D. (2000). *The virtual university: The Internet and resource-based learning.* London: Kogan Page.

Schunk, D. H. (1998). Teaching elementary students to self-regulate practice of mathematical skills with modeling. In D. H. Schunk & B. J. Zimmerman (Eds.), *Self-regulated learning and performance*. Hillsdale, NJ: Lawrence Ehrlbaum.

Tan, S. C., So, K. L., & Hung, D. (2003). Fostering scientific inquiry in schools through science research course and computer-supported collaboration learning (CSCL). In K. T. Lee & K. Mitchell (Eds.), *Proceedings of the International Conference on Computers in Education,* Hong Kong.

Weaver, T. R. (1970). *Unity and diversity in education.* London: Department of Education and Science.

Winne, P. H., & Perry, N. E. (2000). Measuring self-regulated learning. In M. Boekaerts, P. R. Pintrich, & M. Zeidner (Eds.), *Handbook of self-regulation*. San Diego, CA: Academic Press.

Yeung, Y. Y., Cheng, M. M. H., & So, W. W. M. (2003). Supporting the initial training of science teachers in Hong Kong: An effective approach through Web-based peer learning and sharing. In K.T. Lee & K. Mitchell (Eds.), *Proceedings of the International Conference on Computers in Education,* Hong Kong.

Yuen, A. H. K. (2003). Building learning communities through knowledge forum: A case study of six primary schools. In K. T. Lee & K. Mitchell (Eds.), *Proceedings of the International Conference on Computers in Education,* Hong Kong.

Yuen, P. Y., & Cheng, Y. C. (1997). The action learning leadership for pursuing education quality in the 21st Century. *Proceedings of the 5th International Conference on Chinese Education Towards the 21st Century: Key Issues on the Research Agenda,* Hong Kong.

Yuen, P. Y., & Cheng, Y. C. (2000). Leadership for teachers' action learning. *International Journal of Educational Management, 14*(5), 198-209.

Zimmerman, B. J. (2000). Attaining self-regulation: A social cognitive perspective. In M. Boekaerts, P. R. Pintrich, & M. Zeidner (Eds.), *Handbook of self-regulation*. San Diego, CA: Academic Press.

ENDNOTE

1 For details of the concept of contextualized multiple intelligence (CMI) and its development through the new paradigm of learning, please refer to Cheng (2000).

Chapter II
Technologies Challenging Literacy: Hypertext, Community Building, Reflection, and Critical Literacy

Agni Stylianou-Georgiou
Intercollege, Cyprus

Charalambos Vrasidas
CARDET-Intercollege, Cyprus

Niki Christodoulou
Intercollege, Cyprus

Michalinos Zembylas
CARDET-Intercollege, Cyprus

Elena Landone
University of Milan, Italy

ABSTRACT

This chapter describes how new technologies are challenging the traditional concept of literacy and redefining its meaning. New genres of texts change conceptualizations of literacy development and present new challenges for reading and writing. Important consequences for instruction, especially how teachers and students exploit new forms of literacy to enhance teaching and learning, are presented and discussed. A particular emphasis is placed on the role of technology for community building, reflection, and teacher learning. Deconstructing notions of how technology shapes society and the role of critical emotional literacy are also explored.

INTRODUCTION

For many years, a traditional concept of literacy as the ability to read and write print on a page has dominated schooling and adequately served the literacy demands of the society and of the workplace. In the emerging digital economic era, this traditional concept of literacy is being challenged. Traditional reading and writing are but the initial layers of the richer and more complex forms of literacy required in an electronic context. The unique characteristics of reading and writing with computers make us realize that electronic texts represent a substantively new technology that does not readily mesh with the assumptions that have risen from the technology of print. This implies a re-elaboration of techno-space into an infinite, and thus a loss of the finitude of material inscription; consequently, the notion of literacy has to be renegotiated.

Recent theoretical work helps us to better understand the central relationship between literacy and technology. Reinking (1998) provides a transformational perspective of the relation between literacy and technology, observing that technology transforms the nature of literacy. From this perspective, a review of research studies would seek to understand the new forms of literacy possible within new technologies. A critical approach to study the relation between literacy and technology would include studies of how information technologies transform literacy, as well as how educators need to problematize technology as it relates to issues of access, gender, equity, and social justice (Zembylas, Vrasidas, & McIsaac, 2002). Such an approach provides important insights into the many changes currently taking place in the nature of literacy and education around the world.

This chapter addresses four major issues: (1) technology and its impact on the concept of literacy in general and as it applies to language learning; (2) hypertext literacy and the role of navigation in reading comprehension while interacting with electronic texts; (3) the role of technology for community building, reflection, and teacher learning; and (4) deconstructing notions of how technology shapes society and the role of critical emotional literacy in understanding such issues.

TECHNOLOGY AND ITS IMPACT ON THE NATURE OF LITERACY

The Internet and other forms of information and communication technology (ICT) are redefining the nature of literacy. According to Leu (2000) literacy appears to be increasingly *deictic*; its meaning is regularly redefined, as new technologies for information and communication repeatedly appear and new visions for exploiting these technologies are continuously created by the users. New technologies transform contemporary notions of literacy, teacher education, classroom learning contexts, and resources such as text.

A deictic perspective on literacy predicts, according to Leu (2000), that the nature of literacy will change in important ways:

a. *"Strategic knowledge will become even more important to successful literacy activities than it is today. Navigating the increasingly complex information available within global information networks that continually change will require greater and perhaps newer strategic knowledge than is required within more limited and static, traditional texts.*

b. *Literacy will increasingly become a continuous learning task for each of us. Since new technologies and new visions for literacy will regularly ap-*

pear, we will need to continually learn new ways to acquire information and communicate with one another. Social learning strategies will be central to literacy instruction to distribute knowledge about literacy within the global community.

c. *Students should become more aware of the new media and new literacies that are available in digital authoring tools. The new meanings possible by combining multiple media sources create important challenges for the educators as they prepare students for their futures.*

d. *Literacy within global information networks will require new forms of critical thinking and reasoning. Since anyone may publish anything on open networks like the Internet, students should learn how to become more critical consumers of the information they encounter."* (pp. 760-761)

In this chapter we have chosen to elaborate on these issues and we present how technology transforms the notions of teaching and learning to become literate as well as the development of critical emotional literacy. First we present how new genres of texts change conceptualizations of literacy development and present new challenges for reading and writing.

HYPERTEXT LITERACY

Digital forms of expression are increasingly replacing printed forms and, therefore, the way we communicate and disseminate information. Electronic texts that are based on hypertext and hypermedia technologies are now being used in many classrooms to support literacy learning. The new forms of ICT technology pose several questions and challenges to the literacy community. Navigating the increasingly complex information available within global information networks that continually change might require greater and perhaps newer strategic knowledge that is required within more limited and static, traditional texts (Brown, 2000; Leu, 2000). It is still unclear what strategies are actually involved while reading electronic information and whether our fundamental understanding of comprehension can keep pace with the changing nature of text. In this section we describe the unique characteristics of hypertext technology and present the challenges involved in using online information resources.

Hypertext vs. Traditional Texts

How different is navigating an electronic text (hypertext) from reading a printed book? A unique attribute of hypertext systems is the *nonlinearity* of information units. Instead of looking at one unique, predetermined sequence of text, pictures, or graphics, hypertext provides readers the ability to follow multiple reading paths (Landow, 1992; Jacobson, Maouri, Mishra, & Kolar, 1996). The various links in the text (hyperlinks) constitute possible orders in which the different information units could be assembled and read, and each order can produce a different outcome in terms of reading comprehension, message, or problem solving. Unlike books, hypertext has no beginning or ending. Hypertext users can browse through the domain freely and choose when and where to leave the system. They are unconstrained by the linearity of information of print-oriented technologies (Bolter, 2001).

The features in hypertext systems also provide *flexibility* to the reader when compared to reading linear text such as books. Clearly some of this flexibility does exist in books (e.g., table of contents and indexes), but it is not as widely used or exploited (Snyder, 1998). Hypertext permits readers to use these features automatically rather than requiring readers to manually refer to them as needed. This provides additional control to the reader in determining the order that the text is to be read, and allows the reader to follow meaningful links or paths through the document (Jacobson et al., 1996).

Hypertext offers new opportunities for authors and readers to collaborate toward a mutual goal by encouraging readers to take a more *active role* in reading. Readers are required to make decisions and be mentally active while they are interacting with the information they are accessing through the computer. Consequently, learning with hypertext becomes more learner-centered because the emphasis of hypertext is on an active reader. The reader is in control and may use his initiative dynamically. Hypertext readers actively share in shaping the document they read by selecting the links they will follow and the ones that they will ignore. Many hypertext systems allow readers to interact with the system to such an extent that they may become actively involved in the creation of an evolving hyperdocument (Eklund, Sawers, & Zeiliger, 1999). Co-authorship may take a number of different forms, from relatively simple, brief annotations, comments, or modifications of existing material, to the creation of new links connecting material not previously linked (Guzdial, 1998).

Challenges of Navigating Hypertext

The nonlinear representation of the multiple text fragments and the large number of links in a hypertext environment also poses several challenges to readers. Navigating a hypertext system is different from reading a book. It requires more active involvement and metacognitive effort (Jul & Furnas, 1997; Marchionini, 1988). When reading a traditional text, a reader can depend on knowledge of text structure to help create inferences that aid in text understanding (Kintsch & van Dijk, 1978). However, the reader of an electronic text must not only understand the information presented, but must also decide what links to follow and in what order information will be accessed (Lawless & Brown, 1997; Thüring, Hannemann, & Haake, 1995). Each hypertext system has a different structure, and the global coherence (flow of the different sections in a text) depends not only on the authors/designers' decisions but also on the choices that the readers are making while navigating the space of a hypertext. According to Bolter (1998) the reader "seems to be collaborating with the author in the creation of text, in the sense that the choice of links determines what the reader will next see on the screen." Each reader arranges her own unique text, and the paths that she follows might affect text understanding (Hegarty, Quilici, Narayanan, Holmquist, & Moreno, 1998).

The navigational decisions that readers need to make while reading from hypertext may present difficulties and impose a higher cognitive load, especially on readers with low prior knowledge (Charney, 1987; Jacobson et al., 1996; Lawless & Brown, 1997). As a result they may get lost in the hyperspace, unable to identify where they are, what links to follow, and in what order information will be accessed (Conklin, 1987; Marchionini, 1995). Conklin (1987) described this dilemma as 'informational myopia'. In such rich and fluid environments, there is a trade-off between system flexibility and cognitive overload on the user. System flexibility does not guarantee better learning without the learner's active involvement and right decision making. Hypertext readers need to make frequent and important decisions about the selection of relevant links that will enable them to pursue their learning goals. They need to engage in cognitive monitoring, to slow down and take a moment to consider various paths and question why they are considering some paths over others (Charney, 1994).

Since hypertext reading requires readers to make critical navigational choices, our definition and fundamental understanding of comprehension might not keep pace with the changing nature of text. Readers will have to learn how to take advantage of different presentation formats of information, and in doing so they might need to acquire new text comprehension strategies.

Some researchers have suggested that it is likely that text navigation may walk hand in hand with text comprehension. Dillon and Vaughan (1997) advocated that moving through the information space while interacting with hypertext "is frequently the same purpose as the journey, to reach an end point of comprehension—and in this case the journey is the destination" (p. 100). Bolter (1998) claimed that the technology of hypertext makes the process of constructing meaning partly visible. According to the author, "what becomes visible are the choices that the reader makes in following links, as each link followed indicates part of the reader's construction of the meaning of the text." Therefore, studying navigation behavior might help us not only to gain some insights about the strategies that readers employ while reading hypertext, but also to decide what kind of support they need in order to become mindful readers and gain a rich text understanding.

A study conducted by Stylianou (2003) aimed at investigating whether supporting sixth grade students to monitor and regulate their navigation behavior while reading from hypertext would lead to better navigation and learning. *Metanavigation support* in the form of prompts was provided to groups of students who collaboratively used a hypertext system called CoMPASS (Puntambekar, Stylianou, & Hübscher, 2003; Puntambekar, Stylianou, & Jin, 2001) to complete a design challenge while learning about simple machines. The metacognitive prompts aimed at encouraging students to understand the structure of the information space of CoMPASS (how the different text units were related to each other) so that they could make better navigational decisions. The study was conducted in a real classroom setting during the implementation of CoMPASS in sixth grade science classes. Examination of the navigation paths that students followed while using the hypertext environment suggested that providing metanavigation support enabled students to make coherent transitions among the text units and gain a rich understanding of science principles (Stylianou & Puntambekar, 2004).

More research studies need to be conducted to investigate what strategies are involved while reading from nonlinear resources such as hypertext documents. Little is known about how hypertext readers learn or manage learning while interacting with such flexible learning environments. It is not clear whether the metacognitive strategies that apply to traditional expository texts apply to learning from hypertext documents (Puntambekar & Stylianou, 2003). The literacy community needs to investigate whether the new genres of texts require specific skills for reading and writing.

The rapidly changing nature of literacy not only changes conceptualizations of literacy but also has important consequences for instruction, especially how teachers and students exploit new forms of literacy to enhance teaching and learning. In the following section, a particular emphasis is placed on how technology can be used as a tool to support community building, reflection, and teacher learning.

TEACHING AND LEARNING LITERACY THROUGH COMMUNITY BUILDING AND REFLECTION

Networked ICT includes powerful capabilities for information and communication, offering its users the chance to be members of new online cultures and communities on a global scale. Teachers and students are creating and sharing new visions for literacy and learning on the Internet in their classroom Web pages or by collaborating with people around the world through electronic mediums (Leu, in press). The Internet is a very powerful tool that can be used to foster collaborative learning. It enables teachers and students to collaborate and work with people from remote locations and exchange ideas.

Community Building and Web Literacy

An example of technology facilitating collaboration has been introduced by Scardamalia and Bereiter (1994) in a project called CSILE (Computer-Supported Intentional Learning Environment). In this project, technology is used to facilitate the knowledge-building discussion of a community of learners. Knowledge Forum, the second-generation version of CSILE, has been extensively used by a large number of schools from grades K-12 with great success (Scardamalia & Bereiter, 1999). Through the use of this electronic space, a communal database is created by students and their teachers. Students can enter text and graphic notes into the database on any topic their teacher has created. All students on the network can read the notes, and students may build on or comment on each other's ideas. Authors are notified when comments have been made or when changes in the database have occurred. As a teaching and learning medium, Knowledge Forum supports individual and collaborative knowledge building with the emphasis on the creation of knowledge through shared involvement and participation of learners, with the goal of enhancing the individual as well as the whole group's understanding. It enables students to interact actively with others in the process of constructing knowledge. Such pedagogical approaches to the use of technology in learning facilitate the development of collaboration and communication skills as well as Web literacies.

In Finland, an electronic learning space called Netro is also designed to support the development of Web literacy as learners read and write on the Web (Ahtikari & Eronen, 2004). This learning environment integrates Web literacy into university language and communication teaching aiming to raise university students' awareness on aspects of Web literacy. The overall framework of Netro is based on socioconstructivist views of learning, supporting social modes of learning, and collaborative knowledge-building processes (Scardamalia & Bereiter, 1994, 1999). In Netro, learners are encouraged to construct shared knowledge through various activities of reading and writing related to the multimodal content of the Web. The learners are also guided through a virtual tutor to reflect on their individual literacy practices such as how they read online and how they search information, as well as to think about how to use the information found, and how to critically approach and evaluate the multiple modes of representation on the Web.

In light of the developments brought about by rapid technology growth and use in education, language teachers have begun to realize the potential of technology to support reflective language teaching and learning. In the following section, we will briefly describe a technology and literacy project that will illustrate several of the issues raised in this chapter.

Reflection Through the Electronic European Language Portfolio (e-ELP)

The Common European Framework of Reference for Languages underlines the need for the European citizen to have a personal document describing all his/her linguistic experiences and certifications (http://culture2.coe.int/portfolio). It is similar to a language passport, called European Language Portfolio—ELP. So far several ELPs have already been validated and published in hardcopy format. Nevertheless these existing hardcopy editions present several limitations (e.g., difficulties in lifelong updating and maintenance). One of the main objectives of the e-ELP project is to overcome such barriers and take advantage of the affordances of technology in serving the needs of language teachers and students alike.

The traditional paper-based ELP consists of three main parts: language passport (presents at a glance the learners' language proficiencies), language biography (presents documents and information regarding the learner's language learning history), and the dossier (documents and samples of learner's work illustrating language proficiency). The ELP is a crucial tool for European mobility. The specific objectives for this project are:

a. to create an economic, multimedia, and easy-to-deliver e-ELP;
b. to produce an ELP specific for university language students in view of their entrance into the EU job world;
c. to make the ELP available in Italian, Greek, Spanish, and Swedish;
d. to give institutions a tool to monitor the students' language learning process; and
e. to promote a pedagogically correct use of the ELP.

As a pedagogical and assessment tool, portfolios have been used in a variety of ways in education and training. Portfolios interweave several aspects of the education process, including the curriculum, instruction, development of material, evaluation, and assessment of student learning. They have been used extensively in math, literacy, science, language, and art. Activities can be integrated with the use of portfolios in order to encourage students to review their own work; reflect on their learning; analyze their learning strategies, strengths, and weaknesses; and assess their participation within a class context (Calfee & Perfumo, 1993; Glazer & Brown, 1993).

With the growing use of the Internet in education, electronic portfolios have been used extensively by educators. Portfolios, in simple terms, are the collection of selected pieces of work by students. In the context of the ELP, the portfolio includes students' work, documentation, and certifications, as they illustrate the student's language proficiencies. One of the main focuses of a portfolio is the students' reflections on their own work which serve as a record of student learning and growth. When designed and used appropriately following sound pedagogical and design principles, portfolios can serve as a comprehensive document that can be shared with teachers, parents, the community, and future employers, and which demonstrates the learner's learning history and accomplishments. A digital portfolio can include video, audio, text, and images.

The concept of reflection can be transferred to the context of language learners. When learners reflect on their language learning and proficiencies, they engage in those processes essential for deep meaningful learning. Here, portfolios can play an important role. By constructing, rearranging, and evaluating their own digital language portfolio, students can reflect on what they know and have achieved thus far, identify strengths and weaknesses, and establish clear goals for improving their language expertise.

Teacher Professional Development

A key element for the successful implementation of the high-tech educational tools and communication resources in literacy teaching and learning is the professional development of teachers. Determining the most effective ways to support teachers in new electronic worlds is an important challenge for policy makers and educational leaders. Various attempts have been made so that professional development opportunities will be offered to teachers. Teachers must be appropriately trained in order to integrate the power of technology into their teaching. The right training will help teachers enhance their lessons with the meaningful use of interactive multimedia, graphical screen layouts, databases from search engines, and global resources (CEO Forum, 1999, 2000).

Currently, a lot of emphasis is put on enriching pre-service teacher programs so that they will enable prospective teachers to gain knowledge and skills of how to teach the new forms of literacy as a part of their initial certification programs. Case-based instruction is used to provide a rich, authentic context and allow students to observe the actions of experienced teachers. An example of such an attempt is CTELL, a five-year collaborative research project between four research universities in the United States which uses a case-study approach to improve pre-service teacher education and early literacy learning (Teale, Leu, Labbo, & Kinzer, 2002). CTELL (http://ctell.uconn.edu) uses video cases of K-3 classrooms demonstrating research-based principles of effective practice that are delivered over the Internet. The CTELL Project challenges educators to explore how to use Web-based presentation tools to add value to the instructional potential of case-based anchored instruction. Anchor cases (Cognition and Technology Group at Vanderbilt, 1990) involve explorations of classroom instructional scenarios that allow pre-service teachers to understand the kinds of problems teachers encounter and the knowledge they use in their decision making. Video cases become a common anchor for instructors and pre-service teachers to construct knowledge through discussions of theory, research, and practice regarding literacy. Pre-service teachers can explore a classroom by accessing a variety of instructional scenes, samples of children's work, as well as assessment data.

Networked technologies such as the Internet also make it possible to connect in-service teachers to each other, offering opportunities for mentoring, collaboration, formal and informal online learning. Through the Internet, teachers can have access to online professional development opportunities by attending online courses and seminars, follow-up consultations and mentoring, and collaborations with experts and peers. A supportive social structure is one of the key elements for successful online learning. Two notable initiatives that promote professional development of teachers and community building are the Teacher Professional Development Institute (TAPPED IN—http://ti2.sri.com/tappedin/) and the Inquiry Learning Forum (http://ilf.crlt.indiana.edu). What characterizes communities of practice is a shared commitment to a particular practice, which creates a network that enables and promotes knowledge sharing and professional development (Barab, MaKinster, & Scheckler, in press; Hoadley & Pea, 2002; Wenger, 1998; Wenger, McDermott, & Snyder, 2002). For instance, through TAPPED IN educators can participate in online courses, take their own students online, experiment with new ways to teach or conduct research, or participate in community-wide events. The Inquiry Learning Forum (ILF) is another example of teacher communities used for professional development in which mathematics and science teachers collaborate to create inquiry-based environments for their students (Barab et al., in press). The ILF provides opportunities to virtually visit classrooms that are available in the electronic space. It features a large video library of classroom episodes and enables teachers to observe, discuss, annotate, and reflect upon classroom practices as needed.

THE ROLE OF CRITICAL EMOTIONAL LITERACY

Issues of access to technology, equity, gender, and race are discussed within the context of critical education. An important aspect that promotes critical education is developing what we have called "critical emotional literacy" in the context of a "pedagogy of discomfort" (Boler, 1999; Boler & Zembylas, 2003; Zembylas & Boler, 2002; Zembylas & Vrasidas, 2004). A pedagogy of discomfort requires that individuals step outside of their comfort zones and recognize what and how one has been taught to see (or not to see). This pedagogical approach emphasizes two important and interdependent aspects of criticality and literacy in using ICT. The first aspect consists of the capability of questioning cherished beliefs and assumptions, thus opening up possibilities of thinking otherwise. The second aspect refers to the notion that criticality is not only a way of thinking but also a way of feeling and being, i.e., it is a practice, a way of life.

These two aspects of criticality in using ICT in education emphasize a significant but often ignored dimension of literacy in the new emerging digital era. What becomes a central dimension of literacy is that to engage in such a practice, it is not simply a matter of individual abilities or dispositions (Burbules & Berk, 1999), but it requires moving against prevailing valued assumptions, values, and beliefs. It is possible that not everyone can or will become a critically literate individual in using ICT. Nevertheless, a critical education in using ICT recognizes the multidimensional relationship between feeling and thinking, while seeking and promoting new pedagogies that take into account the evolving notions of literacy.

An important step in developing a pedagogy of discomfort includes a realization of the production of new narratives about teaching and learning as a result of the advent of cyberspace. Mestrovic (1997) claims about the "McDonaldization of emotions" may necessitate a fundamental reconsideration of existing pedagogical strategies that educate for a changed understanding of the relationship between reason and the emotions in education (p. xi). While acknowledging the limitations of the use of ICT in education, we agree with Rice and Burbules (1992) that we can educate for critical sensitivities and critical literacy in the use of ICT. Since ICTs are transforming education, key challenges for critical education involve how to analyze such transformations and how to devise conceptual tools and strategies to make use of ICTs that empower traditionally marginalized groups and individuals struggling for democracy and social justice.

Finally, another aspect of developing criticality in the context of a pedagogy of discomfort is acknowledging that criticality is not only "a way of thinking" but also "a way of feeling and being." We are using the word "thinking" in the widest sense possible. It includes believing, inferring, explaining, judging, and interpreting (Gratton, 2001). Criticality is founded on both emotions and reasons (Barbalet, 1998). Thus, critical emotional literacy provides the tools with which to assess the beliefs one holds. This use of criticality in the context of a pedagogy of discomfort empowers educators and their students to problematize their discomfort and emotional challenges and explore their ethical responsibilities.

FINAL THOUGHTS

In this chapter we have suggested that the rapid developments in technology are challenging the traditional concept of literacy and transforming the most basic components of literacy (texts, reading, and writing), the teaching and learning of becoming literate, as well as the development of critical emotional literacy. The unique characteristics of electronic texts present new opportunities as well as challenges for reading and writing and point towards interesting research directions. The literacy community is presented with a major challenge: investigate how learners manage and learn while interacting with networked technologies, and identify what strategies are used to facilitate literacy learning. Another central issue is how to exploit the increasingly powerful, complex, and continually changing technologies for information and communication to support literacy instruction. We presented specific research projects that aim to promote community building to support students as well as teachers to learn from others. Social learning skills are becoming central to literacy instruction, especially as new technologies for information and communication allow teachers and their students to make new connections and view the world in more powerful ways. The use of ICT technologies also has important implications for the development of critical emotional literacy. We argued that it is possible that not everyone can or will become a critically literate individual in using ICT. However, a critical education in using ICT recognizes the multidimensional relationship between feeling and thinking, while seeking and promoting new pedagogies that take into account the evolving notions of literacy. We are confident that the literacy community, both in terms of research as well as practice, is willing to embrace the opportunities of the continually changing technologies of literacies and address the challenges presented with the evolution of literacy.

REFERENCES

Ahtikari, J., & Eronen, S. (2004). *On a journey towards Web literacy—The electronic learning space Netro*. Unpublished Pro Gradu Thesis, University of Jyväskylä, Finland.

Barab, S. H., MaKinster, J. G., & Scheckler, R. (in press). Designing system dualities: Characterizing an online professional development community. In S. H. Barab, S. A. Kling, & J. Gray (Eds.), *Designing for virtual communities in the service of learning*. Cambridge, MA: Cambridge University Press.

Barbalet, J. (1998). *Emotion, social theory and social structure*. Cambridge, MA: Cambridge University Press.

Boler, M. (1999). *Feeling power: Emotions and education.* New York: Routledge.

Boler, M., & Zembylas, M. (2003). Discomforting truths: The emotional terrain of understanding difference. In P. Tryfonas (Ed.), *Pedagogies of difference: Rethinking education for social change* (pp. 110-136). New York; London: Routledge.

Bolter, J. D. (1998). Hypertext and the question of visual literary. In D. Reinking, M. McKenna, L. Labbo, & R. Kiefer (Eds.), *Handbook of literacy and technology* (pp. 3-14). Mahwah, NJ: Lawrence Erlbaum.

Bolter, J. D. (2001). *Writing space: Computers, hypertext, and the remediation of print.* Mahwah, NJ: Lawrence Erlbaum.

Brown, J. S. (2000, March/April). Growing up digital: How the Web changes work, education, and the ways people learn. *Change*, 10-20.

Burbules, N.C., & Berk, R. (1999) Critical thinking and critical pedagogy: Relations, differences, and limits. In T. Popkewitz & L. Fendler (Eds.), *Critical theories in education: Changing terrains of knowledge and politics* (pp. 45-65). New York; London: Routledge.

Calfee, R. C., & Perfumo, P. (1993). Student portfolios: Opportunities for a revolution in assessment. *Journal of Reading, 36,* 532-537.

CEO Forum. (1999). *Professional development: A link to better learning.* Retrieved March 1, 1999, from http://www.ceoforum.org/reports.cfm?RID=2

CEO Forum. (2000). *School technology and readiness.* Retrieved March 1, 1999, from http://www.ceoforum.org/reports.cfm?RID=4

Charney, D. (1987). Comprehending non-linear text: The role of discourse cues and reading strategies. In J. Smith & F. Halasz (Eds.), *Hypertext '87 Proceedings* (pp.109-120). New York: Association for Computing Machinery.

Charney, D. (1994). The impact of hypertext on processes of reading and writing. In S. J. Hilligoss, & C. L. Selfe (Eds.), *Literacy and computers* (pp. 238-263). New York: Modern Language Association.

Conklin J. (1987), Hypertext: An introduction and survey. *IEEE Computer, 20*(9), 17-41.

CTGV. (1990). Anchored instruction and its relationship to situated cognition. *Educational Researcher, 19*(6), 2-10.

Dillon, A., & Vaughan, M. (1997). "It's the journey and the destination": Shape and the emergent property of genre in evaluating digital documents. *New Review of Multimedia and Hypermedia, 3,* 91-106.

Eklund J., Sawers, J., & Zeiliger R. (1999). NESTOR Navigator: A tool for the collaborative construction of knowledge through constructive navigation. In R. Debreceny & A. Ellis (Eds.), *Proceedings of the 5th Australian World Wide Web Conference* (AUSWEB99) (pp. 396-408). Lismore: Southern Cross University Press.

Glazer, M. S., & Brown, C. S. (1993). *Portfolios and beyond: Collaborative assessment in reading and writing.* Norwood, MA: Christopher-Gordon.

Gratton, C. (2001). Critical thinking and emotional well-being. *Inquiry: Critical Thinking Across the Disciplines, 20*(3), 39-51.

Guzdial, M. (1999, April). Teacher and student authoring on the Web for shifting agency. How can CSCL (computer-supported collaborative learning) change classroom culture and patterns of interaction among participants? *Proceedings of the Annual Conference of the American Educational Research Association*, Montreal, Canada.

Hegarty, M., Quilici, J., Narayanan, N. H., Holmquist, S., & Moreno, R. (1998). Designing hypermedia manuals to explain how machines work: Lessons from education of a theory-based design. *Proceedings of the ACM SIGCHI Conference on Human Factors in Computing Systems (CHI'98),* Los Angeles.

Hoadley, C. M., & Pea, R. D. (2002). Finding the ties that bind: Tools in support of a knowledge-building community. In K. A. Renninger & W. Shumar (Eds.), *Building virtual com-*

munities: Learning and change in cyberspace (pp. 321-354). London: Cambridge University Press.

Jacobson, M. J., Maouri, C., Mishra, P., & Kolar, C. (1996). Learning with hypertext learning environments: Theory, design, and research. *Journal of Educational Multimedia and Hypermedia, 5*(3/4), 239-281.

Jul, S., & Furnas, G. (1997). Navigation in electronic worlds—A CHI'97 workshop. *SIGCHI Bulletin, 29*(4), 44-49.

Kintsch, W., & van Dijk, T. A. (1978). Toward a model of text comprehension and production. *Psychological Review, 85*(5), 363-394.

Landow, G. P. (1992). *Hypertext: The convergence of contemporary critical theory and technology.* Baltimore: Johns Hopkins University Press.

Lawless, K.A., & Brown, S.W. (1997). Interacting with multimedia environments: Control, feedback and navigational issues. *Instructional Design, 25*, 117-131.

Leu, D. J. (2000). Continuously changing technologies and envisionments for literacy: Deictic consequences for literacy education in an information age. In M. Kamil, P. Mosenthal, P. D. Pearson, & R. Barr (Eds.), *Handbook of reading research, volume III* (pp. 743-770). Mahwah, NJ: Lawrence Erlbaum.

Leu, D. J. Jr. (in press). The new literacies: Research on reading instruction with the Internet and other digital technologies. In J. Samuels & A. E. Farstrup (Eds.), *What research has to say about reading instruction.* Newark, DE: International Reading Association.

Marchionini, G. (1988). Hypermedia and learning: Freedom and chaos. *Educational Technology, 28*(11), 8-12.

Marchionini, G. (1995). *Information seeking in electronic environments.* New York: Cambridge University Press.

Mestrovic, S. (1997) *Postemotional society.* London: Sage Publications.

Puntambekar, S., & Stylianou, A. (2003). Designing metacognitive support for learning from hypertext: What factors come into play? *Proceedings of the Workshop on Metacognition and Self-Regulation in Learning with Metacognitive Tools at the International Conference on Artificial Intelligence in Education* (AIED), Sydney, Australia.

Puntambekar, S., Stylianou, A., & Hübscher, R. (2003). Improving navigation and learning in hypertext environments with navigable concept maps. *Human Computer Interaction, 18*(4), 395-426.

Puntambekar, S., Stylianou, A., & Jin, Q. (2001). Visualization and external representations in educational hypertext systems. In J. D. Moore, C. L. Redfield, & W. L. Johnson (Eds.), *Artificial intelligence in education, AI-ED in the wired and wireless world* (pp. 13-22). The Netherlands: IOS Press.

Reinking, D. (1998). Synthesizing technological transformations of literacy in a post-typographic world. In D. Reinking, M. McKenna, L. D. Labbo, & R. Kieffer (Eds.), *Handbook of literacy and technology: Transformations in a post-typographic world.* Mahwah, NJ: Lawrence Erlbaum.

Rice, S., & Burbules, N. (1992). *Communicative virtues and educational relations.* Retrieved May 22, 2004, from http://www.ed.uiuc.edu/pes92_docs/rice_burbules.htm

Scardamalia, M., & Bereiter, C. (1994). Computer support for knowledge-building communities. *Journal of the Learning Sciences, 3*(3), 265-283.

Scardamalia, M., & Bereiter, C. (1999). Schools as knowledge-building organizations. In D. Keating & C. Hertzman (Eds.), *Today's children, tomorrow's society: The developmental health and wealth of nations* (pp. 274-289). New York: Guilford.

Snyder, I. (1998). Beyond the hype: Reassessing hypertext. In I. Snyder (Ed.), *Page to screen* (pp. 125-143). New York: Routledge.

Stylianou, A. (2003). *How do students navigate and learn from nonlinear science texts: Can metanavigation support promote science learning?* Unpublished Doctoral Dissertation, University of Connecticut, Storrs, USA.

Stylianou, A., & Puntambekar, S. (2004). Understanding the role of metacognition while reading from nonlinear resources. In Y. B. Kafai, W. A. Sandoval, N. Enyedy, A. S. Nixon, & F. Herrera (Eds.), *Proceedings of 6th International Conference of the Learning Sciences* (pp. 529-536). Mahwah, NJ: Lawrence Erlbaum.

Teale, W. H., Leu, D. J., Labbo, L. D., & Kinzer, C. (2002). The CTELL project: New ways technology can help educate tomorrow's reading teachers. *The Reading Teacher, 55*(7), 654-659.

Thüring, M., Hannemann, J., & Haake, J. M. (1995). Designing for comprehension: A cognitive approach to hypermedia development. *Communications of the ACM, 38*(8), 57-66.

Wenger, E. (1998). *Communities of practice. Learning, meaning and identity*. Cambridge: Cambridge University Press.

Wenger, E., McDermott, R., & Snyder, W. M. (2002). *Cultivating communities of practice: A guide to managing knowledge*. Cambridge, MA: Harvard Business School Press.

Zembylas, M., & Boler, M. (2002). *On the spirit of patriotism: Challenges of a "pedagogy of discomfort."* Retrieved August 27, 2002, from http://www.tcrecord.org

Zembylas, M., & Vrasidas, C. (2004). Emotion, reason and information/communication technologies in education: Some issues in a postemotional society. *E-Learning Online Journal, 1,* 105-127. Retrieved from www.triangle.co.uk/ELEA

Zembylas, M., Vrasidas, C., & McIsaac, M. S. (2002). Of nomads, polyglots, and global villagers: Globalization, information technologies, and critical education online. In C. Vrasidas & G. V. Glass (Eds.), *Current perspectives in applied information technologies: Distance education and distributed learning* (pp. 201-224). Greenwich, CT: Information Age Publishing.

Chapter III
Technological Literacy, Perspectives, Standards, and Skills in the USA

Jared V. Berrett
Brigham Young University, USA

ABSTRACT

This chapter is written to provide a framework for understanding perspectives on technology literacy and how it might be taught in the K-12 setting. Numerous U.S. governmental reports, initiatives, definitions, and professional standards are reviewed. Though there are many fields interested in technological literacy, the argument is made that technology educators may prove to be an excellent resource in meeting the challenge of creating a technologically literate citizenry. A case study of exemplary practice is introduced as a point of investigating how one technology teacher is being successful in teaching technological literacy. It is up to the reader and all those concerned with technological literacy to continue to evaluate and search for best teaching practices for teaching technological literacy in K-12 schools.

INTRODUCTION

Throughout the history of civilization, technology has been an enabling factor in societal change. From a hunter/gatherer to an agrarian-based to an industrial-driven culture, technology advancement has played an essential role (Wright, 1995). Currently our world appears to have moved out of the industrial period into what many are calling the Information Age, where the incredible growth and worldwide acceptance of information and communications technologies seems to be the icon of our era. Though some may argue that the changes society face now are no more significant than those faced in the past, others would suggest that society is facing unprecedented rates of change (Cohen, DeLong, & Zysman, 2000). Of particular concern to this book is the effect these changes are having on the nature of literacy and its relation to technology. This chapter will review numerous perspectives and publications on this topic in an attempt to help the reader formulate an idea of what technological literacy is and to begin thinking about how education in grades K-12 might respond to these changes.

A CHANGING CONTEXT

Innovations in information technology have become so influential in the U.S. that the Under Secretary of Commerce of the United States claimed "the digital economy and digital society are no longer 'emerging.' They are here" (U.S. Department of Commerce, 2000). As the digital economy and the information age continue to grow, individuals, governments, businesses, and organizations are being forced to consider new strategies and guiding principles that focus on knowledge and information to be successful (Drucker, 1992). New ways of thinking about knowledge have also emerged, including a broader acceptance of multiple intelligences (Gardner, 1999) and emotional intelligences (Goleman, 1995) as important elements of an educated workforce. Increasingly, corporations are realizing that in order to adjust and excel in this global economy, a new focus must be directed toward human capital and lifelong learning.

Advances in technologies are not only impacting our economy and business, but they are changing the social fabric of our homes and our schools. Today, more than half of all Americans are using the Internet and more than 65% of the U.S. population uses computers (U.S. Department of Commerce, 2003). It is not uncommon for people to manage investments, make purchases, pay bills, communicate with family and friends, and research services and products online. These are things that would be unheard of 25 years ago because personal computers were not prevalent then.

Though there has been amazing growth in access to technology globally, there is still a serious concern over the digital divide. This gap however is progressively shrinking. In Colombia, for instance, the scientific and technological developments have gained such strength and presence that it can be called a "technological culture," for which the educational system must prepare citizens, making them conscious of its existence and preparing them to approach it (Pena, 1992). Other significant changes in Central and Eastern Europe, including the fall of communist governments, has likewise caused change, need, and opportunity for educating a technologically literate citizenry.

The technological boom our world has been experiencing could be considered in many ways a societal revolution. Such a technological revolution is surely not the first, nor will it be the last. During the Industrial Revolution, John Dewey described the need for schools to keep current with the changing social impacts of a changing and technologically advancing society. "It is radical conditions which have changed, and only equally radical changes in education suffices" (1956, p. 12). If Dewey felt that the changes of his time were so radical and rapid that they required equally radical and rapid changes in the classroom, what should our schools be doing to adapt to the changing nature of our society today? How will technology affect our future? How should we be educating out children differently than we did 5, 10, or 50 years ago?

A CALL FOR CHANGE IN THE U.S.

In the United States, like in other countries, many reports and concerns have given rise to the need for improvement and reform in the education system. In considering several of these reports, it appears that at least part of the answer to education's challenges includes preparing the future workforce to contribute to an advancing technological society. This is surely an important part of a child's education, but there are other things that must be considered as well. For instance, ensuring that students will be able to make effective decisions based on the potential impacts of technologies will be important. Protecting the social fabric of community and family must also be considered as it pertains to being a contributing citizen in a democracy. Other issues might also be considered, but regardless of the perspective, most people would agree that technological advances are forcing our society to reinvent itself, and all members of society should have the opportunity to engage, interact, and live with technology. President Clinton and Al Gore described this challenge as the "Technology Literacy Challenge" (U.S. Department of Education, 1996). They wanted to "energize the nation to make young Americans technologically literate by the turn of the century" (Clinton & Gore, 2000).

A Nation at Risk

Unprecedented attention was drawn to the status of the nation's education system and the need for educational reform when the report *A Nation at Risk* was handed to President Reagan in 1983 (Seaborg, 1991). Among several key issues identified in this report was the concern that American students' general education experience need to prepare them in a wide range of technologies to be globally competitive.

We live among determined, well-educated, and strongly motivated competitors. We compete with them for international standing and markets—not only with products, but also with ideas of our laboratories and neighborhood workshops; knowledge, learning, information, and skilled intelligence are the new raw materials of international commerce. (National Commission on Excellence in Education, 1983, p. 2)

After studying student achievement in America's schools, Paul Hurd concluded: "We are raising a new generation of Americans that is scientifically and technologically illiterate" (National Commission on Excellence in Education, 1983, p. 4). Numerous task forces, reports, studies, and opinion papers have called and continue to call for change in the face of technological advancement and global competition.

Goals 2000

In 1989, President Bush and nearly every governor of the nation agreed to six national goals in education to be achieved by the year 2000 (Seaborg, 1991). This list of goals was disseminated during his State of the Union Address in 1990. Goal #5 reads:

By the year 2000, every adult American will be literate and possess the knowledge and skills necessary to compete in a global economy and exercise the rights and responsibilities of citizenship. (Goals 2000, 1994)

SCANS

Soon after the Goals 2000 objectives were made public in 1990, the Secretary's Commission on Achieving Necessary Skills (SCANS) report was released by the U.S. Department of Labor. This seminal report provided further insight into how the goals espoused in Goals 2000 could to be attained and again drew considerable national attention to education reformation issues in light of a changing society. After spending over a year talking to business owners, public employers, managers, union officials, and other workers whether in an assembly line or at a desk, the SCANS report determined that students needed a new set of competencies consisting of both skill and personal qualities to be applied in the emerging high performance work settings (SCANS, 1991). Schools therefore would be challenged to become high-performance organizations themselves as they prepared their students for their future (see Figure 1). Technology is listed as one of the five essential skills, suggesting the need for students to improve their technological competence by learning how to select, apply, maintain, and troubleshoot technology (p. 2).

Double Helix

The Double Helix of Education & the Economy (Berryman & Bailey, 1992) suggested that K-12, college, and corporate training sectors must address the changes in society specifically through better teaching practice.

Our schools routinely and profoundly violate what we know about how people learn most effectively and the conditions under which they apply their knowledge appropriately to new situations. (Berryman & Bailey, 1992, p. 3)

Common assumptions, learning environments, and outcomes just "don't work" because they do not capitalize on "the point that human beings are inquisitive, sense making animals who learn best when they are fully and actively engaged in solving problems that mean something to them" (Berryman & Bailey, 1992, p. 4). Considering what changes are needed in

Figure 1. Workplace know-how skills established by the SCANS report of 1991

WORKPLACE COMPETENCIES—Effective workers can productively use:

- *Resources:* They know how to allocate time, money, materials, space, and staff.
- *Interpersonal Skills:* They can work on teams, teach others, serve customers, lead, negotiate, and work well with people from culturally diverse backgrounds.
- *Information:* They can acquire and evaluate data, organize and maintain files, interpret and communicate, and use computers to process information.
- *Systems:* They understand social, organizational, and technological systems; they can monitor and correct performance; and they can design or improve systems.
- *Technology:* They can select equipment and tools, apply technology to specific tasks, and maintain and troubleshoot equipment.

FOUNDATION SKILLS—Competent workers in the high-performance workplace need:

- *Basic Skills:* Reading, writing, arithmetic and mathematics, speaking and listening.
- *Thinking Skills:* The ability to learn, to reason, to think creatively, to make decisions, and to solve problems.
- *Personal Qualities:* Individual responsibility, self-esteem and self-management, sociability, and integrity.

education in the 21st century, the idea of "how to teach" as brought out by this report must truly be addressed parallel to the concern with "what to teach" in an increasingly information-driven society.

Technological Literacy Challenge

On February 15, 1996, President Clinton and Vice President Gore announced the Technology Literacy Challenge where they envisioned a 21st Century where all students are technologically literate. "Success as a nation will depend substantially on our student's ability to acquire the skills and knowledge necessary for high-technology work and informed citizenship" (p. 4). Four specific goals were outlined to help achieve technological literacy for all:

1. All teachers in the nation will have the training and support they need to help students learn using computers and the information superhighway.
   a. Upgrading teacher training is key to integrating technology into the classroom and to increasing student learning.
2. All teacher training is key to integrating technology into the classroom and to increasing student learning.
   a. Computers become effective instructional tools only if they are readily accessible by students and teachers.
3. Every classroom will be connected to the information superhighway.
   a. Connections to networks, especially the Internet, multiply the power and usefulness of computers as learning tools by putting the best libraries, museums, and other research and cultural resources at our students' and teachers' fingertips.
4. Effective software and online learning resources will be an integral part of every school's curriculum.
   a. Software and online learning resources can increase student's learning opportunities, but they must be high quality, engaging, and directly related to the school's curriculum.

It is clear that this view of technology is narrowly focused on computers, skills, and access within the schools. Considerable emphasis within the report is placed on technology's role in teaching and learning. The report (see Figure 2) suggested that technology can play a fundamental role in effective pedagogic models of learning. These are some of the same issues identified in the Double Helix report.

Figure 2. The role of technology in education established by the Technological Literacy Challenge in 1996

Technology's Role in Education
• Greater attention is given to the acquisition of higher-order thinking and problem-solving skills with less emphasis on the assimilation of a large body of isolated facts. • Basic skills are learned not in isolation, but in the course of understating (often in a collaborative basis) higher-level "real-world" tasks whose execution requires the integration of a number of such skills. • Information resources are made available to be accessed by the student at a point when they actually become useful in executing the particular task at hand. • Fewer topics may be covered than is the case within the typical traditional curriculum, but these topics are often explored in greater depth. • The student assumes a central role as the active architect of his or her own knowledge and skills, rather than passively absorbing information proffered by the teacher.

The report also suggested that the technology can be a powerful tool for teachers who may use computers and computer networks to monitor and assess the progress of their students, maintain portfolios of student work, prepare materials for use in the classroom, communicate with others, exchange information and consult with others including experts in a variety of fields, access remote information, and further expand their own knowledge and professional capabilities.

Summary: Need for Change

From calls for national educational reform in the United States to the need for more access to information technologies, there is considerable evidence that schools must help students become technologically literate. Many more national and governmental reports have been commissioned, private research studies conducted, and position papers written on the need for such a change and what specific efforts it may entail. Of particular interest in the face of a technologically advancing global society is that "workers in the next century will require not just a larger set of facts or a larger repertoire of specific skills, but the capacity to readily acquire new knowledge, to solve new problems, and to employ creativity and critical thinking in the design of new approaches to existing problems" (PCAST, 1997).

PERSPECTIVES ON LITERACY

This section will focus on how technological literacy can be understood and defined. Central to the definition is the understanding of what literacy is and how it is changing. Therefore, literacy in its simplest form is first defined. Other related terms often confused with technological literacy are then explored, including computer literacy, media literacy, and information literacy. Finally, technological literacy is investigated by first exploring what the term "technology" means, and then how multiple perspectives have made it difficult to understand and define the term.

Literacy

There seems to be an increasing convergence of ideas packed into similar literacy terms making each one difficult to distinguish and define. Part of the confusion is inherent in the foundation term of "literacy." Many have viewed literacy in the past simply as the ability to read and write (Teale & Sulzby, 1986). However, in a recent book entitled *Literacy*, Cooper (1997) suggests that literacy is not just the ability to read and write, but may include the ability to listen, speak, and communicate through technology. With this perspective, literacy is better defined as "the ability to communicate in real-world situations, which involves the abilities of individuals to read, write, speak, listen, view, and think" (Cooper, 1997, p. 7). This definition and view seem reasonable and could be considered inclusive of the technological innovation advancements in technology described as an impetus for changing the views of literacy (Tyner, 1998). These changes appear to support the apparent need for enhancing the skills of all to be able to function in this emerging society.

Computer Literacy

Computer literacy has been defined in diverse ways. Some have described it as a set of computer skills, information on how they are used, and knowledge of their effects, while others urge a less structured approach allowing students to learn about computers through writing, drawing, or composing music (Ragsdale, 1988, p. 160). Numerous courses have been created in schools, colleges, and universities across the nation in order to help students become computer literate. In the late 1980s courses could be found that covered such concepts as the procedure for starting a computer, operating systems like DOS, Windows 95, and the fundamentals of key software like WordPerfect, Quatro Pro, e-mail, and the Internet. (Manini & Cervantes, 1998). Today, legislation from State governing bodies are requiring schools and school districts to teach basic computer literacy classes in an attempt to help students become more literate (USOE, 2004).

Computer literacy should perhaps be considered as an integral part of a general "literacy" definition, however it is limited in perspective and application. This view of literacy typically deals with hardware and software with the intent to increase our understanding of what the machine can and cannot do for us (Horton, 1983). Therefore, one of the challenges to the philosophy of computer literacy as "literacy" is at the very center of the technology itself. Computer software and hardware are changing constantly, thereby making one's training obsolete almost before it has finished.

Media Literacy

Brunner and Tally (1999) use the term media literacy to describe skills that are necessary for teachers that include knowing when and how to use new multimedia technologies. They also claim that teachers must have some basic understanding of how media are constructed, how they are distributed, who owns them, and how they express the values and perspective of their authors in the way they are made. Beyond just teachers' engagement with technology, some of the literature associates media literacy with a broader perspective of visual literacies for all. From this perspective it would be important for teachers and students alike to be able to decode messages, words, and images (Palazon, 2000) which represent different realities around them. Though similar to computer literacy in that students need to be familiar with the technologies that create and deliver media, it differs from it with a focus on the media-saturated world to encourage learners to be producers of effective media messages and critical consumers of ideas and information (Abdullah, 2000).

Information Literacy

Information literacy, as opposed to computer literacy or media literacy, means to raise the level of awareness of individuals and enterprises to the knowledge explosion and how machine-aided handling systems can help identify, access, and obtain data, documents, and literature needed for problem solving and decision making (Horton, 1983). Information literacy has been described simply as "the ability to recognize when information is needed and the ability to locate, evaluate, and use effectively the needed information" (American Library Association, 1989) and may be thought of as inclusive of media literacy. In a recent review of research on the topic, Kuhlthau (1987) states:

Information literacy is closely tied to functional literacy. It involves the ability to read and use information essential for everyday life. It also involves recognizing an information need and seeking information to make informed decisions. Information literacy requires abilities to manage complex masses of information generated by computers and mass media, and to learn throughout life as technical and social changes demand new skills and knowledge. (Kuhlthau, 1987, p. 2)

Information literacy has become of increasing interest to national accreditation agencies, schools, colleges, and universities. (Spitzer, Eisenberg, & Lowe, 1999)

Standards have been created by many states, and recently the American Association of School Librarians (AASL) and Association for Educational Communications and Technology (AECT) have published some in a document called *Information Power: Building Partnerships for Learning* (1998) (see Table 1).

The first three standards focus on accessing, evaluating, and using information which is consistent with the general definition given by the ALA in 1989 and perhaps inclusive of a computer literacy approach. However these standards include two other categories, including independent learning and social responsibility, that are more socially centric rather than technical in nature. Borders of information literacy are blurring from some of the original inceptions and simple definitions of information literacy, evidenced by these standards which include social and independent learning as a central role. This broadening perspective is one that is gaining popularity as will be seen throughout the remainder of the review.

TECHNOLOGY LITERACY

Technological literacy holds the broadest conception of literacy of the current terms used here to describe new skills needed for our advancing society. For many, whether you are in a school or on the street, the first thing that comes to mind when asked "what is technology" is the term "computer" (Rose, Gallup, Dugger, & Starkweather, 2004). Yet, the computer is just one technological innovation that is impacting our society. *Webster's Ninth New*

Table 1. AASL/AECT information literacy standards

Number	Standard
Information Literacy: The student who is information literate…	
1.	accesses information efficiently and effectively.
2.	evaluates information critically and competently.
3.	uses information accurately and creatively.
Independent Learning: The student who is an independent learner is information literate and…	
4.	pursues information related to personal interests.
5.	appreciates literature and other creative expressions of information.
6.	strives for excellence in information seeking and knowledge generation.
Social Responsibility: The student who contributes positively to the learning community…	
7.	and to society is information literate and recognizes the importance of information to a democratic society.
8.	and to society is information literate and practices ethical behavior in regard to information and information technology.
9.	and to society is information literate and participates effectively in groups to pursue and generate information.

Collegiate Dictionary (1986) defines technology as "applied science," "a scientific method of achieving a practical purpose," and "the totality of the means employed to provide objects necessary for human sustenance and comfort." This definition broadens the discussion considerably to all things human made. Aristotle's discussion on *techne* and *logos* (the Greek root of the word technology) reveals that technology is closely linked to science—not just applied science (Barnes, 1990). Paul DeVore states:

It is proposed that technology is the science that deals with the creation, utilization, and behavior of adaptive systems including tools, machines, materials, techniques and technical means and the behavior of these elements and systems in relation to human beings, society and the environment. (1987, p. 8)

Perhaps combining the view of a scientific method and means for satisfying human needs would be a good way to consider the term.

The National Science Foundation (1992) promotes a supportive view that "technology is not an instrument, but a field of study. It involves the application of learned principles to specific, tangible situations" (p. 3). The Science for All Americans document summarizes the perspective that this chapter will promote by stating:

In the broadest sense, technology extends our abilities to change the world: to cut, shape, and put together materials; to move things from one place to another; to reach farther with our hands, voices, and senses. We use technology to try to change the world to suit us better. The changes may relate to survival needs such as food, shelter, or defense, or they may relate to human aspirations such as knowledge, art, or control. But the results of changing the world are often complicated and unpredictable. They can include unexpected benefits, unexpected costs, and unexpected risks—any of which may fall on different social groups at different times. Anticipating the effects of technology is therefore as important as advancing its capabilities. (Rutherford & Ahlgren, 1990)

This broad definition is distinct and compelling because technology affects society every second of every day in a variety of ways—not just through computers. It is important for students to recognize the social responsibility we have for the technologies we develop as a society and individually. It is empowering to teach them that they can impact our society by responsibly using, creating, or designing technology. Not only will this lead to technologically advanced students who can technically compete in today's economy, but they can also be more productive and aware citizens and contributors to our society.

Considering the broad view of technology and the expanding view of literacy, technological literacy is extremely complex, broad, and multifaceted. In Gagel's (1995) dissertation written to clarify the understanding of technological literacy, he states: "Technological literacy draws on the essential themes of both literacy and technology; that is, literacy's knowledgeability and cognitive performance (critical thinking), and technology's knowledge, invention, and praxis" (p. 277). By comparing one's performance to accepted norms of a certain cultural tradition at any given historical moment, one can assess to what extent each of these dimensions are necessary for being considered literate. Technological literacy then must be viewed as fluid, as opposed to constant (Gagel, 1997).

In looking at the word "technology" as it relates to learning and literacy, Chip Bruce identifies three kinds of reasons for the use of technology that are helpful in framing the definition of technological literacy to be used in this chapter. First, there is a need to learn to *use* technology. This would be the focus of those who are intent on using technological literacy to address the economic needs of society. Second, a need is present to be able to learn *through* new technologies. For those using this argument, it is important for individuals to be able to organize, synthesize, and apply information available through the vast resources now available. Third, learning *about* new technologies is important to ensure their use is governed well in a democratic society (Bruce, 2003).

Summary: Technological Literacy

Technological literacy therefore as a construct or concept could be considered an expansion of science in that it uses knowledge of laws in meaningful ways to improve the human endeavor. The real challenge for teachers and students alike then is the fact that technological literacy is constantly in flux. Perhaps by considering our need to use technology, learn through technology, and learn about technology, teachers will be able to help students become more literate in technology.

TECHNOLOGICAL LITERACY STANDARDS

Perhaps the strongest push for technology in today's schools is evidenced by professional organizations producing standards to address what skills and competencies a technologically literate person should possess. Three organizations—the International Society for Technology in Education (ISTE), the American Association for the Advancement of Science (AAAS), and the International Technology Education Association (ITEA)—have had initiatives to write and formally publish these works. By considering these standards, a further in-depth operational understanding of what it means to be technologically literate will be established.

ISTE

The International Society for Technology in Education (ISTE) is dedicated to the improvement of education through the integration of computer-based technology into the curriculum (ISTE, 2000). They have been involved for the past several years in a project to develop National Educational Technology Standards (NETS).

The primary goal of the ISTE NETS Project is to enable stakeholders in Pre K-12 education to develop national standards for educational uses of technology that facilitate school improvement in the United States. The NETS Project will work to define standards for students, integrating curriculum technology, technology support, and standards for student assessment and evaluation of technology use (ISTE, 2001a).

The overall project includes a set of standards for students (NETSS) (see Figure 3), another set for teachers (NETST) developed with the support of the Department of Education and the Preparing Tomorrows Teachers to use Technology Project (PT3) (ISTE, 2001b), and standards for School Administrators (TSSA) (ISTE, 2001c).

The standards for students (NETSS) are broken into six separate broad categories. Within

Figure 3. National Educational Technology Standards for Students and Teachers (NETSS & NETST) written by the International Society for Technology in Education in 1999 and 2000

Student Standards
1. Basic operations and concepts • Students demonstrate sound understanding of the nature and operation of technology systems. • Students are proficient in the use of technology. 2. Social, ethical, and human issues • Students understand the ethical, cultural, and societal issues related to technology. • Students practice responsible use of technology systems, information, and software. • Students develop positive attitudes toward technology uses that support lifelong learning, collaboration, personal pursuits, and productivity. 3. Technology productivity tools • Students use technology tools to enhance learning, increase productivity, and promote creativity. • Students use productivity tools to collaborate in constructing technology-enhanced models, preparing publications, and producing other creative works. 4. Technology communications tools • Students use telecommunications to collaborate, publish, and interact with peers, experts, and other audiences. • Students use a variety of media and formats to communicate information and ideas effectively to multiple audiences. 5. Technology research tools • Students use technology to locate, evaluate, and collect information from a variety of sources. • Students use technology tools to process data and report results. • Students evaluate and select information resources and technological innovations based on appropriateness to specific tasks. 6. Technology problem-solving and decision-making tools • Students use technology resources for solving problems and making informed decisions. • Students employ technology in the development of strategies for solving problems in the real world.

each category the standards are defined. Standard one describes the "basics" or what I described earlier in computer literacy skills. Standard two is unique and interesting as it deals with social impact issues similar to the science and technology standards included in the AAAS Project. The remainder of the standards focus on increasing knowledge about using computer-related technology as an instructional resource or "tool."

AAAS Project 2061

The *National Science Education Standards: Observe, Interact, Change, Learn* (1996) provide a framework for technological literacy within the overarching view of science literacy outlined in the Science for All Americans Project 2061 (AAAS).

As used in the Standards [sic], the central distinguishing characteristic between science and technology is a difference in goal: The goal of science is to understand the natural world, and the goal of technology is to make modifications in the world to meet human needs. (AAAS, 1996, p. 24)

The technology-related standards are framed around two general categories (abilities of technological design, and understanding science and technology) and organized into three different grade levels (K-4, 5-8, and 9-12). Within

Table 2. National Science Education Standards (NSES) content standard E: Science and technology

	Standard	
Number	... abilities of technological design	... understandings about science and technology
	As a result of activities in grades K-4, students should develop the skills to...*	
1.	Identify a simple problem.	Know that science is a way to answer questions and explain the natural world.
2.	Propose a solution.	People invent ways to solve problems.
3.	Implement proposed solutions.	Scientists and engineers often work in teams.
4.	Evaluate a product or design.	Women and men engage in science and technology.
5.	Communicate a problem, design, and solution.	Tools are used.
Standard		
Number	... abilities of technological design	Number
	As a result of activities in grades 5-8, students should develop the skills to...	
1.	Identify appropriate problems for technological design.	Differences and similarities of scientific inquiry and technological design.
2.	Design a solution or product.	Many people and cultures contribute.
3.	Implement a proposed design.	Science informs the development of technology, and technology creates further understanding of scientific principles—they are reciprocal.
4.	Evaluate completed technological designs or products.	Perfect designs or solutions do not exist: must reduce risk and maximize cost, efficiency, and appearance.
5.	Communicate the process of technological design.	Technology designs have constraints.
6.		Technological solutions have benefits and side effects.
As a result of activities in grades 9-12, students should develop the skills to...		
1.	Identify a problem or design an opportunity.	Scientists in different disciplines ask different questions, use different methods, and accept different evidence of support.
2.	Propose designs and choose between alternative solutions.	New technologies often solve challenges, extend scientific knowledge, and introduce new research possibilities.
3.	Implement proposed solutions.	Creativity, imagination, and a strong knowledge base is important.
4.	Evaluate the solution and its consequences.	Scientific inquiry is driven by a desire to understand the natural world, while technology is driven by the need to meet human needs and solve their problems.
5.	Communicate the problem, design, and solution.	Technological knowledge is often kept secret or private because of patents, and so forth.

* *Students in grades K-4 should also develop abilities to distinguish between natural objects and objects made by humans.*

each grade level, specific skills per category are described (see Table 2).

The focus on science is evident throughout the standards. Technology, however, is clearly parallel to science in emphasizing inquiry. Specific to technology, the standards focus on the technological design process, and the differences and similarities of technology and science. It is worth noting that key issues such as diversity and group work are addressed directly.

ITEA

The International Technology Education Association published the *Standards for Technological Literacy: Content for the Study of Technology* (STL) (2000a) as part of the Technology for All Americans Project (TfAAP) (1996) in 2000 (see Table 3). The ITEA standards are broken into five sections, including: (1) nature of technology, (2) technology and society, (3) design, (4) abilities of technological

Table 3. ITEA standards for technological literacy

Number	Standard
Nature of Technology: *Students will develop an understanding of...*	
1.	the characteristics and scope of technology.
2.	the core concepts of technology.
3.	the relationships among technologies and the connections between technology and other fields of study.
Technology and Society: *Students will develop an understanding of...*	
4.	the cultural, social, economic, and political effects of technology.
5.	the effects of technology on the environment.
6.	the role of society in the development and use of technology.
7.	the influence of technology on history.
Design: *Students will develop an understanding of...*	
8.	the attributes of design.
9.	engineering design.
10.	the role of troubleshooting, research and development, invention and innovation, and experimentation in problem solving.
Abilities of a Technological World: *Students will develop abilities to...*	
11.	apply the design process.
12.	use and maintain technological products and systems.
13.	assess the impact of products and systems.
The Designed World: *Students will develop an understanding of and be able to select and use...*	
14.	medical technologies.
15.	agricultural and related biotechnologies.
16.	energy and power technologies.
17.	information and communication technologies.
18.	transportation technologies.
19.	manufacturing technologies.
20.	construction technologies.

world, and (5) the designed world (see Table 3). "It defines what students should know and be able to do in order to be technologically literate and provides standards that prescribe what the outcomes of the study of technology in grades K-12 should be" (ITEA, 2000b).

A unique perspective promoted by the ITEA professional organization is that the standards enable people to develop knowledge and abilities about human innovation in action. They believe "all students can become technologically literate" and that an "effective democracy depends on all citizens participating in the decision-making process"; and since so many decisions involve technological issues, it is important for all to become technologically literate (ITEA, 2000b).

Standards Summary

Even though these professional organizations have a mutual focus on technology and literacy, they have traditionally diverse approaches to achieving their goals. So who is right? What approach to technological literacy should teachers, politicians, or school administrators take? The AAAS and ITEA organizations' definitions and perspectives on technologically literate students are similar, but the ISTE's is much narrower, focusing primarily on computer and information technology. The broader approach to technological literacy—one of all innovations created by humans, and used by AAAS and ITEA—is gaining support but is perhaps too broad.

TECHNOLOGICAL LITERACY K-12

Regardless of the source of standards or slant on technological literacy educators take in defining it, it is clear that there is an increasing emphasis on technological literacy and support for teaching it. Recently the National Academy of Engineering and National Research Council published a book titled *Technically Speaking: Why All Americans Need to Know More About Technology.* The book was produced primarily because, though it is basically understood that all Americans should be better prepared for a highly technological world, "the issue of technological literacy is virtually invisible on the national agenda...Americans appear to be unprepared to engage effectively and responsibly with technological change" (NAE & NRC, 2002). The goal of technological literacy from the NAE's perspective is to provide people with the tools to participate intelligently and thoughtfully in the world around them. Benefits of increasing technological literacy include a better prepared society to make well-informed decisions, improved citizenship, improved economical impacts, and a more abundant supply of technologically savvy workers.

Educators and those involved in education throughout the world face a similar challenge as the U.S. They have a responsibility to improve the technological literacy of their students, but the real challenge lies in "How do you do it?" There is a surprisingly small amount of literature on the actual study of how students become technologically literate, what they are learning, how schools are helping to teach technology literacy, and what programs are doing successfully that could be helpful in transferring this knowledge. Some might argue that technological literacy should be taught within the domain of knowledge. If this approach is taken, a biology teacher might then be looked upon to teach how global positioning systems (GPS) and global information (GIS) systems are revolutionizing farming and crop management around the world. But chances are their required state competency exams are going to allow little time for this. On the other hand, if the approach is taken that technology should be used only as a teaching tool, then a teacher in the field of language arts might embrace the computer as technology and require his class to research a topic on the Internet, then write a paper in a word processor, print it, and submit it. This task just a few years ago would have been done solely in a library and written out by hand or typed on a typewriter. If we only want computer literacy, then this might be an acceptable part of a solution, but are not computers so much more? Will either of these approaches give us students who can make informed decisions as politicians or simple consumers of technology the ability to contemplate the impacts of alternative power and energy supplies for a global transportation system? Or how about the skills to invent, innovate, or design a new communications system that takes advantage of nano technologies? If these skills are important, then a different approach might be required.

TECHNOLOGY EDUCATION: A VIABLE SOLUTION?

Of the three professions that have published standards in support of technological literacy, the field of technology education may be the most well prepared, focused, and organized to help students become broadly technologically literate. In fact, they are the only K-12 organization with teachers dedicated full time to teach technological literacy as a field of study from which their entire curriculum is designed. Not only does this allow technology educators to provide a more inclusive view of technological literacy to their students, but there is some evidence that this approach of technology as subject matter allows for an increased opportunity for students to gain transferable critical thinking, problem solving, and creativity skills like those sought for in the SCANS report (1991). For these reasons a closer look at technology educators must be considered as significant contributors to the overall technological literacy of our students.

The Early Years

Technology education has its roots in what was once known in the U.S. as industrial education. Professionals of this field as early as the 1960s began discussing the changing needs of a new technological age (Lemons, 1988, p. 60). Industry, technology, and education were hotly debated issues among this fields' prominent leaders, especially during the early 1980s at the Mississippi Valley Conferences (MSVs). Innovative leaders suggested that since we no longer lived in the Industrial Age, students needed new skills to prepare them for the new paradigm of life—namely a life of change, adaptation, and learning. Bender, in the 1981 MSV conference, suggested "traditional industrial arts programs have had rigid content that may be no longer valid in an ever-changing technological society if we are to prepare students to cope with life in the future" (Lemons, 1988, p. 67).

In 1985, the American Industrial Arts Association changed its name to the International Technology Education Association (Israel, 1995) to reflect the changing needs of society and emphasize the new priorities and needs of the field. Industrial Arts was defined in 1923 by Bonser and Mossman as "a study of the changes made by man in the forms of materials to increase their values and of the problems of life related to these changes" (p. 5). The ITEA now defines technology education as the "study of technology, which provides an opportunity for students to learn about the processes and knowledge related to technology that are needed to solve problems and extend human capabilities" (ITEA, 2001). The curriculum is problem based and utilizes math, science, and technology principles to promote technological literacy, which is described as "the ability to use, manage, understand, and assess technology" (ITEA, 2001).

The Possibilities

Though the field of technology education has a wonderfully rich perspective on technological literacy and may be poised to provide serious leadership and strength in educating all to become technologically literate, as a profession it has struggled to become unified through transition and change in society.

A concern and goal of the field has been to establish a discipline of technology education and with that to fulfill the goal of creating technological literacy...No discipline has been created and none may ever be able to be created given the nature of people's involvement and use of technology. Disciplines are formed when communities of scholars working together can agree that such a discipline exists. Technology educators are but a small part of the community of scholars who are technologists. (Zuga, 1994, p. 66)

Science education leaders have recognized this challenge and need by claiming that the task ahead is to build technology education into the curriculum, as well as to use technology to promote learning, so that students become well informed about the nature, powers, and limitations of technology (AAAS, 1993).

Though the field of technology education has struggled to realize its potential as a core subject in most of the U.S. schools, there have been many exciting happenings in the field. For instance there has been an increasing recognition from influential organizations like the National Science Foundation, National Research Council (NRC), the Center for Science, Mathematics, and Engineering Education (CSMEE), World Design Foundation, and the American Association for Advancement of Science (AAAS) in technological literacy and the role that technology educators could play. Other challenges still plague the field though, as evidenced by many of its teacher training programs closing down throughout the United States (Custer, 1999) which is contributing to an alarming teacher vacancy rate. In a study conducted by Hoepfl (1999), the average state in the U.S. had 37 technology education positions currently unfilled, and several states reported more than 100. For many states, this meant closing down programs and for others it may mean filling them through emergency certification. Even still, when teaching positions are filled, many of the teachers have high concerns about the curriculum in technology education (Berrett, 1999).

In a way, it is ironic that technology education is struggling to provide leadership and add value to the education of America's schools by teaching technological literacy skills in a time when school reformists continue to call for these skills, and when we live in a society enamored of and driven by technological innovation. William Wulf (2000), president of the National Academy of Engineering (NAE), sums up this quandary of who is to teach it and where do technology education teachers fit best by stating: "The goal of technological literacy requires that the content for the study of technology be delivered by a wide range of teachers...technology education teachers will be called on to advise schools and school districts trying to hop on the technological literacy bandwagon" (Wulf, 2000, p. 12). "It is up to the profession as a whole to step up to the plate, to be the center of the action. 'Is the profession up to the task?'" (p. 12).

A Case Study of Exemplary Practice

There is very little research that describes the learning outcomes of technology education classroom practice. Therefore it is difficult to determine to what degree the profession is being successful in teaching technological literacy. In an attempt to begin to fill this knowledge void, this author conducted a study to investigate the issues of classroom practice, vision, and indicators of student learning in an exemplary teacher's classroom in 2001. The case study was of Brad Thode, a renowned technology education teacher at Wood River Middle School, Hailey, Idaho, USA. Qualitative methods used included daily participant observations over a period of four weeks, 30 hours of interviews with teachers and students, 50 hours of videotaping, document analysis, student sample review, focus groups, and a student survey questionnaire. The original report is more than 300 pages long and includes contextually rich vignettes and data. For the purpose of this chapter however, a brief review of the findings only is presented because they indicated overwhelming success not only on measures of the ITEA, AAS, and ISTE standards for technological literacy, but perhaps most impressively in competencies the SCANS report (1991) identified as being essential for the future workforce in America (Berrett, 2004).

For instance, technology students at WRMS in the sixth through eighth grades demonstrated Workplace Competencies such as the ability to allocate time and resources, work in teams, teach others, lead, negotiate, work with a culturally diverse population, evaluate data, organize and interpret information, monitor and improve systems, and select the right equipment and tools for specific tasks. These skills were evidenced by performing to extremely high levels of competence as they were given the autonomy and responsibility to accomplish a task. They learned that their instructor was not the source of all knowledge and that they could be successful by capitalizing on the things they do best. Students at WRMS also showed high levels of competence in "Foundational Skills" such as basic reading, writing, arithmetic, speaking, and listening skills, along with the ability to learn, reason, think creatively, make decisions, and solve problems, all the while showing responsibility for self, integrity, and high levels of self-esteem.

The findings of this study were so persuasive that this author at times wanted to claim that technology education in fact is *the* way to teach technological literacy. Though this may be the case, it cannot be determined from this single case study since it is not representative of the entire population of technology education classrooms. From a naturalistically generalizable approach however, it is clear that the students in the WRMS program are well on their way to technological literacy and far beyond foundational skills of any other science, math, IT, business, or desktop publishing classroom I have ever been in. What makes technology education and Brad Thode's approach so successful? Thode and Thode (1997) summarize it in the following quote:

It is about opening wide all technology education exploration doors for students; not setting learning limits defined by curriculum organizers. It's about taking advantage of that teachable moment when a student hears about a new technology application and wants to try it right now. It's about never having to say 'that's a good idea, but we can't get to that until next semester.' It's about being a part of the decision making process in the class. By taking an active role in the class, students feel more enthusiastic and excited about the learning process. Technology Education as a subject lends itself easily to this concepts. Few students are passive containers waiting to be thrilled by the vast knowledge of the teacher. The nonlinear approach to curriculum organization not only makes the curriculum come alive for the student, but keeps the teacher excited and enthused as well.

SUMMARY AND CONCLUSIONS

The change and reform literature shows a growing concern and need for technological literacy, as do the increasing changes and new literacies being promoted in all domains of knowledge. This chapter's review of technological literacy and related literature has been written to provide a framework for understanding perspectives on technology literacy and how one field, technology education, may prove to be an excellent resource in meeting the challenge of creating a technologically literate citizenry. With so many people, organizations, governments, and so forth interested in technological literacy, one of the greatest challenges remains in determining best practices in teaching it in the K-12 setting. How should we teach it? This question has yet been well addressed. However, through the literature and within teaching practice there seem to be at least three possible solutions: (1) Technology can be taught as a subject matter treated as its own domain of knowledge as promoted by technology education. (2) Technology can be treated more as a teaching and learning tool as instructional technologists treat it. (3) Technology can be viewed as a component of individual disciplines, leaving the responsibility to teach technological literacy up to the teachers of individual disciplines such as English, Math, and Social Studies. Though arguably each of these perspectives have merit, and undoubtedly all three should be used to some degree in order to be successful—is not technological literacy too important to leave it up to chance or to use just for instructional purposes?

REFERENCES

Abdullah, M. H. (2000). *Media literacy.* ERIC Clearinghouse on Reading, English, and Communication, Indiana University. EDO-CS-00-03.

American Association for the Advancement of Science (AAAS). (1993). *Benchmarks for science literacy.* Retrieved October 2005, from http://www.project2061.org/

American Association for the Advancement of Science (AAAS). (1996). *National science education standards: Observe, interact, change, learn.* Washington, DC: National Academy Press.

American Association of School Librarians & Association for Educational Communications and Technology. (1998). *Information power: Guidelines for school library media programs.* Chicago: Author (ED 315 028).

American Library Association. (1989). *American Library Association Presidential Committee on Information Literacy final report.* Chicago: Author (ED 316 074).

Barnes, J. L. (1990, September 6-13). Clearing up some of the confusion associated with definitions of technology. *Proceedings of the International Design Education Conference.* Edinburgh, Scotland.

Berrett, J. V. (1999). *A study to investigate the acceptance of technology education by technology education/industrial education teachers in Utah.* Unpublished Master's Thesis, Brigham Young University, USA.

Berrett, J. V. (2004, March). Teaching workplace competencies through a nonlinear approach. *Proceedings of the 66th Annual Meeting of the International Technology Education Association,* Albuquerque, New Mexico.

Berryman, S. E., & Bailey, T. R. (1992). *The double helix of education and the economy.* New York: Teachers College, Columbia University.

Bonser, F., & Mossman, L. (1923). *Industrial arts for elementary schools.* New York: The Macmillan Company.

Bruce, B. C. (2003). *Literacy in the information age: Inquiries into meaning making with new technologies.* Newark, DE: International Reading Association.

Brunner, C., & Tally, W. (1999). *The new media literacy handbook: An educator's guide to bringing new media into the classroom.* New York: Anchor Books Doubleday.

Clinton, B., & Gore, A. (2000). *Increasing technology access and innovation.* Retrieved January 2001, from http://www.whitehouse.gov/textonly/WH/Accomplishments/technology.html

Cohen, S. S., Delong, B. J., & Zysman, J. (2000). The next industrial revolution? Information technology makes a difference—finally. *The Milken Institute Review, First Quarter 2000,* 16-23.

Cooper, D. J. (1997). *Literacy: Helping children construct meaning.* Boston: Houghton Mifflin.

Custer, R. L. (1999). Prospects for the future: It's our call. *Journal of Technology Studies, 25*(1), 16.

DeVore, P. W. (1987). Science and technology: An analysis of meaning. *Journal of Epsilon Pi Tau, 13*(1), 2-9.

Dewey, J. (1956). *The child and the curriculum & the school and society.* Chicago: University of Chicago Press. (Original works published 1902 and 1915.)

Drucker, P. F. (1993). The new society of organizations. In R. Howard & R. D. Haas (Eds.), *The learning imperative: Managing people for continuous innovation* (pp. 3-19). Boston: Harvard Business Review.

Gagel, C. W. (1995). *Technological literacy: A critical exposition and interpretation for the study of technology in the general curriculum.* Unpublished Doctoral Dissertation, University of Minnesota, USA.

Gagel, C. W. (1997). Literacy and technology: Reflections and insights for technological literacy. *Journal of Industrial Teacher Education, 34*(3), 6-34.

Gardner, H. (1999). *Intelligence reframed: Multiple intelligences for the 21st century.* New York: Basic Books.

Goals 2000: Educate America Act. (1994). Pub. L. No. 103-HR.1804.

Goleman, D. (1995). *Emotional intelligence.* New York: Bantam Books.

Hoepfl, M. (1999). Alternative routes to certification of technology education teachers. *Proceedings of the 86th Annual Mississippi Valley Technology Teacher Education Conference,* St. Louis, Missouri.

Horton, F. W. (1983). Information literacy vs. computer literacy. *Bulletin of the American Society for Information Sciences,* (9), 14-18.

International Society for Technology in Education (ISTE). (2000). *National educational technology standards for students: Connecting curriculum and technology.*

International Society for Technology in Education (ISTE). (2001a). *National educational technology standards: What is the NETS project?* Retrieved October 2005, from http://cnets.iste.org/

International Society for Technology in Education (ISTE). (2001b). *National educational technology standards for teachers (NETST).* Retrieved October 2005, from http://cnets.iste.org/index3.html

International Society for Technology in Education (ISTE). (2001c). *Technology standards for school administrators (TSSA).* Retrieved from http://cnets.iste.org/

International Technology Education Association. (1996). *Technology for All Americans* (TAA). Reston, VA: Author.

International Technology Education Association. (2000a). *Standards for technological literacy: Content for the study of technology.* Reston, VA: Author.

International Technology Education Association. (2000b). *Standards for technological literacy: Content for the study of technology. Executive summary.* Reston, VA: Author.

International Technology Education Association. (2001). *What is ITEA?* Retrieved October 2005, from http://www.iteawww.org/A6.html

Israel, E. (1995). Technology education and other technically related programs. In G. E. Martin (Ed.), *Foundations of technology education, Council On Technology Teacher Education's 44th yearbook* (pp. 25-117). Peoria, IL: Glencoe McGraw-Hill.

Kuhlthau, C. C. (1987). *Information skills for an information society: A review of research.* Syracuse, NY: ERIC Clearinghouse on Information Resources (ED 297 740).

Lemons, D. C. (1988). Technology education: The culmination of a seventy-nine year quest. Industrial education in transition. *Proceedings of the Mississippi Valley Industrial Teacher Education Conference* (pp. 47-90), St Louis, Missouri.

Manini, C. M., & Cervantes, J. (1998). *Adult basic education basic computer literacy handbook.* Santa Fe, NM: New Mexico State Department of Education (ED 42 480).

Merriam-Webster. (1986). *Webster's ninth new collegiate dictionary.* Springfield, MA: Author.

National Academy of Engineering (NAE) & National Research Council (NRC). (2002). *Technically speaking: Why all Americans need to know more about technology.* Washington, DC: National Academy Press.

National Commission on Excellence in Education. (1983). *A nation at risk: The imperative for educational reform.* Washington, DC: U.S. Department of Education.

National Science Foundation. (1992). *Materials development, research, and informational science education program announcement.* Washington, DC: Author.

Palazon, M. (2000). *The media and transformative learning.* (ED 443 971).

Pena, M. M. (1992). Technology in the general curriculum: A Latin American perspective. In D. Blandow & M. Dyrenfurth (Eds.), *Proceedings of the 1st International Conference on Technology Education: Technological Literacy, Competence and Innovation in Human Resource Development* (pp. 147-152).

President's Committee of Advisors on Science and Technology (PCAST). (1997, March). *Report to the President on the Use of Technology to Strengthen K-12 Education in the*

United States. Washington, DC: Education Technology Initiative.

Ragsdale, R. G. (1988). *Permissible computing in education: Values, assumptions, and needs.* New York: Praeger.

Rose, L. C., Gallup, A. M., Dugger, W. E., & Starkweather, K. N. (2004) The second installment of the ITEA/Gallop poll and what it reveals as to how Americans think about technology. *A Report of the Second Survey Conducted by the Gallup Organization for the International Technology Education Association.* Reston, VA: International Technology Education Association.

Rutherford, F. J., & Ahlgren, A. (1990). *Science for all Americans.* New York: Oxford University Press.

Seaborg, G. (1991). A nation at risk revisited. In *Rebuilding American education for the 21st Century.* Industry Education Council of California, Golden Gate University Press, San Francisco. Retrieved March 2001, from http://www-ia1.lbl.gov/Seaborg/risk.htm

Secretary's Commission on Achieving Necessary Skills. (1991). *What work requires of schools—A SCANS report for America 2000.* Washington, DC: U.S. Department of Labor.

Spitzer, K. L., Eisenburg, M. B., & Lowe, C. A. (1998). *Information literacy: Essential skills for the information age.* Syracuse, NY: Eric Clearinghouse on Information and Technology.

Teale, W. H., & Sulzby, E. (1986). *Emergent literacy; writing and reading.* Norwood, NJ: Ablex.

Tyner, K. (1998). *Literacy in a digital world: Teaching and learning in the age of information.* Mahwah, NJ: Lawrence Erlbaum.

U.S. Department of Commerce, Economics and Statistics Administration, National Telecommunications and Information Administration. (2003, February). *A nation online: How Americans are expanding their use of the Internet.* Retrieved October 2005, from http://www.ntia.doc.gov/ntiahome/dn/index.html

U.S. Department of Commerce, Economics and Statistics Administration, Office of Policy Development. (2000). *Digital economy 2000.* Retrieved October 2000, from http://www.esa.doc.gov /de2000.pdf

U.S. Department of Education. (1996, June). *Getting America's students ready for the 21st century: Meeting the technology literacy challenge.* Retrieved January 2001, from http://www.ed.gov/Technology/Plan/NatTechPlan/

Utah State Office of Education (USOE). (2004). *Educational technology program.* Retrieved October 2005, from http://www.usoe.k12.ut.us/curr/EdTech/default.htm

Wright, T. R. (1995). Technology education curriculum development efforts. In E. G. Martin (Ed.), *Council on Technology Teacher Education 1995 yearbook: Foundations of technology education* (44th ed., pp. 247-285). Peoria, IL: Glenco/McGraw-Hill.

Wulf, W. (2000, March). The standards for technological literacy: A national academies perspective. *The Technology Teacher, 59*(6), 10-12.

Zuga, F. K. (1994). *Implementing technology education: A review and synthesis of research literature information* (Series No. 326). Columbus, OH: ERIC Clearinghouse on Adult Career, and Vocational Education.

Chapter IV
Globalization, At-Risk Students, and the Reconceptualization of Technological Literacy

Leonard J. Waks
Temple University, USA

ABSTRACT

The problem of technology has, since 1970, been radically altered by the global spread of market economies and networked computers. As a result, the notion of technological literacy education that emerged in the 1970s and 1980s must be re-constructed. After reviewing the impact of globalization and the spread of communications networks on the world occupational structure, and the consequent new risks borne by low-tier routine and informal workers, I offer a revision of the concept of technological literacy education that assigns important new roles to non-governmental organizations serving low-tier workers, and that places the hands-on use of networked computers at its core.

INTRODUCTION

In this chapter, I explain in general terms the relationship between globalization, at-risk metropolitan youth, and technological literacy education. I argue that the notion of technological literacy education that emerged in the 1970s and 1980s for middle- and secondary-level education in the developed nations must be reshaped for use in the educational agencies of both developed and developing nations, to take account of changes in the global economy, the rapid urbanization of the developing world, and the migrations of workers from the developing to the developed world. In the first section, I consider the impact of globalization on the emerging world occupational structure, the worldwide movement of jobs and workers, and the risks borne by low-tier routine and informal workers lacking higher order workplace skills.

I then explain why this new situation compels us to reshape earlier notions of technological literacy education, especially for at-risk adolescent learners, and offer some suggestions for revision that assign central roles to links with non-governmental organizations, and hands-on use of networked computers.

GLOBALIZATION

Economic and Technological Globalization

The current stage of *economic* globalization began after World War II, when the economies of Europe and Asia were shattered by the war, and the United States had surplus capital for investment. The world economy was in a state of chronic under-supply, and American firms faced little competition in international trade. They increased their direct foreign investments, creating foreign subsidiaries throughout the world making mass-produced products for global markets. Under these noncompetitive conditions, American firms could set world prices for mass-produced goods and could secure labor cooperation by passing along some part of their excess profits to unionized industrial workers as wages and benefits above world levels.

But by the 1970s European and Asian economies had recovered sufficiently to compete with American firms on a global scale.[1] American firms responded to this competition during the 1970s and 1980s by initiating aggressive antiunion practices, by shifting low-skilled production jobs to nonunion plants in the South or overseas to reduce wages, and by greatly expanding low-wage service sector industries.[2] Western European and Japanese firms also began outsourcing routine jobs to low-wage nations in their regions and beyond. Notions of technological literacy emerging at that time, however, did not pay close attention to those trends.

The globalization of *digital information technology networks* began in the early 1990s. The growth of these networks was accelerated by the need to coordinate the far-flung production and marketing activities of large multinational firms. The first commercial Internet service provider opened for business in 1990, and the World Wide Web was introduced at CERN in 1991. In 1992 the World Bank went online, followed by the White House and United Nations in 1993 (Howe, 2001). Commercial use of the World Wide Web grew rapidly by the mid-1990s.[3]

The driving forces of the growth of digital networks included the rapid technical advances and price declines in computer chips, satellites, and fiber optic cables which facilitated growth in television, telephony, fax, and the Internet, turning the global information grid into a seamlessly integrated resource, "the biggest machine ever made" (Dizard, 1997, p. 1). By the end of 2001 the Internet was growing in the United States at the rate of two million users a month; 143 million Americans were online (54% of the population), an increase of 26 million in 13 months. Schools and colleges also went online, and 75% of 14- to 17-year-olds and 65 % of 10- to 13-year-olds were Internet users by 2002 (NTLA, 2002).

Economic and Technical Convergence: Network Enterprise

The convergence of these economic and technological developments enabled a reorganization of trans-national enterprises: large, vertically organized firms with foreign subsidiaries have been transformed into global 'networks' of downsized flagship firms, small supplier firms, competitor firms, government agencies, and universities in strategic alliances.[4]

These *meta-national* firms, unlike the traditional trans-national firms rooted in a distinct home country, do not take knowledge from the corporate center in the U.S., Western Europe, or Japan, incorporate it into a mass-produced product, produce it in their own subsidiaries, and ship it to markets in *less developed areas*. Rather, they are *true network enterprises*.[5] That is, they depend upon technical and managerial knowledge, experience and insight, and market intelligence from employees and network associates all over the world, connected by in-person and electronic links, to produce small-batch, customized, intermediate, and consumer products for rapid shipment to highly differentiated global markets.[6]

The Global Movement of Jobs and Workers

The globalization process has disrupted traditional patterns of economic activity in developing nations, increased economic inequality, and destroyed traditional rural patterns of livelihood. Economic elites in developing nations have become more tightly linked to their counterparts in the developed nations and more essential to their far-flung enterprises. Meta-national firms draw talent from developing nations to the "global city" planning centers to work on projects that employ their nations' labor or offer products to their markets (Sassen, 1994). Meanwhile the poorest residents of poor nations, undergoing a population explosion and a deterioration of traditional economic roles, are impelled to seek new livelihoods in their own nations' growing urban economies or abroad. Most are unable to find employment in the formal economies of their own nations, and are forced to work in informal economic sectors at home or to migrate (Sassen, 1994; Richmond, 2002).

Reich's (1992) distinction between production workers and in-person service workers explains one important division in workers from developing nations. Sophisticated manufacturing can now be performed throughout the world, and jobs will tend to move to nations with reliable, low-cost workers. The largest proportion of employees and subcontractors of some companies based in developed nations are now found in developing nations. This increases the low-wage routine production sectors of developing nations. But many jobs, both complex and routine, require in-person contact. If you want a brain tumor operated on, or a facility cleaned, you must find an in-person surgeon or facility cleaner. In-person workers across the economic spectrum migrate to wealthier nations with favorable markets for their services.

Globalization and Workforce Reorganization

The formation of global networks and network enterprises competing in global markets has brought about changes in workforce organization in all affected societies (Brown, Green, & Lauder, 2001).

To enhance competitiveness, many firms in developed nations have re-engineered their work systems, employing cross-functional teams of "knowledge workers"[7] using knowledge and information in novel ways for rapid response to global opportunities (Applebaum, Bailey, Berg, & Kalleberg, 2000). They have also cut costs by automating many complex work tasks and relying on low-cost, low-skill routine workers at home and abroad. These workers serve increasingly on a part-time or contingent basis, without job security or health insurance (Cappelli, 1999). In poor nations, the formal economy increasingly has a similar structure of knowledge and routine workers, but also relies heavily on informal workers, for example in food preparation and recycling used parts for machinery.

Four Occupational Categories in the Global Economy

High-Tier Knowledge Professionals

One important group of knowledge workers includes the elite knowledge professionals, labeled "symbolic analysts" by Reich (1991). These workers acquire, permute, and recombine knowledge and information to generate new products and services, while in the process generating new knowledge and information that can be distributed through their networks for "knowledge reuse" (Castells, 1996). Grantham projects a continuing evolution toward a 'team of teams' approach in the creative sectors of the economy, where elite professional knowledge workers come together from throughout the world, "blending interdisciplinary skills focusing on a particular project, completing the project, then disbanding the team as each of its workers move on to other projects" (Grantham, 2000). The most important asset in the productive process is no longer production capital but the knowledge that moves with such workers from one project to another. Elite knowledge professionals thus relate in a new way to the means of production, becoming a distinct "*knowledge class*".

High-Skill Production Workers

The jobs of many front-line production workers in today's networked workplaces have been radically restructured (Applebaum et al., 2000, pp. 3-4). The late 1980s saw modest introduction of self-directing work teams that make and implement decisions without top-down control. The terms *process re-engineering* and *high-performance work systems* refer to such innovations. Employee participation in self-directing work teams in the United states was low in 1990, when among 6.5 million workers in companies responding to a survey, only 300,000 (5%) were in self-directing teams (Levine, 1995, pp. 3-7). But since then, many firms have increased the use of self-managing teams.[8]

The spread of digital information technologies converged in the 1990s with these workplace organization practices. Information technologies are fundamentally different than earlier innovations because information-based tools are highly malleable. They are re-programmable, and the same equipment can perform a wide variety of customized functions. They reduce retooling time and create opportunities for economies of scope. They have a built-in capacity for data collection and analysis to support decision making (Ducatel, pp. 1-4; Applebaum et al., 2000, p. 14).

"Smart" networked environments call for an increase in skills among front-line production workers as a result of both information technologies and worker participation practices. In these environments, the required skills-sets shift from manual and craft skills to higher level cognitive skills including abstract reasoning, because the workers are removed from the physical processes of moving and making. They are system controllers, able to program and maintain their machines, as well as interpret the data read-outs that the machines produce.

Information technologies fit best in settings where workers and managers cooperate directly in production processes (Ducatel, p. 4; see also Ostroff, 1999). Workers in these environments require additional 'soft' skills in communication and group decision making. Applebaum et al. (2000, p. 8) use the term 'high performance work systems' to refer to work situations where new organizational processes and information technologies have converged.

In these systems front-line workers require information processing and communications skills, and must be able to carry out a wide range of tasks. This flexibility presupposes training in the use of networked computers. They also must develop "interpersonal and behavioral skills, to take on supervisory and coordination functions, and to interact effectively with other workers and managers" (Applebaum et al., 2000, p. 208). Prior cross-cultural and cross-class experiences are very valuable in acquiring these soft skills.

Cappelli and his coauthors (1999) have noted that American firms have moved more slowly toward high-skill systems than competitors in Japan and Western Europe, because the larger scale of American production accommodates mass production, long-established traditions of top-down management have been difficult to change, and low-wage minority and immigrant workers are plentiful. Thus the advantages of low-skill mass production using low-cost routine labor remain strong in many segments of American industry.[9] These advantages are also strong in developing nations with large populations of displaced rural workers.

Routine Workers

The direct result of globalization in neo-liberal developed nations such as the United States and the United Kingdom has been a widening income division between high-skills and low-skills segments of the economy, and a relative reduction of the former and expansion of the latter. The routine worker category is growing rapidly while the middle-tier is declining, causing an *hourglass*-shaped distribution, with less room in the middle and more at the bottom. The *routine worker* category contains a high proportion of women, including single heads of households, disadvantaged minorities, and recent immigrants from developing nations (including illegal aliens), in for example material moving, fast food, and low-end retail and healthcare occupations, and as handlers, helpers, and laborers (Amiratimadi & Wah, p. 94). Unlike industrial workers of the 1950s, a large proportion of today's routine workers work in temporary or part-time "contingent" work.[10] From 1970 to 1992 the total payroll of American temporary employees in real terms rose an astounding 3000% (Weinbaum, 1999). While many American routine workers are drawn from poor, native-born black and Hispanic groups, a substantial fraction of both blacks and Hispanics in the American labor force are foreign born. In New York City three-quarters of recent immigrants have come from such poor Third World countries as Haiti, Nigeria, Jamaica, the Dominican Republic, Ecuador, and Columbia. Since 1980 there has been a very substantial increase in the proportion of the new immigrants in lower tier occupations, and a substantial decline in their real and relative earnings (Amiratimadi & Wah, p. 96). These workers form a growing class of *working poor*.

Informal Workers

As subsistence village agriculture collapses in developing nations, villagers flock to the metropolitan regions. In 1950 only one city region, New York, qualified as a "mega-city" with a population in the greater metropolitan area of more than 10 million residents. By 1995 there were 14 mega-cities, and in 2000 there were 19. While the largest actual city in the world, Seoul (10.3 million residents within the city boundary), is in the relatively affluent nation of South Korea, most mega-cities are in relatively poor countries; Mumbai (Bombay), India (second in population with almost 10 million within the city boundaries); San Paulo, Brazil (third, with 9.8 million); and Jakarta, Indonesia (fourth, with 9.4 million) are good examples. In 2000, 14 of the existing 19 mega-cities were in the developing world. It is estimated that more than half of the world's population will live in large cities by 2015, including more than 400 million in 21 mega-cities. Most of these mega-cities lie close to the sea, or are themselves seaports, and are thus situated favorably for both global economic exchange and out-migration. More than half of the world's population lives within 100 miles of such cities and can travel to them in search of work, social services, or migration opportunities.[11] In addition to the mega-cities, by 2000 there were 292 cities in the developing world alone with populations of more than one million people.

Estimates indicate that from 20% to 70% of the mega-urban workforce in the developing world works in the informal economy (the average of various estimates is 50%), altogether outside the government-regulated and taxed system of employment (Simon, 1998). These workers sell hot food or used clothing from makeshift booths, or repair bikes and cars on the streets, or shine shoes, sort and recycle rubbish or used machines, or sell sex services. These informal workers are essential to the economies of these cities, as they feed and clothe a large part of the population, and supply used parts that keep capital equipment operating. They provide a low-cost infrastructure to maintain low production costs in the formal sector of the economy. They comprise a *working underclass* outside the official social system.

Risks and Opportunities in the New Metropolis

The primary risk facing young people throughout the world today is lack of access to work providing a safety net for basic human rights: job security, labor rights, a living wage, unemployment and old age pensions, and access to personal security and health care. Knowledge professionals and high-skill production workers enjoy this basic security. Routine workers in the contingent economy and informal workers, whether in developed or developing nations, do not. The first two work categories require information processing and communications technology skills, as well as 'soft skills' such as working in groups, and communicating across cultural and class divisions.

Young people from poor, minority groups are isolated from middle-class students, in schools earmarked for poor students. Poor young people in the developing world often lack access to middle and secondary school altogether. Under these circumstances it is difficult to form a clear vision of a path to a successful future through education, so these young people have low levels of school attachment.

Many informal urban workers in the developing world are younger than 18. They are subject to unthinkable levels of exploitation, neglect, and violence. For example, according to the non-governmental organization 'War on Want,' between 40,000 and 70,000 children in Jakarta, Indonesia, are illegally trafficked to wealthier nations such as Australia, Taiwan, and Singapore to work at the lowest rungs of industry or as commercial sex workers, and between 200,000 and 300,000 Indonesian children under 18 work as prostitutes.[12]

Urban areas, despite their large risks, actually offer greater opportunities of all sorts today than rural areas. Casual routine laborers in India, for example, are able to find twice as many days of work per year as rural laborers. Young people in urban areas worldwide complete primary, secondary, and tertiary levels of education at much higher rates than rural counterparts. Access to health care and water is better in the cities. As migration accelerates in the developing world, people even have denser family networks in regional cities than in their native villages. These networks provide assistance in locating jobs or setting up informal work regimes, as well as support in times of illness or unemployment. People move to the cities to be near friends and family, to find work, to improve standards of living and opportunities for their children.

Most at-risk students, even in the developed nations, are not acquiring relevant technology and communications skills. In the developing world, few at-risk children remain in school long enough to acquire complex academic knowledge and skills. Technology literacy education must provide ground-floor skills in these areas, and lay the basis for further skill development through further education, on-the-job training, or self-directed learning.

Network technology skills can upgrade the job prospects of routine workers, and facilitate opportunities for college-level occupational training. In surprising ways, these skills can also enhance the work tasks and incomes of informal workers in developing nations, by reorganizing informal work into individual and collective business enterprises. Network technologies can also be useful in delivering knowledge work and entrepreneurial skills even to young people who have left formal schooling for informal work. I discuss these points in the next section.

RECONSTRUCTING TECHNOLOGICAL LITERACY

The "Technology Problem" and the Rise of Technological Literacy

The notion of *technological literacy* education emerged in the United States in the late 1970s and early 1980s, and spread to other developed nations, particularly in the English-speaking world, as a response to specific social and economic factors of that time. Two somewhat conflicting motivations shaped initial technological literacy efforts.

The first was the awareness of a technology problem—the erosion of the ideology of progress through science and technology, which had dominated American ideology in the first half of the twentieth century. This ideology was severely challenged by various techno-shocks after 1950. A selective list would include: the atomic bomb in 1945; the development of the hydrogen bomb in 1952 and its spread to the USSR in 1953; thalidomide babies in the late 1950s and early 1960s; the destruction of habitat due to agricultural pesticides, brought to public attention by Rachel Carson in 1962 in *Silent Spring*; the use of napalm, a jellied form of gasoline produced by the Dow Corporation from 1965-1969 as an incendiary weapon for U.S. forces in Vietnam; and the failure and cover-up at the Three Mile Island nuclear reactor in Pennsylvania in 1979, eerily forecast earlier that year in the thriller *The China Syndrome*.

The steady flow of these shocks prompted deeper and more comprehensive worries about technology and the environment that entered public awareness through anti-war and environmental teach-ins on university campuses during the late 1960s, the publication of the first *Whole Earth Catalog* in 1968, and the first Earth Day in 1970. In response, non-governmental organizations were created to fight the misuse of science and technology, and to promote an environmentally sustainable and peaceful world. These included the Union of Concerned Scientists, established at MIT in 1969; Friends of the Earth, founded by David Brower in 1969 (and successful, two years later, in preventing funding for the proposed American Supersonic Transport); and Greenpeace, organized in 1971 to promote a "green and peaceful planet." Worries about the long-range effects of industrial technologies were sharpened by the computer models of planetary collapse in *The Limits to Growth,* a 1972 report of the Club of Rome, a worldwide group of scientists, economists, businessmen, and political leaders. These developments challenged activists and educators to develop new educational means for controlling technological developments. During the Carter administration years (1976-1980), American educators first sought federal funding for school and college courses in technological literacy, to enable participation by young citizens in the informed resolution of technological issues. Similar courses were also developed in Great Britain, Western Europe, and Australia.

The second motivation for technological literacy education came from the opposite pole on the political spectrum. In the 1970s American industry faced stiff competition in the global marketplace. International comparisons of U.S. and foreign students had also revealed that the former lagged behind their international competitors on measures of science and math achievement. *A Nation at Risk,* issued by President Ronald Reagan's Department of Education in 1983, blamed America's declining dominance over the world economy on its ineffective schools. The report called on American schools to require all students to take demanding courses in science, math, and computer science, and to learn how to use computers as tools in all academic subjects. The root idea of technological literacy embodied in the report was mastery of computer skills for knowledge use in the workplace.

Each of these two conflicting motivations—controlling technological developments to promote peace and protect the environment, vs. mastering technology to dominate world markets—appealed to one of the two dominant political parties. The two were tactically merged in the American technological literacy efforts of the 1980s. New courses in science paid greater attention to technological applications, and employed computers in simulations and as tools for modeling. Technology education replaced 'industrial arts' in general education. Computer literacy programs were introduced. A leading trend, however, was the introduction of courses and course units in science, technology, and society. These aimed to increase scientific and technological sophistication, and to raise student awareness of the central roles of science and technology in society and the potential for controlling these forces through social movements and governments. One great appeal of the science, technology, and society (STS) concept was its use as an heuristic, to generate units in every existing curriculum site: science, social studies, math, language arts, and vocational education. STS thus offered something for every existing teacher, assisted in integrating the curriculum, and demanded no curriculum site of its own.

Initial Models of Technological Literacy Education

Although definitions varied, most accounts of technological literacy education embodying the STS idea offered a learning process that (1) engaged the interest of learners, (2) conveyed scientific and technical knowledge related to identified technology-related problems, (3) guided student investigations of the problems or issues, and (4) facilitated projects in which students planned action steps and 'took action' to address the problems.[13]

The science and technology content already in the school curriculum was accepted as a given background for these units. Technology-related problems were selected to amplify this content and demonstrate its real-world significance. The ozone hole, for example, was a good choice because it related familiar technological objects (refrigerators, aerosol sprays) and familiar health problems (UV radiation and skin cancer) to accessible, yet fascinating, atmospheric chemistry.

Student investigations were seen as akin to discipline-based research projects, and were used as adjuncts to, or replacements for, laboratory experiences in science or projects in social studies. Students linked their textbook knowledge to selected technology-related problems in order to formulate questions and state hypotheses. They then gathered data, subjected it to analysis, drew inferences, stated results, and applied them to the problem that motivated the question for investigation.

Guided student actions were envisioned for the most part as ameliorative individual and collective action steps addressing long-range social problems of technology linked, for example, to environmental protection, occupational health and safety, and population growth. The public was aware of such problems due to efforts of non-governmental organizations, and activated by popular books detailing the "100 things you can do to save the earth." Young people learned to recycle, plant trees, turn down air conditioners, and use bicycles instead of automobiles. Classes engaged in environmental clean-up projects and tested water quality in nearby rivers.

Roadblocks to Implementation

While the STS idea of technological literacy was widely embraced, implemented programs typically neglected the third and fourth stages, issue investigation and action. These activities were unpredictable; they were difficult for teachers to manage and overly challenging for many students. Unlike the first and second stages, awareness and content learning, which fit easily into the standard curriculum format of pre-selected content, learning objectives, and standardized tests, the investigation and social action phases involved unique, creative learning. They did not generate standardized learning outcomes for tests, but took time and effort away from activities that did. They also involved controversial issues, and frequently generated negative feedback from conservative parents and groups insisting that schools 'stick to the facts'.

This technological literacy model also predictably found little support in developing nations. A central concern of the environmental movement was the exponential growth of population, and this issue divided developed and developing worlds. Population growth came to public awareness through the publication of Paul Erlich's *The Population Bomb* by the Sierra Club in 1968, and the establishment of Zero Population Growth, a highly visible non-governmental organization, the same year. Although Erlich focused on population growth in the developed world, with its high-consumption lifestyle, most of the population growth was actually taking place in the developing world. The planetary collapse predictions in *The Limits to Growth* relied upon the spread of industrial technologies and high-consumption urban lifestyles to the developing world. As a result, the anti-technology and environmental movements became vulnerable to charges of "Gringoism"—rich Americans telling people in poor nations to stop having so many children and to slow their pace of industrial development.

Naturally those in the developing nations rejected this message. They saw their growing populations as resources for industrial development, and refused to buy into *limits to growth.* On the contrary, they demanded the transfer of industrial technology and technical knowledge from the developed world to stimulate their economic growth and to free themselves from dependency upon the rich nations. Instead of fostering the continued economic dominance of the richest nations, they sought to increase their own share of domestic and global markets.

The New Problem-Situation for Technological Literacy Education

The global technology problem-situation today is very different than that motivating the initial conceptions of technological literacy, and a new and more relevant conception is needed. Three changes are especially important.

First, the world population in 1970 was approximately 3.7 billion. Today (2004) it stands at 6.4 billion, an increase of 73% in just less than a quarter century. World population growth, however, has dropped from over 2% in 1970 to just over 1% today, and is not, in itself, the pressing issue it was in 1970, though the annual *numerical* increase in population remains more or less constant.

Second, globalization has integrated most nations in Asia, Latin America, and Eastern and Central Europe into the world economy, and digital production and communications technologies have also integrated their populations into the global techno-structure. Seoul, Korea, is not only the largest city in the world, but also the city with the highest level of digital delivery of government services and participation in democratic action (The E-Governance Institute, 2003). Brazil, with its notorious mega-city "favelas" (urban shanty-towns), is also the first nation in the world to establish a completely electronic voting system, with 406,000 electronic ballet boxes (Belos, 2002). As networked enterprises spread, networked computers are found throughout the workplaces of the developing world, and are spreading to schools and youth centers. With concerted effort, networked computers can now be made widely available for education.

Third, non-governmental organizations of the 1970s grew from fringe social movements and were politically marginal. In the global era, however, NGOs play a prominent role at every political level. Many are focused on the specific problems of urban poor and have significant educational outreach programs, including efforts in technology education that make use of networked computers. These organizations are at the heart of global and local civil society, and their role in technological literacy education needs to be expanded.

The initial conceptions of technological literacy have, as a result, three primary shortcomings from a contemporary standpoint.

First, even a quarter-century ago the populations of the largest metropolitan regions were much smaller and less ethnically diverse than today. In global society, migrants from developing nations surge into the metropolitan regions of developed nations. They perform in-person services at every income level from medical specialist to hospital orderly. The cities of the developing nations, in turn, draw migrants from ethnically diverse groups in the surrounding rural areas. The mega-city seaports, with their long histories as trading centers, are particularly diverse. Initial conceptions of technological literacy made passing reference to ethnically diverse students in central city schools of developed nations (e.g., Rubba, 1988; Waks, 1991). Today, soft skills in cross-group communication are understood to be essential in high-technology work settings. A relevant contemporary conception of technological literacy would place concerns of ethnic diversity, and associated problems of stratification and school exclusion and isolation, front and center.

Second, while the initial conceptions certainly encouraged project-based learning, their model projects did not involve networked computers or the manipulation of real-time data. These initial conceptions were forged shortly after the introduction of microcomputers into schools and before the development of the World Wide Web. Computers in most workplaces and schools were weak, stand-alone machines. The economic and technical convergence of globalization had not yet occurred. *Knowledge work* and *networked enterprise* were not yet household words. Today, as a popular slogan says, "the machine *is* the network". An ever-increasing proportion of schools and other educational facilities now have some networked computers, and school projects can gradually approximate the problems faced by high-skilled knowledge workers. An up-to-date conception of technological literacy must take account of this basic fact.

Third, the initial notions of technology literacy education tied technology-related learning to established curriculum sites such as science, math, social studies, language arts, and vocational/technical education. Technological literacy education was essentially a component of the secondary school curriculum. This was unduly restrictive, because poor children in developed nations, especially those from ethnic minority groups, frequently drop out of secondary school, while those in poorer nations simply lack adequate access to schooling at that level. Some at-risk youngsters, however, have opportunities for further learning beyond the official school system. Today metropolitan education can be conceived not merely as an official *school system*, but as a *network* with nodes that include government-funded official and alternative schools, detention centers, and programs operated by non-governmental agencies.[14] A contemporary conception of technological literacy education applicable to 'at-risk' learners, should, without devaluing formal schooling, include educational efforts throughout the network.

Reconstructing Technological Literacy

We can now build on the initial process model of technological literacy education, which, as stated earlier, involved (1) raising awareness and engaging the interest of learners in technology-related problems, (2) conveying scientific and technical knowledge related to those problems, (3) guiding student investigations of the problems, and (4) facilitating student action to address those problems. Each of these components must be reconstructed in light of globalization and the focus on at-risk young people.

1. **Engagement:** When *at-risk* learners are considered, participants in technological literacy efforts should no longer assume to be committed school students. They may instead be disadvantaged youths at risk for dropping out, or part-time students working the informal economies of developing nations (e.g., as trash recyclers or commercial sex workers), or young adult immigrants from poor countries. Recruiting and engaging members of each of these groups pose distinct problems. The problems engaging their attention are rarely abstract problems of environmental pollution, depletion of industrial materials, or protection of endangered species. Often they are problems of personal survival. Nonetheless, they are attracted, for better or worse, by new technologies, such as cell phones, game players, and the Internet, and technological literacy educators can use this attraction as a peg for significant learning experiences.
2. **Conveying Scientific and Technological Knowledge and Skill:** As noted, initial models fit technological literacy into school curriculum sites and accepted the

standard curriculum as background. The problems appropriate for technological literacy education, however, change as we consider new groups of learners and their specific needs. Those on the verge of dropping out of school, or in alternative schools or informal educational settings, are not captive audiences. They only become engaged when lessons address their own pressing problems and interests, as they perceive them, in concrete ways. Technological literacy educators in such settings, as a result, cannot pre-select curriculum content based on standard topics in the secondary curriculum. Instead, content selection requires continuous input and feedback from learners. An example illustrates this point. Homeless trash recyclers in San Paulo formed a cooperative with the assistance of a non-governmental organization, and established a profitable and efficient business by using networked computers to locate trash, organize inventory, and link with markets. The program was able to recruit them and provide technological knowledge and skill because the young workers, from the beginning, helped shape the program to meet their needs.[15]

3. **Guiding Student Investigations and Decisions:** The vision driving 'issue investigations' in early formulations of technological literacy education was overly academic, though few educators at the time were concerned about this. Despite the official goal of "science and technology for all", investigations served primarily to bolster research skills useful for college preparation. Poor students from inferior primary schools lacked the complex academic skills and background knowledge to perform such investigations. As they saw no obvious pathways leading from their present life situation to college and the professional workforce, moreover, many were unable to engage personally in these investigations.
   The root idea of guiding students' use of knowledge and information in inquiries that generate new knowledge and information relevant for practical decisions and actions, however, remains valid, but needs to be conceived in a broader, less exclusively academic way. In today's *knowledge society*, knowledge professionals and high-skilled production workers use complex background knowledge from their educations and their out-of-school experiences in unpredictable ways. They combine it with flows of real-time data and information, and knowledge modules such as online software tutorials, to respond to ill-structured problems and make decisions under conditions of uncertainty. The model of knowledge use in technological literacy education should be that of the *generalized knowledge worker*. Scientific research should be understood merely as a special case of investigation in knowledge work.
   This has three implications. First, problems selected for investigation must be broadened to include those affecting the immediate economic and social challenges of learners (e.g., problems of economic survival, health, safety and shelter, labor rights, exploitation, and trafficking). The educational efforts of some non-governmental organizations are exemplary, and can often serve as models for school-based programs. Second, full weight should be given to the vernacular knowledge of young people. Educators often downplay community-based, real-world knowledge,

using official school knowledge to 'trump' the knowledge young people bring to school. In the process, they unwittingly render learners 'ignorant' in the school setting. But vernacular knowledge ('street smarts') is highly relevant to problems engaging them that can be treated in technology education. Young trash recyclers, for example, know more about the sources of, and markets for, machine parts than schoolteachers. Third, as in the case of the trash recyclers, networked computers should be used as essential tools for gathering, storing, and manipulating information, and for acquiring knowledge on an as-needed or just-in-time basis.

4. **Guiding Student Action:** While participation in the ameliorative projects of non-governmental organizations was given rhetorical support in initial conceptions of technological literacy, it was rarely central to technology literacy education as implemented in schools. 'At-risk' young people, moreover, were too engaged in their own problems to take much interest in abstractions like *the environment* or *future generations*. Today, however, non-governmental organizations often recruit and train young people to take action on their *own* behalf, and the schools should link their students to these organizations. In Indonesia, for example, KOMPAK seeks to increase the awareness of Jakarta's child workers about their labor rights, and to protect them from exploitation and trafficking, through computer games and simulations.[16] In Nicaragua, the Ben Linder cyber café in Managua offers free computer education and Internet access to poor young people, and serves as a hub for community centers throughout the nation.[17]

Networked Learning and Technological Literacy

A greater synergy in metropolitan education can be attained through systematic cooperation between the school and non-government sectors in a broad educational network. Local nodes in the network would include mainstream schools, but also alternative schools, home schools, learning centers, and facilities organized by non-governmental organizations.

Significant learning could be promoted by adopting learning designs incorporating problem-solving tasks gradually approximating adult roles as networked knowledge workers. Today, interactive computer networks have spread to educational and cultural institutions, learning centers, libraries, and homes, making new forms of learning, approximating the conditions of knowledge work, practicable. The networks can be used to connect front-line educators (whether teachers, librarians, parents, or social activists) to a vast world of networked back-line sources of knowledge, information, and expertise. These might include knowledge workers in government, industry, the university, and non-governmental organizations. Networked educational *knowledge workers*, linking students and schools to back-line resources, would reconstruct schools as *network enterprises*.

A *networked school* organization for a metropolitan region would coordinate computer networks of (a) local front-line educational facilities and educators and (b) regional back-line facilities, educators, and resource people. The back-line structure can be used to organize virtual activities connecting young people throughout the region and providing them with many opportunities for hands-on use of computers for problem solving. It could also bring students from different groups in the community together from time to time, as appropriate and possible given the society's goals for interethnic and interclass solidarity, its transport infrastructure, and available resources, for significant in-person project-based learning. These in-person cross-cultural and cross-class experiences in networked-computer environments would be valuable for spreading "soft skills" throughout the region. They could be coordinated with preliminary and follow-up learning at local sites, using network technologies (e-mail, distance learning, list-servs, Web sites) to maintain virtual intergroup contact.[18]

SUMMARY

Globalization refers to the broad integration of economic activity and broad spread of computer networks on an essentially worldwide basis. The convergence of economic and technological developments has enabled a radical restructuring of enterprise, including a reorganization of the global division of labor. A fourfold division of the labor force has resulted, with a growing gap between knowledge professionals and high-skill production workers on the one hand, and routine production and service workers in the contingent economy and informal workers on the other.

Some segments of industry in the developed world rely increasingly on knowledge workers, but many segments, especially in the United States, still compete on the basis of economies of scale in mass market production and use of low-skill, low-cost routine labor. Routine workers are drawn primarily from native-born disadvantaged minorities and recent immigrants from poor nations. They compete for low-wage work without job security or health benefits. Developing nations in the global economy also employ knowledge professionals and highly skilled production workers, but a large part of the workforce exists in the informal economy, outside the government-administered system of employment and taxation.

Knowledge work requires a high level of academic preparation, flexible thinking and problem-solving capabilities, and human relations skills—especially abilities to cooperate in highly diverse work teams. Schools in the developed world with a conventional curriculum offer at-risk students few opportunities for high-level academic achievement, flexible thinking and problem-solving skills, or positive intergroup experiences. As a result, the students are poorly prepared for college and for knowledge work. Their bleak future prospects inhibit attachment to school and academic learning. A large proportion of young people in the developing world lack access to secondary education, and are subject to great risks of exploitation and abuse.

Educational reconstruction would seek to alleviate educational isolation of at-risk students, and engage them in learning tasks approximating those of knowledge workers in the global network economy. Technological literacy education for at-risk students should now be reconceived to emphasize the use of networked computers in addressing complex, poorly structured problems, including the urgent economic and social problems these students face.

Some educational outreach activities of non-governmental organizations offer exemplary models for this kind of education. In many cases they offer the only educational lifeline for child workers lacking access to secondary education, and hence to the academic knowledge and skills required for knowledge work. Some have successfully used networked computers to assist informal workers in converting their work situations into individual or collective business enterprises. Others have provided Internet access to poor young people as a means of bringing them together and providing them with health and education benefits, and to rally them in support of their own basic human rights.

The metropolitan educational organization should be reconceived as a network, rather than merely as a system of official schools. Such a network can support technological literacy through project-based networked learning, arrange activities bringing learners together across ethnic and class lines, and forge a synergy among the metropolitan region's educational providers, including school teachers and social activists in non-governmental organizations.

REFERENCES

Amirahmadi, H., & Wah, T. (2002). New York City: A social profile and alternative economic futures. *Journal of Urban Technology, 9*(1), 85-107.

Applebaum, E., Bailey, T., Berg, P., & Kalleberg, A. (2000). *Manufacturing advantage: Why high performance work systems pay off.* Ithaca, NY: Cornell University Press.

Belos, A. (2002). From jungle to capital the voting is electronic. *The Guardian,* (October 5). Retrieved from http://www.guardian.co.uk/international/story/0%2C3604%2C805092%2C00.html

Brown, P., Green, A., & Lauder, H. (2001). *High skills: Globalization, competitiveness, and skill formation.* Oxford: Oxford University Press.

Cappelli, P. (1999). *Employment practices and business strategies.* Oxford: Oxford University Press.

Castells, M. (1996). *The rise of network society.* Cambridge: Blackwell.

Clegg, D. (2004). *A medical reception on board the mercy ship M.U. Anastasis.* Retrieved from http://www.Healthserve.org/pubs/a0132.htm

Dizard, W. (1997). *Meganet: How the global communications network will connect everyone on earth.* Boulder, CO: Westview.

Doz, Y., Santos, J., & Williamson, P. (2001). *From global to meta-national.* Boston: Harvard Business School Press.

Ducatel, K. (Ed.). (1994). *Employment and technical change in Europe: Work organization, skills and training.* Hants, Aldershot: Edward Elgar.

E-Governance Institute. (2003). *Digital governance in municipalities worldwide: An assessment of municipal Web sites throughout the world.* Newark, NJ: Rutgers University.

Grantham, C. (2000). *The future of work. The promise of the new digital age.* New York: McGraw-Hill.

Gray, M. (1997). *Web growth summary.* Retrieved from http://www.mit.edu/people/mkgray/net/web-growth-summary.html

Hill, P. (1997). A public education system for the new metropolis. *Education and Urban Society, 29*(4), 490-508.

Holusha, J. (1995). First to college, then to the mill. *New York Times,* (August 22), D1.

Howe, W. (2001). *A brief history of the Internet.* Retrieved from http://www.walthowe.com/navnet/history.html

Karrier, T. (1994). Competition and American enterprise. *Challenge, 37*(1), 40-45.

Levine, D. (1995). *Reinventing the workplace: How business and employees can both win.* Washington, DC: Brookings.

Milberg, W. (1994). Market competition and the failure of competitiveness enhancement policies in the United States. *Journal of Economic Issues, 28*(2), 587-96.

NTIA (National Telecommunications and Information Administration). (2002). *A nation online: How Americans are expanding their use of the Internet.* Retrieved from ntia.doc.gov/ntiahome/dn/index.html

Ostroff, F. (1999). *The horizontal organization: What the organization of the future actually looks like and how it delivers value to customers.* New York: Oxford University Press.

Reich, R. (1992). *The work of nations: Preparing ourselves for 21st century capitalism.* New York: Vintage.

Richmond, A. (2002). Globalization: Implications for immigrants and refugees. *Ethnic and Racial Studies, 25*(5), 707-728.

Rubba, P. (1987). Recommended competencies for STS education in grades 7-12. *The High School Journal, 70*(3), 145-150.

Rubba, P. (1988). Integrating STS into school science instruction: Salience for inner-city poor and minority learners. In L. Waks (Ed.), *Technological literacy for the "new majority": Enhancing secondary science education through STS for urban/minority youth.* Washington, DC: U.S. Department of Education.

Rubba, P., & Wiesenmayer, R. (1988). Goals and competencies for pre-college STS education: Recommendations based upon recent literature in environmental education. *The Journal of Environmental Education, 19*(4), 38-44.

Rugman, A., & D'Cruz, J. (2000). *Multi-nationals as flagship firms.* Oxford: Oxford University Press.

Sassen, S. (1994). *Cities in a world economy.* Thousand Oaks, CA: Pine Forge.

Simon, P. (1998). Informal responses to crises of urban employment: An investigation into the structure and relevance of small-scale informal retailing in Kaduna, Nigeria. *Regional Studies, 32*(6).

Waks, L. (1987). A technological literacy credo. *Bulletin of Science, Technology and Society, 7*(1/2), 357-366.

Waks, L. (1991). Technological literacy for the new majority. In Majumdar, Rosenfeld, Rubba, Miller, & Schmalz (Eds.), *Science education in the U.S.: Issues, crises, and priorities.* Philadelphia: Pennsylvania Academy of Science.

Waks, L. (1992). The responsibility spiral: A curriculum framework for STS education. *Theory into Practice, 31*(1), 13-19.

Waks, L. (2004). The concept of a "networked common school". *E-Learning, 1*(2), 317-328.

Weinbaum, E. (1999). Organizing labor in an era of contingent work and globalization. In B. Nissen (Ed.), *Which direction for organized labor?* (pp. 37-58). Detroit, MI: Wayne State University Press.

ENDNOTES

1. Foreign direct investment in the United States, a meager $364 million in 1960, had by 1990 expanded to $48 billion, as foreign-owned companies competed for market share in the large and affluent domestic American economy (Milberg, 1994). In 1960 1.3% of American GDP was produced by foreign firms, while by 1990 they

were producing 8.4% of GDP (Karier, 1994).

[2] The AFL-CIO, which represented 30% of those employed in the non-farm sector in 1954, represented fewer than 12% by the 1990s. Despite these measures, they continued to lose domestic and global market share, and productivity remained stagnant (Milberg, 1994); the annual rate of productivity growth, which had exceeded 4% in the post-war period, dropped to 1.1% in the 1970s and 1980s (Applebaum et al., 2000, p. 227).

[3] In June 1993 there were 130 Web sites, of which only 1.5% had 'dot.com' domain names. By the end of 1993, 4.6% of the 623 sites were commercial; by the end of 1994, 18% of the 10,000; by 1995, 50% of the 100,000; and by the end of 1996, over 60% of the 650, 000 sites (Gray, 1997).

[4] The "flagship-five partners" model is developed in Rugman and D'Cruz (2000).

[5] This term has been in use for about a decade, and has been placed in wide circulation by Manuel Castells (1996).

[6] This is attributed to Doz, Santos, and Williamson (2001). In March 2000 both the *New York Times* and *USA Today* ran cover stories spotlighting *Fast Company,* the trendy new magazine that first captured the spirit of network enterprise.

[7] Peter Drucker originated the term "knowledge worker" in the 1960s to refer to workers holding diplomas from technical institutes and universities. He based the idea of the knowledge worker on the rapidly growing proportion of diploma holders in the advanced economies, and the increasing inputs from these workers in industries previously relying on low-skill labor. Today, however, the term 'knowledge worker" refers more to one with a mindset adjusted to a "smart" work space in a networked environment than to a holder of a specific diploma or bearer of discipline-based scientific or technical knowledge. The primary skills of such workers are knowing how to access, interpret, and apply new knowledge and information to add value to an organization. Drucker's insight has been verified, however, in that a growing proportion of college graduates are found in skilled production jobs (Holusha, 1995).

[8] Lawler compared Fortune 1000 companies in 1987 and 1995, finding that the proportion reporting some use of these teams increased in that period from 28% to 68%. A 1995 survey conducted by the Bureau of the Census and the University of Pennsylvania found that 13% of non-managerial workers in their sample were placed in self-managing teams (Applebaum et al., 2000, pp. 9-10).

[9] Cappelli's work provides economic explanations that supplement the ideology-based account in Brown et al. (2001) for why America, unlike Japan and Germany, depends upon a combination of high- and low-skilled producers.

[10] See Sassen (1994) for a more detailed account of the jobs in the service economy and their role in the workforce of today's 'global' cities.

[11] Interestingly, the access of so many people to these seaport mega-cities eases the delivery of world-class services on a worldwide scale. "Mercy Ships," a fleet of large-scale modern floating hospitals, for example, is able to reach a majority of the world's population quickly and efficiently, through port visits to mega-cities in the developing world for regularly scheduled clinics. Working with local organizations,

they are also able to travel inland to hold clinics and provide health education in nearby towns and villages (see Clegg, D. (2004). A medical reception on board the mercy ship M.U. Anastasis. Retrieved from www.healthserve.org/pubs/a0132.htm).

[12] See the article "Child Trafficking in Indonesia," http://www.waronwant.org/?lid=5034.

[13] The approaches of Rubba and Waks to technological literacy through 'science, technology, and society' were typical of that era (e.g., Rubba, 1987; Rubba & Wiesenmayer, 1988; Waks 1987, 1992).

[14] For one model of metropolitan education as a network, see Hill (1997). I am indebted to Hill's metropolitan network model, but the network I envision here is more inclusive.

[15] Information on the trash recycler program, funded in part by War on Want, can be found at http://www.waronwant.org/?lid=1751.

[16] KOMPAK is also funded in part by War on Want, http://www.waronwant.org/?lid=5034.

[17] The Ben Lindner cyber café receives support from WIRED International, a non-government organization promoting health, education, and inter-cultural dialogue through networked computers. Information is available at http://www.wiredinternational.org/nicaragua.html.

[18] This conception of a networked metropolitan learning Web is outlined in Waks (2004).

Chapter V
ICT in Schools:
What is of Educational Value?

Aidan Mulkeen
National University of Ireland, Maynooth, Ireland

ABSTRACT

This chapter considers how information and communication technology (ICT) can be used to achieve educational value in schools, and encourages teachers to focus on approaches that promote higher-order thinking. It examines the reasons for use of ICT in schools, and argues that clarity of thinking is needed in the face of popular beliefs about ICT. While highlighting the ways that ICT can contribute to important learning objectives, the chapter stresses that many uses of ICT may have little educational value. It argues that the real value of ICT in schools is in enabling more challenging learning activities that develop higher-order thinking, and offers a simple diagram that teachers can use to evaluate their use of ICT. Various ways in which basic technology can be used to promote higher-order thinking are explored. Finally, the chapter considers the factors within a school that are likely to encourage and sustain worthwhile uses of ICT.

INTRODUCTION

Across the world, there is a rush to include information and communication technology (ICT) in schools. In almost every country there is an ICT strategy, and teachers are being trained to use ICT. Teachers and schools leaders are keen to get the best possible benefits from ICT, and wonder how they should be using it. The aim of this chapter is to identify where the benefits lie, and to identify the general principles of how ICT can best be used.

WHY ICT IS IMPORTANT

Before identifying the best uses, we must first be clear about what we are trying to achieve. National ICT strategies often reflect a series of different aims for ICT. In general, four different rationales for use of ICT can be seen. These can be described in general terms as the economic, the social, the pedagogical, and the knowledge society rationales.

The economic rationale is perhaps the most obvious. ICT is increasingly a part of working life for many people. As Negroponte (1995) has so eloquently argued, the development of digital technologies is having an impact on economic structures, changing the way businesses operate, where businesses locate, and even allowing a whole range of new businesses to emerge. ICT skills are seen as important in helping to grow the kinds of business that rely on technology. As a result, developing ICT in schools is a key plank of economic development and competitiveness strategy for many countries.

The social rationale centers on the concern that in the spread of ICT skills, some people will be left behind. Technology is becoming a more important part of everyday life. The numbers of computers are increasing dramatically, as they become cheaper and more powerful (OECD, 2001, pp. 12-13). More and more people use ICT as a main source of their information about the world. By 2001, U.S. Internet users were spending more time online than reading newspapers or magazines, and 45% of Internet users even reported a reduction in their time watching TV as a result of their Internet usage (Jupiter Communications, cited in OECD, 2001, p. 14). As ICT becomes more pervasive and more frequently used, those without ICT skills may become increasingly disadvantaged or marginalized by their lack of skill.

The argument is almost parallel to the argument for literacy. In a society where few people can read, lack of literacy may not be a major barrier, as few things will be written down. Once writing is used by enough people, then those without literacy are at a disadvantage. They may not have the same access to information and may have to rely on others to get information for them. Imagine being illiterate 100 years ago, before radio and television. To find out about jobs or opportunities, or to hear political or public news, you would have to hear from someone else who could read.

The same argument can be applied to ICT. Those with good ICT skills have easy access to information about jobs, economic opportunities, learning opportunities, and news from a variety of perspectives. Of course, those without ICT skills have a variety of other media like newspapers and radio to choose from, but they are at a disadvantage in terms of the speed of access and the range of material they can access. For this reason, ICT is often described as a literacy, and the skill gap is seen as a major cause for concern. As Tom Alexander, then head of the Center for Educational Research and Innovation at OECD, stated:

> *...there are profound concerns now about the gaps opening up between the ICT 'haves' and "have-nots," between those who reinforce their access to, and use of ICT in education what they have and do at home, and those who have little of either. The gaps may become every bit as profound as earlier forms of rigid social and education selection.* (Alexander, 1999)

In this view, the rationale for having ICT in schools is to ensure that all young people get some exposure to ICT. If this does not happen, then there is a risk of a digital divide between those with ICT literacy and those without. Universal experience of ICT while at school is made more important by the uneven patterns of access to ICT at home. In most societies the richest and best-educated groups are also most likely to have home computers (Russell & Drew, 2001; Ginsberg, Sabatini, & Wagner, 2000). As a result, if schools do not act to redress this imbalance, the benefits of the digital age may fall to the already privileged social groups.

ICT as a Teaching Tool

Educators also make a pedagogical case for ICT. This view argues that ICT can be used to support teaching and learning. In this case, the aim is not to teach ICT skills, but to use ICT to teach other parts of the curriculum. ICT as a tool to enhance learning takes a variety of forms. For some people, it means using the computer as a tutor, asking the learner questions, and providing feedback to his/her answers. Examples of this approach can be seen in a variety of educational software, ranging from freeware available for download to sophisticated integrated learning systems (ILS). Others have seen the potential of ICT in the development of more open learning environments where exploratory learning could be facilitated. Examples include the seminal LOGO program, designed to provide an environment for learning mathematical and logical skills (Papert, 1980). This constructivist approach to learning may also be facilitated by use of "content free" software such as word processors (Underwood & Underwood, 1991). Use of such software has been reported to increase motivation and result in student work that is longer, better presented, and of better quality (Stradling, Simms, & Jamison, 1994). More recently, the development of the World Wide Web has brought additional possibilities for education. The vast range of material online can be used to enrich teaching. Perhaps more significantly, the Web can be used to allow students to research topics for themselves (OECD, 2001b, p. 22).

Following the rapid development of the Internet and the World Wide Web, new thinking has emerged that is centered on the concept of a knowledge society. This thinking begins with the idea that pervasive technology will change the kinds of things we need to learn in schools. If we have virtually unlimited information at our fingertips, do we really need curricula so reliant on recall of a set body of facts? In some cases we may even be teaching out-of-date factual material prescribed in a curriculum, while the students have easy access to more relevant and current material online.

Equally, it is clear that access to the Internet does not displace the need for education. In fact, making sense of the material you find online is often quite difficult. In our technological age, we have made information easier to get, but harder to digest. The information available online is difficult to handle for a number of reasons.

First, the material is hidden. The Web can be compared to a huge library, but without an index, and without sections for each topic. Most people either rely on some favorite site that they visit regularly, or they use search engines to find material.

Second, the material is of varied quality. Searching for any given topic, you may find a range of sites. Even when the irrelevant sites are disregarded, you may be left with a wide range of material, some aimed at junior classes in primary schools, and others aimed at postgraduate students. Some may be commercial sites, selling a product. Different sites may reflect different views, different interpretations of history, or different scientific paradigms.

Third, the material on the Web is often read out of context. When you pick up a book, you can read the chapters systematically, or at least look at the contents and see the context in which a particular page is presented. On the Web, search engines often bring the reader to a specific page, presenting just a single item without the sequence or context in which it belongs.

The citizen of a knowledge society may therefore require a different kind of education. They may have less need to memorize large quantities of factual data. On the other hand, to succeed in a world where they are constantly presented with a flood of unstructured and decontextualized information, they may need high levels of other skills. These might include:

- Information-retrieval skills, including searching and refining searches.
- The ability to assess information and select the most relevant, accurate, and appropriate material.
- The ability to interpret, synthesize, and develop a context for information.

These objectives are reflected in the Australian national goals for schooling in the twenty-first century, which includes the expectation that when students leave school they should:

- have the capacity for, and skills in, analysis and problem solving;
- have the ability to communicate ideas and to collaborate with others;
- have qualities of self-confidence, optimism, high self-esteem;
- have employment-related skills and positive attitudes towards lifelong learning; and
- be confident, creative, and productive users of new technologies. (Adelaide Declaration, 1999)

Critics: Risk of Low-Level Thinking

These arguments for the use of ICT in education, or some combination of them, have been widely accepted by governments and education authorities, which continue to invest heavily in ICT. But the drive to put computers into schools has also met with criticism from some writers. One of the frequent themes in the criticism is concern about the quality of work done with ICT. Stoll (1999, p. 4), for example, argues that much of what is done with computers in schools is not very challenging and may be less valuable than what it displaces. Healy (1998, p. 190) also expressed concern. She agrees that the analytical skills outlined above are increasingly important, but she suggests that the use of ICT in schools may actually retard the development of those skills, by providing instant answers and allowing little time for reflection. In addition, working with computers may displace working with real objects (Healy, 1998, pp. 220-221).

These concerns about the quality of work done with ICT raise serious challenges. We cannot simply assume that all uses of ICT in schools are worthwhile. Instead, we must seek the uses that offer educational experiences that are genuinely more valuable than what is displaced.

Some people are so enthusiastic about ICT that they would argue that any use of ICT is valuable, as it provides useful experience of the technology. However, that view does not stand against the rationales for use of ICT identified above. If ICT is being introduced to help develop a competitive economy, then it is true that giving students sophisticated experiences of the technology may help them to develop technology skills. But surely not all uses of the computer will provide anything that could be described as industry-relevant skills. In the most extreme example, if students used a repetition task administered on the computer, how could this be seen as providing experiences of ICT that would help in economic development or employment? Taking a broader view, industry might benefit more from young employees with good mathematical and scientific understanding, a high level of good sense or logic, and the ability to communicate well. If a student has these, surely he/she can learn the basics of word processing very quickly.

Similarly some very primitive ICT training in school may do little to bridge the digital divide. If the key skill for participation in an information society is good information-handling skills, then simply learning to browse the Web is of little benefit.

Types of Use of ICT: The Value Grid

These are not reasons to avoid using ICT in schools. Rather, they present an argument that we should focus our use of ICT on those types of use that promote the really valuable kinds of education. The diagram below shows one representation of where these valuable areas lie.

The bottom of the grid represents different types of uses of ICT in terms of whether they are concerned with learning about the technology, or concerned about using the technology to achieve other educational aims. Thus, learning to operate a particular package would be located on the left side, and using ICT to learn history would be on the right.

The vertical part of the diagram is used to classify uses of ICT in terms of the level of thinking or analysis involved. Uses that demand higher-order thinking skills are at the top of the scale, while uses involving only memory or repetition are at the bottom.

A few examples may help to illustrate these classifications. If a teacher is using ICT to teach the basics of word processing, and teaching this by working systematically through all of the menus one by one, then this is classified as learning about the computer, and involving only low-level thinking, and so is placed on the bottom-left of the graph. If a teacher is using a drill-and-practice program to reinforce basic recall of facts, then this belongs on the bottom

Figure 1. The ICT value hierarchy

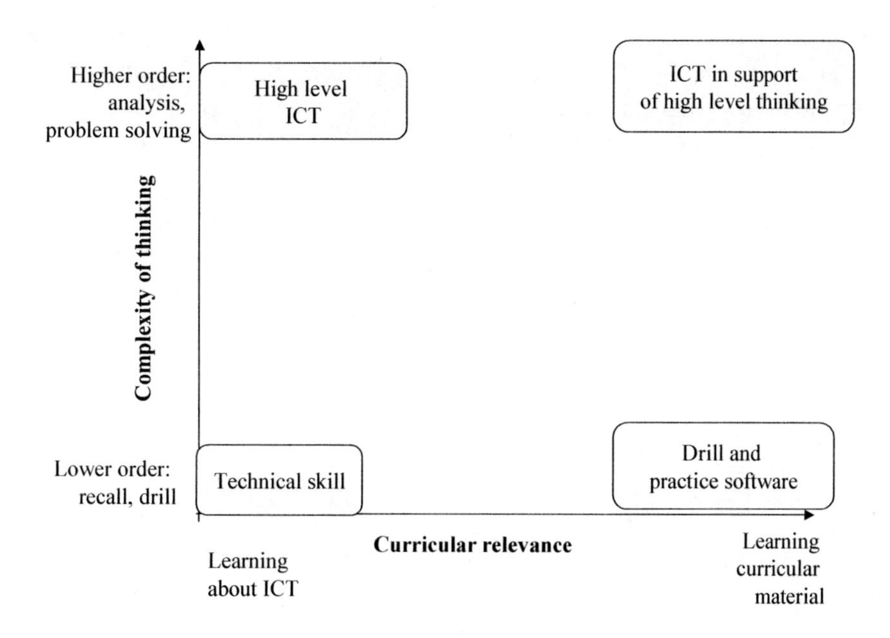

right, involving lower-order thinking but aimed at curricular learning.

If students were doing a project making a complex interactive Web site, this might be classified as involving higher-order skills, but focused on learning about ICT, and so be located in the top-left corner. If they were making a Web site that reported their research in history, that might be located in the top-right corner, as a task that was both challenging and curricular.

It is difficult to make absolute determinations about where each type of activity is located, as the real value of an activity depends a good deal on the subtleties of how it is organized. Take, for example, searching the Web to look for information on a particular topic. This could be a high-level task, involving students in finding material, making assessments of the value of the material, and developing a synthesis of their findings. On the other hand it could be a task involving cutting and pasting without even reading the material found, and have little educational value at all.

So which uses are of value? It depends on why we want ICT in schools. If we are introducing ICT to help build an ICT industry, then obviously the activities in the top-left corner are important, and the top right may be useful in helping to build good thinking skills. If the aim is to bridge the digital divide, then any of the uses along the top, those that involve developing high-level analytical skills, are likely to be useful. If the aim is to use ICT to support learning of other topics, then clearly the activities at the top right are important.

But what of the activities at the bottom of the grid, the activities involving only lower-order thinking. Do they have any place in schools? Let's start at the bottom left of the grid, with the teaching of basic ICT skills. At first glance this seems to be a worthwhile area to include. After all, one might argue that without the basic skills, it is impossible to do the more challenging tasks. But this argument leads to a lot of monotonous teaching of very low-level operational skills in schools, and sometimes the schools never find the time for the more valuable high-level uses. In reality, the level of technical skill needed to be able to engage in challenging activities with ICT is very low. Many challenging activities require only basic word processing or the ability to search the Web. Given the speed with which the operational skills can be learned when needed, and the speed with which the details of packages change, there is little to suggest that spending a lot of school time on basic skills represent a good use of resources.

The other popular use of ICT at the bottom of the grid is the use of ICT as a "drill-and-practice" activity to teach or reinforce basic skills. This may involve rehearsal of factual material, testing of factual knowledge, or doing repetitive mathematical tasks. Teachers and students often find these activities motivating, and believe that they are valuable in teaching or learning factual material. There is no doubt that repetition is an effective way of learning factual material, but it does little to promote understanding or the development of higher-order skills. It therefore does little to serve any of the aims for which ICT is included in schools.

So far, the argument of this chapter has been that ICT is an expensive innovation introduced into schools for economic, social, or pedagogical reasons. To address any of these rationales, it is the higher-order uses of ICT that are needed. Yet sadly, much of the use of ICT in schools is located at the lower end of the grid. These uses may have some value, but they do little to address the reasons why ICT is important in the first place.

So why is so much of the use of ICT directed at these lower-level uses? In part this is because not enough thought is going into why we are using ICT. Often use of ICT is driven by a general uncritical belief that ICT is a good thing. Sometimes this belief is supported by parents, who frequently encourage the use of ICT in schools without a clear analysis of why they believe it is of value (Selwyn, 2001).

Instead of this uncritical acceptance of ICT, we need to be much more demanding of it. We need to recognize that some uses of ICT are actually contra-educational, because they displace more valuable activities. Take for example the teacher who uses a computer drill package to teach the outcomes of an experiment, instead of actually getting the class to perform the experiment. In this case, an educational activity involving higher-order thinking has been displaced by one involving only memory. This is not intended to discourage use of ICT. There are plenty of educationally valuable activities which can be enabled by ICT. However, while recognizing the potential, it is important not to become uncritical enthusiasts. Any teacher considering use of ICT should start with this question: Which uses of ICT will help to move the educational experience into the higher parts of the grid?

WHERE ARE THE REAL BENEFITS?

The preceding section has looked at the reasons that are advanced for the introduction of ICT in schools, and has made the argument that the real benefits lie in uses of ICT that promote higher-order thinking. Assuming that you are convinced by this argument, the next logical question is how ICT can be used to promote the development of these higher-order skills. The answer often lies in the subtleties of how ICT is used.

The basic argument here is that ICT does not have any specific inherent value. It cannot be said that ICT enhances learning or detracts from learning. Instead, the value of ICT depends on the way it is used. The same technology, even the same applications, can be used either to enhance the quality of teaching and learning, or to reduce learning to a memory exercise. Finding the best uses of technology is complex. There are no absolute answers, as the best use depends on the teacher, the context, and the topic. Nevertheless, there are some general principles that can guide the search for the best practices. It seems intuitively obvious that we learn more when we do something to process material, than we do when we sit and passively listen to it. It is similarly obvious that when we are more interested, we put in more effort. We may be more interested in tasks that are real, or have some external purpose, than in tasks that are artificial tasks, invented by the school for no other purpose than to set a task. These basic premises suggest that we should seek teaching activities which involve:

a. students constructing, organizing, or developing the material; and
b. an interesting project or task, preferably an authentic project or task that has a function outside of the confines of the classroom, and preferably a task for which there is a real audience.

There are some characteristics of ICT that lend themselves to these types of useful educational activity. The sections below consider some common applications of ICT, and where they may have a role to play.

Word Processing

Word processing provides a very interesting example of how the same technology can be used in very different ways. Word processing is probably the most frequently used application of personal computers, and in many countries it is the most used software within schools (Mulkeen, 2002). Some of the use of word processing involves learning to work the software, often based on copy-typing material from worksheets or books. This is an effective way to teach the various functions and controls of the package, but does nothing to develop higher-order thinking.

The Quality of Output

Word processing also has characteristics that may be useful to a teacher trying to develop higher-order skills. Firstly, it allows students to produce work that is visually of high quality; this can be helpful in motivating students. For many students, working on a word processor is more motivating that working on paper. Part of this motivation comes from the novelty of working with computers, and as such may be short lived. But a more interesting part of the motivation derives from the ability to produce professional-looking work. Most teachers have probably experienced the satisfaction of seeing a document (perhaps a thesis or project) in printed form and the pride that comes from producing work that looks good. Imagine how much greater the satisfaction can be for those whose handwriting is uneven and messy. Word processing may be particularly helpful in developing writing skills in those with poor handwriting.

This effect has been noted in the U.S. in the ACOT schools (Apple Computers, n.d) and in the UK in the schools in the portables pilot project (Stradling et al., 1994). In these cases, it is not suggested that students use the word processor all the time and avoid learning to write well by hand. But using the word processor allows the skills of composing and spelling to be separated from the task of actually forming the letters. In this way, learners can be encouraged to write more and longer pieces of text without waiting for their handwriting to catch up.

This principle may appear to benefit only learners at the early stages of developing writing skills, but of course the motivational power of presentation can be of use at many levels. Take for example a school that asks each student to write a short story, and then produces these stories as a school magazine for distribution to parents. The task that the students are set is to write a story, and this could equally be done with paper and pen. The extra dimension added by the technology is that it is produced to a high quality, and can then be copied and distributed, providing a real audience for the work.

The Mutability of Text

A second major feature of work produced on a word processor is that it can easily be changed. This characteristic of the changeability or "mutability" (OECD, 2001b) of text opens up a range of possibilities. A few different studies have reported a fascinating pattern, that students who type their work produce longer work and use more evidence to support their arguments (Stradling et al., 1994; Mulkeen, 1998). How could this be? One explanation for this may be that when students type their work, they are able to add new ideas at any point. Thus for students who do not have a clear idea of all of the points they want to make in advance, a word processed document can become an evolving structure onto which ideas are hung, rather than a linear document.

Writing is one of the key skills in a modern society, and many jobs require the ability to produce clear written reports. Skilled writers draft and redraft their work, producing a series of improving versions before the finished product. Yet much of traditional school writing is a one-shot process. In this traditional model, students write material, then the material is marked by a teacher, and the students cast a cursory glance at the marks. It seems logical that if we want to teach students to write well, they should have the experience of reworking their documents in response to feedback. The process of producing the second draft can involve more learning than the first. Even the process of repairing minor grammatical and typographical errors can help to focus students on the accuracy of their work. One very interesting study (Gardner, Morrison, & Jarman, 1993) suggested that students who spent a lot of time using a word processor were more accurate in their spelling and punctuation even when handwriting. This is clearly not a feature of simply using the word processor, but a reflection of a structure where students were required to return to their work and correct these errors, thereby developing greater accuracy. In one project, teachers even deliberately devised tasks that involved students in redrafting texts, with a view to teaching them the skill of developing a document (Gardner et al., 1993).

One of the benefits of the ability to change the text is the ability for a teacher to break a writing task into a series of sub-tasks. A language teacher might, for example, ask students to do a small creative writing task, using the word processor. In the first response to the work, the teacher might consider only the story line and the vocabulary. Once some work had been done on these, a second review of the work might focus on grammar. This type of separation of tasks helps to build student confidence, as they are not presented with all the difficulties at once, and it also allows classes to focus on a particular aspect of the work.

Collaborative Writing

Another dimension to the ability to change text is that it facilitates projects involving collaborative writing. Much of the writing at school is done by individual students, yet we know that in the world outside of education, writing is often done by a group and that some of the best ideas evolve from collaborative activities. Within a school context, collaboration can help students to help each other, building confidence and skill. Writing with pen and paper severely discourages collaborative writing, because the only way that text can be merged or adapted is to rewrite the whole thing. By contrast, word processors make collaborative writing easier. Collaborative writing can take a series of forms. In one model, the students sit around a computer and discuss what they want to say, with one person typing the text as it is agreed. Alternatively, students may work on different parts of a text, and then exchange materials and review the materials, before finally assembling the completed document.

Because documents can be transmitted so easily (and cheaply) using e-mail, the technology facilitates collaborative writing at a distance. Schools can now easily collaborate with schools in other locations to engage in collaborative writing tasks. These provide an external motivation for enhanced effort, by providing an external audience for the work. Working together at a distance can also be a powerful way to build understanding in a cross-community or cross-border context.

Two examples will illustrate the potential. In the Irish Tech Corps, two primary schools decided to use a school magazine as a means to encourage student writing. The schools were located in different cities, 150 miles apart. In each class, each student was asked to do a piece of creative writing on the word processor. Each school then produced a class magazine containing all of the work from their own school, and selected highlights from the other school. The schools each sent copies of the magazine to the parents in their own school, and sent extra copies to the partner school. Thus each child had produced a piece of work that was published and sent home as part of a magazine, and the best writers were further rewarded by seeing their work reproduced in the magazine of another school. This low-cost and very simple project was highly motivational for the children involved, and resulted in great effort being taken with the work (Mulkeen, 1998).

Collaboration at a distance is used in a more formal way in the Dissolving Boundaries project, where schools on each side of the Irish border use ICT to work together on joint projects. The project can be focused on any areas that the teachers and students agree, but the impact is to provide a motivational and authentic task on which the teachers can build real curricular learning while at the same time providing an ongoing contact that helps to build mutual understanding (Austin, Abbott, Mulkeen, & Metcalf, 2003).

Word Processing and Higher-Order Thinking

Word processing, then, is not simply a matter of exchanging handwriting for typing. Instead it offers a new and very different way of writing, writing that can be continuously reshaped and reedited as required. At the same time it produces a very professional-looking product, which is motivating for students. These characteristics of the word processor make it an ideal tool for promoting the higher-level thinking skills identified earlier as so important.

As anyone who writes can attest, organizing your thoughts to produce coherent text is a most challenging task. Writing often forces ideas to be developed with more clarity, and highlights the weaknesses in unclear thinking. Asking students to write their understanding of an issue is therefore a very complex task, which involves making a synthesis of multiple sources, making judgments about what is important, and imposing a structure on the material.

Using a word processor to write allows the writer to experiment with alternative structures, struggle with alternative phrasing, and think more deeply about the choices being made in the writing process. More profoundly, the word processor allows teachers to set tasks that are of a bigger scale, that can be reshaped and modified as new ideas emerge. In short, the word processor makes it easier for students to do project work.

Project work is an attractive approach for a number of reasons. Firstly, it challenges each student to do some creative thinking and to develop ideas and structures for themselves. Secondly, it allows each student to work at their own pace, and at their own level of understanding. Thirdly, the quality of presentation, combined with the ability to improve the document as ideas emerge, means that the output is very rewarding for the student.

This is the essence of the importance of the word processor. It is at its most valuable when used by teachers to delegate more of the responsibility, research, and creativity to the learners. Where this approach works well, it has striking beneficial effects on learners. The ACOT schools, for example, reported increases in the quality of students' work, beyond what the teachers thought was possible. The benefits go beyond simply the quality of work produced. The ACOT schools reported that learners grew in confidence and developed as more autonomous learners (Tierney et al., 1992).

Presentation Software

A very similar set of characteristics make presentation software (such as Microsoft's PowerPoint) useful. Presentation packages allow students to make high-quality presentations with attractive visuals. As with word processing:

a. The quality of the finished product is attractive and motivating for students.
b. The ability to change and develop the work allows complex material to be built up gradually, and allows the teacher to guide development by focusing on one aspect at a time.
c. The ability to change the structure means that different sequences and structures can be tried easily.
d. The software facilitates groups or collaborative work, as students can merge pieces done separately, and can modify work done by their colleagues.

As with word processing, learning to work the package offers relatively little challenge. The real potential of the software is its ability to inspire and enable students to work on challenging projects.

There are two important differences between presentation packages, and word processors and presentation packages. First, the presentation package is primarily intended to accompany a verbal presentation. Verbal communication is an important skill, and one that, in some traditions, has received less emphasis in school than written communication. Yet many of the important events in life (job interviews, meetings, etc.) demand the skill to present ideas cogently in verbal form. Students, particularly the students with poor verbal skills, are often reluctant to practice making presentations verbally. Even confident speakers are often reluctant to speak in a second language, and so this software is of particular relevance for language teachers. Presentation packages provide an easy way to encourage students to make verbal presentations. The students often become engaged with the technology and want to present, just to show off their product. In addition, the sequence of slides can help presenters to make their presentations even when nervous. Finally, the requirement to have a prepared presentation forces students to plan and research a verbal presentation.

The second important characteristic of presentation software is that it is primarily visual. Much school work requires that students express their ideas in words, either written or verbally. Visual presentations give students the opportunity to express ideas in a visual form, using pictures or diagrams.

Teachers can use presentation packages to encourage students to do some research, organize their ideas, and make a presentation to the class. Where this is done, the students may be motivated by the opportunity to make a visual presentation, and by the sense of a real audience for their work. Where this has worked well, teachers have reported high levels of motivation, and benefits in quality of work, verbal skill, and confidence (Mulkeen, 2003b; Austin et al., 2003). As with word processing, the argument is not that the software is inherently valuable or that it produces learning gains. The argument is that the software facilitates and enables a teaching method that involves students doing meaningful and challenging projects for a real audience.

The Internet

One of the most exciting dimensions to ICT in recent years has been the development of the Internet. In most countries the Internet forms a key part of ICT strategy for education. National policies often prioritize increased Internet connectivity for schools. In practice, teachers use the Internet in very different ways, and many do not use it at all. The key uses of the Internet can be divided into three broad categories:

- The Internet as resource library (the World Wide Web).
- The Internet as a communication channel.
- The Internet as a place to publish.

The Internet as Resource Library (the World Wide Web)

The World Wide Web (WWW) can be seen as the biggest library in the world. The number of Web sites is far greater than the number of resources any school library could hope to hold, and the Web has the additional advantage that it is growing and being updated all the time. The Web therefore provides a resource of enormous potential to schools. Students can use it to access more timely, up-to-date materials than in textbooks, and at the same time can have access to a far greater variety of materials than are found within the school.

Following the arguments presented earlier, the key aim of our use of ICT in education should be to promote higher-order thinking skills. Access to a wide variety of sources presents an ideal opportunity to use teaching methods based on students doing research for themselves. Ideally students could find a variety of material, maybe including different perspectives or even opposing views, and make their own synthesis of what they find, based on their analysis and assessment of the material.

However, there are some difficulties. School libraries are generally small collections of carefully selected and classified material seen as appropriate for the students. By contrast, the Web is enormous, but has no central editor, so it is a library without an index, and the material is neither arranged systematically nor selected for its suitability. This presents a series of problems for school use. The first and most obvious is that students can wander into unsuitable material, which often means pornography, but could equally be racist, or material inciting hatred or terrorist acts. There is an enormous volume of pornographic material on the Web, and some of it is specifically designed to present itself to innocent users searching for other things. While there are filtering products available, these are not entirely effective, and schools may be concerned that some unsuitable material may slip through (OECD, 2001b).

Even where relevant material is found, it may not be useful. Material on the Web is at all levels, ranging from material published by primary schools, to leading-edge research. While this is one of the great strengths of the Web, it also means that a student may be presented with material at a variety of different levels, which may be difficult to handle. School texts are carefully written in age-appropriate language and are careful to explain each term when it is first used. Students accustomed to such carefully refined and selected material may find it difficult to adjust to interpreting material in more complex language.

A third difficulty is that the Web presents material out of context. In a book, the argument is built up systematically. It is assumed that the reader can see the chapters in sequence, and concepts that are explained in one chapter can be used in later chapters. The Web presents information out of context. When using a search engine, the reader is often presented with material from the middle of a text, without always having easy access to the preceding pages.

A further difficulty with the Web is that much of the material can be copied and pasted electronically into other documents. While this makes it an ideal research tool for the motivated student, there is always the risk that the less motivated will simply copy material without understanding it.

These characteristics of the Web make it problematic as a resource for student research. Teachers often report difficulties with students:

a. wandering into unsuitable material;
b. getting distracted by irrelevant material;
c. getting overwhelmed by a series of snippets of information and being unable to make a coherent synthesis; or
d. falling into the copy-and-paste trap, where material is copied but not necessarily read or understood.

Teachers respond to these problems in a variety of ways. Some conclude that the Web is not a useful learning tool, and simply avoid using it. These teachers may wonder why all the emphasis on connectivity is justified. Other teachers argue that learning to deal with a huge volume of decontextualized information and make some sense of it is a key skill in an information age, and so it is important that we develop those skills in our students. Of course it is unrealistic to expect that students can simply be "thrown in at the deep end" and expected to develop such skills themselves.

Teachers have tried different ways to make use of the resources of the Web. It may be helpful to see these as a scale with very open and unstructured approaches on one side, and very controlled use on the other. At one extreme there is totally unstructured Web research. In this model, students take a topic and are asked to search for useful material and report their findings. This approach requires that the students do the most thinking for themselves, and so has the greatest potential for learning and especially for the development of those important higher-level thinking skills. However, this approach also has the greatest risk of going wrong, as students can wander into the wrong material or fail to develop a worthwhile understanding of the material.

At the other end of the scale, there are highly controlled ways that teachers can make use of the resources available on the Web. One possibility is that the teacher might search the Web, find useful material, and bring it into the classroom. Alternatively the teacher might direct the students to a specific Web page and ask them to read through the material. These uses are highly controlled and so carry a much lower risk of going wrong. However they also offer much less opportunity for students to be challenged by the material or to have to develop analytical skills for themselves.

In between these two extremes, teachers have tried to develop ways of guiding students to use the Web more effectively, while leaving

Figure 2. Approaches to Web resources

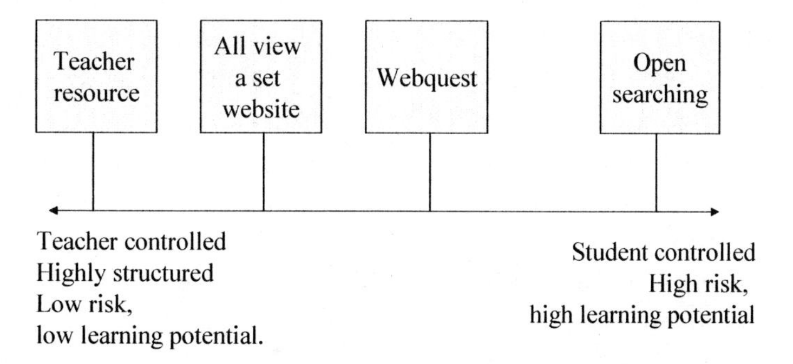

them with the challenge of finding and processing the material for themselves. One of the interesting models is the WebQuest.[1] A WebQuest is a semi-structured research activity, based on the idea that students are given a topic and a series of names sites, and required to develop a verbal presentation on the topic. This structure does not necessarily develop Web searching skills, but it does guide students to the relevant sites, and requires them to read, analyze, and synthesize the information. The model can be used at any level and with any age group. Increasingly, teachers are developing WebQuests and publishing them on the Web, so other teachers can avoid some of the preparatory work and begin by using or modifying an existing WebQuest.

The resources of the Web are not confined to material students can read. Some sites offer interactive activities that can be very engaging. Increasingly, the Web is the mode of distribution for educational software. Where software used to be packaged as a CD-ROM for sale to schools, it is now often available as a download. Much of the material involves a payment before it can be used, but there is still a good deal of material that is available free of charge. This material often falls into one of two main types—tests and simulations. Online texts or multiple-choice quizzes are very popular both with students and with Web developers (Mioduser, Nachmias, Lahav, & Oren, 1999). For the developers, they are one of the easiest kinds of material to produce. For students they are interesting and engaging, and they provide a useful self-check on learning. Students often prefer to be assessed by machines, because the machine is non-judgmental, and to be measured by it is less intimidating than to be measured by a teacher (NCET, 1996). However, if the aim is to develop higher-order thinking, then online tests are often less than ideal. Many are focused on recall of factual information (the easiest kind of multiple-choice test to devise) and are little more than a modern form of rote learning.

Simulations are less common and more difficult to produce. However, when they are well designed, they offer the student the opportunity to adjust different variables and see the impact on an overall system. The ability to "play around" with the variables contributes to the development of students' understanding of the relationship between variables. The weakness of the simulations is often their very open nature. Some students may need guidance or some semi-structured tasks to help them to experiment thoroughly with the simulation and to get the most from it.

The Internet as a Communication Channel

The focus of the previous section was on the use of the Internet as a resource. The Internet can also be used as a means of communication. This may not involve use of the Web, but may instead involve technologies like e-mail, computer conferencing, and videoconferencing. These are most frequently used to communicate with another school, a relationship that provides reciprocal benefits for both schools involved. One of the most obvious uses of inter-school linkage is in the area of language learning, especially where the schools are each learning the language of the other.

But this kind of linkage has a value beyond language learning. First, any kind of link to another school provides an audience for the work of the students. Having an audience is motivational for learners and encourages extra effort to produce better quality work. Second, linkage with another school provides an opportunity for building understanding across divides of culture, geography, or history. In the Dissolving Boundaries project in Ireland, schools in

Southern Ireland are linked with schools in Northern Ireland, and students work together to do curricular projects. The project research suggests that this approach of ongoing interaction over a sustained period is very effective in building a sense of knowing the other students, and a sense that they are "just like us" (Austin et al., 2003). Even when the link is not intended to bridge barriers of historical conflict, there is much to be gained from linkage with another school.

In designing a link with another school, it might be imagined that videoconferencing would be superior to e-mail. Videoconferencing has the advantages of immediate response, moving pictures, and the novelty of a relatively unfamiliar technology. However, e-mail too has its advantages. First, the communication is asynchronous and so is easier to schedule. Second, communication can be based on individuals communicating with individuals, allowing richer communication than a videoconference, which is more likely to involve larger numbers. Third, the written communication, although slower, may allow more space for reflection than a live videoconference (Austin et al., 2003).

The Internet as a Place to Publish

The Internet also has a value as a place to publish student work. While many schools have a Web site, this is often simply produced by teachers as a public statement about the school. From a learning viewpoint, the real power of the Web may be its potential to turn students into publishers. Encouraging students to produce projects that are published online has four attractive features:

1. It provides a real, and potentially very large, audience for the work. As noted earlier, having an audience is an important contributor to motivation and hence to the amount of effort invested in the project.
2. Publishing on the Web also raises the possibility of feedback from readers. It is very easy to ask readers of a Web site to send a message of response. Schools that have produced material that is of interest to others (such as local history) have often received very gratifying feedback from readers that were unknown to them. Messages from unknown readers further contribute to the sense of an authentic audience.
3. It provides a publishing platform that allows multimedia publications. Publication in a school newsletter may be restricted to text, or at best, color pictures. Publication on a Web site can include sound, moving images, animation, and a range of forms of expression. These can of course be distractions, but they can also add to the motivation and effort, as long as the focus remains on the intellectual content.
4. Web publishing does not have to follow the strictly linear form of conventional publication media. Publication on paper usually involves forcing interconnected ideas into a linear structure that is not quite ideal. Publishing on the Web allows ideas to be interconnected in a variety of ways, and so in some ways is closer to the way we think.

Numerous projects in different countries encourage student publication on the Web. Some even offer prizes and publicity for winning entries (see, for example, www.thinkquest.org).

Spreadsheets, Databases, and Datalogging

The discussion up to this point has centered on writing, and on students engaging in projects that are presented in writing or verbally. There are other applications that allow students to develop their understanding of numerical information. The most frequently used are probably spreadsheets, databases, and datalogging. In each case, the key characteristic is the ability to change things and observe the impact of the change. This characteristic of mutability allows students to experiment with numerical information.

With spreadsheets, one of the particularly powerful applications is the ability to draw graphs from numerical data. This can be used to give students an insight into graphs, but also to gain an insight into numerical patterns. Take for example the expression:

$$X^2 + 4X - 8$$

The ability to graph that expression provides an insight into the expression itself. Of course, it could be done by hand, but it is likely that if it is done on a spreadsheet, a greater range of numbers will be included. More importantly, once it is in the spreadsheet, it is very easy to try alternatives. One could ask what would happen if the expression began with "minus X^2"? Or what if it began with X^3? These "what if" questions are an important part of exploring mathematical expressions and are greatly facilitated by the computing power of the spreadsheet.

The graphing power of the spreadsheet also has uses outside of Mathematics. Spreadsheets can be used to make graphical representation of landforms in Geography, such as the cross section of a river, or translating contour lines into a cross section of a land area.

Spreadsheets and databases can also be used to analyze and correlate research data. A very good example of the power of this kind of analysis to facilitate student discovery emerged from the portables pilot project in the UK. Year 6 pupils in a primary school began to carry out a comparative study of the weathering of different types of rock, using the gravestones in two cemeteries. They entered information about the types and age of the stones used, and graded each entry according to how weathered it looked. Subsequently this information was transferred to a spreadsheet and plotted as graphs and charts for comparison with local headstones. While collecting data on the field trip, some pupils noticed the predominance of certain family names on the headstones. This led to a discussion about how to organize databases in order to facilitate searching and sorting. This project went on to involve an exploration of the impact of industrial pollution (by communicating with students in other regions) and an analysis of the age at death, which led to an exploration of social history (Stradling et al., 1994).

This principle of using ICT to facilitate students in doing research is further enabled by the use of datalogging equipment. Datalogging effectively attaches the computer to sensors, and the computer collects readings from the sensors at intervals. These allow students to carry out experiments that would otherwise not be possible, and to achieve more accurate findings than with conventional methods. Examples include:

- measuring the heat in endothermic and exothermic reactions;
- measuring the acceleration due to gravity;
- correlating the light intensity and temperature throughout the day; and
- measuring heat loss in liquids stored in different containers.

A full description of the potential of datalogging is beyond the scope of this chapter, but the interested reader may find more information, and example of good practice, on the Web.

These are only some of the uses of ICT that can support the development of higher-order thinking. The aim of this section has not been to provide an exhaustive list of the possibilities, but to provide enough ideas to illustrate the key messages about how to get the best from ICT. These key messages can be summarized as follows:

- Good use of ICT is use that pushes students towards developing higher-order thinking.
- Good use is not tied to any one application or package, but is dependant on how the package is used.
- Key skills to encourage are evaluation, analysis, synthesis, problem solving, and exploration. By contrast, applications that focus mainly on recall are likely to do little to develop higher-order skills.
- ICT can facilitate approaches that focus more on students doing project work that is educationally challenging. Two key characteristics of ICT were noted consistently:
  - **Mutability:** The ability to change and develop work continually.
  - **Audience:** The ability to present work to an audience outside the classroom.

MAKING IT HAPPEN: HOW SCHOOLS BEGIN TO MAKE GOOD USE OF ICT

This chapter began by looking at the reasons why ICT is important, and focused on the importance of developing higher-order thinking, no matter which rationale is used for the introduction of the technology. The focus then moved to the question of how ICT can be used to develop such skills. The main argument has been that ICT can be used to develop higher-order thinking skills in students. But it is also clear that not all of the uses of ICT achieve that aim.

Reaching those higher-order skills often demands a change in the teaching method. The basis of the shift is a transfer of responsibility to the learner. In each of the example activities described, the use of ICT has not been focused on ICT per se, but has been aimed at using the technology to encourage students to engage in challenging work. The technology was used to motivate and encourage students by providing a real audience for the work done. Students were encouraged to think more deeply about their work by the ability to change and reshape it. Students were encouraged to research material for themselves, and so to evaluate and process the material. Finally, students were enabled to do exploratory research for themselves through spreadsheets, databases, and data logging.

These may be summarized as a shift towards a more student-centered style of teaching, with an emphasis on students taking responsibility for their own learning, and on students constructing and developing their own understanding of the material.

Despite the weight of evidence pointing to the potential of ICT to facilitate higher-order student tasks, and the vast investment in ICT in schools, the reality is that in many countries the progress has been disappointing. Although schools have more equipment than they had a decade ago, and more time is spent in computer rooms, it is not clear that in all schools the technology is being used to promote high-level student thinking. In many cases ICT is being used to teach basic ICT skills or for routine rote learning tasks.

So, what does it take to encourage the use of ICT to achieve those more valuable goals? When teachers are asked why they do not make more use of ICT to support their teaching, one of the most frequent replies is that they do not have enough access to equipment at school. While equipment-rich projects were reporting a change in teaching method (ACOT, 1996) and other schools were reporting that lack of equipment was the barrier, it is not surprising that the idea of ICT as a catalyst emerged, and people began to suggest that provision of ICT could lead to a change in pedagogy.

More recent studies have shown that ICT infrastructure is not the answer. A major international study of ICT in schools showed that the schools that were using a student-cantered approach tended to have higher levels of equipment than average, but that there were also other schools with high levels of equipment that were using very traditional didactic approaches (Pelgrum & Anderson, 1999). A series of OECD case studies of schools innovating with ICT revealed that schools use ICT for very different purposes, and that there was little evidence that ICT had caused a shift in pedagogy. ICT had played a major role in helping to support a change in pedagogy in some schools, but it had not started the change (OECD case studies are available online at www.oecd.org). These studies call into question the idea that ICT is the catalyst for change.

Other studies have focused on the idea that the use of ICT to promote student-cantered learning is primarily a result of teacher beliefs. In the U.S., Becker (1999) reported that the use of ICT by teachers was associated with a "constructivist orientation" in the teacher.

The case studies of ICT as an innovation suggest that getting the best from ICT depends on a complex web of factors. Even with the same levels of equipment, some schools make better use of ICT than others. But there are also differences within schools, as some teachers begin to make use of ICT sooner than others. In most of the case study schools, even where ICT was well developed, some teachers still avoided using it, and the schools often made arrangements to exchange classes to work around them. So it seems that there are some issues at school level that can encourage or discourage use of ICT, and other issues at the level of the individual teacher.

At the level of the individual teacher, some of the key issues are:

- Easy access to ICT;
- A reliable technical infrastructure;
- Sufficient technical skill to feel confident with the technology; and
- A good pedagogical purpose for the technology. This includes believing that the technology is of real value, and having a good idea about how to use it.

Failing to have any one of these may inhibit the use of ICT by an individual teacher.

Making use of ICT takes time, as teachers need time to prepare their work and time to get familiar with the software. Some dedicated and enthusiastic teachers manage to make good use of ICT even when these conditions are not satisfied, but these teachers are unusually proactive, and often feel isolated and unsupported. The more usual and less dynamic teachers are unlikely to start using ICT unless it becomes part of a wider school movement. Where schools have successfully gone beyond the "lone pioneer" stage, this has often been associated with a few key factors at the school level.

First, the school leader plays an important role. The drive provided by enthusiastic school management can allow other difficulties to be overcome. The school principal can facilitate teacher meetings, ensure resource availability, and ensure that timetabling and other arrangements are adjusted to facilitate the adoption of ICT. Perhaps more importantly, the sense that use of ICT is an innovation supported by the management is motivating for teachers, and encourages teachers to explore the possibilities.

Second, teachers are often supported in their first steps with ICT by an enthusiast or "ICT champion" within the staff. This is sometimes a formal role, perhaps that of ICT coordinator. In other cases a teacher assumes the role voluntarily. Either way, the presence of a mentor that is supportive and encouraging can greatly facilitate teachers in the early stages of adoption of ICT. This mentor serves a dual role, by providing a point of contact for discussing the pedagogy to be employed, and by solving minor problems with the technology that might otherwise stop the innovation entirely.

Third, teacher adoption of ICT is often encouraged by participation in professional development activities in the school, with colleagues. In most countries, teachers are offered or required to attend training courses in ICT. This kind of training has some benefits, but often fails to address the specific context of the individual school. By contrast, training for the staff conducted in the school has the double advantages of:

a. addressing the context (infrastructure, etc.) of the school, and
b. providing an opportunity for staff to come together to consider the possibilities of ICT as a group. This has often helped to provide a "critical mass" of interested teachers.

CONCLUSION

This chapter began with a question; the question was why ICT is important in schools. It is clear that there is a good deal of ambitious talk about ICT, and that much of this may be overstated. ICT is justified by a series of rationales, based generally on the economy, society, and pedagogy. In each case, it was argued, ICT is only of value if it reaches the higher-order thinking skills.

Using ICT in the classroom does not automatically address these higher-order skills. Many applications of ICT involve learning low-level operational skills or memorizing facts. But there are ways in which ICT can be used to develop higher-order thinking. Many of these involve encouraging students to engage in project work, or work which requires them to discover, assess, and construct information.

Getting the higher-order benefits is not a question of technology, but one of pedagogy. ICT can help. Many of the ideas described would be much more difficult to implement without the technology. But it is also clear that the technology is not the driver. Making the best use of ICT depends on a series of factors, but one central issue is the teacher's beliefs about teaching. If a teacher is convinced that students can learn well through the kinds of student-centered work described here, then the teacher is much more likely to find a way to implement it. By contrast, where teachers cannot see the value of this approach, any amount of infrastructure can be used to support more traditional didactic practices.

This has a series of implications. For teachers, the key is to focus on how ICT is used, rather than on how much or how often it is used. Quality of use does not depend on the sophistication of the task, but on the sophistication of the thinking involved in completing it. Using the illustration in Figure 1, teachers should seek to move their use of ICT further towards the top of the graph—towards the higher-order thinking.

For schools, the key is to encourage teachers towards the appropriate uses. If there is not a school-wide interest in ICT, then it is difficult for individual teachers to feel supported to experiment with the possibilities. If the general view of ICT is as a technical skill area, then it is difficult for teachers of other subjects to begin to use it as a pedagogical support. One key task is to build a school vision of the potential of ICT in achieving the important aims; leadership, ICT champions, and whole-school work seem important steps in this direction.

REFERENCES

ACOT. (1996). *Teaching learning and technology: A report of 10 years of ACOT research.* Apple Computers.

Adelaide Declaration on National Goals for Schooling in the Twenty-First Century. (1999). Retrieved from http://www.curriculum.edu.au/mceetya/nationalgoals/

Alexander, T. (1999). New OECD initiatives on ICT, education and learning. *Proceedings of the Dissolving Boundaries, ICT and Learning in the Information Age Conference,* Dublin, Ireland.

Apple Computers. (n.d.). *High school writing. Apple K-12 effectiveness report no. 5.* Retrieved from www.apple.com/education/k12/leadership/effect5.html

Austin, R., Abbott, L., Mulkeen, A., & Metcalf, N. (2003). Dissolving boundaries: Cross national cooperation through technology in education. *The Curriculum Journal, 14*(1), 55-84.

Becker, H. (1999). *Internet use by teachers: Conditions of professional use and teacher-directed student use.* Teaching, Learning and Computing: 1998 National Survey Report #1, Center for Research on Information Technology and Organizations, USA.

Gardner, J., Morrison, H., & Jarman, R. (1993). The impact of high access to computers on learning. *Journal of Computer Assisted Learning, 9*(1), 2-16.

Ginsberg, L., Sabatini, J., & Wagner, D. (2000). Basic skills in adult education and the digital divide. *Learning to Bridge the Digital Divide.* Paris: OECD.

Healy, J. (1998). *Failure to connect.* New York: Simon and Schuster.

Mioduser, D., Nachmias, R., Lahav, O., & Oren, A. (1999). *Web-based learning environments (WBLE): Current pedagogical and technological state.* Research Report No. 54, Science and Technology Education Center, Tel-Aviv University, Israel.

Mulkeen, A. (1998). *Irish Tech Corps—External review.* Report of the External Evaluation of Irish Tech Corps.

Mulkeen, A. (2002). *ICT school census.* A Report to the National Center for Technology in Education.

Mulkeen, A. (2003b). *SchoolSat: Satellite Internet access for schools. Pedagogical evaluation report.* Report to the European Space Agency and the National Center for Technology in Education.

NCET (National Center for Educational Technology). (1996). *ILS phase II evaluation report.* NCET.

Negroponte, N. (1995). *Being digital.* London: Hodder and Stoughton.

OECD. (2001). *E-learning, the partnership challenge.* Paris: OECD.

OECD. (2001b). *Learning to change.* Paris: OECD.

Papert, S. (1980). *Mindstorms: Children, computers and powerful ideas.* New York: Basic Books.

Pelgrum, W. (1993). Attitudes of school principals and teachers towards computers: Does it matter what they think? *Studies in Educational Evaluation, 19*(2), 101-127.

Pelgrum, W., & Anderson, R. (1999). *ICT and the emerging paradigm for life long learning: A worldwide educational assessment of infrastructure, goals and practices.* Amsterdam: International Association for the Evaluation of Educational Achievement.

Russell, N., & Drew, N. (2001). *ICT access and use.* Research Surveys of Great Britain, DFEE Research Brief No. 252.

Selwyn, N. (2001). Promoting Mr. Chips: The construction of the teacher/computer relationship in educational advertising. *Teaching and Teacher Education, 17,* 3-14.

Stoll, C. (1999). *High tech heretic.* New York: Doubleday.

Stradling, B., Simms, D., & Jamison, J. (1994). *Portable computers pilot evaluation report.* National Foundation for Educational Research, NCET.

Tierney, R. J., Kieffer, R., Stowell, L., Desai, L., Whalin, K., & Moss, A. (1992). Computer acquisition: A longitudinal study of the influence of high computer access on students' thinking, learning, and interactions. *Apple Classrooms of Tomorrow Research Report No. 16,* Apple Computer, California.

Underwood, J., & Underwood, G. (1991). *Computers and learning : Helping children acquire thinking skills.* Oxford: Blackwell Publishers.

ENDNOTE

[1] The WebQuest model was developed at San Diego State University. Details are available at http://edweb.sdsu.edu/webquest/.

Chapter VI
Web-Based Technologies, Technology Literacy, and Learning

Wan Ng
La Trobe University, Australia

ABSTRACT

This chapter consolidates information regarding the role of technology in K-12 education, including policies related to it at a global level, the current status of its use in the classroom, and its impact on student learning. Its main focus is on the World Wide Web where its rich source of information and educational tools remains largely untapped in many K-12 classrooms. The chapter provides an illustration of how some Web-based technologies can be used to promote constructivist learning and foster the development of technology literacy in K-12 students. It also informs researchers and educators of issues and challenges faced by teachers and students in the classrooms in using these Web-based technologies and resources as pedagogical tools for concept development and promoting technology literacy.

INTRODUCTION

The shaping of today's contemporary society can be recognised in the quick pace of the integration of technological development in all sectors of life. In the business world, the impact of technology has seen many routine tasks, both in administration and production, streamlined and reduced in the amount of time required. Employees in the 21st century are focused on tasks that require higher-order thinking skills (Weil, 2002). The transformation in the activities of the workplace from the industrial era to the current information/knowledge era has meant that a different set of skills is required of the workforce of today. In the industrial era, the workplace was very much hierarchically structured so that workers were told what to do, jobs were routine and stable, and workers dominantly used the same set of skills throughout their careers. Re-training and lifelong learning were not common features of that era. However, in the 21st century:

Organizational structures need to change. Hierarchies need to be broken down and networked organizations developed. Successful organizations are flexible and able to adapt quickly. Group and team working, often cross functional, geographically displaced and changing frequently, stimulates creativity and innovation, enhances communication and knowledge sharing, and utilizes the best skills and experience on every task. (Oxbrow, 2000, p. 2)

To prepare students for these trends in society, there is a need for them to be reflected in the teaching and learning in the classroom. As part of this preparation, there is a need to be aware of the divergent aspects of technology—from the mechanical thoughtless machine that takes over routine tasks, to the sophisticated and partially automated machine that requires the user to have a complex understanding of the task being done and to be able to critically evaluate sometimes contradictory information and options that the technological process produces. Both of these dimensions are relevant, but in different ways, to education and work.

At the school level, students need to be prepared not only to enter the workforce, but also to become informed and active citizens who are able to critically evaluate information reported by the media, information found in books, magazines, and journals, as well as on the World Wide Web, so that they can shape their participation in social life. They need to become critical thinkers who are able to make decisions about matters that will affect their own lives and those of people around them. This trend in education of moving towards process knowledge (learning to learn and think) and critical thinking development has been discussed by Good (1999). In his paper "Future Trends Affecting Education," Good describes the evolving nature of education and also sees future trends where schools will be results-driven and striving towards high achievement as they become more accountable and are increasingly required to compete for students and funds. As a consequence, Good argues, there will be more emphasis on process (rather than content) knowledge and the development of students' critical thinking skills. The presence of technology will drive changes in the education system by shifting the balance between the two sets of conditions listed in Table 1 (Good, 1999, p.12).

Table 1. Evolving nature of education (Good, 1999, p. 12)

From...	***Toward...***
School time	Learning anytime, anyplace
Teacher-centered	Student-centered
Textbook funds	Education resource funds
One pace for all	Different rates and styles of learning
Buildings	Multiple access points for learning
Mass instruction	Personalized instruction

The increasing presence of more sophisticated technologies in life, including in schools, has meant that since the mid-1990s, across the globe, policymakers have made technology an explicit priority in educational policies and associated planning documents (Meredyth, Russell, Blackwood, Thomas, & Wise, 1999, cited in VCAA, 2002). These documents include:

- In the United States, *Getting America's Students Ready for the 21st Century* sets out the long-range, national goal to improve student achievement through the use of technology in education (p. 25).
- In the United Kingdom, the revised *National Curriculum* requires that information technology be more consistently integrated into the curriculum (p. 25).
- Denmark's new *Primary and Secondary School Act* identifies information technology as an educational priority (p. 25).
- Norway is now implementing a comprehensive *New Information Technology* program as part of its *National Plan* for 1996-1999 (p. 25).
- Information technology has recently been made part of the basic general education in Finland (p. 25).
- Information technology education has been identified as a major goal in New Zealand's *Education 1997-1999 Government Strategy and Education for the 21st Century* (p. 25).
- Singapore's *Masterplan for Information Technology in Education* (MOE, 1997) sets out strategies for achieving national milestones for the integration of information technology in education, along four key dimensions: curriculum and assessment, content and learning resources, physical and technological infrastructure, and human resource development (p. 25).
- In Thailand, the National Information Technology Committee developed a policy for information technology in schools that was approved by the Cabinet in 1996. The policy seeks to create prosperity and social equity among all segments of the population through use of information technology (p. 26).

In Australia, similar school charter priorities have been identified, for example, in Victoria, the *Learning Technologies in Victorian Schools 1998-2001* statement (DoE, 1998) outlined a visionary stance explaining objectives and targets to be achieved in schools. One of these targets included the education of students to become routine and competent users of learning technologies in their daily programs in schools. In the Asia Pacific region, the ideology is that technology will enable students and teachers to get better access to information and assists with bridging the digital divide in education. However, in terms of technology use in education, there is at present a wide disparity in the region (UNESCO Report, 2003). Countries that have launched schoolnets at the primary, secondary, and tertiary levels include Australia, New Zealand, China, South Korea, Japan, India, Malaysia, and Thailand. However, other countries in the region are still far behind this level of implementation.

ISSUES IN THE USE OF TECHNOLOGY IN CLASSROOM EDUCATION

In the same way that technology has transformed and revitalised businesses in many ways, though not necessarily always positively, education policymakers in developed nations had desires and expectations in relation to the mirroring of this transformation in education. They attempted to realise these expectations by investing vast amounts of money in getting schools and tertiary institutions connected. The pace of change in schools as a result of this presence of technology has been progressive, albeit at a less dynamic rate than in the tertiary and business sectors. While the impact of technology on K-12 education is currently relatively minor in comparison, and the issues surrounding this assertion will be discussed later, its potential and impact in the long run could be profound, complex, and exciting (Good, 1999). Not only is the pace of change at the school level slow in comparison with the business sector, it is well behind that seen at the students' personal levels and in the home (Australian Bureau of Statistics [ABS], 2000). The ABS report was based on a 1998 study which investigated the types of information technology skills students and teachers were using in the classroom. Teachers and students from the three major Australian education sectors—government, Catholic schools, and independent schools—were surveyed. Data received from 220 school principals, 1,258 teachers, and 6,213 students in the final year of primary school and the final year of junior secondary school found that most of the computer-based skills that students possess were learned in the home, and students' computer skills in many advanced areas such as Web authoring and creating presentations, and use of communication tools on the Internet such as chat forums, far exceed those of their teachers.

The full potential of technology in learning at the school level continues to be debated. Sceptics have cast doubt on the potential of technology in transforming education, arguing that technological advancements decades ago did not result in television or radio having much impact on education (Good, 1999). Such arguments are somewhat skewed and ill-informed. Educational technology such as the computer has a component of 'intelligence' in it that most other technologies do not possess. Learners can be actively engaged with the computer in ways that they cannot be with other technologies, because computers provide a means of responding to learner needs and control by accessing information within seconds and there is provision of instant feedback on a task performed. At a cognitive level, however, where the impact of computer-based technology on learning is still being researched and investigated, there is continuing debate. While advocates (Salomon, Perkins, & Gloerson, 1991; Bork, 1992; Lockhard, Abrams, & Many, 1994; Jenkins, 1999) argue that technology enhances learning, others are more apprehensive about the changes that it brings to the classroom (van Dusen & Worthen, 1993; Settlage, 1995; Cummings, 1996; Wellington, 2003). Lankshear, Snyder, and Green (2000), cited in VCAA (2002), state that technology in education is "...yet a further form of social control or enforced consumption, which promotes the interests of state and corporate sectors" (p. 1).

While millions and even billions (Fulton, 1999; McKenzie, 1999) of dollars have been invested into getting schools connected in different countries, questions have been raised regarding the huge level of investment and expenditure and the benefits it brings to students' learning in the classroom. At the global level, there is insufficient detailed research in this area and there is little definitive research data to support the claim that technology in the classroom indisputably leads to improved learning (Johnson, 1996; Fulton, 1996; McKenzie, 1999; Hardin & Ziebarth, 2000). Evaluating and measuring the effectiveness of technology in teaching and learning is complex and complicated by a number of variables such as the teachers' technical competency and/or instructions, the student's ability, the quality of the equipment used and the user-friendliness of the software, and the contribution of the social environment to the learning in a setting which is technologically enhanced. New methodologies to measure the full potential of technology in the learning process need to be developed.

Apart from the lack of effective methodologies to measure the contribution of technology to the learning process, the implementation of technology usage itself in the classroom is plagued by a number of issues (Mckinsey, 1997; Stevenson, 1997; Ofsted Report, 2001; Jenkins, 1999; Tebbutt, 2000; Ng & Gunstone, 2003). The use of technology in teaching is *ad hoc*, ranging from enthusiasts who have explored and put into place strategies for its frequent use in the classroom, to those with varying degrees of enthusiasm and skills for these technologies, many of whom have reacted only in response to non-pedagogic pressures such as the necessity to include in their performance review the use of technologies in their teaching, or because the use of technologies is in the school charter. These sorts of issues relate to obstacles that prevent the frequent use of computer-based technologies in the classroom. Fulton (1999) asserts that the obstacles relate to access, teacher professional development, and school support. This is supported by findings of Ng and Gunstone (2003), who identified the six main obstacles to the greater use of technology in the area of Science teaching and learning as:

- the difficulty in getting access to computers and associated equipment,
- the lack of time to investigate and plan computer-based teaching and learning activities,
- the lack of financial support to purchase equipment and software,
- the lack of teachers' skills and knowledge of appropriate applications using computer-based technologies,
- the lack of suitable software programs at the school level, and
- management issues in terms of student behaviour related to lack of equipment and technical skills in using the hardware or software programs

These obstacles are not separate entities such that when one obstacle is addressed, the implementation of technology usage automatically falls into place and running technology-based activities becomes smooth and trouble-free. These obstacles are interdependent like a 'chain' where a broken link at any one point would produce negative effects for the entire chain. The support from administration, both at the government and local level to ensure adequate equipment is available for teaching, is necessary as is the need to support teachers with preparation time and professional development opportunities to develop skills and time afterwards to plan and practise.

Research into the effectiveness of using computer-based technology in the teaching and learning processes is a priority area if we are to realize the full potential of its use. However, methodologies for measuring effectiveness in these environments so that the researcher is able to distinguish between learning achieved as a result of the use of the technology or other pedagogical support such as teacher instruction or the social milieu need to be very carefully designed.

While the impact of technology on students' learning at the cognitive level is somewhat inconclusive, the motivational impact is quite clear. There are numerous reports over an extended period (e.g., Kromhout & Butzin, 1993; Dywer, 1994; Pedretti, Mayer-Smith, & Woodrow, 1998; Forcheri & Molfino, 2000; Mistler-Jackson & Songer, 2000; Ng & Gunstone, 2002; Wallace, 2002) in the literature that support the motivational advantage that computer technology brings to the classroom. Students are motivated to learn with technology as they have ownership and control over their learning in terms of pace and choice of Web site content (Ng & Gunstone, 2002). However, issues of weaker and less confident students appearing to be aided by technology but not actually learning, as well as those associated with students from a non-English-speaking background, need to be addressed (Ng & Gunstone, 2002).

LEARNING IN A TECHNOLOGY-ENHANCED ENVIRONMENT

While there are still many issues to be addressed in terms of 'effective' learning using technology, at the theoretical level, learning that is mediated by technology has been discussed by many authors, among them Semple (2000). She argued that the application of learning theories to the use of technology in the classroom is a "matter of fitness for purpose" (p. 27), and depending on the learning theory that is dominant at the time, appropriate learning environments can be created for meaningful learning to take place. In this chapter I will discuss the contribution of the constructivist learning theory to a technology-enhanced learning environment. Constructivism has different meanings for different people and takes many forms (Matthews, 1998), but this chapter will be concerned with educational constructivism and the theories behind it.

Constructivism is a theory of knowledge that offers particular explanations of how we come to know what we know. As a learning theory, it has been highly influential in Western education in the last three decades and a vast amount of literature has been accumulated on it, particularly in the area of Science education (Matthews 1993, 1998; Driver, Squires, Rushworth, & Wood-Robinson, 1994; Fensham, Gunstone, & White, 1994; Solomon, 1994; von Glaserfeld, 1995; Phillips, 1995, 2000). The two notable theorists quite separately associated with educational constructivism were Jean Piaget and Lev Vygotsky. Piaget's (1955, 1972) constructivism is also known as personal constructivism and is based on his cognitive developmental theories, which propose that concept formation in the individual follows a clearly defined set of stages that must be experienced sequentially by that individual. Piaget's constructivism is also called cognitive constructivism. The underlying principle in cognitive constructivism is that knowledge resides in individuals and that it cannot be given or transmitted complete to them by their teachers. Learners must construct their own knowledge in their minds and build upon the knowledge through experiences. Real learning can only take place when the learner is actively engaged in the process, either at the operational level where the learner is engaged in physical manipulations or at the cognitive level where (s)he is mentally processing information or stimuli.

Vygotsky's (1962, 1978) social constructivism makes similar assertions to Piaget's cognitive constructivism on how learners learn—that is, that knowledge is progressively built up and continually re-interpreted, but Vygotsky places more emphasis on the social context of learning. In his theory, the learning process involves interaction with other individuals where culture and society will influence the learning. A difference between cognitive and social constructivism is that in the former, the teacher plays a limited role, acting as a facilitator, whereas in the latter, the role of the teacher is active and involved in helping students to grasp concepts by guiding and encouraging group or other analytic work.

Educational constructivism, as applied in many classroom situations, draws on the cognitive and social theories of both Piaget and Vygotsky, and sees learning as a dynamic and social process. A technologically enriched environment is potentially consistent with the underlying assertions of constructivism in that the learner can be an active participant in the construction of knowledge, and that existing knowledge and a socially interactive environment are factors that affect this construction of knowledge. The interrelationship between constructivism and technology as revealed by empirical research has been discussed by Nanjappa and Grant (2003). Their discussions based on studies in a variety of settings, including teacher education, online learning, and K-12 education, looked at constructivism in light of collaborative and cooperative learning methods, engaging in critical and reflective thinking, evaluation through electronic portfolios, and a critical look at emerging teacher roles within constructivist paradigms.

In a technology-mediated learning environment, the interactive, open, and non-linear nature of learning requires learners to be actively analysing, evaluating, and making decisions while manipulating the information at hand in order to construct new knowledge or solve a problem. They will constantly have to compare their own prior knowledge of a body of information with that presented in the learning environment and seek means of re-confirming their prior knowledge or to de-construct and re-construct new meanings. In such an environment, collaborative learning is possible at both local and global levels, and the flexible learning that is possible with technology provides among other things self-paced, individual learning pathways for learners to proceed along.

Blended into constructivist learning theory is the theory of constructionism. In his books *Mindstorms* (1980) and *The Children's Machine: Rethinking School in the Age of the Computer* (1993), Papert, a student of Piaget's, linked constructivism to technology. He devised the programming language Logo for children to develop their cognitive skills in mathematics by constructing "turtle" graphics or mathematical models on the computer screen. It was from studies of children using Logo that he developed the concept of constructionism. According to Papert (1991) students are engaged in learning when constructing a public artefact that others will see, critique, and use, be it a sand castle or a theory of the universe. Constructionism is linked to constructivism in the students' active learning and learning by doing, all of which takes into consideration students' prior knowledge and experiences. For instance, Anderson and Witfelt (2003) reported on a project where students design a 'game' to connect computer-based and experimental learning environments. The project involved 12- and 13-year-old students studying home economics and working in groups to solve problems of survival by learning about food of various kinds, the principles of cooking, and concepts of energy and nutrition via an interactive game prior to cooking in the kitchen. The students also engaged in "storylines pedagogy where they used a MOO-environment as the software genre. A MOO is an Internet-based multi-user environment that allows multiple users to share a community of rooms, virtual spaces to explore specific concepts or engage in particular kinds of interactions and objects based on text, and to interact with each other (Holmevik & Haynes, 1998). Such an environment offers students the opportunity to undertake problem-solving tasks where they explore, investigate, and create narratives based on their own ideas and experiences. A MOO environment encourages teamwork and promotes integration of learning areas across the curriculum.

For example the "storyline" could be along the theme of throwing a party (Anderson &

Witfelt, 2003). The students take up employee roles in various sections of an imaginary ecological restaurant, such as the economic department, the production planning department, and the kitchen. They design a business plan, investigate the cost of production of different cooked products, and physically experiment with various recipes to find out about the cheapest and best way of producing each recipe. Skills in science, mathematics, English, and those associated with home economics are developed in this particular example of a storyline. As a result of this evaluative task-oriented groupwork, the MOO prototype promotes teamwork and higher-order thinking skills in students. In this environment, where active construction of new knowledge and learning at an individual and a social level is taking place, and where students are developing both technical and cognitive skills, they are developing to be technologically literate people. The concept of technology literacy is the focus of the next part of this chapter.

DEFINING TECHNOLOGY AND TECHNOLOGY LITERACY

As indicated at the beginning of this chapter, the term technology in itself has a very broad meaning in education, as in other contexts. It embraces any piece of equipment that assists with the teaching and learning processes. The pen and paper, whiteboard and texta pen, the overhead projector, the video/DVD machines, and the television are a few examples that can be called technology. In learning about food and material technologies, the kitchen knife, the cooking and eating utensils, the sewing machine, the metal sheet cutter, and the drill are tools that are consequently also technology. However, at the K-12 levels of education, when talking about technology, the term usually refers to the hardware and software associated with displaying and/or processing data or information on the computer. 'Computer' as defined by its ability to process information would include the mobile phone, e-book, laptop, PC tablets, and PDA (personal digital assistants or, simply put, hand-held computers). These devices and their usage are commonly referred to as information and communication technology (ICT). In the context of this chapter, the term technology has the more specific meaning of computer-based technology or ICT, and these latter terms will be used interchangeably.

Technology Literacy

Parallel to the elaboration of the relationship between handwriting and spelling on the one hand and literacy, meaning the ability to critically engage with the meaning of texts on the other hand, the rapid development in computer-based technology has been accompanied by the development of a range of terms relating to its literacy, such as ICT literacy, net literacy, online literacy, and digital literacy (Oliver & Tomie, 2000). The evolution from the more narrowly defined computer literacy to the more broadly encompassing term of ICT literacy has been reviewed by Oliver and Tomie (2000). It is generally accepted that technology literacy cannot be defined solely as the acquisition and mastery of technical skills. Instead Oliver and Tomie (2000) state that other important key skills outside the domain of traditional curriculum areas should be included. These key skills, viewed as lifelong learning skills, include communicative skills (reading, comprehension, and writing); skills to search for and gather information from multiple sources; and the ability to make critical judgements about the authenticity, relevancy, currency, and accuracy of the information and to use the information in a meaningful and critical manner. Consistent with

Oliver and Tomie's views of technology literacy, the U.S. State Educational Technology Directors Association (SETDA) National Leadership Institute Toolkit has defined technology literacy as:

...the ability to responsibly use appropriate technology to communicate, solve problems, and access, manage, integrate, evaluate, and create information to improve learning in all subject areas and to acquire lifelong knowledge and skills in the 21st century. (http://www.setda.org/content.cfm?sectionID=24)

In expanding the above views of technology literacy by embracing social, conceptual, and ethical attributes, the International Society for Technology in Education (ISTE) has a set of six broad considerations for the use of technology by students (http://cnets.iste.org/students/s_stands.html). These are:

1. **Basic Operations and Concepts:** Students are able to access resources and utilize them in daily work.
2. **Social, Ethical, and Human Issues:** Responsibility and citizenship are an essential consideration as students learn with technology.
3. **Technology Productivity Tools:** Technology plays a pervasive role in the knowledge construction of student work.
4. **Technology Communication Tools:** Effective communication is enriched through the use of technology.
5. **Technology Research Tools:** Students leverage learning opportunities by utilizing technology for research.
6. **Technology Problem-Solving and Decision-Making Tools:** Problem solving is a valued-student skill that can be amplified through the use of technology.

Based on the views stated above, a technologically literate student should be able to demonstrate an understanding of the nature and operation of technology systems and selectively use these tools for learning, communication, and research purposes. At the operational level, the student will need to have mastery of technical skills in using computer-based tools such as the spell-check and grammar check functions for writing, entering data into cells on a spreadsheet and using the appropriate graphing tools to produce a chart, using a search engine in seeking information, and saving Web pages and graphics/animations into the hard disks or bookmarking the sites. At a cognitive level, a technologically literate student needs to be able to examine and make judgements based on critical analysis of what the software produces as a result of an entry. For instance, if the student is unsure of which word to use given a list of words in the spell-check, (s)he would draw on other resources such as an online dictionary to find the meanings of the words. The student should not accept responses provided by the software at face value, but should think about what is on the screen and whether the information provided is appropriate in the context of the task that (s)he is carrying out. (S)he needs to be aware of the fact that the spelling and grammar checking software is unable to pick up words that are correctly spelt but incorrectly used in the context of the sentence, for example, *form* and *from* or *trial* and *trail*. In doing a presentation, the student should be able to decide if a PowerPoint presentation or one based on Web pages or an alternative technology is the better way to get across ideas that (s)he would like to convey. For instance, the student might like to show the number of and the different food chains in a food web. Animations would be useful where one chain at a time shows up on the screen. This could be done using PowerPoint or Flash animations. Where there is a need to show a table that cannot fit into a slide of a PowerPoint presentation, using a Web page would be better, as the student could scroll up and down showing a complete table. Being technologically literate combines both physical manipulation (skills) and mental processes where understanding of both the benefits and the limitations of computer-based tools or software is necessary in order to use them effectively to support learning or to demonstrate understanding.

The worldviews that students bring with them at the different stages of their schooling play a major role in determining their degree of technology literacy. These worldviews, similar to those identified as significant in the constructivist learning theory, consist of their preconceived views of the subject matter. These views will vary according to the depth of knowledge of and how they have been taught to use the different computer-based tools. These factors are important in the students' ability to make judgements about whether what is on the screen is valid and to determine what other resources they could draw upon to help them make those judgements.

It is the purpose of the rest of this chapter to apply the views discussed above in terms of technology literacy and learning in a technology-enhanced environment to the use of selected resources that are available on the World Wide Web (WWW) for K-12 classroom education. Issues and challenges associated with research on the WWW and its subsequent use of information by students will be a major focus, as Web research is the most common technological practice in classrooms.

THE WORLD WIDE WEB: A LARGE FILING CABINET OF EDUCATIONAL RESOURCES

The terms 'Internet' and 'World Wide Web' are often used interchangeably, as if they mean the same thing. They are, however, different from each other. The Internet is the infrastructure that houses and transports material, and is a global system of linked networks of computers facilitating communication and the transfer of data via systems such as electronic e-mail or the WWW. The WWW is the material that we seek to access. It is a collection of pages of information on the Internet written in a coding system called hypertext markup language (html). These pages may be interactive and contain texts, graphics, images, sound, video, and/or animations. In order to access the information on the WWW, an Internet connection and a Web browser are needed. The WWW expands when additional pages are created, while the Internet expands when additional computers are connected.

As the pages on the WWW have continued to proliferate, the amount and variety of resources available to students at the primary and secondary school levels has increased dramatically over the last few years. With vast amounts of textual material and information being available on all school-taught topics, project work that relies on computer-based research investigations has become very popular in the classroom (Becker, 1999). This has meant that there has also been a dramatic expansion in the technology literacy that is required to sift through the material and to make wise choices about what material to use and how to use it. Apart from information, there are a variety of resources available on the WWW, some of which are cost free for teachers and students to use, leading to the same technology literacy challenges for teachers as there are for students. These Web-based resources include lesson plans for classroom use, homework and online tutoring services, shareware, and multimedia and simulation resources. Advantages in using these WWW-based tools and resources include the fact that many of them can be used at no cost, and teachers and students are able to access them outside the school environment, making independent research and learning more possible for a wider number of learners. However, these resources and tools bring with them challenges and issues that educators have to focus on in order to bring about meaningful learning for their students.

THE WWW AS A RESEARCH TOOL: ISSUES AND CHALLENGES

The WWW contains both enormous riches and profound traps. It is a rich source of up-to-date information for school-based research projects at the upper primary and secondary levels. It provides a learning environment where students can carry out independent research and learning in ways that will assist them with constructing understandings of new topics or concepts. In using the WWW as a research tool in their learning, students are actively engaged with (processing and sorting out) information on the screens of their computers. An advantage the WWW offers is that the vast amount of information available appears in varied styles of writing and presentation, for example some sites have only text and others have pictures, diagrams, flow charts, and multimedia materials. These different forms of presentation mean that students from different backgrounds, with different learning styles or at different stages of (particularly second) language development, have a better chance of finding something that makes sense to them than they would have if only one version of the material were presented by the teacher in only one way. However, the variety of material means that there is a commensurate sophistication required in the accessing, comparison, and evaluation of the content of the pages.

The challenges to technology literacy are compounded because the construction of knowledge using the WWW as a learning tool does not follow a systematic, structured path, and factors such as the different document structure on the WWW have been shown to affect how students retain information that they have viewed (Brown, 1998). Furthermore, conducting research by using the WWW is a quite divergent process where students can go off in unexpected directions. The unstructured nature of this type of learning requires that students navigate strategically in a hypermedia environment. Attempting to both monitor and absorb the available information while also monitoring and assessing the strategic appropriateness of the learning path could lead to cognitive overload (Warschauer & Kern, 2003). In addition, students need to reflect on and link information from one site to another in order to construct understanding in a manner that they feel comfortable with and that will lead to a cohesive interpretation. The type of unstructured and open, text-based learning that is possible when using the Web would pose more problems for students who are not proficient in the English language, such as English as Second Language (ESL) students and students whose literacy skills are somewhat lacking (Ng & Gunstone, 2002). In addition to these obstacles, the complexity of using the WWW as a research tool (Ng & Gunstone 2002; Wallace, 2002) includes:

- the time-consuming nature of doing a search,

- difficulty in locating appropriate information,
- understanding the information on the Web, and
- using the information found to construct understanding of the concepts under study.

As the number of resources and educational sites on the WWW grows at an astounding rate, using the WWW as a research tool becomes more and more of a challenge. Research has shown that students learn ineffectively when asked to search the Web for information in an open and unstructured manner (Krajcik, Blumenfeld, Marx, & Soloway, 1994; Ng & Gunstone, 2002; Wallace, Kupperman, Krajcik, & Soloway, 2000; Hoffman, Wu, Krajcik, & Soloway, 2003). If students are to be prepared as lifelong learners, they will need to learn skills to handle the WWW appropriately. These skills include, among other things, distinguishing between the different search engines and using the more 'suitable' search engine for a particular purpose; narrowing down the search and hence reducing the number of hits that they have to look at; searching for images; and being able to critique the contents of Web sites that they have found in terms of accuracy, reliability, and level of difficulty.

Much has been written about the use of critical thinking skills to evaluate Internet resources and the problems and issues related to teaching these skills. A bibliography on evaluating WWW resources can be found at http://www.lib.vt.edu/research/libinst/evalbiblio.html. While students may not be able to evaluate the accuracy of the content on a Web site without further assistance from teachers, to use WWW resources effectively they should be able to judge the reliability of the source by noting the authorship/producer of the Web site and the currency of the Web site such as when it was created and last updated. Information coming from reputable tertiary and research institutions or specialised commercial centres should contain information that is more accurate than content from sites written by sole authorship with no identifiable credentials. The students will need to learn how to make these kinds of judgements by learning where to look for these details and how to seek further information about them.

Other skills students need to learn to handle the WWW appropriately are being able to make sense of the information displayed in front of them to construct meaningful knowledge in their minds, bookmarking and/or downloading and storing the information for later perusal. Drilling these skills out of context will be of little use, but practising them in ways that demonstrate their significance in making choices is essential. Hence the time invested in the classroom to develop these skills and the opportunity to put them into practice would be invaluable to both the personal and academic development of the students. These skills elaborate the wider sense of technology literacy outlined earlier and underscore the close relationships between learning to make use of 'technology' and general educational purposes.

Wallace et al. (2000) and Hoffman et al. (2003) have carried out extensive studies on how middle school students use the Web and Web-based technologies to carry out inquiry-based assignments. The studies of Wallace et al. (2000) showed that while students could use Web technologies easily, they did so with little exploration and evaluation, and lacked the skills to develop understanding of the content obtained through the use of the Web. This suggests that a narrower view of computing skills had been taught without a complementary grounding in the wider sense of technology literacy. Hoffman et al. (2003) reported that an important point with students using online resources for understanding of content is the level of engagement with inquiry strategies—search and assess. Students who are highly engaged would critically examine the information at a site before doing another search in the 'assess-search' pathway to content understanding. This integration of engagement and critical evaluation highlights some of the tensions involved in using ICT. As I indicated earlier, ICT is highly motivating, but the motivational enthusiasm can be seductive. Students must simultaneously develop a critical distance from the experience so that they can maintain both an intellectual and a technical control over it. This is where the teacher is central.

Similar to the arguments presented above, McKenzie (1998) summarises what a technology-literate student is in his article, "Grazing the Net, Raising a Generation of Free-Range Students." He describes a student using the WWW as an "infotective" who is "a student thinker capable of asking great questions about data (with analysis) in order to convert the data into information (data organized so as to reveal patterns and relationships) and eventually into insight (information which may suggest action or strategy of some kind)" (p. 27).

McKenzie's "infotectives" are students who are problem solvers, applying higher-order thinking skills in searching, sorting, and evaluating their way through piles of often fragmented information and rearranging them until a picture emerges. McKenzie argues that the decision-making skills in carrying out these actions are the same sorts of skills that students will use in making up their minds about important issues concerning their own lives.

If students are to use the WWW for research purposes, it is necessary to provide them with a substantial amount of time to do that. When students are provided with only one to two periods in the computer laboratory to conduct research on a project (as is often the case in schools), the result is that little construction of knowledge takes place. There is a tendency for students to print out or download the information from a couple of Web sites and compile them into some form of presentation in an attempt to complete the project (e.g., Krajcik et al., 1994). This tendency is, therefore, not to evaluate either the information itself or the locations of that information. In consequence, the students end up regurgitating the words of others rather than creating their own informed understandings. When this happens, they do not learn.

Role of the Teacher

The role of the teacher in supporting learning in a technology-enhanced environment is crucial and multi-faceted. In order to be able to undertake this role effectively, the teacher needs to possess competent WWW search and evaluation skills and to have a good knowledge of the topic under study. Unlike overtly structured classroom instruction where students are directed according to the lesson plan for the period, when appropriately adopting the role of a facilitator, the teacher needs to be able to come in at different points of the students' learning since it will be inevitable that the students will be working at different paces. The important role a teacher plays in a WWW-based learning environment has been outlined by Wallace et al. (2000), who concluded from their studies that students benefited in their learning when their teacher provided clear expectations for classroom activities and provided scaffolding activities to assist them with developing strategies for searching, evaluating, and synthesising Web-based information. Tasks such as WebQuests are one way in which a teacher could overcome the difficulties faced with the open and unstructured way of finding information for a research assignment.

WEBQUESTS: STRUCTURED USE OF THE WWW AS A RESEARCH TOOL IN K-12 CLASSROOMS

WebQuests are Internet-based activities where students explore online a large body of information in a content area and complete a research task. They are student-centred but teacher-designed, inquiry-oriented activities in which students interact with information that comes primarily from the Internet (Dodge, 1997). WebQuests are useful tools in K-12 classrooms to aid with development of technology literacy skills in students and to promote constructivist learning in a technologically enhanced environment. This assertion is further discussed below.

WebQuests are created by teachers according to a standard format (Dodge, 1997) and usually uploaded to the school's intranet or server. The design of a useful WebQuest requires considerable time, as the teacher is required to pose the problem and find relevant Web sites to direct students to. The six components in a standard WebQuest design (Dodge, 1997) are:

1. *Introduction,* where background information is provided.
2. *Task,* where the finished product that is expected from students is described.
3. *Resources,* where primarily Web-based resources are listed.
4. *Process,* where the WebQuest process is broken up into steps providing a framework for students to follow.
5. *Evaluation,* where the criteria for the evaluation of the product, usually in the form of an evaluation rubric, is outlined.
6. *Conclusion,* which brings closure to the quest and students reflect on what they have learnt and propose future directions where appropriate.

WebQuests present students, individually or in groups, with authentic, usually real-world situations where a problem has to be solved. They are often created with a multidisciplinary approach to extend students' learning beyond simple fact-finding about a topic. A well-written WebQuest requires students to analyse a variety of resources and use their creativity and critical-thinking skills to derive genuine and reasonable solutions to a real-world problem (Yoder, 1999). For example, in a group work task, students take on such different roles as that of a biologist, a chemist, an environmentalist, and an economist to investigate environmental, health, and cost issues associated with acid rain. Technology-based interdisciplinary learning of this type relates closely to students' daily experiences (Raizen, Sellwood, Todd, & Vickers, 1995) and fosters higher-order thinking skills when they have to apply their various skills in the different discipline areas to analyse, evaluate, synthesise, and communicate information in order to solve the problem given in the task. The communication aspect of carrying out WebQuest tasks further enhances students' technology literacy. In communicating their work using technology, students need to be considerate in their choice of words, be it in an e-mail to an expert making enquiries about a matter that (s)he is unsure of or joining in a chat forum that takes place within or without the school settings. Students often communicate their product upon completing the WebQuest task by posting it on the WWW. In displaying their products to an audience that could range from peers of their own age to well-informed professionals, students need to ensure that their information is well analysed and presented where ethical and social issues are considered.

In developing WebQuests, teachers provide students with a structured learning framework in which scaffolding and resource support are available. However, the learning is self-directed as students are allowed the freedom to access different resources on the Internet (such as Web sites and e-mailing a scientist/institution) and printed materials to construct their own understanding, beliefs, and values out of their experiences. It is this teacher-mediated but self-directed learning that assists in the development of the richer sense of technology literacy. The teacher can choose different moments to intervene to pose the critical questions that will focus the students on the choices and decisions that they need to make in evaluating the material that they are working with. The teacher can also open up new directions that the students will have the option of exploring in their searches for possible answers to the teacher's questions.

WebQuests as Useful Pedagogical Tools for Concept Development and Promoting Technology Literacy and Constructivist Learning

Instructional strategies incorporated into WebQuests and the supply of resources such as pre-screened Web sites provide students with guidance on what to focus on and where to look for information on the World Wide Web. This directed approach narrows the search that has to be carried out by students on the Web. This enables the students to focus on developing their cognitive skills rather than challenging their Web search skills. However, students still make use of Web-based search skills to navigate and evaluate information on the recommended sites and those linked to them. The provision of pre-screened Web sites provides an additional advantage in ensuring a student-safe online environment. In WebQuest tasks, the students have some degree of freedom to think and learn in ways that suit them and to present the final product in a style that is in accordance with their own creativity. One of the concerns associated with using technology-based activities like WebQuests in the classroom is that the students will lose their focus on learning (for example, Science) content due to their inability to connect information from different Web pages together or because they become distracted with the technical skills of using technology in researching and creating a presentation. The role of the teacher would be to guide and assert that specific knowledge be demonstrated by the student in their final products, and as part of that process to direct students to the ambiguities and contradictions present in their work. A well-designed WebQuest will challenge the students to use their understanding of the discipline area to solve the problem established in the teacher's scenario.

WebQuests enable students to self-direct and have ownership of their own learning under the broad direction and mediation of the teacher, hence increasing their motivation to learn. WebQuest tasks that require students to solve a problem that has relevance to their way of living further increase their motivation to learn. Apart from independent study, WebQuest tasks also promote the development of critical and higher-order thinking skills. One of the pioneers of WebQuests, March (1998) in his article "WebQuest for Learning" states:

A WebQuest forces students to transform information into something else: a cluster that maps out the main issues, a comparison, a hypothesis, a solution, etc. (http://www.ozline.com/webquests/intro.html)

A comprehensive review of the literature pertaining to critical thinking has been discussed by Vidoni and Maddux (2002). They have used Weinstein's (2000) critical thinking framework to discuss the fostering of critical thinking in students undertaking WebQuest tasks. The distinction between the different forms of thinking—critical, creative, constructive, and higher-order thinking—is explained by Thomas (1998) at the Centre for Studies in Higher Order Literacy, University of Missouri-Kansas City who stated:

Critical/creative/constructive thinking is closely related to higher-order thinking; they are actually inseparable. Critical/creative/constructive thinking simply means thinking processes that progress upward in the given direction. First, one critically analyzes the knowledge, information, or situation. Then they creatively consider possible next-step options, and then finally, they construct a new product, decision, direction, or value. (http://members.aol.com/MattT10574/HigherOrderLiteracy.htm)

As these comments indicate, when properly constructed, WebQuests require learning that is indistinguishable from the kinds of learning envisaged in both 'technology literacy' and 'literacy' in general. In this way, the critical thinking or higher-order thinking skills that are called for are the same as the higher-order thinking skills that education in general seeks to promote. They result in a citizenry able to make informed decisions on the basis of potentially contradictory inputs by reference to evidence and the students' own sense of judgement. However, in order for this to occur, teachers must be in a position to create WebQuests that are appropriate to such goals and affordable/accessible in their particular contexts. An issue with designing and creation of WebQuests is that it requires a considerable amount of time on the teacher's part. There are, however, hundreds of ready-made WebQuests for K-12 students on the World Wide Web posted by educators who are willing to share them. Teachers could adapt these to suit the context and environment in their own classrooms.

TECHNOLOGY LITERACY AND OTHER WEB-BASED RESOURCES

There are other Web-based resources, such as sharewares, that are useful in K-12 classroom for promoting constructivist learning and technology literacy. Shareware is a marketing method where software is distributed without payment. However, not all of it is free. There is, therefore, a distinction between shareware and freeware. The software is copyrighted and can be downloaded from the WWW or copied from one user to another. Most shareware programs are demonstration versions of the full program, so that they can be downloaded for trial before purchasing. Fully commercial software packages such as the concept mapping software *Inspiration* (http://www.inspiration.com), graphics software for drawing *Smartdraw* (http://www.smartdraw.com/), Web authoring tool *Macromedia* (http://www.macromedia.com/software/dreamweaver/), and the science- and mathematics-based *Crocodile* series (http://www.crocodile-clips.com/index.htm) have demonstration versions which could be downloaded for trial within a timeframe before purchasing. There are shareware programs that are free, such as the downloading of images from the Clipart Web site at http://classroomclipart.com/cgi-bin/kids/imageFolio.cgi?direct=Clipart. Other freeware programs on the WWW are simulation programs such as java applets; these will be discussed in more detail in the next section of this chapter.

The use of shareware or freeware requires teachers and students to be technologically literate: technical skills are needed in downloading the programs, extracting (such as unzipping) and installing them into the computer, and cognitive skills required for evaluating and manipulating the software in order to construct new knowledge. In addition, the teacher needs to decide if scaffolding support is required to assist students with the learning.

Two educational shareware programs that are useful across all subject areas in K-12 education are *Hot Potatoes* (http://www.halfbakedsoftware.com/hot_pot.php) and the Web-based concept mapping tool called *Cmap Tools* (http://cmap.ihmc.us/). *Hot Potatoes* allows the user to create interactive Web-based exercises easily by just entering texts, questions, and responses. The finished product can then be uploaded onto a server where, if it is in a public domain like geocities.com, other users would be able to access it and provide feedback, hence creating a collaborative and social environment. The five basic programs in the *Hot Potatoes* suite are:

1. the *JQuiz* program, which creates question-based quizzes, such as multiple-choice and short-answer quizzes;
2. the *JCloze* program, which creates gap-fill exercises;
3. the *JCross* program, which creates crossword puzzles that can be completed online;
4. the *JMix* program, which creates jumbles-sentence exercises; and
5. the *JMatch* program, which creates matching exercises.

Most of these do not challenge the outer bounds of the concept of technology literacy that were addressed by WebQuests, but they do reinforce some of the literacy skills that complement the information evaluation dimensions of literacy that are highlighted above. The next package, *Cmap Tools,* moves closer to the evaluative/transformational understanding of literacy.

Concept mapping is a visual, at-a-glance representation of the inter-relationships between key concepts or ideas. Concept mapping tools such as *Inspiration* and *Kidspiration* provide flexible means for creating concept maps, allowing students to organise ideas in a logical but not rigid manner, and they can build on the maps as they progress through their learning.

An interesting Web-based concept mapping software (free for non-commercial use), called *Cmap Tools,* can be used to construct stand-alone, individual maps, but the maps can also be shared with other Web users. The *Cmap Tools*, developed by the Institute for Human and Machine Cognition (IHMC), is made up of servers and the *Cmap Tools* program for building concept maps. The servers can be anywhere on the Internet and users can share, browse, and search other concept maps and collaborate in the manipulation of other maps. Specific collaboration features of *Cmap Tools* include:

- synchronous collaboration where two or more users can be editing the same Cmap at the same time and simultaneously chatting,
- collaboration by the addition of new discussion threads (forums) to the concept map building,
- post-it notes to comment on others users' Cmaps, and
- sharing of claims and propositions derived from users' concept maps in relation to a domain of knowledge that is under study.

Research into the collaborative use of Cmap tools and concepts is relatively new, and a number of publications can be found at the IHMC Web site: http://cmap.ihmc.us/Publications/. In a paper describing a project called Project Quorum, Cañas et al. (2001) suggest that meaningful and collaborative learning can take place in students within classrooms, and between schools through the software system called the *Knowledge Soup* (a feature of *Cmap Tools*), where sharing of claims or propositions derived from students' concept maps can take place.

Both *Hot Potatoes* and *Cmap Tools* promote constructionist modes of learning and allow users to develop their technology literacy attributes at the same time. In getting students to create exercises from anyone of the programs from the *Hot Potatoes* suite or construct a concept map, they are actively engaged in the learning where demonstration of understanding is translated into an artefact created as a result of critical selection of content and expressions. The open nature of the tasks undertaken with these software programs and the making of choices about programs in using *Hot Potatoes* mean that the students have higher levels of ownership and control over their own learning.

Simulations

Another useful Web-based resource that fosters self-paced learning and promotes both technology and scientific literacy in K-12 students is simulation programs. Simulations are software programs that attempt to replicate a real or imagined phenomenon or concept in a dynamic and often interactive way. It is a non-linear and manipulable model (Thomas & Hooper, 1991), and has inbuilt features for the learner to track his/her progress, providing feedback in realistic forms (Alessi, 1991; Barab, Bowdish, & Lawless, 1997). Simulations are multimedia resources, where elements of graphics, text, audio, video, and animations are integrated to make the investigation more interesting and concepts easier to understand. In addition, with multimedia resource support, it is possible that students learn more information more quickly compared to practices such as traditional classroom lectures (Najjar, 1996). Hargrave and Kenton (2000) reported that simulations designed for and used prior to formal instructions can change students' learning significantly. Pre-instructional simulations are exploratory environments where students are able to manipulate variables to explore their own thinking and to generate questions about concepts that they experience difficulty with. The teacher's preparation task changes as a result of being provided with such information regarding students' learning in this type of environment.

Small simulation programs called *applets* are quite readily available without cost on the WWW, particularly for Mathematics and the different Science discipline areas. Applet is a name for a computer animation created using Java script or other appropriate programming language. Applets are interactive, and most are intended for use over the WWW and are linked to Web pages, for example, the virtual frog dissection at http://curry.edschool.virginia.edu/go/frog/Frog2/. Web browsers (for example, Internet Explorer, Netscape Navigator, and Mozilla) will run the applets, provided the appropriate 'plug-in' software has been installed for the browser. The Java Virtual Machine (free from Sun Microsystems—http://java.com/en/index.jsp) is required in order to use Java applets and Flash Player (free from Macromedia—http://www/macromedia.com) for using Flash animations.

Simulations provide the learner with close to real learning situations, and are particularly useful for learning science and Mathematics where some of the concepts are very abstract. With simulations, students are able to visualise the abstract nature of concepts such as through the use of models, hence assisting with their understanding. For example, secondary students learning about projectile motion can access the Java applet at http://www.walter-fendt.de/ph14e/projectile.htm and experiment with different variables such as mass and angle of travel. This will overcome the difficulty with carrying out real-time projectile motion experiments due to a lack of school equipment to measure the motion of an object. These models also mean that teachers do not need to worry about students who are "creative and dedicated" in exploring the potential motion of projectiles within the classroom, even at the risk of hurting other students. Students have found the visual aspect of simulations such as this both motivating and useful in helping them learn projectile motion concepts because:

> *You would not be able to get the precise exact things in a prac because in projectile (experiments) you have to throw something obviously and you can't pause it in the middle of it to see how it is travelling whereas on a computer it shows the path [and] the applets made it easier (to learn) and more fun. I enjoyed the applets, if you are enjoying what you are doing you are going to absorb more information.* (Ng, in press)

In manipulating the parameters on the computer monitor, students are critically examining and evaluating the information and making decisions that will assist with the construction of their own understanding.

CONCLUSION

Computer-based technology has an enormous potential to bring about new ways of learning in K-12 classrooms. It provides a medium for a rich source of knowledge as well as for conducting dialogue. It facilitates an inclusive environment for students where individual learning pathways are possible. The interactive, self-paced learning environment that technology offers raises the self-esteem of students and provides them with a sense of power in being able to control and be responsible for their own learning. Learning with technology is not constrained to the classroom, and with information easily available anytime and anywhere where there is an Internet connection, lifelong learning and skills to do that are essential components of today's school education.

In order to be independent learners with technology, students need to learn basic technical skills to operate the computer and use software programs appropriately, including those found on the WWW. They also need to develop higher-order thinking skills to enable them to critically evaluate the information displayed in front of them and to use the information appropriately in solving a problem or learning a concept. In developing these skills, students become technologically literate and are well prepared to become responsible citizens who are able to make choices, based on informed decisions, on matters that will affect their own lives.

For the teacher, in a technologically enhanced classroom, teaching practices need to change as (s)he faces different challenges to the traditional methods of teaching and learning. The role of the teacher becomes one of supporter and facilitator of learning where (s)he is likely to spend more time in supporting individual students and less on teaching whole classes (Jenkins, 1999). In order for teachers to make an impact on students' learning using computer-based technologies, they need to be provided with support to be professionally developed in using them. The beginning sections of this chapter highlighted the disparity and issues associated with teachers' use of technology in the classroom. Ng and Gunstone (2003) reported that despite the huge amount of money invested in getting computers into the classroom, the state of use of computer-based technologies in Victorian schools' Science classes can be summed up in the saying, "three-miles wide, half an inch deep." Without the proper support, teachers will take on board the technology and give it up after a short period of time.

In concluding, this chapter has sought to address the meaning of technology literacy and why it is important for students at the K-12 levels to be technologically literate. It discusses the use of a variety of Web-based resources to foster the development of technology literacy in students and how these resources engage students in active construction of knowledge. In order to understand the impact of Web-based technologies on students' learning and teachers' practices, extensive research carried out in technology-enhanced learning environments are required. Some research questions that need to be addressed include:

- What kind of thinking processes go through students' mind in searching, selecting, and evaluating information on the WWW?
- What kinds of strategies do students use in navigating in a hypermedia environment?
- What learning strategies do students employ to construct understanding of concepts with simulations?
- What kinds of methodologies are effective in assessing students' learning in a technology-enhanced environment?
- How do we distinguish learning as being an outcome of interactions with computer-based technologies and not other variables associated with the teaching and learning processes inside the classroom?
- How does the role of the teacher change in a technology-enhanced environment? Does it change what they teach and how they teach it?

REFERENCES

Alessi, S. M. (1988). Fidelity in the design of instructional simulations. *Journal of Computer Based Instruction, 15*(2), 40-47.

Anderson, K., & Witfelt, C. (2003). Educational game design: Bridging the gab between computer-based learning and experimental learning environments. *Conference Proceedings of the 6th International Conference on Computer-Based Learning in Science* (vol. 1, pp. 402-413).

Australian Bureau of Statistics. (2000). Australia now—A statistical profile, communications and information technology. *Real Time: Computers, Change and Schooling.* Retrieved May 12, 2004, from http://www.abs.gov.au/Ausstats/abs@.nsf/0/d34a3b2e9ed5bc12ca2569de0028de8f?OpenDocument

Barab, S. A., Bowdish, B. E., & Lawless, K. A. (1997). Hypermedia navigation: Profiles of hypermedia users. *Educational Technology Research and Development, 45*(3), 23-42.

Becker, H. J. (1999). *Internet use by teachers: Conditions of professional use and teacher-directed student use* (Report #1). Irvine, CA: Center for Research on Information Technology and Organizations, University of California, Irvine, and the University of Minnesota. Retrieved April 13, 2004, from http://www.crito.uci.edu/tlc/findings/internet-use/text-tables.pdf

Bork, A. (1992). Learning in the twenty-first century: Interactive multimedia technology. In M. Giardina (Ed.), *Interactive multimedia learning environments: Human factors and technical considerations on design issues* (pp. 2-18). Berlin: Springer-Verlag.

Brown, I. (1998). The effect of WWW document structure on students' information retrieval. *Journal of Interactive Media in Education, 98*(12). Retrieved May 4, 2004, from http://www-jime.open.ac.uk/98/12/brown-98-12-01.html

Cañas, A. J., Ford, K. M., Novak, J. D, Hayes, P., Reichherzer, T. R., & Suri, N. (2001). Using concept maps with technology to enhance collaborative learning in Latin America. *The Science Teacher, 68,* 49-51.

Cummings, L. E. (1996). Educational technology—A faculty resistance view. Part II: Challenges of resources, technology and tradition. *Educational Technology Review, 5,* 18-20.

Dodge, B. (1997). *Some thoughts about Webquests.* Retrieved May 4, 2004, from http://edweb.sdsu.edu/courses/edtec596/about_webquests.html

Driver, R., Squires, A., Rushworth, P., & Wood-Robinson, V. (1994). *Making sense of secondary science: Research into children's ideas*. London: Routledge.

Dwyer, D. (1994). Apple classrooms of tomorrow: What we've learned. *Educational Leadership, 51*(7), 4-10.

Fensham, P. J, Gunstone, R. F., & White, R. T (1994). Science content and constructivist views of learning and teaching. In P. J. Fensham, R. F. Gunstone, & R. T. White (Eds.), *The content of science* (pp. 1-8). London: The Falmer Press.

Forcheri, P., & Molfino, M.T. (2000). ICT as a tool for learning to learn. In D. M. Watson & T. Downes (Eds.), *Communications and networking in education* (pp. 175-184). Boston: Kluwer Academic.

Fulton, K. (1999). *How teachers' beliefs about technology and learning are reflected in their use of technology: Case studies from urban middle schools.* Thesis, University of Maryland. Retrieved May 12, 2004, from http://learn.umd.edu/fulton-thesis.html

Good, D.G. (1999). Future trends affecting education. *Denver, CO: Education Commission of the States.* Retrieved April 13, 2004, from http://www.ecs.org/clearinghouse/13/27/1327.htm

Hardin, J., & Ziebarth, J. (2000). Digital technology and its impact on education. Retrieved April 14, 2004, from http://www.ed.gov/Technology/Futures/hardin.html

Hargrave, C. P., & Kenton, J. M. (2000). Preinstructional simulations: Implications for science classroom teaching. *Journal of Computers in Mathematics and Science Teaching, 19*(1), 47-58.

Hoffman, J. L., Wu, H. K., Krajcik, J. S., & Soloway, E. (2003). The nature of middle schools learners' science content understandings with the use of on-line resources. *Journal of Research in Science Teaching, 40*(3), 323-346.

Holmevik, J. R., & Haynes, C. (1998). *High wired. On the design, use and theory of educational MOOs.* Ann Arbor: University of Michigan Press.

Jenkins, J. (1999). *Teaching for tomorrow: The changing role of teachers in the connected classroom.* EDEN Open Classroom Conference, Balatonfured. Retrieved May 25, 2004, from http://www.dlab.kiev.ua/edl/JENKINS. HTM

Johnson, D. (1996). Evaluating the impact of technology: The less simple answer. *From Now On, The Educational Technology Journal, 5*(5). Retrieved May 20, 2004, from http://www.fno.org/jan96/reply/html

Krajcik, J. S., Blumenfeld, P. C., Marx, R. W., & Soloway, E. (1994). A collaborative model for helping middle grade science teachers learn project-based instruction. *The Elementary School Journal, 94*(5), 483-497.

Kromhout, O., & Butzin, S. (1993). Integrating computers into the elementary school curriculum. *Journal of Research on Computing in Education, 26*(1), 55-69.

Lankshear, C., Snyder, I., & Green, B. (2000). *Teachers and technoliteraty: Managing literacy, technology and learning in schools.* New South Wales: Allen & Unwin.

Lockhard, J., Abrams, P. D., & Many, W.A. (1994). *Microcomputers for 21st century educators* (3rd ed.). New York: Harper Collins.

March, T. (1998). Webquests for learning. Retrieved May 20, 2004, from http://www.ozline.com/webquests/intro.html

Matthews, M. R. (1993). Constructivism and science education: Some epistemological problems. *Journal of Science Education and Technology, 2*(1), 359-370.

Matthews, M. R. (Ed.). (1998). *Constructivism in science education: A philosophical examination.* Dordrecht: Kluwer.

McKenzie, J. (1998). Grazing the Net. *Phi Delta Kappan, 79*(September), 26-31. Retrieved May 20, 2004, from http://www.fno.org/text/grazing.html

McKinsey & Company. (1997). *The future of information technology in UK schools.* London: McMinsey and Company.

McKenzie, J. (1999). Scoring high with new information technologies. *From Now On, The Educational Technology Journal, 8*(7). Retrieved May 20, 2004, from http://www.fno.org/apr99/scoring.html

Meredyth, D., Russell, N., Blackwood, L., Thomas, J., & Wise, P. (1999). *Real time: Computers, change and schooling.* Canberra: Commonwealth Department of Education, Training and Youth Affairs.

Mistler-Jackson, M., & Songer, N. B. (2000). Student motivation and Internet technology: Are students empowered to learn science? *Journal of Research in Science Teaching, 37*(5), 459-479.

Naiiar, L. J. (1996). Multimedia information and learning. *Journal of Educational Multimedia and Hypermedia, 5*(2), 129-150.

Nanjappa, A., & Grant, M.M. (2003). Constructing on constructivism: The role of technology. *Electronic Journal for the Integration of Technology in Education, 2*(1). Retrieved May 25, 2004, from http://ejite.isu.edu/Volume2No1/nanjappa.htm

Ng, W. (in press). Web-based resources for learning: A case study of Year 10 students' perceptions of learning concepts of *Motion* using Java applets. *International Journal for Learning, 11.*

Ng, W., & Gunstone, R. (2002). Students' perceptions of the effectiveness of the World Wide Web as a research and teaching tool in science learning. *Research in Science Education, 32*(4), 489-510.

Ng, W., & Gunstone, R. (2003). Science and computer-based technologies in Victorian government schools: Attitudes of secondary science teachers. *Journal of Research in Sci-*

ence and Technology Education, 21(2), 243-264.

Ofsted Report. (2001). *ICT in schools.* Retrieved April 14, 2004, from http://www.ofsted.gov.uk/publications/index.cfm?fuseaction=pubs.displayfile&id=1043&type=pdf

Oliver, R., & Tomei, L. (2000). Information and communications technology literacy: Getting serious about IT. *Proceedings of ED-MEDIA 2000. World Conference on Educational Multimedia, Hypermedia and Telecommunications,* Virginia. Retrieved May 25, 2004, from http://elrond.scam.ecu.edu.au/oliver/2000/emict.pdf

Oxbrow, N. (2000, October). Skills and competencies to succeed in a knowledge economy. Information Outlook. *Washington, DC: Special Libraries Association.* Retrieved April 14, 2004, from http://www.findarticles.com/cf_dls/m0FWE/10_4/66276583/p1/article.jhtml

Papert, S. (1980). *Mindstorms.* New York: Basic Books.

Papert, S. (1991). Situating constructionism. In I. Harel & S. Papert (Eds.), *Constructionism.* Norwood, NJ: Ablex Publishing.

Papert, S. (1993). *The children's machine: Rethinking school in the age of the computer.* New York: Basic Books.

Pedretti, E., Mayer-Smith, J., & Woodrow, J. (1998). Technology, text and talk: Students' learning in a technology enhanced secondary science classroom. *Science Education, 82,* 569-589.

Phillips, D. C. (1995). The good, the bad, and the ugly: The many faces of constructivism. *Educational Researcher, 24*(7), 5-12.

Phillips, D. (Ed.). (2000). *Constructivism in education: Opinions and second opinions on controversial issues.* Chicago: University of Chicago Press.

Piaget, J. (1955). *The construction of reality in the child.* London: Routledge & Keegan Paul.

Piaget, J. (1972). *Psychology and epistemology: Towards a theory of knowledge.* London: Penguin University Books.

Raizen, S., Sellwood, P., Todd, R., & Vickers, M. (1995). *Technology education in the classroom.* San Francisco: Jossey-Bass.

Salomon, G., Perkins, D. N., & Gloerson, T. (1991). Partners in cognition: Extending human intelligence with intellectual technologies. *Educational Researcher, 20,* 2-9.

Semple, A. (2000). Learning theories and their influence on the development and use of educational technologies. *Australian Science Teachers' Journal, 46*(3), 21-28.

Settlage, J. (1995). Children's conceptions of light in the context of a technology-based curriculum. *Science Education, 79*(5), 535-553.

Solomon, J. (1994). The rise and fall of constructivism. *Studies in Science Education, 23,* 1-19.

Stevenson, D. (1997). *Information and communications technology in UK schools. An independent inquiry.* London: The Independent ICT in Schools Commission. Retrieved May 20, 2004, from http://rubble.ultralab.anglia.ac.uk/stevenson

Thomas, M. (1998). *Higher-order thinking strategies for the classroom.* Retrieved May 20, 2004, from http://members.aol.com/MattT10574/HigherOrderLiteracy.htm

Thomas, R., & Hooper, E. (1991). Simulations: An opportunity we are missing. *Journal of Research on Computing in Education, 23*(4), 497-513.

Tebbutt, M. (2000). ICT in science: Problems, possibilities...and principles? *School Science Review, 81*(297), 57-64.

UNESCO Report. (2003). Retrieved May 20, 2004, from http://www.unescobkk.org/education/ict/v2/info.asp?id=11002

Van Dusen, L. M., & Worthen, B. R. (1993). Factors that facilitate or impede implementation of integrated learning systems. In G. D. Bailey (Ed.), *Computer-based integrated learning systems.* Englewood Cliffs, NJ: Educational Technology Publications.

VCAA (Victorian Curriculum and Assessment Authority). (2002). *E-merging: Perspectives on information and communications technology—Teaching and learning in the VCE.* Victoria: VCAA.

Vidoni, K. L., & Maddux, C. D. (2002). Webquests: Can they be used to improve critical thinking skills in students? *Computers in Schools, 19*(1-2), 101-117.

von Glasersfeld, E. (1995). *Radical constructivism: A way of knowing and learning.* London: Falmer Press.

Vygotsky, L.S. (1962). *Thought and language.* Cambridge, MA: MIT Press.

Vygotsky, L. (1978). *Mind in society: The development of higher psychological processes.* Cambridge, MA: Harvard University Press.

Wallace, R., Kupperman, J., Krajcik, J., & Soloway, E. (2000). Science on the Web: Students on-line in a sixth grade classroom. *Journal of Learning Sciences, 9*(1), 75-104.

Wallace, R. M. (2002). The Internet as a site for changing practice: The case of Ms. Owens. *Research in Science Education, 32*(4), 465-487.

Warschauer, M., & Kern, R. (Eds.). (2000). *Network-based language teaching: Concepts and practice.* Cambridge: Cambridge University Press.

Weil, L. (2002). Successful strategies: Technology in education. *New York: Business Coalition for Education Reform.* Retrieved April 14, 2004, from http://www.bcer.org/projres/Technology in education3.pdf

Weinstein, M. (2000). A framework for critical thinking. *High School Magazine, 7*(8), 40-43.

Wellington, J. (2003). Has ICT in science teaching come of age? *School Science Review, 84*(309), 39.

Yoder, M. B. (1999). A productive and thought-provoking use of the Internet. *Learning and Leading with Technology, 26*(7), 6-11.

Chapter VII
New Media Pathways:
Navigating the Links between Home, School, and the Workplace

Helen Nixon
University of South Australia, Australia

Stephen Atkinson
University of South Australia, Australia

Catherine Beavis
Deakin University, Australia

ABSTRACT

This chapter uses the case of students enrolled in the Multimedia Pathway offered by Harbourside High School to discuss the tensions and contradictions inherent in the views that: (a) school curriculum and pedagogy have much to learn from young people's informal and leisure-based learning; and (b) school-based courses in new media are important because they increase student retention and the chance of success in post-school employment. We draw on literature about the "new work order" (Gee, Hull, & Lankshear, 1996) to explore the nature of these students' learning about and with ICTs and show that the students' knowledge exists "in a network of relationships" (Gee, 2000) that bridge the formal and informal learning divide. Finally, we discuss the parts played by their in- and out-of-school engagements with ICT in their becoming the kinds of portfolio people supposedly required by the new capitalism.

INTRODUCTION

The terms literacy and technology remain highly contentious within the field of education. What is meant by literacy and the methods used to measure it vary quite markedly in educational and historical contexts across the world. Similarly, while there is a shared concern to research the potential impact of new information and communication technologies (ICT) on patterns of teaching and learning, there are major discrepancies about which aspects and uses of these technologies should be incorporated into formal learning environments and how this can be accomplished. While government policymakers tend to regard ICT in relation to ideas of *smartness*, *efficiency*, and the *knowledge* (or *new*) economy, educators and educational researchers promote them as offering new tools for learning and critical thinking, and the development of new literacies and sociocultural identities. This clearly has ramifications for the ways literacy is taught and conceptualised throughout the years of schooling, in K-12. Outside school, meanwhile, students engage with ICT on another level entirely, as tools for the maintenance of social networks for leisure, and for learning and participating in the cultures of their peers. Whatever the differences in perspective, it remains the case that a society's dominant understandings about literacy and technology will have significant implications for the development of school curriculum.

We begin this chapter by introducing some of the theoretical work being done within Literacy Studies on the relationship between literacy and technology, and consider some of the broad issues and challenges this raises for curriculum development and pedagogy. The main body of the chapter addresses those issues and challenges specific to the secondary curriculum as the personal computer becomes ubiquitous and new media pathways open up at work, in the home, and at school. To illustrate this, we use brief case studies from a recent research project conducted in a suburban, publicly funded Year 8-12 comprehensive secondary school in which both literacy and ICT are high curriculum priorities.

The Literacy-Technology Interface in Literacy Studies

The increasing pervasiveness of computers has focused the attention of literacy educators on the literacy-digital technology interface. The argument has been made that information and communication technologies and the new media are changing what it means to be literate (e.g., Cope & Kalantzis, 2000; Kress, 2003; Lankshear, Snyder, & Green, 2000). This is more than a question of changing what it means to be *functionally* literate. Rather, it has been suggested that more complex models of literacy, and theories of meaning-making or semiosis, are required in an age in which the image, sound, and hypertext have become integral to the modes and media of representation and communication (Kress, 2003).

From this perspective it is not tenable to try to incorporate ICT into conventional literacy frameworks. Instead, literacy scholars and educators are faced with the challenge of responding to the changing world of literacy-technology in informed and systematic ways. Australian scholars have been at the forefront of this theoretical work in literacy-technology studies which begins from the theoretical position that literacy is a social practice (Street, 1995). This work assumes that "emerging technology-mediated literacy practices...can be understood only when they are considered within their social, political, economic, cultural, and historical contexts" (Snyder, 2002, p. 5) because it is within those contexts that literacy and meaning are embedded. Bill Green, for example, has argued the need for a "3D model of l(IT)eracy" in which l(IT)eracy learning has operational, cultural, and critical dimensions. This work received national attention in *Digital Rhetorics* (Lankshear, Bigum, Durrant et al., 1997b), the first major document bringing together literacy and technology in a report to the Australian federal government on literacy and IT in primary and secondary schools (see also Lankshear et al., 2000). Similarly, Peter Freebody, Allan Luke, Carmen Luke, Colin Lankshear, and others have argued the need for people to develop a complex ensemble or repertoire of literate practices which includes engagement with ICT as critical consumers and producers (see Cope & Kalantzis, 2000; Durrant & Green, 2000, Lankshear et al., 2000; Luke & Freebody, 1999; Luke, 2002). Elsewhere in the Anglo-American context, literacy educators have increasingly argued the need to bring together the terms literacy, ICT, and new media (e.g., Alvermann, 2002; Kress, 2003; Lankshear, Gee, Knobel, & Searle, 1997a).

A socio-cultural approach to literacy learning contends that literacy technologies change, that ICT provide new ways of 'doing' literacy, that different contexts of social practice 'embed' different forms of literacy, and that what literacy means is always changing along with "new modes of human practice and ways of experiencing the world" (Snyder, 2002, p. 26). Green's coinage of the term "l(IT)eracy" encapsulates the view that literacy and technology are moving closer together to form what he calls a "literacy-technology nexus." Scholars who adopt this position maintain that the literacy-technology interface poses significant challenges to the fields of both literacy education *and* technology education. It is argued that formerly taken-for-granted concepts in both fields—including literacy, language, text, authorship, and information—require radical reassessment at a time when the technologies and industries of print publishing, broadcast media, and computing are converging (Durrant & Green, 2000; Kress, 2003; Lankshear et al., 1997a).

An early response to the challenges to literacy presented by technology in the context of libraries, in particular, was the development of 'information literacy'. As a field, information literacy ostensibly presents the context in which to address these changes, and there has been a proliferation of 'information literacy' courses which seek to equip students with skills seen to address the demands of the changing literacies associated with new technologies. However, older approaches to the links between literacy and technology as information literacy which characterise such courses are themselves inadequate to represent or comprehend these interactions. As Boyce notes, consequences of the introduction of digital communications in the library hold true elsewhere: they have the effect of "destablising the ground of literacy by changing the alignment of text, technology and literacy and bringing established practices and alliances into sharp relief" (Boyce, 2001, p. 125). Not only is literacy reconfigured in this context, but so too are understandings about the kinds of uses that might be made of technologies (as distinct from just the mastery of skills packages) and the mutually constitutive effects of technologies on literacy, texts, and users. Idealised constructions that simplistically equate technological proficiency with higher-order thinking and workplace opportunity fail to address complex relations between literacy and technology; pedagogy, creativity, and engagement; and real-world politics, global economic markets, and employment opportunity.

New Media, New Literacies, and New Learning

A second important theme in the research literature about school-aged children and ICT suggests that today's educators are faced with the new task of teaching the "Nintendo generation" (Green, Reid, & Bigum, 1998) or the "electronic generation" (Buckingham, 2002). The argument is made that young people now routinely engage in "new literacies" and "new kinds of learning" in Internet cafes, videogame arcades, and as they use personal computers and game consoles in their bedrooms and living rooms (Buckingham, 2000, 2003; Gee, 2003). In order for school-based curriculum and pedagogy to be stimulating and relevant for the "digital generation," they need to be informed by research into children's and young people's new media use outside schools (see Sefton-Green, 2004). Further, young people's experience of digital culture and communication in the online context shapes their expectations about texts, interaction, agency, and (digital) literacy, and foregrounds the limitations of a focus on print-based texts and literacies within the school curriculum.

One implication of these propositions is that as everyday literacies become multiple and complex, schools may need to redesign the literacy curriculum to enable students to use the new media to explore and reflect on these different literacies, and to provide opportunities for students to utilise the affordances of the new media in their construction and analysis of text. Similarly, in all curriculum areas an acknowledgment of the power and presence of new media and digital culture in young people's lives, and a recognition of media culture as both "context and resource" for teaching (Durrant & Green, 2000), is needed in order to re-energise school subjects and cast them into a more contemporary form.

While not arguing for the direct utilisation or duplication of digital culture texts and environments within the school, Gee (2003), Lankshear and Knobel (2004), and others nonetheless propose that schools might learn from identifying and in some ways replicating or capitalising on the elements that contribute to the highly efficient ways in which learning takes place in and around digital culture. Gee (2003) points out that computer games would not be the commercial success they are without powerful inbuilt learning principles guaranteeing continued patronage and sales. Schools and education systems, he argues, would do well to build such elements into their own learning environments, pedagogies, and resources. He identifies 36 features or 'learning principles' in games, characterised as "semiotic domains," which facilitate their immense popularity and take up. These principles include such things as an *active critical learning principle*—in which "all aspects of the learning environment...are set up to encourage active and critical...learning" (p. 207); a *design principle* (learning about and coming to appreciate design principles as core to the learning experience); and a *semiotic principle* entailing "learning about and coming to appreciate interrelations within and across multiple sign systems (images, words, actions, symbols, artefacts, etc.) as a complex system is core to the learning experience" (p. 207).

Once again, neither Gee nor Lankshear and Knobel are suggesting that schools adopt wholesale the features of computer-related learning that takes place outside them. Lankshear and Knobel (2004), for example, caution against assuming that parallels can readily be created between in- and out-of-school experiences of multimodal texts and literacies. However, like Gee they argue the importance of "looking for principled ways of appropriating roles, ways, tools, and uses that young digitally 'at home' learners already have into grounded, academically and scholastically sound practices that articulate well to the world beyond the school" (Lankshear & Knobel, 2004, p. 19). Schools seeking to prepare their students for active participation in an ICT-saturated workplace, then, need to find ways to provide not just training in specific skill packages, but the opportunity for the sorts of agency, ease, and ownership in and around technology that the 'digitally at home' experience as they inhabit and make use of their digital cultural worlds.

The Literacy-Technology Nexus and the School Curriculum

Micro-computers have been a feature of many Australian schools since the 1980s. However, only relatively recently have government policy and public opinion begun to assume that schools should integrate or embed computers and digital media into the curriculum and everyday classroom practice. Following a worldwide trend, both state and federal governments in Australia have devoted significant proportions of public money to first hard wiring schools, and then providing professional development for teachers in how best to use computers to develop appropriate curriculum to educate workers for the knowledge economy. Much of the impetus for this direction in government policy has come from economic and cultural globalisation which has, somewhat paradoxically, forced nations to consider how they might simultaneously reinforce their sense of national identity *and* their international competitiveness in a global cultural economy (Nixon, 2004).

In 2001 the Australian Ministerial Council for Employment, Education, Training, and Youth Affairs (MCEETYA) endorsed *Learning in an Online World*, an agreed national framework for change, with the following overarching goals:

- All students will leave school as confident, creative, and productive users of new technologies, particularly information and communication technologies, and understand the impact of those technologies on society.
- All schools will seek to integrate information and communication technologies into their operations, to improve student learning, to offer flexible opportunities, and to improve the efficiency of their business practices. (Education Network Australia, 2001)

Meanwhile, each state has developed its own IT policies and implementation plans. In South Australia, the Department of Education, Training, and Employment developed a plan "Creating the Information Society" and allocated AU $85.6 million to the DECS*tech* 2001 project designed to amplify, extend, and transform student learning. The key curriculum strategy of this project was a Learning Technologies Project, which included the establishment of the Discovery Schools and Discovery Teachers programs. Like the Navigator Schools, which had preceded them in Victoria, the South Australian Discovery Schools had a brief to showcase their work as exemplar schools which had accessible models of new learning environments, and to share their work with others and thus create a premium professional development resource for teachers (Pearson, 2003). The rationale of developments like these was that:

Schools must increasingly equip students with ICT literacy and fluency skills to both grow the pool of skilled labour to support the ongoing development of an internationally recognised ICT industry and to support the evolution of a knowledge economy characterised by sophisticated users and consumers of ICT in all areas of the society. (Education Network Australia, 2001, p. 25)

However, despite this emphasis in government policy on ICT literacy, there are few details about what this new literacy entails or how it might best be developed. Nor is there any explicit connection made between the teaching of literacy and the teaching of technology in the school curriculum. In the minds of literacy studies scholars, this is a major oversight. Despite the increasing presence of computers in schools, it remains the case that most learning is textually, if not now exclusively, verbally based—a form of logocentrism alluded to by Green (1993) as "the insistence of the letter." Even as the technology of computers becomes as widespread as the technology of pen and paper, literacy and written texts of various kinds are both the medium and the product of instruction. Associated changes to literacy have broad implications for text-based pedagogy and curriculum in all subject areas. The school curriculum authorised by some state education departments has established expectations in this regard. In Queensland, for example, the *Literate Futures* (Education Queensland, 2003) statement describes literacy as:

the flexible and sustainable mastery of a repertoire of practices with texts of traditional and new communications technologies via spoken language, print and multimedia. (p. 3)

Attempts to account for the multiple forms of semiosis entailed in IT-based text and communication have popularised the notion of "multiliteracies" (New London Group, 1996), and foregrounded the need to find ways in which to analyse, think, and talk about digital/electronic texts. Textual analysis and critical reflection, long-time core elements of English/literacy curriculum—and recently given new life in the broader context of literacy-based curriculum as "critical literacy" (Luke, 2000)—would appear more essential than ever in relation to the use and analysis of technologically generated multimodal texts. As Luke (1996) argues, what are required in tomorrow's workforce are people who are highly skilled at reading and manipulating symbols. Such facility with symbolic forms of representation is not the focus of most traditional IT and Computers in Education approaches, nor of Vocational Education and Training (VET) competencies in which 'computer literacy' or ICT literacy has largely meant the acquisition of certain identifiable and measurable skills such as knowing how to use the software and hardware that industries and governments use. Moreover, as Gee et al. (1996) argue, in the workforce of the "new world order," simple skill mastery is not sufficient to provide agency, creativity, and innovation. Fulfilling the promise of VET education to position young people to take their place as active and productive members of a networked economy and society requires attention to more than the operational uses of technology. Attention must also be paid to matters of text and literacy, representation and expression, meaning-making and design, concerns traditionally at the core of the English/literacy curriculum.

Literacy and Technology at Harbourside High School

...a comprehensive school catering for students from Year 8 to Year 13. We are designated as an Adult Re-entry school, a school for the Engineering Pathways, one of six Discovery schools in SA, as well as being one of a very small number of schools around Australia that are official Microsoft certified

Professional Training providers. We provide a range of pathways to ensure students are successful when leaving school in gaining employment, training or entry into TAFE or university. Harbourside High School is one of the leaders in Vocational Education and Training (VET) in South Australia and has well-established links with Business and Industry. (2004 Middle School Course Book, Harbourside High School, p. 1)

When we embarked on a recent study of teenagers' informal use of computers for Web authoring and computer games playing[1], we recruited some of our teenagers from Harbourside High, a secondary school located in a relatively low socio-economic area with a long industrial working class history. The school began as a boys' technical school whose original focus was on training students for occupations in the skilled and semi-skilled manual trades. Since the mid 1970s, however, in accordance with government efforts of the time to break down the class distinctions that such schools seemed designed to perpetuate, Harbourside, like many other technical schools, became co-educational and comprehensive.

Although all Australian secondary schools are now comprehensive, they are nonetheless diverse and unequally resourced. We cannot claim, therefore, that Harbourside is a representative state secondary school. However, it is of particular interest here because of its commitment to linking together literacy and ICT and embedding computers across the curriculum. Historically, this represents a fundamental shift away from teaching *about* the use of technology as a discrete area of expertise, to an exploration of the potential of teaching *with* technology. Rather than seeing computers from a technicist perspective, as a feature of hard scientific development and a set of technical skills to be mastered, the school's leadership team is aware that much research into the application of new technologies in schools has emphasised the need to focus instead on the meaning and uses of technology and the ways these influence, and are shaped by, the socio-cultural context of the user. Digital media offer new ways to create and interpret text, sound, and image, and to engage with new narratives and new concepts of authorship and composition.

Despite the espoused emphasis on literacy and technology across the curriculum at Harbourside High, the senior secondary school curriculum is more traditional than the rhetoric might suggest. The provision of skilled school leavers and a commitment to applied, vocational learning remain key elements of Harbourside's stated mission. To support this, the school offers vocational education and training (VET) courses at a senior secondary level that specifically deal with information technologies, hardware and software, and multimedia production, as well as a Microsoft-Certified Professional qualification. The Year 10 curriculum prepares students for senior secondary school where the South Australian Certificate of Education pattern, and students' career aspirations, guide subject choice. According to their choices in Year 10, students enter one of a number of possible curriculum "pathways" that lead into Years 11 and 12. Three of these pathways are designated as "vocational education pathways:" an engineering trades pathway, an IT pathway, and a Design and Multimedia pathway. The learning programs and assessment for courses in the IT and Design and Multimedia pathways are national rather than state based; they count as Australian Qualifications Framework credentials, and allow entry into post-secondary studies accredited through Technical and Further Education (TAFE). At the same time, they are also accredited through the Senior Secondary Assessment Board of South Australia, thus making university entry also possible for students who successfully complete them.

Knowledge Workers and the New Capitalism

VET courses make explicit links between compulsory schooling and the workplace, and are based on a utilitarian view of education. Yet, unlike the skills taught in other VET pathways, vocational education and training in the manipulation of digital technologies has a vastly elevated cultural and social (class) status and the associated promise of greater financial rewards in the job market. They are essentially the skills of an educated middle class, hence their crucial importance to the conception of the knowledge economy and new capitalism. On another level, the shift in emphasis from trades in the old technical schools to vocational training in IT and Multimedia in contemporary comprehensive schools is based squarely on transformations in industry, the economy, and as a consequence, the skills required by the local workforce.

Economic globalisation and the free flow of capital have led to a more expansive worldwide search for cheap labour by the manufacturing industries. Local employment opportunities in many trades in the local community and Australia more generally have dwindled, along with the vast bulk of skilled and unskilled positions that these trades supported. Many manufacturing positions have disappeared offshore to be absorbed by the industrialised working classes of poorer nations where massive unemployment and restricted rights to organise keep workers plentiful, cheap, and compliant. ICT industries have also begun to outsource many of their activities, especially to sub-contractors based in India, but despite claims that this undermines employment opportunities for local IT graduates, the sector continues to be seen as a local growth industry.

How work, identity, community, creativity, teaching, and learning are being transformed by new technologies is an important part of this discourse. In addition to knowledge, a key concept for new capitalism is mobility, the ability to refocus, adjust, and adapt skills to different tasks, places, and projects, "like a dynamically flowing liquid":

> *There are no discrete stable individuals, only ensembles of skills stored in a person, assembled for a specific project (to be reassembled for other projects), and shared with others within "communities of practice."* (Gee, 2000, p. 414)

As Gee argues:

> *If the old capitalism had a deep investment in creating standardized, stable identities (Leach, 1993), the new capitalism has a deep investment in creating what we might call 'shape-shifting portfolio people', at least for more privileged rungs of the new capitalism. Shape-shifting portfolio people live to fill up their portfolios with attributes, achievements, and skills that they can flexibly arrange as things change and new contexts demand that they redefine themselves.* (p. 414)

However, it is also clear from this that not all young people are destined to become equal in the hierarchy of knowledge workers. Rather, Gee continues, there are likely to be three basic "slots" for people in the new capitalism:

a. symbol analysts, the professionals who design, implement, and transform networks and systems and can reap large rewards because they have high-level sociotechnical knowledge;

b. enchanted workers, the front-line (product or service) workers who are paid lean wages, though they may get bonuses for productivity, but who must, nonetheless, collaboratively and proactively understand and continually redesign their work as a whole process; and
c. backwater workers, who work, often part-time or on demand, for very minimum benefits and who do the remaining old capitalist work, low-level service jobs, or jobs that require brute strength, but little (value-adding) knowledge. A pressing issue is how schools are equipping or sorting children for these three slots. (Gee, 2000, pp. 414-415)

Considering the role that education systems play in preparing their students to become self-reflexive and critical knowledge workers—and perhaps shape-shifting portfolio people—is, therefore, an important political as well as educational issue.

Curriculum Pathways and Imagined Life Trajectories

There is a good deal of quantitative evidence that children and young people use personal computers in homes, cafes, recreation centres, schools, and workplaces for very diverse purposes (e.g., Australian Bureau of Statistics, 2003; Centre for Media Education, 2001; Livingstone & Bovill, 2001). One goal of our Cyberkids study was to bridge a gap in the qualitative research in literacy education about the literacy or meaning-making practices teenagers engage in as they use ICT in their leisure time. We were also interested in how they represented themselves as they participated in off-line and online communities associated with their ICT use. Adolescence or the early teen years (between Years 6 and 10 in Australian schools) is a time in which particular aspects of identity-production come to the fore. It has been suggested, for example, that people of this age are actively exploring questions about who they are as individuals, as well as developing a sense of the social communities they want to participate in and belong to (Chandler & Roberts-Young, 1998). Both computer game-playing and Web site production—particularly the making of personal Web sites—provide opportunities for expressing a range of feelings, and constructing, mediating, and trying out versions of self. Overseas studies have shown that young people have quite different 'profiles' as computer users (Chandler-Olcott & Mahar, 2003; Upitis, 1998) and represent themselves quite differently on their Web sites (Abbott, 2001; Stern, 1999). We wanted to explore how this international research applied to contemporary Australian teenagers.

Over six months we interviewed fifteen 15- to 16-year-olds in South Australia and Victoria who used the World Wide Web and played computer games. First they were given access to a digital camera and asked to create "digital portraits" titled "Me and My Cyberworld" that represented themselves in relation to digital media or cyberspace. We then talked to them individually about their portraits using a loosely structured interview protocol which included prompts that enabled them to talk-through their experiences of making the portraits and using digital media. These methods of data collection elicited information about teenagers' patterns of media use, including their use of digital media as part of peer socialisation. These methods also informed us about their experiences as producers of digital texts, including the vocabularies they were developing with respect to the digital literacies and digital aesthetics associated with multimedia textuality and composition.

Eight of the young people we interviewed were Year 10 students who attended Harbourside High School and had chosen a curriculum that led to either the IT or Design and Multimedia pathway in Years 11 and 12. Several of the students said that they had chosen their secondary school on the basis of its IT facilities and VET pathway programs, and hoped in the future to secure jobs in graphic design or computer programming. Although it had not been an original focus of the study, increasingly we became interested in the ways in which these young people imagined and talked about their educational and social trajectories as they moved into the senior secondary school and workplace contexts. What 'pathways' were they travelling, and how did they connect their anticipated success on these educational and life trajectories with their use of computers inside and outside school? In what follows we briefly consider the cases of Karen, Nick, David, and Mark, four participants in the Cyberkids project who attended Harbourside High.

Karen: Career Pathway in Art and Design

Karen had a computer at home and had had some prior experience of Web authoring at school in Years 7 and 9. She was interested in painting and drawing with pen and paper, but also in learning how to use computers for graphic design. With a friend, Karen also attended an animation academy at the Technology School of the Future in Adelaide over a period of several months for five hours each Saturday. There she was learning to use Flash software to create animations, and she expressed pleasure that she was "learning how to use Flash properly."

Karen was excited at the potential opened up by computers for creative work in graphic design. Her particular interest was in creating logos and designing Web backgrounds. These skills were at the time being taught in her Design and Multimedia class, but she also spent many hours working on these things at home. As part of her Year 10 curriculum Karen studied Visual Arts as well as courses in the VET Design and Multimedia pathway. She saw the two as complementary and mutually informing, noting that "visual arts... is developing my drawing skills, which can then help me with graphic design because I can draw and scan." She had combined the two subjects in her curriculum pathway in the expectation that this would give her an advantage in a future career pathway, saying that "when I got to high school, I realised that I wanted to do graphic design when I leave." She added that she had deliberately combined drawing with computer-aided design "so [that] hopefully I can get somewhere with it." She was aware that it was important to build up a portfolio of different design skills in order to increase her chances of success in both post-school study and the employment market, and she devoted both school time and leisure time to pursuing this objective.

At the time we spoke with Karen, her goal was to learn how to work with Photoshop and Flash software. Because she was not yet fluent using either program, her digital portrait for the study was produced using PowerPoint, although many of its effects were created in Photoshop. For Karen, the means by which she learned the skills she was seeking to develop, and the locations in which she practised them, were of little importance. At home she downloaded trials of software used in her Design and Multimedia classes so that she would have more opportunities to "try things out and learn more." Having a trial of Photoshop loaded at home, for example, allowed her to experiment and "muck about with different brushes and filters and opacity." If she had any difficulties she would ask her classmates for assistance, either from home by using Internet chat, or else in person at school the following day. Thus there was a constant flow between her in-school and out-of-school uses of ICT, not only in learning how to use software and create digital graphics, but also in her social interaction with peers.

Both inside and outside school, Karen was being inducted into the culture of public display associated with visual art and graphic design. At the animation academy she attended outside school, she was preparing for 'showcase' of her work to be shown to an audience at the end of the course. The concept of the showcase has also been adopted by Harbourside High School where assemblies and other public events provide platforms for students to display their work to wider audiences. Assessment practices in computer-based classes at her school require students to assemble digital portfolios of their work in much the same way as professional graphic artists do. These portfolios can also be accessed for various public 'showcases', and much of Karen's digital portrait for the Cyberkids project effectively showcased what she had recently learned to do using the filters and other effects available in Photoshop software.

Nick: Web Builder and Technological Aesthete

When we met 15-year-old Nick, he had been interested in computers for several years and—with some assistance from his father—had recently bought his own computer at a government auction. Nick's teacher considered him to be one of the most talented students in her Year 10 Design and Multimedia class, someone who already knew most of what the curriculum was designed to teach. Nick confirmed this judgement when he told us that "I know most of the stuff, so I'm not actually learning much, but it's good to do." Like many of his peers, he had for several years been playing computer games and participating in Internet chat. What made him of particular interest to the project, however, was his devotion to Web design and the production of computer graphics. His digital portrait, produced using Flash, was the most technically accomplished of those produced for the Cyberkids project and the only one to integrate sound. Nick explained that he had "not used the digital camera much because... it felt just like taking pictures of me using a computer." Instead, he integrated screen shots of some of the Web sites he visited and the online activities he participated in during his leisure time.

Nick had high status among his friends and peers because of his skills in Web design. He was also admired because he had bought his own dot.com, a Web site that publicised a Web hosting and design business that he hoped to have fully established some time in the near future. Nick spent hours designing different homepages for this site, which regularly changed in appearance. As he told us, "I mainly make banners and stuff for Web pages. I really like making layouts of stuff. I make lots of layouts... I just make them and sometimes delete them." The site itself was not fully functional and was variously described on its homepage as being "under remoulding" or "under construction." The frequency with which Nick changed the site's design, together with its permanent state of incompletion, suggested that for him the utility of the site was not what mattered. Rather, his pleasure lay in designing the new graphics and layouts and changing the look of the site. In short, Nick could be described as a 'technological aesthete' (Abbott, 2001), someone for whom designs were created for their own sake, and as something to be admired.

In relation to his future working life, Nick told us that he did not have a lot invested in the

Design and Multimedia VET pathway. Making computer graphics was fun, and Design and Multimedia classes merely provided more time in which he could develop this interest. As he put it, "I haven't got my eye set on Design and Multimedia as my future. I'm more into maths and science so I sort of use this as a hobby." Although he was unsure exactly what vocational path he wanted to take once he left school, Nick told us that he would probably pursue a professional career in law or medicine.

David: Technological Aesthete and Portfolio Builder

David was Karen's and Nick's classmate and friend. He had been involved with computers since he was about 10 years old, even though his family had only recently purchased a home computer. Prior to that, David had accessed computers through his primary and secondary schools and in the homes of his friends. At the age of 15, not only did his parents assist him in buying what he needed for his computer, David had also managed to find a part-time job and was therefore able to purchase even more software and peripherals for the home computer. Further, the fact that his best friend's father owned a computer company meant that David was in a very good position to satisfy his desire to continually augment and update his computer equipment. He took great pride in the look and functionality of his customised desktop, carefully explaining its features, pointing out how its organisation increased his efficiency, and anticipating the changes he would make to it when he had a new computer and broadband access.

When asked to explain how he had learned what he knew about computers, Web site building, and graphic design, David pointed to a number of sources, citing free tutorials on the Internet and purchased software programs, examples of other people's work found on the Web, and tips from his classmate Nick and other friends he had met in online chatrooms. He explained that he and Nick made a particularly compatible pair when it came to teaching and learning from each other. In his view they had together been able to achieve more accomplished outcomes than they might have done had they each been working on their own. He added that much of this peer teaching was carried out online during out-of-school hours rather than in school.

I only know him through school but I've got him on my msn [Microsoft Network] messenger and I just talk to him a lot, like he's a great guy...I'm like, 'Hey, how do you do this?' and he shows me—because I'm more into the java scripting, the scripting side—and he's more into the Web design, like the pictures and graphics and...yeah we do pretty well together so it's all right.

David shared with Nick many of the qualities of the technological aesthete. He enjoyed Web designing and creating new kinds of computer graphic imagery, and gained great pleasure from describing in detail how he had achieved certain kinds of effects. He was obviously familiar with the current trends in professional design circles as discussed in the online forums he frequented. For example, he described particular effects used in his graphics and layouts as being "a very widely known technique nowadays" or "the latest craze going around now, like for Web pages," and sometimes passed his own judgements on these trends by saying things like "it's the latest Web craze going on from the image designers. I like it. I think it looks quite good." Thus David conveyed a sense of being an apprentice to the experts; a novice in pursuit of the skills and knowledge required to be able to display or perform a certain fully realised design aesthetic.

David was an active participant in a number of online communities. He frequented sites devoted to graphic design to which both he and Nick posted their own work as well as feedback to other graphic artists. When he showed us a favourite site, he explained very thoroughly how particular effects had been generated and what he appreciated about them. Unlike Nick, however, David was also what Abbott (2001) described as a "community builder." He showed a good deal of excitement about and commitment to the sense of 'community' he finds online, and indeed used that term to describe the way in which people helped each other to learn new skills. He acted as moderator/host to a Web site devoted to a role-playing computer game which he no longer played but which many of his friends continued to play. David not only spent several hours a week maintaining and moderating the site, he also assisted site participants to learn new skills and carried out what were in effect forms of "consultancy" and brokerage by making changes to the site requested by the site host and introducing people he knew from different online communities who might be of assistance to each other.

David told us, "I decided I wanted to have a computer job when I'm older." He had chosen to attend Harbourside High School because of its high level of computer resourcing and provision of IT-related curriculum pathways. He had also asked his friends who attended the school whether its publicity about computing was realised in practice before he made his final decision. David had not been disappointed in his school choice, noting: "I like it because I get lots more opportunities." David seemed more consciously aware than either Karen or Nick of the value of building and demonstrating a portfolio of skills. His digital portrait, produced using a number of programs, took the form of a Web site. Although it was less graphic and more verbal than his interests and skills might suggest, his portrait not only displayed some of his computer skills, but also represented himself as particular kind of aspirational and employable young man. It was a vehicle for David to "perform an identity" (Merchant, 2003) associated with a well-planned and clearly projected successful future. Interestingly, David even spoke of his involvement in the project as an achievement, referring to his digital portrait as a "portfolio," and thanking the university and his teacher for "choosing" him to participate in the research.

Mark: Literary Stylist and Flash Fan

Mark had a long history with computers. His father had worked with computers when Mark was very young, and in the early 1990s his family had been the first on their street to have an Internet connection. Mark could talk at length about the history of his engagement with computers, noting that he had "learned to use computers on DOS on Magic Desk" and had begun playing with *Commander Keen* computer games at the age of four. He described his progression from "arcade games on the old 486 and the Commodore later" through to the first "flight sims on eight discs" to more sophisticated games made possible by first Windows 95 and then Pentium computers. At the age of 15, however, he used the computer for chat, music, and graphic design rather than games.

Mark was a Flash enthusiast. He admired work produced using Flash software and frequented Web sites devoted to it. Originally he had been working on a PowerPoint portrait for the study, but because PowerPoint annoyed him, he had decided to use Flash even though he was not very proficient with it. Although he enjoyed doing Flash animations and clay animations, he noted, "I'm not very good." He was very apologetic about his portrait's level of sophistication, telling us that "I had to do it on a Saturday night and stuff. I was busy all Saturday so I had to cobble it together." His desire to use Flash—despite his lack of proficiency in using it—underlines the symbolic importance of Flash, and the peer status attributed to it, among a particular group of young males in his class.

Among his English teachers Mark was considered to be a gifted student with a talent for writing. He wrote poetry and short stories, frequented Web sites devoted to writing, and had begun to establish a Web site for writers at Harbourside High that the English subject coordinator was hoping would provide a place for young writers to post and discuss their work. Although Mark was also proficient using computers, he emphasised that in his view, English should retain its emphasis on the printed word and leave multimedia production to other curriculum areas. Nonetheless, he did not consider computers to be as insidious as television, which he held in low regard as a passive and crassly commercial medium. Rather, he told us that since he had learned more about multimedia production, he considered aspects of the process to be creative and of considerable aesthetic value:

It has the same effect as painting a picture I suppose, so I don't feel like it's like watching TV. If you're on Photoshop I still feel like you get the same satisfaction, you get the same kind of feelings as if you're painting a picture.

Of the 15 digital portraits produced in the study, Mark's was the only one that consciously engaged with the metaphoric possibilities of the title "Me and My Cyberworld;" this perhaps reflected his literary leanings. For example, he wrote of his cyberworld as being "a different place," "a world to escape into," a world which you "sign into" and use to "open up" new possibilities. He also saw his online world as a place in which he could interact with like-minded others and, therefore: "I'm not alone in my world." As he explained it:

I suppose what came to mind first was just like I'd decided to make something to show that it's a world that you're in, because it's important as having an escape, a world to escape into.

Is that how it feels to you?

Yeah. I don't know if it feels better or worse, it's just a completely different place, which I suppose is what I like about it, it's just a completely different place that you can go into, not like the real world, and I suppose like this is more and more evident, especially with like the world changing so fast that everyone needs something that they're going to escape into as well...

Mark explained that the portrait he had actually produced was a poor imitation of what he had imagined and hoped for. In part he attributed this to a lack of facility with the medium, noting that when he had tried to put some of his ideas into practice, "It didn't do what I had in mind." We sensed that Mark may still have felt more 'at home' using written language to express poetic ideas even though he remained open to the possibilities of the new media. Overall, the way Mark represented himself in his digital portrait and in the interviews suggested a young man who had an artistic rather than a vocational interest in computing.

Continuing Challenges

Although Harbourside High's catchment area for students encompasses many poorer suburbs—the school has a commitment to meeting the special needs of students from low-socio-economic backgrounds as well as those from indigenous and non-English-speaking households—in recent years teenagers from more prosperous suburbs have begun to make up more of the overall student population. Karen, Nick, and David, for example, live in a comparatively new and affluent residential area, a marina enclave which is a conspicuous departure from the more basic housing stock and working-class associations of suburbs in the immediate vicinity of the school. In this social climate, the introduction of VET pathways has been an important strategy used by the school to reorient the curriculum in order to attract and retain a diversity of students and funding sources. However, a strategy of emphasising the vocational utility of school curriculum—which has become increasingly common in Australian schools—introduces new philosophical and practical challenges in relation to curriculum and pedagogy.

One challenge arises from the fact that nationally accredited VET courses bring with them a ready-made curriculum and form of assessment that has not traditionally been used in secondary schools. Assessment of student achievement in VET is made according to "elements of competency and performance criteria" that define the national qualifications framework. For example, when making their digital portraits, Karen, Nick, David, and Mark demonstrated national competency CUFIMA01A/03 as they "incorporated digital images into a multimedia sequence." This is described in the national framework as one element of competency in the "skills and knowledge required to produce and manipulate digital images for a multimedia production within the cultural industries" (National Training Information Service, 2004). This particular element of competency includes:

- create graphics that incorporate the principles of design using the designated software;
- edit, enhance, amend, and save digital images into a designated multimedia sequence;
- combine digital images into a designated multimedia sequence;
- integrate digital images into a designated multimedia sequence; and
- evaluate the outcome for visual impact, effectiveness, and fitness of purpose. (National Training Information Service, 2004)

From a literacy educator's point of view, it is precisely at this point that any analogy between literacy and competency loses its utility. Nationally accredited element of competence skill descriptions are necessarily generic. However, there is a complete failure here to adequately describe the *repertoire* of literacy practices required to produce a multimedia product that has the required "visual impact, effectiveness, and fitness of purposes" in particular circumstances. This is a serious omission, not only in relation to assessment, but also in relation to teaching and learning about multimedia production. There is little or no elaboration, for example, about how to teach young people to discern and reproduce indicators of "visual impact" or "effectiveness and fitness of purpose" in a multimedia text. In any competencies framework these things are taken as read.

We would argue that for the literacy-technology nexus to be adequately addressed in the school curriculum, different questions raised by literacy studies about multimedia textuality and composition need to be addressed from the earliest years through to "vocational" and other programs in the final years of schooling. This requires, in turn, further research into the meaning of "style and genre" (O'Hear & Sefton-Green, 2004) in new media texts, including the "multimodality" (Burn & Parker, 2001) of texts constructed using words, image, and sound. It requires more research into principles and grammars of meaning-making or "semiosis" (Gee, 2003; Kress, 2003) in order to enable teachers and children to develop a shared language with which to explore the formal properties, possibilities, and constraints of commercial software and multimedia texts as well as texts that they produce themselves. Finally, principles of "active, critical learning" (Gee, 2003)—which include the exercise of such intangible things as creativity and imagination—are crucial to any adequately 3D conception of l(IT)eracy learning. Such complex issues now beginning to be foregrounded within literacy education continue to complicate the neat precision of many "ICT literacy" competency frameworks.

Further complicating this picture are the misgivings of some educational researchers about the ability of schools and curricula in their current state to take advantage of the full range of potential offered by the incorporation of ICT into formal educational contexts (Bigum, 2002; Lankshear & Knobel, 2004; Sefton-Green, 2004). Part of this concern is driven by the observation that a dominant feature of the use and usefulness of ICT outside schools is their ability to strengthen and multiply communicative relationships—between ideas, especially as a feature of non-linearity and hypertext, as well as between people—and that while information is an important element of information and communication technologies, it is their facilitation of access and exchange rather than the provision of information itself that has been most revolutionary. The gradual vernacular and conceptual shift over the last half decade from IT to ICT following the inclusion of communication is telling in this regard: ICT, unlike IT, can no longer be defined primarily by their ability to store and compute data and content. They are fundamentally about connections, connectivity, and relationships between and amongst individuals, groups, communities, corporations, institutions, government departments, and so on, and between information and the processes of knowledge creation. These aspects, too, need to be taken into account in re-imagining K-12 literacy education. Schools, in general, have been slow to engage with this important feature of ICT, and therefore participate in and contribute to the communicative networks that have become such an integral part of ICT in wider society.

As our case studies have suggested, students' use of ICT outside school, as well as being self-directed and recreational, are often expressly about networks geared towards the exchange of skills and knowledge for mutually beneficial ends. Furthermore, while schools may be resistant to allowing students' out-of-school activities into the classroom, some young users of ICT are, it seems, already adept at drawing on the skills and knowledge acquired outside school for the demonstration of competencies required by the formal educational curriculum. In this way their practices and identities are reminiscent of Gee's idealised workplace and his "shape shifting portfolio people" in their emphasis on teamwork, and the flexible appropriation and application of knowledge to accomplish a diverse array of tasks and projects. School assignments—and those imposed by visiting educational researchers—were similarly treated by these students as tasks among others to be carried out within a wider context of endeavour that included the demands of formal education and training, leisure-time pursuits, and after-hours and weekend courses. In short, as they use the new media, some young people are not only following school-designed curriculum pathways, but are also forging their own pathways between home and school, and towards their desired workplaces of the future.

REFERENCES

Abbott, C. (2001). Some young male Web site owners: The technological aesthete, the community builder and the professional activist. *Education, Communication and Information, 1*(2), 197-212.

Alvermann, D. (Ed.). (2002). *Adolescents and literacies in a digital world.* New York: Peter Lang.

Australian Bureau of Statistics. (2003). *Children's participation in cultural and leisure activities, Australia (Pub. No. 4901.0).* Retrieved from http://www.abs.gov.au/ausstats/abs@.nsf/0/0b14d86e14a1215eca2569d70080031c?OpenDocument

Bigum, C. (2002). Design sensibilities, schools and the new computing and communication technologies. In I. Snyder (Ed.), *Silicon literacies: Communication, innovation and education in the electronic age* (pp. 130-140). London; New York: Routledge/Taylor & Francis.

Boyce, S. (2001). Engineering literacy in the library. In C. Durrant & C. Beavis (Eds.), *P(ICT)ures of English: Teachers, learners and technology* (pp. 125-142). Adelaide: Wakefield Press.

Buckingham, D. (2000). *After the death of childhood: Growing up in the age of electronic media.* Cambridge: Polity Press.

Buckingham, D. (2003). *Media education: Literacy, learning and contemporary culture.* Cambridge: Polity Press.

Burn, A., & Parker, D. (2001). Making your mark: Digital inscription, animation, and a new visual semiotic. *Education, Communication & Information, 1*(2), 155-179.

Centre for Media Education. (2001). *Teen sites.com: A field guide to the new digital landscape.* Retrieved from http://www.cme.org/teenstudy/index.html

Chandler, D., & Roberts-Young, D. (1998). *The construction of identity in the personal homepages of adolescents.* Retrieved from http://www.aber.ac.uk/media/Documents/short/strasbourg.html

Chandler-Olcott, K., & Mahar, D. (2003). "Tech-savviness" meets multiliteracies: Exploring adolescent girls' technology-mediated literacy practices. *Reading Research Quarterly, 38*(3), 356-385.

Cope, B., & Kalantzis, M. (Eds.). (2000). *Multiliteracies: Literacy learning and the design of social futures.* Melbourne: Macmillan.

Durrant, C., & Green, B. (2000). Literacy and the new technologies in school education: Meeting the l(IT)eracy challenge? *The Australian Journal of Language and Literacy, 23*(2), 89-108.

Education Network Australia (EdNA). (2001). *Learning in an online world: The school education action plan for the information economy.* Ministerial Council for Education, Employment, Training and Youth Affairs (MCEETYA). Retrieved from http://www.edna.edu.au/edna/file12663

Education Queensland. (2003). *Literate futures: The teacher summary version.* Retrieved from http://education.qld.gov.au/curriculum/learning/literate-futures/pdfs/lf-teacher-summary.pdf

Gee, J. P. (2000). Teenagers in new times: A new literacy studies perspective. *Journal of Adolescent & Adult Literacy, 43,* 412-420.

Gee, J. P. (2003). *What video games have to teach us about literacy and learning.* New York: Palgrave Macmillan.

Gee, J., Hull, G., & Lankshear, C. (1996). *The new work order: Behind the language of new capitalism.* Sydney: Allen & Unwin.

Green, B. (Ed.). (1993). *The insistence of the letter: Literacy studies and curriculum theorizing.* Pittsburgh, PA: University of Pittsburgh.

Green, B., Reid, J., & Bigum, C. (1998). Teaching the Nintendo generation? Children, computer culture and popular technologies. In S. Howard (Ed.), *Wired up: Young people and the electronic media* (pp. 19-41). London: UCL Press.

Kress, G. (2003). *Literacy in the new media age.* London/New York: Routledge/Falmer.

Lankshear, C., Bigum, C., Durrant, C. et al. (1997b). *Digital rhetorics: Literacies and technologies in education. Current practices and future directions.* Canberra: Commonwealth of Australia, DEETYA.

Lankshear, C., Gee, J., Knobel, M., & Searle, C. (1997a). *Changing literacies.* Buckingham/Philadelphia: Open University Press.

The digitally at home. *Proceedings of the American Education Research Association (AERA) Annual Meeting,* San Diego. Retrieved from http://www.geocities.com/c.lankshear/roles.html

Lankshear, C., Snyder, I., & Green, B. (2000). *Teachers and techno-literacy: Managing literacy, technology and learning in schools.* Sydney: Allen and Unwin.

Lankshear, C., & Knobel, M. (2004, April 15). Livingstone, S., & Bovill, M. (Eds.). (2001). *Children and their changing media environment.* Mahwah, NJ: Lawrence-Erlbaum.

Luke, A. (1996). Text and discourse in education: An introduction to critical discourse analysis. In M. Apple (Ed.), *Review of research in education 1995-1996* (pp. 3-48). Washington, DC: American Educational Research Association.

Luke, A. (2000). Critical literacy in Australia: A matter of context and standpoint. *Journal of Adolescent and Adult Literacy, 43*(5), 448-461.

Luke, A., & Freebody, P. (1999). *Further notes on the four resources model.* Retrieved from http://www.readingonline.org/research/lukefreebody.htm

Luke, C. (2002). Re-crafting media and ICT literacies. In D. Alvermann (Ed.), *Adolescents and literacies in a digital world* (pp. 132-146). New York: Peter Lang.

Merchant, G. (2003). "Imagine all that stuff really happening:" Narrative and identity in children's on-screen writing. *Proceedings of the British Educational Research Association (BERA) Annual Conference.*

National Training Information Service. (2004). Retrieved from http://www.ntis.gov.au/cgi-bin/waxhtml/~ntis2/unit.wxh?page=80&inputRef=23438 &sCalledFrom=std

New London Group. (1996). A pedagogy of multiliteracies: Designing social futures. *Harvard Educational Review, 66*(1), 60-92.

Nixon, H. (2004). Cultural pedagogies about ICTs and education in a globalised cultural economy. In M. Apple, J. Kenway, & M. Singh (Eds.), *Globalising public education: Policies, pedagogies and politics* (pp. 45-60). New York: Peter Lang.

O'Hear, S., & Sefton-Green, J. (2004). Style, genre and technology: The strange case of youth culture online. In I. Snyder & C. Beavis (Eds.), *Doing literacy online* (pp. 121-144). New York: Hampton Press.

Pearson, J. (2003). Information and communications technologies and teacher education in Australia. *Technology, Pedagogy and Education, 12*(1), 39-58.

Sefton-Green, J. (2004). Literature review in informal learning with technology outside school. Retrieved from http://www.nestafuturelab.org/research/reviews/07_01.htm

Snyder, I. (2002). *Silicon literacies: Communication, innovation and education in the electronic age.* London/New York: Routledge/Taylor & Francis.

Stern, S. (1999). Adolescent girls' expression on Web home pages: Spirited, sombre and self-conscious sites. *Convergence: The Journal of Research into New Media Technologies [Special Issue: Children, Young People and Digital Technology], 5*(4), 22-41.

Street, B. V. (1995). *Social literacies: Critical approaches to literacy in development, ethnography and education.* London: Longman.

Upitis, R. (1998). From hackers to luddites, game players to game creators: Profiles of adolescent students using technology. *Journal of Curriculum Studies, 30*(3), 293-318.

ENDNOTE

[1] The project, *Cyberkids and Cyberworlds: New Literacies, Identities and Communities in Formation,* was funded in 2003 by an Australian Research Council Discovery Grant. Chief investigators were Helen Nixon from the University of South Australia and Catherine Beavis from Deakin University, Australia. Research assistants were Stephen Atkinson and Sandy Muspratt.

Chapter VIII
Responsible Technologies and Literacy:
Ethical and Legal Issues

Elizabeth A. Buchanan
University of Wisconsin-Milwaukee, USA

Tomas A. Lipinski
University of Wisconsin-Milwaukee, USA

ABSTRACT

This chapter presents a case study of research conducted in the state of Wisconsin, USA, on the awareness of and knowledge surrounding ethical and legal uses of technology by primary teachers, administrators, and technology coordinators. The authors use the term responsible technologies to define the concept of ethical and legal awareness; the chapter reports on the findings from the pre- and post-in-service surveys, and makes recommendations for greater awareness of the ethical and legal implications surrounding technology use in general, and surrounding copyright in particular.

INTRODUCTION: RESPONSIBLE TECHNOLOGIES AS PART OF TECHNOLOGY LITERACY

Preparing future citizens for the responsibilities of full participation in the information society means more than just imparting an understanding of the technical uses of the information and communication technologies (ICTs) that support society. There is a level of literacy that receives less attention than keyboarding, word processing skills, or programming skills. This level of literacy entails an understanding of the *responsible uses of technology* in general and of *computing* in particular. A sense of computing responsibility is a global issue, not relegated to any particular country or area of the world; this responsibility relates to ethical and legal frameworks and guidelines. But, where do our young citizens learn this responsibility? Moreover, how and what do we teach when we speak of responsible uses of computing?

In the United States in particular, educators, policymakers, industry leaders, and many more people are beginning to ask these very important questions of K-12 education, as this realm of schooling seems to be the most appropriate setting in which to instill legal and ethical understanding of technology use. For instance, Slind-Flor (2000) asks: "Should we be teaching reading, writing, and copyright?" The authors believe the answer to this question is a straightforward yes, and to this end, the responsible technologies (RT) project was funded by the University of Wisconsin System. Responsible technologies, as a concept, is the understanding, knowledge, and uses of technologies in ethical and legal ways. The concept of RT begins with teachers, media specialists, technology coordinators, and school administrators, and is then passed down to their students; a systemic approach is necessary. RT thus becomes a continuum, a process by which all stakeholders in K-12 education embrace the idea of using technology in legal and ethical ways. Therefore, school children and young adults are instilled with a set of skills for present and future technology use—and an attitude of responsibility from which to use those skills as informed citizens in society.

The necessity for the responsible use of ICT is becoming widely recognized, especially as part of the primary and secondary educational experience. Many factors contribute to this growing necessity: the developing case law (e.g., *Chicago School Reform Board of Trustees vs. Substance, Inc.*, 354 F. 3d 624 (7th civ. 2003), cert. denied 125 S. ct. 54 (2004)), as well as the increased awareness of copyright issues among educators, in part due to the publicity surrounding P2P (Peer-to-Peer) litigation (*A&M Records, Inc., v. Napster, Inc.*, 239F.3d 1004 (9th cir. 2001); *In re Aimster Copyright Litigation*, 334 f 3rd 643 7th (cir. 2003); *Metro-Goldwin Studeios, Inc. v. Grokster, Ltd.*, 125 S. ct. 2764 (2005)). Piracy is, in fact, *growing*, as recent surveys indicate (Business Software Alliance, 2004):

> *A majority of U.S. children continue to download songs, despite acknowledging they know it is illegal. According to the survey, 88 percent of kids between the ages of 8 and 18 know that most popular music is copyrighted, but 56 percent download music files anyway. Survey participants said they were generally more concerned about downloading viruses in music files than being prosecuted for copyright violations.*

Further reasons include the exponential growth in technology use in school settings; a growing mandate from states and departments of public instruction for standards which include a component of legal and ethical competence in technology use, demonstrates that legal and ethical issues are gaining prominence in educational settings but have far to go in making their environments both compliant in terms of the law and committed in terms of ethical uses of technology, combining the legal and ethical issues into a compound concept of responsible technologies.

A number of U.S. states have implemented model academic standards for the use of technology in the classroom that include the introduction of these concepts into the primary and secondary curriculum. In Wisconsin, the locus of the present study, the Wisconsin Department of Public Instruction issued the Wisconsin's Model Academic Standards for Information & Technology Literacy (hereafter, WMASITL) (Fortier, Potter, & Grady, 1998). While the majority of the WMASITL concern minimum technological standards and learning

(Part A: Media and Technology, Part B: Information and Inquiry, Part C: Independent Learning), as might be expected, Part D focuses on the learning community and reflects the necessity of education as a prerequisite to effective participation in the "workplace of the 21st century." Implementation of the standards requires that "[s]tudents in Wisconsin will demonstrate the ability to…use information and technology in a responsible manner, respect intellectual property rights." A series of performance standards are presented that suggest the given level of knowledge a member of a K-12 learning community should possess:

- *By the end of grade 4 students will… Use information, media, and technology in a responsible manner, by the ability to demonstrate use consistent with the school's acceptable use policy, understand concepts such as etiquette, defamation, privacy, etc. in the context of online communication… Respect the concept of intellectual property rights, by the ability to explain the concept of intellectual property rights, describe how copyright protects the right of an author or producer, to identify violations of copyright law as a crime….*
- *By the end of grade 8 students will… Use information, media, and technology in a responsible manner by the ability to describe and explain the applicable rules governing the use of technology in the student's environment, demonstrate the responsible use of technology, recognize the need for privacy and protection of personal information… Respect intellectual property rights by the ability to explain the concept of fair use, and that the application of the concept may differ depending on the media format, relate examples of copyright violations, explain and differentiate the purposes of a patent, trademark, and logo….*
- *By the end of grade 12 students will… Use information, media, and technology in a responsible manner by the ability to assess the need for different informational polices and user agreements, understand concepts such as misrepresentation and the need for privacy of certain data files or documents… Respect intellectual property rights by the ability [to] explain why fair use is permitted for educational purposes but not for profit situations, and the conditions under which permission must be obtained for the use of copyrighted materials….* (WMACITL, 1998, pp. 14-15)

Unfortunately, as previous research has demonstrated, educators are ill prepared to teach legal and ethical concepts regarding the uses of new technologies in the classroom (Green, 1993; Carter & Rezabek, 1993; Patterson, 1996; Monts, 1998). Teacher education programs typically do not include coursework in technology law and ethics, but instead focus on curricular integration and use of technologies. Moreover, administrators also receive little training and place legal and technical issues regarding new technologies low on the concern list (Rice, 1991). Until faced with the actual threat of impending litigation, educational organizations in the PK-12 environment offer ineffective guidance either by doing little to change the behavior of those in its employ, such as teachers and other staff, or to mold the development of the charges in their care, the students. Copyright violations are a significant area of concern.

As a result, students learn first hand (by observation) not only how to infringe copyright and abuse the rights of others online, for example, but also to take such actions without any consideration or understanding of intervening concepts such as fair use. Copyright infringement in particular is now second nature among the upcoming generation of information society participants. It is no wonder, unfortunately, that the current educational system seems more designed to turn out the next generation of teenage hackers than it is law abiding or conforming "netizens." For this reason, copyright law and ethics were a major consideration of this study, over other equally important issues, such as plagiarism or computer theft.

K-12 schools are particularly at risk for violating ethical principles and setting legal precedents, as teachers are often pressed for time, money, and resources, but mainly lack of knowledge about technology law and ethics. Previous educational rhetoric, policy, initiatives, and funding have focused decidedly on getting new technology into the classroom with little thought on the legal and ethical implications of those decisions. Perhaps it was assumed that the adoption of responsible technologies would be a given, but an ongoing research study indicates that this is simply not the case. The researchers propose that a systematic program of responsible technologies is necessary to directly assist teachers, administrators, and ultimately all students in making ethically sound and legal decisions regarding technology use in schools.

THE WISCONSIN RT PROJECT: GOALS, PROCEDURES, OUTCOMES

The authors set out to design a selective research and education project surrounding the idea of responsible technologies, with a particular emphasis on copyright law and ethics. The population of the study included participating school districts located throughout Milwaukee, Ozaukee, Waukesha, and Washington counties, in the state of Wisconsin, and was carried out with the cooperation of the Cooperative Education Service Agency, District #1 (CESA 1). Funding for the project came from the University of Wisconsin System. This chapter should therefore be considered a case study of a particular group and setting, where the data were collected and the education component conducted throughout Fall 2001. While the results are limited in generalizability, the conceptual approach and theoretical, not statistical, significance should be considered in a global perspective. While cultural norms and of course laws differ greatly, the authors believe the overall concept of infusing a systemic approach to teaching educators, administrators, and students about responsible use (noting that the actual laws and ethical norms introduced may differ across cultures) is an appropriate design in assisting educators and preparing students for computing use.

The original goals and objectives of the RT project as articulated in the grant proposal included:

1. to assess the current state of knowledge and perceptions of a sample of K-12 teachers, media specialists, technology coordinators, and administrators from a five-county population in Southeastern Wisconsin surrounding legal and ethical implications of technology use in the classroom;
2. to teach this sample about legal and ethical uses of technologies in the classroom;
3. to work with participants to develop resources about legal and ethical uses of technologies in the classroom and make these resources widely available; and

4. to assist K-12 educators in becoming aware and knowledgeable about law and ethics in order to better instruct their students so they become responsible users and consumers of technologies.

The initial phase of the project consisted of an assessment of participant readiness to implement the WMASITL. This was accomplished by administering both a pre- and post-survey that included an extensive Copyright Battery Questionnaire. Training sessions and the formation of feedback focus groups were also used to impart additional skills training to the participants and to gauge remaining problems such as institutional barriers. Case study methodology relies on a triangulation of data sources, thus the surveys, focus groups, and school documents were considered in data analysis.

A total of 27 participants completed both the pre- and post-survey. Results were reviewed to compare differences between the pre-survey and the post-survey responses. Comparisons were also reviewed between the following groups (see Table 1): those who taught library or technology-related subjects and those who did not; those who taught primary grades (K-6) and those who taught secondary (7-12) (in the instances where a teacher taught in both categories, he/she was placed in the category that he/she taught more—i.e., a teacher who taught grades 4-8 was placed in primary since grades 4, 5, 6 represent a majority of the grades); and those who taught greater than 15 years and those who taught less than 15 years.

The pre- and post-survey was done in two sections: Opinion Questions, and Practical Copyright Law Questions (the Copyright Battery). Each section is further broken down into four parts: Pre vs. Post; Lib/Tech vs. Non Lib/Tech; Years of Experience; and Primary vs. Secondary. A criterion for what was considered a significant difference is given in each section. (Parenthetical references refer to survey part and then to question number within that part, i.e., (2.01) is question number one from part 2 of the survey.)

Regarding the levels of agreement with the school's and teacher's knowledge of, enforcement of, and support of legal and ethical use of technology, the post-survey responses indicated a decrease in confidence. In the post-survey,

Table 1. Survey group participants, by comparative categories

Library/Technology Subject	15
Non-Library/Technology	12

Primary	13
Library/Technology Subject	11
Administrator	3

Years Teaching/Administrating	1-15 Years	16+ Years
	18	9

less participants indicated that their school had a copyright policy (73.1% to 63.0%) (2.01), that they were familiar with the content of such a policy (66.7% to 40.7%) (2.02), and that their school administrator was supportive of responsible use (88.5% to 66.7%) (2.04). This may be due to the observation that many participants scored low on the Copyright Battery in the pre-survey measure, yet assumed when taking the pre-survey measure that their knowledge of the law and the response of their school to copyright issues was adequate, then after participating in training mechanisms, came to realize that their level of knowledge regarding copyright law and its application in practice was not as high as it was originally thought. It is notable that more non-library/technology teachers disagreed that they were familiar (i.e., were not familiar) with their school's copyright policy (75.0%) than library/technology teachers (50.0%) in the post-survey (2.02).

A notable finding regarding the support of the school was that none of the participants agreed that their school as a whole punishes teachers for the infringement of copyright in either the pre- or post-survey (2.03), which conflicts with the fact that the majority of them (greater than 50%) said that their administrator was supportive of responsible use (2.04). During the follow-up focus group discussions, this anomaly of sorts was rather simple to explain, and common among participants: administrators pay lip service to copyright issues in schools. While supporting, at least in policy formulation, copyright compliance, administrators are reluctant to take action against violators of those policies. This is also consistent with an earlier study, and reflects the fact that administrators assign low importance to copyright issues (Rice, 1991). This was also reflected by the number of participants who agreed that their administrator was supportive of responsible use (2.04)—a reduction of approximately 22% (88.5% to 66.7%) of participants. Again, this may reflect the experience of the participant-educators that upon return to their respective schools, armed with more accurate copyright information, response to the implementation of more appropriate compliance measures was met with some reticence from administration. On a related issue, a majority (greater than 50%) strongly believed (a scoring of a 4 or 5) that their school as a whole punishes students for unethical or inappropriate use of technology (3.07). This again may reflect a tendency to operationalize general rules of behavior (e.g., a prohibition on sending harassing messages), far more readily, through an acceptable use policy for example, than articulating hard and fast rules regarding copyright.

A number of participants reported a greater understanding of ethical uses of technology, the meaning of intellectual property, and the definition of an acceptable use policy (AUP) in the post-survey. A large number of people reported that they enforce AUPs in their classroom in the pre-survey, 80.7% chose level 3, 4, or 5, but it is worth pointing out that in the post-survey, 100.0% chose level 3, 4, or 5 (3.06). There was also an increase in the levels participants gave regarding what they teach their students (why plagiarism is wrong (3.08), why copying software is wrong (3.09), respecting others in online environments (3.10), privacy rights (3.11), and using technology responsibly (3.12)). It is worth mentioning, however, that the levels of agreement in the pre-survey were high (more than 50% reporting levels 3, 4, or 5) in these areas as well. It seems that more participants learned the need to also teach the social implications of technology: 44.4% indicated a level of 1 or 2 in the pre-survey; 88.8% indicated a level of 3 or 4 in the post-survey regarding social implications. This does not appear inconsistent with the findings regarding the external classroom environment, as the participant educators gained new knowledge of the variety of issues surrounding the responsible use of technology, and attempted to implement such knowledge into the classroom (one of the overall objectives of the three-year project), yet the support structure external to the implementation of the educator-student (classroom) environment remains underdeveloped or was inadequate, as indicated by questions regarding institutional practices and thus might have prevented an even higher level of application in the classroom and elsewhere in the school.

Not many participants were aware of the "Ten Commandments of Computer Ethics" (Barquin & Computer Ethics Institute, 1992) before the in-service training sessions (63% indicated level 1), and no participant indicated level 4 or 5. However, in the post-survey, 48.1% indicated a level of 4 or 5 (3.15). More people indicated a higher level of confidence in teaching technology ethics in their classroom (3.18). Before and after the in-service, participants felt that technology ethics should not be taught distinct from the regular curriculum (greater than 70% indicated level 1 or 2) (3.19), and that it is appropriate to teach K-12 students (greater than 60% indicated level 1) (3.20). However, in both of these areas, the number of people indicating a level of 4 or 5 jumped from 3.7% to 14.8% and 0.0% to 7.4%, respectively.

Significant differences (greater than 30%) between the number of people who gave a level of 4 or 5 (in cases where an agreement level of 5 is considered ideal) were found when comparing those who were library or technology teachers with those who were not. When compared with teachers who taught subjects other than library or technology, more library/technology teachers knew, practiced, or felt the following:

- Their school had policy about ethical uses (64.3% compared with 33.3%) (3.01)
- Are familiar with ethical uses of technology (92.9% compared with 58.4%) (3.02)
- Discuss ethical uses with their class (64.3% compared with 25.0%) (3.03)
- Enforce AUPs (85.7% compared with 50.0%) (3.06)
- Teach students to respect others in online environments (78.5% compared with 41.7%) (3.10)
- Teach students about privacy rights and technology (64.3% compared with 25.0%) (3.11); about censorship (64.3% compared to 33.3%) (3.13); and the social implications of technology use (64.2% compared to 25.0%) (3.14)
- Familiar with "Ten Commandments of Computer Ethics" (64.3% compared with 25%) (3.15)
- Explain filtering to students (57.1% compared with 16.6%) (3.17)
- Feel confident about teaching technology ethics (78.6% compared with 41.7%) (3.18) and ethics and copyright and other legal ownership or property issues (64.3% compared with 25.0%) (3.21)

There were not many differences in responses between teachers who had 1-15 years of experience and those who had greater than 15 years of experience. The only significant difference found when looking at the opinion questions was regarding the view of the school administrator's support of responsible use of technology: 88.9% of teachers with greater than 15 years experience agreed that the school administrator is supportive, 55.6% of teachers with less than 15 years experience agreed (2.04).

Only one significant difference was found between the responses of primary and secondary teachers. The question to which there was greater than 30% difference between the groups' responses (percentage of people who gave a level of 4 or 5 when an agreement level of 5 is considered ideal) regards teaching students to respect others in online environments (76.9% of primary teachers compared to 45.5% of secondary) (3.10). This may reflect a perception, albeit mistaken, that at the secondary level, education in proper online behavior is not as important because teachers assume students have had such training at the primary level. This of course may not be a correct perception—that is, the student may not have received adequate training in responsible uses at the primary level, or at least may still be in need of reinforcement in those areas. Moreover, the WMASITL envisions that responsible use of technology education including intellectual property should be a continuing and evolving process, with not only reinforcement, but also expansion of concepts parallel to a student's cognitive and technical abilities. This of course suggests that the WMASTIL are not fully implemented in a number of schools as represented by the survey participant responses.

The Copyright Battery

A few cases of correctly applying copyright law were clearly understood by the participants in both pre- and post-surveys. (Please note: the wording of a statement does not reflect the actual wording of the question as it appeared in the copyright battery, but rather is an attempt to convey the idea behind the question, i.e., the specific factual context presented in a given question may be absent.) The questions to which more than 70% of the responses were correct in both surveys were the following (u: unacceptable, and a: acceptable):

- Every personal use a fair use (4.02) (u)
- Every educational use a fair use (4.04) (u)
- Showing purchased video to class on relevant topic (4.20) (a)
- Showing purchased video again next semester (4.21) (a)
- Reading Faulkner to English class (4.26) (a)

A greater number of people correctly answered the copyright law questions in the post-survey. The questions to which an additional 25% or more answered correctly in the post-survey are as follows:

- Making a copy of CD using Napster for classroom distribution and use (4.03) (u)
- Making a copy of a chapter of a book for use in preparing for class (4.05) (a)
- Copying the two most important pages of a chapter for each student (4.08) (a)
- Requiring students to bring copy of required chapter to class (4.13) (u)
- Making copy of test from workbook for each student (4.14) (u)
- Taping PBS episode off of TV and showing to class during the following week (4.15) (a)
- Placing off-air copy of magazine-format news program in library for individual use (4.19) (a)
- Showing purchased video at school's annual science fair (in cafeteria or gym) (4.22) (u)
- Showing purchased video as reward (4.23) (u)
- Broadcasting teacher's dramatic performance to class in neighboring district (4.25) (u)
- Reading Faulkner to pass time in Math class (4.27) (u)

- Student scanning National Geographic photo to include in term paper (4.29) (a)
- Student using 20-second sound byte of 3:38-minunte song in presentation (4.30) (a)
- Student using entire CD in presentation (4.31) (u)
- Student showing that presentation to 10th grade school assembly (4.33) (u)
- Teacher creating work with 20 seconds of music and 2 minutes of video (4.36) (a)
- Using 2 minutes from off-air recording of movie last year (4.37) (u)
- Allowing students to upload/download music using MP3.com or Napster (4.41) (u)
- Referring to Web site containing infringing material (4.42) (u)

There were questions concerning the copyright law that participants were unsure of both during the pre- and post-surveys. The questions to which less than 35% knew the correct answer both before and after the in-service are as follows:

- Making a copy of the same chapter for each teacher in curriculum group (4.06) (u)
- Making copy of chapter of book for each student (4.07) (u)
- Recording same program off-air as last semester and using it within 10 days (4.16) (u)
- Taping a program from cable-only station and using it next week (4.17) (u)
- Student using 2 minutes of Smashing Pumpkins music video in presentation on American Authors of Romantic Period (4.34) (u)
- Teacher using same presentation this year as last, using 20 seconds from purchased CD and 2 minutes from video borrowed from library (4.38) (a)
- Broadcasting that presentation to neighboring school (4.39) (a)
- Teacher using that presentation at upcoming conference (4.40) (a)

Teachers who taught library or technology gave the same responses as those who taught other subjects except in a few areas. The following questions were answered differently (greater than 30% difference in number of people who answered correctly):

- Making copy of chapter of book for use in preparation (preferred response of Yes: 50.0% lib/tech; 85.7% non lib/tech) (4.05) (a)
- Making copy of entire book for use in preparation (preferred response of No: 25.0% lib/tech; 57.1% non lib/tech) (4.12) (u)
- Making copy of a test from workbook for each student (preferred response of No: 11.7% lib/tech; 85.7% non lib/tech) (4.14) (u)
- Performing entire score/script of Oklahoma to drama class (preferred response of Yes: 25.0% lib/tech; 57.1% non lib/tech) (4.24) (a)

Unfortunately, the library or technology specialists in these four questions provided the less preferred response in the first and last scenarios in the previous aforementioned series, opting instead for curtailing a use that is acceptable (under fair use guidelines and under the statute, respectively) and in supporting a use that is not fair (or unacceptable under the classroom fair use guidelines, i.e., 4.12 and 4.14). While the questions were not necessarily technology dependent or technology intensive, the copyright library/technology educators still missed the mark, perhaps revealing that old habits, or in this case copyright misconceptions "die hard."

Teachers with more than 15 years of experience gave the same responses as those who taught less than 15 years except for two cases. The following questions were answered differently (greater than 30% difference in number of people who answered correctly):

- Copying two most important pages from a chapter for each student (preferred response of Yes: 88.9% 1-15 yrs; 55.6% 16+ years experience) (4.08) (a)
- Student using 2 minutes of Smashing Pumpkins video in presentation (preferred response of No: 33.3% 1-15 yrs.; 77.8% 16+ years experience) (4.34) (u)

A possible explanation for the discrepancy might reflect the fact that the correct answer is contained in the various copyright fair use guidelines that have been created over the years. While the fair use guidelines are not law, their reflection in the survey, exhibits a reality among participants that many schools have formalized use of the guidelines within a policy framework. For example, the latter question references the portion limitations of the Fair Use Guidelines for Educational Multimedia (reprinted in Bielefield & Cheeseman, 1997, pp. 92-102). New teachers may be unaware of the existence of the guidelines, much less a working knowledge of its content; perhaps most copyright information is learned post-service, through repeated exposure via continuing education.

When comparing the responses of primary teachers and secondary teachers, several differences were found. The following questions were answered differently (greater than 30% difference in number of people who answered correctly):

- Making a copy of a chapter for use in preparation (preferred response of Yes: 84.6% primary; 36.4% secondary) (4.05) (a)
- Student using 20 seconds of 3.38-minutes Smashing Pumpkins song in presentation (preferred response of Yes: 84.6% primary; 45.5% secondary) (4.30) (a)
- Student using 2 minutes of Sleepy Hollow in presentation (preferred response of Yes: 92.3% primary; 45.5% secondary) (4.32) (a)
- Teachers using 20 seconds of music and 2 minutes of video when creating similar work (preferred response of Yes: 84.6% primary; 45.5% secondary) (4.36) (a)
- Verbal chastisement for infringing uploading/downloading enough to qualify for "immunity" under section 512 (preferred response of No: 53.8% primary; 90.9% secondary) (4.43) (u)

As suggested earlier, primary educators were more knowledgeable of copyright issues than their secondary counterparts, and may reflect the attitude that such issues, though contrary to WMASITL, are not as important at the secondary level, even when, as four of the above questions reflect, the incorrect error by the secondary educators was on the side of caution—that is, mistakenly believing that the law or guidelines prevented them from making a "fair" use of the material. This underscores the concept that the benefit of understanding the responsible uses of technology, including copyright, includes not only the pejorative or the "thou shall not," but the ability to maximize the rights of users such as educators and students under the law and interpretive guidelines.

While the impact of the in-service training sessions and focus group resulted in a great awareness of responsible technology issues, especially copyright, there remain areas of confusion. (The participants believed that additional follow-up or in-service sessions would help.) In general, there appears to be a substantial return in terms of knowledge, considering the limited hours of actual time contributed by the participants (i.e., 10 to 12 hours of in-service, though a potential for additional hours of self-study, review, and reinforcement through the Web site learning resources existed). This statement is underscored by the observation that the responsible technologies portion of the WMASITL represents a fourth and final portion of a framework for implementing the standards, yet in the experience of the authors and from the commentary of the focus group participants, the pre-service or in-service training educators receive or the administrative or curricular environment of the a school rarely devotes a proportional time to learning or teaching about responsible technologies.

RECOMMENDATIONS

As K-12 educational institutions struggle to meet various literacy standards, we recommend that a component of responsible use of technologies be included. As noted earlier, many states of the United States are including this component. How such standards of responsible use are enforced is a challenging issue.

Legal

Legal recommendations are based on a combination of legal statutes, legislative history, court decisions, regulation, and various so-called fair use guidelines. These recommendations speak to the copyright issue in particular.

Schools should establish a copyright compliance officer as the go-to person. In order to have effective responsibility for copyright issues, this person should have knowledge of copyright and licensing practices within the district, and the administrative authority to impact practices within the district (read discipline or enforcement). The person functions not only as an information resource with respect to copyright law and practices under the applicable licenses, but also as an arbiter of practices with the authority to intervene when abuse is discovered. This person can also fulfill the designated agent requirement under the notice and take-down provisions of Section 512 of the copyright law that offers online service providers liability limitation.

Second, a copyright committee should be established not only to deal with major issues as they arise, but also to make an annual review of compliance efforts and to recommend future responses, such as in-services, policy changes, and so forth (Lipinski, 2003b).

Further, a district copyright policy must be established or reviewed. The policy can serve as a best practices document and informational tool, as well as evidence, assuming it is enforced, of the district's sincere desire to encourage compliance with the copyright law. Also, recent changes to the copyright law require the establishment of copyright policies by accredited nonprofit educational institutions engaging in distance education (Lipinski, 2003a).

Districts should also consider a broader copyright compliance program including information (signs, brochures, newsletters, postings, in-services, etc.) and warning notices regarding copyright law. Again this is required by recent changes to the copyright law for accredited nonprofit educational institutions engaging in distance education. Other provisions of the copyright law require the use (posting) of warning notices as well: library reproducing equipment under section 108, ILL under section 108, library circulation of software under section 109 (Lipinski, 2005).

When schools, their teachers, and administrators themselves embrace an attitude of compliance with the law, and a responsibility to society at large, their students should follow.

Ethical

Kizza (1996) has suggested that "teaching computer ethics early in children's education should be seen as a gradual creation of a culture among the children, a culture they may live with and pass on" (p. 49). Similarly, Buchanan (2004) has recommended computer ethics education must complement the push towards technological literacy in K-12 education in order to instill a sense of civic responsibility surrounding technology use. Thus, we must implement a normative technology ethics education into schools when children begin keyboarding, so that a natural respect of and adherence to principles of social responsibility in regards to technologies ensues seamlessly. What are technology ethics? They are not as clear cut and dry as the aforementioned legal recommendations, as philosophical understandings of ethics is complex and varied, and different philosophical theories posit different approaches.

However, Barquin and the Computer Ethics Institute (1992) have offered a "Ten Commandments of Computer Ethics," which serves as an efficacious starting point from which educators can instill a sense of responsibility to students and among themselves.

1. Thou Shalt Not Use A Computer To Harm Other People.
2. Thou Shalt Not Interfere With Other People's Computer Work.
3. Thou Shalt Not Snoop Around In Other People's Computer Files.
4. Thou Shalt Not Use A Computer To Steal.
5. Thou Shalt Not Use A Computer To Bear False Witness.
6. Thou Shalt Not Copy Or Use Proprietary Software For Which You Have Not Paid.
7. Thou Shalt Not Use Other People's Computer Resources Without Authorization Or Proper Compensation.
8. Thou Shalt Not Appropriate Other People's Intellectual Output.
9. Thou Shalt Think About The Social Consequences Of The Program You Are Writing Or The System You Are Designing.
10. Thou Shalt Always Use A Computer In Ways That Insure Consideration And Respect For Your Fellow Humans.

These guidelines can be interpreted along a continuum of complexity, as students become more sophisticated technology users. The commandments embrace the most significant issues revolving technology use, from theft/copyright violations, to hacking, to plagiarism, to the larger social questions about technologies in general and the role of technology in society. These commandments can be instilled throughout the curriculum, into core educational areas such as language arts, social studies, and history. Based on the RT project, this research suggests that students learn about technology ethics most effectively through example. This can include examples from current events, or through role playing and case studies. Such activities can be implemented across grade levels, with varying levels of complexity and sophistication. Importantly, however, technology ethics must not be presented as a standalone lesson; students must see the interconnections.

CONCLUSION

While this case study represents a particular group of educators, it is the model of RT that holds importance for educators and policymakers in the U.S. and worldwide. It is the hope of the authors that as computing technology is embraced and implemented, these are done with an awareness of and attention to legal and ethical discourse.

In conclusion, it is critical for primary schools to integrate the concept of responsible technologies into staff development and student curriculum *systemically*. Both elements are necessary, until teacher preparation programs teach more effectively about legal and ethical uses of and issues in technology; and secondly, until K-12 educators understand RT, they cannot be expected to teach students about it and how to embrace technology legally and ethically.

REFERENCES

Barquin, R., & Computer Ethics Institute. (1992). *The Ten Commandments of Computer Ethics.* Retrieved May 2000 from http://www.brook.edu/its/cei/overview/Ten_Commanments_of_Computer_Ethics.htm

Buchanan, E. (2004). Ethical considerations for information professionals. In R. Spinello & H. Tavani (Eds.), *Readings in cyberethics* (pp. 613-124). Boston: Jones and Bartlett.

Business Software Alliance. (2004). *Survey finds U.S. kids continue downloads.* Retrieved May 20, 2004, from http://www.washington post.com/wp-dyn/articles/A37231-2004May18.html

Carter, A., & Rezabek, L.L. (1993). The awareness of copyright issues by preservice teachers. *International Journal of Instructional Media, 20,* 20.

Fortier, J., Potter, C., & Grady, S. (1998). *Wisconsin model academic standards for information technology literacy.* Madison: Wisconsin Department of Public Instruction.

Green, D. W. (1993). *Copyright law and policy meet the curriculum: Teachers' understanding, attitudes, and practices.* ERIC Doc. #ED 364 946.

Kizza, J. M. (Ed.). (1993). *Social and ethical effects of the computer revolution.* NC: McFarland.

Lipinski, T. (2003a). Legal issues in the development and use of copyrighted material in Web-based distance education. In M. G. Moore & W. G. Anderson (Eds.), *Handbook of American distance education* (pp. 481-505).

Lipinski, T.A. (2003b). The climate of distance education in the 21st century: Understanding and surviving the changes brought by the TEACH (Technology, Education, and Copyright Harmonization) Act of 2002. *Journal of Academic Librarianship, 29,* 362.

Lipinski, T.A. (2005). *The Complete Copyright Liability Handbook for Librarians and Educators.* New York: Neal-Schuman.

Monts, D. R. (1998). *Student teachers and legal issues* ERIC Doc. #ED 428 039.

Patterson, F., & Rossow, L. (1996). Preventative law by the ounce or by the pound: Education law courses in undergraduate teacher education programs. *National Forum of Applied Educational Research Journal, 9,* 38+.

Rice, R.L. (1991). *Behavior opinions and perceptions of Alabama public school teach-*

ers and principals regarding the unauthorized copying and use of microcomputer software. ERIC Doc. #ED 340 703.

Slind-Flor, V. (2000). Students flunk IP rights 101. *The National Law Journal*, (March 13), B6.

APPENDIX A

This appendix includes the survey protocol as administered in Fall 2001.

Completing this survey indicates that I am at least 18 years of age and I am giving my informed consent to be a subject in this study.

Responsible Technologies Teacher Assessment (© Buchanan and Lipinski, 2001)
Center for Information Policy Research
School of Information Studies, University of Wisconsin-Milwaukee

Please answer the following.

Part One:

1. Grade you teach:
2. Subject you teach:
3. Male/Female:
4. Years teaching:
5. District:
6. Did you receive instruction in copyright law as it applies to the educational environment as part of your undergraduate teacher education? YES/NO
   If YES:
   Within that course or training, estimate the number of hours devoted to the topic of copyright? Hours _____
   How long ago was the training/course? Year _____
7. Did you receive instruction in copyright law as it applies to the educational environment as part of a graduate course of study? YES/NO
   If YES:
   Within that course or training, estimate the number of hours devoted to the topic of copyright? Hours _____
   How long ago was the training/course? Year _____
8. Have you attended a continuing education training course on the topic of copyright law? YES/NO
   If YES:
   Within that course or training, estimate the number of hours devoted to the topic of copyright? Hours _____
   How long ago was the training/course? Year _____

Part Two:

Indicate whether or not you agree with the following statements:

1. My school has a copyright policy.
   Agree ____ Disagree____
2. I am familiar with the content of my school's copyright policy.
   Agree ____ Disagree ____
3. My school as a whole punishes teachers for infringement of copyright.
   Agree ____ Disagree ____
4. My school administrator is supportive of responsible (legal and ethical) uses of technology.
   Agree ____ Disagree ____

Part Three:

On a scale of 1-5, with 1 being the weakest and 5 being the strongest, rate your agreement with the following statements:

	Weakest Strongest
1. My school has a policy about ethical uses of technology (computers, World Wide Web, software, etc.).	1 2 3 4 5
2. I am familiar with ethical uses of computer technology.	1 2 3 4 5
3. I discuss ethical uses of technology with my classes.	1 2 3 4 5
4. I understand what intellectual property is.	1 2 3 4 5
5. I understand what an acceptable use policy is.	1 2 3 4 5
6. I enforce acceptable use policies in my classroom.	1 2 3 4 5
7. My school as a whole punishes students for unethical or inappropriate use of technology.	1 2 3 4 5
8. I teach my students why plagiarism is wrong.	1 2 3 4 5
9. I teach my students why copying software or committing other copyright violations is wrong.	1 2 3 4 5
10. I teach my students to respect others in online environments.	1 2 3 4 5
11. I teach my students about privacy rights and technology.	1 2 3 4 5
12. I teach my students to use technology responsibly.	1 2 3 4 5
13. I teach my students about censorship.	1 2 3 4 5
14. I teach my students about the social implications of technology use.	1 2 3 4 5
15. I am familiar with the "Ten Commandments of Computer Ethics."	1 2 3 4 5
16. I feel filtering technology is appropriate for my school.	1 2 3 4 5
17. I explain filtering and its functions to my students.	1 2 3 4 5

continued on the following page

18. I feel confident teaching about technology ethics within my regular curriculum.	1 2 3 4 5
19. Technology ethics should be taught distinct from the regular curriculum.	1 2 3 4 5
20. Ethics is too impractical to teach to K-12 students.	1 2 3 4 5
21. I feel confident teaching ethics and copyright and other legal ownership or property issues within my regular classroom.	1 2 3 4 5
22. Copyright or other legal concepts are too impractical to teach to K-12 students.	1 2 3 4 5

Part Four:

Please select the best answer to the 44 questions below.

Assume that no previous specific permission or license exists to govern these situations. Base your answer upon knowledge of copyright statues, regulations, "fair use guidelines," and case law.

Copyright Basics and User Rights. Under Section 107...

1. Can a teacher make a copy of a pre-recorded music CD or videocassette for personal use?
   Yes ___ No ___ Possibly ___ Probably Not ___ I Don't Know ___
2. Is there a blanket exception for personal reproductions of copyrighted material, in other words is every personal use a fair use under the copyright law?
   Yes ___ No ___ Possibly ___ Probably Not ___ I Don't Know ___
3. Can a teacher make a copy of a pre-recorded music CD or videocassette using Napster or DeCSS technology for classroom use?
   Yes ___ No ___ Possibly ___ Probably Not ___ I Don't Know ___
4. Is there a blanket exception for educational reproductions of copyrighted material, in other words is every educational use a fair use under the copyright law?
   Yes ___ No ___ Possibly ___ Probably Not ___ I Don't Know ___

Classroom Use of Copyrighted Material. Under Section 107 and the "Classroom" Guidelines...

5. Can a teacher make a copy of a chapter from a book for use in preparing his or her class?
   Yes ___ No ___ Possibly ___ Probably Not ___ I Don't Know ___
6. Could the teacher make a copy of the same chapter for each teacher in her curriculum design group for today's planning meeting?
   Yes ___ No ___ Possibly ___ Probably Not ___ I Don't Know ___
7. Can a teacher make a copy of a chapter from a book (20 pages long, about 8,000 words) for each student in the class?
   Yes ___ No ___ Possibly ___ Probably Not ___ I Don't Know ___

8. What if the teacher just copied the two most important pages from the chapter, could he/she make a copy for each student in the class?
   Yes ___ No ___ Possibly ___ Probably Not ___ I Don't Know ___
9. Could the teacher do this on a weekly basis throughout the entire semester, make a copy from the same book of the two most important pages from successive chapters in successive weeks for each student?
   Yes ___ No ___ Possibly ___ Probably Not ___ I Don't Know ___
10. Could the teacher, on a weekly basis throughout the entire semester, make a copy of the most important page or two from a different book each week of the semester for each student in the class?
    Yes ___ No ___ Possibly ___ Probably Not ___ I Don't Know ___
11. Can the teacher make a copy of an entire book for use in preparing an upcoming unit for his or her class?
    Yes ___ No ___ Possibly ___ Probably Not ___ I Don't Know ___
12. Does it make any difference if the book is out of print; e.g., if it is out of print, could the teacher then make a complete copy of the book for use in preparing for his or her class?
    Yes ___ No ___ Possibly ___ Probably Not ___ I Don't Know ___
13. Instead of the library or the teacher making copies for each student, could the teacher require students to bring a copy to class of the required chapter, on the assumption that students would retrieve the book or chapter from the reading or electronic reserve, copy the chapter or print it out, then bring it to class?
    Yes ___ No ___ Possibly ___ Probably Not ___ I Don't Know ___
14. Can the teacher make a copy of a test from a workbook for each student in the class?
    Yes ___ No ___ Possibly ___ Probably Not ___ I Don't Know ___

Off-Air Taping and Use. Under Section 107 and the "Off-Air" Taping Guidelines...

15. A teacher taped off of their TV at home an episode of the *American Experience on the Dust Bowl*, it aired on PBS last evening; can the teacher show it in American History class next week, the unit is on the Depression?
    Yes ___ No ___ Possibly ___ Probably Not ___ I Don't Know ___
16. Can the same teacher record the same Dust Bowl off-air tape when it re-airs again next semester, and then use it in the depression unit next semester, within ten schools days of its re-taping?
    Yes ___ No ___ Possibly ___ Probably Not ___ I Don't Know ___
17. If the program that is taped off-air is from a cable-only station, like CNN or MSNBC, and is also a topical "magazine-format" news program, could the teacher use it next week?
    Yes ___ No ___ Possibly ___ Probably Not ___ I Don't Know ___
18. Assuming Section 107 allows the copying, under Section 108 could a teacher place a copy (made from an off-air recording) of the nightly NBC news program to have as a resource in the library for individual student research use, *not* for classroom use?
    Yes ___ No ___ Possibly ___ Probably Not ___ I Don't Know ___

19. Assuming Section 107 allows the copying, under Section 108 could a teacher place a copy (made from an off-air recording) of a topical "magazine-format" CNN or MSNBC news program as a resource in the library for individual student use, *not* for classroom use?
    Yes ___ No ___ Possibly ___ Probably Not ___ I Don't Know ___

Using Pre-Recorded Videos and Other Copyrighted Works in the Classroom. Under Section 110...

20. If the teacher purchased from a video catalog a copy of a NOVA episode on the Yellowstone Fire, can the teacher show it in his/her ecology class?
    Yes ___ No ___ Possibly ___ Probably Not ___ I Don't Know ___
21. Could a teacher use the purchased copy of a PBS NOVA episode on the Yellowstone Fire next semester in his/her ecology class?
    Yes ___ No ___ Possibly ___ Probably Not ___ I Don't Know ___
22. Could the teacher play the same NOVA Yellowstone Fire tape at the school's annual science fair exhibit, displays are usually housed in the cafeteria or gymnasium?
    Yes ___ No ___ Possibly ___ Probably Not ___ I Don't Know ___
23. Since the school has won the regional science fair with their exhibit on fires and firefighting, can the teacher show the John Wayne classic *Hell Fighters* about a team that fights oilrig fires off the Louisiana coast, or the more recent *Firestorm* starring ex-football player turned actor Howie Long as a reward?
    Yes ___ No ___ Possibly ___ Probably Not ___ I Don't Know ___
24. Can a teacher perform (sing and act out) the entire score/script of Rodgers and Hammerstein's *Oklahoma* to his or her drama class?
    Yes ___ No ___ Possibly ___ Probably Not ___ I Don't Know ___
25. Could the teacher broadcast his or her dramatic performance of *Oklahoma* to a drama class in the neighboring school district?
    Yes ___ No ___ Possibly ___ Probably Not ___ I Don't Know ___
26. Can a teacher read a Faulkner short story to an English class?
    Yes ___ No ___ Possibly ___ Probably Not ___ I Don't Know ___
27. Can a teacher read the same Faulkner story in order to pass the time to a Mathematics class?
    Yes ___ No ___ Possibly ___ Probably Not ___ I Don't Know ___
28. Could the teacher (Faulkner to an English class) stream (via the Internet) the reading to an English class in the neighboring school district, assuming the class is studying the same or similar unit of material?
    Yes ___ No ___ Possibly ___ Probably Not ___ I Don't Know ___

Creating Multimedia Works. Under Section 107 and the "Multimedia" Guidelines...

29. Can a student scan a photograph from *National Geographic* into a word processing document and use it as an illustration in a class term paper?
    Yes ___ No ___ Possibly ___ Probably Not ___ I Don't Know ___

30. Could the same student take a twenty-second sound byte from a three-minute–and-thirty-eight-second Smashing Pumpkins song and use it as backing music in a Mass Media class PowerPoint presentation?
    Yes ___ No ___ Possibly ___ Probably Not ___ I Don't Know ___
31. Could the same student use the entire Smashing Pumpkins CD as backing music in the same presentation?
    Yes ___ No ___ Possibly ___ Probably Not ___ I Don't Know ___
32. Could a student take two minutes of the Johnny Depp movie *Sleepy Hollow* and use it in a multimedia presentation in a class on American authors of the Romantic Period?
    Yes ___ No ___ Possibly ___ Probably Not ___ I Don't Know ___
33. Could the student show the same Johnny Depp *Sleepy Hollow* American authors of the Romantic Period multimedia presentation to a tenth grade school assembly during Fine Arts Week?
    Yes ___ No ___ Possibly ___ Probably Not ___ I Don't Know ___
34. Could a student use two minutes of a Smashing Pumpkins music video in the same multimedia presentation in a class on American authors of the Romantic Period?
    Yes ___ No ___ Possibly ___ Probably Not ___ I Don't Know ___
35. Could the student keep the video project with the twenty-second Smashing Pumpkins song and the two-minute *Sleepy Hollow* video clip and use it in her portfolios next year when applying for a scholarship to the School of Visual and Performing Arts at Syracuse University, Syracuse, New York?
    Yes ___ No ___ Possibly ___ Probably Not ___ I Don't Know ___
36. Could a teacher create a similar work on American authors of the Romantic Period, with twenty seconds of music (commercial recording) and two minutes of video (theatrical film) and use it in class?
    Yes ___ No ___ Possibly ___ Probably Not ___ I Don't Know ___
37. Could the same teacher incorporate the two minutes of the *Sleepy Hollow* film from a copy that was taped off-the-air last year when it aired on HBO?
    Yes ___ No ___ Possibly ___ Probably Not ___ I Don't Know ___
38. Could the teacher use the multimedia work (twenty seconds of music from a CD (personal copy purchased from Target) and two minutes from a pre-recorded tape (free rental copy obtained from the local public library)) on American authors of the Romantic Period next year when covering the same unit?
    Yes ___ No ___ Possibly ___ Probably Not ___ I Don't Know ___
39. Could the same teacher broadcast the multimedia presentation on American authors of the Romantic Period to students in a neighboring school district studying the same unit using a secure electronic network connection in real time?
    Yes ___ No ___ Possibly ___ Probably Not ___ I Don't Know ___
40. Could the teacher keep the video project with the twenty-second Smashing Pumpkins song and two-minute *Sleepy Hollow* and use it at an upcoming NEA-sponsored conference demonstrating the use of new media technology to teach a particular topic?
    Yes ___ No ___ Possibly ___ Probably Not ___ I Don't Know ___

Liability Issues. Understanding Case Law

41. Under Section 512, could the school allow students to upload and download music or video files using MP3.com, Napster, or similar technology?
    Yes ___ No ___ Possibly ___ Probably Not ___ I Don't Know ___
42. Under Sections 512 and 1201, could a teacher refer in class to a Web site that is likely to contain infringing material, such as one that has over one thousand theatrical videos downloadable for free, or one that has available for downloading anti-circumvention technology on it such as DeCSS that would allow users to "crack' a protection code on a DVD?
    Yes ___ No ___ Possibly ___ Probably Not ___ I Don't Know ___
43. Under Section 512, if a school notices a pattern of infringing uploading or downloading by a student or a teacher and responds with a verbal chastisement, is this a sufficient response (in order for the school district to escape liability)?
    Yes ___ No ___ Possibly ___ Probably Not ___ I Don't Know ___
44. Under Section 1201, could the school media curriculum team circumvent a technological protection measure (encryption, scrambling, etc.) placed on a CD-ROM encyclopedia resource in order to determine whether or not the school district should license the resource for student and staff use?
    Yes ___ No ___ Possibly ___ Probably Not ___ I Don't Know ___

Chapter IX
Internet Chatrooms:
E-Space for Youth of the Risk Society

Cushla Kapitzke
University of Queensland, Australia

ABSTRACT

This chapter uses Ulrich Beck's (1992) concept of Risk Society to contextualize the current 'youth problem' and the emergence of the techno-genre, Internet relay chat (IRC), in advanced capitalist societies. It argues that unsympathetic social policies combined with increased levels of surveillance in physical environments have contributed to the uptake of virtual space and online chatrooms as a means of social contact and engagement for youth. To the uninitiated, 'chat' is an ungovernable space of indecipherable codes, virtual skulking, and suspect subcultures. The chapter begins with a description of the rhetorical conventions of chat and a review of extant literature on it. It examines adult responses to teen chat through investigation of their representation in newspapers and compares this with text from 100 chatrooms. The purpose of this was to investigate whether adult prohibitions about chat are justified. Data showed that chat is a discursive space with highly regulated protocols and social mores, and that its delegitimation can be construed as an exercise in social control and governance over the textualities and sexualities of youth.

For both progressives and traditionalists the youth problem described and symbolized a period of acute transformation...In the highly dramatic perceptions of a dramatic era, youth was either damned or beautiful. (Fass, 1977, p. 16)

Youth is a material problem; it is a body...that has to be properly inserted into the dominant organization of spaces and places, into the dominant systems of economic and social relationships. (Grossberg, 1994, p. 34)

INTRODUCTION

"Damned or beautiful." This provocative phrase encapsulates the intensity of adult anxiety about youth in America during the turbulent 1920s. Yet these signifiers pertain equally well to the ambivalence and moral dread adults harbor about young people today. Forces of change that society faced at the beginning of the twentieth century—industrialization, urbanization, the changing status of women, and the desire for personal fulfillment through leisure—continue unabated at the beginning of the twenty-first century. As youth then were a focus for the strains of a new century, so too public sensibilities in this late modern moment remain on high alert about the attitudes and activities of those who are defined primarily by their age (cf., Lee, 2001).

Most current discourses of youth—both popular and scholarly—are typically essentialist, conceptualizing *adolescence* as an unproblematized biophysical phenomenon, and framing *adolescents* as psychologized individuals deemed to be less than adults in terms of age and maturity. By contrast, I consider the concept of youth to be a discursive construction, one that functions to universally position them as *other* to the construction of *adult*. In order to avoid this mutually oppressive and unhelpful binarism, I focus instead on the sociological concept, Risk Society, as a means of contextualizing what is uncritically termed, the "youth problem." This is a deliberate strategy aimed at shifting the burden of critical scrutiny from the young to those with social responsibility and a duty of care towards those who, though young chronologically, comprise a physically, culturally, and emotionally complex and differentiated social group. More specifically, the chapter examines issues of youth and risk through a focus on the communications form, Internet relay chat (IRC).

As social and cultural practices, new literacies emerge from particular sociohistorical contexts. A raft of economic, political, and technological developments have recently afforded the widespread use of chat as a means of social interaction for youth. A key reason for choosing chat as data was that this socially significant genre remains relatively under-researched and under-theorized. To the uninitiated, chat is an ungovernable space of indecipherable codes, virtual skulking, and suspect subcultures (Valentine & Holloway, 2001). This chapter debunks that myth through a technical description of the chat genre and a consideration of the reasons why so many young people are turning to online environments for social contact and engagement. It describes the responses of some adults to this techno-textual phenomenon through examination of the representation of chat in newspapers and, through analyses of chatroom exchanges, considers whether adult concerns and prohibitions are justified.

INTERNET RELAY CHAT: SOCIAL E-SPACE

Internet relay chat is a form of social interaction that allows groups of people to converse in real time by typing messages on a computer screen. IRC is one of the many new technosocialities—along with instant text and video messaging, e-mail, networked game-playing, and blogging—afforded by computer networks, tele-existence, and cyber-corporeality. Because the rapid-fire text of chat screens can seem incomprehensible to the uninitiated, a description of the codes and rhetorical protocols that render it different from conventional print text follows.

Chatrooms—sometimes called channels—are created and "owned" by individual users through the simple process of typing "/join #chatroomname" within a chat software program. The creator of the chatroom then registers the room with the chat server by using a command assigning the creator—and those with access to the registration password—a set of restricted commands. These commands might include setting the chatroom topic, banning a person from re-entering the chatroom, or giving special privileges to certain users. The first challenge for chatroom creators is to convince others to enter the chatroom for social engagement. One way of doing this is through cues provided by the name of the chatroom, which signify conversational markers such as a shared interest. Thus, the room #Singapore might contain people interested in talking to others from Singapore, whereas #Metalheads signifies discussion for heavy metal music fans. Different chatrooms open as separate windows and tile for ready access like the windows of the Microsoft Office suite.

Chat screens consist of sequential messages on a continually scrolling screen, with messages and turns positioned on new lines like dialogue in a transcript. Messages sent to a chatroom are visible within seconds of being typed, and are conveyed to others using the Enter key. Delay in the amount of time it takes to send or receive messages is referred to as lag and is considered disruptive by participants. The immediacy and synchronicity of interaction makes participants view IRC as talk (Du Bartelle, 1995). Each message is prefixed with the sender's alias placed between brackets at the beginning of a line.

While thousands may share the same IRC network at any given moment, individuals join specific rooms with targeted populations of users. This reduces the number of messages being sent and displayed to a manageable level. For example, there might be 13,000 users connected to the IRC network called *Dalnet* at the same time, but only 20 people in the room called #chatzone. Chatroom users have several interactional alternatives. Users can participate in a number of chatrooms simultaneously, or chat privately with one other person. Alternatively, they may choose not to participate actively, opting instead just to read messages. This is called lurking.

If participants are willing to be identified, they use the same alias—a nickname registered under a password—each time they enter a room. An alphabetic list of the aliases of all participants (i.e., the user list) is displayed in a column on the right side of the screen. An @ sign in front of the names at the top of the column indicates users with access to restricted operational commands within the chatroom. A message prefixed by three stars represents users who are performing a command such as joining or quitting the room. Users leave chatrooms by typing specific commands, or by using the command toolbar at the top of the screen interface. For example, if a user named Alleycat types the "/leave" command to quit a room called Funfactory, then the following message would appear in the message scroll:

"***Alleycat (Internet port address here) has left #Funfactory"

Once a person leaves the room, their name disappears from the current user list at the side of the screen. Other shared signifying devices to convey meaning include the use of color, upper and lower case, bold text, non-alphabetic and non-numeric symbols, and numerals.

Research on Internet Relay Chat

Some adults, including teachers, consider online chat a genre unworthy of serious academic attention. Yet statistical information shows that online chatting is an important social activity for large numbers of young people. A survey of children in the United States aged 7-12 years found that close to 13 million teenagers used instant messaging[1] to communicate with friends, and roughly half of the sample population visited chatrooms (Lenhart, Rainie, & Lewis, 2002). Figures from other countries show similar trends. The Australian Bureau of Statistics found that 95% of Australian children aged 5 to 14 years had used a computer in the previous 12-month period, and almost half (47%) had accessed the Internet (ABS, 2000). Of note is that the most common Internet activities for children at home were using e-mail, browsing the Web for leisure, and engaging in chatroom activities. Other research conducted by the Australian Broadcasting Authority (2001) reported that 15- to 18-year-olds dominate chatroom usage. While 42% of respondents aged 18 to 24 years had visited chatrooms, only 9% of respondents over 50 years of age had used them. The incongruity between those using chat (i.e., young people) and those eschewing it (the age bracket of large numbers of teachers) is a cue for cultural and hence curricular disjuncture within K-12 education.

A growing corpus of research testifies to the importance of chat as a communicative medium for youth. Nevertheless, because of a difficulty in researching users by demographic variable, most research has focused on chat participants as a unitary group. That is, researchers have approached chat as an emergent communications medium and not differentiated user groups by age or gender (see Danet, Ruedenberg-Wright, & Tamari-Rosenbaum, 1997; ten Have, 2000; Hutchby, 2001; Mar, 2000; Reid, 1996; Werry, 1996). Because chat studies draw from logged public chat text created by anonymous or near-anonymous users, it is difficult to ascertain the age of participants reliably (Orthmann, 2000). This dilemma is compounded by the logistical and ethical challenge of accessing chatrooms designed specifically for young people. Programming devices such as fee-based memberships and invitation-only chatrooms that are used to keep undesirables out, keep researchers out as well. Furthermore, the few studies that focus on younger chat users present contradictory accounts. For example, in a report for the Crimes Against Children Research Center, Finkelhor, Wolak, and Mitchell (2000, p. 33) found that while the quantity of online sexual solicitation of children is "potentially alarming," elsewhere the same authors stated that "few youths reported bad experiences with online friends" (Wolak, Mitchell, & Finkelhor, 2002, p. 1).

The similarities that have been found between the discursive features of teen homepages and chat indicate a blurring of boundaries between conversation and publishing (Abbott, 1998). This blending of literacies is an area that teachers and university-based researchers could explore further, considering recent work that is being conducted on the benefits of computer game-playing for learning (cf., Gee, 2003; Fromme, 2003). Nevertheless, education does not rate highly as a topic of conversation in chatrooms. Roberts, Foehr, Rideout, and Brodie (1999), for example, found that chatrooms visited most frequently by 8 to 18-year-olds could be organized into three categories: Entertainment (music and celebrities); Sports, gaming, and hobbies; and Lifestyle issues and relationships, in that order of popularity.

Another area of related studies is the use of chatrooms for formal learning done online (see Albright, Purohit, & Walsh, 2002; Alvermann & Heron, 2001). Most schools, however, bar chatroom use. This value position is operationalized through school policies and codes of practice such as the installation of surveillance programs that log chat server contact, filtering utilities that block access to particular sites or domains, and direct supervision. Stated reasons for this are student safety and duty of care obligations. While it is understandable that administrators, teachers, and parents are cautious about the uses and content of some chatrooms, is it possible that blanket prohibitions on chat are motivated by social and political ideologies? Thousands of young people meet regularly in chatrooms for social contact with peers, much as their forbearers did at dances or movies on a Saturday night. Why, then, are adults so negative about youth chat? Is this anxiety always justified?

Most of the aforementioned research is instrumentalist in theoretical and methodological approach. That is, their conceptual frameworks lack an historical and social framing. By contrast, an analysis of the conditions of possibility for the emergence of IRC within a particular social, cultural, and political matrix will contextualize its popularity and use. Ulrich Beck's (1992) concept of the Risk Society (*Risikogesellschaft*) is used because it enables the integration of seemingly disparate trends like social policy developments, school failure, the emergence of a youthful 'underclass', youth crime, and other unhelpful categorizations associated with school to work transitions of younger people in fast capitalist societies.

THE RISK SOCIETY

Like Giddens (1991), Beck argues that social relations of nation-state societies—what he calls "first modernity"—were based on territorial and communal principles. But global integration and local fragmentation have 'radicalized' social life in this stage of 'second' or late modernity. Economic and cultural globalization and the erosion of collective identity through the obscuring of social class and community ties have reduced the certainty and predictability of the past. In the face of environmental degradation, famine, economic meltdown, the threat of nuclear war, and terrorism, governments are no longer primary sources of security for their citizenries. Consequently, the pursuit of protection and safety from these anxieties and stresses has replaced the primary concern of industrial societies, which was preoccupation with the creation and distribution of wealth.

Beck's theory links increased levels of social anxiety to new forms of social governance. Within a context of pressures from a free market environment and transnational economic competition, nation-states are predisposed to adopt neoliberal social policies that protect the productive and wealthy at the cost of caring for the needs of the disadvantaged. This has entailed the privatization of public utilities and the dismantling of welfare systems. While it is true that the advent of modernity weakened traditional family ties and obscured the social structures of exploitation, recent developments have exacerbated these effects, particularly for the young. Paradoxically, while conventional structures of class, race, and gender seem to have lost their significance in a world characterized by complexity and plurality, research shows that an individual's life chances remain contingent upon them.

Destabilized cultural norms and values have recast social boundaries and made once dominant groups marginal. In turn, this has created high levels of uncertainty over social membership and identity. Risk, responsibility, and trust have become 'disconnected' from their sources. To whom can responsibility be entrusted if neither government nor corporation is accountable for the ungovernable effects of "turbo capitalism" (Luttwak, 1999)? Theorists use the term *organized irresponsibility* to describe these uncontrollable flows and configurations of social and economic constituents. The lack of national, structural, and cultural guides has meant that individuals become more personally reflexive, or self-conscious, about the micro decision-making processes of their personal lifeworlds (Giddens, 1991). These processes occur increasingly in isolation as new technologies mediate human activities and as human interaction occurs less through face-to-face contact. As governments, employers, and schools espouse discourses of personal accountability, individuals perceive difficulties and problems as deriving from personal inadequacies and failings. In effect, the political is made personal to the extent that anxiety has become a national health risk (Culpitt, 1999; Wilkinson, 2001). What are the implications of these changes for youth who, while increasingly isolated, remain dependent in so many ways upon adults for their welfare?

Youth and the Risk Society

The mass media typically represent young people as self-centered, materialistic, unmotivated, amoral, and cynical (for a discussion of media moralities, see Aitken, 2001, p. 151). Davis (1999, p. 7) notes that some go so far as to accuse them of "destroying the American Dream." Yet, an array of endemic structural hindrances such as changes in the labor market, chronic under-employment and unemployment, limitations of vocational education programs, and the casualization of the workforce currently diminish the life chances of many socially disadvantaged young people. The response of social policymakers to these challenges are viewed by many as less than helpful. A corpus of cross-national research shows that public policies have, in large part, marginalized youth. In Europe and Australia, for example, the formulation of youth policy has moved away from strategies of consent to those of compliance through coercion (cf., Jamrozik, 2001; Williamson, 1997). Increasingly punitive interventions in the lives of the young include the withdrawal of social security entitlements, workfare welfare programs, the removal of workplace protection rights, curfews, school exclusion, and the extension of custodial provisions for young offenders (Bessant, 1995; Dwyer & Wyn, 2001). Excluded as they are from decision-making processes, youth remain objects of policy creation and implementation.

Many remain confused by contradictions between the loss of agency and control they experience in their lives and society's expectations of them. Not surprisingly, some become unresponsive to government-mandated strategies. The point here is that, whereas governments and policymakers pay considerable attention to the financial fallout of globalization with respect to the "risky" behavior of youth (cf., Gruber, 2001), they frequently overlook the impact of these developments for their well-being. While I seek not to essentialize or slate *adults* as a unitary category, it is notable nevertheless that the clamor about the youth *crisis* sits awkwardly beside an emergent body of international literature that is expressly critical of those with public responsibility for providing young people with safety and a future (see Davis, 1999; Dwyer, 1996; Evans, Behrens, & Kaluza, 2000; MacDonald, 1997). Heightened levels of governance, surveillance, and moral panic are poor substitutes for material prospects and social resources of hope.

Further to this, ways in which certain groups are constructed as "at risk" constitute ethical and political values, interests, and agendas. Technologies of public surveillance play a role in identifying and tracking such groups (see Aitken, 2001). Similar to religious discourses of "sin" in the pre-modern era (Douglas, 1992), the concept of risk functions to categorize and normalize ruptures in the social order. As processes of globalization recast boundaries of social inclusion and exclusion, so the notion of risk is used as a means of managing the anxieties, uncertainties, and disputes over social membership. Within this context, the identities and lives of young people are constituted, disciplined, and tracked by a web of texts such as birth certificates, school entry and attendance records, examination results, census data, employment profiles, purchasing habits, and so on. Demographic databases are instances of Foucault's "grids of specification" that position and "call up" the multiple and fragmented identities of the postmodern subject. Ericson and Haggerty (1997) argue that this proliferation of textual records is a response to the management of risk and the concomitant need for governmentality and security. Access to such data is crucial to the risk assessment procedures of social institutions like education, employment, health, and welfare.

Young people know that their social conduct is subject to closer scrutiny and supervision than that of adults, and are aware that, despite legislation prohibiting the sharing of personal data, inter-agency cooperation plays an important role in this process (Decker, 2002). They may not be cognizant of the full extent to which this occurs, but in this "post-privacy" era, young people understand that few spaces of their lives are free from prying "electronic eyes" (Lyon, 1994). In industrial economies, young people used their wages to buy 'space' of their own in sporting events, rock concerts, and videogame arcades. These social locations provided a measure of privacy and autonomy away from the purview of adults. But events and public places like street corners and shopping malls of today are increasingly surveyed, and the wont of youth to "loiter" in such places renders them easy targets for those who, in the current climate of instability and uncertainty, seek scapegoats for the strain of social change. It is within this context that online environments provide a space for social exchange that is relatively free of adult intrusion.

DATA COLLECTION AND ANALYSIS

Data for this study of youth and chatrooms were collected from two sources: adult accounts of teen chatrooms in newspaper articles and the content of teen chatroom logs.

Media Treatment of Teen Chatrooms

To better understand adult accounts of teen chatrooms, a search of reputable newspapers was conducted. While the limitations of newspapers as valid representations and sources of reliable information is acknowledged, they do nonetheless provide a means of understanding one powerful medium that shapes and, conversely, is shaped by adult perceptions of teen chat. To explore this source of information, the descriptors *teen* and *chatroom* were used to search the online database LexisNexis for articles published in the previous six months. The search retrieved 111 articles, 100 of which were randomly selected and analyzed for keyword and thematic content. Those selected were from national newspapers such as *The New York Times*, *Daily News* (New York), *The Washington Post*, *The Toronto Sun*, *The Guardian* (London), *Chicago Tribune*, *The Times of India*, and *The Weekend Australian*.

The articles were overwhelmingly biased in their representations of chat. Seventy-five of the 100 articles represented teen chatrooms negatively and as having adverse effects on participants. Forty-seven referred specifically to criminal behavior of some kind; 43 discussed illicit sexual activity occurring on or through chat; 35 focused on some aspect of pedophilia; and 12 dealt with pornography. Two common complaints were that students are "online all night, until 3 or 4 [o'clock]" and that chat adversely affects literacy levels. The following lament from an interview of a teacher about chat is representative:

We're losing a generation of readers to this [i.e., chat], because kids go to schools where they can choose between doing homework and playing video games, and they will not make the right choice. (Terzian, 2002)

Only eight of the 100 articles introduced positive or productive aspects of chatroom use. I would argue that this partial coverage fails to provide a balanced and open discussion on the benefits and drawbacks of online teen chat.

Chatroom Logs

To gauge the extent of risk associated with chat, the content of chat text from three major chat servers—America Online, MIRC, and MSN—was examined. One hour of text was logged from 100 discrete chatrooms, totaling 100 hours of chat text. Shorter chunks of more chatrooms rather than longer ones from fewer chatrooms were chosen in order to provide a broad coverage of categories and conversation topics. From the range of chatrooms that were available, 10 categories of rooms were developed, and 10 chatrooms for each of the categories were randomly selected. The categories were Race and Gender, Geographic Location, Teens, Lifestyle, Hobbies, Music, Romance and Relationships, Celebrities, Games and Software, and Generic Chat.

Content analyses of text showed that a high proportion of the logged material was at shallow levels of both content and contact. Because many individuals entered and left the chatrooms without contributing to the exchange, much of the text comprised *join and quit* messages. Most of the exchanges could be categorized into three forms of interchange: greetings and farewells between acquainted chatters, invitations to private messaging, and other 'small' talk. The latter frequently took the form of chatters describing what they were doing at the keyboard, or what they had been doing that day or that week. The following excerpt, found in #beginners, illustrates how participants—four in this case—talked about what they were eating or drinking while chatting, or about food in general. The text shows how the shared topic acts as a basis for jokes and further talk such as *storying* on the topic.

"* Tazzy2 has a sour puss look on my face...eating sour gummy worms
<gervy> rofl Tazzy2 aren't those great tho
<Tazzy2> yummmmy
<gervy> right now i would kill for chocolate tho
<VeL_VeT> I'm kinda thinking chips:*)
<Tazzy2> oh gervy...Angel is selling the best chocolate for school...wanna buy some??? hehehehehe
<VeL_VeT> hmmm...barbecue doritos:*)
<gervy> rofl Tazzy2 mailing choc doesn't work well
* gervy mailed chocolate to australia and it cost a fortune rofl
<gervy> oh man Tazzy2 yes he sent me some fantastic candy...oh man my mouth is watering
<Tazzy2> he sent me a care package from PA to CA...& it didn't melt
<gervy> yeah he sent me one too but was in cool weather
<WolfChylde> I send candy to germany and its fine
(From #beginners)"

Tazzy2 opens the conversation with a wry comment describing how her face looked because of the tart sweet s/he was eating. Gervy responds affirmatively with the shorthand term rofl, an acronym meaning "rolling on the floor laughing," and Tazzy2 continues the light banter with the onomatopoeic, "hehehehehe." Tazzy2 and gervy find some common ground in that they have purportedly received chocolates from a male acquaintance. Interjections by VeL_VeT and WolfChylde are ignored by the pair of interactants. Thus, shared information about mundane activities provided participants with a topic of conversation and a subject for joke making. Nearly every chatroom visited contained playful talk and jokes based on prior chat, sometimes around a single word. The social function of much of the chat in the logs was as a discursive, playful, and fun-seeking means of social activity through teasing and other forms of repartee.

Further analyses showed that the style and content of chat was similar across chatroom types: that is, the talk from both generic and topic-specific chatrooms was largely indistinguishable. There was only slight variation in content between categories like Celebrities, Lifestyle, Teens, and Romance and Relationships chatrooms. In some topic-specific chatrooms (#hockey, #Africanamerican), roughly half of the talk was topical and half was chit-chat on roommates, television, music, friends, work, and so on. In others (#astrology) nearly all of the talk was devoted to everyday chat, greetings, and insulting banter such as name-calling. It was difficult to differentiate generic chatrooms like #Chatzone and #Chatterz, as the talk consisted of greetings and farewells, invitations to message privately, inducement to Web sites (mostly pornography), verbal insults, and other more mundane talk.

By contrast, exchanges of chatrooms in the category Race and Gender displayed greater variation in style. This occurs because participants construct their text to approximate the languages and dialects of their real-world speech communities. As illustrated in the two examples below, the differentiated use of slang, spelling, and grammar demonstrates discursive competence and legitimacy to belong to particular chatrooms.

"Guest_eL_JoKeR_y_sU_Ramfla : DONDE ESTA LA RAZA
AncientPowerfulGunner : who is talking about you baby
?pequitas12054 has left the conversation.
Guest_eL_JoKeR_y_sU_Ramfla : BIEN VATO Y TU?
lovieluv20 : *cop still*
Classybutton21 : **u dont need to take it**
buky_818 : muy bien bracias
(From #Hispanicamerican)
Lady_Saw82 : **THA BLACK MAN GOT DISCONNECTED FROM THA BLACK WOMAN**
Mainerd3 : **LADY U RIGHT!!!!!**
ReAtrice_Inc1 : **YALL LADIES BE PRESSED FOR A MAN TO MARRY YOU**
4_A_GENTLE_MAN_ONLY : ***THEY DO BUT IT TAKES A STRONG BLK MAN TO UNDERSTAND ONE***
FamiliarKatteyez420 : true
Mainerd3 : **SO WHEN CAN WE GET BACK???**
CHOCOLATE101 : ***I'M CATCHING HECK........LIVING HERE ALONE.....I NEVER REALIZED HOW MUCH YOU MEAN TO ME.....***
(#From #Africanamerican)"

Note the reliance on the bolding of fonts for emphasis in the #Africanamerican room. Chatrooms here were providing participants with space to demonstrate membership of and competence within social groups *of their choosing*. This means that those who may be ascribed a range of identities in face-to-face conversation (e.g., "adolescent," "basketballer," "church member," "shoplifter," or "school truant") are able to choose which of these to selectively talk into being in the disembodied space of a chatroom. Chatrooms offer teens—and adults masquerading in teen chatrooms—a space for assuming and performing preferred identities. In this data, participants employed a range of registers and subdiscourses to construct text-based identities. These included *homie* talk and *geek* talk, that is, talk that showed familiarity and competency with chatrooms and computers (see below).

"Hurried_Frog59 : jus livin in da country no 1 got money 4 present or calender we lose track of time
(From #12-17Nzteens)

<'kinteK-> can anyone help me with splintercell for xbox msg me~
(From #VAgamers)"

Many teens would also participate in nonteen chatrooms like Goths, metalheads, sportsfans, punks, hackers, gamers, and sexually identified groups. Thus, in analyzing the chat of teens, it is difficult to make generalizations about the category as a whole because so many different sub-registers, discourses, and identities are used within the exchanges.

Nevertheless, across categories of chatrooms, teens deployed racist and anti-racist discourses, as well as normative discourses of sexism, hegemonic masculinities, and extreme femininities. In the data, these discourses were especially common in two rooms, #eminem and #p-daddy3, as illustrated below.

"skateforska17: I hate the fack that its cool to be black now a days. You see all the upper class white kids istning to rap and wering street cloths and talkin all that bull

Marco is a Polo: yeah eminem is pretty dumb

KidHacker2K: u guys are a lil girls comin in the eminem chatroom saying he sux

SLIMSHADEYSBABE1: u all need to shut up and stop bein racist

(From #eminem)"

In contrast to the racial hostility of this interaction, chat collected from the #brissyandcoastteens room displayed more *preppie* and upper middle class talk about money and leisure pursuits. Some participants talked about school, mostly in terms of which school they were attending and what subjects they were studying. Biographical details like locality or employment were used as conversational resources between unacquainted participants and, possibly, as a way of checking that the other person really was a teen.

"GOdZiILaR32 : **wat school**
GOdZiILaR32 : **?**
conspiracy : **glendale tech high**
GOdZiILaR32 : **wat u study?**
conspiracy : **- subjects:**
business studies
ancient history
software design and tech
english
maths
ipt (information process tech)
GOdZiILaR32 : **sounds good**"

These textually constructed identities are used as shared resources to locate likely partners for ongoing conversation and also to identify suitable topics of conversation. However, while participants have the option of choosing an identity for conversational purposes, they are also aware of their obligation to display a certain level of competency in performing the identity, otherwise they risk having their credibility and incumbency questioned.

As already noted, much of the data across categories of chatroom contained talk about social competency and identity. The extract from #teentalk below shows how talk is concerned with establishing one's own identity and that of others through information on age, sex, and location—a discursive genre commonly signified as "asl." Participants provide clues to their age, sex, geographical location, and interests through rhetorical devices such as the design of their nicknames (e.g., SXC_MAN_MELB_84), questions such as "Any chicks from Port Macquarie in here?," and condensed biodata such as the "22/m/vic" format (22-year-old male from Victoria).

"SpHinX82 : **any chicks from port macquarie in here?**

GHETTO_GIRL_4_U_247 : ***HEY HAS JAYCEE2003 BEEN IN HERE***

?UnbiasedDifference has joined the conversation.

yindi_2 : **im back**

BrandedSinger : syd westy girl's where r u

?Beer_drunk_Punx_Oi has joined the conversation.

SXC_MAN_MELB_84 : asl everyone ?

?GHETTO_GIRL_4_U_247 has left the conversation.

?Princess12323 has joined the conversation.

\\ĶeЯvO_tнa†š_Яi†• _•• RVO : ***22/m/vic***"

A high proportion of talk was devoted to challenging participants in terms of their moral or mental adequacy, chatroom competency, literacy skill, or claim to a particular identity incumbency. Participants seemed to enjoy using the chatroom as a space for ascribing identity—usually unflattering—and trading insults through a battle of wits. The following extract illustrates this combative feature of chat.

"IamSpecial_1 : **Rotar....Is there a gibberish translator in the house? I can't make head nor nail of that uber-babble you flung onto the screen during your latest spasmodic seizure. Try learning elementary grammar before attempting to inflict your next literary abomination**

ΞXρO□˝• D••+•••Gαï• «««©: **special, DO U HAVE DOWNSYMDROME?**

IamSpecial_1 : **NOPE CANT SAY THAT I DO**

MarshallMathersisdeadICP : she has more than down syndrome

IamSpecial_1 : **ARE YOU A KLINFELTER**

InsultingEevilWeevil : boring sh*t

Host Sysop_NineMSN kicked InsultingEevilWeevil out of the chatroom: Do not use profanity

IamSpecial_1 : **I THINK YOU ARE**

MarshallMathersisdeadICP : are u a bovine dendrophiliac special, Its ok to be one

(From #teentalk)"

Despite the distinctive conventions of abbreviated and symbolic chat language, participants frequently corrected the spelling, punctuation, and grammar of their own and other's texts.

"HRT_Babe69 : **learn how to spell fatface**

(From #hotttchix4hotttguys)

Mr_Chewie__ : wat a comment comming form a POO

«˜°¤vîxøη¤°˜» : form?

Mr_Chewie__ : my bad

«˜°¤vîxøη¤°˜»: i think u mean from
Mr_Chewie__ : from
«˜°¤vîxøη¤°˜»:
Mr_Chewie__ : that its
Mr_Chewie__ : thanks
(From #the_night_club)

Copulaterking1 : copulater look it up dumy astec
Aztec_Arsenal_19 : **its**
Aztec_Arsenal_19 : **Aztec**
Aztec_Arsenal_19 : **with a z**
(From #hispanicamerican)"

The following excerpt illustrates the current craze for the genre of trivia quiz games.

"Mã°terMind : **13. Who found the long lost explorer david livingston?**
Mã°terMind : **Here's a hint: henr_ ________**
•••©HêêkŸßöÝ™••• : henry lawson
¤•°°£o££ie°°•¤™: **henry**
•2BorNot2B75 has joined the conversation.
• ©дЯψ§д®B€дЯ : **herny scotts**
Mã°terMind : **Hey i see you changed your name 2BorNot2B75 , or should i say bigv75**
•2BorNot2B75 has left the conversation.
• ©дЯψ§д®B€дЯ: **henry larson**
••• ıłd'Яĕψψ'11•• : **henry no last name**
•Freoguyinmelb1 has left the conversation.
¤•°°£o££ie°°•¤™: **henry winkler**
• ©дЯψ§д®B€дЯ : **who has henry?**
••• ıłd'Яĕψψ'11•• : **lol**
•Niña-Bonita36 has joined the conversation.
nazral1 : ***lol***
•silkboxers_melb has joined the conversation.
Mã°terMind : **Time's up! The answer was: henry stanley**
•Freoguyinmelb1 has joined the conversation.
¤•°°£o££ie°°•¤™ : **wb freo**
Freoguyinmelb1 : **ta babe**
•Ąйa• : **wb freo**
Mã°terMind : **14. Cervix of Uterus belongs to which system in the human body?**
Freoguyinmelb1 : **ty sexy**
•••©HêêkŸßöÝ™••• : reproduction
•Ąйa• : **lol**
Mã°terMind : **Here's a hint: Repr________**
°ßlõ• d• °ßõmb°Sh• ll° : reproductive
Mã°terMind : **Winner: °ßlõ• d•°ßõmb°Sh•ll°, Answer: Reproductive, Time: 11.467, Streak: 1, Wins: 4, WPM: 12, Rank: 435th**
(From #a Melbourne room)"

One can imagine teachers approving of this informal learning about the language of history and biology. In some ways, the text is reminiscent of the Socratic method used in classrooms, except that here participants are voluntary and are using the Internet to find answers. Speed and skill in searching are required to find an answer in the shortest possible time to beat other players. As the quote below shows, some also use chatrooms as a resource for homework tasks.

"Toyota_Supraz_RuleToyota_Supraz_Rule (this iz homework) ***Any 1 no who the dutch navigator who discovered australia in 1642
(From #12-17Nzteens)"

Thus, while most adults view chatrooms primarily as a space for social interaction and leisure, language learning and literacy practices valued by schools can occur.

However, it must be conceded that questions along the lines of: "n e hot chix wanna chat?" were more common than questions about explorers. In every type of chatroom outside of gay channels, the large number of turns seeking a "whisper," or private messaging, with partners of the opposite sex indicates that adults and teens viewed chatrooms as a space to meet and converse with potential romantic/sexual partners. As well, the focus on location by those seeking partners online suggests that they intended using locality as a conversational resource and/or furthering the relationship offline. Irrespective of whether their intentions were to find a romantic/sexual partner, or merely to indulge in flirtation and/or sexual talk, it is evident that participants viewed chatrooms as a space to communicate with "strangers" for social purposes. As illustrated below, participants in teen chatrooms sometimes framed their invitations to "whisper" by complaining about "boredom."

"donotnowatminiknameis: ne 1 wanna whisper cause im boared
(From #teentalk)

Rach: this room is still very boring
(From #teengothchat)"

Sexual innuendo, verbal abuse, advertising, and swearing were more prevalent in some chatrooms than in others across all categories, and distribution of this behavior appeared to depend on different rules and levels of moderation in the individual chatrooms. It should be mentioned here that since the data from celebrity sites were collected mostly from AOL (America Online[2]), the content of AOL chatrooms contained significantly higher amounts of advertising for pornographic sites, sexually explicit and abusive language, and various types of solicitations than other types of chatrooms. From the number of advertisements received in AOL chatrooms, it was apparent that some users had set up pornbots, which are electronic scripts or programs designed to send advertisements as messages to other users. Consequently, because of the number of advertisements like the following, the chatroom #BritneySpears contained more offensive content than did the chatrooms #childslavesex or #transsexualloungeroom.

"GurlWithWEBSITE4: 17/f/FL/ Come Watch Me And My BI GirlFrineds take off all of our clothes execpt our cute perfect smelling snow white panties! Grade A jackoff material!
(From #britneyspears)"

I would argue, then, that this constitutes a form of seepage of textual genres. That is, the genre of pornography is reframing what is supposedly a forum for talk about "tweenie" pop music. It is also possible to speculate that pornography may become emasculated and lose its taboo power through such ready availability. Participants' complaints and blasé comments about the preponderance of pornographic messages provided evidence of this trend.

To investigate this issue and the extent of danger from the "perverts" and "sickos" referred to by the newspaper articles, I visited adult rooms devoted specifically to sex talk. This examination revealed that content of these rooms was no more graphic or titillating than that of erotic fiction available from the local library. For example, one hour of text taken from chatrooms with names suggesting hardcore content (e.g., #Transsexual Lounge Room, #Humiliation Palace, #Child Slave Sex, #Erotic) was analyzed. Talk in the #Transsexual Lounge Room comprised mostly conversation on makeup, greetings, and music. Similarly, chatters in the #Child Slave Sex room were either shy or wary because talk was slow in coming, and what was there consisted mostly of greetings (e.g., "Hi"). Moderators of the #Erotic room regulated the trade of porn and banned both underage users and advertising. It too consisted overwhelmingly of greetings and small talk. #Humiliation Palace was dedicated to S&M role-play but was nevertheless heavily controlled by the host, resulting in many "kicks." While some talk was of an explicit sexual nature, the bulk of it comprised idle chat such as greetings and users' asl profiles.

"brooke28fem2 : hi celestine
BonierElk : hi celedtine
Host MissPennyC kicked 2_flex out of the chatroom: for giving our phone numbers (Access ban set for 24 hours)
hulkinggiant : hi miss penny
?r8ted_r has left the conversation.
?Hamptons-finest1 has left the conversation.
celestine_cd : ahhhhh the lovvvvvvvvvvvverly broooke pur
brooke28fem2 : lol
?jaimielynn has left the conversation.
(From #Transexual Lounge Room)"

Undoubtedly, the Internet is used as a medium for social activities that can be termed "sleazy" and to gain access to and possibly harm young people. Nevertheless, this small piece of research suggests: (1) that those who hold negative and narrow views about chat need to be informed that media representations are not necessarily balanced, and (2) that there are positive aspects about chat for young people. Perhaps those who are dubious and fearful should venture online to see for themselves.

CONCLUSION

The small scope of the present study can do little more than pose questions for others to explore. It is well established that teen chatrooms afford young people a significant social space for forming relationships and for framing multiple identities through symbol and language. Material presented here shows that virtual disembodiment enabled chatters to select a wider range of subjectivities than that offered by the constraints of face-to-face contexts. Indeed, if chat environments are dangerous places, are they also spaces where teens can engage in risky behavior like talking to strangers, inviting rejection by posting personal advertisements, or challenging others with relatively few adverse effects? Is it possible that chatrooms may be one of the 'safer' spaces for youth of the risk society because they provide opportunity to experiment socially from the security of home—if home is, in fact, a safe place? As well, far from being anarchic, many of the chatrooms were highly regulated discursive spaces characterized by strict forms of governance and censorship. Some chatters adopted authoritarian roles themselves. After all, having been schooled, schoolspeak is part of their discursive biographies and repertoires of practice.

Why, then, are adults so reserved and even hostile about youth chatrooms? These are questions of social control and governance of youth around new textual forms and their sexualities (see Luke & Luke, 2001). Indeed, the subtext here is that the social currency and value of text and its uses has increased because of the shift from material commodities to symbolic goods in the knowledge society (Stehr, 2001). It seems that societal and school responses to youth chat are based on principles similar to those of the pre-modern era when owning certain books was deemed a crime worthy of a flogging or being burned at the stake. While the present flogging of youth may be metaphorical, it illustrates nevertheless that literate practice remains an issue of ideological restraint and power over identity, agency, and cultural practice.

Even where chat has entered the K-12 curriculum as a legitimate genre within literacy-based curricular activity, it seems that adults find it hard to relinquish their positions of authority and control (see Kapitzke & Bruce, 2005)

Young people deserve the right and the space to speak from their lived experience, no matter how alternative or threatening these subcultures and practices appear to adults. In pre-digital times, the body was the surface upon which youth explored and inscribed their exploration of agency and action, risk and resistance, pain and pleasure. Agency was conducted through face-to-face interaction with peers and others in material, public places. Yet, the range and possibilities of these places for social, cultural, and economic engagement are diminishing at an alarming rate as democratic, public spheres are being eroded. In the face of this loss, Giroux (1997) goes so far as to claim that society is "demonizing" and "destroying" youth because "millions more kids are abused by silence than by leering pedophiles...In the nationwide discussion about protecting kids from the sickos who prey on them, the kids are missing" (Eurydice, cited in Giroux, 1997, p. 31). Perhaps, then, the real question here is: Who, in fact, are the sickos preying on kids? Is it possible that as well as the predators of chatrooms, some are operating out of corporate boardrooms?

Unprecedented access to a range of people, information, and ideas renders problematic the historical ascription of innocence and dependence of children and adolescents upon the superior experience and knowledge of adults. Indeed, because young people excel at social interaction in the de-materialized spaces of e-culture, adults have lost one element of power over them. Could this diminished control be linked to the alarmist newspaper reportage on chatrooms that is picked up by concerned parents and responsible teachers? Are adults engaging honestly and productively with the reality that young people function in a different social spatiality, one that is free of place-bound constraints? Would it be more useful for adults to cease decrying the depravity of youth chat forums, and instead view them as potentially valuable social spaces, literacy practices, and pedagogic sites that constitute less peril for teens than some public places, products of poor social and economic policy?

Transnational research has shown that school literacy programs that exclude emergent and socially significant literacies are not of optimal educational value (Goodson, Knobel, Lankshear, & Mangan, 2002; Lankshear & Knobel, 2003). As the introductory quotation by Grossberg (1994) so clearly articulates, youth remains a material problem because adults are unsure how to locate them within new forms of space and place, and within emerging systems of economic and social relations. If parents, youth workers, and teachers want to retain any semblance of credibility, influence, and affiliation with youth, they would do well to acknowledge their penchant for online social pursuits and to meet with them in forums like chatrooms. A good place to start is in K-12 literacy programs where technologically competent students can assume pedagogical and technical roles by assisting teachers to use chat.

ACKNOWLEDGMENT

I would like to acknowledge the contribution of Rhyll Vallis, who helped with the description of IRC (Internet Relay Chat), and the logging and analysis of data.

REFERENCES

Abbott, C. (1998). Making connections: Young people and the Internet. In J. Sefton-Green (Ed.), *Digital diversions: Youth culture in the age of multimedia* (pp. 84-105). London: UCL Press.

ABS (Australian Bureau of Statistics). (2000, August). *Use of the Internet by householders, Australia.* Cat. No. 8147.0. Media release 8147.0. Retrieved January 16, 2003, from http://www.abs.gov.au/

Aitken, S. C. (2001). *Geographies of young people: The morally contested spaces of identity*. London: Routledge.

Albright, J., Purohit, K., & Walsh, C. (2002). Louise Rosenblatt seeks QtAznBoi@aol.com for LTR: Using chatrooms in interdisciplinary middle school classrooms. *Journal of Adolescent and Adult Literacy, 45*(8), 692-705.

Alvermann, D., & Heron, A. (2001). Literacy identity work: Playing to learn with popular media. *Journal of Adolescent and Adult Literacy, 45*(2), 118-23.

Australian Broadcasting Authority. (2001). *Internet@Home*. Retrieved January 16, 2003, from http://www.aba.gov.au/internet/research/home/exec_summ.htm

Beck, U. (1992). *Risk society: Towards a new modernity* (M. Ritter, Trans.). London: Sage.

Bessant, J. (1995). *Youth, unemployment and crime: Policy, work and the "risk society.* Melbourne: University of Melbourne.

Culpitt, I. (1999). *Social policy and risk*. London: Sage.

Danet, B., Ruedenberg-Wright, L., & Tamari-Rosenbaum, Y. (1997). "Hmmm... where's that smoke coming from?" Writing, play and performance on Internet Relay Chat. *Journal of Computer-Mediated Communication, 2*(4).

Davis, N. (1999). *Youth crisis: Growing up in the high-risk society*. Westport, CT: Praeger.

Decker, S.H. (2002). *Gangs, youth violence and community policing*. Belmont, CA: Wadsworth.

Douglas, M. (1992). *Risk and blame: Essays in cultural theory*. London: Routledge.

Du Bartelle, D. (1995). Discourse features of computer-mediated communication: "Spoken-like" and "written-like." In B. Warvik, S. Tanskanen, & R. Hiltunen (Eds.), *Organization in discourse* (pp. 231-239). Turku: Turun Yliopisto.

Dwyer, P. (1996). *Opting out: Early school leavers and the degeneration of youth policy*. Hobart: National Clearinghouse for Youth Studies.

Dwyer, P., & Wyn, J. (2001). *Youth, education and risk: Facing the future*. New York: Routledge.

Ericson, R. V., & Haggerty, K. D. (1997). *Policing the risk society*. Toronto: University of Toronto Press.

Evans, K., Behrens, M., & Kaluza, J. (2000). *Learning and work in the risk society: Lessons for the labour markets of Europe from Eastern Germany*. Basingstoke: Macmillan/Palgrave.

Fass, P. S. (1977). *The damned and the beautiful: American youth in the 1920's*. New York: Oxford University Press.

Finkelhor, D., Wolak, J., & Mitchell, K. (2000). *Online victimization: A report on the nation's youth*. Alexandria, VA: National Center for Missing and Exploited Children.

Fromme, J. (2003). Computer games as a part of children's culture. *Game Studies, 3*(1). Retrieved May 16, 2003, from http://www.gamestudies.org/0301/fromme/

Gee, J. P. (2003). *What video games have to teach us about learning and literacy*. New York: Palgrave Macmillan.

Giddens, A. (1991). *Modernity and self-identity: Self and society in the late modern age*. Cambridge: Policy Press.

Giroux, H.A. (1997). *Channel surfing: Racism, the media and the destruction of today's youth*. New York: St. Martin's Griffin.

Goodson, I., Knobel, M., Lankshear, C., & Mangan, J. M. (2002). *Cyber spaces/social spaces: Culture clash in computerized classrooms*. New York: Palgrave Macmillan.

Grossberg, L. (1994). The political status of youth and youth culture. In J. S. Epstein (Ed.), *Adolescents and their music: If it's too loud, you're too old* (pp. 25-45). New York: Garland.

Gruber, J. (Ed.). (2001). *Risky behavior among youth: An economic analysis*. Chicago: University of Chicago Press.

Hutchby, I. (2001). *Conversation and technology: From the telephone to the Internet*. Oxford: Polity Press.

Jamrozik, A. (2001). *Social policy in the post-welfare state: Australians on the threshold of the 21st century*. Frenchs Forest: Longman.

Kapitzke, C., & Bruce, B.C. (2005). The arobase in the libr@ry: New political economies of children's literatures and literacies. *Computers and Composition, 22*, 69-78.

Lankshear, C., & Knobel, M. (2003). *New literacies: Changing knowledge and classroom learning*. Buckingham: Open University Press.

Lee, N. (2001). *Childhood and society: Growing up in an age of uncertainty*. Philadelphia: Open University Press.

Lenhart, A., Rainie, L., & Lewis, O. (2002). *Teenage life online: The rise of the instant-message generation and the Internet's impact on friendships and family relations*. Retrieved from http://www.pewinternet.org/reports/toc.asp?Report=36

Luttwak, E. (1999). *Turbo-capitalism: Winners and losers in the global economy*. London: Orion.

Luke, A., & Luke, C. (2001). Adolescence lost/childhood regained: On early intervention and the emergence of the techno-subject. *Journal of Early Childhood Literacy, 1*(1), 91-120.

Lyon, D. (1994). *The electronic eye: The rise of surveillance society*. Minneapolis: University of Minnesota Press.

MacDonald, R. (Ed.). (1997). *Youth, the "underclass" and social exclusion*. New York: Routledge.

Mar, J. (2000). Online on time: The language of Internet Relay Chat. In D. Gibbs & K. Krause (Eds.), *Cyberlines: Languages and cultures of the Internet* (pp. 151-174). Melbourne: James Nicholas.

Orthmann, C. (2000, December). Analysing the communication in chatrooms—problems of data collection. *Forum Qualitative Sozialforschung/Forum: Qualitative Social Research, 1*(3).

Reid, E. (1996). Communication and community on Internet Relay Chat: Constructing communities. In P. Ludlow (Ed.), *High noon on the electronic frontier: Conceptual issues in cyberspace* (pp. 397-410). Cambridge, MA: MIT Press.

Stehr, N. (2001). *The fragility of modern societies: Knowledge and risk in the information age*. London: Sage.

ten Have, P. (2000). Computer-mediated chat: Ways of finding chat partners. *M/C—A Journal of Media and Culture, 3*(4).

Terzian, P. (2002). Reaching adolescents. *Newsday,* (October 30).

Valentine, G., & Holloway, S. L. (2001). Online dangers? Geographies of parents' fears for children's safety in cyberspace. *Professional Geographer, 53,* 71-83.

Werry, C. C. (1996). Linguistic and interactional features of Internet Relay Chat. In S. C. Herring (Ed.), *Computer-mediated communication: Linguistic, social and cross-cultural perspectives* (pp. 47-63). Amsterdam: John Benjamins.

Wilkinson, I. (2001). *Anxiety in a "risk society."* New York: Routledge.

Williamson, H. (1997). *Youth and policy: Contexts and consequences. Young men, transition and social exclusion.* Burlington: Ashgate.

Wolak, J., Mitchell, K., & Finkelhor, D. (2002). Close online relationships in a national sample of adolescents. *Adolescence, 37*(147), 441-455.

ENDNOTES

[1] Instant messaging differs from chat in that only two parties are present in message rooms, whereas anywhere up to 300 users may occupy a chatroom.

[2] America Online is a large Internet server that also operates a chat server. Its size and success is, in large part, because the Windows applications software (95 and onwards) was configured with a connection/sign-up to America Online. However, because of the level of spamming and the perception that it is used by those who are less able to find a better/cheaper Internet and chatroom service provider, AOL chat is despised by hardcore chatters of other servers.

Chapter X
Re-Schooling and Information Communication Technology: A Case Study of Ireland

Roger Austin
University of Ulster, Northern Ireland

John Anderson
Education Technology Strategy Management Group, Northern Ireland

ABSTRACT

This chapter examines the place of information communication technology (ICT) in re-schooling; the authors discuss what re-schooling means and use Northern Ireland as a case study to explore how an entire school system is starting to be transformed through a combination of political change, curriculum development, and the integration of a managed ICT service. They argue that where ICT is being used to build social cohesion and social capital in addition to developing pupils' employability and academic performance, it is helping schools to become learning organisations. The authors provide an analysis of the social, economic, and political context of Northern Ireland, but argue that the lessons that have been learned about the relationship between ICT and re-schooling have resonance on a global scale.

INTRODUCTION

The central focus of this chapter is an examination of what the term re-schooling means and what role information and communication technology (ICT) and digital literacy might play in this process. Northern Ireland is chosen as a case study in the analysis of the question: What is re-schooling?

The work of David Istance and his team (2004) in the Organisation for Economic Cooperation and Development (OECD) has focussed attention on major challenges facing education and schooling systems across the world. In the

Schooling for Tomorrow Project, they have developed six scenarios based around three clusters of ideas; the first cluster is based around the notion of "attempting to maintain the status quo." This is characterised as teachers working as they do at the moment, largely in isolation and fearful of what change might mean for their status; a more extreme version of this outlook is the "meltdown scenario" caused primarily by an extreme shortage of teachers through loss of morale, coupled with an inability to recruit sufficient numbers of young graduates who would have a far wider range of career options open to them.

The second cluster of scenarios have been termed re-schooling; the first example of this is called "schools as core social centres," where the school becomes the key social and community institution, and school walls would come down so that substantial community activity took place within the school. The second instance of this model is the "school as a focused learning organisation," "revitalised around a strong knowledge agenda in a culture of high quality, innovation, experimentation, and diversity." Teachers would be part of a wider local, national, and global community, aided by the opportunities created from digital connectivity, and there would be extensive networks linking the school to further and higher education. There would be a high premium on teacher professional development within schools as organisations that justified the use of the term-focused learning organisations.

The third cluster of scenarios is based around ideas of 'de-schooling', with schools no longer central to the education of young people, and government playing a less important role compared to a range of religious and commercial providers. In effect market forces would drive the shape of educational provision, and schools as we know them would cease to exist.

Istance reports that at a recent conference 85% of participants thought that the "school as a focused learning organisation" was highly or rather desirable while 81% supported the "school as core social centre." However, at a second conference to ascertain where 20 countries thought they actually were on a matrix showing all six scenarios, it was clear that OECD countries are "pretty much stuck in the bureaucratic status quo." While it was acknowledged that reform was taking place on a piecemeal scale, there were substantial barriers to change, not least of which are the attitudes of middle class parents who are described as "risk averse" and "wanting their children to succeed in terms that they know and did reasonably well by themselves."

The term re-schooling shares common ground with the language of transforming learning, which has been used to talk about applications of ICT that go beyond automating learning and enhancing learning; in this chapter, an analysis is offered of the ways in which ICT has started to 'transform learning' in Northern Ireland and to suggest that this is an illuminative case study of one of the necessary but not sufficient conditions in re-schooling.

THE NORTHERN IRELAND SCHOOLING SYSTEM

Any discussion of what re-schooling might mean needs to take account of what is already in place; what are the most significant characteristics of the education system in Northern Ireland? First, the system is segregated to a high degree with 95% of children from a Protestant background attending controlled, state schools, while those with a Catholic upbringing generally attend maintained schools which subscribe to and often display a Catholic ethos. Only 5% of children attend integrated schools.

All children, except those with severe learning difficulties, attend primary schools from age 5 to 11, and most then take an externally set examination at the age of 11 to determine whether they will attend a grammar school or a high school. Students may leave school at 16, but many choose to continue their studies until the age of 18 either at their secondary school where they take academic or vocational examinations or by transferring to one of the current 16 Colleges of Further and Higher Education which offer vocational courses. These colleges are all mixed in terms of gender and religious affiliation. Students from school and the Further and Higher Education Colleges may apply for places at University at the age of 18; the quality of education is frequently invoked to attract inward investment.

The curriculum, compulsory for all pupils from 5-16, is shaped by the Council for the Curriculum, Examinations, and Assessment (CCEA), which also plays a key role in assessing student learning. Until 2004, it could be said that the curriculum had a strong 'subject' focus, particularly for pupils aged 8 and above; up to the age of 11, pupils in primary schools are taught most of their work by a single teacher, and in this sense there is some flexibility about how subject knowledge is embedded within broad themes such as literacy and numeracy. There is, however, a strong focus on the core subjects of math, English, and science which are assessed at the age of 8 and 11. At the age of 11 however, when students transfer to a post-primary school, their day is timetabled on the basis of subject disciplines, with periods of time, ranging from 30 minutes to an hour for each lesson. At the age of 16, most students are entered for an external examination, the General Certificate of Secondary Education (GCSE). Those who continue their studies after 16 take either the General Certificate of Education at Advanced level or the more vocational Higher National Diploma. Research from Gallagher and Smith (2000) shows that examination results for the most academically able students in Northern Ireland are at least as good if not better than elsewhere in the United Kingdom, but that there is also a substantial "tail" of young people who leave school at the age of 16 with few or no qualifications. This group is higher than the UK average.

In addition to the conventional academic and economic purpose of schooling, there have been various attempts to address the problems that have arisen from the nature of schooling in a deeply divided society. A range of initiatives, some identified with the Department of Education, have been summarised by Hagan and McGlynn (2004). They point to the generally limited impact such initiatives have had, with the exception of the integrated school movement which has, according to other research (McGlynn, 2004), had a "very significant" long-term effect on the ability of 55% of those who attended such schools to mix with people of a different background to their own.

Before analysing the broader historical, political, and social context within which schooling is located, what conclusions can be made about the schooling system? Its mainly segregated character is reinforced by the arrangements for initial teacher training; some 50% of all those who enter teacher training attend either a mainly Catholic university college or a university college whose student body is predominantly Protestant. The remaining 50% of places are in 'mixed' university courses. In-service support for teachers is provided through one of the five Education and Library Boards whose advisory staff receive government funding to provide professional development. Accredited courses for teachers are run by the Higher Education institutions which generally receive no funding from government for this work; fees are paid by teachers or their schools.

There is considerable pride in the current system of education in Northern Ireland; it has provided a source of social stability during difficult times, has steadily improving standards according to the Educaton and Training Inspectorate (2003), and is accepted as having a highly innovative approach to ICT.

THE SOCIAL, ECONOMIC, AND POLITICAL CONTEXT OF NORTHERN IRELAND

What is it that makes Northern Ireland an appropriate entity to use as a focus for such a study? First, its relatively small size with a population of 1.8 million people and its relatively autonomous educational administration make it possible to effect changes to the education system that impact on all the children in its schools. Although Northern Ireland is constitutionally part of the United Kingdom and is currently governed directly by Westminster-appointed ministers, in practice it has been able to develop educational policies that reflect the particular needs of the region. In this respect, it is not unlike Scotland and Wales, again both part of the United Kingdom but with varying degrees of autonomy in policymaking. In other words, there is a concentration of administrative and financial power in Northern Ireland that makes it possible for any changes in the nature and organisation of schooling to be managed on a system-wide basis.

Second, unlike Scotland and Wales, whose people have a strong sense of national identity, the citizens of Northern Ireland are divided about whether they are Irish, British, or Northern Irish. The issue of political identity has a significant overlap with religious affiliation, with the 44% Catholic population more likely to think of themselves as Irish and the Protestant population inclined to associate themselves with Great Britain. The schooling system in Northern Ireland reflects these differences with over 95% of schools effectively segregated along denominational lines. Only 5% attend 'integrated' schools where children of different religious affiliation are educated together. This level of segregation is in some respects exacerbated by housing patterns; in a housing census in 2001, 18.6% of the population was recorded as living in publicly owned housing. Most of such properties are in housing estates where the display of either loyalist/protestant or republican/nationalist identity through wall murals, flags, and painted curb stones is highly visible. Although the stock of public sector housing has declined from an estimated peak of around 38% in 1971 (Poole, 2004), what remains are estates where, in some cases, paramilitary control is exercised through punishment beatings and manipulation of the drugs trade. This might not be of any special significance were it not for the recent troubled history of the region, which has endured 30 years of inter-communal violence. Although the level of disorder has reduced significantly since the so-called 'Good Friday' agreement of 1998, Northern Ireland is still a deeply divided society.

The divisions are both reflected and reinforced through the political party structure; all of the four main political parties in Northern Ireland draw their support mainly from one side of the community or the other. Despite being part of the United Kingdom, the Conservative, Labour, and Liberal Democrat parties that make up the bulk of the Members of Parliament in the House of Commons at Westminster do not campaign in Northern Ireland. Electors in this part of the United Kingdom are invited to vote for parties that do not represent standard 'class' differences; instead they are predominantly canvassed on different versions of loyalty to the union with Great Britain or different degrees of support for the idea of a united Ireland.

Readers of this chapter who are unfamiliar with the history of Ireland might wonder how these constitutional and political arrangements emerged. Great Britain had effectively controlled Ireland since the sixteenth century, but in the face of growing demands by the nationalist, Catholic population of Ireland from the end of the 19th century, independence was granted to 26 out of the 32 counties of Ireland in 1921. The six counties that were excluded were those in the north east of the island, counties with a predominantly Protestant population that wanted to remain part of Great Britain. The creation of the Republic of Ireland and Northern Ireland at this time solved an immediate problem but left a legacy of suspicion on both sides and, for many nationalists on both sides of the border, a sense that there was unfinished business. In effect, since 1921 the status of Northern Ireland and the issue of whether there should be a reunification of the island have dominated political debate on the island of Ireland and to a much lesser extent in Great Britain.

For the purposes of this chapter, it should be noted that since 1921, the education and schooling systems in Northern Ireland and the Republic of Ireland differ significantly in terms of control, organisation, and curriculum. The schooling system in Northern Ireland is sometimes seen as part of the problem and at other times as part of a possible solution. In other words, when we consider what the term re-schooling might mean in practice, this needs to embrace the broad question of whether this includes a values dimension. What kinds of attitudes and behaviours might re-schooling involve both in terms of community relations within Northern Ireland and indeed between Northern Ireland and the Republic of Ireland?

A third reason for considering Northern Ireland as a suitable case study is to examine the relationship between schooling and the economy. This is of particular importance because one of the reasons most frequently invoked to justify high levels of government spending on educational ICT infrastructure and training is related to employment trends and required skills. It is claimed in the 2001-2002 annual report of the global digital divide that "the use of ICT contributed close to 50% of the total acceleration in U.S. productivity in the second half of the 1990s." Further:

Industrialised countries with only 15% of the world's population are home to 88% of all Internet users. Internet access on its own does not ensure economic growth and prosperity; nevertheless it seems clear that those without connectivity and skills will be disadvantaged from participation in the information society.

Barber (2001) has claimed that one of the functions of education is to meet the needs of a knowledge-based and global economy. Northern Ireland's geographical position on the geographical periphery of Europe places it at a disadvantage when it comes to the manufacture and transportation of goods. The recent "troubles" have also discouraged outside private investment; one consequence of this is that the public sector provides 60% of adult employment, a position that the Northern Ireland administration is keen to change. Their strategy is to encourage a much more vigorous entrepreneurial dimension in the economy with a particular emphasis on tourism, and an ICT-based service sector. Information processing and data distribution using a skilled workforce and 21st century communications is thought to offset any disadvantages due to geographical location.

Since the introduction of compulsory education in the 19th century, one of the functions of schooling has been to prepare young people for working life; that working life is expected to be very different from the pattern set towards the end of the twentieth century with relatively stable employment often spanning an entire life with a single employer. In the current Northern Ireland curriculum, business studies as a subject has been introduced as an examination subject at 16 and is growing in popularity. The Colleges of Further and Higher Education have a specific remit to address essential skills shortages. But what is it that employers want in terms of work-related skills and in terms of the understanding and use of ICT? In the context of re-schooling, what vocationally relevant skills are most likely to be important and to what extent are such skills being valued in the emerging curriculum in Northern Ireland? ICT is identified as a key skill in the curriculum and employers want this as a key asset, but further work needs to be done to look at the specific applications of ICT that are most needed.

How can we best summarise the predominant characteristics of Northern Ireland? It is emerging from a period of considerable unrest, and the ways it is seeking to resolve internal conflict are often seen as a model for other parts of the world. Traditional support structures, including the family and the church, are strong but evolving. As we shall see in the next section, while some parts of Northern Ireland have strong roots in the past, in other respects it has ambitious plans for the future. A key part of the future is the development of e-learning.

TRIGGERS FOR CHANGE

The Curriculum

The importance of the 1998 Peace Agreement should not be underestimated; although it has detractors, the principles of respect for diversity and the commitment to the development of an inclusive society are starting to bring about change. A revised curriculum, particularly for those aged 11-14, has a far more central commitment to the development of skills and values rather than the acquisition of inert knowledge.

To take just one illustrative example of this, the teaching of history is now to be grounded in "inquiry, evidence, and multiple perspectives." The Curriculum Council for Examination and Assessment's advice (2003) gives teachers a framework for developing pupils' knowledge, understanding, and skills, and claims that there are three key objectives in teaching history. It makes a distinction between the role of history in "developing pupils as individuals," "developing pupils as contributors to Society," and "developing pupils as contributors to the economy and environment." Under each of these headings, CCEA offers further guidance; for example, under the first heading it suggests that this includes the investigation of "how history has been used by individuals and groups to create stereotypical perceptions and to justify views and actions." It goes on to say that this extends to "a willingness to challenge stereotypical, biased or distorted viewpoints with appropriately sensitive, informed and balanced responses and take responsibility for choices and actions." This is part of a wider commitment to the development of what the document calls "moral character," which is described as the capacity to "show fairness and integrity in dealing with others," being "reliable and committed to tasks," and taking "responsibility for choices and actions." One reading of this is that students should have a far stronger sense of how history has affected their own lives and how it has affected their "personal identity."

The role of history in contributing to society draws on the concepts of citizenship, cultural understanding, media awareness, and ethical awareness; in a bold departure from previous practice, this is now to include the investigation of "some ethical issues in history or historical figures who have behaved ethically or unethically in relation to, for example, persecution, slavery, the use of the atom bomb, the decision to declare war, the partition of countries." The role of history in contributing to the economy and environment refers to employability, economic awareness, sustainable development, and environmental responsibility. In other words, what we have here is an explicit set of values even though that term is not used in this official document.

Taken overall, the new curriculum framework is forward-looking and challenging; its orientation is towards the re-schooling notion of "school as a focused learning organisation," "revitalised around a strong knowledge agenda in a culture of high quality, innovation, experimentation, and diversity." Current plans to provide professional development for staff in the move from a curriculum based on content to one based on skills supports the re-schooling proposition of schools as focused learning organisations.

These deep changes in the curriculum need to be understood in light of a further significant policy development. The Northern Ireland administration has decided that academic selection at the age of 11 will stop in its present form in 2008. This will have major implications for primary schools and for those that cater for young people aged 11-18; it may involve much closer links between secondary schools and Colleges of Further and Higher Education. E-learning is one of the ways in which such cooperation might occur through the joint design and delivery of vocational courses to broaden curriculum choice for pupils.

THE ICT INFRASTRUCTURE

We noted earlier that part of the definition of re-schooling included the concept that "teachers would be part of a wider local, national and global community, aided by the opportunities created from digital connectivity and there would be extensive networks linking the school to further and higher education." In this section of the chapter, we examine how an ICT infrastructure is emerging in Northern Ireland that is creating the framework for re-schooling.

The aspiration to connect secondary and further education requires great confidence in having widespread, reliable access to mission-critical infrastructure and the connectivity fit to deliver substantial elements of curriculum content to large numbers of pupils simultaneously, together with teachers who know how to develop high-quality e-learning courses and teach and assess effectively online.

How Northern Ireland finds itself able to venture into such an aspiration goes back to an analysis undertaken in 1997 into the failure of the integration of ICT in the curriculum on account of the major obstacles facing schools and teachers in their attempts to do so. In 1989, a UK-wide upheaval of the school curriculum resulted in the creation of a statutory curriculum for children of 4 to 16 years of age. Northern Ireland's version of the national curriculum was unique in a number of ways. The new curriculum contained a number of so-called "educational (cross-curricular) themes"—including information technology (IT)—which were intended to be delivered through the context of the programmes of study throughout compulsory schooling. However, in the years that followed, reports from the Department of Education's Education and Training Inspectorate indicated that, while there was *some* use of IT in teaching as a consequence, it was relatively patchy, never continuous, and often wholly token. Use eventually reached a plateau from which it did not progress, notwithstanding some excellent and exciting innovation in practice.

The survey conducted in 1997, indicated that:

- more than 70% of computers in schools were "old";
- most of them could not support modern multimedia;
- only 1% of primary schools had networks that could support ICT in the curriculum; and
- when teachers were asked what were the main constraints in moving forward with effective use of ICT, the items top of their list were:
    - poor access to appropriate technology provision in the classroom,
    - weak teacher competence and confidence in ICT,
    - lack of training and support, and
    - the financial constraints under which schools operated.

Furthermore, purchase of computer systems and software were wholly local decisions, taken individually by schools and by each of the five local education boards with the result that there were as many as five or six incompatible systems in use in classrooms—sometimes even in the same school and the same classroom. Without a common, or even coordinated, approach, investment fluctuated wildly from school to school, resulting in great inequalities in access and educational experience for young people, depending where they lived and which school they attended. Nor was it possible to begin to imagine a plan for systemic change in the role of ICT in delivering education.

The 1997 survey provided a diagnosis, to which the Department of Education in Northern Ireland responded rapidly with a prescription in the form of an Education Technology Strategy, which ventured to implement a wholly integrated and comprehensive policy solution. Although the language of re-schooling had not yet been developed, the vision expressed through the policy was clearly designed to transform learning and teaching; it set out some 50 targets applying to everyone in the education service across the four dimensions of:

- **Learners and Their ICT Skills:** Covering standards of pupil competence, curriculum integration, accreditation and assessment for learners, homework policy, special education needs, and policy on acceptable use of the Internet.
- **Teachers and Learners:** Setting targets for teacher competence in ICT, the use of ICT in the professional development of teachers, quality assurance by the schools' Inspectorate, the role of the senior management in schools, and access to personal computers for teachers.
- **Schools and Resources:** *D*ealing with links with school libraries, the home, the community, and setting expectations for accommodation and equipment needs.
- **Implementation and Support:** To address and resolve the issues of financial resourcing, levels of expected use of computers, technology refresh, communication and network tariffs, and user training and support.

Shortly afterwards, in November 1998, all four government education departments in the UK (England, Wales, Scotland, and Northern Ireland) published "Open for Learning; Open for Business," which challenged the ICT supply industry to put forward proposals for 'competing managed services' as a new model for supplying ICT to schools. Uniquely in the British Isles, the five education boards in Northern Ireland set out, through the ET Strategy, to treat the entire school service of one country as a "joint purchasing group" to procure a single managed service solution. In doing so Northern Ireland moved the entire school education service in response to the government's challenge to:

Consider the possible advantages in providing high-speed infrastructure and large-scale data-storage across a range of local authority service areas, and the range of procurement models available for such services, including the Private Finance Initiative and other forms of public private partnership. (And to) consider the potential for forming purchasing consortia with other local institutions or authorities, and operating joint local support arrangements for the use of the Grid in schools, libraries and elsewhere, including public/private partnerships.

The demonstrable benefits of a centralist approach to a managed computer service for school administration offered a foundation for a 'prototype' National Grid for Learning (NGFL) service. The Education Technology (ET) Strategy committed the Department of Education in Northern Ireland to "consider how the various elements of ICT, curricular and administration/ management systems can be brought together as a coherent whole." In essence, it was recognised that the effective exploitation of ICT in education represented a national challenge, regionally articulated through the ET Strategy, which now needed to be implemented in Northern Ireland in a way which was *centrally coordinated and locally delivered.* It was this bold move that has made Northern Ireland's ICT solution for schools a matter of considerable interest to other jurisdictions around the world.

While the journey to implement the strategic plan from 1998 to 2001 is a story in its own right, schools in Northern Ireland now have, through a project called Classroom 2000 (C2k), a high-quality, sustainable infrastructure, connectivity, and resources which meet strategic targets. The ICT issues of affordability, value for money, and sustainability, which challenge many education authorities around the world, have been resolved through a centrally funded managed service approach. In other words, re-schooling has become possible for all children, irrespective of whether they are attending Catholic, Protestant, integrated, special, or Irish medium schools. Furthermore, they will be part of this enriched learning environment throughout the years of compulsory education, from 5 to 16.

Grant-aided schools receive, at no cost to themselves, a core entitlement, based on their pupil numbers, including:

- an infrastructure of 60,000+ networked computers with broadband connection to the Internet and linked to schools' existing hardware (legacy systems);
- access to a wide range of content and services to support the Northern Ireland Curriculum and the teachers' professional development;
- an integrated suite of services for school administration and management;
- broadband connection of schools' networks into a single wide-area education network, connected to the network linking all public libraries in the province;
- tools to facilitate the development of online teaching and learning; and
- first-line support through a central help desk.

C2k works with a wide range of partners from both private and public sectors to deliver an integrated and supported service, installed, maintained, and upgraded by specialist providers.

Figure 1. Layout of ICT infrastructure

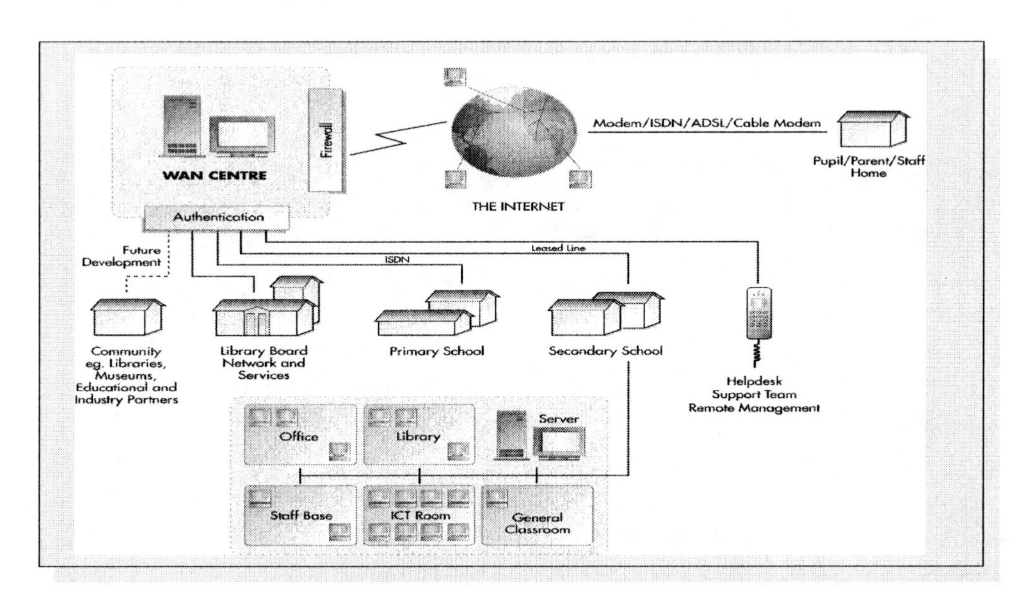

Digital Content Delivered on Local School Networks

Teachers and pupils welcomed the regional software licences, which allow them to run most of the C2k curriculum software packages (80 titles in primary schools and 120 in post-primary schools) on school laptops and other legacy systems. Many of the titles are supported by specialist education consultants employed locally by the software companies.

Learning NI: Online Digital Content and Services

Learning NI, a new online learning environment, is funded by the Department of Education and built, serviced, and maintained by Hewlett Packard for all education users. Initial content has been licensed from educational publishers. Local adaptation of additional digital learning resources can be supported through the partnership with the publishers. Over time, schools, teachers, advisers, librarians, professional officers, assessors, examiners, and anyone who creates content, resources, services, and courses will to able to develop local content and services, working collaboratively whenever possible. In Learning NI, many activities, whether hosted by organisations, interest groups, or individuals, can take place in safety. Learning NI will be one *software platform* which may house various *learning environments*, accessible to the registered users. In practice, schools, teachers, and others will define how it is used and how it will grow.

Learning NI will provide:

- all teachers and pupils with Internet connection, each with personal, secure e-mail;
- a range of digital resources from a variety of sources, including subscriptions to online libraries of curriculum content, which can

be drawn down and modified for local use in the Northern Ireland curriculum;

- online tests and formative assessment tools;
- collaboration and sharing (teacher-teacher, pupil-pupil, teacher-pupil) through e-mail, online classrooms, text and videoconferencing (educators will be able to share best practice, pupils will be able to work on joint projects);
- easy access to learning resources and personal work files from inside and outside the school network;
- personal homepages and calendars for teachers and pupils;
- a flexible means of setting and submitting homework;
- tracking, recording, and reporting the use made of the content by individuals;
- one-to-one feedback between pupil and teacher; and
- access to the school's administration system.

Once it is fully developed, Learning NI will support best practice in:

- creating resources and courses once, adapting them, and delivering them flexibly—anywhere, anytime;
- tracking and recording the use of content by the learner;
- assessing to aid differentiated, diagnostic, and individualised approaches to teaching;
- providing online courses to support professional development;
- facilitating collegial cooperative networks and the building of communities of practice;
- supporting critical reflective practice in initial teacher education, as well as in early and continuing professional development;
- enabling the integration of curriculum/administrative applications for all teachers; and
- involving parents more closely in their children's learning.

This high level of service has been completed just in time, at a time when the statutory curriculum is being revised with personal skills

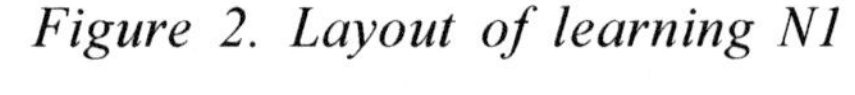

Figure 2. Layout of learning NI

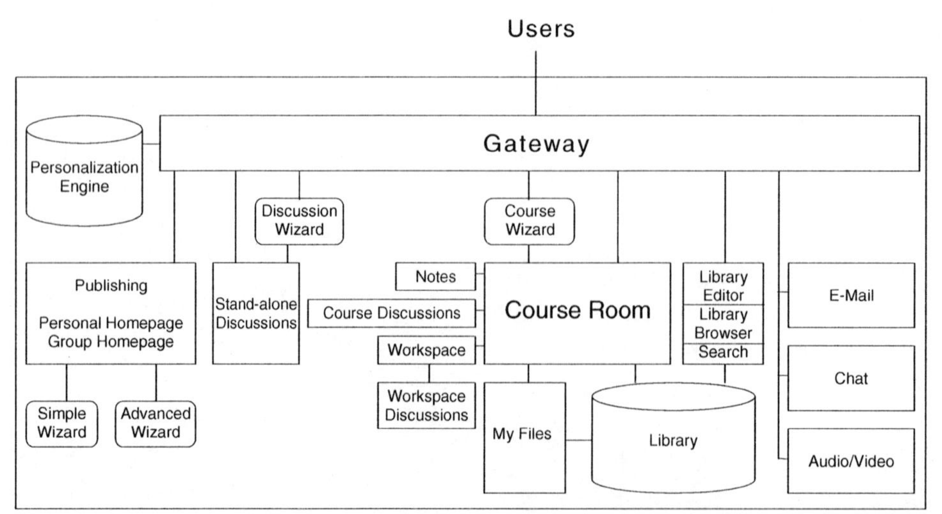

and competencies, such as critical and creative thinking skills, moving centre stage in teaching, to better address the needs of young people for living and working in the 21st century economy.

Once the revised curriculum has been implemented—a seven-year task—our curriculum will look significantly different. Competence in ICT will be embedded, using sample teaching plans, with critical and creative thinking skills, and personal and communication skills. Alongside literacy and numeracy, ICT competence will be monitored annually. Formative assessment, including online adaptive testing, will provide rich feedback for pupils on individual strengths and weaknesses, and replace key-stage testing at 8, 11, and 14. Many of the new changes are already being piloted and independently evaluated through the Northern Ireland E-Learning Partnership, a strategic umbrella arrangement which facilitates a variety of over 20 organisations responsible for the school service to work together, evaluate what they are doing, share their learning, build a common understanding of quality and standards in e-learning, and avoid reinventing the digital wheel.

Futuristic scenarios envisage learners routinely submitting work on the school network, taking part in group sessions through videoconferencing when not in school, borrowing tablet PCs from the school if they do not have their own, and taking assessments online just when they are ready. Teachers should be able to assign coursework threads for their pupils through personal digital devices wirelessly connected, working in online teams with local and international experts to create new resources and courses, and having computers on school premises accessible more than 12 hours a day.

These visions are represented in a new strategy—emPowering Schools—setting out milestones for 2008, and envisaging education in 2020. The strategy builds ambitiously on the £250 million injection in infrastructure and connectivity by C2k, and aims to see practice transformed "so that all young people will learn, with, through and about the use of digital and online technologies."

E-learning pilots (the "e" stands for *enhanced* learning) demonstrate that e-learning means more than providing enriched content, important though that is. We place the emphasis on the interaction between teacher and learners, and we see innovation in practice range from using the Internet during a lesson, through to taking part in online courses. Over the past 18 months, the Northern Ireland E-Learning Partnership has brought all the stakeholders together to learn collaboratively about e-learning, to build models of effective practice, to kite-mark standards for online provision, and to identify professional development needs.

The pilots range over the school curriculum—science, AVCE (Advanced Vocational Certificate in Education), ICT, citizenship, geography—and from work with pupils ages 8 to 17. They illustrate how ICT may play a part in the reorganisation of secondary schooling, as part of the current government review to replace the transfer test for 11-year-olds.

The pilots also address professional development—for students, beginning and serving teachers, advisers, head teachers, and professional development tutors. They are building capacity and creating a real grasp of issues in design, pedagogy, assessment, and mentorship, and the effective blending of online and face-to-face environments to create collegial communities of practice.

Achieving the vision will require more infrastructure innovation: wireless, personal access, better bandwidth, changes in school layouts, as well as innovation in school organisational arrangements. As in the past, these changes will happen because of a single, integrated approach, locally supported and implemented across the province.

THE PROCESS OF CHANGE: AN ILLUMINATIVE CASE STUDY

Hiebert, Gallimore, and Stigler (2002) among many other commentators have stressed the critical role of the teacher in making change happen in the classroom. Any notion of re-schooling, even with a 21st century ICT infrastructure, is unlikely to get very far without teacher ownership of the process. Cradler and Bridgforth (1996), commenting on a range of studies on the impact of ICT, noted the findings of these and other studies generally and consistently show that technology alone does not have a significant effect on teaching and learning. Technology is a tool that when used with tested instructional practices and curriculum can be an effective catalyst for education reform. When we ask the question about what benefits we might expect from an investment in ICT, the most frequent answers, at least in the United Kingdom, relate to improved academic performance. Evidence from the Impact2 project showed that the properly managed application of ICT had the effect of accelerating pupil performance by up to a term. We wish to argue that there are other criteria that might be used to measure the value of ICT through an analysis of the use of ICT for links between schools, and that this might be part of the wider process of re-schooling.

The Dissolving Boundaries Project

As we have noted, Northern Ireland is the place where Ireland and Great Britain meet; its citizens are pulled in different political directions, and on the island of Ireland itself there are barriers, not least of which is the boundary that separates the North from the South. There is still a substantial amount of suspicion by people and in some cases a reluctance to cross the border; although there are very few police checks since the Good Friday Peace Agreement of 1998, the border is still present both psychologically and politically.

In 1998, the two government departments of Education in Belfast and Dublin agreed to support a project that would link student teachers across the border using ICT; after their time studying at the University of Ulster and Maynooth, the students carried their new ICT skills into the schools where they were doing teacher practice so that their pupils could work together to create a joint Web site based on curricular collaboration. This project, called This Island We Live On, evolved into a more ambitious scheme called Dissolving Boundaries through Technology and Education.

There are now 121 schools in total, half in Northern Ireland and half in the Republic of Ireland, working as partners on a range of educational topics. Pupils and teachers are using both real-time videoconferencing and asynchronous computer conferencing to work together, and the results of their collaborative activity are often presented through PowerPoint slides or the construction of a Web site. The project team has evaluated the work of this project since 2001 using a combination of questionnaire, semi-structured interview with teachers and students, and analysis of comments expressed through computer conferencing. Six key conclusions have emerged from the data.

The first is the role of videoconferencing; all schools have been provided with ISDN lines and video-phones so that teachers and pupils can communicate synchronously. As the schools move towards broadband, connectivity is likely to be secured through IP-based videoconferencing, which will bring costs down significantly. The impact of videoconferencing has been very strong in both primary schools with children ages 8-12 and in special schools. Part of this effect is certainly related to the attractiveness of the visual dimension for this type of pupil, and it has made the link across the border 'more real' for them. Wenger (1998) has argued that learning is "a fundamentally social phenomenon" and that the primary focus of the theory is on learning as social participation. In other words, when teachers from two schools on different sides of the border in Ireland agree to help their children to work collaboratively together on some aspect of history or any other topic in the curriculum, they are engaged in learning that is authentic and meaningful. The presence of a distant audience of peers gives the pupils the chance to express their understanding through both real-time links in videoconferencing and through computer conferencing. How utterly different this is compared to the normal practice of asking pupils to write individually something which will only be read by the teacher.

The second conclusion relates to the organisation of work between the linked schools; it has become clear that the optimal conditions for effective interaction and learning occur when the children in each school are formed into groups of four to six children. Each group is linked to a group in the other school for the purposes of planning their work, sharing information, and working towards an end product. This arrangement means that children are not overwhelmed by having too many superficial contacts, but neither are they limited to a one-to-one link. The focus of learning is on collaboration within each group and between the two distant groups working together. As we have said, communication takes place through videoconferencing and asynchronous computer conferencing. Within the computer conference, each linked group has their own folder with two discussion areas, one called "pupil café" and the other "work in progress'\." It is striking, though not surprising, that the majority of messages exchanged are in the "pupil café" rather than the more serious "work in progress" discussion area. The project team members sometimes worry that outsiders may regard these messages as trivial and unrelated to the real business of studying history or whatever subject is being studied. However, taking Wenger's point about learning being centrally concerned with social participation, we might rather celebrate the willingness of these young people to ignore the border between them and simply get on with chat about everyday life.

The third conclusion that can be drawn from the work of the Dissolving Boundaries programme is the way that ICT has enabled links to be made between young people in special schools and with those in mainstream education. In one link, pupils in two special schools on two sides of the border are using videoconferencing to communicate. The benefits of this according to the teachers involved were:

- It helped to develop social skills. The children need to develop the skills of talking to a child they do not know, and taking turns talking.
- They felt they got to know each other during the conferences. When they met for a face-to-face, they had already decided who their friends were, and they sought out specific people.
- They got to know about Belfast—there was a lot of looking at maps and at the other school.
- The project was very popular with parents and with the board of management. Parents do not expect that a child with special needs will be videoconferencing with another school.

In another link, pupils from a deaf school in the Republic of Ireland are linked to pupils in a mainstream school in Northern Ireland where they are using computer conferencing to investigate imperialism. The teachers involved commented:

...we had our first conferencing session with our partner school...It worked [the] first time and both schools found it very interesting. It was good to see the girls we are currently working with and especially to understand and communicate with them. My students and I had learnt our names using the iris mantel alphabet and the Dublin girls were able to make them out, it was great. We are currently negotiating a face to face meeting and have agreed in principle that we will do one, probably in Dublin in February.

And, from the other teacher:

I was delighted with today's conference. I must admit that I was apprehensive and worried that we wouldn't be able to understand each other. It was a huge relief and very exciting for us. I was very impressed at how well the girls communicated—I didn't have to intervene at any stage. Looking forward to the next conference.

The fourth conclusion relates to the impact of the project on the teachers involved; this project presented considerable difficulties in terms of classroom management and the need to work cooperatively with colleagues in a different jurisdiction. Yet, the majority of teachers involved chose to continue with their links even after the first year of external support was reduced. What was it that prompted teachers to stay with the project? There are three significant pointers. The first is that this work was embedded in the curriculum and in that sense was not seen as an additional burden. Second, the very positive impact of the links on pupils' interest and enthusiasm generated a positive buzz in many classrooms and schools which teachers found motivational. Third, it is clear that the project gave many teachers the opportunity to develop their own ICT skills and confidence; they had a real incentive to make use of what they had learned and were able to put new skills to immediate use. At least one reason for this was that another teacher and another class relied on work being sent or messages being posted. One teacher was reported as saying in the 2002 report:

When you're working independently, I suppose if you meet difficulties you might choose to go back to a more reliable method, but when you're working collaboratively, there is pressure on you to make it a success for your own benefit and for the pupils.

Teachers, particularly in primary schools, commented on how the ICT skills that children were learning in the Dissolving Boundaries were being transferred to other areas of learning. They also commented on how their own role was changing; one of them said:

The project has improved my teaching, especially group work. It has made me much more aware of the different dynamics within and between groups...and of the different personalities. I have integrated ICT a lot more in all subjects.

Another commented on how the project was reinforcing a move to a more facilitative role:

I'm more facilitative and it's filtering through everything. I can sit the children down and they can get on without me. They're well aware of their time limits and they work out how much time to spend on each bit to make sure they get finished.

These comments are important in the light of work by Harris (2002) who argues that in order for ICT to be effective in the classroom, teachers need to change their existing practice, particularly to give pupils more opportunity for independent learning. Teachers in the programme found that more of their time was devoted to developing pupils' talking and listening skills because it was essential that pupils understood what was being said on videoconferencing. Again, because of the emphasis on collaboration through the use of ICT, teachers reported having to look at new ways to engage with their class. More discussion amongst pupils took place, and many teachers regarded themselves as "facilitators" in the class; as one put it, this was a "a welcome change from chalk and talk!" Teachers found themselves listening to suggestions from pupils as to how to organise videoconferences. Teachers also listened more to suggestions from pupils as to which way to proceed in the collaborative tasks, and which way to present information to their partner schools. Pupils themselves compiled quizzes based on their curricular topic to exchange with their partner school. All these activities gave pupils a new independence in the classroom.

Flexibility was a word commonly used by teachers as they attempted to explain their new role. Some of this flexibility involved time management—teachers had to be spontaneous in their use of computer suites when they became available. Student-led activities were a new feature in many classrooms. One teacher described this as "being able to stand back and let children lead the way." Another described it as "pupil driven and pupil centred." Flexibility also related to working with groups and changing the make-up of groups according to the different strengths of the pupils. Groups for the programme were often different than groups in math, English, and so forth. Many teachers reported more emphasis on cooperative group work within the classroom. In short, what we can observe from this is clear evidence that ICT applications of this sort are enabling a major shift to take place in teachers' sense of their role.

Fifth, what effect is the programme having on children's attitudes and values? Based on teachers' assessment of the effect of the project on their pupils in 2002, the project had most effect on pupils' "motivation" and ICT skills. They also commented on how preparing work for a distant audience had helped to improve their children's sense of self-esteem and confidence. Cultural awareness came fifth in importance from a set of 10 possible aspects. In the 2003 report however, which included more specific data from the pupils themselves on this question, pupils commented explicitly on increased cultural awareness emerging through the project. One male primary school pupil in Northern Ireland, for example, used space at the end of a questionnaire to write: "It shows to everyone not just to say I don't know her so I hate her, but to talk and learn about someone." Another in a similar school in the Republic of Ireland wrote: "I think Dissolving Boundaries helped me to get to know people more and not judge them by their history and appearance. Thanks." Analysis of preliminary results from work being carried out in 2003-2004, to be published in 2004, is confirming this evidence: 67% of teachers questioned said that the programme was having either a "significant" or "very significant" effect on north-south understanding. One primary teacher from Northern Ireland said that her pupils gained "greater awareness of other children's area, experiences and culture and showed an awareness of similarities rather than differences." In fact, several teachers commented that their pupils saw their partners simply as "other children" and realised that "life down south is similar to life up north." Many pupils, particularly from the South, had never travelled across the border, but the programme "brought North nearer and children got different viewpoints." One primary school teacher, having travelled with pupils to the North to meet their partner school, gave an insight into the minds of her pupils: *"Some of my pupils had not been to the North before, and even thought that the coach might be stopped at the border!"*

The final conclusion about this programme is its strong focus on a particular type of learning: this learning is characterised by its investigative and authentic nature, its collaboration with others, and the high levels of creativity arising from the need to use technology to present findings. This is a long way from passively absorbed information transmission unconnected to the real world of the children and their environment.

In what sense can the project be said to have transformed learning or in any way contributed to re-schooling in the sense of providing a better-defined purpose for learning? We suggest it does this through an emphasis on those ICT applications that place a premium on cooperation, collaboration, and inter-cultural understanding. We assert that its use of ICT to promote inclusive practice, to break down barriers between young people who have grown up in a climate of mistrust and its drive towards unlocking creativity through investigative study, are core qualities in the context of re-schooling.

CONCLUSION

There are many instances of good practice in ICT in Northern Ireland, and the analysis provided here of one particular instance of this should be seen only as an illustrative case study. So what conclusions can be drawn about the place of ICT in re-schooling? If we recall one of the definitions of re-schooling, it is the "school as a focused learning organisation," "revitalised around a strong knowledge agenda in a culture of high quality, innovation, experimentation, and diversity." Teachers would be part of a wider local, national, and global community, aided by the opportunities created from digital connectivity, and there would be extensive networks linking the school to further and higher education. There would be a high premium on teacher professional development within schools as organisations that justified the use of the term "focused learning organisations."

The ambitious ICT infrastructure discussed in the chapter and the illustrative case study of the Dissolving Boundaries programme are beginning to get close to this definition. Carr and Hartnett (1996) have argued that one of the main functions of schooling is to help young people work together for the common good. In this sense, the promotion of social cohesion through rebuilding social capital can be seen as a key priority for policymakers interested in the re-schooling agenda. This is particularly important given the very small percentage of children in Northern Ireland who attend integrated schools and the future difficulties that will lie ahead in terms of the cost of building new schools at a time when demographic data shows a declining school population. One response to this is to claim that all schools, irrespective of their Protestant or Catholic intake, should use school development plans to focus on core values and to redefine their purpose in terms of what we have called some of qualities associated with re-schooling. A central element of this will be to examine the values associated with the ways that ICT is used in the school and to recognise that ICT can be used in ways that can transform learning.

Outside Northern Ireland, what broader implications can be made from this case study? We suggest that there are three. The first is that government policy in regard to investment in ICT and education needs to examine the potential function of ICT as a tool not simply to improve the quality of academic and vocational learning but as an aid to social cohesion. While there may be specific historical, political, and cultural reasons why this matters in Northern Ireland and on the island of Ireland, this chapter has provided evidence to show that communications technology can offer real benefits when children and their teachers are empowered to work collaboratively across geographical boundaries. The second is the value of the concept of re-schooling as a touchstone for guiding ICT policy and evaluating its impact; we have argued in this chapter that the benefits of ICT as an enabling mechanism in re-schooling should be provided on a completely inclusive basis, for all children. The third implication is that the use of ICT is not value-free; decisions about the level of investment, the architecture of the learning environment, and the learning that emerges from this are all heavily charged with judgments about the role of schooling in society. Policymakers, academics, teachers, and all those involved in education need to be clear whether the purpose is reproduction of existing culture or transformation through re-schooling.

REFERENCES

Annual Report of the Global Digital Divide Initiative. (2001-2002). *World economic forum.* Retrieved June 30, 2004, from http://www.weforum.org/pdf/Initiatives/Digital_Divide_Report_2001_2002.pdf

Austin, R., Abbott, L., Mulkeen, A., & Metcalfe, N. (2002). *Dissolving boundaries in the north and south of Ireland: Cross-national cooperation through ICT in education.* Retrieved September 1, 2004, from www.dissolving boundaries.org

Austin, R., Abbott, L., Mulkeen, A., & Metcalfe, N. (2003). *The global classroom: Collaboration and cultural awareness in the north and south of Ireland.* Retrieved September 1, 2004, from www.dissolvingboundaries.org

Austin, R., Smyth, J., Mallon, M., Mulkeen, A., & Metcalfe, M. (2004). *Transforming learning through dissolving boundaries.* Retrieved October 29, 2004, from www.dissolving boundaries.org

Barber, M. (2001). *Teaching for tomorrow.* Retrieved July 31, 2004, from http://www.oecdobserver.org/news/printpage.php/aid/420/Teaching_for-tomorrow.html

Carr, W., & Hartnett, A. (1996). *Education and the struggle for democracy.* Buckingham/Philadelphia: Open University Press.

Classroom 2000. (2003). Retrieved April 28, 2004, from http://www.c2kni.org.uk/corp/corporate. htm

Cradler, J., & Bridgforth, E. (1996). *Recent research on the effects of technology in teaching and learning.* Retrieved July 30, 2004 from http://www.wested.org/techpolicy/research.html

Dissolving Boundaries through Technology in Education. (1999). Retrieved April 5, 2004, from www.dissolvingboundaries.org

Educational Technology in Schools Survey. (1997). *KPMG for DENI.* Retrieved May 15, 2004, from http://www.deni.gov.uk/strategy/etr/index.htm

Empowering Schools Strategy. (2004). Retrieved June 16, 2004, from http://www.empowering schools.com/

Entrepreneurship and Education Section of the Economic Development Policy Web Site, Department of Enterprise, Trade and Invest-

ment. Retrieved June 30, 2004, from http://www.detini.gov.uk/cgi-bin/get_builder_page?page=94&site=3&parent=45

Gallagher, A., & Smith, A. (2000). *The effects of the selective system of education in Northern Ireland. Main report, volumes 1 and 2.* Bangor: Department of Education for Northern Ireland. Retrieved September 1, 2004, from http://www.deni.gov.uk/parents/transfer/d_transferadvice.htm

Hagan, M., & McGlynn, C. (2004). Moving barriers: Promoting learning for diversity in initial teacher education. *Journal of Intercultural Education, 15*(4).

Harris, S. (2002). Innovative pedagogical practices using ICT in schools in England. *Journal of Computer Assisted Learning, 18*(4), 449-458.

Hiebert, J., Gallimore, R., & Stigler, W. (2002). A knowledge base for the teaching profession: What would it look like and how can we get one? *Educational Researcher, 31*(5), 3-15.

Istance, D. (2001). *What schools for the future?* Paris: OECD.

Istance, D. (2003). *Networks of innovation: Towards new models for managing schools and systems.* Paris: OECD.

Istance, D. (2004a). *Knowledge management in the learning society.* Paris: OECD.

Istance, D. (2004b). *Innovation in the knowledge economy: Implications for education and learning systems.* Paris: OECD/CERI.

Keane, M. (1990). Segregation processes in public sector housing. In P. Doherty (Ed.), *Geographical perspectives on the Belfast region* (pp. 88-108). Newtownabbey: Geographical Society of Ireland.

Mallon, M., Kappe, F., & Neale, B. (2003). *Towards a restructuring of the learning environment for Northern Ireland.* Retrieved September 1, 2004, from http://www.aace.org/conf/elearn/speakers/kappe.htm

Matchett, M. (2003, April 28-29). Teacher education in a climate of change—the inspectorate's view. *Proceedings of the Department for Employment and Learning Conference—Teacher Education in a Climate of Change,* Limavady, Northern Ireland.

McGlynn, C. (2004). Integrated education in Northern Ireland in the context of critical multiculturalism. *Irish Educational Studies Journal, 22*(3), 11-28.

Northern Ireland Census 2001 Output. (2001). Retrieved May 20, 2004, from http://www.nisra.gov.uk/census/Census2001Output/index.html

Northern Ireland E-Learning Partnership. (2002). Retrieved August 10, 2004, from http://www.elearningfutures.com/templates/template3.asp?id=1

Open for Learning, Open for Business. (1998). *The governments' national grid for learning challenge.* London: HMSO.

Pathways. Proposals for Curriculum and Assessment. (2003). Retrieved July 31, 2004, from http://www.ccea.org.uk/

Poole, M. (2004, May 25). E-mail correspondence with the author.

Wenger, E. (1998). *Communities of practice. Learning, meaning and identity.* /New York: Cambridge University Press.

Chapter XI
Transforming the K-12 Classroom with ICT:
Recognizing and Engaging New Configurations of Student Learning

Colin Baskin
James Cook University, Australia

ABSTRACT

This chapter begins with four very public examples of how K-12 education providers across Australia are attempting to assimilate new teaching and learning technologies into existing teaching and learning structures. The transition as predicted is not altogether smooth, and questions are raised as to where and how the discourses of literacy, education, and technology converge in the information and communication technology classroom. The discussion presents a layered case study that brings together the practical discourse of the teacher, the new discourses of literacy, teaching and learning confronting our students, and the challenge these provide to the management discourse of school administrators. In doing so, it points conclusively to the fact that new configurations of learning are at work in our online classrooms.

INTRODUCTION

Using a convenience sample of middle school SOSE students, the discussion draws on quantitative as well as qualitative methods to explore and document the educational, social and information literacy outcomes of students (and their teacher) in their first experience of online learning. The emerging community of practice is the crucial node at which technology-in and technology-and education is aligned, and its members organised and merged. This situated account describes how this merging is taking place, and how allegiance to the practice of learning both reengineers and re-orients the very roles, relationships, and distributed knowledge of the school community. In particular, the chapter offers a *gendered* account of how students mediate online learning, how new literacies are appropriated in learning exchange, and how ICT-enhanced teaching challenged one teacher's classroom practice. The chapter urges that before we discount the transformative values of ICTs in teaching and learning, we need to consider the case for a new definition of student learning that focuses on "the demands of the new world environment" (Blasi & Heinecke, 2000, p. 5).

NEW LEARNING TECHNOLOGIES AND EDUCATIONAL REFORM: TRANSFORMING LEARNING

This discussion is framed by the unique and very public circumstances facing schools and school communities across Australia as they endeavour to get up to speed with new learning technologies. In a recent edition of the *Sydney Morning Herald,* an article ambitiously entitled "Experience the Power of E-Learning" (Wilson, 2002) described how the "horse and buggy days of education" were numbered. It went on to announce the transformation of an entire educational system as part of a "learning revolution" capable of "turning education on its head." Educational reform born of new configurations of learning is walking steadily towards our schools along the information superhighway. Such bold expectations reflect a deeper belief that technology will improve learning, but as yet there remains some gap between the pervasiveness of this discourse and the actual progress of information and communication technologies (ICTs) in adding value to teaching and learning efforts. The following scenarios capture something of this impasse.

Scenario One

Under Australian political structures, the constitutional responsibility for public as distinct to private education provision falls to each of the six state and two territory governments. In 2001, the New South Wales (NSW) State Treasurer announced a State Government plan to commit $21 million over four years to provide an e-mail account for every teacher and student in NSW. In doing so, he proclaimed that, "In years to come, I believe people will look back on this as the year we began a revolution in NSW schools." The Treasurer continued to state that creation of such e-mail accounts "would transform how children learn" (*Sydney Morning Herald,* 2001). Later, when asked to elaborate on how such lofty outcomes could be realised through e-mail provision, he passed the question to the Minister for Education for a more judicious response. A follow-up article reported somewhat tongue-in-cheek:

It turned out [the Treasurer] had little idea how e-learning accounts would work or how they differ from the Internet and extensive education intranets and free e-mail accounts that exist. Instead, he kept insisting the media should ask the... Minister for Education. 'I'm one of those people who still writes his Budget speech with a pen in hand and a piece of paper so I think I will leave the details to John Aquilina's people' Mr. Egan said modestly. And, er, no he doesn't have an e-mail account at home himself. (Hewett, 2001)

Scenario Two

In one of our Southern states, Victoria, the Federal Member for Murray, Dr. Sharman Stone, has publicly criticised the State Labour Government for its lack of action in addressing rising Internet costs to state educational providers in the rural regions of Victoria. This criticism was delivered against a backdrop where the state government had just awarded 50 state secondary schools across Victoria a $5,000 share of state grants to foster the further development of information and communication technology skills.

Scenario Three

The New South Wales State Government has been publicly applauded by parent groups for providing free e-mail services to students across the state (see Scenario One above). The same political administration is simultaneously lauded by local school administrators for providing a quota of only one ISDN line per school, making widespread use of the e-mail system both impossible and impractical (Parker, 2002).

Scenario Four

In Queensland, State Education Minister Anna Bligh has reacted strongly to the recently discovered knowledge that the targeted 'critical mass' of teachers with information technology skills has not emerged with the political velocity she would have liked. Teachers in our Northern most state have been slow to embrace technology in the classroom. As a result the minister is examining the value of withholding funding for technology-based projects within "recalcitrant" schools (Johnstone & Fynes-Clinton, 2002).

RECOGNISING NEW CONFIGURATIONS OF LEARNING BEFORE YOU KILL IT!

This discussion raises the possibility that ICTs in teaching and learning can be both rewarding and confronting for all parties involved—school administrators, policymakers, educators, school communities, and their students. The scenarios presented above suggest a widespread belief (at least in political circles) that ICTs will enhance student learning in the K-12 schooling sector. Yet in each scenario (above), as we race to meet this learning revolution, our school administrators create a new casualty on the information superhighway. These same scenarios are indicative of the gap between the pervasiveness of the school reform mantra and the inchoateness with which people address the very real question of how learning will be improved by incorporating ICTs into school curriculum.

What do we mean when we talk about how ICTs might improve learning and teaching? Does, for example, improved learning in terms of ICTs act as a proxy for a known outcome statement? Should this be expressed in terms of higher test scores, or pedagogically as models of *deeper* compared to *shallower* learning? Or, as with the field of Information Literacy, should we document 'learning' against the standardised observations common in ICT effectiveness studies that code, measure, and compare observations against sets of industry standards?

A recent U.S. report demonstrates the failure of ICTs to achieve these kinds of results. The National Assessment of Educational Progress test scores for 1999 showed no significant change between 1994 and 1999 in reading, mathematics, or science for any of three age groups: 9-year-olds, 13-year-olds, and 17-year-olds (Campbell, Hombo, & Mazzeo, 2000). Technology has transformed many social practices, but the millions of new computers in schools have delivered no real improvement in aggregated literacy, numeracy, and science benchmark scores in this U.S. study.

An alternate approach might suggest there is something to be gained in pursuing the new literacies, new knowledges path, where "improved learning" is interrogated as a process wherein new learning dynamics, processes, and outcome states are currently in the act of production. If what-counts-as-learning is taken as already existing and known, then "improved learning" can only mean higher steps on the existing ladder of learning outcomes; learning that is more of the same, but incrementally better than the same. If on the other hand new forms of learning are emerging, then "improved learning" describes a form of learning that is different to what has gone before. This discussion posits that recognition of this different learning in an ICT setting poses problems for classroom practitioners in the K-12 setting and beyond.

For example, it is unlikely in Scenario One that even the then NSW Education Minister could give an adequate response to the question of how and where ICTs are transforming 'horse and buggy' schools, because there is very little documented evidence to date to suggest that educational technologies are being used to transform aggregated learning in any meaningful way. Certainly a few resourced schools are able to publicise stories about their use of laptops across the curriculum, document expensive hardware provision, and project this to host communities as self-promotion. Yet, this is not the same as broad scale curriculum innovation that draws on rich constructivist learning sites with embedded rich media. Nor does it deliver, as Scenario Four suggests, the critical mass of ICT skilled teachers required to sustain the 'learning revolution' within our schools. Stiles (2000) in his UK-based study of the implementation of ICTs for learning, concluded that failure in the use of technology in learning was based on several factors:

- the failure of ICT-enhanced curricula to engage the learner in authentic learning tasks;
- the propensity of ICT instructors to mistake "interactivity" for engagement;
- the bias in ICT-enhanced teaching and learning towards focusing on content rather than outcomes;
- the futile attempts of teachers in ICT-enhanced settings to mirror traditional pedagogical approaches onto new learning technologies; and
- an across-the-board failure to recognise the social nature of learning.

Stiles' (2000) fourth point is particularly relevant here. In general, ICT uptake in schools has to date focused primarily on trying to overlay new technologies on traditional forms of teaching, without making substantive changes to the character of teaching (Campbell et al., 2000). As each of Scenarios One, Two, Three, and Four suggest, there exists an attitude that computers remain separate from, rather than integral to established educational process. School communities are ambitious about technology-and education, but are less adept when it comes to technology-in education.

It is conceivable therefore that many teachers will not take full advantage of computer technology in their classrooms (evident in Scenario Four) as many teachers know and understand that the greatest benefits of ICT-mediated learning cannot wallpaper traditional modes of schooling. For here, the teacher rather than the learner remains in control of learning and predetermines many aspects of computer (class) interactions, how the machines will fit into the overall learning agenda, what the surrogate computer will teach and what the teacher will teach, and how much class time will be allocated to computer activities. This is not a unique view. A parallel U.S. study concludes that despite considerable rhetoric to the contrary, new technologies were hardly being used at all in schools (Cuban, 2001) with Fiske (1998, p. 11) arguing that "the only significant technological innovations of the 20th century to find a secure place in schools are the loudspeaker, the overhead projector, and the copy machine." To suggest that local Scenarios One, Two, Three, and Four (presented above) are out of context with general trends in the schooling sector would be incorrect. Clearly, the trends repeat in Australia; it is only the setting that varies.

When we consider the more qualitative questions of how and why actual ICT uptake in schools fails to match the uptake of the ICT hype of politicians, perhaps this 'trend' is symptomatic of the dwindling capacity of state educational providers and administrators to meet and greet the information age head on. At the heart of Scenarios One, Two, Three, and Four lays a central and important question—how do we as an educational community transform classroom practice using new technologies? In doing so, how do we develop pedagogy and curriculum that engages the interests and (prior) experiences of the student, whilst opening access to future learning options to them? As early as 1993, Seymour Papert urged the academic community to re-examine how the relationship between children and computers affects the traditional learning culture of schools. In some ways, the attempts of Australian policymakers nationwide reflect this engagement, with technology now a designated 'key learning area' in all state curricula. Computer competence amongst teachers is also articulated in policy; in our Northern most state Queensland, the Education minister has developed teacher competency measures and a performance continua to include benchmarks that feature changing a printer cartridge, word processing operations, and knowledge of e-mail and the World Wide Web (Johnstone & Fynes-Clinton, 2002). Policy rhetoric, though at times inconsistent with current examples of practice, clearly underlies a reformist move towards integrated information literacy, technology, and education. New learning technologies (ICTs) aim to facilitate "rich curriculum tasks," but as yet stop short of umbrella status as pedagogy in their own right (Lankshear & Knobel, 1995), and remain on the periphery of classroom practice. From the limits-to-technology-growth argument's annotated list of critics, Perelman's (1992, p. 23) key point is that schools are "no longer the primary modellers of information processing and knowledge transmission." This position is elaborated somewhat by the Queensland Minister of Education, Anna Bligh, in her affirmation that schools will now have to "prove their teachers can use IT equipment" before they receive it (Johnstone & Fynes-Clinton, 2002).

DIVERGENT DISCOURSES: CONVERGING PRACTICES

Clearly, there is a widespread climate of expectation about the new technological era. Schools and school systems are grappling with these expectations and competing as to what they try to put in place. It is a period when many people (from parents to bureaucrats) really want to know what is meant by improved schooling and therefore what constitutes good educational practice. Some indication of what this may be is provided in the Queensland School Reform Longitudinal Study (QSRLS), located within the literature on effective schools, and which has wide appeal for policymakers, administrators, and parents alike. It attempts to identify pedagogies, school contexts, and external conditions that are linked to improved student outcomes (Lingard, Hayes, & Mills, 2002;

Newmann & Associates, 1996). Potentially, 'improved learning' under such a model sees schools, teachers, and students using ICTs in ways that are intellectually demanding, in learning environments that are socially supportive, that value difference, and that have connections beyond the immediate confines of the classroom.

At a very instrumental level, Scenarios One, Two, Three, and Four offer a rough consensus to the technology/education problem. If school reform is stalling as each of these scenarios imply, and if ICTs are not the umbrella answer to improved learning as the lack of teacher uptake (Scenario Four) reflects, then the question becomes one of convergence. At what point (if any) do the discourses of information literacy, education, and technology converge and reconcile? How does this intersection better align the practical discourse of the teacher and the management discourse of school administrators such that it opens space for navigating new educational, social, and technological outcomes? The community of practice that comes to occupy this emerging space is the crucial node at which technology-in and technology-and education is aligned and its peoples (students, teachers, parents, administrators, and policymakers) implicated and merged.

"Rich description" is the tool of analysis commonly used to examine these types of connections. Recent research debates have made clear that "we do not so much *describe* as inscribe *in discourse*" (Lather, 1991, p. 90). What we see and how we name things is shaped by the conceptions and agendas we bring to it. There is a difference providing rich descriptions from familiar guiding agendas (disadvantage, words and what they construct, explicit and inexplicit pedagogical devices), and ones where, to some extent, it is not clear what aspects of the ICT-enhanced classroom we should even be considering. The next part of this discussion brings to life a situated account of how the merging of 'technology-in and technology-and' education is taking place within a local North Queensland school, and of how allegiance to the practice of learning has reengineered and re-oriented the very roles, relationships, and distributed knowledge of the classroom. It begins with a quantitative analysis of student learning behaviours in their maiden online experience, and uses their teacher's qualitative account of the online classroom as a tool to help guide reflections on changing classroom practices. In this case, the crucial node at which technology and learning align is in an early high school Studies of Society and Environment (SOSE) classroom; it is also the site at which emerging learning identities are formed.

IF YOU GO DOWN TO THE WOODS: A MIDDLE SCHOOL STUDY OF RAINFORESTS ONLINE

The essential impact of learner engagement with online learning environments is an emerging sense of learner control over the learning experience (Baskin, 2001). To the teacher in a K-12 setting, this represents substantive curriculum change, but not merely in terms of teaching and delivery. Resource-based learning and the shift from teacher-centred to learner-centred environments require that teachers rethink the fundamentals of their teaching role. To effectively weave usage of new information literacies, resources, and ICTs into the curriculum and culture of an early secondary school classroom requires improved understandings of learning theory. This may involve a questioning of:

- What are the essential characteristics of the classroom environment in terms of stimulating learning? Can these be replicated online?
- Furthermore, what is meant (for example) by a 'leaner-centred approach' to an early high school SOSE program?

The Blackboard® Learning Management System (LMS) was the e-learning platform featured in this classroom. The sample presented here is a Year 8 early high school class, engaged in a semester-length integrated SOSE classroom. The unit of study featured in this analysis was intended to occupy a timetabled block of five weeks; its focus was on local, national, and global rainforest issues. This unit was chosen to be developed and taught online as it contained pre-existing:

- peer-to-peer learning activities (paired and group collaborations);
- cross-disciplinary collaborative interactions (science, technology, health sciences, commerce, sociology, and administration);
- situated problem solving (task and interpersonal components); and
- stimulus for learners to engage with a variety of new literacies (learning objects, databases, e-paedias, virtual tours, and resource repositories).

The aim of offering the subject online was to extend the borders of the classroom to enable learners to have increased access to information resources for simulations, group work, and problem solving. A second focus of the online delivery of the Rainforests Unit was to enable interactions that promote a sense of belonging to a wider and *richer* learning environment than the traditional classroom. To achieve these dual aims, a heavy reliance was placed on the communication and collaboration suites embedded in the host learning management system (Blackboard®) to extend student learning into new spaces and places.

CHALLENGING EXISTING TEXTUALITIES

One feature of current educational practice in the SOSE field is the codifying of knowledge into existing subject matter, for example the diversity of topics woven into a representative Year 8 SOSE textbook and course of study. The resultant package is an anthropological and theory-laden textual construct, what Callon (1986) calls "obligatory" passage points of discourse around what "counts as social education." Its treatment of the Rainforests Unit warrants mention. Its textual device attempts to enrol students in "established" understandings of rainforest problems, while at the same time convincing them of the indispensability of existing (and at times ideological) solutions to rainforest issues. This kind of textual engagement at some level obviates the need for the student to participate in the search for active solutions, perhaps to the extent that "knowledge is lifted out of practice" (Wenger, 1998, p. 265). In this light, the teaching of Rainforests Unit does not necessarily cause learning about rainforest issues.

To the extent that teaching and learning are linked in practice, the linkage is not one of cause and effect, but more of resources and negotiation. Unlike a traditional K-12 classroom where everyone is learning the same thing, participants in an online setting contribute in a variety of interdependent ways to the learning of the class and to engaging with others around that purpose. The online setting becomes a way of organising learning, while providing the technology context in which learning can be demonstrated through active participation. The role of information literacy in the online setting lies in its capacity to enable teaching and learning about rainforests to interact so that each becomes a structuring resource for the other. Perhaps the sharing of some observations may illuminate this point in more detail.

Something Different Happened: Time Zones in a Rainforest

Data presented here is a summary of the learning experiences of 33 Year 8 students (*n =17 female, n=16 males*) and their teacher. These participants were drawn together as a convenience sample, surveyed and interviewed after completing a five-week online unit on "*Rainforests*" as part of their Year 8 SOSE program. The first sign that something different had happened in the ICT classroom was that the planned five-week unit was much shorter than anticipated. In fact the teacher completed it in two weeks, most of the student groups in three, and one group of boys failed to complete it in the five allocated weeks. The class was both divided and together in its online experience of learning about rainforests. Not only did the online setting separate student/teacher and student/student in learning *time and space*, but the online medium also showed that learning time does not align with teaching time, and that learner perceptions of both are a powerful influencer on learning experience and outcomes. When the traditional stop signs of the classroom—the bells and timetabled periods—gave way to 24/7 orientations to learning in an "already there" online setting, students began to adopt individual and self-paced approaches to learning by freely accessing learning materials. This increasingly happened outside timetabled classroom activities; the teacher's original reaction was one of dismay: "What about my lesson plans?"

LEARNING DESIGN AND LEARNING ARCHITECTURE

The Rainforests host subject site is heavily constructivist, and the aim of its design is to stimulate candidate membership by recruiting learners *vis-a-vis* Wenger's (1998, p. 270) three-component design infrastructure of "engagement, imagination, and alignment." For many students, the option to work online was not just a curriculum delivery experience, but also an opportunity to connect, relate, and have serious fun through an established learning resource. The learning architecture (LMS) supports the pervasiveness of Wenger's (1998) mantra of engagement, imagination, and alignment by providing opportunities for access to:

- **Communication suites/tools/places to promote and expand asynchronous engagement:** This involved group architecture that featured small learning groups each comprising six students, as well as class-level discussion boards, file exchange, and e-mail facilities.
- **Web-mounted materials and experiences (i.e., Virtual Amazon tour):** These enable students to negotiate and construct an image of themselves in their world that is not timetable nor teacher-centric, and which is not reliant on *right-answerism* for its confirmation.
- **Simulations and interactions to practice and form practices about how to critically interact with the rainforest:** These include, but are not limited to, embedded learning resources, featured Web sites, and online threaded discussions. Peer review and collaboration was also a feature of the rainforests learnscape, in so far as student's were asked to act as both reviewers and producers of texts.

Figure 1. Rainforests online

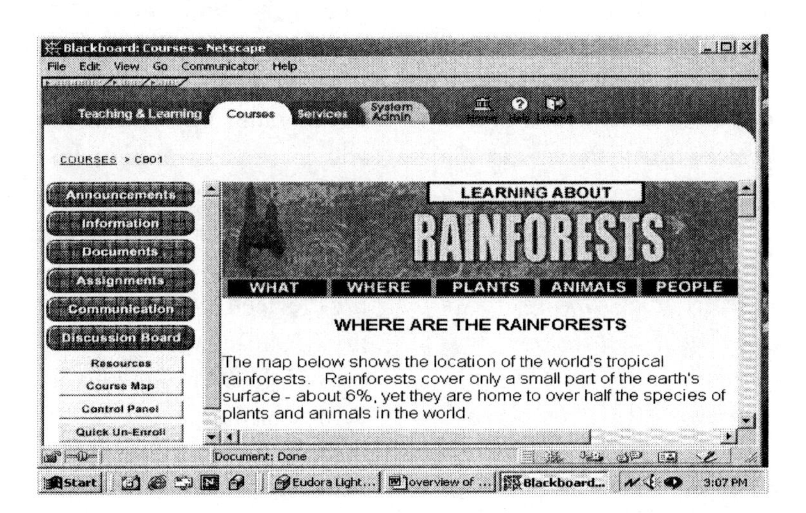

For the learner, the lure of technology is both obvious and profound; the short-run rewards in supporting the teaching and learning process and its management are significant. Information and communication technologies open pathways to the creation of learning environments that are able to support the diverse needs of learners whilst widening traditional access to learning opportunities. Despite the novelty of the subject site and method, the degree to which students appropriated online learning behaviours varied incredibly. Some students identified strongly with new learning approaches; others did not at first. Some reacted to the online environment with suspicion, some with resistance. Some saw it an opportunity to participate, whilst others saw it as an invitation for peripheral lurking.

Teacher appropriation was similarly varied. The LMS offers a variety of features and new literacy tools to enhance the delivery of subject content materials and activities. These include a conferencing system, online chat, student progress tracking, group work organisers, student self-evaluation, grade maintenance, access control, navigation tools, auto-marked quizzes, e-mail, course calendar, student homepages, digital drop boxes, and embedded search engines. The choice for uptake by the classroom teacher is comprehensive and the technologies broad. Initially the magnitude of the pedagogical divide confronted the host teacher.

...we used to spend a week on this or a week on that and now they (students) are all over the place...I worry that they don't spend enough time on the important bits, and I'm buggered if I know how to test them? (Teacher)

The teacher, like his students, reported that the traditional teaching pyramid had "been inverted" and felt at times at the "bottom of the technological totem pole." Despite initial feelings of inadequacy, by the end of the Rainforests Unit, learning relations had significantly reformed, as teacher and students began to engage in new ways of learning through the online learning resources. A new set of learning relations slowly began to evolve.

We learned a lot…even about Rainforests, but mostly about how to learn. I would like to say we met on a level playing field, but the kids were way ahead of me. But (we) swapped ideas and traded skills and before long I was in about 12 discussion groups, and was able to start pointing and linking these together. The silence in the classroom was deafening…the noise in the discussion boards was huge. It was like unleashing a monster…I wondered how these things stop but then remembered we still have the bell, thank God! (Teacher)

A METHODOLOGY FOR CAPTURING STUDENT FEEDBACK

When a classroom teacher states that we "learned a lot…but mostly about how to learn," I am intrigued to know more about the nature of this learning, how it is (re)distributed within this particular classroom, and what kind of learning improvements or costs accrue. In keeping with the online setting, an authenticated Web survey featuring 20 items (a CGI form) was generated and posted to the host subject Web site. Raw data from the survey was treated by placing the data (ranked 1-5 in nominal format) into a frequency distribution to view comparative differences of median scores across and between participating students. There were 33 out of 58 valid learner responses, valid in so far as they contained completed student data sets.

Additionally, factor and multiple regression analyses were conducted to locate and measure the learning behaviours students identified as the most 'relevant' to their learning in an online environment. The data collection process ensured anonymity for all participants. The aggregated data confirmed that the online setting facilitated an authentic learning context for the study of rainforests by opening access to relevant forms of learner participation. The subject design and architecture emphasised active, learner-centred learning paradigms based on constructivist principles (Beaty, Hodgson, Mann, & McConnell, 2002), including the need for peer interaction and collaboration (Ramsey, 2002). The data set shows that the online environment was able to enrich student learning, and led to a more informed teacher perspective about what-counts-as-learning in a Year 8 SOSE classroom.

Learning-In and Learning-Through Technology

As previously stated, a feature of the Rainforests Unit was the host course management (LMS) software, which provided opportunities for embedding and networking collaborative learning groups. Learning group activities conducted over (a planned) five weeks featured student collaborations in solving situated problems related to understanding rainforests and rainforest management. Online student meetings consisted of sharing information, dissecting course materials, environmental site analyses, collecting project data, collating project data, interpreting data, as well as publishing results from virtual field trips and projects (e.g., virtual Amazon). The summative assessment for the unit involved students designing, developing, and testing their own board game simulation entitled *"Rainforests."* Formative assessment involved a range of progressive online quizzes (using the LMS quiz generator) that often directed students to the archived and published work of other students as a point of reference and debate. The tabled results of the CGI form survey (Table 1) indicate that students endorse

Table 1. Summary statistics—student perspectives online learning

Individual perspective	**Agree** No.	(%)	**Disagree** No.	(%)	**Unsure/DK** No.	(%)
Learnt a lot about Rainforests	28	[85]	2	[6]	3	[9]
Made new friends & connections	24	[73]	2	[6]	7	[21]
Improved my computer skills	28	[85]	0	[0]	5	[15]
Felt at risk at first	15	[45]	7	[21]	11	[34]
Found it easier to speak on-line in a group	26	[79]	5	[15]	2	[6]
Learned a lot through the experiential exercises	28	[85]	0	[0]	5	[15]
Learnt to have confidence in other students	24	[73]	4	[12]	5	[15]
Learnt to use on-line communications	29	[88]	1	[3]	3	[9]
I took control of my own learning	27	[82]	3	[9]	3	[9]
I was able to relate materials to real world issues	25	[77]	5	[15]	3	[9]
I felt comfortable giving/receiving feedback	18	[54]	9	[28]	6	[18]
Group perspective						
Showed up immature students	16	[49]	10	[30]	7	[21]
Saw how my behaviour affects others	24	[73]	3	[9]	6	[18]
Fun	30	[91]	1	[3]	2	[6]
Learnt to include quiet people	24	[73]	3	[9]	6	[18]
The Rainforests Project						
Unorthodox/unusual way of learning	17	[52]	12	[36]	4	[12]
Learnt to manage learning	27	[82]	4	[12]	2	[6]
Learnt to manage myself	27	[82]	1	[3]	5	[15]
Learnt to manage others in group work	22	[67]	8	[24]	3	[9]
Forced me to manage my time	25	[76]	4	[12]	4	[12]

the online environment as an appropriate and fun (91%) forum for learning.

The survey questions attempt to capture the what (quantitative) as well as the why (qualitative) factors underlying student feedback about online learning at the individual, collaborative, as well as curriculum levels. Reported learning transfer is high (85%), as is the reported increase in computer (85%) and communication tool skills (88%) and software applications. Completion of the online learning activities clearly required learners to apply new literacy and communication tools as a means of actively negotiating learning resources. Learner responses also reflect a high level of acceptance of (and a corresponding shift to) self-directed learning (82%), and increased self- (82%) as well as time-management (76%) opportunities. Some 90% of the survey cohort accessed the learning resources outside of scheduled class time, indicating a readiness to extend (and in some ways challenge) the limitations of the timetabled classroom. This is strong evidence of engagement with and acceptance of a new learning context, one that supports learners to take learning beyond the pedagogical intentions of the traditional classroom setting.

A climate of active learning exchange (73%) was evident as students moved between formative assessments using subject discussion boards. Learners were able to apply information literacy to form connections between ideas (association), create meaningful patterns of ideas (integration), test consistency between new patterns and old ones (validation), and own new information in a way that allows it to inform their practice (appropriation) as resource managers. Students valued the opportunity for learning to learn through both global and local materials and activities (77%), for the inbuilt modelling of learning behaviours (85%) and self-reflection and feedback (54%).

Students also reported a sense of engagement in realistic challenges that mediated their study of the environment, their interactions with peer group members, and the available ICT learning resources. Learning was not just confined to learning about technology, but encompassed learning-in and learning-through technology. Feedback on aspects of member participation indicate an increased awareness of how "my behaviour affected others" (73%), of the need to manage group processes (67%), and how to include and accommodate others (73%) in collaborative learning tasks. The online environment was clearly able to stimulate authentic experiential and interpersonal challenges for students in a Year 8 SOSE course of study. In the words of one female student: "The boys aren't nearly as loathsome online" (Georgie).

The 'identities of participation' that emerge through these classroom interactions point to a learning community that is closely connected by knowledge resources, whose membership is locally differentiated (by skill, computer orientations, exposure, preferences, proximity), yet one that remains locally connected through learning. This is more significant given the steady drift (over five weeks) away from 'timetabled' classes as a critical factor in determining when learning about rainforests would take place. In terms of induction to literacy in technology, some students saw the concept of online learning as somewhat unorthodox (52%) at first, with nearly half the students reporting that they initially felt at risk (45%) in what was an unfamiliar environment. Beyond the initial difference in perceptions, a virtual learning environment does pose a unique set of challenges for the novice user. It immediately enlists the learner in a process of having to decide where to go within the site, what counts-as-learning, and which territories of knowledge need to be claimed and labelled. The traditional classroom setting demands less immediacy from the learner; the initial contrast is the shrinkage of response time, wherein new relations and forms of learner membership are negotiated, owned, and enacted. The feedback from students about their maiden experience of online learning is indeed glowing, and assigns a significant potential to ICT and resource-based learning in SOSE classrooms of the future.

What or Who Helped Me to Learn?

If, as the survey data suggests, ICT delivery can improve learning effort, questions of "when, where, and how" this value adding takes place become important. In order to identify "what-or-who-helped-me-to-learn," a principal components factor analysis with *varimax* rotation was conducted to examine which (if any) structure of variables (see Table 1) students attribute to "better learning experiences." Six principle factors with *eigenvalues* greater than one were extracted using SPSS. In other words, a good online learning experience boils down to patterns of learner participation (at a range of levels) within the learning community (Wenger, 1998). In particular, the emerging factors included

1. how learners manage themselves,
2. how learners manage their learning,
3. how they use communication tools and processes,
4. how they organise online collaborations,

5. the degree to which learners seek and incorporate teacher feedback and evaluation, and
6. the gender of the learner.

From these results, a multiple regression analysis was used to examine associations between these six factors (e.g., which, if any, factors above can be used to explain or account for student learning outcomes). The regression analysis showed a strong association (r^2=.802) between *learner attitude and approach*, *learner self-management*, *learner's use of Blackboard communication tools*, *gender*, and *collaborative learning relationships*, with *teacher feedback and evaluation* held constant as the dependent variable. In all, 80% of all variance (that is what-or-who-helped-me-to-learn) has been accounted for by these five independent variables.

For the teaching practitioner, this is very good news indeed. What-counts-as-improved-teaching in the traditional classroom still counts as much (if not more) in the online setting. Teaching with ICTs still requires an ability to generate enough excitement, energy, relevance, and value to attract and engage learners, and to communicate this through feedback and evaluation. What is needed to make the transition to online teaching successful is a translational pedagogy that is able to situate the teacher and their students within more contemporary (read ICT-enhanced) learning systems. As stated previously:

- *The first goal of the "Rainforests subject site is to articulate the internal direction, character, and energy of the classroom. This already is in the classroom; hence the online classroom is built on pre-existing personal networks and clear curriculum statements.*
- *A second goal of the subject site is to open a dialogue between insider and outsider perspectives, whilst making space for different levels of learner participation. Learning is not always direct and declarative.*

These goals advocate greater learner control over the learning experience. The learner control afforded to users of the subject site is best evidenced in its space labelling properties.

- GroupWare enabled easy transition between private and public spaces (subject-wide and small group spaces), shifting the learning focus from the macro (class) level to the micro (small learning group) level through the simple selection of a navigation icon.
- E-mail provided a conduit for one-on-one networking for the sharing of information with limited clusters of people, and back channel group discussion pages (small group pages) helped orchestrate the public space before students go public with publishing their work and/or ideas.
- ICTs were therefore able to add value to learning by raising individual awareness, and in the longer term developing a systematic body (memory or archive) of knowledge that can be easily accessed by each learner within the classroom. This is a powerful characteristic of new literacies.

As the Rainforests Unit progressed, students settled into a pattern of Web site use built around the functionalities of the site. The mix of idea-sharing forums and tool-building projects fostered both casual classroom connections as well as facilitated learner outcomes. The combination of whole class, as well as small learner group gatherings created a balance between the familiarity of teacher-centred interactions, and the buzz students describe from working and playing in a distributed learning environment.

My cyber-dentity was Baby Spice, but all the kids christened me Mrs. Doubtfire. I thought I could trick them and just merge into the group but it was not that easy...they were on to me as much online as they were in the classroom. What was different was that they started a discussion thread called ask Mrs. Doubtfire and I suddenly realised how ridiculous Baby Spice seemed. (Teacher)

The Relevance of Gender

The very fact that we have a middle-aged male teacher using ICTs to adopt the *cyberdentity* of Baby Spice is evidence that something different in learning is at work here. Here is evidence that subjectivities and relationships are mediated by literacy in technology, and are associated with shifts in power relationships. There is a long literature on inequalities and schooling, with some longstanding stabilities (class or SES) and some changing forms (gender), and each has been an ongoing issue of interest in relation to the organisation of schools and school systems. The introduction of expensive technology and possibly new forms of literacy, competence, skill, and knowledge display potentially change some existing relationships with which education is instrumentally familiar.

Harding (1997) points to the fact that technology is "gendered space" within the school curriculum. Although it is "dominated" by males, this dominance is not based on competency or learning outcomes (Kirkpatrick & Cuban, 1998; Durnell, Glissov, & Siann, 1995; Cockburn & Arnold, 1985). Girls and women are very pragmatic and confident users of computers. To some degree, this is represented in the higher relative weighting females assign to communication tools as a component of the rainforests learning environment (see Figure 2).

The uptake of communication and learning technologies by female students is evidence that learning is changing and that what counts as knowledge is changing. What is more important however is the issue of what the fundamental knowledge, both content *and* competencies, is that learners now need for vocational success (read access), to new worker identities and opportunities. Gee, for example, argues that the new economy calls up a "shape shifting portfolio person," and that skills in design and collaboration therefore warrant and acquire new significance (Gee, 1999).

Given the outcomes in Figure 2, it is no surprise that female respondents report a higher level of enjoyment of the online learning experience than their male counterparts (see Figure 3). In fact, all female students reported enjoying the online learning experience more than they did their recent unit of study in the traditional chalk-and-talk SOSE classroom. The pragmatic adoption of online communication processes by female students also signalled a change in the nature of learning relationships within the online classroom. Short, Williams, and Christie (1976) hypothesised that users of communication media are in some sense aware of the degree of social presence of each medium and tend to avoid using particular interactions in particular media. Specifically, users avoid interactions requiring a higher sense of social presence in media which lack such capacity. Social presence, they contend, varies among different media; it affects the nature of the interaction and it interacts with the purpose of the interaction to influence the medium chosen by the individual who wishes to communicate. In this context, a gendered pattern of ICT uptake is emerging through the data;

Figure 2. Use of communication tools

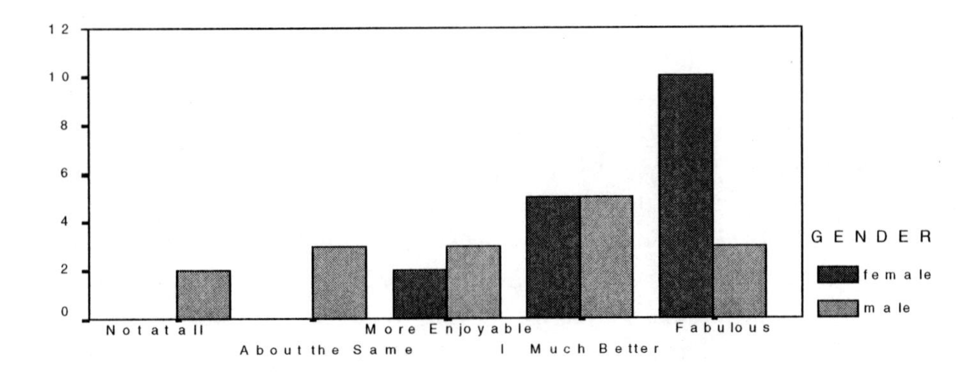

Figure 3. Enjoyed online learning

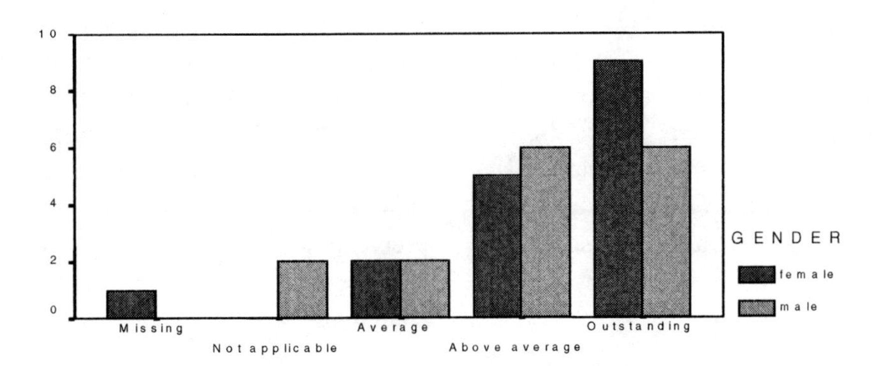

subjectivities and relationships are mediated by technology in so far as female students are transforming their online learning experience by transforming the role of technology to enhance what Stiles (2000) termed the "social" aspect of learning.

In any learning context, knowledge construction hinges mainly on an *understanding* of the *relationships* between different components of knowledge; understanding is demonstrated in the ability to interpret, integrate, and create new knowledge (Resnick, 1987) actively, purposefully, effectively, and strategically. When mediating learning, from a constructivist viewpoint the learner has to do the thinking, relating, constructing, integrating, and interpretation. Knowledge is, after all, personally constructed (Toohey, 1999, p. 56). However, in any teaching context, the teacher needs to make the learning platform available; they must explicate its dimensions and processes, and provide the scaffolding for writing, thinking, and constructing to take place purposefully and effectively. Teaching and learning thus have a context of practice.

To this end, Anderson (1995) describes different types of knowledge, each of which can be linked to rainforests. Declarative (also known as representational or propositional) knowledge (Anderson, 1995) is knowledge about facts, beliefs, things, and events usually expressed as theories, principles, rules, and frameworks (knowing what). When this theoretical knowledge is converted into action, it becomes procedural knowledge (Anderson, 1995), realised as the skills and techniques related to knowing how to perform cognitive tasks. When both the theoretical (propositional) and practical (procedural) knowledge are enacted, it becomes conditional knowledge—that is *knowing* when, where, and how cognitive tasks are to be performed and applied. When the theoretical (propositional), practical (procedural), and applied (conditional) knowledge are used as deliberate learning strategies, strategic knowledge comes into being, meaning the learner is negotiating the when, where, how, and why of learning. There is clear (directional) taxonomy between these four categories of knowledge and evidence within this case study that new improved forms of social learning are emerging among female students.

Most student responses identified that traditional teaching and learning relationships and roles had changed in an ICT-enhanced classroom (see Figure 4).

Female students seemed more aware and indicated a stronger desire for supported online learning than did male students—all female respondents deemed additional peer and teacher support as helpful, important, or necessary (see Figure 5) as a strategy for consolidating learning by balancing information literacy competence with the rhythm of the ICT classroom.

Female students reported enjoying the opportunities for self-directed learning, and in turn using these opportunities to orchestrate and structure whole community and small community communications to support their learning. For all participants, the mix of idea-sharing forums and tool-building projects enabled informal (as well as formal) learning channels to develop in a way that supported interpersonal development. Its culture promoted an increased sense of responsibility for learning, for self- and time-management as a learner, and was able to create dialogue between the learning materials and the context of the learner. The data presented here also suggests that the online class-

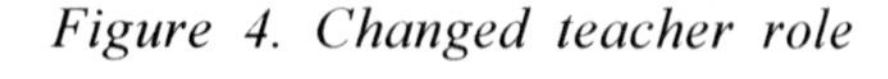
Figure 4. Changed teacher role

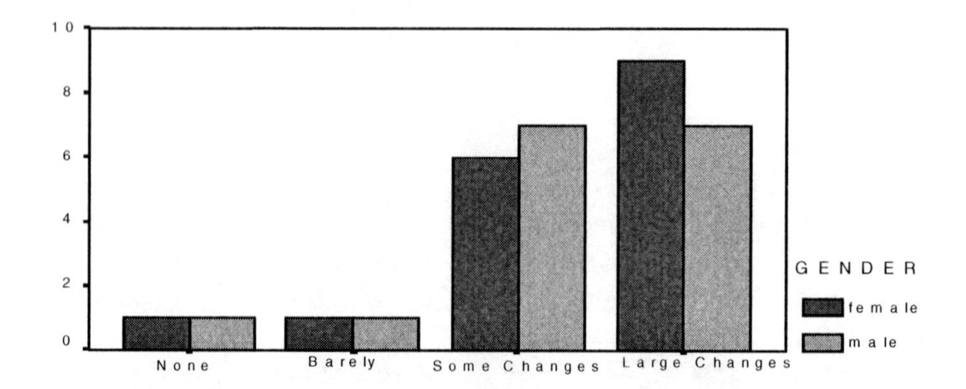

Figure 5. Need for learner support

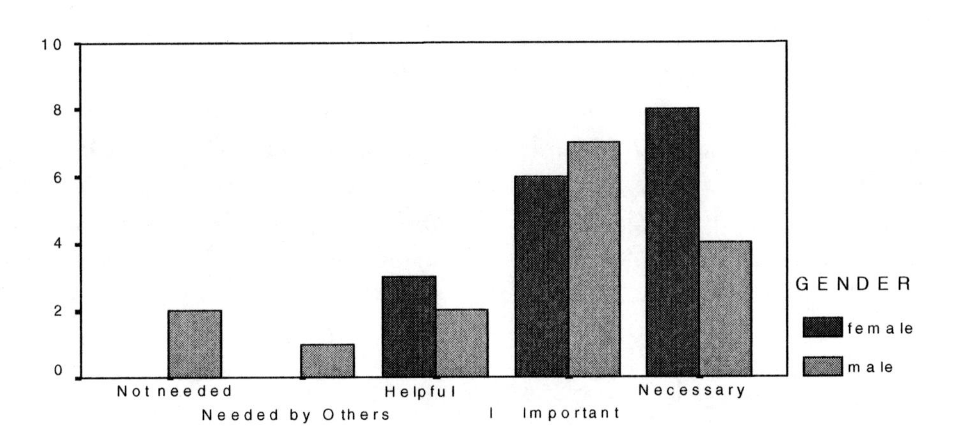

room is less extreme in its representations of 'real-world' classroom identities and dysfunction, and also points to evidence that learning misbehaviour is more transparent when embedded in ICTs. The ICT framework also laid the groundwork for a more explicit examination of what the traditional classroom enacts as (hidden) largely tacit forms of classroom organisation. Data presented here suggests that ICTs help teachers and learners explicitly declare:

- *who is the authority, and who is expected to listen and is enabled to speak;*
- *how activities, groupings, and transitions are determined;*
- *who judges performance, and how this is done;*
- *what counts as knowledge; what constitutes performance;*
- *what links are made across the curriculum; which have currency beyond it;*
- *where control is located;*
- *where knowledge is produced;*
- *when communication is enabled and for what purpose; and ultimately*
- *why computers are used.*

THE TEACHER'S PERSPECTIVE

As educators we are traditionally encouraged to focus on creating structures, systems, and roles within our classrooms that achieve relatively fixed (sometimes banded, sometimes hierarchical) goals that enable our students to fit well into other school-based or systemic structures and processes. To most teachers this challenge presents itself in the form of strategies and techniques for classroom management. The interview data presented here brings to a focus the challenges facing teachers and students when interfacing between two textually mediated delivery or classroom management contexts. The teacher in this study reports a range of challenges in a variety of areas: technology, literacy, logistics, organisation, and delivery (Dabbagh, 2001). What emerged from his observations was a sense of dissonance, a fragmentation of teaching practice across two conflicting platforms.

Two roles…on the one hand me, the constructivist, the facilitator moving in and around the knowledge construction processes of the student. They expect me to be their peer, their mentor...I am supposed to contribute equally to the subjective and unstructured as well as the structured discussion within the class. On the other hand a different me...the assignment marker...bringing the lower end of the class closer to the top end...the expert who will ultimately be expected to pass judgement on the rigor of student work in the most objective way possible. This conflict means the roles have to be performed independently—this results in a huge increase in my workload. (Teacher)

After analysing the activity within the subject-wide discussion forum, over half the discussion threads were generated by the teacher, and more than 50% of the total responses were directly attributable to the teacher. Most of the teacher's discussion threads were attempts to "set (discipline) the collaborative agenda" for the class, including setting up activities, assigning groups, and indicating useful resources. As the Rainforests Unit progressed, more and more of the teacher's online time was spent on 'weaving' the student discussions towards an outcome. In the words of the contributing teacher, this was "heavy reflective work," the "very stuff" of good teaching.

One task had students using e-mail to prepare and submit a summary report of their board game...(you know) the final assessment piece for the unit. This created huge response pressures. Even the suggestion of "e-mail contact" raises the expectation that I am permanently on tap for feedback, and that feedback will be needed yesterday rather than today. Great!...so one Monday I lose my spare period when I would normally send out e-mails and for the rest of the week I am apologising to kids for my tardiness. They got very clever..."Hey Mr. _______, I can't do this assignment until you have approved my concept." The system had turned on me...it was (like) trial by media...make that multi-media (hah). And then...yes there is a then...you (interviewer) told me to use the technology to work for me, you remember...to copy and paste responses in e-mail rather than type it all. So the kids start to compare my feedback comments...and what do you know...they accuse me of sending out the same rotten e-mail. I have never felt more under the microscope. (Teacher)

The conflict between face-to-face and online literacy and learning processes was marked. In the teacher's words, "It is normal bread and butter practice to close (a lesson) by pulling together the key themes of a lesson." However, the demands of new literacies and their synchronous and asynchronous properties shift the responsibility for mediating discussions (read closure) to the facilitator. As the same teacher concludes, "Closing an online discussion helped me to demonstrate effective modelling and synthesising strategies, but it took me far too long (time) to achieve this."

At the administrative level, the organisational and logistical aspects of online learning seemed also to challenge the response capacity as well as the mindset of school administrators. It was difficult to "get a working computer lab," and even more difficult to "mediate the tribal practices of the IT and multimedia teachers" to secure server access and "some form of ongoing help." Assessment also appeared problematic in an online environment. The teacher felt compelled to be able to 'feedback' to students and parents about the quality of each student's participation, but felt he lacked the 'repertoire' (read time and means) to determine:

...which student contributions actually enhanced the debates; who was original

and who was responsive in discussions; how to deal with "lurkers" and non-participation, and how to educate about attribution of ideas and resources. In short...I felt the collective was engaged seriously in learning but I found it hard to say the same for each individual.

As to the degree to which technology added value to the classroom practices of this particular teacher, we must borrow from "Tina's" experiences.

Tina ________ just talks a lot. She is incessant. In class I would speak to her all day everyday if she had her way. I added up all the words I had typed to her over the last four weeks...about 1,800 words. Stay with me...I have a point. Now, if I speak at...say 160 to 170 words a minute this means that in four weeks I have spoken to Tina for the equivalent of about 10-12 minutes. You tell me...is that enough?

The risk in over-focusing on the ICT experiences of one teacher is that we may tend to over-identify with a very singular and idiosyncratic episode of teaching, vis-a-vis a sample of one. If we ask, looking through the wide-angle lens, "What does literacy in technology do?" a surprisingly focused answer comes back. It recaptures the expressivity of youth cultures which school texts and teacher-generated learning materials exclude. This creates a different sense of self and society: literacy allows us to access human society in the formal terms denied to an oral culture that just narrates its ongoing drama. New literacies enable both sets of rules to speak with equal voice and equal authority. ICTs in this way are transformative tools; the fact remains that this online classroom had an intrinsic quality that led to questioning (both productive and less than productive) about the dual modes of teaching delivery. Despite the glowing response from students to the new forms of literacy and learning, two enduring principles of the 'trial by multimedia' exist for our more circumspect teacher.

- online teaching leads to an increase in teacher workload, and
- it can also lead to dissatisfaction (at least ambiguity) with the quality of the teaching experience.

ENGAGING NEW CONFIGURATIONS OF LEARNING

Education has always had at its core the desire and the requirement to prepare students for the world in which they live. Much has been written before this chapter about the challenge of shifting educational practice to reflect the demands of contemporary society, resulting in suggestions of new skills and new understandings of how such education can be achieved. A growing number of researchers and educationalists (individuals and groups) describe variations on a common range of desirable attributes. Students need to be able to motivate and direct their own learning, engage in creative, productive activities that are relevant beyond the classroom. They need to be well equipped to think critically, make decisions, and be sensitive to multiple cultural and world views. Invariably the ability to creatively and responsibly use information and communication technologies appears as a requirement.

The findings of this project suggest that a learner's digital literacy skills will emerge in ways that reflect his/her local circumstances; in a Year 8 classroom, this means how they manage themselves, their learning, how they use online communication tools, how they form networks of collaboration, how they appropriate teacher feedback, and how they challenge aspects of gendered practice. The findings also highlight issues impacting social, school, and assessment policy and practice. Revitalised schools are something many of us strive for: staff, students, and communities alike. One problem facing the ICTs in schools movement at the enterprise level is that staff, students, and administrators in our schools experience technology differently. This is reflected in all four opening scenarios, and captured within the student and staff data presented in the body of this chapter.

Because experiences of new learning technologies are different, so too will be the range of outcomes. Technology transforms by inclusion as this project shows: the transition to a better social and learning opportunity for the individual (and for groups) depends largely on where and how you are positioned in relation to information literacy access. If you are a mid-career teacher, a disenfranchised male high school student in a Year 8 SOSE class, or a teacher in a school that for social, cultural, or economic reasons does not meet the IT state benchmarks, then the social pact of ICT-enhanced education offers at best diminishing marginal returns.

The failure of ICTs in K-12 schooling is that we ask existing teachers to scaffold new learning processes in a new generational context, using old generation theories, approaches, and concepts. This impasse provides the case for a new definition of student learning that focuses on "the demands of the new world environment" (Blasi & Heinecke, 2000, p. 5). The challenge presented here is to find better ways to respond to a clear identification of the need for more relevant and meaningful standards or new definitions of student achievement.

Parallel to this are significant methodological issues for how we might channel ICTs into a study of current K-12 learning practices, and how this could build upon existing research models. Current ICT effectiveness studies are designed to code, measure, and compare observations against sets of standards to draw common classroom observations about improvements in student learning. A more productive outcome might be to place the pedagogical practices of teachers under scrutiny as this project has done, to question what technology is used for, how learning is assessed, and what performances are valued. A new research emphasis on learning recognises that not all learning may be observed by changes in behaviour, and that some learning may well be unplanned and therefore unanticipated (Wenger, 1998). It is the latter, that which Wenger (1998) terms "emergent learning," that holds much promise for ICTs in education, for here new configurations of learning are at work within our classrooms. That this happens despite our accidental attempts to 'kill' them off is a profound statement of their pedagogical potential and productivity.

REFERENCES

Anderson, J. R. (1995). *Cognitive psychology and its implications* (4th ed.). New York: W. H. Freeman and Co.

Baskin, C. (2001). The Titanic, Volkswagens, and collaborative group work: Remaking old favourites with new learning technologies. *Australian Journal of Educational Technology, 17*(Spring).

Beaty, E., Hodgson, V., Mann, S., & McConnell, D. (2002). *Towards e-quality in networked e-learning in higher education.* Retrieved from http://www.csalt.lancs.ac.uk/esrc/manifesto.htm

Blasi, L., & Heinecke, W. F. (2000). From rhetoric to technology: A transformation from citizens to consumers. In R. A. Cole (Ed.), *Issues in Web-based pedagogy: A critical primer.* London: Greenwood Press.

Callon, M. (1986). Some elements of a sociology of translation. In J. Law (Ed.), *Sociological review*. London: Routledge & Keegan Paul.

Campbell, R., Hombo, C., & Mazzeo, J. (2000). *NAEP 1999 trends in academic progress: Three decades of student performance* (NCES Publication No. 2000-469). Washington, DC: U.S. Department of Education, Office of Educational Research and Improvement, National Center for Education Statistics. Retrieved January 23, 2001, from http://nces.ed.gov/nationsreportcard/pdf/main1999/2000469.pdf

Cockburn, T., & Arnold, E. (1985). *Machinery of dominance: Women, men and technical knowledge.* Boston: Boston University Press.

Cuban, L. (1998). High-tech schools and low-tech teaching: A commentary. *Journal of Computers in Teacher Education, 14*(2), 6-7.

Dabbagh, N. (2001). The challenges of interfacing between face-to-face and online instruction. *TechTrends, 44*(6).

Durnell, A., Glissov, P., & Siann, G. (1995). Gender and computing: Persisting differences. *Educational Research, 37*(21).

Fiske, E. R. (1998). Computers at the crossroads. *Technos, 7*(1), 11-13.

Gee, J. P. (1999). New people in new worlds: Networks, the new capitalism and schools. In B. Cope & M. Kalantzis (Eds.), *Multiliteracies: Literacy learning and the design of social futures.* London: Routledge.

Harding, J. (1997). Gender and design in technology education. *The Journal of Design and Technology Education, 2*(1).

Hewett, J. (2001). Pen and paper treasurer who can't dot com the I's. *Sydney Morning Herald,* (May 30), 1.

Johnstone, C., & Fynes-Clinton, M. (2002). Teachers face the tech test. *Courier Mail,* (April 3, North Queensland Edition), 1.

Kirkpatrick, H., & Cuban, L. (1998). Should we be worried? What the research says about gender differences in access, use, attitudes and achievement with computers. *Educational Technology,* (July-August).

Lankshear, C., & Knobel, M. (1995). Literacies, text and difference in the electronic age. *Critical Forum, 4*(2).

Lather, P. (1991). *Getting smart: Feminist research and pedagogy with/in the postmodern.* New York: Routledge.

Lingard, B., Hayes, D., & Mills, M. (2002). Developments in school-based management: The specific case of Queensland, Australia. *Journal of Educational Administration, 40*(1).

Newmann, F. & Associates. (1996). *Authentic achievement: Restructuring schools for intellectual quality*. San Francisco: Jossey Bass.

Papert, S. (1993). *The children's machine: Rethinking school in the age of the computer.* New York: Basic Books.

Parker, M. (2002). Parents' fears cast a wide net. *Daily Telegraph,* (February 27), 23.

Perelman, L. (1992). *School's out: The radical new formula for the revitalization of America's educational system.* New York. Avon Books.

Ramsey, C. (2002, April 8-10). Using virtual learning environments to facilitate new learning relationships. *Proceedings of the BEST 2002 Annual Conference,* Edinburgh. Retrieved from www.business.ltsn.ac.uk/events/BEST%202002/Papers/c_ramsey.PDF

Resnick, L. B. (1987). *Education and learning to think*. Washington, DC: National Academy Press.

Short, J., Williams, E., & Christie, B. (1976). *The social psychology of telecommunications.* London: John Wiley & Sons.

Stiles, M. (2000a). Developing tacit and codified knowledge and subject culture within a virtual learning environment. *IJEEE, 37*(1), 13-25.

Toohey, S. (1999). *Designing courses for higher education.* Buckingham, UK: The Society for Research into High Education and Open University.

Wenger, E. (1998). *Communities of practice: Learning, meaning and identity.* Cambridge: Cambridge University Press.

Wilson, E. (2002). Experience the power of e-learning. *The Sydney Morning Herald,* (April 23), 11.

Chapter XII
The Complexities of Measuring Technological Literacy

Marcie J. Bober
San Diego State University, USA

ABSTRACT

Evaluating the impact of technology infusion is fraught with challenges. In this chapter, the author argues that the variance in evaluation rigor and quality about which so many complain depicts definitional confusion about technological literacy—the central premise underlying nearly all technology initiatives. She offers strategies for improving how we operationalize technological literacy as a construct—in part by drawing on the best of the many standards systems proffered by well-respected professional associations and educational agencies. She closes the discussion with a brief examination of other evaluative complications that exacerbate measurement/assessment—to wit, the criticality of engaging stakeholders; timely evaluator selection; and robust, up-front evaluation design and planning,

TECHNOLOGY INITIATIVES: EVALUATIVE QUANDARIES

Schools committed to technology infusion—those that actively develop infrastructure, acquire hardware and software, and provide professional development—generally seek *sponsors* to fund their efforts (e.g., federal, state, or local governments; private foundations, often tied to large corporations). Evaluating the impact of these often large-scale, multi-year initiatives is fraught with challenges (Ertmer, 2003; Rockman, 2004). One confounding factor is the pilot or demonstration nature of so many programs and projects—a *fluid* condition that breeds idiosyncrasies that thwart comparative analysis and rigorous measurement. Two federally funded initiatives—*Technology Innovative Challenge Grants* (TICG, launched in 1995) and *Preparing Tomorrow's Teach-*

ers to Use Technology (PT3, launched in 1999)—aptly demonstrate this situation. As their Web sites detail,[1] many projects—especially those awarded in each initiative's later years—were led by consortia with members whose level/nature of involvement, commitment, and backgrounds varied dramatically. Some projects were fairly localized; others, however, featured a broad stakeholder/constituency base. Though overarching goals and outcomes were fairly common across projects, enabling and terminal objectives tended to be tailored/particularized to institutions and the oversight state agencies and/or professional associations with which they were affiliated. As important is that few projects were static, either administratively or procedurally. Turnover in personnel was the norm rather than the exception, as were changes in project scope and direction (Ertmer, 2003; Johnston & Toms Barker, 2002).

Despite fluidity and other complexities just described—factors that clearly affect how project assessment unfolds—evaluators have long been interested in educational technology and its potential to positively affect instructional processes. Robust assessment of technology integration/infusion (including use and access, professional growth, and influence on academic and other performance indicators) dates back to the Apple Classrooms of Tomorrow (ACOT) project[2] where—over a 13-year period (1985-1998)—researchers studied how the "routine use of technology by teachers and students [in seven classrooms, representing a cross-section of primary and secondary schools] might change teaching and learning."

The past decade, in particular, has been witness to a virtual flood of evaluative research on technology—conference sessions, technical reports, white/position papers, journal articles, and texts—both at the K-12 and university levels. Unfortunately, not all investigations have been methodologically stellar; few, in fact, have attempted to replicate or even extend the processes that ACOT meticulously advocated and modeled.[3] By 1998, staff at the Office of Educational Research and Improvement (OERI)[4] were sufficiently concerned about evaluation quality *and* accuracy to fund invitation-only institutes (offered at the University of Michigan in 1999 and 2000) where evaluators assessing grants associated with such prominent federal initiatives as TICG, Star Schools[5] and PT3 could share ideas and experiences (best practices, lessons learned) and build a repository of evaluation tools for colleagues to adopt or adapt. OERI also sponsored publication of a *how-to guidebook* for the "nonevaluator" (generally, a building-level resource teacher or technology coordinator) and a *sourcebook*[6] whose contributors focused on:

- the theoretical constructs associated with measuring growth, impact, or change in specific areas (e.g., learner outcomes in the cognitive or affective domains; pedagogical changes among teachers, technology integration);
- current evaluative practices in each of these several areas;, and
- different (but promising) measurement approaches.

Linda Roberts (2004, pp. viii-x), Director of the Office of Educational Technology[7] during the Clinton Administration, elaborates on specific OERI concerns. She details an array of measurement shortcomings that include evaluator inexperience or naïveté *as well as* a tendency among evaluators to target only a few classrooms or schools, explore short- rather than long-term effects, narrowly focus on individual experiences that cannot easily be generalized to other groups or settings, omit critical details about students' actual technology-supported learning experiences, and over-rely on anecdotal or perceptual evidence. Each of these problems is nontrivial—given the increased emphasis on scientifically based research and experimental or quasi-experimental designs that *No Child Left Behind*[8] demands (Rockman, 2004).

Means and Haertel (2004) also point to somewhat obvious but often downplayed difficulties that underlie technology assessment—among them, selecting the "sheer number of technologies and technology uses" (p. 3) that might be examined; determining "true" issues of investigative interest; deciding how best to account for differences in student characteristics as well as variance in teachers' backgrounds, teaching philosophies/styles and experiences, and training; and establishing methods for measuring complex constructs—for instance, higher-order and interpersonal skills that include creative problem solving, decision making, analytical thinking, collaboration/teaming, and constructive criticism.

In sum, then, this author argues that the variance in evaluative rigor and quality about which so many complain depicts *definitional confusion about technological literacy*—the central premise underlying nearly all technology initiatives. Because there is no one definitive or authoritative meaning or explanation to which evaluators may turn, measurement too often explores isolated application skills ("comfort" with software features/functions); frequency of technology use and the "regularity" of access to computers and peripherals; and—occasionally—changes in curricular reform that feature tasks/activities/projects in which computers and Internet use are prominent aspects. Berrett (2003), in fact, argues, that the evaluative research conducted to date has been little more than "rhetorically, philosophically, or logically based commentary"—premised on inadequate designs that fail to account for real classroom practices. Simply put, because technological literacy is poorly defined as a construct, the evaluation community cannot adequately measure its attainment, describe its many manifestations or permutations, or offer specific techniques to speed its development and ensure that "learner" growth is sustained.

LITERACY: A BROAD LOOK

Like technology, *literacy* defies easy explanation—and this section provides only a glimpse at the many ways it has been characterized. Some argue against any single definition, suggesting that literacy is a "socially constructed combination of skills and contextual understandings that are particularly appropriate/effective for a specific time and a specific space."[9] Others quarrel over the tendency to equate *literacy* (at least semantically) with skill, ability, or competence (International Technology Education Association, 2000). Until the late 1800s, a *literate* person was one *familiar with literature, well-educated,* or *learned.* Not surprisingly, the term was often linked to one's years of formal schooling or the degree/diploma attained. Today, according to the 4th edition of the *American Heritage Dictionary of the English Language* (2002), the term suggests one's ability to read and write—to *systematically process* and *manage* information—a description of prowess with *data* strikingly similar to the outcomes of most K-12 technology initiatives. In that vein, then, one who is *functionally* literate can "manage"—operate successfully—in society...but barely. He or she can read, write, speak, and *deal with copious amounts of information* just well enough to conduct everyday (basic) transactions at work, at home, and in the community.[10]

Educators and researchers now tend to take a wide-ranging and more sophisticated stance toward literacy, arguing (in the main) that it incorporates the abilities "to think, assess, deconstruct, critically analyze, synthesize, create and communicate across a variety of media including text" (Virtual Teacher Center, 2002). They recognize that a host of conditions—the global face of today's economy (including a trend among U.S. employers to manage labor costs through outsourcing), sustained military efforts to combat both domestic and international terrorism, and an activist political agenda—all validate efforts to define literacy more precisely—to deconstruct this large and somewhat abstract term into its many facets. In other words, we are beginning to favor the notion that literacy *implies being knowledgeable in a particular subject or field* (cultural literacy, environmental literacy, visual literacy, etc.).[11]

It is fair to suggest, then, that *technological literacy* is far more than one's ability to use computers and the Internet (Bugliarello, 2000; Bybee, 2003).

The issues in our everyday life for which we need technological literacy go beyond knowing [about] devices...They are issues that affect how we go about making personal...as well as community decisions, such as how to vote on a proposed incinerator or on a bond issue to build a bridge that may be of greater immediate benefit to a neighboring community. They are issues of risk, safety, cost-effectiveness, standards, and trade-offs—all interwoven. (Bugliarello, 2000)

Clearly we shortchange the *transformative* power of technology when we focus exclusively on hands-on skill with devices. Gagel (1997), too, leans toward a definition that can accommodate rapid changes in specific hardware and software (both design and use). One who is technologically literate understands technology development—and its historical and cultural contexts. He or she can easily cope with technological change, creatively solve technological problems, use technology to make decisions and navigate complex processes, and assess technology's impact on human life (Bugliarello, 2000; Dugger, 2001; Gagel, 1997; Wonacott, 2001). Such a position well aligns with a view of technology itself as the "way people modify (invent, innovate, change, alter, design) their natural environment to suit their own purposes" (Dugger, 2001, p. 514).

Technological literacy is often seen as an extension of the language arts. Rafferty (1999), for example, organizes technological literacy into three distinct subgroups:

- **Text-based/alphabetic literacy** encompasses three core abilities: *reading well* and *reading to learn* (narrative and expository aptitudes, respectively) as well as *reading to do* (document aptitude), e.g., "interpreting and using information from different kinds of nonprose formats—forms, charts, graphs, maps, and other visual display—in which information is not arranged in sentence or paragraph form."
- **Representational literacy** is the ability to analyze information and understand how meaning is created; it implies competence with media, visualization, and messaging (sometimes termed alternative or nontraditional texts) and how each may be used to attract a "reader's" attention and present varying points of view. One who is representationally literate can create or construct personal meaning from symbols, including "alphabetic characters, icons, photographs, or other visual images."

- **Tool literacy** is the ability to use hardware devices and software applications to read, interpret, evaluate, critique, and use information.

The American Library Association (ALA), on the other hand, suggests that technological literacy is *interwoven with and supports information literacy* (2000, p. 3). Its argument is that *fluency* with information technology "may require more intellectual abilities than the rote learning of software and hardware associated with *computer literacy*, but the focus is still on the technology itself" (p. 3). Information literacy is the more robust construct, according to the ALA. It is "an intellectual framework for understanding, finding, evaluating, and using information—activities [that may call for technology savvy and sound investigative methods, but which are mainly accomplished] through critical discernment and reasoning" (pp. 3-4). ALA therefore *embeds* technology into the performance indicators/outcomes associated with each of its six comprehensive information literacy standards (for higher education). Standard Two (*the information-literate student accesses needed information effectively and efficiently*) is reprised in Table 1 in its entirety, clearly illustrating the ALA's integrative emphasis.

But many researchers point to the interdisciplinary nature of technological literacy—that it incorporates cultural and social competence, scientific and mathematical proficiency, financial astuteness, business savvy, and communication know-how (Virtual Teacher Center, 2002a, 2002b). Berrett (2003), in fact, bemoans that preservice educators—in particular, those specializing in instructional or educational technology—continue to artificially separate technology education from the disciplines on which their graduates will actually focus (science, social studies, English/language arts, mathematics, and the arts). He reminds us that the American Association for the Advancement of Science (AAAS) has long taken the position that those who are *scientifically literate* "know how technology connects to society and have an understanding of the elements of the designed world including agriculture, materials, manufacturing, energy, communication, health, and computer technologies and [their] implication for the human enterprise."

Without a doubt, those charged with evaluating technology initiatives face an uphill battle. They must wade through a morass of perceptions and beliefs about the prime construct under study as they assess how well the interventions and/or activities capture its tenets—all while contending with a complex sociopolitical, and often volatile environment that features *implicit* and *explicit* stakeholders who have varying information needs, levels of investment, and interest (Mohan, Bernstein, & Whitsett, 2002).

OPERATIONALIZING TECHNOLOGICAL LITERACY: STANDARDS SYSTEMS

There is little doubt that lack of definitional consensus about technological literacy compromises the evaluator's ability to characterize the literate technology user (and distinguish him/her from other users), adequately assess/measure what it means to use technology in innovative ways, and depict what technology infusion or integration "looks like."

One compensatory mechanism has been the proliferation of "standards" systems. Since the early 1990s, several well-established (and well-connected) professional groups have developed (and actively promoted) comprehensive metrics or scales by which technological literacy may be progressively measured. Some

Table 1. ALA performance indicators and outcomes for literacy in Standard Two

Performance Indicator	Outcomes
Indicator #1: The information-literate student selects the most appropriate investigative methods or information retrieval systems for accessing the needed information.	Outcomes: (a) identifies appropriate investigative methods (e.g., laboratory experiment, simulation, fieldwork) (b) investigates benefits and applicability of various investigative methods (c) investigates the scope, content, and organization of information retrieval systems (d) selects efficient and effective approaches for accessing the information needed from the investigative method or information retrieval system
Indicator #2: The information-literate student constructs and implements effectively designed search strategies.	Outcomes: (a) develops a research plan appropriate to the investigative methods (b) identifies keywords, synonyms, and related terms for the information needed (c) selects controlled vocabulary specific to the discipline or information retrieval source (d) constructs a search strategy using appropriate commands for the information retrieval system selected (e.g., Boolean operators, truncation, and proximity for search engines; internal organizers such as indexes for books) (e) implements the search strategy in various information retrieval systems using different user interfaces and search engines, with different command languages, protocols, and search parameters
Indicator #3: The information-literate student retrieves information online or in person using a variety of methods.	Outcomes: (a) uses various search systems to retrieve information in a variety of formats (b) uses various classification schemes and other systems (e.g., call number systems or indices) to locate information resources within the library or to identify specific sites for physical exploration (c) uses specialized online or in-person services available at the institution to retrieve information needed (e.g., interlibrary loan/document delivery, professional associations, institutional research offices, community resources, experts, and practitioners) (d) uses surveys, letters, interviews, and other forms of inquiry to retrieve primary information
Indicator #4: The information-literate student refines the search strategy if necessary.	Outcomes: (a) assesses the quantity, quality, and relevance of the search results to determine whether alternative information retrieval systems or investigative methods should be utilized (b) identifies gaps in the information retrieved and determines if the search strategy should be revised (c) repeats the search using the revised strategy as necessary
Indicator #5: The information-literate student extracts, records, and manages the information and its sources.	Outcomes: (a) selects among various technologies the most appropriate one for the task of extracting the needed information (e.g., copy/paste software functions, photocopier, scanner, audio/visual equipment, or exploratory instruments) (b) creates a system for organization of the information (c) differentiates between the types of sources cited, and understands the elements and correct syntax of a citation for future reference (d) records all pertinent citation information for future reference (e) uses various technologies to manage the information selected and organized

are independent frameworks, meant to inform instructional practices or gauge readiness to move forward; others are contractually or otherwise connected to accreditation or credentialing requirements.[12] A few are prescriptive—organized around grade levels, curriculum/lesson plans, and performance assessments.

Not surprisingly, each features its own terminology—but there is considerable overlap among them. Also common is the consistent message that hardware and software savvy *do not* alone portend technological competence. Nonetheless, the simpler frameworks tend to focus on self-assessed skill with different software and hardware (*novice to mastery*), frequency of technology use (*rare to always*), awareness of technology-infused teaching techniques or practices (*unfamiliar to highly familiar*), and perceptions of support or access to technical assistance (*no support to 24/7*). Those more comprehensive in scope and scale imply that true proficiency goes well beyond isolated skills to the ways in which knowledge, skills, and values converge or meld to affect the teaching and learning dynamic and contribute to pedagogical excellence. They point to the behavioral and attitudinal shifts that result from consistent exposure to technology and promising techniques for successful classroom integration (Roblyer, Edwards, & Havriluk, 1997)—specifically, a focus on learners rather than the instructor, projects and tasks that feature both independent and team opportunities (hence, cooperation and healthy competition), curriculum with an interdisciplinary and open-ended flavor, and an emphasis on learners pursuing creative and appropriate solutions rather than right ones.

Six of the most prominent systems are briefly depicted in the subsections that follow. *Interesting to note is that for most, measurement of technological literacy is implied or inferred rather than explicit; the naïve reader may not immediately realize that system elements, whether individually or collectively considered, distinguish the* literate *technology user*.

NETS

Many in the K-12 arena look to the highly touted *National Educational Technology Standards* (NETS) project[13] to fashion their definition or description of technological literacy. Developed by the International Society for Technology in Education (ISTE)[14] with funding from the PT3 initiative, NETS is, in fact, the benchmark by which a growing number of funded technology-infusion efforts are assessed. Since 1999, several state agencies and professional associations affiliated with teacher preparation have wholly adopted one or more of its several versions.[15] The allure of NETS—whether focused on K-12 students, preservice candidates, veteran teachers, or school administrators—is its breadth of vision.[16] First, performance standards are couched in terms of *essential conditions* that must be in place institutionally and programmatically if technology is to flourish. To illustrate, the conditions associated with *teacher preparation* include: *a shared vision for technology*; *equitable access to hardware, software, and telecommunications*; *financing, tenure/promotion, and other administrative policies which suggest that technology is valued administratively*; *skilled faculty who are encouraged to grow professionally and committed to disciplinary expertise—fully aware and capable of modeling both general and disciplinary-specific technology use*; *availability of technical assistance*; *rigorous and relevant content standards as well as sound curricular sources*; *a commitment to student-centered approaches; continuous and rigorous assessment practices to ensure technology use is effective;* and *active support (resources, expertise) from outside the school (including the business and other communities).*[17]

All NETS competencies, regardless of audience (students, teachers, administrators) are organized around *clusters*. For students, indicator outcomes (termed *profiles*) are leveled by grade range (PreK-2, 3-5, 6-8, 9-12); for teachers, they are structured around the four phases that characterize teaching training and early practice: general preparation, professional preparation, student teaching/internship, and first-year teaching.

While not a *curriculum*, per se, NETS does provide more curricular support than the other frameworks showcased in this section. Example "lessons" are fairly explicit[18] a searchable database in the *NETS for Teachers* area of the Web site allows one to locate materials according to such criteria as *lesson title*, *teacher profile* (e.g., first year or student teaching/internship), *subject area* (including educational foundations), and *grade range*.

Seven Dimensions

The Milken Family Foundation for Educational Technology promotes its *Seven Dimensions for Gauging Progress*.[19] First published in 1998, the document helps policymakers, educators, and technology directors determine "the conditions that should be in place for technology to be used to its greatest educational advantage in any classroom." Central to the companion piece, entitled *Technology in American Schools: Seven Dimensions of Progress—An Educator's Guide*, is a continuum of progress indicators for each dimension organized around three stages of progress: *entry*, *adaptation*, and *transformation*. Transition steps guide an educator from one stage to the next.

Each of the framework's dimensions (Dimension 1: Learners; Dimension 2: Learning Environments; Dimension 3: Professional Competency; Dimension 4: System Capacity; Dimension 5: Community Connections; Dimension 6: Technology Capacity; and Dimension 7: Accountability) is relatively independent, comprising core areas that stakeholders (most often, within a K-12 environment) should consider as technology and telecommunications are deployed.

For example, fundamental to Dimension 3: Professional Competency are core technology fluency; curriculum, learning, and assessment; professional practice and collegiality; and classroom and instructional management. Teachers interested in generating a status report/profile of their knowledge, skills, and attitudes in these four areas complete the *Professional Competency Continuum Online Assessment Tool.* The General Assessment provides a competency overview, while Detailed Assessments in the four major areas or strands generate customized advice and resources.

Especially appealing about the Seven Dimensions is that the assessment scales (in particular, those associated with professional development) signify receptivity to change and innovation (not merely personal skill, comfort, or frequency of use/application) *and* encourage group-level (not merely individual) participation (helping to ensure that program transformation is about *us*...not merely about *me*).

ITEA's *Standards for Technological Literacy*

The *Standards for Technological Literacy: Content for the Study of Technology*—published in 2000—is an outgrowth of the International Technology Education Association's (ITEA's) *Technology for All Americans Project*.[20] The *Standards,* which explicitly address what every student in grades K-12 should know and be able to do to be technologically literate, were closely reviewed by the National Research Council (NRC) and the National Academy of Engineering (NAE) prior to publication—and the framework reflects their influence in terms of design, flow, and organization.[21] Currently, it is endorsed by the above referenced organizations as well as the American Association for the Advancement of Science's (AACE's) *Project 2061*,[22] the National Council of Teachers of Mathematics (NCTM), and the National Science Teachers Association (NSTA). The *Standards* differ from other frameworks to emerge in recent years because they define the study of technology as a discipline.[23] As such, they are far more than a "checklist for the technological facts, concepts, and capabilities that students should master at each level" (Dugger, 2001, p. 514). One chapter, in fact, advocates for technology as an integral part of the school curriculum (detailing the importance of preparing students to live in a technological world), while another calls on educators, the business community, and others to vigorously support the effort.

The standards themselves, 20 in all, are of two distinct but complementary types. The *cognitive* standards attend to basic knowledge about technology—how it works and its place in the world. The *process* standards target specific student abilities or competencies. Organizationally, the standards fall into five distinct themes:

1. Those associated with the nature of technology theme call for students to acquire knowledge of the characteristics and scope of technology; the core concepts of technology; and the relationships among technologies; and the connections between technology and other fields.
2. Those associated with the technology and society theme task students with learning the cultural, social, economic, and political effects of technology; the effects of technology in the environment; the role of society in the development and use of technology; and the influence of technology on history.
3. Those allied with the design theme focus on students understanding the attributes of design; engineering design; and the role of troubleshooting, research and development, invention and innovation, and experimentation in problem solving.
4. Those featured in the abilities for a technological world theme call for students to apply the design process; use and maintain technology products and systems; and assess the impact of products and systems.
5. Those associated with the designed world theme focus on students selecting and using a broad range of technologies: medical, agricultural and biotechnological; energy and power; information and communication; transportation; manufacturing, and construction.

Coupled with each standard are *benchmarks*—which specify its fundamental content elements. Simply put, "benchmarks are statements, organized around grade ranges (K-2, 3-5, 6-8, and 9-12), that describe the specific knowledge and abilities that enable students to meet a given standard" (Dugger, 2001, p. 515). Finally, the document features examples (vignettes) that illustrate how a teacher might plan, implement, and assess instruction to help students attain mastery of a specific standard.

Like NETS, the ITEA *Standards* are attractive—comprehensive, developmentally appropriate, precise without being overly prescriptive, coupled with peripheral guidance (e.g., assessment and professional development standards) that makes implementation feasible, well-connected to the sciences, and thoughtfully presented.

CTAP

The state-funded *California Technology Assistance Program* (CTAP)[24] provides assistance to schools and districts integrating technology into teaching and learning. Effective use of technology is promoted through regional coordination of support services based on local needs organized around five core areas: staff development, technical assistance, information and learning resources, telecommunications infrastructure, and funding.

CTAP features two proficiency profiles: *preliminary* and *professional.* Important to note is that the profiles *roughly* represent the technology proficiencies integrated into California's two-level credentialing structure[25]; on the whole, they speak to technological competence and comfort.

Each proficiency profile is organized into discrete sets or areas (factors) that themselves are classified into larger dimensions or facets of the teaching experience (e.g., Dimension 1: Communication and Collaboration; Dimension 2: Preparation for Planning, Designing, and Implementing Learning Experiences; and Dimension 3: Evaluation and Assessment). Furthermore, each factor (regardless of profile) represents a *general* or *specific* knowledge/skill. Two skills associated with preliminary profile illustrate the distinction: *Demonstrates knowledge of current basic computer hardware and software terminology* is a general knowledge/skill (Dimension 2) while *Uses computers to communicate through printed media* is deemed specific (Dimension 1).

Once registered in the CTAP2 network, users complete one or more of the surveys associated with each proficiency set/area, view their results, and then select professional development opportunities "customized" to identified weaknesses and strengths. Ongoing skill assessment is encouraged; higher scores on the surveys imply the effectiveness of the activities instructors have opted to attend/complete. Important to note, however, is the behaviorist orientation of the survey items and the simplistic measurement scale (introductory, intermediate, proficient).

StaR Chart

The CEO Forum on Education and Technology offers two *School Technology & Readiness (StaR) Charts*[26]—one for the K-12 community, the other for colleges/universities offering teacher preparation programs. Each features a comprehensive assessment that organizes results into one of four profiles. The *Early Tech* label suggests that an institution, college/school, or teacher prep program offers little or no technology *or* promotes use that is—at best—perfunctory and low level (e.g., linear tutorial or drill/practice). *Target,* on the other hand, describes an institution, college/school, or teacher prep program that serves as an innovative model for others to emulate—where staff organize students in innovative ways; offer robust, challenging tasks/activities that build decision-making, problem-solving, and communication/interpersonal competence—and improve academic performance; and target capacities that reflect the work settings students will face upon graduation.[27]

Each of the four profiles is fairly complex, composed of several component parts. For example, the *Digital* element of the K-12 chart is itself composed of five factors:

1. format,
2. role of education and degree to which digital content is infused into the curriculum,
3. how students use digital content to enhance their learning,
4. percentage of students with access to digital content—as well as frequency or regularity of access, and
5. percentage of the budget allocated for purchase/acquisition of digital resources.

Institutions reportedly use the ratings data in several ways: to set benchmarks and goals (e.g., around infrastructure/capacity or professional development)—and then monitor progress toward their attainment, to identify technology needs for which grant/award applications may be written, to determine how best to allocate technology funds already available, and as the basis of statewide technology assessments. The idea is to make institutions focus on what are dubbed the five key building blocks for student achievement in the 21st century: assessment, alignment, accountability, access, and analysis.

The assessments are decidedly lean—especially when compared to the detail of the charts themselves. Nonetheless, the framework is simple to deploy and offers immediate results and next-steps prescriptions that, though generic, are easy to interpret.

enGauge

The enGauge framework—jointly produced by the North Central Regional Educational Laboratory[29] and the Metiri Group[29]—is designed to help schools and school districts use technology more effectively and transform themselves into high-performance organizations.

Like NETS, it is structurally organized around *essential conditions* (reprised here),[30] each featuring several indicators that attend far more to beliefs about innovative teaching/learning and social/digital equity than proficiency with specific hardware and software.

- **Forward-thinking, shared vision**—a systemic approach which ensures that students are prepared to learn, work, and live successfully in a knowledge-based global society.
- **Effective teaching and learning practices**—how well the vision is being executed via well-designed (and technology-integrated) instructional settings premised on sound field-based research.
- **Educator proficiency with effective teaching and learning practices**—the extent to which educators can proficiently implement, assess, and support a variety of teaching and learning strategies.
- **Digital-age equity**—how well *all students* and *unique/specialized student needs* are being addressed in educational tasks that align with/reflect the vision.
- **Robust access anywhere, anytime**—the extent to which students and staff have the access they need to support effective instruction.
- **Systems and leadership**—how well the system has transformed itself into a high-performance learning organization.

Like the CEO Forum's StaR Chart, enGauge includes assessments that generate profiles (or snapshots) of the institution or survey-taker's progress toward "effective" views about and uses of technology; the output also depicts how the institution or respondent prioritizes "21st century skills" and values technology integration in various subject areas. enGauge generates a menu of "high-impact" technology-based resources matched to an institution and/or individual's profile.

The enGauge output is lengthy and detailed—unwieldy to transform or convert into concrete plans of action. The framework's system-wide view, though laudable, calls for many vantage points or positions to be consolidated (from teachers to district technology coordinators)—an immense task for which many school personnel are not well suited or properly trained. enGauge also does not offer much guidance about how to get started and ensure success; in general, the fairly trite advice features five steps: *Learn* (investigate the rationale and history behind the 21st century skills); *Advocate* (set a goal worth striving for); *Focus* (find the fit for your school; make the commitment); *Activate* (try things; make necessary system changes; get everyone ready); and *Impact* (implement with integrity).

IMPLICATIONS OF DEFINITIONAL CONFUSION

It is evident that definitional confusion about technological literacy undermines—indeed, compromises—rigorous evaluation. While the previous section presented an array of conceptual/classification schemes to which evaluators might turn to focus their efforts, they are often unable to do so. Practically speaking, evaluators are often selected *after* project funding—and must "comply" (at least in part) with the designs on which original awards were predicated. Unfortunately, those not at liberty to cast technological literacy in the light they see fit may find that all phrases of the evaluation process are negatively affected—design, implementation, analysis, reporting, and next-steps decision making (Fitzpatrick, Sanders, & Worthen, 2004; Russ-eft & Preskill, 2001). Promising programs or projects may be modified or even terminated without cause or reason, while initiatives with severe limitations or shortcomings may be encouraged to continue without change.

Thus, this chapter closes with a brief discussion of two critical areas that may be seriously affected by definitional confusion about technological literacy—issues about which the educational community (policymakers, financial auditors, school staff, parents and students, vendors, future employers) should be fully aware.

- **Stakeholder Engagement:** The research suggests that stakeholder involvement is key to assessment quality. The focused evaluator recognizes and empowers those affected by or vested in technology initiatives, and actively seeks ways to avoid their misidentification. Whether charged with determining a program or product's merit or worth (the *summative* stance), suggesting program or product improvements (the *formative stance*), or providing general knowledge about evaluation practices and the importance of systematic data collection (the *informative* stance), the well-intentioned evaluator cannot proceed unless key stakeholders (those with verifiable interest in the initiative under investigation, and able to take advantage of the results) are on board regarding the issues to investigate and the methods for doing so.
  Definitional confusion about technological literacy can cause even astute evaluators to misidentify or ignore groups with critical input , and/or to misinterpret the context/environment in which the initiative unfolds—and so misjudge such dimensions as *formality*, *complexity*, and *information need* (Mohan et al., 2002). They may also "miss" or downplay the ethical complexities that underlie their efforts—including shifting priorities; inad-

vertent exclusion or stereotypic portrayals of certain individuals and groups; costly activities with controversial or ill-defined outcomes; high-profile staff whose futures depend on success (positive results); and "unnatural" ties to projects already in place or on the drawing board (Fitzpatrick et al., 2004). Finally, they may overlook critical organizational dynamics and fail to establish trusting and respectful relationships with key individuals and groups.

- **Evaluation Planning and Design:** By and large, evaluators agree that successful studies are guided by established frameworks or approaches—not by data collection methods (Russ-eft & Preskill, 2001). The choice of approach is predicated on several factors—conceptual, logistical, operational, managerial, technical. Most technology initiatives are long term in nature—forcing the astute evaluator to select a single or mixed approach that attends to programs/projects that unfold in phases or life stages *and* accommodates both formative and summative issues of interest. The evaluator must recognize (and address) potential constraints—including limited access to resources/information sources, tight timelines, a limited budget, potential interference with day-to-day functioning, the extent to which anonymity and confidentiality can be assured, and so forth (Fitzpatrick et al., 2004).

  Definitional confusion about technological literacy can lead to critical errors in judgment about evaluation design; in addition, it can spur problems with sampling; selection of instruments or collection methods; when data should be collected and how often or at what point(s) in time; selection of analytical methods; how best to triangulate perceptions (beliefs, values, attitudes, convictions) with data more focused on knowledge or skill attainment; data control; and report generation and dissemination (Fraenkel & Wallen, 2003).

REFERENCES

Association of College & Research Libraries, American Library Association (2000). *Information literacy competency standards for higher education.* Retrieved May 27, 2004, from http://www.ala.org/ala/acrl/acrlstandards/standardsguidelines.htm

Berrett, J. (2003, April). Technological literacy: Researching teaching and learning in the K-12 setting. *Proceedings of the 2nd AAAS Technology Education Research Conference,* Washington, DC. Retrieved May 26, 2004, from http://www.project2061.org/meetings/technology/tech2/Berrett.htm

Bugliarello, G. (2000). Reflections on technological literacy [electronic version]. *Bulletin of Science, Technology & Society, 2,* 83-39.

Bybee, R. W. (2003). Fulfilling a promise: Standards for technological literacy [electronic version]. *The Technology Teacher, 62*(6). Retrieved June 12, 2004, from Academic Search Elite (EBSCO) database.

Dugger, W. E. Jr. (2001). Standards for technological literacy. *Phi Delta Kappan, 82*(7), 513-517.

Ertmer, P. (2003). Transforming teacher education: Visions and strategies. *Educational Technology Research & Development, 51*(1), 124-128.

Fitzpatrick, J. L., Sanders, J. R., & Worthen, B. R., (2004). *Program evaluation: Alternative approaches and practical guidelines* (3rd ed.). Boston: Allyn & Bacon.

Fraenkel, J. R., & Wallen, N. E. (2003). *How to design and evaluate research in education* (5th ed.). Boston: McGraw-Hill.

Gagel, C. W. (1997). Literacy and technology: Reflections and insights for technological literacy. *Journal of Industrial Teacher Education, 34*(3). Retrieved June 8, 2004, from Academic Search Elite (EBSCO) database.

International Technology Education Association. (2000). *Standards for technological literacy: Content for the study of technology.* Retrieved June 6, 2004, from http://www.iteawww.org/TAA/Publications/STL/STLMainPage.htm

Johnston, J., & Toms Barker, L. (2002). Introduction. In J. Johnston & L. Toms Barker (Eds.), *Assessing the impact of technology in teaching and learning: A sourcebook for evaluators* (pp. 1-7). Ann Arbor: Institute for Social Research, University of Michigan.

Means, B., & Haertel, G. D. (2004). Introduction. In B. Means & G. D. Haertel (Eds.), *Using technology evaluation to enhance student learning* (pp. 1-8). New York: Teachers College Press.

Mohan, R., Bernstein, D. J., & Whitsett, M. D. (2002). Editors notes. *New Directions for Evaluation, 95*, 1-4.

North Central Regional Educational Laboratory, & Metiri Group. (2003). *enGauge: 21st century skills for 21st century learners.* Retrieved June 4, 2004, from http://www.ncrel.org/engauge

Rafferty, C.D. (1999). Redefining literacy: Literacy in the information age [electronic version]. *Educational Leadership, 57*(2), 22-25. Retrieved June 15, 2004, from http://www.ascd.org/

Roberts, L.G. (2003). Forward. In G.D. Haertel & B. Means (Eds.), *Evaluating educational technology: Effective research designs for improving learning* (pp. vii-x). New York: Teachers College Press.

Roblyer, M. D., Edwards, J., & Havriluk, M. A. (1997). *Integrating educational technology into teaching.* Upper Saddle River, NJ: Merrill.

Rockman, S. (2004). Preface: Positioning evaluation and research within PT3 projects. *Journal of Technology and Teacher Education, 12*(2), i-vii.

Russ-eft, D., & Preskill, H. (2001). *Evaluation in organizations: A systematic approach to enhancing learning, performance, and change.* Cambridge, MA: Perseus Publishing.

Virtual Teacher Center. (2002a). *About literacy.* Retrieved June 3, 2004, from http://www.virtualteachercentre.ca/literacy/default.asp

Virtual Teacher Center. (2002b). *Literacy definitions.* Retrieved June 3, 2004, from http://www.virtualteachercentre.ca/literacy/Resources/definitions.html

Wonacott, M. E. (2001). *Technological literacy* (ERIC Digest No. 233). Columbus, OH: ERIC Clearinghouse on Adult Career and Vocational Education. (ERIC Document Reproduction Service No. ED459371).

ENDNOTES

[1] The Technology Innovation Challenge Grants (TICG) program supported innovative uses of technology in the K-12 setting. For the most part, funded projects focused on *infrastructure*, *professional development*, and/or *curricular reform.*

See http://www.ed.gov/programs/techinnov/index.html for further details. Preparing Tomorrow's Teachers to Use Technology (PT3) was designed to ensure that *newly credentialed* teachers were well prepared to use technology in innovative ways. Among the "techniques" or strategies employed were faculty development, course restructuring, certification reform, alternative assessment methods, and unique student/veteran teacher configurations. Eligibility for funding was *relatively* similar (though not identical) to the requirements imposed for TICG. See http://www.ed.gov/programs/teachtech/eligibility.html for further details.

2 See http://www.apple.com/education/k12/leadership/acot/.

3 See http://www.apple.com/education/k12/leadership/acot/history.html.

4 On November 5, 2002, President Bush signed the Education Sciences Reform Act into law, which produced a new organization—the Institute of Education Sciences; see http://www.ed.gov/about/offices/list/ies/index.html?src=mr.

5 The comprehensive Star School initiative was one of the first technology programs ever funded by the Department of Education (see http://www.ed.gov/programs/starschools/index.html). Although limited now to continuation grants only, the original program funded eligible telecommunications partnerships that—via distance learning technologies—provided: (a) instructional programs (in mathematics, science, foreign languages, literacy, and vocational education) to underserved students, (b) professional development to their teachers, and (c) technical assistance. *Underserved* students included those who were socially or economically disadvantaged, illiterate/non-readers, limited English proficient, and learning or otherwise disabled. More than one million students and their teachers in all 50 states and territories have participated in a Star School program.

6 Johnston, J., & Toms Barker, L. (Eds.). (2002). *Assessing the impact of technology in teaching and learning: A sourcebook for evaluators.* Ann Arbor, MI: Institute for Social Research, University of Michigan.

7 See http://www.ed.gov/about/offices/list/os/technology/index.html.

8 A key tenet of *No Child Left Behind* (NCLB) is *accountability*; legislation specifically states that schools: "...must describe how they will close the achievement gap and make sure all students, including those who are disadvantaged, achieve academic proficiency." As part of that process, they must produce state and school district report cards (annually) that discuss progress toward well-articulated goals and objectives. Schools that do not attain their progress benchmarks "must provide supplemental services, such as free tutoring or after-school assistance; take corrective actions; and, if still not making adequate yearly progress after five years, make dramatic changes to the way the school is run."
Associated with accountability is scientific rigor. Simply put, instructional programs/interventions must be grounded in *scientifically based research*; there must be *reliable evidence* that a program or practice is effective. Experimental or quasi-experimental research designs—specifically those that feature randomized assignment to groups—are preferred. In fact, NCLB encourages researchers and practitioners to test educational practices much as scientists might "assess the effectiveness of medications." See http://

www.ed.gov/nclb/landing.jhtml for further details.

[9] See http://english.mansfield.ohio-state.edu/writing.

[10] See http://www.curriculum.wa.edu.au/support/cd/CF/fwk05a.htm.

[11] See the American Heritage Dictionary of the English Language, 4th edition (2002) at http://dictionary.reference.com/search?q=literacy.

[12] See http://cnets.iste.org/ncate/, which details the complex relationship between the International Society for Technology in Education (ISTE)—which promotes its National Educational Technology Standards (NETS)—and the National Council for Accreditation of Teacher Education (NCATE)—the accrediting body for more than 80% of teacher-preparation programs.

[13] See http://cnets.iste.org/teachers/.

[14] See http://cnets.iste.org/.

[15] See http://pt3.org/stories/index.html for examples of NETS infusion into accredited credentialing programs—targeting both preservice candidates and teacher educators.

[16] See http://cnets.iste.org/nets_overview.html.

[17] See http://cnets.iste.org/teachers/t_esscond.html.

[18] See http://cnets.iste.org/teachers/pf/pf_achieve-equit-access.html, for example.

[19] See http://www.mff.org/publications/publications.taf?page=158.

[20] According to its Web site, ITEA is the largest professional educational association and information clearinghouse devoted to enhancing technology education (across the K-12 spectrum). The Association's Technology for All Americans Project (TfAAP) was largely funded by the National Science Foundation (NSF) and the National Aeronautics and Space Administration (NASA)—and unfolded in three phases. Phase 1 focused on the philosophical underpinnings for the study of technology in K-12 classrooms, while Phase 2 resulted in standards for the study of technology. Phase 3 focused on operationalizing the standards—addressing professional development, program enhancement, and assessment. See http://www.project2061.org/default_flash.htm and http://www.iteawww.org/TAA/Publications/STL/STLMainPage.htm for further details.

[21] See http://www.nap.edu/books/0309053269/html/index.html, for example.

[22] See http://www.project2061.org/default_flash.htm.

[23] Coupled with the recently published Advancing Excellence in *Technological Literacy*. The standards tackle key facets of the instructional process: student expectations, assessment strategies, professional development guidelines, and program standards.

[24] See http://ctap.k12.ca.us/whatis.html.

[25] Eligibility for the Level 1 credential is predicated, in part, upon successful completion of an accredited teacher preparation program. It has a five-year limit and is nonrenewable. [Other terms and conditions apply—including minimum scores on different qualifying assessments.]
The Level 2 credential is permanent, but it must be renewed every five years. Renewals are predicated on the candidate meeting specific conditions or qualifications in several areas (or having them waived). No qualifying assessments are associated with the Level 2 credential.

[26] See http://www.ceoforum.org/.

[27] Three key or guiding questions serve as an advance organizer for the StaR Chart (both versions): *Is technology (at the school or district level) being used in ways that ensure the best possible teaching and learning? What is a school or district's "technology profile?" What areas (at the site/district level) should be targeted to ensure effective integration?*

[28] NCREL is one of 10 Regional Educational Laboratories (RELs). With support from the U.S. Department of Education, Institute of Education Sciences (formerly the OERI), the Laboratories work as vital partners with state and local educators, community members, and policymakers in using research to tackle the difficult issues of education reform and improvement. NCREL specializes in the educational applications of technology, providing research-based resources and technical assistance to *any* interested educators, researchers, and policymakers (but focusing, in particular, on communities in Illinois, Indiana, Iowa, Michigan, Minnesota, Ohio, and Wisconsin). See http://www.ncrel.org/info/about/ and http://www.relnetwork.org/about.html for further details.

[29] See http://www.metiri.com/.

[30] See www.ncrel.org/engauge/framewk/index.htm for full descriptions of each essential condition and its associated indicators.

Chapter XIII
Systemic Innovations and the Role of Change–Technology: Issues of Sustainability and Generalizability

Chee-Kit Looi
National Institute of Education, Singapore

Wei-Ying Lim
National Institute of Education, Singapore

Thiam-Seng Koh
National Institute of Education, Singapore

Wei-Loong David Hung
National Institute of Education, Singapore

ABSTRACT

There has been increasing interest in issues of sustainability and scalability concerning educational innovations and reforms. In Singapore, the Ministry of Education in the first IT MasterPlan for Education has equipped all schools with adequate IT infrastructure and teacher training. After seven years into this investment, the "take-up" of technology for learning has been generally at the basic level. Three challenges lie ahead in the effective integration of IT into the curriculum for meaningful student learning: (1) how can schools embrace technology where school practices, curriculum, pedagogy, teachers' and students' beliefs are aligned to its effective use; (2) how can the process of change be facilitated systematically in schools such that these changes leverage on IT as a catalyst to enhance learning; and (3) how can policy-wide initiatives be set in place to enable schools, teachers, and teacher-training institutes to be fully aligned in order to enact systemic innovation change? This chapter discusses these issues as part of the research efforts arising from the Learning Sciences Lab (LSL).

BACKGROUND AND MOTIVATION

In the past year, centers for the science of learning (e.g., www.learnlab.org, http://life-slc.org/) have been launched in the United States. These centers funded by the National Science Foundation seek to produce knowledge on effective ICT-enabled educational practices, through varying research foci such as capacity building and technologies that facilitate pedagogical changes. Besides pursuing research to understand how learning occurs in formal settings, informal settings, and even at the implicit level such as at the neurological plane, most centers are also concerned with the sustainability and scalability of effective educational practices.

The background to such efforts in the U.S. and elsewhere is that the investments in learning and how technology engages learning through innovations have not been widespread. A quick sweep across the world indicates that only a small percentage of teachers embraces technology effectively and uses it as a means for deep or engaged forms of learning. In other words, the numerous conferences, journals, and manifold dissemination of "how technology enables learning" do not measure up to the scale of implementation in practice.

There is thus a united call for greater reform—how we can be more effective as researchers in seeking transformation in ICT-enabled pedagogy which are more pervasive in our schools. What are some of the conditions and issues faced, and how can we leverage on the current successes to chart future directions? We recognize that technology is only one tenet in the complex system of education that would include societal needs, policies, curriculum, pedagogy, practices, epistemic beliefs, skills, and others. It is not just a technological enabler in learning, but the only one tenet that affords the catalytic effect to trigger change due to its ability to be adapted and enacted across cultures and contexts. For the purpose of this chapter, the term *change-technology* will refer to technologies with such catalytic effect, differentiating it from the general usage of the word technology.

In Singapore, the Ministry of Education has invested millions of dollars into the IT MasterPlan for Education. In the first MasterPlan for Education, all schools have been equipped with the appropriate networked-based IT infrastructure to enable IT-based learning in the schools. After five years into this investment, the take-up of technology for learning has been generally at the basic level, with only a small number of schools experimenting and innovating with the use of IT in teaching and learning. The real challenge in going ahead still lies in whether the IT adoption in schools leads to meaningful engaged learning. Thus, there are three challenges that need to be addressed: (1) how can schools in Singapore embrace technology where school practices, curriculum, pedagogy, teachers' and students' beliefs are in full alignment to the effective use of technology; (2) how can the process of change be facilitated systematically in schools such that these changes leverage on IT as a catalyst to enhance learning; and (3) how can the larger scheme of things such as policy-wide initiatives (at the Ministry level), community-parental concerns, be set in place to enable schools, educational professionals, and practitioners to be fully aligned in order to enact this innovation for sustainability?

This chapter describes the efforts of the Learning Sciences Lab (LSL) in the National Institute of Education in tackling the challenges and issues faced concerning systemic innovation in schools which are enabled by technology. The aim of this chapter is to describe and understand some of the challenges underlying sustainability and scalability in technology innovations in the context of LSL's efforts. LSL is proposed as an experimental lab where ideas and concepts related to learning interactions and teaching pedagogies can be prototyped and implemented in classrooms and schools (Looi, Hung, Bopry, & Koh, 2004, p. 92).

At this stage, while we recognize that a systematic approach has to be adopted if change-technologies are to be taken up meaningfully, it remains a challenge as to how such innovations can be scaled up, potentially extending across schools over a sustained period. The issue of scalability and sustainability is an arduous and complex one. Alongside the U.S. centers in tackling these common issues, LSL hopes to make a contribution to the field, particularly in the context of the Asia-Pacific region where Singapore is situated. The above background serves as the motivation for the rest of this chapter.

SYSTEMIC INNOVATIONS FOR EDUCATIONAL REFORM

From a business prospective, innovation is about pursuing radical new business opportunities, exploiting new or potentially disruptive technologies, and introducing change into the core concept of a business (Wolpert, 2002). Fishman, Marx, Blumenfeld, Krajcik, and Soloway (2002), on the other hand, describe innovation as one that emerges from design and design experiment research with technology and pedagogy in classroom settings. It is about developing technologies that are integral in core school curriculum with linkages to the professional development of teachers. We are particularly interested in technology-enabled systemic innovation in LSL.

In our opinion, systemic innovation is a combination of the above two perspectives. It is about the enduring pursuit of seeking effective scalable ICT-enabled pedagogies, and our main approach is by introducing change in the way research is being conducted. Systemic innovations as a result of design experiments should bear the potential of being adapted, enacted, and reenacted across contexts with the goal of producing deep learning. It is a change-technology that reflects the cultural, political, and economic forces of their times, bearing close relevance to the reform efforts and the learners' needs (Cuban, 2001). The key issue here is the coherence of the epistemic orientation of the innovation with the reform efforts. The underlying philosophy of innovations must support and foster the newer kinds of pedagogies such as inquiry-based, problem-solving, and student-centered learning.

In tandem with the contemporary researches conducted in the learning sciences, change-technologies facilitating this process need to be grounded in theoretical underpinnings such as situated cognition, distributed cognition, and constructivism for them to be sustained. Studies of innovations have shown that successful change is always bottom-up, middle-out, and top-down (Dede, 1998). The drivers for bottom-up change are the learners themselves. Typically, students would generally be more motivated when given the opportunity to learn by doing, to collaboratively construct knowledge, and to experience relationships with others in the sociocultural setting such as counterparts from other institutions and experts in the field. Bottom-up change also comes from enlightened and insightful teachers who know when and how learners are experiencing a very different kind of learning experience such that it is apt for them to inculcate good skills, attitudes, and beliefs.

The broad educational goal arising from the 1970s and 1980s of equipping our young with the necessary skills and knowledge to be a productive workforce for industrialization is no longer as relevant as it was before. There is now a heightened need for customized learning, in helping learners to excel in what they can do best, and in equipping them with the ability to think and discern, to be innovative and enterprising for the new economy. While the community at large, school leaders, and practitioners recognize this societal need, the notions of how learning take place and the strategies to be adopted to bring about the change remain diverse. Some believe in increased rote learning, providing additional application exercises and practice, sometimes with the use of multimedia to capture and maintain learners' attention. Others speak the constructivist jargons, by having groupwork, learning by doing without a change in their underlying belief towards teaching and learning. The lack of internal coherence between practice and belief in this case, resulted in frustrations and confusion, lending the classroom back to its original didactical form. Yet others simply refuse to change.

Educational researchers, in their efforts to generalize their findings for adoption across varying school contexts, often overlook the need to explicate their theoretical underpinnings, philosophy, and epistemic orientations on how learning takes place (Davis & Sumara, 2003). As such, practitioners in an attempt to take on constructivist approaches, without an in-depth understanding of the learning philosophy, often find themselves in an instructional dilemma when faced with a classroom with all its complexities. The resultant learning process is often unsatisfactory, coupled with phrases such as "I try to run a constructivist classroom" to describe the phenomenon (p. 129).

At LSL, we do not seek to obtain a unified acceptable notion of learning. Rather, in acknowledgment of the divergent perspective, we see three sets of pertinent issues that need addressing: (1) in the designing for learning, teachers must be equipped with the right epistemologies: (2) meaning-making for learners should be facilitated with learning technologies; and (3) the success in scalability and sustainability largely hinges on the alignment with school polices and practices (Looi et al., 2004).

ISSUES OF SUSTAINABILITY

Currently, at the top of the agenda in the setting up of the lab is how to go about conducting research such that ICT-enabled pedagogies enjoy a reasonable sustained period in schools. In the past, funded research projects, for instance the investigation of the impact of CSCL technology (Knowledge Forum or KF) on students engaging in scientific inquiry, was implemented with well-thought-out customized strategies across schools, coupled with comprehensive training for the teachers involved. However, like with many other projects, they dealt little with the larger scheme of things such as impact on formal students' assessment, schools' core curriculum, and the ongoing professional development for teachers (Ibrahim & Tan 2004; Tan, Hung, & So, 2005; Tan, Yeo, & Lim, 2005; Teo, Chiam, & Ng, 2004). Thus, without an expansive long-term approach to implementation, such meaningful use of technology often does not enjoy a sustained period in schools after the completion of the research projects. Typically the technology innovations are left to be run only on the school's off-peak periods, and are fully dependent on the teachers' enthusiasm to sustain it. So what must we do differently from the past?

Our sense of it is that it is an endeavor faced by many like-minded colleagues around the world summarized in Sabelli and Dede's writing (n.d.):

Decades of funded study that have resulted in many exciting programs and advances have not resulted in pervasive, accepted, sustainable, large-scale improvements in actual classroom practice, in a critical mass of effective models for educational improvement, or in supportive interplay among researchers, schools, families, employers, and communities.

Researchers have repeatedly experienced the "death" of their ICT integration effects resulting from their withdrawal from the project. Reasons for a limited shelf life are multifaceted and perplexing. Evidences of repeated letdowns remain largely visible. Teachers' pedagogies in the classroom are still fairly didactic, occasionally spiced up by the use of multimedia. Policymakers and school leaders' key performance indicators remain at a statistical level.

This dilemma faced by all who engage in the work of systemic reform resides in the tension between the desire to scale effective practice on the one hand, and issues of localization and fit on the other (Honey & McMillan-Culp, 2000). The range of issues to consider can have widespread impact—from the running of the school's organizational policies and procedures, to practice in the classroom, modes of assessment, and even management of parental expectations. To bring about such a change requires strong leadership and shrewd management. Such daunting tasks may not seem palatable at the outset to many school and policy leaders.

According to Cohen and Ball (1999), this challenge can be analyzed in terms of *specifications* vs. *development*. Specification is the degree of details in which an innovation is described for school take-up. Some initiatives are general, consisting of goal statements that suggest a general direction. Others are more specified being clear about curriculum, intended teaching practices, and desired learning goals. Development refers to the provision of resources required to enact innovations such as learning philosophy, curriculum materials, professional development, and examples of teaching practice that can be educative for others. Specification without development requires teachers to figure out how to enact the innovation in their local setting, which can be a barrier to all but the master teachers. Development without specification provides resources for improvement, but not a clear picture as to what goals are to be attained. A take-away for us at LSL is to find the balance between these two tensions. Researchers need to be aware of both the complexities of the actual classroom conditions that practitioners face, and the epistemic beliefs, pedagogic skills, and the level of mastery of the teachers. This balance is not a stone-cast rule in schools' implementations. It requires customization to the school's sociocultural context. Hence, the direction of being in close partnership with the Ministry of Education, bridging the lab with practitioners in conducting research, indicates a start in the right direction.

There is a need to differentiate between maintenance and sustainability. Sustainability, unlike maintenance, has the element of self-rejuvenation—the ability to evolve. This is an important criterion for innovations to be sustained over time. It is a common phenomenon for innovations to 'fade away' when the initial hype subsides or when changes in sociocultural conditions such as leadership or policies come into effect. Leaders and stakeholders must find ways to adapt innovations to new conditions without compensating the underlying intent and learning philosophy of the original design. One of the major contributions of the research conducted by Century and Levy (2002), drawn from a study of nine communities with K-6 science education programs, is the definition of sustainability and the contextual factors that affect it.

According to Century and Levy (2002), changes must be contained within which the core intent and philosophy of the original reform is still reflected. They have thus defined sustainability to be:

The ability of a program to maintain its core beliefs and values and use them to guide program adaptations to changes and pressures over time. (p. 4)

To do so, consideration has to be given to the influential factors (see Figure 1) at any levels of the school system in the progression of each innovation, from establishment, maturation, to evolution. At each of these three phases, an innovation has particular goals and particular strategies for achieving those goals. To be sustained, goals must be realized at different levels of the school system, which require multiple strategies to be employed simultaneously. Thus at any point in the development of an innovation, attention must be given to the influential factors at any of the levels in the system (see Figure 2) such that the stakeholders' orientation is accurate.

The Letus community (www.letus.org) offers another perspective into the issue of sustainability. If an innovation is not usable, adoption is unlikely, not to mention sustainability or scaling up of technologies. In tandem with other literature, Blumenfeld, Fishman, Krajcik, Marx, and Soloway (2000) put forth that the three dimensions within school systems—school culture, capability of practitioners, and policy/management—are the key aspects to look into,

Figure 1. Factors summarized from Century and Levy (2002)—sustaining change: A study of nine school districts with enduring programs

Factors Related to the Concrete Element of a Program/Innovation	Factors Related to Influences on a Program/Innovation and Outcomes that Result from the Program/Innovation
Accountability	Implementation and Adaptation
Instructional Materials	Philosophy
Leadership	Critical Mass
Money	Perception
Partnerships	Quality
Professional Development	History—Origins and Longevity

Figure 2. Abstracted from Century and Levy (2002, p. 6)

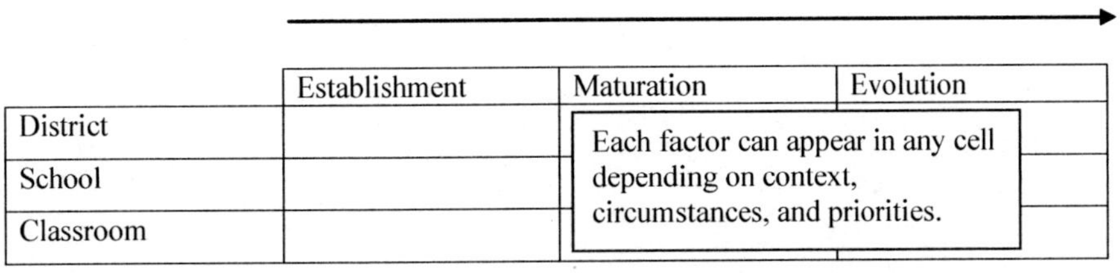

	Establishment	Maturation	Evolution
District		Each factor can appear in any cell depending on context, circumstances, and priorities.	
School			
Classroom			

thus the conceptualization of the usability framework. Together, the three dimensions—when arranged in the form of three axes originating form a common point—form the three-dimensional space known as the usability cube. An innovation can hence be placed within the space for gap analysis. Such an analysis helps researchers attempting to do such work in understanding the opportunities and processes within which one needs to look for sustained success in schools.

Transplanting an initiative that has been successful in one school to another is difficult. Tyack and Cuban (1995) remind us in their historical examination of school reform: despite more than a century of reform efforts, "educators have variously welcomed, improved, deflected, co-opted, modified, and sabotaged outside efforts at reform" (p. 7). If it were easy to transfer effective practices from one setting to another, we would have done so long ago. Each school has its own unique culture, and when innovation and change-processes are implemented carelessly, it can potentially run into the danger of becoming entirely transformed—diverging sharply from the initial intentions.

A strategy proposed by Hargreaves and Fink (2000) is to use model successes to reculture as well as to restructure schools—adopt the philosophy and intentions into the existing culture of adventurous volunteer schools before enacting the structures for change. The goal of systemic innovations must first be established in the larger cultural context such that it becomes an "entire continent of change" (Hargreaves, Earl, & Ryan, 1996). In fact, they argued that educators should go beyond the policymakers by making their practice and improvements visible to the public such that a broad social movement for large-scale, deep, and sustainable transformation can be created (Hargreaves, in press). Concomitant to this notion, Cuban, in an interview with John O'Neil (2000), highlights that the innovations that have the best chance of sticking are those that have constituencies growing around them. An example he cited is that of special education. When reform efforts reflect some deep-rooted socio concern such as preparing autistic children to lead fulfilling lives, it will be widely accepted over a sustained period of time.

Figure 3 is an attempt by Century and Levy to map out the complexity of sustainability and

Figure 3. Layered complexity of innovation (abstracted from Century & Levy, 2002, p.18)

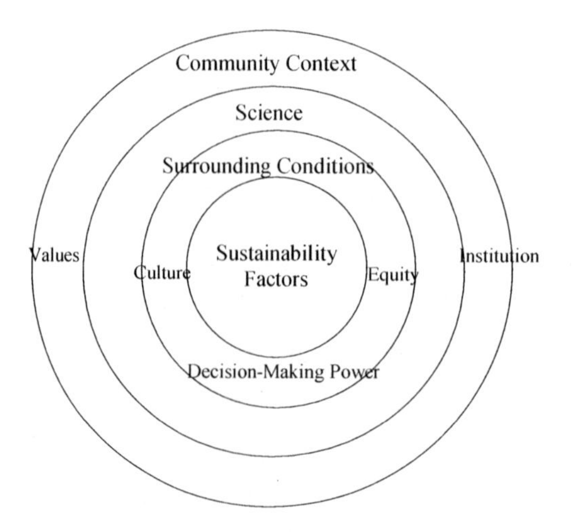

Figure 4. Five "P" levels of changes

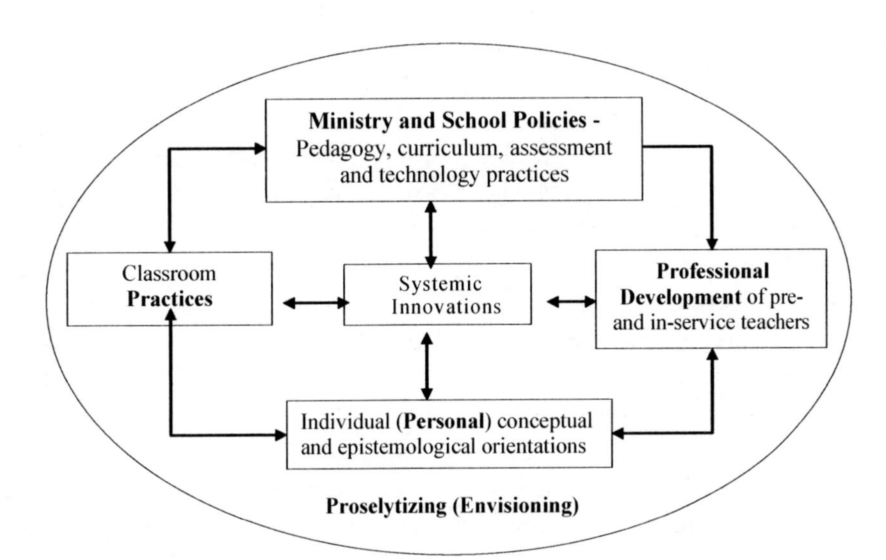

scalability in a way that appropriately represents the layered nature and complexity of their interactions. In the center sits the innovation and its influential factors. These factors are embedded in the larger surrounding conditions such as the school culture, decision-making power, and equity issues. The next ring represents the content area (which is science in their scope of study) and other correlated issues such as pedagogy, content knowledge, status of science learning, beliefs of teachers, and so forth. Finally, in the broadest ring, there is the community and its corresponding sociocultural issues exerting pressure on the innovations and the stakeholders' decisions.

OUR PERSPECTIVES IN APPROACHING SUSTAINABILITY AND SCALABILITY

It is clear that the research work we do at LSL cannot be undertaken in linearity but planned and implemented from a systems perspective, factoring in all pertinent components of the reform system with a design-research methodology. Under this perspective, there must be a rich interplay between the macro scheme of things and the micro details, bringing together a design focus and the critical assessment of the design elements. It is in the hope that repeated design experiments, implemented in real-life settings, can help inform the theory of learning in-situ and improve educational practices in schools. Towards this end, we know we have to work with the curriculum people, the assessment people, school educators, and the teachers at different levels in order to overcome the issues highlighted earlier in this chapter.

In every reform effort, the catalytic innovation forms the core of the system. It could be a result of the research efforts by the change-agents with support from the ministry, with the goal of achieving deep learning in our learners. The innovation, when implemented, should facilitate change-processes at four dimensions.

First, this change occurs at the sociocultural plane: change in school policy practices, social interactions, learning relationships within online and face-to-face settings, and other patterns of collaboration. Second, this change must be seen at the level of classroom practices, strategies, and pedagogies where possibly transitions are seen from teacher-centered to learner-centered approaches. The third, which is related to the second, concerns changes at the professional development training-education programs of pre- and in-service teachers. Fourth, change occurs at the person or individual plane: conceptual and epistemological change, problem solving and meta-cognitive dispositional change, and change in personal beliefs about how learning and teaching occurs through technology. We plan to leverage on our advantage in bringing alignment along all four dimensions to have a better chance of real systemic innovations occurring in the Singapore school system.

Our strategy in facilitating such change-processes lies in our tight collaboration with the Educational Technology Division (ETD) of the Ministry of Education, which we believe is one of the necessary requirements for systemic and sustained interventions, and in building up capacity of our people to spread the meaningful use of technology in the classroom. In partnership with the Ministry, appropriate policy and decisions can be made that can have a direct bearing on the schools' sociocultural environment. Such informed policy-wide initiatives, once set in place, enable schools, teachers, and teacher-training institutes to be fully aligned to effect change. At the same time, with an effective professional development model in place, changes can slowly be made with regards to classroom practice as well as personal epistemology of the practitioners.

We therefore see that a very important objective of LSL is capacity building, inculcating the right beliefs and dispositions in our pre-service and in-service teachers, heads of departments and school leaders, and graduate students, to have expansive mindsets towards students' learning. They would recognize that while they stick to what works for several years in preparing our students to do well in examinations, they are also cognizant of the competencies and skills that students need to engender to live in a very different world, often characterized by complex, ambivalent, and relativist attributes. In order words, educators are more willing to invest in inquiry and reflection when they see it as relevant to problems of practice and the realities of their schools and classroom. This reflection often leads to self-initiated changes in how they support their students' learning (Goldman, 2005).

When these "change agents" go out or back to schools, or go out to inspire and train more teachers, we hope to see a profusion effect, albeit slow in building, that will see a significant percentage of them pursuing their ideals and dreams about educational innovation, setting up good practices, and inspiring others. This process, according to Goldman, builds a culture of collaborative inquiry that gradually spreads throughout the system, expanding its capacity as a whole. In other words, change is initiated, sustained, and carried through systems by people (2005, p. 71).

The four recommendations which we have highlighted above can be summarized with five "P" levels in changes, namely (see Figure 4):

1. proselytizing,
2. policy,
3. practices,
4. professional development, and
5. personal.

These five levels of changes presume that change and reform is necessary to begin with. This presupposes a vision for change, and this vision or proselytizing arises from the need to transform pedagogies in the schools because of the kinds of dispositions and competencies required in the 21st century. This vision then leads to required changes in policy structures such as how teachers are rewarded and how students are assessed for their competencies. As such, classroom pedagogies and practices need to emphasize skills, knowledge, and dispositions that develop our learners for the future. But teachers need capacity building, which is translated into professional development for such work. And as we foster a social-cultural milieu befitting such learning, personal epistemologies, and beliefs of school leaders, teachers and learners begin to gradually change towards that overarching vision. That overarching vision may have been a clear target in the beginning, but as all stakeholders tread on this journey, the vision is made clearer—"the path is made by walking." In a truly social constructivist fashion, the community involved in this effort clarifies the vision for themselves.

CONCLUSION

In this chapter, we look at sustainability and generalizability issues of innovation enabled by ICT, and explore how we in LSL would pursue a research agenda that can inform educational reform and lead to better chances of sustainability and scalability. It is acknowledged that such transformation in the educational landscape requires enduring effort and perseverance. It incorporates diverse, multidisciplinary skills and knowledge with no clear starting point as elements in the system are inter-linked. It is thus important that the goals remain clear in the midst of this perplexing complexity. As Hargreaves and Fink summarized:

> *Ultimately, only three things matter about educational reform. Does it have depth: does it improve important rather than superficial aspects of students' learning and development? Does it have length or duration: can it be sustained over long periods of time instead of fizzling out after the first flush of innovation? Does it have breadth: can the reform be extended beyond a few schools, networks or showcase initiatives to transform education across entire systems or nations? Successful school reform is a Picasso, not a Rembrandt. It approaches change not in one or two dimensions, but, like a cubist painter, views it from all three. (Hargreaves & Fink, 2000,* p. 30)

Thus, leveraging on the progress we made thus far in education and educational research in Singapore, having a national IT agenda (IT Masterplan I and II) supported by close LSL-ETD collaboration, and creating the capacity in our education workforce and leaders to appreciate the breadth and depth of learning from multidisciplinary perspectives rather than be fixated on traditional views of learning, are moves set in the right direction. Not only do they address the three issues raised earlier in this chapter, together with an alignment of efforts with various stakeholders at different levels of educational organization, we hope to see the fruition of transferable workable processes and technologies supported by research findings to at least a few schools and continuous improvement through design research on the conditions for fostering sustainability and scalability.

REFERENCES

Blumenfeld, P., Fishman, B., Krajcik, J. S., Marx, R. W., & Soloway, E. (2000). Creating usable innovations in systemic reform: Scaling-up technology-embedded project-based science in urban schools. *Educational Psychologist, 35*(3), 149-164.

Century, J. R., & Levy, A. J. (2002, April 3). Sustaining change: A study of nine school districts with enduring programs. *Proceedings of the Annual Meeting of the American Educational Research Association.* Retrieved April 10, 2005, from http://cse.edc.org/work/research/rsr/default.asp

Cohen, D.K., & Ball, D. L. (1999). *Instruction, capacity, and improvement* (CPRE Research Report Series No. RR-043). Philadelphia: University of Pennsylvania Consortium for Policy Research in Education. Retrieved March 30, 2003, from http://www.cpre.org/Publications/rr43.pdf

Cuban, L. (2001). Answering tough questions about sustainability. *Proceedings of the 1st Virtual Conference on Local Systemic Change.* Retrieved April 10, 2005, from http://sustainability.terc.edu/index.cfm/keynote

Davis, B., & Sumara, D. (2003). Why aren't they getting this? Working through the regressive myths of constructivist pedagogy. *Teaching Education, 14*(2), 123-140.

Dede, C. (Ed.). (1998). Learning with technology. *1998 ASCD Yearbook.* Alexandria, VA: Association for Supervision and Curriculum Development.

Fishman, B., Marx, R., Blumenfeld, P., Krajcik, J.S., & Soloway, E. (2002). Creating a framework for research on systemic technology innovations. *Journal of the Learning Sciences.*

Goldman, S. R. (2005). Designing for scalable educational improvement. In C. Dede, J. P Honan, & L. C. Peter (Eds.), *Scaling up success: Lessons learned from technology-based educational improvement.* San Francisco: Jossey-Bass.

Hargreaves, A. (In press). *Beyond anxiety and nostalgia: Building a social movement for educational change.* Phi Delta Kappan.

Hargreaves, A., Earl, L., & Ryan, J. (1996). *Schooling for change.* London: Flamer Press.

Hargreaves, A., & Fink, D. (2000). The three dimensions of educational reform. *Educational Leadership, 57*(7), 30-34.

Honey, M., & McMillan-Culp, K. (2000). Scale and localization: The challenge of implementing what works. In M. Honey & C. Shookhoff (Eds.), *The Wingspread conference on technology's role in urban school reform: Achieving equity and quality* (pp. 41-46). Racine, WI: The Joyce Foundation, The Johnson Foundation, and the EDC Center for Children and Technology.

Ibrahim, A., & Tan, S. C. (2004). Computer-supported collaborative problem solving and anchored instruction in a mathematics classroom: An exploratory study. *International Journal of Learning Technology, 1*(1), 16-39.

Looi, C. -K., Hung, D., Bopry, J., & Koh, T. -S. (2004). Singapore's Learning Sciences Lab: Seeking tranformations in ICT-enabled pedagogy. *Educational Technology Research & Development, 52*(4), 91-99.

O'Neil, J. (2000). Fads and fireflies. The difficulties of sustaining change. *Educational Leadership,* (April), 6-9.

Sabelli, N., & Dede, C. (n.d.). Integrating educational research and practice: Reconceptualizing the goals and process of research to improve educational practice. Retrieved March, 2005, from http://

www.virtual.gmu.edu/SS_research/cdpapers/integrating.htm.

Tan, S. C., Hung, D., & So, K. L. (2005, forthcoming). Fostering scientific inquiry in schools through science research course and computer-supported collaborative learning (CSCL). *International Journal of Learning Technology.*

Tan, S. C., Yeo, A. C. J., & Lim, W. Y. (2005). Fostering scientific inquiry skills with computer-supported collaborative learning: A case study. *Journal of Computers in Mathematics and Science Teaching.*

Teo, C. L., Chiam, C. L., & Ng, F. K. (2004). Fostering knowledge building among low achievers through technologies: A perspective from SingaporeFostering Knowledge Building among Low Achievers through Technologies: A Perspective from Singapore. *Teaching and Learning, 25*(1), 89-102.

Tyack, D., & Cuban, L. (1995). *Tinkering toward utopia: A century of public school reform.* Cambridge, MA: Harvard University Press.

Wolpert, J. D. (2002). Breaking out of the innovation box. *Harvard Business Review,* (August), 77-83.

Section II

Teaching and Learning with Technology

Chapter XIV
Digital Video in the K-12 Classroom:
A New Tool for Learning

Christopher Essex
Indiana University, USA

ABSTRACT

This chapter describes how digital video (DV) production can be integrated into K-12 education. It describes how recent technological developments in digital video technology provide an exciting new way for teachers and students to collect, share, and synthesize knowledge. It argues that DV can provide tangible, real-world benefits in student learning, as it requires that students work actively and collaboratively on authentic real-world tasks. Furthermore, DV projects can be tied to technology literacy and curriculum standards. The reader is guided through the stages of the DV production process, and specific K-12 projects are described. Guidelines for choosing hardware and software are provided. Parent and administrative concerns about the use of DV are discussed. The goal of this chapter is to provide K-12 teachers and administrators with the information they need to integrate digital video production into the curriculum.

INTRODUCTION

Until recently, video production had little place in K-12 education, except for the videotaping of sporting events and theatrical productions, and those tasks were generally reserved for the adults—teachers, media center directors—and perhaps a trustworthy secondary student or two with special training. Video equipment was scarce, fragile, and expensive, and the learning curve for mastering the process was a steep one. As the 21st century begins, however, advancements in video and computer technology have brought digital video (DV) production into the K-12 classroom, and teachers integrating this new technology into their curriculum can provide a powerful new tool for student learning. Digital video production is rapidly becoming an important element of the technological literacy curriculum for K-12 students.

This chapter will provide K-12 teachers and administrators with the background necessary to integrate digital video into the curriculum. I will begin with a discussion of recent technology developments that have led to the current availability of high-quality, low-cost DV hardware and software. A discussion of how digital video can affect the teaching and learning process follows, with a special focus on meeting technology literacy and other curriculum standards. The reader will then be guided through the four stages of the digital video production process: planning, shooting, editing, and delivery. I will provide guidance in the selection of DV hardware and software for the K-12 classroom. A number of real-world K-12 DV projects will be described. Finally, concerns related to implementing DV projects will be addressed. At the end of the chapter, three appendices will provide useful hints and strategies for creating high-quality DV presentations.

TECHNOLOGY DEVELOPMENTS

Since the 1990s, there has been a huge increase in the number of computers in K-12 classrooms. In 1995, the average school in the United States used 72 computers for instructional purposes; by 2001 the number of computers had nearly doubled, increasing to 124 per school (National Center for Education Statistics, 2003). This growth is not happening in the U.S. alone, either; in the UK, for example, the average number of computers per public school grew from 27 in 2000 to 34 in 2001 (British Educational Suppliers Association, 2002). Similar changes can be seen in Korean schools; there, the government has provided one personal computer for every five students (Korea Education & Research Information Service, 2003). Of course, not all of these computers are capable of displaying digital video (DV), but this number is growing; 67% of U.S. schools in 2003 had computers with DVD drives, a tremendous increase from just 5% in 2001 (Market Data Retrieval, 2003).

Camcorders and handheld video cameras are also becoming common in K-12 schools. A total of 63% of Maryland schoolteachers, for example, reported that they have access to camcorders (ORC Macro, 2002). A quarter of UK schools have at least one digital camcorder, and the cameras are being used. Nearly 50% of the schools stated that the camcorders were used at least once a week (British Educational Suppliers Association, 2002). Overall, access to the hardware required for digital video production is becoming increasingly available in K-12 schools.

The days when video editing required expensive specialist equipment and extensive training and practice are now behind us. Computer software now makes these activities, which were once the realm of professionals, literally child's play. Software such as Apple's iMovie and Microsoft's Windows Movie Maker 2, both of which come without additional cost as part of their respective maker's operating systems packages, simplify the video production process to such a degree that it can be taught in just a couple of sessions. Using this type of software, K-12 students can trim video clips; rearrange them; add music, narration, and sound effects; choose transitions between scenes; and create titles and subtitles.

Advances in camcorder technology have led to increased picture resolution, better color definition, the ability to record images at low light levels, and the ability to lessen the "shakiness" common to handheld videos. The availability of reasonably priced digital videotape formats (Digital8 and MiniDV) has allowed for video to be easily copied from one tape to another without any loss of quality, as was common when editing analog video formats such as VHS and Beta. In the past, even when there was good quality raw footage, the copies made after editing would show serious quality degradation. Because these digital tape formats are much smaller than previous analog formats, camcorder size and weight have been reduced, which makes the equipment more accessible to younger students. Firewire connectivity, a recent technical advance, allows video to be quickly and easily transferred from the camcorder to the computer through a specific type of cable and peripheral card. Finally, because of the rapid pace of technological developments, the cost of camcorders has been coming down, making them more affordable for budget-conscious school districts.

In the past, distributing video was a slow, costly, and laborious process. Duplicating videotapes, even when high-speed duplication and/or multiple VCRs were available, took a large amount of time, and the resulting videocassette copies were fragile, heavy, and costly to mail. Now, video productions can be distributed through sturdy, lightweight, and inexpensive CDs or DVDs, which can be mailed for little more than the price of a regular letter. Video can also be streamed over the Internet, essentially for free, if the school already has Internet connectivity, giving a nearly infinite number of viewers access to the video on a worldwide basis. Video productions can be shared with family and friends, wherever they might be located, or with other schools.

In summary, digital video production technology has finally matured to the level at which the hardware and software are available and appropriate for use at the K-12 level. Delivery technology has also kept pace, making it possible to share student work to a degree that has never before been possible. The fact that an elementary, middle, or secondary school class can create a video that can be seen by people all over the world is something that could not have occurred at any previous time in history.

DIGITAL VIDEO TECHNOLOGY AND LEARNING

Researchers from the North Central Regional Education Laboratory (Jones, Valdez, Nowakowski, & Rasmussen, 1995) argue that today's students learn best when they are involved in "engaged, meaningful learning and collaboration involving challenging and real-life tasks; and [using] technology as a tool for learning, communication, and collaboration" (p. 1). DV production activities are an excellent way of meeting these demands for a new type of instruction for today's learners.

Using DV tools, students can actively describe elements of the world around them and arrange these elements in a structure. Duffy and Cunningham (1996) noted that "video has moved out of the 'professional production' limitation and out of the television and movie theaters to become a general medium available for the viewing and analysis of any event" (p. 18). It can be argued that digital video production is a good fit with constructivist learning models. Jonassen, Peck, and Wilson (1998) state that video production tools require learners to be active producers of information, not just passive consumers. They must also be intentional learners and must work cooperatively to construct information using this media.

The student in a DV production project is an active learner, constructing audiovisual documents that represent information and ideas and the relationships between them. These relationships can be altered and explored, and information can be constantly added, deleted, or revised as the project develops. Unlike a textbook, or even a commercial videotape or DVD, a student-created video production requires the active participation of the learner, as he/she creates a visual structure to describe some event, idea, or narrative.

Video production provides an opportunity for students to participate in collaborative learning, and the taking on of authentic tasks. In the real world beyond the classroom, video productions involve a wide range of people taking on numerous roles, such as writer, producer, sound engineer, camera person, on-camera performer, and so forth. These professional roles can be emulated in the design of K-12 DV productions. Students can also be rotated among the various roles to gain the full variety of skills associated with the process. Students must work together collaboratively, use good communication skills, and practice their teamwork and decision-making skills in order to complete the project.

Video projects also provide an opportunity for students to actively gain media literacy skills. K-12 students are surrounded by audiovisual messages in today's culture, and too often they are passive receptors. By understanding the ways in which these types of communications are created, they are better prepared to deal with these messages appropriately.

Some specific benefits of digital video production activities are that students:

- author or coauthor knowledge, rather than being passive consumers;
- actively select from real-world sources of information;
- extend and build upon knowledge from textbooks, classroom discussion, and other traditional sources of course information;
- design a product that can be an archivable, reusable learning object:
- design a product that is not static, but can be revisited and revised at any time, based on new knowledge;
- collaboratively work together to produce a product, taking on a variety of roles; and
- can be expected to be highly motivated and engaged, because of the active nature of the process and typical student interest in audiovisual media.

DIGITAL VIDEO TECHNOLOGY AND TEACHING

Digital video technology allows teachers exciting new ways of communicating to and motivating their students. Teachers with DV skills and tools can enhance their teaching in a number of ways. The following list of opportunities for teacher projects is adapted from Baugh (2003):

- **Inspiration and Stimulus:** The teacher can show the students a short video clip to spark off conversation, or to give students inspiration for creative work. For example, recording a local poet reading her work might be a wonderful way to encourage English students to try to write a poem of their own.
- **Delivery of Content:** Today's tools make it easy for teachers to create DV presentations for their students that would be difficult to replicate in person. For example, an art teacher could scan in various works of art and use the pan and zoom features available in some DV editing tools to highlight various elements of an artwork. He or she could also record narration to go along with the visuals. While there is a fair amount of work in the

beginning, the DV project could be shown to students for years afterwards.

- **Recording of Classroom Events:** Most teachers have exciting educational events throughout the year such as spelling bees, science fairs, and reenactments of historical events, speeches, plays, musical presentations, and others, which deserve recording. These recordings can also be used to provide future students with a model for their work.
- **Assessment and Evaluation:** Video can record students at work, and provide evidence of how they interact with each other, including how much each participates in a project. It can also provide valuable input for measuring the effectiveness of a lesson or instructional strategy.
- **Recording Best Practices:** Teachers can record their classroom activities to share with other teachers, or with student teachers, to demonstrate best practices or new instructional strategies. These videos could also be part of a teaching portfolio.
- **Sharing with Parents:** Encouraging parental involvement in their children's education is difficult when, as it is in many families, both parents work during the school day. Video provides an excellent "window on the classroom" for parents, and helps them to feel included in the school community.

The above are just a few ways in which DV projects can be utilized by teachers in their daily work. Of course, it is important to remember that classrooms do not exist in a vacuum. Teachers need to address curriculum standards developed by others as they choose their classroom activities.

Curriculum Standards

The importance of technological literacy to today's K-12 curriculum is being increasingly recognized by school districts. Forty-nine of 50 U.S. states have utilized the National Educational Technology Standards for Students and Teachers (NETS) as part of their efforts to integrate technology in their K-12 curricula (Use of NETS by State, 2004). These standards were developed in "a unique partnership with teachers and teacher educators, curriculum and education associations, government, businesses, and private foundations" (ISTE Standards, n.d.)—in other words, important stakeholders in the education of the nation's children. Now that school districts have adopted technology standards, they are requiring that students create technology-based projects as part of their graduation requirements (Groton Public Schools..., 2004; 2005 Graduation Requirements..., 2004).

While much of the focus on implementing these standards has been Internet-based (e.g., the creation of educational Web sites, keypad activities, and Web-based research), digital video production is another computer-based technology that can be used to meet these standards. For example, NETS Technology Foundation Standard for Students 4b, "Students use a variety of media and formats to communicate information and ideas effectively to multiple audiences," and 5a, "Students use technology to locate, evaluate and collect information from a variety of sources" (NETS for Students: Technology..., 2004) are a perfect match for student DV projects.

The NETS guidelines also provide Performance Indicators for Technology-Literate Students based on these standards. Among these performance indicators are:

- Use technology resources (e.g., puzzles, logical thinking programs, writing tools, digital cameras, drawing tools) for problem solving, communication, and illustration of thoughts, ideas, and stories. (Grades PreK-12)
- Use technology tools (e.g., multimedia authoring and presentation, Web tools, digital cameras, scanners) for individual and collaborative writing, communication, and publishing activities to create knowledge products for audiences inside and outside the classroom. (Grades 3-5)
- Design, develop, publish, and present products (e.g., Web pages, videotapes) using technology resources that demonstrate and communicate curriculum concepts to audiences inside and outside the classroom. (Grades 6-8)
- Collaborate with peers, experts, and others to contribute to a content-related knowledge base by using technology to compile, synthesize, produce, and disseminate information, models, and other creative works. (Grades 9-12) (NETS for Students: Profiles..., 2004)

The above is just a sampling, as other performance indicators could also be aligned with digital video production projects. The availability of digital video production hardware and software in the schools allows for a much wider variety of options for meeting these standards.

Other Reasons

Of course, meeting standards related to technological literacy is just one possible reason for developing a K-12 DV project. When teachers design DV production activities, like any other K-12 classroom activities, they need to have specific educational objectives in mind. Digital video production-related activities can be used to meet any number of educational objectives, of a wide variety of types:

- Meeting subject-area standards
- Improving different aspects of literacy (reading, writing, listening, speaking)
- Providing creative ways to present knowledge and to communicate
- Developing media literacy and becoming more critical consumers of media
- Providing opportunities for independent learning and creative problem solving
- Encouraging cooperation, team building, and social skills
- Differentiating instruction for different achievement levels and learning modalities (Teacher Planning, 2002)

Teachers should not think of DV production as a stand-alone topic, just for students who are interested in going on to media-related careers. DV production should instead be integrated throughout the curriculum. DV projects, and their related writing, shooting, and editing tasks, can be a wonderful way for students to collect, organize, and present information on any content area topic.

One may immediately think of digital video as being a part of activities in the school's music, drama, English, and physical education classes, but teachers should also consider how video could enliven student projects in the sciences, math, and art, among others. For example, in History class, students could develop a presentation similar to the American TV series of the fifties, "You are There," in which actors dramatized historical events in the format of a modern television news broadcast. Another idea would be for early elementary teachers to have their students videotape the various animals in their neighborhood and make a video presentation on the various species and breeds that they discovered. Further examples and links to online descriptions of real-world classroom DV projects are provided later in this chapter.

THE DIGITAL VIDEO PRODUCTION PROCESS IN THE K-12 SETTING

Digital video projects are created in a four-stage process. The stages are: planning, shooting, editing, and delivery. Regardless of the content, whether the video project is of a dramatic or artistic performance or a record of a real-life event, the DV production process will require these four stages. In order to get a full overview of the process, and to have the most efficient and effective project design, readers should review the entirety of the four stages below before beginning a project.

While it is not a specific stage of the process, most K-12 teachers will want to follow up a DV production activity with an evaluation of its success. An example of an evaluation rubric can be found at http://www.groton.k12.ct.us/mts/compcurr/rubricvp.pdf.

Planning

The first stage of a DV production involves choosing or writing a script. Having a script that puts as many students as possible in front of the camera is important: think crowd scenes, and teams and partners rather than individual, standalone roles. A script that involves just two people may not be appropriate for a 35-student class. Parents will be disappointed not to see their children on the screen.

Once the script is ready, students should create a storyboard for each DV project (a template for storyboards can be found at http://www.mediafestival.org/downloads.html). Storyboarding means creating a series of cartoon-like sketches that show each shot or scene in a production. This helps everyone to visualize what the final product will look like. Without such a storyboard, one risks shooting extra, unnecessary footage, or forgetting to shoot important material.

Storyboarding is an authentic task for students, as most if not all professionally made movies are developed using this strategy. Almost any professional DVD of a motion picture has a "behind the scenes" feature that includes shots of the director reviewing storyboards as part of planning out the movie, and this can be shown to the students as a model. The artwork involved does not need to be of graphic-novel quality, but it should give a sense of the type of shot chosen (close-up, mid, or wide angle). Microsoft PowerPoint can be used to make storyboards. Students can create their images using the full-screen Slides view, and then print them out using the Handouts view. A text description of each shot is helpful, too, though not required. The storyboard can be taped up on the wall, so that everyone can easily see the full scope of the project.

It is also important at this stage of the project that all the student participants understand their roles in the production, whether they are on camera or behind the scenes. Obviously, in a dramatic production, actors will need to memorize their scripts, but others, such as lighting and prop assistants, will also need to practice their jobs. If it is possible to allow students to act in multiple roles in the production, that will improve their understanding of the overall process.

If students are to operate the camera, they need to be trained to use it. Some teachers may be concerned about students actually operating the camcorder during shooting. Of course, this is a fun and glamorous task, and students typically clamor for this job. However, the camcorder is an expensive and delicate instrument, and cannot be quickly and easily repaired if something bad happens, so it is tempting to designate it as off limits to student operation. If indeed it is the only camcorder for the school, this might be a wise policy, but if there are other cameras available, I would recommend instead creating a "driver's license" program for the camera, much as some schools have a "driver's license" for the Internet (Emerson J. Dillon..., n.d.; Welcome to the Modesto..., n.d.). The student would work through an online or print handbook about the features of the camcorder and take a quiz on its use before being allowed to operate it.

Even if one student is operating the camcorder, it is possible to arrange things so that other students can view what the camera sees. The fact that camcorders have large (sometimes four inches or more, measured diagonally) LCD viewscreens allows a limited number of students who are not operating the camcorder to view what is being recorded, though it may get crowded if there are more than a couple of observers. One could also connect the camera to a television to allow a larger group to view the proceedings.

Overall, the DV production process should be structured as a cooperative learning activity, in which each student is an active participant in constructing the final video. This will help to ensure high student motivation and "buy-in" to the project, as well as decreasing classroom management problems.

Shooting

Once the full production process has been carefully planned out, a script written, storyboards made, production roles distributed to students, and skills practiced, the shooting stage can begin.

The foundation of any DV project is a well-shot raw videotape recording. There is an old saying in the computer world, "Garbage in, garbage out" (GIGO). The GIGO effect is true in regards to digital video as well. While editing programs such as Apple's iMovie and Windows Movie Maker 2 provide some basic adjustments that can be used to fix video, such as tint, brightness, and contrast correction, they can only do so much if the recording was poorly shot in the first place. For example, if there was a backlighting problem and the main subject appears as only a silhouette, there is nothing that can be done to fix it. Similarly, if the camera operator forgot his/her jacket and was shivering during the handheld videotaping, nothing can edit out the shakiness in the resulting video. Appendix A provides "Twenty-Five Tips for Shooting Digital Video," which I developed based on my years of experience shooting video in K-12 and higher education settings.

Teachers should allow time to show digital "dailies" to the students after each day's videotaping. Students enjoy seeing the immediate results of their work, and observing what has been recorded can provide valuable feedback for the performers and camera operators. The occasional "blooper" can be a source of great hilarity, and if the editor(s) has time, they may want to edit together a short "blooper reel." However, one does not want to over-emphasize the amusing mistakes made; it is important to make sure that the students realize that the purpose of the project is to create a professional-looking production, not a laugh-filled blooper reel; otherwise the situation may degrade to the point of students purposefully blowing lines, dropping props, and so forth, to garner the laughter of their friends.

Editing

As stated before, it is important to let the students see, more or less immediately, the results of their work. It is also important that students understand how editing is done. A modern, basic DV editing program such as Apple iMovie or Windows Movie Maker 2 is certainly well within the capabilities of a late elementary school student to master. In fact, teachers may even have students who have already created their own DV productions at home. I remember well shooting Super 8 movies with my sixth-grade friends in the seventies; we could only dream of the kinds of tools sixth graders have available to them today.

Every student should get a chance to work with video using DV editing software. Even if the students have not shot any video yet, many programs include sample video files that students of fourth grade and up can manipulate. If sample files are not available, the teacher can create a set of files and copy them onto multiple machines or onto a central server for the students to access. It is useful for students to see how different edits of the same footage can give very different effects. This might be a useful time for the teacher to discuss how editing is used in TV advertisements, news broadcasts, and "reality television," and how editing can effect a viewer's interpretation of televised events. An understanding of how video productions are edited can improve a student's critical thinking abilities and overall media literacy.

The task of editing a DV project can be complex and time consuming, so make sure there is enough time in the schedule for it. Editing at 10:00 in the morning on a video that has to be shown to the entire school at noon is no fun! Sorting through multiple takes for the "best" version of a scene can be challenging, as one performer's best take can be another's worst. Sometimes there is a first take with enthusiastic performers but a few missteps and/or forgotten lines, and a later take with fewer mistakes but less energy all around; which does one choose? Sometimes combining the best parts of one take with another is possible.

Thankfully, today's DV software makes these editing tasks much simpler than they were in VHS-only days. But even with today's easy-to-use software, before you begin the editing phase of your project, you will want to review the specific tips I have provided in Appendix B.

Delivery

There is a multiplicity of options available for delivering a DV project to its audience. Here are a few possible ways to deliver a video to K-12 students, parents, and/or relatives:

- Big-screen TV display to students in a classroom
- Broadcast over local cable community access TV channel
- Broadcast over school TV system
- Streaming video over the Internet
- Downloadable video file on the Web
- DVDs played on students' home players
- CD-ROMs played on students' home computers
- VHS videotapes played on students' home VCRs
- Attached to a PowerPoint presentation and projected on a screen

Before choosing from these options, one should first define the audience for your project and consider whether you may have multiple audiences. Who will be watching the video? What technology will they have available to them? Making a video in a format that your audience cannot view is going to be a waste of your students' valuable school time.

Even though delivery is the final stage of the DV development process, the choice of delivery method(s) needs to be decided as early as possible, as it will have implications on how the video is shot and edited. For example, if one were planning for delivery only through streaming video (video that does not need to be fully downloaded to begin viewing), one would want to have more close-ups of people speaking, since the frame size of the video will necessarily be small, and the detailed features of wide-shots would be lost. One would want to avoid flashy transitions, since they will not display well in the smaller format, and use larger titles for improved readability. Also, few schools would have access to a high-speed VHS dubbing system, so if the project is to be distributed on VHS, it should be as short as possible so that it could be duplicated for a large number of people quickly. DVD- and CD-based projects can be copied on consumer equipment at rates many times faster than the length of the recording, and a longer project would be more feasible. Since no duplication is needed at all for streaming Web delivery, the project could feasibly be hours and hours long, or even be a live Web stream from the school.

Multiple Delivery Options

It is, of course, possible to have more than one delivery method. For example, the teacher could show the video to the class and also make copies for students and their friends and relatives on VHS, CD-ROM, and/or DVD. This could be a great fundraising opportunity for the school if the videos were packaged nicely with printed or hand-drawn covers. It is important to consider multiple audiences, and future multiple audiences, as soon as possible in the DV project development process. If this decision is made too late in the process, it may be difficult to implement. For example, if one has produced a DVD and then decides to stream the video over the Internet too, and the original project files have been deleted, the process of creating the streaming video file would be much more complicated than if both delivery methods had been planned for in the beginning. An even worse case scenario would occur if a streaming file was created, the original files deleted, and then a DVD version was proposed. Unless the raw footage was available, it would be impossible to create a DVD of acceptable quality. Thus, if one decides on delivery media early in the process, the video editing software will allow one to quickly export the DV project to any number of delivery formats.

Cost

When choosing among various delivery methods, cost is going to be an important factor for a K-12 school district. One should first review existing delivery hardware and software, as well as networking infrastructure. Some school districts may have already installed video-on-demand delivery systems, in which the video files are stored on a central server (Masullo & Brown, 1995; Video On-Demand, n.d.) and streamed to individual classroom TVs via a high-speed network connection. This is the most costly option, unfortunately, if the networking infrastructure is not already in place.

The least-expensive option is to deliver the final product via VHS tape. Videocassette playback machines should be available in even the poorest school districts, as they have been around for decades without any major technology change. DVD players are now available quite inexpensively too, though, and the much higher quality and ease of use of a DVD would make this a better option with only a minimal increase in cost.

Streaming video over the Internet, as it does not require the purchase of any blank cassettes or disks, or stamps and packaging to mail them in, can be another lower-cost option, assuming that the streaming video server has already been provided by the school and that computers that receive the video will have high-speed connections.

Again, when choosing delivery options, the first and most important thing is to know your audience and your own existing technology infrastructure, and to make your delivery decisions based on that knowledge at the beginning of the DV development process. In Appendix C, I have included some additional tips for choosing among delivery options.

CHOOSING HARDWARE AND SOFTWARE

The proper choice of equipment and computer software for a DV project makes a big difference in the quality of the end result, the digital video presentation, and the ease and efficiency with which it is created.

Computer Hardware

When considering computer hardware choices for a digital video project, more is better, which means that one should aim for the highest-specification computer that one can afford. Churning out the gigabytes of data that DV consumes is a processor-intensive task, and so the faster the processor, the faster the final, compressed video will be rendered. One should also have as much RAM as possible installed, at least 512 MB but ideally 1 GB or more. Digital video also requires a large amount of hard drive space, and a second internal or external hard drive should be considered as well. A computer with less than 40 GB of hard drive space is insufficient, unless the videos to be created are very short, and there is no need to archive the raw files.

There needs to be some way to pass the digital video from the camcorder on to the computer, and this is done by way of a video capture card with a Firewire cable connection. Firewire is sometimes called IEEE 1394 or iLink. All Apple and most higher-end Windows machines come with video capture cards and their associated Firewire ports as standard equipment. If not, a card will need to be purchased and installed. There is a huge number of capture cards put out by different companies; you will want to do some research on the Web before making your purchase, as performance and cost vary a great deal.

If the project is to be delivered using CD-ROMs or DVDs, then the computer will also need to have a drive capable of creating ("burning") them. These also can be purchased separately if they are not installed on the computer. You may wish to consider buying an external "burner," with its own case, as it can be shared among multiple computers. It is also worth noting that DVD burners can create CDs, but not vice versa. Finally, there are two DVD formats: DVD-R and DVD+R; while there are some technical differences, both will create disks that can be played in standard DVD players. In order to avoid purchasing two types of blank media, it is best to standardize within a school or school district, or to purchase a burner that can use both types of disks.

Finally, it is worth mentioning that it is possible to create video DVDs without a computer at all. Some camcorders, as described below, record directly onto small DVDs. One can also purchase standalone DVD recorders that are similar to a home VCR except that they record onto DVD instead of VHS tape. These machines can be a lifesaver if all one needs to do is to get video onto a disc immediately, but editing the recorded video is essentially impossible, and one cannot create an attractive menu with custom graphics. Also, some of the cheaper models only have analog audio/video inputs, as opposed to digital, which will cause some loss of quality, and so are to be avoided.

Camcorders

There are hundreds of camcorder models available, with widely varying specifications. In this section, I will discuss the most important factors in camcorder choice for K-12 use, including (in order of importance): media type, viewscreen size, number of charge-coupled devices, zoom factor, audio and video inputs and outputs, image stabilization, loading mechanism, and miscellaneous lesser features.

Media Type

The days of big, bulky VHS camcorders are over, and there are a number of options in terms of the media camcorders record on. Even the smaller VHS-C camcorders, which accept the same type of tape in a smaller cartridge, are obsolete. These types of analog tape only record 250 lines of resolution, while Digital 8, MiniDV, and MicroMV can record twice that, and thus provide far superior picture quality. VHS is also subject to generational loss—the loss of quality resulting from copying one tape onto another; digital tapes can be copied without any lost information or degradation of video quality. In the analog days, editing video would start off with a reasonably acceptable quality VHS tape of the raw footage, but by the time the best takes of the scenes were edited onto a master tape and copies for the students were made, the resulting video quality was but a fuzzy ghost of the original. With digital media, each student can have a copy that looks as clear and colorful as the master.

In terms of specific types, MiniDV has become the standard for semiprofessional digital video work. There is a broad range of MiniDV camcorders from many manufacturers to choose from. Blank tapes are readily available, should be for many years, and all of the editing and authoring software is compatible with this format. Standalone MiniDV players and MiniDV/VHS dual decks for dubbing from one format to the other are also available through online sources, though rarely seen in brick-and-mortar stores.

Sony's Digital8 format is based on its analog Hi8 format, and its predecessor 8mm, and uses a larger cassette than MiniDV (though still considerably smaller than VHS). Digital8 camcorders use the same cassettes as Hi8 but record a digital signal, which means that half the tape time (60 minutes on a 120-minute blank) is lost, but what is recorded is of much higher quality than an analog camcorder would provide. This format is used in Sony's lower-end DV cameras. The Digital8 camcorders are larger and bulkier, because of the greater size of the tape cassettes, and the cameras do not have all the features of the pricier MiniDV models. However, the video quality can be just as good as MiniDV, and, if one happens to have an archive of old 8mm and Hi8 analog tapes, one may want to choose a Digital8 camcorder, as it can play these tapes and thus digitize them. If the project has only a limited budget, Digital8 may be the best solution.

There are two other, less common options to consider: MicroMV and DVD-R. The primary advantage of MicroMV is the tiny size of the tape cartridge. This miniaturization of the tape allows for very compact camcorder designs, camcorders that could literally fit in a coat pocket. However, the format is so new that these camcorders are not widely available, nor are blank tapes, which are also more expensive than other blanks. The video format that MicroMV uses is also incompatible with current editing and authoring software.

Camcorders that record directly onto DVD blanks are very convenient in some instances, as the disks can be put into a set-top player immediately after recording a session and played back. If all one wants is to quickly get raw unedited video onto a DVD, this may be the best choice. However, the camcorders themselves are typically heavier and bulkier, due to the size of the disks used. A number of recording quality levels is available, but recording at the best quality will give only a limited recording time. Editing the resulting video may also be challenging, and there is limited compatibility with existing software.

To sum up, the MiniDV format is best suited as the media of choice for K-12 digital video projects. It is the standard format for consumer-level digital video editing, provides great quality at a reasonable price, and hardware and media should be easily available for a long time.

Viewscreen

All camcorders have a small, traditional viewfinder, not unlike that of a still camera, but most of the time, using the larger pull-out LCD display (or viewscreen) is more convenient and enjoyable. Typically, these monitors are in the 2.5- to 4-inch range. I recommend choosing the largest viewscreen possible, unless one is looking for the absolute smallest and lightest camera. Students love seeing themselves on this bright, colorful TV display, and it will provide an excellent idea of what the video, when presented on a big-screen TV or a computer monitor, will look like. The viewscreens can usually be flipped around so that the person recorded can see him/herself.

CCDs

CCDs, or charge-coupled devices, are internal camcorder components that sense light and turn it into a digital signal. The larger the number of the CCDs, the better the picture you record. In a 3-CCD camcorder, each CCD can focus on a different range of color, instead of having one sensor to capture them all, thus giving a fuller, more realistic range of hues, which is important given the colorful nature of many K-12 productions. Unfortunately, 3-CCD camcorders can still be fairly expensive, and many entry models come with only one CCD. However, even a 1-CCD camcorder will provide an image of a quality that far outperforms analog models.

Zoom

Another area where "more is more" in a camcorder is its ability to "zoom" in on a distant object or person. This is specified by degrees of magnification, for example 3x or 10x. The higher the number here, the better. A camcorder with a 10x optical zoom will make the distant object seem 10 times closer. This can, of course, be a major advantage when shooting action that is distant from you, or in getting close-ups of people. It is important to note that one should ignore the digital zoom numbers. These are generally much higher than the optical zoom numbers, but using the digital zoom quickly degrades the quality of the video signal. Zoom in enough and all that is visible are fuzzy blocks of color. If one opens up a still image in a program like Photoshop and keeps zooming in on the image, eventually all that is visible are the individual pixels, the small dots of color of which the image is composed. That is what happens when the digital zoom is used.

Audio Inputs and Outputs

Most camcorders have microphone and headphone jacks. An external microphone is a must in K-12 environments. Relying on the microphone built into the camcorder will ensure that the camera operator's own breathing and coughing are captured more clearly than the comparatively distant teacher and student voices. Wearing headphones provides the camera operator with a sense of what the camcorder is hearing. Listening to the room through headphones will provide a better sense of how noisy the environment is and whether or not the viewer will be able to make out the conversation being recorded. Wearing headphones is a must during videotaping sessions, in case the battery for the external microphone cuts out or the microphone gets unplugged. I have had a situation where a student-teacher was videotaped in front of her classroom teaching a lesson, and her lapel mike stopped working. The camera operator was not wearing headphones, so he did not notice the mike failure. The end result was a nicely shot hour's worth of video with no sound at all—totally useless for our purposes, which were to document the content of the student-teacher's interactions with the children.

Video Inputs and Outputs

There are a number of cable inputs on the side and/or back of most camcorders. Sometimes these inputs are hidden by a plastic cover. Most digital camcorders have Firewire connections to allow you to connect your camcorder to a computer or a standalone DVD recorder. An analog video in/out jack is another useful option. This allows one to plug the camcorder directly into a television or VCR. You will want to make sure that the AV connection works in both directions, so it will allow you to input video to the camcorder from an analog VCR as well as output video to a television and/or a VCR.

Image Stabilization

You should also look for an image stabilization feature when shopping for a camcorder. While the use of a tripod is recommended, sometimes there is no choice but to shoot video in a handheld mode, and this feature will help cut down the shakiness inherent in this type of footage.

Loading Mechanism

It is important to consider where the tape is loaded into the camcorder. For example, if the tape door is close to the built-in microphone, this means that the noise from the tape mechanism may be picked up while recording. You also want to avoid bottom-loading camcorders. If one mounts the camera on a tripod (as should usually be done), the camera will have to be detached from the tripod each time the tape runs out, which could be inconvenient and could cause one to miss part of a live event.

Lesser Features

There are also features that are of much less importance when choosing a camcorder, or of no relevance at all for the K-12 environment. First, I would also recommend ignoring the capability of shooting still pictures with the camcorder. For one thing, most video editing programs will allow one to export stills of video frames, so when videotaping, one is already shooting thousands if not millions of potential still photos. Also, the resolution of the still photos generated by reasonably priced camcorders is well below the quality provided by a still camera in the same price range. Today's digital still cameras can shoot images of 8 megapixels and higher, something no camcorder can yet approximate.

Similarly, some camcorders provide the capability of shooting MPEG movies. These movies are generally very short (around 30 seconds), have low resolution, and are really only suitable for sending via e-mail or putting on a Web site. If that type of video is what is desired, it can be easily created from your digital videotape using desktop video software, as described in the next section.

Computer Software

Two types of software are needed to create digital video productions: editing and authoring software. The features of both types of software are described in this section, and various software packages are recommended.

Video Editing Software

Editing software allows users to perform the following tasks:

1. Capture video clips (import them into the computer)
2. Trim and arrange video clips
3. Add transitions between clips
4. Apply video effects to clips
5. Add titles (opening titles, credits, subtitles)
6. Add audio (sound effects, narration, music)
7. Adjust audio (volume, balance, tone)

On the Apple platform, there are four primary options for editing software: Apple's iMovie, Final Cut Pro HD, Final Cut Express, and Avid's Free DV. The first is the default, free program that is installed on every new Apple Macintosh, or can be inexpensively purchased as part of the iLife suite. As the name suggests, Final Cut Pro HD is a professional-quality tool with a professional-level price (around $1,000). Final Cut Express costs about $300 and is designed to fit somewhere between the consumer-level iMovie and the professional Final Cut Pro HD. Avid's FreeDV is another no-cost option, a cut-down version of Avid's professional-quality video editing software.

On the Windows platform, there is also a free tool, Windows Movie Maker 2, which is part of Windows XP. Avid FreeDV, mentioned above in the Apple section, is also available for Windows and is a free program with basic editing features. Adobe Premiere, a professional tool, is expensive but pretty much the industry standard. There are also innumerable other video editing tools available, such as MainActor, Screenblast, ShowBiz, Studio8, Vegas Video, Visual Communicator, Xpress DV, and many others. Some come bundled with video cards.

For both platforms, K-12 educators should start with the free products, iMovie and Movie Maker 2. These tools can handle all of the editing tasks listed above, and the learning curve is not nearly as steep as for the professional-level tools. The additional expense and learning time required for the pro tools would only be worthwhile if the educator finds that the free tools cannot meet his/her needs.

DVD Authoring Software

DVD authoring tools take the video files that are created by the editing software, convert them to DVD specifications, add a menu, and then burn the projects to disc. If all one needs to do is to create streaming video files, or files to be included on a CD-ROM or as part of a PowerPoint presentation, then an authoring tool is not needed. Any video editing tool should export to at least one suitable video format (Quicktime, Realvideo, or Windows Media). Since the tool may not export in all three formats, checking with the server administrators at one's school district for compatibility with their servers is important before choosing video editing software if one is going to stream the video.

On the Apple platform, there is a free, consumer-level tool and a professional-level DVD authoring tool. The free tool is called iDVD and, like iMovie, comes installed on new Macs, and is also available in the inexpensive iLife suite. Windows, unfortunately, does not offer a free DVD authoring tool. Sonic's DVDit is a good, inexpensive, beginner-level tool, though, with attractive, customizable themes similar to those in Apple's iDVD. Adobe Encore is an excellent pro-quality DVD authoring package.

It is also worth mentioning that there are some software packages available for Windows that are a compilation of both editing and authoring tools, such as Sonic's MyDVD Studio. These tools generally offer fewer features than standalone programs.

SAMPLE K-12 PROJECTS

Over the years, I have worked with K-12 students on several digital video projects. I will briefly describe two of them, and then cite further K-12 projects that can be viewed online.

The first project was an exercise in digital storytelling with fourth through eighth graders. The students were involved in every stage of the production, from the initial brainstorming of story ideas to participating in the final cut of the movie, which was shown to the rest of the school with great approval. The students were led through the DV production process as described earlier: first, they chose the overall plotline, which involved a mad scientist and his creation, a monster that terrorized the school. Then they wrote the individual scenes, which they turned into storyboards; chose props, costumes, and sets; and acted out the scenes in front of the MiniDV camcorder. So far, this process was not too dissimilar from other typical school theatrical projects the students had been involved in, but the fact that the production was videotaped and captured on computer meant that the very best take of each scene, or part of a take, could be used. Student actors typically are extremely variable from scene to scene, in terms of quality of performance, and this ability allowed for each actor to appear at his/her best in the final production. Students could view their performance objectively on the computer screen, and help to decide their best takes.

The editing software used (in this case, iMovie) also allowed for special video effects such as fire, lightning, and earthquakes that would have been impossible or dangerous to replicate in real life. Other video effects could be added to the software via downloadable plug-ins, some available for free. The software even allowed for variations in speed, so that a scene's pace could be increased or decreased to a greater or lesser degree. Actions could even be shown in reverse, something difficult or impossible to produce in the past. Recorded narration was added over the original soundtrack, as was music and sound effects. The final product was shown to the other students in the school, and then distributed via DVD to the student participants. The final DVD product was very professional in appearance, with animated menus and scene selection. It was created with Apple's iDVD program, which provided a large number of predefined themes that could be customized for each individual production.

The second project was part of a larger thematic unit on civilization building. In this activity, which encompassed content related to history, politics, sociology, ecology, and other subjects, students were put into teams that were located in a specific environment (rainforest, plains, mountains, etc.). The goal for each team was to survive and improve conditions for their people in their given environment. As a culminating activity, the students created a news report on the events in their society. This allowed them to reflect on their experiences while sharing them with the other students. Digital video software allowed us to emulate the appearance of a TV newscast, with titles, subtitles, and appropriate music. We were also able to emulate the fast-paced editing of a typical newscast. Again, the final production was shown to the students, and a DVD was created so that the students could share the newscast with family and friends, and have a permanent record of their activity.

Other K-12 Digital Video Projects

The two projects described above are just a small sample of the kinds of video projects that are possible. The range of potential K-12 video projects is endless. Extensive lists of K-12 school projects can be found online at these Web sites:

- http://www.people.memphis.edu/%7Eaimlab/imovie/handout.htm
- http://www.schoolhousevideo.org
- http://www.springfield.k12.il.us/movie/index.html
- http://education.apple.com/education/ilife/

In the following paragraphs, I will highlight just a few of the outstanding K-12 DV projects found in the above resources and elsewhere.

Digital video production skills can lead to various career paths, and it is important that today's children get opportunities to explore possible future employment as camera operators, electronic graphics designers, sound mixers, script writers, lighting technicians, and other occupations by working through authentic tasks that these professionals undertake in their day-to-day jobs. Waimalu Elementary School in Hawaii has established a video studio in which fourth through sixth graders are allowed to practice their skills as they create school newscasts, announcements, student profiles, game shows. and other DV projects (Waimalu Elementary School..., 2002).

In the Capistrano Unified School District in California, fourth graders learned the DV production process and utilized their new skills to document the school science fair. This allowed them both to share the various science principles demonstrated at the fair with those who could not attend, and to archive the exhibits for future reference. Along the way, the activity helped their teacher meet academic standards in Communication, Language Arts, and Technology (Capistrano—Digital Movies, n.d.).

Many students find traditional book reports to be uninteresting to create and uninvolving to listen to. In the Minnesota Public School system, elementary students are creating video-based "Book Talks" to share their favorite reads with other students. The ease of use of today's tools—such as, in this case, iMovie _makes it easy for students to record their spoken thoughts and add appropriate graphics to create an appealing DV presentation that they can then burn to CD to also share with parents and relatives. (Bill Bierden's Web Site, n.d.)

In Wisconsin, high school students analyze TV commercials and use what they have learned to create their own DV commercials, which are shown on the school's TV system. This project is undertaken within a joint English, speech, and technology education course, and media literacy and writing and presentation skills are emphasized. These students came away from the activity better able to defend themselves against the distortions and manipulations of the advertising that they are constantly exposed to (Teaching Through Technology Program, n.d.).

In North Dakota, high school students made personal connections with U.S. history by interviewing war veterans and editing a video in which the ex-soldiers shared their wartime experiences. This gave historical events and concepts a personal impact that merely reading about them would not have done (Apple—Education…, 2004).

Wired News profiled a number of Maine schools that have integrated DV into their instruction (Dean, 2002). A creative writing class wrote screenplays which they filmed using digital camcorders and edited using iMovie. As a part of the activity, the teacher had the students review scripts of movies they were familiar with, in order to learn about how to construct a video story. "You guys are warehouses of information for movies because you see a lot of 'em," he told the students. "And now you know how movies are structured" (Dean, 2002, p. 1).

These types of DV projects are empowering students to learn about their world in new ways and to share what they have learned with others. Traditional skills like reading, writing, and math are not being thrown away but are instead given a new technological dimension that allows modern, media-saturated students to organize, structure, and create information in ways that their parents were not able to imagine. Today's students are becoming as comfortable with the DV camcorder and the personal computer as their parents were with the blackboard and the typewriter, and they should be allowed to use these tools as they work on projects throughout the K-12 curriculum.

Other Uses

There are many other additional uses for digital video in the K-12 environment that are not purely academic. For example, in these days of heightened concern for student safety, the administration of a school or district might want to prepare a video showing what students should do in a number of dangerous scenarios. Of course, video also offers an ideal way to showcase student achievement in athletics; not just to preserve high school football games but also as a tool to enhance student performance. For example, videotaping student attempts to swing a baseball bat or golf club could help them to improve their form.

CONCERNS

Parents, teachers, and administrators often voice concerns about the use of digital video in the K-12 environment. This section discusses issues related to copyright, student safety, privacy, and presenting DV material in the classroom.

Copyright

It is important that K-12 teachers make sure that their students are aware of copyright and privacy concerns. While the chances of actually getting sued are minimal, students should not wear shirts or sweatshirts with trademarked logos or characters in them when in front of the camera. In music videos and other professional productions, these are often "fuzzed out." It is also best to avoid the use of copyrighted music in DV productions. Even a song such as "Happy Birthday" is copyrighted (Mikkelsen & Mikkelson, 2002). If copies of a video with copyrighted music on it were sold to parents, it could invite legal problems.

Student Safety

Similarly, it is also wise to have a written agreement from each performer in a video—in the case of a child, that child's parents, especially if the video is to be shown outside of the classroom, such as before a school assembly or on a local public-access cable network. An example of a school district student video release form can be found at http://www.altoona.k12.wi.us/schools/Pedersen/Registration/PhotoVideo.pdf and http://www.cms.k12.nc.us/resources/documents/photoRelease.asp. A Google search on the terms "Student Video Release Forms" will provide additional examples.

If students are to be shooting video "in the field," it is important to discuss proper behavior beforehand. Some schools require students to fill out an explicit agreement about what constitutes acceptable behavior during a video shoot before allowing them to have access to equipment. An acceptable behavior form is available at http://www.olejarz.com/arted/digitalvideo/appropriateness.pdf

Privacy and Other Content Issues

The small size of today's DV camcorders allow for the easy invasion of others' privacy. Students could potentially record material that was embarrassing to others. Offensive or illegal behavior can also be recorded with a video camera. Pornography could be edited into a video presentation. Finally, students can cause harm by creating content that "makes fun" of another student, teacher, or administrator. Just like with any other school project, student behavior needs to be monitored during all stages of the DV production process. Requiring adherence to the storyboarding procedure described earlier in this article, and reviewing the "daily rushes" that the students create, should help to minimize the potential of bad behavior during these projects.

Presenting Video in Class

Using video in the classroom has gained a certain stigma in some parent and administrator minds because some teachers have used TV as a way to fill in class time and avoid coming up with other types of activities. It is important that teachers create the proper classroom environment and atmosphere when presenting DV projects. The following tips (adapted from Teacher Resources Video Programs, n.d.) should be kept in mind when planning DV presentations to K-12 students:

- **Preview the Video:** If you are not totally familiar with the video content, or have not viewed it for a while, preview the video before showing it to the students.
- **Show Video in Short Segments:** Help students to digest the information in a video by breaking it up into small segments of a few minutes and discussing the content between segments.
- **Leave the Lights on:** Turning off all of the lights encourages students to relax and even drift off, perhaps even into sleep. Leaving the lights on reminds students we are still in school and that this is an active, intellectually engaging activity. Also, if the lights are on, students can more easily take notes on the video, encouraging them to be "active viewers."
- **Make the Video part of a Larger Sequence of Activities:** Do not just show the video as a standalone activity. Present it in the context of other types of related activities. Presenting an activity or reading dealing with the content of the video before presenting it will help the students put the video information in context.
- **Provide an Advance Organizer:** Before the students view the video, provide them with a list of things to look for and focus on.

Keeping these guidelines in mind should allow teachers to create and present DV productions in class without facing controversy.

CONCLUSION

The potential of computer-based digital video production for expanding and extending our K-12 students' capability to learn, analyze, and communicate is tremendous, and we are just beginning to explore the infinite numbers of possible educational implementations of this new technology. Today's advanced but affordable desktop video production hardware and software are providing students with opportunities to enhance their learning in creative and innovative ways that were impossible in the past. Digital video is a medium that allows students to combine moving and still images, text, sound effects, music, and dialogue in ways that no other media can. A student involved in a DV project is an active learner, constructing and organizing knowledge through this new medium, not just a passive receptacle of knowledge.

Digital video production should not be seen as a standalone activity, designed just for those future techies who will go on to create their own dot.coms, but instead should be a part of every content area curriculum. Making use of DV in the content-area classroom offers students a new way to explore the facts and concepts of a field in a way that is both highly active and highly motivating. The potential to share this new knowledge with other students, nearly anywhere else in the world, is another development that has only recently become available to K-12 students. Digital storage, such as CD-ROMs, DVDs, and streaming video over the Web, allows these productions to be shared with vast audiences at little expense. Because of these newfound capabilities, digital video production should be a part of every K-12 teacher's technological literacy curriculum.

REFERENCES

2005 Graduation Requirements. (2004). Retrieved December 10, 2004, from http://www.sumner.wednet.edu/arounddistrict/pages/Plan%202005/2005GradReq.html

Apple—Education—Stories of the American Experience. (2004). Retrieved May 27, 2004, from http://education.apple.com/education/ilife/project_template.php?project_id=104&subject_id=1

Baugh, D. (2003). *iLife—Developing the Creative Curriculum.* Retrieved October 24, 2004, from http://www.denbighict.org.uk/dv/ilife.html

Bill Bierden's Web Site. (n.d.). Retrieved on May 24, 2004, from http://www.users.ties.k12.mn.us/~wbierden/booktalk.htm

British Educational Suppliers Association. (2002). *Information and communication technology in UK state schools: 2001 summary edition.* Retrieved January 20, 2004, from http://www.siia.net/sharedcontent/divisions/GLOBAL/besaexecsumm1101.pdf

Capistrano—Digital Movies. (n.d.). Retrieved on May 24, 2004, from http://www.ocde.k12.ca.us/showcase/districts/cusd/promect6/index.html

Dean, K. (2002). *Maine spawns budding Kubricks.* Retrieved December 10, 2004, from http://www.wired.com/news/school/0,1383,56246,00.html

Duffy, T. M., & Cunningham, D. J. (1996). Constructivism: Implications for the design and delivery of instruction. In D. H. Jonassen (Ed.), *Handbook of research for educational communications and technology.* New York: Macmillan.

Emerson J. Dillon Middle School Internet driver's license tutorial. (n.d.). Retrieved December 10, 2004, from http://www.phoenix.k12.ny.us/ejd/info/computer/tutorial/driver.htm

Groton Public Schools—Information and Technology Literacy K-12. (2004). Retrieved May 2, 2004, from http://www.groton.k12.ct.us/mts/compcurr/itlfshsgrad.htm

ISTE Standards. (n.d.). Retrieved April 2, 2004, from http://www.iste.org/standards/

Jonassen, D. H., Peck, K., & Wilson, B. G. (1998). *Learning with technology: A constructivist perspective.* Columbus, OH: Prentice-Hall.

Jones, B. F., Valdez, G., Nowakowski, J., & Rasmussen, C. (1995). *Plugging in: Choosing and using educational technology.* Retrieved December 30, 2004, from http://www.ncrel.org/sdrs/edtalk/newtimes.htm

Korea Education & Research Information Service. (2003). *Adapting education to the information age: A white paper.* Retrieved December 10, 2004, from http://www.keris.or.kr/english/pdf/2003-WhitePap.pdf

Market Data Retrieval. (2003). *New Market Data Retrieval report finds growth in leading-edge technologies amid decreased technology spending.* Retrieved January 26, 2004, from http://www.schooldata.com/pdfs/Tech_in_Ed_PR.pdf

Masullo, M. J., & Brown, D. (1995). *Pioneering video on-demand projects in K-12 education.* Retrieved October 11, 2004, from http://ianrwww.unl.edu/EDUPORT/R-ISDL95.HTM

Mikkelson, B., & Mikkelson, D. P. (2002). *Happy birthday, we'll sue.* Retrieved May 24, 2004, from http://www.snopes.com/music/songs/birthday.htm

National Center for Education Statistics. (2003). *Digest of education statistics, 2002: Learning resources and technology.* Retrieved January 20, 2004, from http://nces.ed.gov/programs/digest/d02/ch_7.asp

NETS for students: Profiles for technology-literate students. (2004). Retrieved May 1, 2004, from http://cnets.iste.org/students/s_profiles.html

NETS for students: Technology foundation standards for all students. (2004). Retrieved May 1, 2004, from http://cnets.iste.org/students/s_stands.html

ORC Macro. (2002). *State of the state survey: Access to technology.* Retrieved January 10, 2004, from http://www.thinkport.org/6d9a650b-aca7-4acf-9701-09375c273bc3.asset

Teacher planning. (2002). Retrieved May 27, 2004, from http://edtech.guhsd.net/video/Plan.htm

Teacher resources video programs. (n.d.). Retrieved October 24, 2004, from http://www3.iptv.org/exploremore/ge/Teacher_Resources/video_tips.cfm

Teaching through technology program 15: 9-12 Projects. (n.d.). Retrieved May 27, 2004, from http://www.ecb.org/ttt/program15.htm#video

Use of NETS by state. (2004). Retrieved May 30, 2004, from http://cnets.iste.org/docs/States_using_NETS.pdf

Video on-demand. (n.d.). Retrieved October 11, 2004, from http://homework-hotline.org/pages/vod.html

Waimalu Elementary School Video Studio. (2002). Retrieved May 4, 2004, from http://www.k12.hi.us/~waimalu/video.htm

Welcome to the District 108 Internet driver's license tutorial. (n.d.). Retrieved May 4, 2004, from http://www.pekin.net/pekin108/tech/tutor/

APPENDIX A. TWENTY-FIVE TIPS FOR SHOOTING DIGITAL VIDEO

1. Learn how all the little buttons and menus work on your camera before you start shooting with it. The manual is your friend!
2. Be aware of the time. Most camcorder batteries last for less than two hours, and even less if you use the LCD display screen.
3. Start shooting before the action begins, and let the camera run a bit afterwards. Editing will be a lot easier if you have extra footage (at least 5 seconds) before and after the event you are trying to capture.
4. Look at what you are shooting through the LCD display at least in the beginning, to check colors and focus. Make sure your view of the screen is exactly straight-on—looking at the screen at an angle can provide an inaccurate view.
5. When framing a shot, remember the rule of thirds. Imagine that your LCD display is divided into three equal columns and rows, like a tic-tac-toe game. Then imagine the four spots where the horizontal and vertical lines meet. Those four imaginary spots mark the location where the most interesting parts of your subject should be. (Do not worry, you do not have to hit all four spots.)
6. Use an external microphone if at all possible. Your actions (and your breathing) are louder than you think. Heavy breathing in the background of your video can provide a disturbing, Darth-Vader-like effect. Also, make sure you turn off the microphone when it is not being used, so that the battery does not run down.
7. If you do not have an external microphone, get in close to your subject, so that your camera's built-in microphone can capture what they are saying. A zoom lens from across the room may work well, in terms capturing the visuals, but the audio will be low in volume and hard to understand.
8. Wear headphones so that you can ensure that the microphone is working.
9. Use fresh videotape. Even though the signal is digital, it is still recorded on good old magnetic tape, which can degrade during use, and really should not be reused more than three times.
10. Use a tripod if possible. Practice moving the camera smoothly with the tripod beforehand. If you cannot use a tripod, lean against something sturdy or place the camera on something.
11. Do not "spray paint" or "yo-yo" your shots. Do not quickly pan over here to something interesting, then pan over there to something else. Avoid excessive zooming in and out.
12. Instead of using the digital effects on your camera, save them for editing, where they are undoable if you do not like the end result.
13. Similarly, avoid the use of in-camera titles. They are permanent and you will have more font, color, and size options if you add them on the computer.
14. Consider the lighting in the environment. Be careful to avoid backlighting whenever possible. Pushing the backlight button to compensate often leaves you with a washed-out background and a generally

unpleasant image. If indoors, get as much light on the subject as possible to avoid dark, dingy-colored footage.

15. Make sure you have not pushed the Nightvision button! If you have, everything will have a sickly green tint.
16. Make sure Autofocus is on, unless you are planning to manually focus each shot.
17. Always do a white balance before shooting to ensure the purity and vividness of the colors in your video. This usually just involves focusing the camcorder on a white sheet of paper and pushing the White Balance button.
18. If you are planning to mix footage from two camcorders, make sure they are both the same quality, ideally the same camcorder model and same tape brand. Do not try to mix Hi8 and MiniDV shots in the same video, for instance.
19. Simplify your background. Too much detail is distracting.
20. Get close. With online video, closer shots will seem sharper, since more pixels will be devoted to the subject. It will also tend to blur the background, meaning less data to process there. Step closer rather than zooming in, as a zoomed-in shot (especially without a tripod) tends to exaggerate camera jiggle.
21. During videotaped interviews, encourage people to speak succinctly and in shorter sentences, to make editing easier.
22. Look for "B-roll" footage, sometimes known as "cutaways." "A-roll" footage is your main event, the focus of the video content. But what else is going on? If there is an audience, shoot some footage of their reactions. Capture other aspects of the event. This will make your video more varied and lively.
23. Let us know where we are. Shoot some establishing shots of the location.
24. Record some room noise (nobody talking) on the videotape, for use in audio editing later.
25. Have fun! If your video is boring to you, it will be boring to the audience.

APPENDIX B. TEN TIPS FOR EDITING DIGITAL VIDEO

1. Unless you have unlimited hard drive space, capture only the parts of the video that you actually need. There is no reason to fill up your hard drive with footage that will only be trashed later.
2. If you will be selling the video, or if you envision the possibility of doing so, make sure that none of the elements (graphics, sound, video) that you add during editing are copyrighted. It will be much harder to replace those elements later.
3. Capture full clips and then edit later. It is easier to shorten a lengthy clip than to recapture or recover footage that has been either excised or not captured. I would recommend assembling an initial version of your project without concern for time length, and then trimming the individual clips down.
4. In most situations, editing should be as invisible as possible to the viewer. You want the audience to pay attention to the content of the video, not your ability to use transitions, filters, sound effects, and so forth. Choose simple transitions and avoid flashy effects, unless your goal is to create a music video or similar project.
5. Assemble your clips and give them an initial edit, before adding your sound effects, transitions, and so forth. Otherwise, your effects may become displaced or may need to be redone.

6. Pay attention to your audio. Make certain that both stereo channels are audible and of similar volume. Make sure that clips are of similar volume. Make sure that your soundtrack is loud enough—you do not want to have to require your viewers to increase their television volume above normal to hear your presentation.
7. Preview the "rough cut" of your video with people who are not involved in the production. Note when they seem to be interested and when they seem to drift off. Minimize the boring sections.
8. Keep all your raw editing files until you are absolutely sure that the editing phase is over, and that you will not need to produce a version in another format.
9. Even if your final product is a DVD, export your project onto MiniDV for archiving purposes. If you later decide you need to change or update the project, it is easier to capture it from MiniDV than it is from a DVD disk. Archiving onto a hard drive is another option, though a tape is cheaper than a hard drive.
10. Shorter is better. No matter whether the purpose is to educate or entertain, or a little of both, your audience will appreciate a short, sweet video presentation. This is true whether your audience is made up of squirmy kids, busy teachers, or exhausted parents.

APPENDIX C. THREE TIPS FOR DELIVERING DIGITAL VIDEO

1. Consider your audience's technology. Who will be watching your video and how will they access it? Obviously, handing out DVDs to parents who do not have players will not work; similarly, streaming high-bandwidth video to parents with a dial-up modem connection to the Internet will be a failure as well.
2. Consider the fact that you may have multiple audiences. Perhaps you need to choose multiple delivery methods for different segments of your audience. Perhaps a DVD for classroom use and VHS for parent copies. Or a Quicktime version for relatives to view on the Internet.
3. Consider future audiences. Is this a project you may want to show 10 years from now? If so, that small-frame CD-ROM version of your project may seem pretty low quality and antique. You may also have to run around to find a computer that still has a CD-ROM drive. Making high-quality copies of your projects in a number of formats will help you beat the built-in accelerated obsolescence of computer technology.

Chapter XV
Technology as a Classroom Tool: Learning with Laptop Computers

Ann E. Barron
University of South Florida, USA

J. Christine Harmes
James Madison University, USA

Katherine J. Kemker
University of South Florida, USA

ABSTRACT

Laptops and other ubiquitous devices in the classroom provide powerful opportunities to integrate technology as a classroom tool that supports student learning. However, effectively using laptops to achieve learning outcomes can be a daunting task for teachers and students—at least initially. This chapter examines the research on one-to-one computers and outlines learning outcomes that can be achieved when technology is used as a classroom tool. Sample lesson plans are provided to illustrate specific learning outcomes and highlight technology literacy issues, for both teachers and students participating in wireless classrooms. The integration of laptop computers into the curriculum can create collaborative, student-centered learning environments and increase student and teacher technology literacy.

INTRODUCTION

"Whether it's called a laptop program, one-to-one computing, ubiquitous computing, or 24/7 access, schools and school districts around the country are exploring the benefits and challenges of what happens when every student has a laptop computer" (Rockman, 2003, p. 24). One of the challenges of integrating laptop computers as classroom tools is the impact on the technology literacy of both teachers and students. As technology is merged with the curriculum, the definition of literacy evolves to new levels.

This chapter examines the research on one-to-one computers in the classroom, describes learning outcomes that can be achieved when technology is used as a classroom tool, and outlines technology literacy for students and teachers in a laptop environment. Sample lesson plans with specific learning outcomes are included to provide authentic examples of using technology as classroom tools.

TECHNOLOGY AS A CLASSROOM TOOL

Taylor (1980) categorized the use of computers in classrooms as tools, tutors, or tutees. In the 1980s, very few classrooms had computers for student use. A common configuration was to have a computer lab where students could work with computers once or twice a week as *tutors* (such as integrated learning systems) and *tutees* ("teaching" the computer by writing programs).

In the past 10 years, the number of classrooms with computers and Internet access has increased dramatically—due to a decrease in the price of computers, the E-rate program, and the proliferation of technology in our society. "In fall 2001, 99 percent of public schools in the United States had access to the Internet. When NCES first started estimating Internet access in schools in 1994, 35 percent of public schools had access" (Kleiner & Farris, 2002, p. 3). Although Taylor's categories are still useful today, the emphasis in many schools is shifting towards using computers as classroom *tools*.

As a tool, the computer can be used in classrooms to save time, access information, communicate effectively, and engage students in critical thinking. As Jonassen stated: "Carpenters use their tools to build things; the tools do not control the carpenter. Similarly, computers should be used as tools for helping learners build knowledge; they should not control the learner" (1996, p. 4). When using technology as a tool, students can focus their energies on higher-order tasks, and teachers can focus on facilitating learning.

STUDIES RELATED TO LAPTOP COMPUTERS

For technology to be an effective classroom tool, it must be available when needed—just as other classroom tools (such as pencils and paper) are accessible within a classroom. Providing laptop computers for students in a classroom is one way to address this issue, and laptops and other wireless technologies are being purchased by more and more school districts. The Quality Education Data report that was published in 2004 stated that during the 2004-2005 school year, "about half of the districts in the country will have portable wireless labs on carts (or COWs) to bring technology directly to students in their classrooms" (Quality Education Data, 2004, p. 1). This report was based on 493 surveys that were submitted by randomly selected public schools in the Quality Education Data's National Education Database (a registry of over 4.3 million educators). Using computers as a classroom tool has finally become a feasible option for many schools.

In the past few years, several large-scale implementations of laptop computers (such as Microsoft's Anytime Anywhere Learning Program and Maine's Learning Technology Endowment) have taken place (Lemke & Martin, 2003; Rockman et al., 2000). In addition, smaller laptop initiatives can be seen in schools and districts throughout the world.

Based on these laptop initiatives, several studies have been conducted to assess the educational benefits of laptops in the classroom. A variety of issues have been investigated with regard to the ubiquitous technology, including motivation, writing skills, student achievement, absentee rates, and perceptions. A synopsis of the following evaluations is presented in this section: Beaufort County School District (South Carolina); National Anytime, Anywhere Learning Initiative; Walled Lake Consolidated Schools (Tennessee); School District No. 60 (British Columbia); and Maine Learning Technology Endowment. These projects represent large, systematic studies that took place between 1997 and 2003.

Beaufort County School District (South Carolina, 1997-1999)

The Learning with Laptops program at Beaufort County School District in South Carolina began with 300 sixth graders in 1996 and expanded to sixth-, seventh-, and eighth-grade students by 1999. The students were provided with laptop computers and used them throughout the school day as electronic notebooks. In the report for the third year of the study, Stevenson (1999) provided results based on surveys completed by 105 teachers and 689 students. In addition, the achievement scores on a standardized test (1999 Metropolitan Achievement Test) were analyzed to determine if students who used laptops scored significantly better or worse than students who did not participate in the laptop project. The findings include (Stevenson, 1999, pp. 1-2):

- Both teachers and students responded positively regarding the impact of the laptop computer project. However, third-year student participants were less positive than the first- and second-year respondents had been.
- Students using the laptops continued to score well on standardized achievement tests. The third-year users in particular maintained their scoring advantage over non-users.
- Laptops were most often used in English/language arts, history/social studies, and science. They were not used to any extent in mathematics classes. Amount of use was also dependent on grade.
- Two of the biggest problems identified with the project were dispersion of laptop students (mixing laptop and non-laptop participants in the same class) and the mechanical reliability of the laptops. Teachers also indicated that lack of keyboarding skills among students was a problem, though students disagreed.
- Among students who had traditionally not found success in schools, those who participated in the laptop project performed better than those who did not. Free/reduced lunch students using laptops scored approximately the same on standardized achievement tests as students not on free/reduced lunch who were not laptop participants.

Microsoft/Toshiba (Anytime, Anywhere Learning, 1998-1999)

In 1996, Microsoft and Toshiba began the Anytime, Anywhere Learning project with 19 pioneer schools in the United States (Rockman et al., 1998). A series of studies was conducted by an independent research company and published as the project was expanded over the years. Using surveys, computer logs, writing samples, interviews, observations, and standardized test scores, the researchers documented the impact laptop computers had on students' achievement, behavior, and perceptions. In the 1999 study (the third in a series of studies), more than 400 students and 50 teachers participated in the evaluation process. The major findings include (Rockman et al., 2000):

- Laptop teachers showed significant movement toward constructivist teaching practices.
- Laptop teachers showed significant gains in how often they used computers for specific academic purposes.
- Laptop students performed better on the writing assessment than non-laptop students.
- Standardized test score comparisons were inconclusive.
- Comparison groups of laptop and non-laptop students showed less clear differences in some areas than the study conducted the previous year.
- Laptop students rated their confidence in computer skills more highly than non-laptop students.

Walled Lake Consolidated Schools (Tennessee, 2001)

Walled Lake Consolidated Schools in Tennessee also participated in the Anytime, Anywhere Learning project. Researchers from the University of Memphis and Wayne State University conducted evaluations for the implementation of laptops in fifth-, sixth-, and seventh-grade classrooms (Ross, Lowther, & Morrison, 2001). Teachers who participated in the project received over three weeks of training that centered around lessons designed to "engage the students in critically examining community and global issues, while strengthening student research and writing skills" (Lowther, Ross, & Morrison, 2001, p. 1). The analyses for Year 2 (2000-2001) included both quantitative and qualitative measures to collect data from eight schools (with approximately 250 students using laptops). "The data set for the evaluation included classroom observations, student achievement writing and problem-solving test scores, student surveys and focus groups, teacher surveys and interviews, and parent surveys" (p. 4). Classrooms with laptop computers were compared with computer-extended (CE) classrooms (which had 1 to 5+ computers). Findings include (Ross et al., 2001):

- The writing performance of laptop students was significantly better than that of CE students.
- Significant advantages for the laptop group were found in all aspects of problem solving.
- Perhaps the finding of greatest importance from the past two years is that full access to computers in the classroom, whether or not there is access at home, is what drives curriculum and learning most substantively (p. 97).

The third year of the project included a study that compared classrooms with laptop computers with classrooms that had a laptop cart. The student/computer ratio was one-to-one in both cases. Findings include (Ross, Lowther, Wilson-Relyea, Wang, & Morrison, 2003):

- There were relatively few differences in teaching methods between laptop and cart classes.
- The writing performance in the laptop classes was significantly higher than in the cart classes.
- Problem-solving scores in the laptop classes were significantly higher in five out of seven categories.

Teaching and Learning Initiative (Henrico County, Virginia, 2002-2003)

Henrico County Public Schools in Richmond, Virginia, has deployed over 27,000 wireless laptops in the past few years. High school students and staff received laptops in 2001-2002, and middle schools entered the program in 2002-2003. The goals of the project were to improve test scores, enhance quality of educational content, and raise graduation rates (Henrico County Public Schools, 2004). Students use the wireless laptops throughout the school, and they can take them home at night. Teachers post their assignments to the Web, enabling students to organize their resources and access them from any location. The following results have been reported:

- Higher levels of academic achievements were evident in core subject areas. For example, the scores in the Standards of Learning (SOL) tests showed the highest countywide performance ever in all four content areas.
- Dropout rates are at an all-time low (down to 1.52%).
- SAT scores rose by 13 points.

School District No. 60 (British Columbia, 2002-2003)

School District Number 60 is located in northern British Columbia, Canada. The district includes 5,600 students, served by 20 schools in rural, semi-rural, and urban areas. In February of 2002, the school district implemented the Wireless Writing Project by providing wireless technology in five classrooms (approximately 120 sixth- and seventh-grade students). The primary objective was to evaluate the potential for improving students' writing abilities. Using a variety of data collection methods (including impromptu writing samples based on BC Performance Standards for Writing; teachers' in-class assessment of student achievement; surveys of students, teachers, parents, and administrators; classroom visits; and student portfolios), the following results were reported (Jeroski, 2003, p. i):

- Improvements in writing achievement, as measured on controlled writing assessment and in-class assessments were strong and consistent.
- Teachers, parents, and students were all extremely enthusiastic about the use of iBooks and their impact on student achievement, motivation, and attitude.
- Teachers, parents, and students described positive changes in other aspects of achievement, most notably on technology skills, student attitudes, motivation, and work habits. Students appeared to be better organized, more responsible, and more confident.

Maine Learning Technology Endowment (Maine, 2002-2003)

In the Fall of 2002, every seventh-grade student in the state of Maine received a laptop computer to use at school. In some of the districts, the students were able to take the laptop home at night. In 2003, the program expanded to include both seventh and eighth graders, and involved 33,000 students and 3,000 teachers. Policymakers funded this program for three primary reasons: "increased economic viability for the state, higher levels of academic achievement, and a desire to close the digital divide" (Lemke & Martin, 2003, p. 2). The analysis of this program is still underway; however, researchers have reported the following, unanticipated results (Lemke & Martin, 2003, p. 3):

- Students are becoming respective, responsible "ambassadors" of the program.
- Teacher skepticism is down—and student retention is up.
- Parent-student communication is improving.

THEMES FROM LAPTOP STUDIES

Although it is difficult to synthesize the findings from these studies because of the substantial differences in research designs, implementation techniques, and available software, several themes emerge. With regards to academic achievement, the Rockman et al. (1998) study found comparisons of standardized test scores to be inconclusive; however, advantages for laptop users were noted on standardized tests in the Beaufort County (Stevenson, 1999) study. In addition, investigations of potential correlations between laptop use and writing achievement were found in school districts in Walled Lake and British Columbia (Jeroski, 2003; Ross et al., 2001). In addition, significant differences in problem-solving ability were noted in the Walled Lake report (Ross et al., 2001), and Henrico County is reporting increased test scores in all content areas (Apple Computer, 2003). Of particular interest were the findings related to the academic achievement of free/reduced lunch students in the Beaufort County study. One of the summary statements (referring to the standardized achievement data) was:

Most telling, however, may be the fact that the students who were on free or reduced lunch in 1995/96 as fifth graders, but have been three year laptop participants in middle school program, have continued as eighth graders to perform at the level of non-laptop students not qualifying for free or reduced lunch. When both groups were fifth graders, the non-free/reduced lunch group had significantly outperformed the free/reduced lunch group. (Stevenson, 1999, p. 26)

Another theme in the studies relates to an increase in technology literacy for both students and teachers. The Anytime, Anywhere Learning project reported that laptop students rated their confidence in computer skills higher than non-laptop students (Rockman et al., 2000). A similar finding was reported at the Walled Lake Schools (Ross et al., 2001). Likewise, teachers in the Anytime, Anywhere Learning project showed significant gains in how often they used computers for academic purposes (Rockman et al., 2000), and teachers involved in laptop projects in Maine and Tennessee also felt their computer skills had improved (Great Maine Schools Project, 2004; Lowther et al., 2001).

A few unexpected findings also emerged. Two of the multi-year studies indicated that there were fewer differences between laptop and non-laptop users in the more recent years. Rockman et al. (2000) reported, "When we first examined comparison groups of Laptop and Non-Laptop students in the 1997-98 school year, differences between the two groups were often stark. While group differences are still large for many measures, such as computer use and computer skills, in other areas differences have diminished" (p. 7). Stevenson (1999) found that although the student and teacher attitudes continued to be positive, the participants in the most recent year were more negative than participants in the first two years. This "narrowing of the gap" might be the result of increasing access to technology for all students (both at school and at home), or it could be that the Hawthorne effect provided a "halo" for early participants and initial results in the laptop classrooms.

LITERACY ISSUES FOR TEACHERS AND STUDENTS

Implementing laptops in the classroom has broad implications for the technological literacy of teachers and students. While these tools have great potential for improving teaching and learning, successful implementation requires professional development workshops and continued support for teachers. Teachers need integration models, technical support, and ideas for lesson plans. While it is not necessary that teachers become experts in the use of each software or hardware tool, they need to be skilled enough to recognize appropriate applications of the tools and to assist the students with their media projects. Providing training that includes modeling of technology integration, along with sample lessons, can be helpful strategies.

Many of the studies described in this chapter include factors related to the impact of the laptops on teachers' technological literacy. "For any professional development activity, teachers need time to plan, practice skills, try out new ideas, collaborate, and reflect on ideas" (North Central Regional Educational Laboratory, 2000, p. 5). Recognizing the need to prepare teachers, many of the laptop programs provided computers for the teachers six months to one year prior to deploying the laptops to the students. In addition, training sessions were available to prepare teachers for both the technological and the pedagogical aspects of the initiatives.

Students' literacy is also an important issue. First of all, students must have basic computer skills to operate the laptops. To address this issue, Henrico County required that students complete a basic training course that included the use, care, and troubleshooting techniques for the laptops—prior to deploying the laptops (Henrico County Public Schools, 2004). Secondly, an environment must be created in the classroom that allows students to develop both academic skills and technological literacy. In order to prepare students for the workforce, they must be provided the proper tools to be able to use technology effectively and confidently in the real world (Pearson & Young, 2003).

An important issue for laptop initiatives is the selection of software programs/tools for the computers. The majority of laptop computers come equipped with basic, tool-based software, such as word processing, spreadsheets, and presentation programs. However, when the laptop computer is equipped with more extensive tools, such as graphic organizers, digital video editing, databases, and multimedia authoring, the learning outcomes and literacy skills can be extended beyond basic skills. If students are to use computers to develop cognitive skills, they need access to the following resources (Apple Computer, 2003; Laptops for Learning Task Force, 2004):

- **Word Processing Programs:** Word processors can be used as classroom tools for students to develop their communication and basic thinking skills. Teachers can use word processors to create worksheets, templates, or digital documents for students.
- **Spreadsheet Programs:** Spreadsheets are another basic cognitive tool, often used to store and manipulate numerical data. Spreadsheets can be extremely versatile, requiring higher-order thinking skills, and allowing students to graphically display relationships to construct and communicate concepts.

- **Presentation Software:** An important part of technology literacy focuses on the ability to communicate ideas, thoughts, and concepts. The development process for presentations requires students to access, frame, and synthesize information so that it can be presented in a logical manner, incorporating text, video, and graphics as appropriate.
- **Graphic Organizer Programs:** Concept mapping is a strategy that requires learners to construct visual maps of mental models for others to understand or "see" the problem. Higher-order thinking occurs as students create models that visually interpret, collect, and utilize data (Jonassen, Howland, Moore, & Marra, 2003).
- **Database Programs:** Databases allow students to gather information, then organize and present the information in various formats. Constructing a database involves analyzing, synthesizing, and evaluating information.
- **Web and Multimedia Authoring Programs:** Authoring multimedia programs and creating Web sites provide students the opportunity to become content experts as well as designers and producers. Learning outcomes for authoring include communication, creative thinking, and critical thinking as students collaborate to develop a product that will effectively communicate a message or concept.
- **Video and Sound Production and Editing Programs:** A digital camera allows students to communicate a story or concept visually through pictures. Students are able to capture moments in time and bring them into the classroom to share with their peers. Like the invention of written language, the potential of using desktop movies in the classroom, in which the students are able to create their own movies, seems to represent a major landmark in human history (Forman, & Pufall, 1988).

LEARNING OUTCOMES

Means and Olson (1995) predicted that technology could expand learning opportunities for all students when it is implemented as a tool in instructional activities. However, technology alone cannot improve teaching and learning. Studies have found that success ultimately depends on how effectively the computer is integrated with the curriculum (Baker, Herman, & Gerhart, 1996; Glennan & Melmed, 1996; Schacter, 1999).

There is a wide range of ways that technology can be integrated as a curricular tool—from using a word processor for a report to recording video for an oral history. There are also many potential learning outcomes from technology-enriched activities. For the purposes of this chapter, we will classify learning outcomes into 5Cs:

1. Content/Basic Thinking
2. Communication
3. Collaboration
4. Critical Thinking
5. Creative Thinking

Based on the concepts of Jonassen's (2000) Integrated Thinking Model, the 5Cs provide a useful framework for developing learning activities for laptops.

Content/Basic Thinking

Content/basic thinking outcomes focus on core subject matter (in Science, Mathematics, Social Studies, etc.) and include the skills, attitudes, and dispositions required to learn and recall basic academic content and general knowledge. Activities that could be enhanced with technology might include having students label a map in a graphics program or create an animation explaining how to measure perimeter. Word processors, spreadsheets, and graphic organizers are especially effective in this category.

Communication

Communication skills involve the ability to convey ideas and information and to interact with others. Technology can be used to enhance and encourage communication skills by allowing students to illustrate ideas via photographs, sound, or video. Word processors, presentation software, digital video, graphic organizers, and authoring tools provide excellent avenues for conveying ideas.

Collaboration

Collaboration skills include the ability to distribute roles, accept responsibility, and interact effectively. Technology is a great tool for enhancing cooperation and social negotiation. Activities that would fall into this category might include collaborating on a digital video project or using a graphic organizer to brainstorm an idea. Designing and developing Web sites and multimedia projects also provide opportunities for students to work in groups and build social skills.

Critical Thinking

Critical thinking involves three critical skills—analysis, evaluation, and making connections (Jonassen, 2000). Analysis involves the ability to identify individual parts of relationships, conduct comparisons, and classify objects into categories. Synthesis focuses on activities that require students to integrate, combine, or connect ideas into a plan or project. Evaluation involves assessing the reliability, usefulness, and accuracy of information, on the basis of specific standards or criteria. Activities that would fall into this category might include using spreadsheets or databases to collect data, compare options, and make predictions. Editing video and producing multimedia projects also enhance critical thinking skills.

Creative Thinking

Creative thinking focuses on the generation of new knowledge and involves skills such as synthesizing, elaborating, and imagining (Jonassen, 2000). Elaborating involves the ability to modify, extend, and hypothesize. Imagining focuses on being able to visualize, speculate, and predict outcomes based on specific circumstances. Activities that would fall into this category include writing a story using a word processor, constructing a presentation, producing a video, or creating a multimedia program.

SAMPLE LESSONS

A great way to get started with a laptop initiative (from the teacher's perspective) is to examine case studies and sample lessons. There are several showcases of "laptop lessons" available on the Web. The lessons in this section were selected to illustrate a range of software programs/tools, subject areas, and learning outcomes. Each example outlines specific activities, literacy skills, and learning outcomes. These sample lessons were synthesized from detailed lesson plans that are available on the Web (corresponding URLs are provided for each lesson).

Body Length Experiment (Learning Outcomes)

In the lesson described in Figure 1, the basic content focuses on the use of centimeters for measurement. Critical thinking is required for comparing the length of specific body parts to the entire length of the body and for using data to support the conclusions and comparisons. Communication skills are necessary for creating tables and graphs to share the results of the data collection and for writing concluding paragraphs to convey the findings.

Body Length Experiment (Literacy Requirements)

Students need to use spreadsheets for data collection and creation of tables and graphs. They use a word processing program to write their concluding and comparison paragraphs. Teachers need to create templates for data collection and graph creation, as well as examples in a spreadsheet program.

Seaside Science (Learning Outcomes)

The lesson described in Figure 2 involves a field trip to the beach. Before going to the beach, students study basic concepts such as the difference between univalve and bivalve shells, the phyla of sea life, and the basics of the scientific method. At the beach, student work in collaborative groups to gather and classify shells and sea life. After collecting their specimens, students use their critical thinking skills to first classify the shells into categories and then

Figure 1. Lesson description for Body Length Experiment

Body Length Experiment
Author: Florida Center for Instructional Technology
Web site: http://etc.usf.edu/wireless/plans/lp0007.htm
Lesson description: Students in third through fifth grades work in pairs to measure the entire length of their bodies and individual different body parts. Each pair of students uses one worksheet (a template created by the teacher in a spreadsheet program) to enter their data. Students then create bar graphs based on their data. Finally, the students write a paragraph summarizing their results and a concluding paragraph to describe the relationship between the length of specific body parts and the length of their entire bodies, using data to support their conclusions.

Figure 2. Lesson description for Seaside Science

Seaside Science
Author: Submitted by Rhonda Bajalia at Toshiba/NSTA Laptop Learning Challenge
Web site: http://www.nsta.org/programs/laptop/lessons/e1.htm
Lesson description: Students take a field trip to the beach and observe sea life. They work in groups of four to gather and classify shells as univalve or bivalve, and they classify sea life according to phyla. Groups then use this data to create frequency tables and graphs. The teacher performs a desalinization experiment following the scientific method, and students form hypotheses and draw conclusions before submitting a final report.

create a pie graph for the entire class showing the ratio of univalves to bivalves. Using these same skills, students then classify the various sea life into phyla, and create a frequency table and bar graph to communicate the distribution of sea life by phyla. Critical thinking is necessary for interpretation of the desalinization experiment. Communication skills are required for writing up a final report, including presentation of data in charts and graphs.

Seaside Science (Literacy Requirements)

This activity involves the use a spreadsheet program to create templates for frequency tables, pie graphs, and bar graphs. Students would need to be able to work with spreadsheet templates, entering data that would create charts and graphs. In order to take notes and write a final report, they need to be able to use a word processing program. Teacher literacy requirements include creating spreadsheet templates and supervising students in the creation of charts from data.

Baffling Biomes (Learning Outcomes)

Basic content in the lesson that is described in Figure 3 includes identification of the major biomes of the world and their characteristics such as animal life, climate, and vegetation. Production of the multimedia project requires creative thinking—the only specifications given are that the project must be a movie that includes the physical features, wildlife, vegetation, and climate of the biome. Critical thinking is needed as students must also identify and explain potential problems that their biomes may be facing. Collaboration is key as students work in cooperative groups to research their biome and create the final project. The communication outcome is covered by students using the multimedia project to convey their ideas to the rest of the class.

Baffling Biomes (Literacy Outcomes)

In order to complete this lesson, the teacher and students must be familiar with Internet re-

Figure 3. Lesson description for Baffling Biomes

Baffling Biomes
Author: Submitted by Dorothy E. Henson at Apple Learning Interchange
Web site: http://ali.apple.com/ali_sites/ali/exhibits/1000871/
Lesson description: In this lesson, students begin by studying biomes of the world. Students are placed into groups of five, and each student draws from a hat to determine which biome he or she will study. Each group must create a multimedia project. They begin their project by using books and recommended Internet sites to research their biomes. Then they create storyboards for the multimedia project, which are submitted to the teacher. Once the projects have been planned, the class visits a zoo and an aquarium, where students can capture the video and audio necessary for their projects. Students use iMovie or a similar program to produce their projects and present them to the entire class.

Figure 4. Lesson description for Ecosystem

Ecosystem
Author: Florida Center for Instructional Technology
Web site: http://etc.usf.edu/wireless/plans/lp0016.htm
Lesson description: In this example, fifth-grade students were studying about the ecosystem at their school. The teacher sectioned off small plots of ground on both the sunny side of the school and the shady side of the school. Students were grouped into pairs, and each pair was assigned a one-foot-by-one-foot plot of ground—half of the class on the sunny side of the school, the other half in the shade. They were instructed to count the number of living (such as plants and animals) and non-living (such as dirt and rocks) things, and enter their data into a spreadsheet. They did this for three days, examining a different plot each day. When they were finished with data collection, they created a bar graph from their data. They then wrote a paragraph about each individual ecosystem and a paragraph comparing the three ecosystems. In addition to using spreadsheets to record and synthesize data, students took examples of the living and non-living things they found in their plots of land and scanned them into digital images. As a culminating activity, students created a poster that included data, a digital picture of what they scanned, and a written report.

search, using a digital camera, and editing digital video and digital audio.

Ecosystem (Learning Outcomes)

In the Ecosystem lesson (described in Figure 4), basic content outcomes include identifying the plants and animals living in various environments in the community. Another core concept is using quarter percents to figure the amount of dirt in each plot of land. Students are collaborating as they work in pairs to complete the project. Using communication skills, students must create a poster with representations of data and digital pictures to convey their ideas to their classmates. Another communication outcome is a written report based on the data and their observations. Critical thinking outcomes include estimating the amount of grass in a whole plot, and solving a problem by generating, collecting, organizing, displaying, and analyzing data using a bar graph.

Ecosystem (Literacy Requirements)

In order to complete this lesson, students would need to be literate in basic computer operations, as well as the productivity tools of spreadsheets and word processors. Scanners and accompanying graphics software would also be needed. In addition to being comfortable with graphics software, teachers need to be able to create spreadsheet templates that the students can use for entering data.

Flight Project (Learning Outcomes)

The interdisciplinary Flight Project lesson (see Figure 5) showcases how content/basic thinking skills can be achieved in several areas, such

Figure 5. Lesson description for Flight Project

Flight Project
Author: Submitted by Laura Richter at Maine's Learning Technology Initiative
Web site: http://www.msad54.k12.me.us/MSAD54Pages/SAMS/cedarsite/Lindbergh/connections.htm
Lesson description: This interdisciplinary lesson for middle school students is based around the travels of a suitcase, as it follows the route that Charles Lindbergh took on his flight to Paris. Over a two-year period, the suitcase travels this route and at each stop items are added to the suitcase. Once the suitcase returns to the school, teachers from several subject areas use it as a springboard for lessons in geography, history, physics, geometry, and language arts. For example, the science teacher created a lesson on the Bernoulli Principle and other principles of flight. The social studies lessons focused on geography along the route and social issues of the time period, and math lessons involved plotting the circular route of flight and navigation. As a final, culminating project, students create a documentary, performance, WebQuest, exhibit, or picture book that includes elements from each subject area. Throughout the project, students are involved in activities such as using e-mail to communicate, conducting Internet research, and creating Web sites to track the progress of the suitcase.

as physics, history, and geography. Students work in collaborative groups on various aspects of this interdisciplinary unit. Communication skills are essential, as students use e-mail and listservs to communicate with pilots and schools along the route to track the progress of the suitcase. The culminating projects involve creative thinking skills in that students are required to synthesize content from all of the disciplines and create a product. Critical thinking is also necessary to examine relationships across the subject areas.

Flight Project (Literacy Requirements)

Teachers would need to be literate in the use of e-mail, listservs, Web authoring, word processing, multimedia presentations, and video editing. They would also need to have Internet research skills and be able to provide guidance for students in search strategies and techniques for critiquing information obtained from Web sites. In order to complete the projects, students need skills in word processing, multimedia presentations, and video editing. In order to communicate with outside experts, skills in using e-mail, listservs, and Web authoring tools are also required.

CONCLUSION

By examining case studies and "lessons learned" from other schools, teachers can prepare themselves and their students to truly integrate technology as a classroom tool. As evidenced in the sample lessons and the matrix presented in Figure 6, laptop lessons can target several learning outcomes and increase technology literacy for both teachers and students.

Figure 6. Matrix of sample lessons, learning outcomes, and literacy requirements

Lesson	Learning Outcomes	Technology Literacy
Body Length Experiment	Content/Basic Collaboration Communication Critical Thinking	Spreadsheet Word Processing
Seaside Science	Content/Basic Collaboration Communication Critical Thinking	Spreadsheets Word Processing
Baffling Biomes	Content/Basic Communication Collaboration Critical Thinking Creative Thinking	Internet Research Digital Video Digital Photographs
Ecosystem	Content/Basic Communication Collaboration Critical Thinking	Spreadsheet Graphics Editing Word Processing
Flight Project	Content/Basic Collaboration Communication Critical Thinking Creative Thinking	Internet Research Web Site Creation Digital Photographs Digital Video Word Processing

Providing computers for students to use as tools in the classroom is now a feasible option for many schools. However, successful learning activities depend on more than simply having access to technology. By focusing on learning outcomes, providing the software tools necessary, ensuring that teachers and students have the requisite skills, and integrating technology when appropriate, an effective learning environment that is conducive to using technology as a tool can be achieved.

REFERENCES

Apple Computer. (2003). *Profile in success: Henrico County Public Schools.* Retrieved May 31, 2004, from http://www.apple.com/education/profiles/henrico2

Baker, E. L., Herman, J. L., & Gearhart, M. (1996). Does technology work in schools? Why evaluation cannot tell us the full story. *Education and technology: Reflections on computing in classrooms.* San Francisco: Jossey-Bass.

Forman, G., & Pufall, P. B. (Eds.). (1988). *Constructivism in the computer age.* Hillsdale, NJ: Lawrence Erlbaum.

Glennan, T. K., & Melmed, A. (1996). *Fostering the use of educational technology: Elements of a national strategy* (MR-682-OSTP/ED). Santa Monica, CA: RAND. Retrieved April 8, 2002, from http://www.rand.org/publications/MR/MR682/

Great Maine Schools Project. (2004, February). *One-to-one laptops in a high school environment. Piscataquis Community High School study: Final report.* Retrieved June 1, 2004, from http://www.mitchellinstitute.org/Gates/finalLaptopreport.doc

Henrico County Public Schools. (2004). *Teaching & learning initiative.* Retrieved May 31, 2004, from http://www.henrico.k12.va.us/ibookm

Jeroski, S. (2003, July). *Wireless writing project: Research report phase II. School District 60, Peace River North.* Retrieved June 1, 2004, from http://www.prn.bc.ca/Wireless_Writing_Program.html

Jonassen, D. H. (1996). *Computers in the classroom: Mindtools for critical thinking.* Englewood Cliffs, NJ: Prentice-Hall.

Jonassen, D. H. (2000). *Computers as mindtools for schools: Engaging critical thinking.* Columbus, OH: Prentice-Hall.

Jonassen, D. H., Howland, J., Moore, J., & Marra, R. M. (2003). *Learning to solve problems with technology: A constructivist perspective.* Upper Saddle River, NJ: Prentice-Hall.

Kleiner, A., & Farris, E. (2002). *Internet access in U.S. public schools and classrooms: 1994-2001.* (NCES 2002–018). U.S. Department of Education, National Center for Education Statistics. Washington, DC: U.S. Government Printing Office.

Knapp, L. R., & Glenn, A. D. (1996). *Restructuring schools with technology.* Boston: Allyn and Bacon.

Laptops for Learning Task Force. (2004, March). *Laptops for learning: Final report and recommendations of the Laptops for Learning Task Force.* Retrieved June 2, 2004, from http://etc.usf.edu/L4L/

Lemke, C., & Martin, C. (2003, December). *One-to-one computing in Maine: A state profile.* Retrieved May 22, 2004, from http://www.metiri.com/NSF-Study/ME-Profile.pdf

Lowther, D., Ross, S. M., & Morrison, G. R. (2001, June). Evaluation of a laptop program: Successes and recommendations. *Proceedings of the National Educational Computing Conference,* Chicago, IL (ERIC Document Reproduction Service No. ED462937).

Means, B., & Olson, K. (1995). *Technology and education reform: Technical research report. Volume 1: Findings and conclusions.* Menlo Park, CA: SRI International.

North Central Regional Educational Laboratory. (2000). *Pathways to school improvement: Providing professional development for effective technology use.* Retrieved June 2, 2004, from http://www.ncrel.org/sdrs/areas/issues/methods/technlgy/te1000.htm

Pearson, G., & Young, A. T. (Eds.). (2003). *Technically speaking: Why all Americans need to know more about technology.* Washington, DC: National Academies Press.

Quality Education Data. (2004). *K-12 technology budget is projected at $7.06 billion for 2004-2005 school year.* Retrieved November 16, 2004, from http://www.qeddata.com/TFPPressrelease.htm

Rockman et al. (1998, September). *Powerful tools for schooling: Second year study of the laptop program.* Retrieved March 21, 2004, from http://rockman.com/projects/laptop/laptop2exec.htm

Rockman et al. (2000, June). *A more complex picture: Laptop use and impact in the context of changing home and school access.* Retrieved March 21, 2004, from http://rockman.com/projects/laptop/laptop3exec.htm

Rockman, S. (2003, Fall). Learning from laptops. *Threshold,* 24-28. Retrieved May 22, 2004, from http://www.rockman.com/articles/LearningFromLaptops.pdf

Ross, S. M., Lowther, D. L., & Morrison, G. R. (2001, December). *Anytime, anywhere learning. Final evaluation report of the laptop program: Year 2.* Retrieved January 19, 2004, from http://www.nteq.com/Research/Laptop%20Yr2%20Final%2012-10-01.pdf

Ross, S. M., Lowther, D. L., Wilson-Relyea, B., Wang, W., & Morrison, G. R. (2003, December). *Anytime, anywhere learning. Final evaluation report of the laptop program: Year 3.* Retrieved November 19, 2004, from http://www.nteq.com/Research/Laptop_Report_Yr3.pdf

Schacter, J. (1999). *The impact of education technology on student achievement.* Santa Monica, CA: The Milken Exchange on Education Technology. Retrieved April 9, 2004, from http://www.mff.org/publications/publications.taf?page=161

Stevenson, K. R. (1999). *Evaluation report—Year 3: Middle school laptop program.* Beaufort County School District: Beaufort, SC, laptop project. Retrieved January 18, 2004, from http://www.beaufort.k12.sc.us/district/evalreport3.htm

Taylor, R. P. (Ed.). (1980). *The computer in the school: Tutor, tool, tutee.* New York: Teachers College Press.

Chapter XVI
Tapping into Digital Literacy: Handheld Computers in the K-12 Classroom*

Mark van 't Hooft
Kent State University, USA

ABSTRACT

This chapter describes the integration of handheld computers in K-12 classrooms and its impact on digital literacy. Following a brief description of this new technology for education, teacher stories are used to illustrate what types of educational activities are possible above and beyond what is possible with available technology, what pedagogical changes need to be made to effectively integrate handheld technology in K-12 classrooms, how handheld devices can be adapted to harness their full potential as ubiquitous devices for teaching and learning, and how digital literacy skills influence and are being influenced by this technology. The ultimate goal of the author is to show that handheld computers have the potential to have a tremendous impact on teaching and learning, given the right context.

INTRODUCTION

When it comes to technology, the world in which we live today is very different from the one that existed 10 or 20 years ago. New developments and inventions occur on a daily basis, including phenomena such as hybrid automobiles, human cloning, and nanotechnology, changing the ways in which we go about our lives. Education is affected like any other field through the continuous introduction and integration of new tools such as digital imaging and video, the Internet, wireless technologies, and more recently, personal technologies like mobile phones and handheld computers. These new tools have the potential to fundamentally change teaching and learning when integrated appropriately and under the right conditions.

The development of handheld devices can be traced back to the 1970s, starting with Xerox PARC's research into the Dynabook concept, a highly mobile, notebook-sized computer with artificial intelligence capabilities. This was followed by the development of related devices such as the Psion I (1984), GRiDPaD (1988), Amstrad's PenPad and Tandy's Zoomer (1993), the unsuccessful Apple Newton, which was in development and production for about 10 years (1993-1995), and the eMate (1997-1998). However, while others struggled, US Robotics (bought in 1997 by 3Com) introduced the Palm Pilot in 1996, featuring a graphical user interface, text input using *Graffiti* handwriting recognition software, and a cradle for data exchange with a desktop computer. This device became the forerunner of several generations of devices powered by the Palm OS, ranging from the Palm Pilot 1000 to current handhelds like the Tungsten E and Zire72 (Bayus, Jain, & Rao, 1997; Williams, 2004), and a plethora of peripherals. During the same time, Microsoft also actively pursued the development of a portable device, modifying its Windows operating system to fit on handhelds produced by such companies as HP, Dell, and Compaq. This development did not have a real impact on the mobile computing market until Microsoft's release of Windows CE 2.0 in 1997, and the Handheld PC Professional and Windows Mobile 2003 Operating Systems (HPC Factor, 2004).

Handheld computing enthusiasts have been advocating the use of these small and portable devices in classrooms in an effort to get closer to a truly ubiquitous computing environment. The term "ubiquitous computing" was defined in 1991 by Mark Weiser from Zerox PARC as an environment in which "a new way of thinking about computers in the world...allows the computers themselves to vanish into the background" and become indistinguishable from everyday life (p. 94). Weiser emphasized that ubiquitous computing in this sense does not just mean portability, mobility, and instant connectivity, but the existence of an environment in which people use many computing devices of varying sizes (which he described as tabs, pads, and boards) that interact with each other, combined with the aforementioned change in human psychology to the point where users have learned to use the technology well enough that they are no longer consciously aware of its presence and do not have to be. While the change in our knowledge and use of a wide variety of computing devices is not yet at the level that Weiser envisioned more than a decade ago, we are much closer to reaching the technological requirements: "cheap, low-power computers that include equally convenient displays, a network that ties them all together, and software systems implementing ubiquitous applications" (Weiser, 1991, p. 99).

Weiser's vision of ubiquitous computing fits well with current visions of technology integration in education and its potential impact on teaching and learning. Academic research has shown that computer use and student learning gains are "closely associated with having computers accessible to all students in teachers' own classrooms" (Becker, Ravitz, & Wong 1999; see also Marx et al., 2000; Norris & Soloway, 2001; Soloway et al., 2001). A 1:1 student-to-computer ratio is needed to make computing in schools truly personal and meaningful, but for many school districts, attaining this ratio is a financial impossibility when desktop or laptop computers are considered (Norris & Soloway, 2001). Handheld devices seem to provide a more realistic alternative for integrating technology into the classroom to create a ubiquitous computing environment and meeting the challenges of improving student achievement, because of their small size and comparatively low cost in acquisition and ownership (Hennessy, 1997, 2000; Robertson et al., 1996; Sharples, 2000a). As a result, handhelds are starting to make their way into classrooms, often supplementing the existing technology infrastructure. Some scholars have defined the resulting learning environment as "handheld-centric," with students having access to and actually using a variety of equipment besides handheld computers, including networked PCs, probeware, and digital cameras. The value of a handheld-centric classroom thus becomes "providing all students with access to valuable resources on a shared but timely basis," where each tool has been earmarked for its intended use (Norris & Soloway, 2004; Tatar, Roschelle, Vahey, & Penuel, 2003).

Because they have the potential to create a truly ubiquitous computing environment and change the role of existing technology in classrooms, handheld devices are also altering the nature of technology integration in teaching and learning, and can act as catalysts for radical changes in pedagogical practices. Fung, Hennessy, and O'Shea (1998) describe this changing role of technology as a paradigm shift, comparing it to the historic shift in reading from initially being done as an elitist activity in centers of learning such as monasteries and universities to an integral part of everyday life. In the case of handheld computing, the fundamental difference from the more traditional desktop computing environment lies in the fact that in addition to the more traditional uses, users "interacting with a mobile system interact with other users [and] interact with more than one computer or device at the same time" (Roth, 2002, p. 282; see also Cole & Stanton, 2003; Danesh, Inkpen, Lau, Shu, & Booth, 2001; Mandryk, Inkpen, Bilezkjian, Klemmer, & Landay, 2001). Therefore, handheld computers lend themselves well for both individual and collaborative learning if used appropriately. Roschelle and Pea (2002), for example, highlight three ways handheld devices have been used to increase learning collaboratively—classroom response systems, participatory simulations, and collaborative data gathering—and suggest there are many more such uses (see also Danesh et al., 2001; Mandryk et al., 2001; Roschelle, 2003; Roschelle, Penuel, & Abrahamson, 2004).

Moreover, because of their small size, handheld computing devices no longer constrain the user like desktops and laptops do. As such, they enable students to take the initiative and explore, allowing for a more authentic and deeper immersion in technology, not as a separate subject of study, but as an integrated part of the whole curriculum. Taking this idea beyond the classroom, handhelds encourage the use of technology in everyday activities and enable students to understand the computer as a lifelong-learning tool anywhere, anytime (Inkpen, 2001; Sharples, 2000b), eventually leading to the type of ubiquitous computing that Weiser envisioned.

Paired with the increasing integration of technology in K-12 education is a need for expanded literacy skills for teachers and students. Traditionally, being literate meant that one could access, evaluate, and use information from a variety of sources. Therefore, an information-literate person recognizes the need for accurate and complete information for decision making; decides what information is needed and where to obtain it; accesses, evaluates, and organizes this information; integrates new information into existing knowledge; and uses it for critical thinking and problem solving (Doyle, 1992; Langford, 1999).

The introduction of technology into the classroom has put a virtual flood of information in the hands of teachers and students. Literacy skills as originally defined, without technology, are still essential. With the rising popularity of the Internet, mobile and wireless technologies, and the explosion in data collection, processing, and storage, there is a more pressing need for educators to teach students how to find, sift, process, and analyze data, and make meaning of it all. Therefore, it is now more important than ever for teachers to help students acquire the necessary literacy skills to deal independently and effectively with massive amounts of information (Fitzpatrick, 2000; Rice & Wilson, 1999; Risinger, 1998; Saye, 1998). These skills include learning how to use traditional literacy skills to deal with new forms of information, and new digital literacy skills to amplify existing skills with technology tools. For example, according to Gilster (1992), digital literacy skills address the fact that information is no longer limited to text but also includes still images, video, sound, and interactive Web pages. Second, information retrieval has changed from being mostly book based to being more Internet based, requiring increased information construction from multiple sources. Finally, digital literacy is interactive and multidimensional, and requires the ability to read, evaluate, integrate, and use resources from multiple sources, and communicate these newly constructed pieces of knowledge to others.

In sum, handheld devices possess certain characteristics that allow for frequent and immediate access to a wide variety of tools and information sources for teachers and students. This requires more and increasingly refined information and digital literacy skills, especially when considering Weiser's vision of ubiquitous computing, where the focus is no longer on the tools but on their use. Within this theoretical framework, let us now turn to some specific examples of handheld use for teaching and learning, in order to explore:

- what types of educational activities handheld devices make possible above and beyond what is currently possible with available technology to improve teaching and learning;
- what pedagogical changes need to be made to effectively integrate handheld technology in K-12 classrooms;
- how handheld computers can be adapted to harness their full potential as ubiquitous devices for teaching and learning; and
- how literacy skills are influencing and are being influenced by this new technology.

The teachers who tell the stories that follow are participating in an ongoing investigation by the Research Center for Educational Technology (RCET, http://www.rcet.org) at Kent State University (Ohio, USA) into the use of different types of handheld devices in K-12 education and their impact on teaching and learning. The research has its origins in RCET's participation in the Palm Education Pioneer project (http://www.palmgrants.sri.com/background.html) in 2001-2002 (see also Vahey & Crawford, 2002; van 't Hooft, Diáz, & Andrews, 2003; van 't Hooft, Diáz, & Swan, 2004). RCET researchers monitored the use of about 280 Palm™ IIIc devices in the classrooms of 11 K-12 teachers in 10 local schools, teaching in a variety of (subject) areas including computer science, math, science, language arts, social studies, and special education. Since then, they have also engaged in research related to the use of Alphasmart's® Dana™ in science education, as well as the integration of Texas Instrument's TI-83 Silver Edition graphing calculator in preservice secondary social studies education. While the stories describe the use of handhelds at particular grade levels, this use could easily be adapted for students at different age levels and abilities, making them a good testimony for the flexibility of handheld computers and their potential for integration into the curriculum.

The first two stories were written by an elementary special needs teacher who integrates a variety of technology tools in her teaching, including desktops, laptops, handheld computers, and digital cameras. As she describes below, the main advantage of having a 1:1 student-handheld ratio for her and her students has been the ability for the technology to be adapted to *individual* students and their particular disabilities. She teaches grades 1-4 at a suburban elementary school; her class size ranges from 10-15 each year, and includes students with a variety of socioeconomic/ethnic backgrounds and learning disabilities.

THE CHAIR

—by Karen McClain, Stow, Ohio, USA

Each time I walked into the school office, the old wooden armchair in the corner of the room caught my eye. It had curved arms and beautiful turned spindles for legs. I was curious about the history of the chair and about the stories the chair could tell if it could talk. Our school was built 64 years ago, and I suspect the chair was a piece of office furniture purchased at the same time.

Daily, children would sit patiently in that chair, waiting to see the principal. Sometimes they had good news to share. Sometimes they were waiting because they broke a school rule. Children would sit with tears in their eyes waiting for a parent to take them home because they were ill or had missed the school bus. Some children sat in the chair because their parents had important business with the principal. Seldom did I see an adult occupying the chair. The chair was certainly large enough for an adult, but it was assumed to be a chair reserved for children.

I decided to ask the principal if my class, a special education class, could paint the chair to make it a little brighter for the children who regularly occupied it. When she gave her approval, I began to write a plan that would make the project meaningful for my students. After much discussion about the chair, the class chose to incorporate book titles into the overall theme of the chair. The students wanted to ask every student in the school to nominate a favorite book title for the chair. The librarian agreed to allow the class to survey the student body during library checkout time. Armed with handheld computers loaded with *WordSmith*, the *Memo Pad,* and foldable keyboards, the fourth-grade resource room students went to the library with every class in the school. At the end of the week, they had collected over 150 book titles.

The children sorted the book titles by grade level and created ballots with only the top three or four titles from each grade level. They returned to the library the next week and asked each student to vote for his or her favorite book title. The ballots were sorted by classroom and the results were entered into a student-created spreadsheet on the handhelds.

The initial data was collected on the handhelds using the spreadsheet application in the *Documents-to-Go* software suite. The children transferred the information from their handhelds to laptops using a USB cradle and opened their data in *Microsoft Excel.* They had to organize the spreadsheet to show each classroom's favorite book titles and create a spreadsheet for each grade level. They were able to copy and paste each room's totals into a new spreadsheet that reflected the appropriate grade level, enter the SUM function, and have the total book title counts for the grade level. Using this data, the students created a series of charts and graphs. A bar graph was created for each classroom. Grade level results were illustrated in a pie chart. All of the graphs were displayed in the hallway and eventually given to each classroom.

Finally it was time to paint the chair. We researched houses on a Web site specifically featuring painted lady houses, discussing the colors that were used and the way the painters often painted each section of a spindle a different color. The children selected a base color for the chair and five additional pastel colors. All of the resource room students, grades one through four, participated in the painting. The first step was to sand the chair and arrange for the custodian to glue the loose joints. Then the children took digital black and white pictures of all sides of the chair. They used colored pencils to color their digital photographs. Over the next two weeks, the children took turns painting the many different parts of the chair. When the children were satisfied that they had fully painted the chair, we enlisted two parent volunteers to paint the titles of 10 favorite books on the flat surfaces of the chair.

Since I always try to encourage the children to visit the local public library during the summer months, I arranged for the chair to be placed on display in the children's room of the library. Periodically, I stopped by to check on the chair. The librarians told me that my students visited the chair regularly as well. They always made it a point of explaining which part of the chair they had painted.

At the end of the summer, the chair was returned to its familiar spot in our school office. Every day children sit in the chair waiting for the principal, the nurse, or a parent. I always smile when I see them reading the book titles. My hope is that a title catches a child's eye and that that child becomes curious enough about the book to check it out from the library and read it!

THE GINGERBREAD VILLAGE

—by Karen McClain, Stow, Ohio, USA

Prior to the winter holiday, I always try to teach the geometry math unit. The shapes of holiday decorations easily lead to discussions about geometric shapes and figures. The children are interested in the holiday and a culminating activity that involves the creation of a decorated gingerbread house using geometric shapes and figures leaves an imprint on each child's memory. I have found that my students easily recall the math concepts after the winter break, and believe that they are still able to recall the geometry unit when taking the state-required assessments in math.

To get started, the children drew a floor plan of their favorite room at home using handheld computers and the drawing application that is on the *Print Boy* menu. They estimated the actual length of each wall and each piece of furniture, and entered their estimates into the *Memo Pad* application on the handheld. The next day, each student took a six-foot tape measure home and measured the actual length of the walls, windows, doors, and furniture in their favorite room. I asked them to compare the actual size of each measurement with the estimates on their handheld sketches. From this discussion, they were able to better understand how important it is to have actual measurements with a ratio to build from a blueprint. The second- and third-grade students then used the handheld calculator to calculate the perimeter of their rooms. The fourth graders calculated the perimeter and the square footage of their rooms.

Next, my students used the *Memo Pad* application to list as many geometric shapes and figures as they could find in the classroom. They earned certificates for finding the most unusual figure, the most figures, congruent figures, symmetric figures, and so on. We played a form of "I Spy" to find shapes and figures around the school. I also encouraged the children to use a variety of senses to discover the shapes and figures. They used their visual sense by just looking for a shape or figure. The tactile sense was used by tracing objects with their fingertips. The kinesthetic sense required tracing a larger object using their hands and arms. This usually involved a door, window, or a piece of furniture. Finally, my students used their auditory sense by listening for a sound that drew them to an object, such as the ticking of a clock. For some children, the aroma of dinner cooking led them to a shape in the kitchen. In this case, they were actually using the olfactory sense to discover more shapes and figures.

After several lessons that centered on shapes and figures in and around the house, it was time to work on the culminating activity, which was to make a gingerbread house. I found pre-made houses made of a synthetic material which simulated gingerbread at a local craft store. I was able to get four different styles, so that we had a variety of homes when the activity was completed. I recruited my sister as an extra hand for this activity. First, students selected a house to decorate. They drew the footprint of the house on the drawing application on the handheld, then took digital pictures of their houses using the *Kodak PalmPix®* camera attachment. Next, students went back to their notes on the *Memo Pad* and determined a ratio that would be an appropriate scale for the gingerbread house to a real house. Each child was certain that his/her gingerbread house would be a 'mansion' if it were enlarged to the ratio they selected. They incorporated walls, doors, windows, and eventually all of the furniture important to a child. Some drawings had fireplaces and others had entertainment centers complete with video games. All of the children had televisions (big screen!) and computers in their dream houses.

Finally it was time to mix up the icing recipe used to cement the candy to the houses and begin adding real candy to each pre-fabricated house. We had strict rules about how much candy could be eaten during the project! We made sure we had a large variety of candy in many shapes and colors. During the decorating process the children took lots of pictures, using either the handheld cameras or a regular digital camera. We saw lots of sharing of ideas and heard wonderful conversations. Without prompting, the children called the candy by shape. "I need a rectangular prism for my window," said one child. Another student added a flat round piece of candy to his chimney and told me, "The circle on my roof is my satellite dish!" When we asked a second grader if we could help him stand up the snowman candy that was laying flat in front of his house, he promptly told us that his snowman was "making a snow angel!" That quickly led to a demonstration on the classroom floor and a discussion about symmetry.

After the houses were complete and arranged on a table to represent a village, the children wrote stories about living in a gingerbread village. They completed the stories on *Word-To-Go*, the handhelds enabling them to take their stories and write outside of the classroom. Students shared their stories using the infrared beaming capabilities on the handhelds to beam them to each other. When they took their mansions home for the winter break, they also took their handheld computers, containing 15 stories written by their classmates to read during the holidays. The stories were rich with description. The vocabulary the children used included the math terms they had studied. The excitement they shared with each other was contagious.

WADING INTO SCIENCE WITH HANDHELD COMPUTERS

—by Shawn Jones, Kent, Ohio, USA

This fifth-grade teacher started working with us about a year ago. He teaches at a local elementary school that provides educational services to students from working class/ university neighborhoods. As a result, the student population tends to be heterogeneous, with class sizes of about 20-25 students. Even though he was a novice when it came to handheld devices, he and his students became proficient in their use very quickly. Their story, as told by him here, is evidence of that. — by Mark van 't Hooft

As part of a water quality monitoring project of the Ohio Department of Natural Resources (ODNR, http://www.ohiodnr.com/dnap/monitor/default.htm), the County Soil and Water Conservation District, and Kent State University's Desktop Video Conferencing Project, my fifth-grade students have been monitoring the water quality in a local stream for almost two years. It has enabled me to provide students with authentic and meaningful learning experiences. Because of their small size and mobility, Palm OS-based handheld devices and Pasco science probes have played a huge role in this project, especially with respect to on-site data collection, and the ability to carry digital data back to the classroom for immediate analysis and feedback.

At the site, my students have been collecting information about the stream by turning over rocks in the water and catching the fleeing organisms in a waiting net. These organisms are then counted, recorded, and returned to the stream. The amount and variety of organisms found provides us with an indication of the quality of the water. In addition, students record characteristics of the water itself using probes attached to handheld computers. The devices and probes we are using allow us to measure up to three different variables at once, in this case water temperature, pH levels, and force of the water flow. Because the devices are small and easy to carry, even when the probes are connected to the handhelds, kids have been able to go wherever they needed to go to collect data, including the middle of the streambed. In addition, the data they have been collecting is recorded straight into *Data Studio*, a data analysis program on their handhelds, and can be viewed numerically or graphically as it is being collected or transferred to a desktop computer afterwards.

While we really only needed the temperature and pH probes to measure the water quality, I decided to take the force meter as well. The reasoning behind this is that we were studying force in science and students were having a difficult time understanding how much force a Newton really represents. In order to give students a real-life example, we connected a force meter to a handheld and tied a small parachute made out of string and plastic to the meter. Students took turns holding the parachute in the stream and letting the water drag it, while looking at the readings on the handheld. The students could actually feel and see at the same time how much force is exerted at varying amounts of Newton. Another nice feature about having the handheld computer was that if we had to do any calculations or conversions (like Newtons to grams to pounds), we had the calculator right there to help us with that.

Back at school, we transferred the data to our desktops, and projected the graphs onto a large screen for collaborative analysis and discussion of our findings. We compared our findings from the fall to the spring as well and created a double bar graph to compare the two trips. In addition, using IP-based videoconferencing we have been able to compare our findings with other classrooms in Ohio and the ODNR, as well as consult natural resource experts at Kent State University.

Even though the water quality project is great in itself, the mobile technology has added another dimension to it. Students were excited to see how a piece of technology that they had mainly been using in the classroom, like a desktop computer, could be taken into the field to collect data and be used for scientific research. They were amazed at how our little handhelds could do so much and in so many different ways. I am excited about the fact that besides learning the science content, my students have developed new skills to collect and use a variety of data formats (text, graphs, images) from a variety of sources (science probes, handhelds, the Internet, videoconferencing, and e-mail) in a variety of ways (data collection and analysis, making comparisons, understanding new concepts, communication with others).

LEMONADE! GET YOUR LEMONADE!

—by Kadee Anstadt, Hudson, Ohio, USA

One area that is often overlooked when it comes to teaching with technology is social studies. However, in a society that is becoming increasingly dependent on computer technology, it is essential that social studies educators teach with and about the latest technology to give their students "the knowledge, skills, and attitudes required...to be able to assume 'the office of citizen'" (NCSS, 1994, p. 3). Besides the literacy skills students need to deal with the glut of information available on the Internet (the favorite technology tool of social studies teachers), the nature of the content that they teach should cause social studies educators to think about the impact of technology on society (including their students). In fact, "as learning technologies become more sophisticated, so too must our critical assessment of their impact on our lives" (Ross, 2000). This example of handheld use shows how a fourth-grade teacher used technology to provide all of her students with an interesting yet meaningful experience and teach about some fairly complex concepts at the same time (see also van 't Hooft & Kelly, 2004, for another example of the use of handhelds in social studies). She teaches an average of 40-45 students a year, divided in two classes, in an upper-middle-class, suburban school. She describes her unit as follows. — by Mark van 't Hooft

The unit I created for this purpose is an economics unit which falls under the topics "Scarcity and Resource Allocation" and "Production, Distribution, and Consumption" strands in the fourth-grade *Academic Content Standards for Social Studies* in the state of Ohio (2002). According to the *Standards*, my fourth graders should be working toward understanding "why entrepreneurship, capital goods, technology, specialization and division of labor are important in the production of goods and services" (Benchmark B, p. 34), and "how competition affects producers and consumers in a market economy and why specialization facilitates trade" (Benchmark C, p. 34). On the surface these terms can be foreign to the average fourth grader, but can be made more accessible if learning about them is done in meaningful, integrative, value-based, challenging, and active ways according to the standards for teaching excellence of the National Council for the Social Studies (1994).

I wanted my students to come to understand these concepts in a way that would make the most sense to them, and a good way to do that is through personal experience in the form of a simulation. Doing a simulation on the desktops in my classroom simply was not an option, because students would have had to work in groups of about five students per computer. In that scenario somebody always misses out due to lack of involvement or engagement, which is crucial in an activity that requires direct personal experience to have an effect on student understanding of social studies concepts. However, my students had been using handheld computers in language arts, and each student had access to one while in class; therefore I knew that they knew how to use this technology. I decided to use two tools on the handhelds for the economics unit: a business simulation called *Lemonade Tycoon*, and *FreeWrite*, a word processor for journaling.

For the simulation, each student set up a lemonade stand, making a variety of business decisions along the way. They had to acquire the appropriate resources to begin, including ingredients, supplies, tools, and a venue to run their businesses. Students had to make business decisions including how many supplies to buy; what recipe to use; where to set up the stand based on the weather forecast, news reports, and cost of operation; how much to charge per cup; and whether or not to advertise. Each school day, students ran their businesses for one business day and put their individual results in a spreadsheet. At the end of the week, students were responsible for graphing the profit-and-loss pattern for the week along with daily sales and temperature.

Throughout the simulations, students kept a daily journal. In each entry, they were to discuss some basic facts, including the weather, news events, the recipe they used, and where they set up. Next, they were to describe what happened during the business day. Finally, and most importantly, my students reflected on the decisions they made at the start of the day, whether or not they worked, and how they would adjust their business practices for the following day. In addition, students used their handhelds to reflect on what they had learned about economics that day and how it applies to their own lives.

PATTERNS IN NATURE

—by Jan Kelly, Mogadore, Ohio, USA

One advantage of teaching elementary school is that the curriculum is not as compartmentalized and concepts can be taught across subject areas. Cross-curricular connections can help students in their ability to understand the inter-relatedness of the disciplines as they apply to their lives. This elementary school teacher is a perfect example of how interdisciplinary teaching can have a lasting impact on what students learn, as she describes her experiences using handheld technology in a predominantly lower-middle-class school. She teaches about 25 students per year, coming from a variety of academic and socioeconomic backgrounds. — by Mark van 't Hooft

One of the content areas in our elementary curriculum consists of a study of the relationships between plants and animals and what each needs for survival. Our unit of study was entitled "The Importance of Living Things." Standards developed by the state of Ohio (USA) require students to analyze plant and animal structures and functions needed for survival, and describe Earth's resources including minerals, soil, water, air, animals, and plants, and the ways in which they can be conserved. I incorporated elements from other disciplines to help them make concrete connections to each of those discipline areas. In an effort to connect math, science, and language arts, this unit focused on patterns.

The interdisciplinary unit began as an investigation into the geometric patterns found in nature. I introduced my students to different types of polygons in math, and initially had them graph coordinate planes on graph paper. There was an interesting dialogue as the students were asked to discuss why they were given this activity. They came to the conclusion that in order to accurately duplicate a shape, graph paper and coordinates were important, if not essential, tools.

Next, I used a poem to illustrate the use of shapes in daily American life. Cloverleaves were explored in relationship to highways, and aerial views of communities were found on the Internet to show how we have boxed off, squared off, and divided property, parking lots, and neighborhoods. Students were encouraged to write similar poems of their own, at the same time satisfying some of the state language arts standards.

Following the writing activity, my students took their handheld computers and *PalmPix* cameras into the school yard to find and photograph polygons in nature. They digitally captured common objects such as pine cones, leaves, rocks, tree bark, and landforms. Flowers were photographed, as well as salt and sugar crystals. Each student remarked about the large numbers of polygons they were able to find, never having realized how patterns played such an important part in the world around them. They produced collages of their polygons in nature and hung them prominently in the school hallways.

To more deeply investigate the concept of patterns, we explored the idea of tessellating some of the patterns we had found. An easy way in which to do this is by using *Tessellation*, a handheld application that allows students to draw a design and immediately see it in tessellated fashion. The students were amazed to see the continued patterns and the beauty of their finished designs. Because the handheld devices enabled students to quickly tessellate a shape, they were able to experiment with many different ones.

With each student having access to a handheld device, there was an added sense of ownership of the content in this unit of study as the students were able to individually gather photos for their collages. They were able to develop their own style of expression in their artwork and poetry. They made sense of the connections between math and science as they explored the world outside the four walls of the classroom, and reflected on this along the way. This was made possible with the use of the handheld devices, the *PalmPix* cameras, and a variety of handheld applications. The portability and immediate access of the tools they used helped to enrich this unit of study far beyond what I would have been able to do without the technology.

GOT ROOTS?

—by Ric Hughes, Ravenna, Ohio, USA

A couple of years ago, Ric Hughes and about 20 of his eighth-grade students worked with Kent State University's Research Center for Educational Technology and the local public television station on a social studies/ language arts project entitled "Speaking of History: Doing Oral History Projects" (see http://www.pbs4549.org/HISTORY.HTM for more information about this project and http://www.pbs4549.org/HISTORY/BROWNMID/GOTROOTS.HTM for Ric's project). Originally this project was to focus on local history, but the events of September 11, 2001, made students realize the importance of family, and their focus shifted accordingly. Ric's students learned by doing that history is much more than remembering facts, and that oral history records feelings and impressions of people as they lived their lives in the past. In addition, this project is an excellent example of meaningful learning in a ubiquitous computing environment that includes handheld devices. Students took advantage of the many tools available to them during a six-week learning experience in the SBC Classroom at Kent State University (see http://www.kent.edu/rcet/classroom/SBC Ameritech Classroom.cfm and www.kent.edu/rcet/classroom/SBCAC-Archives-2001-2002.cfm#Roots). Technology used included handhelds, voice recorders, digital still and video cameras, a document camera, and desktop computers to create multimedia presentations. Ric, a former language arts teacher, recounts his experiences as follows. — by Mark van 't Hooft

In today's global world in which increasing numbers of people are connected by technology such as the Internet, e-mail, and instant messaging, it is more important than ever that students develop solid communication skills, both written and oral. The oral history project we engaged in was a perfect opportunity for my students to do just that. Before collecting actual data from immediate family and relatives, I set aside plenty of time for students to become familiar with some of the skills they would need, such as listening, writing, and note taking, as well as skills related to the analysis of primary sources. In addition, this practice time gave them a chance to become familiar with the technology they were going to use throughout the project.

My students started out by interviewing each other (using their handhelds to take notes and pictures) and presenting their interviews to class using multimedia presentations. This simple activity taught them how to plan for and ask good questions, how to prepare for and act during an interview, and how technology could aid them in the process. To teach students the importance of correctness in oral communication and detail in recording information, I put them in teams of three, had one student look at and explain a graphic design to another, and had the third student draw the design on paper, based on what the second student had told him/her. Students then compared the original design with their version of it.

The second important element in preparing my students for collecting their family histories was teaching them how to deal with and interpret historical source material. Again, I started with something my students were familiar with, and had them create a timeline of their own lives using the software *Timeliner*. Students brought in all kinds of materials to include, such as pictures, diplomas and awards, and family heirlooms. All artifacts were digitally captured with either a scanner or digital camera and inserted in the timeline. Next, I brought in several primary sources, and had students guess what they were and write about them using their handhelds.

Third, I had students practice the skills they had mastered up to this point by introducing the Springer family, whose life has been captured on the Web site, "You Be the Historian" (http://www.americanhistory.si.edu/kids/springer). Students worked in pairs and investigated the lives of Thomas and Elizabeth Springer as they lived in New Castle, Delaware, about 200 years ago. Using their handhelds, students took notes on what they discovered, and wrote a narrative as a fictional member of the Springer family. They used the infrared beaming function on the handhelds to share and peer-edit their writing.

Now that they had been exposed to the data collection and analysis skills, students practiced with the technology tools they would be using during the actual interviews. We spent several days to practice, using handhelds, digital audio recorders, and digital still and video cameras to capture information. Students also used this time to write and refine interview questions and practice their oral communication skills.

The real interviews took place a few days later. Some students visited relatives and talked to them at their places of residence; others were interviewed at the SBC Classroom. All interviews were either audio or video recorded. Students took copious notes on their handhelds and took pictures with digital cameras. In addition, students got access to a variety of other primary resources, including photo albums, artifacts, and family records. Many students learned new things about their families, but most of all the interviews reemphasized the importance of family to them. Based on the information my students gathered, they created multimedia presentations, producing historical knowledge in the process. When students shared their final products with their classmates, many of the family members who had been interviewed were present to see them as well.

In retrospect I have to say that my students learned more from the experience than they probably realized at the time. Many of the skills they picked up or improved upon are skills they use in their daily activities. I also believe that many of them learned to appreciate technology as lifelong tools for learning, because of the authentic context in which they learned how to use different types of devices and software for a variety of tasks.

DISCUSSION

In a nutshell, existing research in the area of ubiquitous/portable computing indicates that handheld devices provide immediate access to a variety of computing tools in a mobile and affordable package that allows for anywhere, anytime computing for all students. There is not a single use that could be considered as the *killer* application of handheld integration, and each case study exemplifies this in its own way. Instead, it is the nature of handheld devices (small form factor, portability, ease-of-use, versatility) that is going to cause them to have a substantial impact on teaching and learning as we know it today.

A common thread in the case studies is that handheld computing in schools requires a new technology literacy on the part of students and teachers in order for it to become truly ubiquitous, so that the technology becomes as transparent as Weiser envisioned almost 15 years ago. While schools are still a ways away from that kind of ubiquitous computing in teaching and learning, the case studies show that many teachers and students are well on the way toward reaching this lofty goal. Let us return to the questions that guide our investigation into ubiquitous computing in this chapter, and use them to put the individual case studies into the larger context of technology integration and digital literacy in K-12 education.

What types of educational activities do handheld devices make possible above and beyond what is currently possible with available technology to improve teaching and learning?

It should be obvious by now that handheld technology enables a wider variety of learning activities than was previously possible. For one, while handheld devices offer students more personal and private space when using technology as evidenced in the Factors of Production unit (provided there is a 1:1 ratio), the infrared or wireless capabilities built in to the hardware encourage more student collaboration through the sharing of work or the use of multiple interconnected devices to achieve a common learning goal. In this respect, a variety of grouping formats are possible, such as student pairs in the Patterns in Nature unit, and larger groups such as those illustrated in the Water Quality and Chair units.

Second, almost all of the examples of handheld use described in this chapter take learning beyond the classroom walls to places such as the school yard, a local stream, and home. The great advantage in this respect is that student learning can be made more active, authentic, and meaningful, as illustrated in one way or another by almost all of the case studies. Students were involved in a variety of activities that necessitated the use of technology outside of the classroom for conducting polls (Chair unit), capturing real-life data using probes and digital cameras (Water Quality and Patterns in Nature units), and comparing estimates to reality (Gingerbread Village unit). As many of the teachers noted, students seemed to reach a deeper understanding of the concepts learned.

Third, because of the immediate availability of a variety of tools in one small package, students are now able to represent what they have learned in lots of different ways. One way in which this is illustrated is through either individual (Lemonade Tycoon unit) or group representations (Water Quality unit) of concepts learned. Another way in which multiple representations of learning can be demonstrated is seen in the Patterns in Nature unit. Students provided proof of understanding through digital images, drawings, poems, and tessellations, using handheld devices where appropriate. In the Gingerbread Village unit, students produced a sampling of digital sketches, images, lists, calculations, and stories as proof of their understanding of geometric shapes.

Fourth, handheld technology enables students to collect data using different tools and sources. In the Chair unit, students used their devices to administer a survey and use digital images to plan for the painting of the chair. In the Gingerbread Village unit, students made digital sketches to create estimations of room dimensions which they used as the basis for measurements of the real thing later on. In the Water Quality unit, students used several probes attached to handhelds to collect scientific data in numerical and graphical formats. Students used their computers in the Lemonade Tycoon unit to gather data for a fictitious business which they used to essentially analyze their business performance. Finally, the last case study describes how students gathered an array of patterns in nature.

In sum, mobile and affordable technologies that present all students with immediate access allow opportunities for a wider range of learning activities. These activities involve both individual and collaborative learning which can take place inside and outside of the classroom, an assortment of representations of concepts across subject areas, and using a variety of data collection tools and sources. As a result, students tend to be more motivated for and engaged in learning tasks. Moreover, these types of activities lead to active and challenging learning that is meaningful and authentic, with the potential that handheld devices will become lifelong learning tools for their users.

What pedagogical changes need to be made to effectively integrate handheld technology in K-12 classrooms?

Integrating technology to the extent that handheld devices make possible does require some changes in teacher pedagogy. However, instead of encouraging teachers to completely rewrite their existing curricula, it is important to remember that promoting a radical change like this is often counterproductive. In our case, evolution, not revolution, is the key (Soloway, 2004). Most of the teachers who shared their stories in this chapter started from existing curriculum and took one or two activities in which they integrated handheld technology. From there, the rest was history.

The initial PEP research uncovered that handhelds have a great potential in educational settings. Teachers and students agreed that "accessibility of a computer for each student is the greatest benefit. Students are able to collect, store, and organize data. They can research, calculate, write, and share information…The Palms enhance student collaboration and encourage students to use higher level thinking skills." Handhelds also enable students to take the initiative and explore, allowing for "a more authentic and deeper immersion in technology, not as a separate curriculum, but as an integrated part of [the] whole curriculum." Moreover, handhelds encourage "the use of technology in everyday activities and enable students to understand the computer as a tool" (van 't Hooft & Diáz, 2002; see also Vahey & Crawford, 2002). It is within this context that teachers and curriculum specialists should approach pedagogical changes.

When it comes to the actual pedagogy, there is plenty of research that indicates that a more constructivist approach to teaching in which knowledge is constructed rather than transferred works better when using technology (e.g., Cajas, 2001; Doolittle & Hicks, 2003; Jonassen, 2000a, 2000b; Salomon, Perkins, & Globerson, 1991; Saye, 1998). It has not been proven whether a constructivist (as opposed to a more traditional) approach leads to increased technology use. Instead, it seems that technology tends to be a catalyst for more constructivist ways of teaching (e.g., Rice & Wilson, 1998). Therefore, it is my belief and the belief of many of the teachers I work with on a daily basis that an influx of technology in the classroom will eventually lead to teaching approaches that are more in step with the information that teachers and students have shared with us in previous research. The case studies presented here should exemplify this as well.

One area that has not been mentioned in the case studies but that will have a tremendous impact on teaching and learning is assessment. The immediacy and portability of a handheld device allows for a shift from summative to formative assessment. This type of assessment has added benefits in that it tends to be more precise and enables a teacher to make curriculum adjustments on the fly in order to address gaps in student learning. Finally, handheld-based assessment creates opportunities for student self-assessment, peer assessment (e.g., peer editing of written work), and group assessment (e.g., the use of TeamLab to assess group processes).

How can handheld computers be adapted to harness their full potential as ubiquitous devices for teaching and learning?

The initial PEP research studies (Vahey & Crawford, 2002; van 't Hooft et al., 2003, 2004) also indicated that handheld devices still had a ways to go when it comes to seamless adaptation in K-12 classrooms. Teachers and students ran into a variety of hurdles when using their palm-sized computers, some of which we still see today. The main issues include text input, issues of logistics, durability, and to a lesser extent screen size, availability of affordable software, and Internet access.

Currently, the most pressing issue by far is text input. Without the use of an external keyboard (which decreases mobility), text input is limited to an onscreen keyboard or handwriting recognition tools such as the *Graffiti* application on PalmOS devices. For younger users, fine motor skills are often not far-enough developed for problem-free usage, while older users often get frustrated because they cannot keep up when taking notes, for example. In addition, teachers in the early elementary grades have expressed their concerns when it comes to *Graffiti*, because its alphabet is different from the regular alphabet that students are taught for writing. One way in which this matter has been resolved is through the introduction of devices such as AlphaSmart's Dana™, a portable device that runs PalmOS and sports an integrated, full-size keyboard. The trade-off is that its larger size decreases portability to some extent.

A close second are issues related to the logistics of setting up and maintaining a set of handheld devices in a classroom. While 1:1 computing resolves issues of access, a teacher is now responsible for the successful implementation of 25-30 devices instead of the five desktop computers that are fairly standard in American elementary and secondary classrooms. High on the list are having enough chargers around to keep the batteries charged and getting students in the habit of remembering to charge their devices on a regular basis, finding ways to quickly and efficiently transfer and/or back-up handheld data on classroom computers, and distributing handheld software and files. For most teachers 5-10 minutes of class time is simply not available on a regular basis. Instead, teachers have come up with a variety of often innovative syncing and beaming schemes which can be easily implemented and take up very little instructional time.

Third, durability has been and still is an issue. While actual hardware failures of handhelds have decreased over the years due to new product development and heightened user awareness, it still happens and seems to be more of an issue in an environment where 1:1 ratios of computers to students has become commonplace. Any time you hand technology to a bunch of children, stuff is going to break; the key seems to be to teach students some basic maintenance skills right from the start and to involve parents as well. Our experiences have been that teachers who spent some time teaching students how to take care of the handhelds, and who have created parent awareness of handheld use by students, have dealt with far fewer hardware problems. Technical and administrative support play an important role here as well.

Minor issues of screen size, affordable software, and Internet access should be mentioned as well. Interestingly enough, screen size seems to be mostly a problem for teachers. Students tend to be very comfortable with the handheld form factor, a phenomenon which I attribute to the Gameboy Syndrome and the fact that digital kids are growing up with ever-smaller computing devices that are not necessarily limited to gaming.

The availability of affordable software has become an issue in that once teachers realize the potential of handheld devices, they often push the envelope of handheld use. There is a long list of handheld software available on the Internet in a variety of formats: freeware, shareware, and commercial ware. It is often difficult to find good educational freeware, simply due to the sheer volume of what is downloadable, and not all of it is good. When it comes to shareware or commercially available products, financial resources are often an issue. Even though individual licenses for handheld products tend to be reasonable, one needs to remember that licenses for handheld software are usually bought in bulk, and the price of a classroom set of 25-30 licenses is all of a sudden not so reasonable anymore.

Internet access has been somewhat of an issue in the past and is likely to become more of an issue as wireless capabilities such as Bluetooth and 802.11.b are becoming more standard. This creates the need for increased wireless network infrastructures and support in school buildings, as well as Internet sites that are reformatted to fit on smaller screens. For now, it appears that wireless capabilities are used primarily for printing and transferring data across devices.

Finally, there are some legal and ethical concerns that surface anytime you put technology into the hands of students, especially when this technology is pervasive and nearly invisible, as handhelds tend to be. Many critics of portable devices for students have argued that because of their size, handhelds are easily stolen or lost. Reality tends to prove them wrong. For example, out of the 280 handhelds that were issued to the teachers in our original handheld research project, four were either stolen or lost, and out of those four, three were eventually returned. Research projects in the United States comparable to ours have reported similar ratios. Different districts deal with this issue in their own ways, but for the most part parents sign some kind of form at the beginning of the year stating that their children will be using a handheld and that they are responsible for care and replacement if needed. This policy is no different from others with regards to use of school property by students (e.g., textbooks, music instruments, or athletic equipment).

The other issue deals with student rights to privacy. Being in an environment with technology that is small yet has capabilities to capture images, create messages, and send them without anybody knowing can create problems in that students can and will find ways to use technology in inappropriate ways. The key to solving this problem has not been to punish, but to educate. Organizations like the International Society for Technology in Education have acknowledged that ethical use of technology is an area that should be addressed by educators, and have included it in their standards for teachers and students (ISTE, 1999). Our experience has been that unethical use of handheld technology has been sparse, and definitely not higher than unethical use of other types of technology, and that teachers have been able to effectively deal with issues on a case-by-case basis. A common example is students beaming each other notes with inappropriate content.

How are digital literacy skills influencing and are being influenced by this new technology?

New technologies usually require new and/or improved literacy skills. In the case of handheld technology, the traditional literacy skills as described for example by Doyle (1992; recognizing the need for accurate and complete information for decision making; decides what information is needed and where to obtain it; accesses, evaluates, and organizes this information; integrates new information into existing knowledge; and uses it for critical thinking and problem solving) are more important than ever before, because ubiquitous computing exposes more students to more information from more sources more often. Therefore, the increase of technology in classrooms requires more teaching and learning about the process of dealing with vast amounts of information. Handheld technology has the capability of helping students amplify these skills.

The latter is especially the case in our contemporary society that is heavily data driven. The use of technology as described here requires an additional (sub)set of literacy skills usually described as digital literacy skills. These skills include learning how to deal with information that is no longer limited to text but also includes still images, video, sound, and interactive Web pages. Second, information retrieval has changed from being mostly book based to being more Internet based or technology based (e.g., digital images, handheld surveys), requiring increased information construction from multiple sources. Finally, digital literacy is interactive and multidimensional. It requires students to be able to read, evaluate, integrate, and use resources from multiple sources and communicate these newly constructed pieces of knowledge to others. Handheld devices are a prime example in this respect, especially when it comes to sharing information with others. As stated, handheld devices allow learners to provide evidence of learning by way of different forms of representation. Examples of student work in the case studies include poems, stories, digital images and drawings, graphs, and calculations.

CONCLUSION

Handheld computers have the potential to have a tremendous impact on teaching and learning given the right context. Through our research and daily interactions with teachers, we have seen the most successful implementations in environments where there is enough technical and administrative support to add handheld computers to the existing technology infrastructure, where teachers are willing to integrate this technology through a process of evolution, and where there is an increased focus on (digital) literacy skills. In addition, the case studies show that successful implementation does not necessarily equal high levels of handheld technology use on a daily basis. Integration works best when the technology is almost invisible, yet available when appropriate and needed, and creates an environment of meaningful and authentic use. This, in return, increases student levels of interest, motivation, and engagement, resulting in knowledge construction and a deeper understanding of concepts learned.

Within the larger context of ubiquitous computing—that is, the type of environment Weiser had in mind when he put forth his ideas in the early 1990s—handheld computers play a prominent role. Even though Weiser and his colleagues focused on the integration of ubiquitous devices for everyday applications, the leap to educational settings is not that difficult to make. It should also be easy to see why increased levels of literacy, including the digital variety, are so important in a society where technology is becoming more invisible, yet all the more important, especially in the things we do that we take for granted. Handheld computing should help us all to become more aware of that.

REFERENCES

Bayus, B. L., Jain, S., & Rao, A. G. (1997). Too little, too late: Introduction timing and new product performance in the personal digital assistant industry. *Journal of Marketing Research, 34*(1), 50-63.

Becker, H., Ravitz, J. L., & Wong, Y. (1999). *Teacher and teacher-directed student use of computers and software*. Report #3, Teaching, Learning, and Computing: 1998 National Survey. Irvine, CA: Center for Research on Information Technology and Organizations, University of California, Irvine.

Cajas, F. (2001). The science/technology interaction: Implications for science literacy. *Journal of Research in Science Teaching, 38,* 715-729.

Cole, H., & Stanton, D. (2003). Designing mobile technologies to support co-present collaboration. *Personal and Ubiquitous Computing, 7,* 365-371.

Danesh, A., Inkpen, K., Lau, F., Shu, K., & Booth, K. (2001). Geney™: Designing a collaborative activity for the Palm™ handheld computer. *Proceedings of CHI, Conference on Human Factors in Computing Systems,* Seattle, Washington.

Doolittle, P. E., & Hicks, D. (2003). Constructivism as a theoretical foundation for the use of technology in social studies. *Theory and Research in Social Education, 31,* 72-104.

Doyle, C. (1992). *Outcome measures for information literacy within the national education goals of 1990.* Final Report to the National Forum of Information Literacy. Summary of Findings. (Eric Document Reproduction Service, ED 351 033).

Fitzpatrick, C. (2000). Navigating a new information landscape. *Social Education, 64,* 33-34.

Fung, P., Hennessy, S., & O'Shea, T. (1998). Pocketbook computing: A paradigm shift? *Computers in the Schools, 14,* 109-118.

Gilster, P. (1992). *Digital literacy*. New York: Wiley Computer Publishing Co.

Hennessy, S. (1997). *Portable technologies and graphing investigations: Review of the literature*. CALRG Technical Report 175. Milton Keynes, UK: The Open University, Institute of Educational Technology.

Hennessy, S. (2000). Graphing investigations using portable (palmtop) technology. *Journal of Computer Assisted Learning, 16,* 243-258.

HPC Factor. (2004). *A brief history of Windows CE: The beginning is always a very good place to start.* Retrieved October 12, 2004, from http://www.hpcfactor.com/support/windowsce/

Inkpen, K. (2001). Designing handheld technologies for kids. *Personal Technologies Journal, 3,* 81-89. *Proceedings of CHI, Conference on Human Factors in Computing Systems,* Seattle, Washington.

International Society for Technology in Education. (1999). *National education technology standards.* Retrieved May 18, 2004, from http://cnets.iste.org.index.htm

Jonassen, D. H. (2000a). *Computers as mindtools for schools: Engaging critical thinking.* Upper Saddle River, NJ: Merrill.

Jonassen, D. H. (2000b). Transforming learning with technology: Beyond modernism and post-modernism or whoever controls the technology creates the reality. *Educational Technology, 40*(2), 21-25.

Langford, L. (1999). Information literacy? Seeking clarification. *School Libraries Worldwide, 4*(1), 59-72.

Mandryk, R. L., Inkpen, K. M., Bilezkjian, M., Klemmer, S. R., & Landay, J. A. (2001). Supporting children's collaboration across handheld computers. *Proceedings of CHI, Conference on Human Factors in Computing Systems,* Seattle, Washington.

Marx, R. W., Blumenfeld, P., Krajick, J., Fishman, B., Soloway, E., Geier, R., & Tal, T. (2000). *Inquiry based science in the middle grades: Assessment of student learning in the context of systemic reform.* Unpublished Manuscript, University of Michigan Center for Learning Technologies in Urban Schools, USA.

National Council for the Social Studies. (1994). *Expectations of excellence: Curriculum standards for Social Studies.* Washington, DC: National Council for the Social Studies.

Norris, C., & Soloway, E. (2001, June). Towards realizing the potential of palm-sized computers in K-12. *iMP Magazine.* Retrieved May 18, 2004, from http://www.cisp.org/imp/june_2001/06_01soloway.htm

Norris, C., & Soloway, E. (2004). Envisioning the handheld-centric classroom. *Journal of Educational Computing Research, 30*(4), 281-294.

Ohio Department of Education. (2002). *Social studies academic content standards.* Columbus, OH: State Board of Education.

Rice, M. L., & Wilson, E. K. (1999). How technology aids constructivism in the social studies classroom. *The Social Studies, 90,* 28-33.

Risinger, F. C. (1998). Separating wheat from chaff: Why dirty pictures are not the real dilemma in using the Internet to teach social studies. *Social Education, 62,* 148-150.

Robertson, S. I., Calder, J., Fung, P., Jones, A., O'Shea, T., & Lambrechts, G. (1996). Pupils, teachers, and palmtop computers. *Journal of Computer Assisted Learning, 12,* 194-204.

Roschelle, J. (2003). Unlocking the value of wireless mobile devices. *Journal of Computer Assisted Learning, 19,* 260-272.

Roschelle, J., & Pea, R. (2002). A walk on the WILD side: How wireless handhelds may change computer-supported collaborative learning. *International Journal of Cognition and Technology, 1*(1), 145-168.

Roschelle, J., Penuel, W. R., & Abrahamson, L. (2004). The networked classroom. *Educational Leadership, 61*(5), 50-53.

Ross, E. W. (2000). The promise and perils of e-learning. *Theory and Research in Social Education, 28,* 482-492.

Roth, J. (2002). Patterns of mobile interaction. *Personal and Ubiquitous Computing, 6,* 282-289.

Salomon, G., Perkins, D. N., & Globerson, T. (1991). Partners in cognition: Extending human intelligence with intelligent technologies. *Educational Researcher, 20*(3), 2-9.

Saye, J. W. (1998). Creating time to develop student thinking: Team-teaching with technology. *Social Education, 62,* 356-362.

Sharples, M. (2000a). *Disruptive devices: Personal technologies and education.* Educational Technology Research Paper Series 11. Birmingham, UK: University of Birmingham.

Sharples, M. (2000b). The design of personal mobile technologies for lifelong learning. *Computers and Education, 34,* 177-193.

Soloway, E. (2004, May). Why handhelds? *Proceedings of the America's Future Classroom: Advancing Learning with Handhelds Conference.* Oklahoma City, Oklahoma.

Soloway, E., Norris, C., Blumenfeld, P., Fishman, B., Krajcik, J., & Marx, R. (2001). Log on to education: Handheld devices are ready-at-hand. *Communications of the ACM, 44*(6), 15-20.

Tatar, D., Roschelle, J., Vahey, P., & Penuel, W. R. (2003). Handhelds go to school: Lessons learned. *IEEE Computer, 36*(9), 30-37.

Vahey, P., & Crawford, V. (2002). *Palm education pioneers: Final evaluation report.* Menlo Park, CA: SRI International.

van 't Hooft, M. A. H., & Diáz, S. (2002). *Palm education pioneers: Final report.* Unpublished report, Kent State University, USA.

van 't Hooft, M. A. H., Diáz, S., & Andrews, S. (2003). Byte-sized learning: Handhelds in K-12 classrooms. *Learning Technology, 5*(2), 37-38.

van 't Hooft, M. A. H., Diáz, S., & Swan, K. (2004). Examining the potential of handheld computers: Findings from the Ohio PEP project. *Journal of Educational Computing Research, 30*(4), 295-311.

van 't Hooft, M. A. H., & Kelly, J. (2004). Macro or micro: Teaching fifth-grade economics using handheld computers. *Social Education, 68*(2), 165-168.

Weiser, M. (1991). The computer for the 21st century. *Scientific American, 265*(3), 94-95, 98-102.

Williams, B. (2004). *We're getting wired, we're going mobile, what's next?* Eugene, OR: ISTE Publications.

ENDNOTE

* Parts of this research have been funded by Palm Inc., through the Palm Education Pioneer Project, the friendly folks at GoKnow who provided us with PAAM, Rubberneck, and HLE, as well as an AT&T research grant through the Research Center for Educational Technology. I would also like to thank Karen McClain, Jan Kelly, Kadee Anstadt, Shawn Jones, and Ric Hughes, and all the other teachers and students who generously shared their time, stories, classrooms, and handhelds!

Chapter XVII
Digital Literacy and the Use of Wireless Portable Computers, Planners, and Cell Phones for K-12 Education

Virginia E. Garland
The University of New Hampshire, USA

ABSTRACT

Wireless technologies have transformed learning, teaching, and leading in K-12 schools. Because of their speed and portability, laptops, planners, personal digital assistants (PDAs), and cellular telephones are major components of digital literacy. In this chapter, current international trends in the educational uses of portable technologies will be discussed. The implications of newer hardware specifications and educational software applications for laptop computers will be analyzed, including inequities in student access to the handhelds. Next, the role of planners and PDAs as more recent instructional and managerial tools will be evaluated. This study also includes a review of the current debate over whether or not cell phones, especially those with photographic capabilities, should be allowed to be used by students in schools. Finally, potential uses of wireless technologies for interactive learning and collaborative leadership on a global basis will be investigated.

OVERVIEW OF DIGITAL LITERACY AND WIRELESS TECHNOLOGIES

Language acquisition and mathematics skills are the core elements of traditional views of literacy. In the new millennium, technology has become another basic skill for K-12 students across the globe. Wilhelm wrote about the focus of a 2002 Berlin conference on technology literacy in the 21st century and paraphrased the words of German Chancellor Gerhard Schroeder:

Shroeder boldly asserted that digital literacy should now take its place as a basic literacy alongside reading, writing, and arithmetic, recognizing the pivotal role of information and communication technologies in underwriting lifelong learning, economic productivity, and democratic engagement. (p. 297)

Not only in Europe, but also in Asia and the United States has there been a drastic paradigm shift towards the integration of new technologies in learning. Other nations are beginning to explore the digital advantage shared by the more technologically advanced countries.

The widespread use of wireless computers and handhelds has already had a positive impact on many schools throughout the world. Laptops and handhelds are efficient tools for individualizing instruction. Teachers and students are using them to maximize learning, especially in mathematics, language, and science instruction. Educational administrators are finding wireless handhelds to be very effective in the areas of teacher supervision, budgeting, and school safety.

What is Wireless Technology?

Wireless communication networks date back millennia. In the Roman Empire, for example, geese were used to alarm residents of fires. Napoleon implemented the semaphone signals on poles, and during the Battle of Waterloo, the Rothschilds sent carrier pigeons to London with word that Wellington was going into battle. Congolese have traditionally used tam-tam drums to communicate complex signals to other tribes in the central Congo. In the 1970s the United States established wireless satellite data links to Europe. By the late 1990s, technologically advanced countries were using wired base stations for laptops, cell phones, and personal digital assistants (PDAs).

Today, there are two primary types of wireless technology: RF and Bluetooth. Radio frequency (RF) is used as another term for 802.22a and b, which need a wired base station. Wireless does not work in RF unless there is a wireless access point in the near area. If, for example, a portable computer has a wireless card installed, it can be connected to the Internet without having to have an ethernet cable. Wireless is for a short period of time, then a power source is needed for charging. Some school campuses use 802.11b Wi-Fi wireless LAN for wireless laptops.

With the second type of wireless technology, "Bluetooth," adapters and USB ports are used. Popularized since 2000, the primary uses of Bluetooth are for PDAs and combination cell phone/planners, the Palm Zire 71, printers, keyboards and a mouse, and headsets. There is a direct beam between two pieces of hardware, such as between two Palm Zire 71s, in which a photo can be taken and then transmitted from one Palm and received by the other.

Honan (2004) compares these RF and Bluetooth configurations:

Wireless products fall into two categories, Bluetooth and 27MHZ radio frequency (RF). Both have advantages and disadvantages. RF devices are less expensive, more widely available, and far more compatible with older Macs than their Bluetooth counterparts. Bluetooth devices have a much longer range, can transmit encrypted signals, and aren't prone to interference from other devices operating on the same frequency. Furthermore, RF products rely on wired base stations that plug into your Mac's USB port, while Bluetooth products can use either your Mac's built-in Bluetooth receiver or a comparatively small USB adapter that connects directly to the port without any wires. We prefer Bluetooth technology for wireless input devices. (p. 28)

Although Bluetooth needs fewer configurations, it tends to be slower and cover shorter distances than RF.

There are some potential difficulties with using wireless technologies. In Oak Park, Illinois, three sets of parents of elementary students are suing the district to have the wireless computer networks removed because "a bit of Internet research turned up some alarming studies purporting to show that radio frequency radiation similar to that used in wireless networks could break down DNA or damage the protective barrier surrounding the brain. An appendix to the complaint lists 28 such studies" (Sanchez, 2004, p. 12). Other wireless network issues, such as inconsistency of signals, bandwidth problems, logistical and security concerns are important, but are more suitably addressed in the more technical texts than in this chapter.

How Are Schools Using Wireless Networks?

With either RF or Bluetooth wireless configurations, there are practical benefits to educators. Newer Mac and PC laptop specifications of RAM and CPU speeds and memory have enhanced wireless communication. The benefits of wireless connectivity over a wired infrastructure are analyzed by Gonzales and Higby (2003):

An explosion of newer mobile devices such as laptops, cell phones, and personal digital assistants (PDAs) has fueled the recent high demand for wireless mobility...A primary advantage of wireless communication involves reduced costs as compared with the expense of wired installations...Many computer manufacturers are developing mobile devices with built-in capabilities of wireless connectivity that support 802.11b and 802.11g. (p. 33)

For older schools where new wiring would be difficult and costly, such as those in the United Kingdom, Finland, Japan, and the United States, wireless broadband is the best option. It may also be beneficial to plan on wireless and avoid the expenses of hardwiring for newer school facilities.

How much are educational leaders investing in the newer hardware? Current wireless trends in the United States for 2002-2003, the most recent year for which data is available, indicate five significant areas of growth: first, $776 million was spent on wireless technologies in public schools; second, it was found that handhelds are becoming more affordable than computers, making computer labs outdated; third, the favorite handheld is the Palm (33%), followed by the Alpha Smart (8%); fourth, Apple remains the public schools' favorite computer (30%), with Dell in second place (22%); and, lastly, laptops are taking an increasing share of the middle school and (46%) and high school (48%) markets (McLester, 2003). Given the legal power of state and local control over education in the United States, there are limitations in the teachers' and administrators' choice of either PC or Mac systems in the American K-12 schools.

In other countries, where public education may be more centralized, there is opportunity for more standardization in the types of hardware selected. Iceland is a good example of technology integration at its best. According to Trotter (2004):

Iceland has more subscribers to Internet service, per capita, than any other nation...technology is considered a basic tool for real-world preparation. Teachers use it to help run their classes and bring resources into the learning process, as well as to encourage students to learn in deeper ways and express themselves better...And schools all have access to a nationwide centralized system for student information. (pp. 42-43)

For Icelanders, wireless technologies are already fully integrated in their educational experiences.

Some schools in Europe and Asia have opted for partnerships with technology corporations. In the United Kingdom, the East Manchester Education Action Zone ("The Zone") was established from grants from Microsoft:

Each school has it own wireless LAN network with high-speed broadband access. Schools in The Zone have access to hardware, including 1,000 Toshiba Satellite and Tecra laptops, interactive whiteboards, personal digital assistants and Portege tablet PCs. (Bentley, 2004, p. 55)

According to educators in this urban city, "The Zone" students now have opportunities to gain digital literacy skills and engage in more meaningful learning activities with their wireless technologies.

Microsoft is also having an impact on schools in Thailand, Malaysia, and Indonesia. In August 2004, the Thai-language Microsoft Windows XP Starter Edition operating system for first-time computer users was officially unveiled. In addition, Microsoft has the Starter Edition in Malay and Indonesian languages to help new PC users (*The Nation* (Thailand), 2004). Governments in Southeast Asia now have the opportunity to apply non-English-based Windows software in educational settings.

Wireless configurations in any country can enable students, teachers, and administrators to access the Internet anywhere in the school building. COW (computers on wheels) carts allow laptops to move from room to room without having to be concerned about Internet plug ins. In individual classrooms, a wireless tower can be used to beam the Internet connection on only one computer to an entire classroom of laptops. It becomes cost efficient for schools to use these approaches because additional space and wiring is unnecessary.

However, Vail (2003) points out some security concerns with wireless networks:

But with Wi-Fi—as the techies call high-frequency wireless networking—the good news of easy connectivity is bad news for security, You don't have to be a sophisticated hacker to tap into a nearby wireless network, so users suggest limiting Wi-Fi to the classroom. Keep the school server and other vital systems plugged into the wall. (p. 37)

School technology staff should therefore be aware of the disadvantages of "Wi-Fi" in terms of security for student grades and other confidential records.

A school's wireless network infrastructure can support a vast array of portable technologies. Three categories of wireless devices have emerged as valuable educational tools: (1) portable computers/laptops, (2) planners and personal digital assistants (PDAs), and (3) cellular telephones (cell phones). What follows is an analysis of the effectiveness of each of these three wireless technologies in K-12 education.

PORTABLE COMPUTERS/ LAPTOPS

Portable Computer Hardware

Portable computers include laptops, tablets, convertible tablets, and desktops. Schools are increasingly using these technologies because of their portability in school and at home as well as their wireless connectivity. Popular hardware choices in American schools include AlphaSmart's Dana Wireless ($429), a two-pound laptop powered by Palm OS, which incorporates built-in Wi-Fi technology (802.11b) and was released in 2003. According to the editors of *T.H.E. Journal* (2003), this Dana wireless "offers a screen larger than the standard Web-enabled PDA and an integrated full-size keyboard, which makes it perfect for K-12 and higher education students and teachers" (p. 27). Other laptops recently purchased by schools include the Compaq Tablet PC TC 1000 from Hewlett Packard and the Acer TravelMate C110 convertible Tablet PC.

There has been controversy about giving students in some U.S. schools access to laptops, such as the Apple iBook, with Internet capability. Wildstrom (2003) evaluates the distinct advantages of AlphaSmart's Dana over other laptops and handhelds:

> *Dana's biggest deficiency is its lack of Internet connectivity...Schools that have given laptops to students have had a constant struggle with kids installing their own software and downloading pornography, music, and other prohibited material from the Web. Henrico County has called in its more than 11,000 Apple iBooks twice to tighten security settings...Dana nicely fills the gap between inadequate handhelds and overkill laptops* (p. 26)

Without filtering devices and teacher supervision, some laptops are being inappropriately used by students.

International Models of Laptop Use

On the K-12 level, there are three general models of laptop use: (1) each student having his or her own laptop for use anywhere, (2) a classroom set of computers being shared by teachers, and (3) a desktop computer set in each classroom. The most effective model, despite the possibility of inappropriate use, appears to be a laptop for every student to use both in and out of the classroom.

In Asheboro City, North Carolina, the school district used wireless technology to address a serious deficiency in writing ability in the elementary schools:

> *The district proposed a project to put an AlphaSmart portable word processor into the hands of every third-grade student to focus on writing improvement...they purchased 400 AlphaSmart 2000 devices and...conducted staff training...It wasn't long before the AlphaSmarts were being used during science observations, on social studies projects, and more.* (Bigham, 1999, p. 90)

In this project, the AlphaSmarts were generally not allowed out of the classroom setting.

The Essex, England, E-Learning Foundation, a charity supported by Intel, recently gave 200,000 laptops to students and teachers, "as part of a scheme to encourage the use of IT in education and set up the nation's first county-wide wireless network" (Lloyd, 2004, p. 11). Full implementation of the project, which involves significant staff training, will take three years. Similarly, the Malaysian Education Ministry has a five-year technology integration plan. In 2004, approximately 35,000 elementary and secondary teachers in Malaysia were trained and given Intel Centrino notebooks. The Malaysian IT plan includes equipping all schools with WiFi capability (*New Straits Times* (Malaysia), 2004).

The administrators of Bishop Hartley High School, a private Catholic school in Ohio, gave an entire student class their own personal Compaq Tablet PCs, which they could use anytime and anywhere. According to Barton and Collura (2003), the tablet PCs have distinct advantages for improving writing skills of students:

For the eternally disorganized student, the tablet PC is the breaker of old habits and the initiator of new ones. By using a note-taking program such as the Microsoft Windows Journal, students can now not only type their notes into the computer, but also handwrite them on the screen using an electronic pen…The tablet PC software also allows students to convert their handwriting to typewritten text, so those poorly organized students will never again have to look beyond their Windows Journals files to find lost notes. (p. 39)

Because American public schools seem to have more financial, security, and safety issues than their private-school counterparts, there are few cases of public high schools giving Tablet PCs to each student.

Portable Computers and Digital Literacy

Constructivist learning with concept mapping, text creating, Web searching, and data mining is very effective with the laptop as a tool. In the United States, the National Science Teacher's Association cosponsored a Laptop Learning Challenge with Japan's Toshiba Corporation in 1999, with a focus on science and mathematics education:

Some award-winning ideas showed students using laptops to facilitate group work, to analyze data immediately during a lab exercise, or to conduct scientific investigations in the field rather than in the classroom…Other uses for laptops include creating spreadsheets to solve math homework problems; creating book reports that inspire student creativity with presentation software such as PowerPoint or HyperStudio. (Belanger, 2001, p. 2)

Teachers can more easily evaluate student work from laptops if they are connected to the school network, where their assignments can be saved in a central file server for the teacher to grade and leave for the student to retrieve.

In a wireless networked school or classroom, the interaction between teacher questioning and student response is faster and more effective than in the traditional classroom:

The students enter their responses into a personal computing device, such as a graphing calculator, a palm-size computer, a laptop or even a special-purpose device known as a response pad, which is similar in appearance to a TV remote control. The next step is crucial: The teacher's desktop computer collects the students' work, processes it into a meaningful graphic that the teacher and students can quickly interpret, and displays this graphic to the whole class. From a pedagogical point of view, this rapid accumulation of student feedback enables the teacher to adjust instruction as needed. (Roschelle et al., 2004, p. 52)

This individualizing of instruction is made possible by the speed and efficiency of the new wireless computers.

However, Paul Gilster (1997), author of *Digital Literacy,* claims that there are "core competencies" for evaluating online resources. He claims that the "globe-spanning information network" can be both "misleading and deceptive" to the user. According to Gilster, "critical thinking about content" is the most valuable digital literacy. Training teachers to evaluate the reliability of Web pages, newsgroup postings, and Internet search engines is key to classroom use of the newer technologies.

Case Study of One Laptop per Student

The Maine Learning Technology Initiative provides a laptop for every one of the state's 34,000 seventh and eighth graders (Schachter, 2004). Belanger (2001) discusses the advantages of this concentrated model, in which "teachers are free to integrate technology fully into instruction as well as assignments, since all students have access to a computer for homework, study, and projects. In the class set and dispersed models, teachers are free to integrate laptops during the school day; however, there may still be students within the same class who lack access to a computer in the home, so integration options are more limited" (p. 2). Access to a computer at any time, in school or at home, is considered essential to the students involved in the Maine Learning Technology Initiative.

The achievement results in one Maine school district rose as a result of connecting all of the state's seventh and eighth graders and their teachers to a wireless network. Schachter (2004) evaluates the impact laptops have had on one rural school:

Since the start of the program, their scores in reading, writing and math have improved enough to remove Pembroke from Maine's list of underperforming schools. After school detentions have almost become obsolete. "The laptops are integrated in the classroom all day long, and the students have become totally self-directed, independent learners." (p. 33)

The one-on-one access to wireless laptops has also personalized instruction in the more populated and wealthier Maine school district of Freeport. In that district, middle schoolers are "using their laptops as combination textbooks, writing tools, reference libraries (students can connect to a database of newspapers, periodicals and encyclopedias) and multimedia vehicles for creating class projects" (p. 33). However, there are some critics of the Maine initiative. In the property-poor district of Buckfield:

Students typically end their daily school routine by sliding their iBooks onto the shelves of metal carts, which are wired to recharge the computers. The laptops recharge while students go home and complete their homework. Teachers and students agree it's unsatisfying and unproductive when "anytime, anywhere learning ends at 2:20 p.m. at the schoolhouse door. (Trotter, 2004, p. 30)

Consistency among districts in implementing the goals of the Maine laptop initiative is crucial to success. In property-poor areas of the United States and other countries, more funding should be made available for teacher training and hardware support.

Wireless Computer Applications for School Administrators

Paperwork reduction is realized with the rapid exchange of information on wireless laptops. Principals can observe classes and collect data on their laptops, send the preliminary evaluation to the teachers via the laptop, and set up post-conference interviews. Teachers and administrators can more easily monitor student behavior in and outside the classroom with portable, wireless computers. In addition, wireless laptops can be indispensable in the budget process. With the use of Excel software, spreadsheet data can be beamed from one laptop to another during collective bargaining negotiations or budget meetings. Managing by walking around has become a great deal more effective with the laptop, planners, and cell phones now readily available to the school leader.

PLANNERS AND PDAS/HANDHELDS

The most popular and recognizable planners and PDAs currently in use in the schools of Europe, Asia, and the United States and Canada are Palm Pilots, iPaqs, Axims, and the X5. These wireless handhelds have distinct advantages over the larger laptops and stand- alone computers, cost being the most notable:

Districts are buying more of the devices each year, finding that more handhelds provide students tremendous computing and learning power at about a 10th of the cost of a regular computer. "You can buy wireless laptops for $2,000 or Palm handheld computers for $200, Where's the dilemma?"...with handhelds, the manufacturers claim, teachers and students can accomplish about 80 percent of the things they can with a regular computer. (Joyner, 2003, p. 42)

In overcrowded and older schools, handhelds are replacing outdated computer labs because they are cost-effective space savers.

Palms are now making huge inroads in the K-12 market in the United States:

In 2000, higher education made up 70% of Palm's education market; now K-12 schools account for 70% of the company's education sales...IDC, a Framingham, Mass., research firm, projects that $40 million will be spent on handhelds in the 2003-04 school year. But that amount will skyrocket to $175 million in 2004-05 and to $300 million by 2005-06. (Joyner, 2003, p. 42)

Although specific cost figures are unavailable, it is widely believed that the American PDA and cell phone markets lag behind those of Finland and Japan.

Handheld Applications for Teachers and Students

Texas Instruments (TI) was one of the first companies to manufacture a handheld instructional tool: the calculator. Fourteen years ago TI developed the TI-81, a graphing calculator that helped students visualize and understand complex geometry and other mathematical equations. Recently, TI incorporated some of the functions found in the PDAs to its graphing calculator, introduced in 2003 as the TI-Navigator. It is a wireless tool that allows teachers to network the class's graphing calculators, to send assignments to students, and to immediately review homework.

Some other student uses of the handhelds include learning activities from the humanities as well as the sciences. With these versatile, wireless handhelds, students in Kentucky are downloading e-books such as *Romeo and Juliet* onto their Casio E-11 PDAs. In Illinois students use the Vivonic Fitness Plan software on their handhelds to keep a diary of their exercise and health records. Science students in Oklahoma attached special probes to their Palms and tested water quality. One other teacher used a checkbook program on a handheld to teach money management (Joyner, 2003).

In a 2003 study of PDA use by 22 students in a pilot program, Wangemann, Lewis, and Squires (2003) found that American students' learning benefited from the use of their PDAs:

54% agreed that technology gives each individual more control over his or her own learning, and 50% agreed that technology helps one to be more productive...67% of student teachers felt that the PDAs increased student motivation and interest, 71% felt that they increased the ability to collaborate and communicate, and 80% of the student teachers increased their productivity by having a portable and accessible learning tool. (p. 27)

Gateway is another company which is leading in handheld sales to schools. In 2003 it marketed its first handheld, the 100X Professional PDA, a "pocket PC." When it was tested in the schools, teachers were found using the Gateway PDAs to "take roll, make lesson plans, organize schedules, and track students in special education classes" (Joyner, 2003, p. 43). Teachers are also using PDAs to increase parental contact. In one project, Motorola pagers were issued to teachers and parent participants. Using Palm Pilots in the classroom, teachers recorded "notable student behaviors" and sent pager messages to parents about the student conduct issues (Strom & Strom, 2003, p. 14). Pagers can also be used to contact parents about positive student behaviors, changes in scheduled activities, or emergencies.

Handheld Applications for School Administrators and Librarians

Principals and supervisors are finding handhelds to be extremely effective organizers. Districts are providing their administrative teams with planners like the Palm Pilot for scheduling meetings and for collecting data on students and teachers. Using programs like UDT Solutions Mobile Administrator, principals can walk around their buildings with student records, schedules, parent information, and other school data readily at hand. In collective bargaining and school budget negotiations, some administrators are loading Excel software on their Palm Pilots in order to quickly make key fiscal decisions.

A newer use of the Palm Pilot Zire 71 for teacher supervisors is in the teacher evaluation process. Principals are using their Palms to record data on classroom observations on wireless keyboards, which are easily attached to the Palm device. In addition, with teacher and student permission, supervisors can photograph classroom activities as a supplement to the narrative accounts. Thus, the Palm Zire 71 is making the teacher evaluation process more data driven and more meaningful.

Library staff have also found professional uses in PDAs. Already, e-books can be downloaded onto handhelds in one of two formats, Palm OS/Win CE and HTML. For librarians and media specialists, the features which are standard issue on Palm Pilots "allow staff members to leave their desks and work instead from the library's stacks, classrooms, meetings or wherever else they are needed. The e-mail function may be especially relevant to libraries as they move further into e-mail and live chat reference services" (Embrey, 2002, p. 25).

The instant messaging capabilities of most PDAs today are invaluable tools to administrators, librarians, and teachers alike in times of school emergencies. For instance, without vocalizing a concern about a possible intruder in a school building in front of her students, a teacher could instant message the principal from his or her PDA. In lockdown situations, administrators can contact outside police and fire authorities for assistance with their PDA instant messaging,

PDA Issues and Challenges

All handhelds have the clear disadvantage of being prime targets for theft or loss because of their small size and mobility. Realistic policies must be put in place to provide educators with training in preventive theft measures as well as to provide for possible losses. The controversial issue of Palm Pilots with photographic capabilities, such as the Palm Zire 71, will be discussed in the next section on cell phones.

CELLULAR TELEPHONES

Of all the wireless handheld devices discussed in this chapter, cell phones are the most controversial. Without proper guidelines, some students will use them inappropriately or even illegally. Cell phones have been used by students in schools of the United States and Ireland to interrupt classes, to cheat on tests, to make drug deals, and to take pornographic pictures of unsuspecting students in lavatories or locker rooms. On the other hand, students in Colorado's Columbine High School used cell phones to properly alert outside authorities during that violent tragedy. Many students use them to contact their parents about changes in school activities and schedules. The current debate centers around the types of policies needed to create the most beneficial use of these ubiquitous handheld wireless phones.

The newer cell phones have voicemail, paging, e-mail, instant messaging, news, cameras, music, and planners. AT&T and Verizon are two of the top cell phone competitors in the United States. In one corporate/school partnership sponsored by the AT&T Learning Network and Ericsson, called the Safe Schools program, cell phones were made available to school personnel in several urban centers. Each selected school received two Ericsson Digital PCS phones and a digital control channel for enhanced conversations and extended battery life. Adults found that they could quickly call for assistance in emergencies on the school grounds, injuries were dealt with efficiently, and rival gangs disappeared when they saw school personnel with the phones (*Inside School Safety*, 1999).

Some schools in Europe, the United States, and Canada are replacing the intercom system with cell phone communication for teachers and administrators. In cases of individual student names being called over the intercom, interrupting the entire class, and creating privacy issues as well, instant messaging or cell phone messages from the office to the teacher are positive communication alternatives.

In one California program, cell phones were:

...programmed to connect directly to local law-enforcement agencies so school officials can report emergencies...At Laguna Creek High School in the Elk Grove, Calif., district, the building is only six years old and each classroom already had a phone, Still, school officials believe the donated cell phones will enhance their communications and provide better security to the staff and 2,500 students. "We plan to use them at student events where

we have no access to phones," says Chris Hoffman, an assistant principal. "At a football game, you might have 5,000 or 6,000 people, and if something happens, you want to be able to make emergency phone calls" (Kennedy, 2000, p. 38)

On the other hand, one school is banning cell phone use by students:

Security expert Kenneth Trump said having hordes of students making emergency calls at the same time could actually hinder the sending of emergency messages by jamming phone lines. (Current Events, 2002, p. 3)

In addition, police in Europe, Asia, and the Americas have found that "walkie talkie" and cell phone use during bomb threats can actually trigger the detonation of bombs.

Some school districts are banning cell phones and pagers because they were going off during classes. Recently, administrators in Ireland, the United Kingdom, and the United States are establishing policies to remove all camera phones and handhelds from school grounds. The newest cell phones can snap and send digital pictures in an instant:

Fears range from students taking shots of tests like junior James Bonds to shooting photos of unsuspecting classmates as they change in the locker room. (Curriculum Review, 2004, p. 4)

According to Carroll (2004), there are serious legal implications to students' unauthorized use camera phones:

...a picture-taker invades someone's privacy when he or she snaps a picture of a person. However, if the photographer deletes the picture and no one else ever sees it, the privacy issue is removed. Problems arise...when the image is transmitted to another person, via e-mail or directly to another camera phone, or by posting it on the Web. (p. 8)

In the American legal system, there is a clear invasion of personal privacy in the transmission of unauthorized digital photographs.

The California School Boards Association recommends two options on the issues of cell phone and pager use:

Option One: "Students may possess or use electronic signaling devices, including but not limited to pagers, beepers, and cellular/digital telephones, provided that such devices do not disrupt the educational program or school activity. Electronic signaling devices shall be turned off during class time and at any other time directed by a district employee. If disruption occurs, the employee may direct the student to turn off the device and/or confiscate the device until the end of the class period, school day or activity." Option Two: "Except for prior consent for health reasons, possession or use of electronic signaling devices, including but not limited to pagers, beepers and cellular/digital telephones, is prohibited." (Danforth, 2003, p. 32)

More school districts are taking the moderate approach, due to parental pressure, of adopting the first policy.

FUTURE ISSUES

Wireless technology has a wealth of potential to improve learning, instruction, and school operations. Training is needed to demonstrate how these portable tools can be applied to the classroom. For many students, wireless handhelds provide new opportunities for interactive learning, collaboration with peers, and data mining and research. Teachers need support in their acquisition of new professional skills in using the wireless technologies which are now available in some schools. In the future, we may even use scanned beam technology, available through Microvision in the next five years, to take the moral high ground on teaching biology without unnecessary pain and death:

Imagine students in your school dissecting a frog in biology class. Without slicing a millimeter of the frog's skin, the students see images of the animal's veins, heart, and fatty tissue hovering over the specimen. This isn't fantasy, according to Matt Nichols, spokesman for Microvision. It's called scanned beam technology. (Pascopella, 2003, p. 25)

Instead of wireless handhelds, we will soon have wireless headbands, which can generate images from any data or video source to the eyes directly from the headband.

Despite the vast potential for the wireless technologies, the "digital divide" between the "high-tech" and "low-tech" schools is a serious issue. This gap exists between school districts in the United States (Garland & Wotton, 2002), as well as between school regions in China and between countries across the globe. According to Borja (2004):

In China's high-tech nerve centers such as Shanghai, Beijing, and the verdant southern coastal province of Guangdong, many middle and high school students surf the Internet on the latest wireless-enabled laptop computers, send text messages to their friends on personal cellphones, and download class lessons from their school's Web sites. But in China's rural west, technology is scarce. Schools often lack enough teachers, supplies, and up-to-date textbooks—never mind computers and the Internet's "information highway." (p. 24)

There is also a huge disparity between the technologies available in other Asian countries. Borja states:

Countries such as South Korea, Singapore, and Taiwan have implemented far-reaching national "master plans" to install high-speed computers in schools, train teachers to bolster their lessons using technology, and encourage students to conduct online research, build Web sites, and develop online projects. But other, poorer countries—such as Vietnam, Laos, and Mongolia—lack such blueprints. (p. 25)

The level of technology integration, if it exists at all, is directly correlated to the wealth of the country. Some efforts, such as the pilot projects sponsored by the World Bank in China (Garland & Yang, 1994) and UNESCO in Laos, are attempting to close the digital divide.

In conclusion, the globalization of wireless technologies has incredible benefits for educational collaboration and cross-cultural understanding. With adequate teacher training and financial support, digital literacy has the potential to provide meaningful, interactive learning in schools around the world.

REFERENCES

Barton, C., & Collura, K. (2003). Catalyst for change. *T.H.E. Journal, 31*(4), 39-42.

Belanger, Y. (2001). Laptop computers in the K-12 classroom. *ERIC Digest* (ED440644).

Bentley, R. (2004). East Manchester school uses wireless LAN, broadband and PDAs to motivate learning. *Computer Weekly,* (February 3), 55.

Bigham, V. S. (1999). Going portable. *Curriculum Administrator, 35*(10), 89-90.

Borja, R. (2004). Asia. *Education Week, 23*(35), 24-28.

Carroll, C. (2004). Camera phones raise whole new set of privacy issues. *Education Week, 23*(23), 8.

Colgan, C. (2003). School boards have policy options. *American School Board Journal, 190*(7), 30-33.

Current Events. (2002). Calling cell phone bans into question. *101*(19), 3.

Curriculum Review. (2004). Technology update. *43*(5), 4-6.

Embrey, T. (2002). Library applications in support of the needs of students and teachers. *Teacher Librarian, 29*(5), 24-28.

Garland, V. E., & Wotton, S. E. (2001-2002). Bridging the digital divide in public schools. *Journal of Educational Technology Systems, 30*(2), 115-124.

Garland, V. E., & Yang, D. (1994). East meets west: Technology training in the People's Republic of China. *Technological Horizons in Education, 21*(6), 50-53.

Gilster, P. (1997). *Digital literacy.* Hoboken, NJ: John Wiley & Sons.

Gonzales, R., & Higby, C. (2003). Let's go wireless. *Tech Directions, 63*(4), 30-36.

Honan, M. (2004). Wireless input devices. *Macworld, 21*(3), 28.

Inside School Safety. (1999). Cell phones provide direct line to safer schools. *3*(10), 6-7.

Joyner, A. (2003). A foothold for handhelds. *American School Board Journal, 190*(9), 42-45.

Kennedy, M. (2000). Lifelines to the office. *American School and University, 72*(10), 38-39.

Lloyd, T. (2004). Essex schools to hand over 200,000 laptops. *ITTraining,* (May), 11.

McLester, S. (2003). A studied look at wireless. *Technology and Learning, 23*(10), 4.

The Nation (Thailand). (2004). Microsoft XP now in Thai. (August 12).

New Straits Times (Malaysia). (2004). Notebooks for 35,000 teachers. (April 6).

Pascopella, A. (2003). High-tech trends: Tablet PCs, multimedia labs and digital video are proving their worth across the country. *District Administration,* (December), 22-25.

Roschelle, J., & Penuel, W. (2004). The networked classroom. *Educational Leadership, 61*(5), 50-55.

Sanchez, J. (2004). WiFoes. *Reason, 35*(8), 12-13.

Schachter, R. (2004). A tale of two laptops. *District Administration, 40*(3), 32-36.

Strom, P., & Strom, R. (2003). Teacher-parent communication reforms. *High School Journal, 86*(2), 14-22.

Trotter, A. (2004). Digital balancing act. *Education Week, 23*(20), 29-32.

Trotter, A. (2004). The Viking journey. *Education Week, 23*(35), 42-48.

Vail, K. (2003). School technology grows up. *American School Board Journal, 190*(9), 34-38.

Wangemann, P., Lewis, N., & Squires, D. (2003). Portable technology comes of age. *T.H.E. Journal, 31*(4), 26-31.

Wildstrom, S. (2003). A Dana for every schoolkid? *Business Week,* (3829), 26.

Wilhelm, A. (2002). Wireless youth: Rejuvenating the Net. *National Civic Review, 91*(3), 293-303.

Chapter XVIII
Using WebQuests to Support the Development of Digital Literacy and Other Essential Skills at the K-12 Level

Susan E. Gibson
University of Alberta, Canada

ABSTRACT

This chapter introduces the WebQuest as one means of addressing effective technology use for developing digital literacy skills at the K-12 education levels. It argues that technology use that promotes constructivist learning principles has been found to have the greatest effect on learning. Furthermore, the WebQuest and its extension, the Web Inquiry Project, exemplify strategies that promote constructivist learning principles when they are designed to encourage student-directed learning, problem solving, higher-level thinking, perspective taking, real-world authentic issues, and collaboration. The author hopes that by providing specific examples of each of these strategies, readers will be better able to envision effective, constructivist-based technology use for their classrooms.

WHAT IS LITERACY?

These days literacy has taken on a whole new meaning. The current buzzword in educational circles is digital literacy. According to the North Central Regional Educational Laboratory (NCREL) Web site (http://www.ncrel.org/engauge/skills/agelit.htm), digital literacy encompasses a large number of literacies, including basic literacy such as language proficiency and numeracy, scientific literacy, economic literacy, technological literacy, visual literacy, information literacy, multicultural literacy, and global awareness. In addition to these literacies, NCREL calls for an emphasis in K-12 schooling on other skills such as higher-level thinking—including critical and creative thinking and problem solving—as well as communication skills in order for students to be successful in the 21st century. This is a major undertaking for schools and teachers as they struggle with how to address all of these aspects of digital literacy in their teaching. Computer technologies can be used to assist and support teachers in their endeavors.

This chapter begins by reviewing what we currently know about effective computer use to support and enhance teaching and learning. Constructivism is then examined as a promising theoretical framework for that use. The remainder of the chapter looks at WebQuests and their extension, Web Inquiry Projects, as approaches that have the potential to effectively address constructivist learning principles and digital literacy skills, as well as essential higher-level thinking, problem solving, and communication skills.

WHAT DOES RESEARCH TELL US ABOUT WHAT MAKES EFFECTIVE AND MEANINGFUL TECHNOLOGY INTEGRATION?

Before examining ways to address literacy skills in teaching, it is important to review what we know about effective technology use. Computers are becoming more readily available in many K-12 schools worldwide, and the Internet is often hailed as an innovation with unprecedented potential for the improvement of teaching and learning.

However, teachers are at varied levels of awareness about the possibilities for employing these technologies in effective and efficient ways to enhance teaching and learning. Repeatedly the research has found that computer technologies have had "only isolated, marginal effects on how and what children learn in school, despite early champions of their revolutionary educational potential" (Roschelle, Pea, Hoadley, Gordin, & Means, 2000, p. 77). Consequently, "many computers in schools, even up-to-date multimedia computers with high-speed Internet access, are not being used in ways that significantly enhance teaching and learning" (Kleiman, 2000, p. 8). The foremost problem seems to be that teachers tend to use computers as add-ons to the ways they have always taught which often is modeled after a traditional, transmissionist approach. As noted by Howard Gardiner (2000), "When the [computers] are plugged in, they are all too often simply used to 'deliver' the same old 'drill-and-kill' content" (p. 33). As McAdoo (2000) asserts:

> *The issue of equity centers not on equality of equipment but on quality of use. The computers are there, yes, but what is the real extent of access? And are schools able to raise not just students' level of proficiency but also their level of inquiry, as advanced use of technology demands?* (pp. 143-144)

The key to best use is not the fact that computers are being used, but how they are being used. Computer use needs to go beyond low-level tasks such as students being able to demonstrate understanding of how to operate the various technologies with proficiency, to tasks that encourage more advanced learning with computers. Over a decade of research indicates that the most effective uses of computer resources in schools occur when the technology is used by students as an information processing and productivity tool to achieve a task. Ravitz, Becker, and Wong (2000) identified effective uses as those practices that encourage student engagement in learning, the release of agency from teachers to students, and collaborative knowledge-building around authentic or ill-defined problems.

When computers are used as tools to complete projects in which students apply new learnings, there is greater potential to promote cognitive development by stimulating the development of intellectual skills such as inquiry, reasoning, problem solving, and decision-making abilities, as well as critical and creative thinking. Tools such as databases, spreadsheets, multimedia, e-mail, software, and the Internet are used to complete authentic projects, requiring students to use information to solve problems, promote cognitive, and social development, as well as a positive attitude toward learning. These computer tools have the power to stimulate the development of intellectual skills such as inquiry when the emphasis is more on student understanding than on getting the right answers. Using an inquiry approach to learning with computers places less emphasis on acquiring and presenting information and more on constructing knowledge, making meaning, drawing on personal life experience, and taking responsibility for learning. Learning experiences with technology that are active, cooperative, constructive, intentional, and authentic are more meaningful for students (Jonassen, Howland, Moore, & Marra, 2003). Making use of real-world contexts to anchor the learning, visualization and analysis tools, scaffolds for problem solving, and opportunities for feedback, reflection, and revision add to the meaningfulness of the learning experience (Means, 2001).

Where the greatest challenge for teachers lies is in thinking differently about teaching and learning. As noted by Girod and Cavanaugh (2001), using computer technologies in the ways outlined above:

...asks teachers to view not only learners, but also learning tasks in new ways. Rather than asking students to complete predetermined and well-defined tasks such as worksheets, step-by-step lab experiments, and projects designed with a single goal in mind, teachers must embrace learning activities that are ill structured, ill defined and open-ended. (p. 1)

Computer technologies can help teachers to develop new approaches to teaching and learning, but teachers need to be exposed to these new understandings and new capabilities that are possible through the use of technology as a tool for the learner They also need to determine where technologies fit into their philosophy of teaching:

Otherwise schools overt curriculum may change, but its hidden curriculum is likely to keep targeting conformity instead of creativity, individual-competitive rather than collaborative learning skills and tradition rather than innovation. (Laferriere et al., 2001, p. 1)

As noted by Doolittle and Hicks (2003), "A philosophical and theoretical foundation provides answers to the questions of why and how specific pedagogy, including the application of technology, should be employed" (p. 76). Where success has been most apparent has been in cases where teaching is transformed through the use of computer technologies and where learning is happening in ways that were impossible or difficult without the use of these technologies.

HOW CAN CONSTRUCTIVIST LEARNING THEORY HELP TEACHERS TO DESIGN MEANINGFUL, COMPUTER-ENHANCED LEARNING ENVIRONMENTS?

The merging of technology and constructivism offers many possibilities for designing innovative learning environments. This is perhaps the most significant change for teachers as a result of the technologically driven society—the need to shift from teachers as transmitters of knowledge to students as constructors of knowledge. In the past the school was one of the primary agents for dispensing academic and social information to young people, but that is no longer the case. Teachers need to recognize that one person or one textbook can no longer hold all the important knowledge. In constructivist learning environments, "learners actively create their own knowledge, rather than recapitulating the teacher's interpretation of the world" (Jonassen, 1996, p. 12). Teachers who support this view recognize the importance of the active involvement of their students in learning and the need for a learning environment that encourages students' independent exploration of ideas.

Constructivism is a theory about how people learn. Learning is viewed as an active and social process in which the learner constructs new knowledge and rethinks prior ideas. The new learning always begins with and builds upon the learners' previously stored knowledge; as the learners elaborate upon and interpret the new information, their initial ideas are reshaped. Learning occurs most effectively when it is situated in experiences that are authentic and meaningful to the learner. Working collaboratively, students are given support to engage in task-oriented dialogue with one another. They are routinely asked to apply knowledge in diverse and authentic contexts, to explain ideas, interpret texts, predict phenomena, and construct arguments based on evidence (Windshchitl, 2002).

Technology use that is shaped by constructivist learning principles supports a more student-centered, inquiry-oriented approach to teaching. Such constructivist uses of computer technologies provide learning opportunities based on authentic tasks and allow for exploring and reflecting. However, teachers need to remember that the technology does not teach students, but rather the students only learn when they construct their own knowledge and think and learn through their experiences. Teachers also need to recognize that the four classroom walls no longer bind places of learning. Every classroom has the potential to be a global learning environment. In this way, computer technologies can help to bridge the gap between the artificial world of school and the dynamic needs and interests of young citizens by engaging students in projects that investigate real-world issues, draw on multiple perspectives, and encourage collaboration. We need to engage students actively in collaborative activities in which they interact with peers in classrooms around the world to promote understanding and appreciation of multiple perspectives (Githiora-Updike, 2000). Teachers too can be collaborators with each other both locally and globally as they plan and share suggestions via electronic communication networks.

What is needed in classrooms are technology uses that help to develop students' higher-order thinking and problem-solving skills by providing opportunities for them to think critically and analytically about information and represent their new understandings in multiple ways (Staley, 2000). The inquiry should begin with students' prior background knowledge and experience, and engage them in creatively applying the resultant new knowledge. This learning environment should represent as much as possible the complex real world of problem solving; however, students need to be taught the skills to work in such environments. This is where a more structured type of learning environment such as a WebQuest can provide initial assistance in developing the requisite skills by providing a guided process.

HOW IS A WebQuest AN EXAMPLE OF A CONSTRUCTIVIST LEARNING APPROACH TO TECHNOLOGY USE?

The WebQuest is one example of how to design Internet-based learning experiences for the K-12 level that promote digital literacy as well as the development of essential higher-level thinking, problem solving, and communication skills. There is a growing body of literature on the value of WebQuests as an instructional approach to integrate structured inquiry and the use of technology (Hicks, Sears, Gao, Goodmans, & Manning, 2004). A WebQuest is an inquiry-oriented activity in which most or all of the information used by learners is drawn from the Web. WebQuests are designed to efficiently make use of learners' time by focusing on using information rather than looking for it, and to support learners' thinking at the levels of analysis, synthesis, and evaluation (Dodge, 1996). Through a WebQuest, students can actively explore issues and questions from a number of different perspectives, as well as searching for solutions and making moral and ethical decisions about real contemporary world problems. In an authentic WebQuest there is no single correct answer. While engaged in the inquiry through a WebQuest, students are constructing their own personal meaning about the issue under investigation.

The rationale for using a structured inquiry approach such as a WebQuest design at the K-12 level can be traced back to Bruner's cognitive development theory. For Bruner, the outcome of cognitive development is thinking, "knowledge is a process, not a product" (Bruner, 1966, p. 72). Bruner's discovery learning and inquiry teaching methods envision the learners creating their knowledge by "rearranging or transforming evidence in such a way that one is enabled to go beyond the evidence so assembled to additional new insights" (Bruner, 1961, p. 22). This requires an activity structure that scaffolds learners' experience so that they must move beyond simply finding information to using that information to think through and resolve a problem or issue. "The thinking skills associated with WebQuests include comparing, classifying, inducing, deducing, analyzing, constructing, abstracting, and analyzing perspectives" (Norton & Wiburg, 2003, p. 180).

WebQuests can also enhance students' communication skills as many involve working in cooperative groups and role-playing. Working either independently or in groups, the students explore an issue or problem in a guided and meaningful manner. Some WebQuests have the students take on roles that help to make the group work together more efficiently and effectively. These roles can include a group leader, recorder, communicator, encourager, and evaluator, among others. Other WebQuests have the learners assume the roles of particular players in a role-playing setting where they access, analyze, and synthesize the information provided from the perspective of that player.

The most authentic WebQuests engage students in *perspective taking* on a particular problem or issue. Students investigate the context and the issue from an individual's perspective in order to build a better understanding of the person, the event, and the setting. The goal is for students to use the information collected to construct an argument based on evidence. They then publicly share their findings with the class and the class tries to come to some kind of resolution to the problem under investigation. This resolution may mean coming to class consensus, or if there is a conflict of resolutions, then agreeing to disagree. Role-playing can be particularly beneficial for teaching students the importance of perspective taking when problem solving. Here is where WebQuests have the greatest potential for addressing the multicultural literacy aspect of digital literacy mentioned on the NCREL Web site. Investigating problems from a number of different cultural perspectives can help learners to better understand the wide diversity of views on any one issue, as well as the important cultural foundations of those views. This can lead to learning to respect and appreciate diversity.

How is a WebQuest Structured?

Usually a WebQuest consists of the introduction, task, process, resources, evaluation, and conclusion. The first part, the introduction, lays out the task or the problem to be investigated, provides some background information, and acts as a motivator to get the students interested in the activity. The task outlines the overall challenge the students will be engaged in and explains what they will be doing to represent what they have learned from completing the WebQuest. The task also usually provides the focus questions that frame the investigation and facilitate the learning process. The process provides a description of what needs to be done in order to accomplish the task in a step-by-step fashion. Here, students are usually assigned roles or provided with differing perspectives on the issue or problem being investigated. The resource section provides information sources that are needed for solving the task. Most of the resources used for the inquiry are other Web sites that have been vetted by the teacher and linked directly to the WebQuest. Many WebQuests provide direct access to individual experts, current news sites, and searchable databases for information sources.

The evaluation section provides information for students on how they will be assessed. The assessment tool often included is a rubric for providing feedback on the outcome of the inquiry. Other formative types of assessment can be used throughout the inquiry including personal reflective logs, skills checklists, and self and group feedback on the effectiveness of their group work. The conclusion brings closure to the Quest by reviewing and summarizing the learning from the experience, and often challenges learners to extend their learning in new ways.

WebQuests can be either *short term,* on the average one to three classes, with the goal of acquiring and making sense of new information, or *longer term,* in which a student analyzes a body of information, transforms it in some way, and demonstrates an understanding of that information in a public way. Longer-term WebQuests can take anywhere from a week to a month (Norton & Wiburg, 2003). Throughout the Quest, the teacher acts as the facilitator checking to see that students understand the role that they are to take and that they stay on task.

Where Can WebQuests Be Found?

WebQuests appropriate for the K-12 level can be chosen from a series of pre-designed WebQuest collections (see http://webquest.sdsu.edu/ or http://www.kn.pacbell.com/wired/bluewebn/), or the teacher can create it to address a specific topic of study. The latter allows for more active student involvement in deciding what problem they might like to inves-

Figure 1. WebQuest Web site on "Does the Tiger Eat Its Cubs?"

Figure 2. WebQuest Web site on "DNA for Dinner?"

tigate and in designing an interesting and relevant learning experience around that problem. One example of a WebQuest for upper elementary, titled "Does the Tiger Eat its Cubs" (http://www.kn.pacbell.com/wired/China/childquest.html), explores the way children in orphanages in China are treated (Figure 1).

The question that students investigate is, "What's the truth about how children are treated in China?" Students are directed to investigate the information from a number of perspectives. They are divided into three teams. One team reads international news reports, another reads responses from the Chinese people, and a third examines the government of China's position as stated in China's One Child Policy. The class then comes back together and discusses their findings with the challenge of making a consensus decision on the issue. The culminating activity is to write a letter to the government expressing their opinion on what they feel should be done about the situation.

In another pre-designed, secondary-level WebQuest called DNA for Dinner (http://www.angelfire.com/ma4/peacew/dna.htm), students examine the issue of genetically altered food (Figure 2).

They are to investigate the issue, "Should genetically engineered food crops be specifically labeled for consumers and why?" and draft a law based on their investigation of the issue. They are then encouraged to e-mail a representative in the federal government detailing their investigation and their concern over the issue, and explaining their proposed solution. A WebQuest such as this one is an example of how the learning activity can be designed to increase students' motivation to want to learn by connecting what is learned in school to real-world experiences (Watson, 1999).

A similar WebQuest only for elementary aged children that is focused on a real-world issue is the EnergyQuest (http://www.geocities.com/brookwebquest1/webquest.html). This WebQuest projects students into a futuristic world where they must work together to solve the issue of depletion of non-renewable energy resources (Figure 3).

Working in groups, each student selects an energy source to become an expert on, including solar, wind, hydro, geothermal, or bioenergy sources. The group then prepares a presentation to the UN Secretary General on which energy source appears to be the best solution to the energy crisis.

Even younger children can be involved in WebQuests that help them to better understand the importance of protecting the environment and to see that they too can impact their environment. "Paper or Plastic: An Internet WebQuest on Recycling" (http://oncampus.richmond.edu/academics/education/projects/webquests/paper/top.htm) is designed for the Grade 3-level learner (Figure 4).

Students working in teams take the role of either a paper advocate, a plastic advocate, a person who thinks that there are too many problems associated with recycling for it to be worth the time and money, or a concerned citizen who just wants to see what the plastic and paper industries are doing to help the environment. After researching their viewpoint, the group comes back together to decide whether to choose paper, plastic, or neither and then creates a poster to be hung in grocery stores based on their findings.

WebQuests can also be a powerful way for students to be immersed in historical events and work with historical documents. In the "Scrooge for Mayor" WebQuest (http://www.coollessons.org/Dickens.htm), students work in

Figure 3. WebQuest Web site on "Depletion of Non-Renewable Energy Resources"

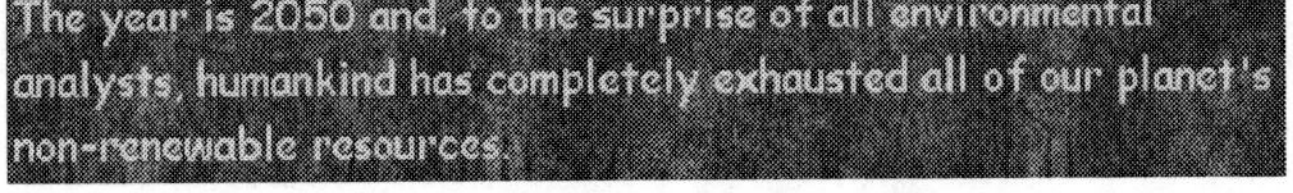

Figure 4. WebQuest Web site on "Paper or Plastic"

Figure 5. WebQuest Web site on "Scrooge for Mayor"

Figure 6. WebQuest Web site on "Kyotoquest"

This WebQuest will allow you to explore the Kyoto Protocol and its effects on Alberta's people, economy, environment, and government. The Kyoto Protocol is an international agreement created by the United Nations. Its goal is for developed countries to reduce greenhouse gas emissions by an average of 5.2% below 1990 levels, by 2012.

teams to develop a campaign proposal for Scrooge using information about labor, education, industrialization, and quality-of-life issues of Dicken's representation of England in *A Christmas Carol.*

Each campaign team is made up of a team manager, research analyst, public relations person, and political strategist. Students are directed to focus on how Scrooge's viewpoint on daily life in London will need to change, and what solutions to London's problems and programs he will need to support in his run for mayor. Each person on the campaign team is responsible to write an article for a newspaper describing what they found out, including what life was like in the area in the 1840s, the conditions that make it necessary to bring about change, what changes are proposed, and how these will better things; they also must write an editorial on the topic, "Is the industrial revolution a good thing?" The team must create a campaign poster, a pamphlet, and a PowerPoint presentation that are to be used to communicate their ideas to Scrooge. This WebQuest is an excellent example of how WebQuests can be used to integrate various subject areas in meaningful ways. This particular WebQuest addresses the learning outcomes of social studies, reading, language arts, and science.

Teachers can also design WebQuests to meet their own personal needs. For example, Kyotoquest (http://www.geocities.com/mteichrob) is a WebQuest that has been designed to actively engage Canadian students living in the province of Alberta in constructing an understanding of the Kyoto Protocol and its political, environmental, humanitarian, and economic effects on their province (Figure 6).

The curriculum specific to the province of Alberta was used as the framework for this design, and the issue of the Kyoto Protocol was chosen as the focus of the WebQuest because of its relevance and currency for the students in the province. As outlined in the provincial curriculum document, the Quest emphasizes the interrelationships between people and their environment, as well as the impact people have on their environment. Students examine a variety of perspectives supporting each of the diverse ways that the Kyoto Protocol will affect Albertans. The issue that students are to investigate is: *Should the Kyoto Protocol be ratified?*

In teams of four, students select either the role of citizen, environmentalist, provincial government representative, or federal government representative. The citizen is to explore family and job-related issues and perspectives pertaining to the Kyoto Protocol. The environmentalist is to study scientific facts and issues related to the environment. The provincial government representative is to address Alberta's role and position on the Kyoto Protocol, and the federal government representative examines Canada's role and position. After individually examining each viewpoint, the team members then write a collective newspaper article and send an e-mail to their local government representative stating their views and concerns. As students study this significant and relevant topic, they learn that they can influence change by both being informed and by taking action on important issues.

Numerous templates are available to help K-12-level teachers in their construction of their own WebQuests (see http://webquest.sdsu.edu/LessonTemplate.html or http://www.ozline.com/templates/webquest.html).

Older students may also be encouraged to try developing their own WebQuests and sharing them with classmates. A database of sample student-developed WebQuests can be found at the ThinkQuest Library site (http://www.thinkquest.org). Having students create their own WebQuests challenges them "to explore a topic, summarize what the most important events or facts are in relation to the topic, and then put together the links and questions for other students to follow" (Whitworth & Berson, 2003, p. 480). When students engage in creating their own WebQuests, it can also enhance the development of their critical, creative, and higher-level thinking skills. The two Web sites noted previously provide templates that students can use for creating their own quests.

What Are Some Limitations of WebQuests?

The WebQuest approach is intended to capitalize on the possibilities provided by the Internet for guided inquiry learning while eliminating some of the disadvantages such as time wasted looking for resources, accessing inappropriate resources, and lack of sufficient experience with the research process (Milson, 2002). There are some limitations to using WebQuests, however, that teachers need to be aware of. One of these limitations is that not all WebQuests encourage higher-order thinking. Many WebQuests are designed merely as fact-finding exercises that do little to engage students in problem solving. No attempt is made to involve students in role taking or learning to view problems from multiple perspectives. Others lack clear directions, and this can detract from taking control of the learning experience.

There are a number of Web sites that provide rubrics for determining the quality of WebQuests (see, for example, http://www.ozline.com/webquests/rubric.html, http://bestwebquests.com/bwq/matrix.asp, http://webquest.sdsu.edu/webquestrubric.html, or http://www.todaysteacher.com/WebQuestIntroduction/assessing_webquests.htm). The criteria included in these assessment include: engaging opener; clear question and task; roles match the issues and resources; higher-level thinking is required; opportunities for feedback are built in; and the conclusion ties in to the introduction, makes the students' cognitive tasks overt, and suggests how this learning could transfer to other domains/issues.

Another limitation of WebQuests is that students are most often removed from the process of selecting resources on which to base their investigation. As current information becomes easily accessible online, it is increasingly important that students have the opportunity to develop their critical analysis capabilities (Mason et al., 2004). Also educators are warned not to simply rely on Internet filtering software, but rather to focus on teaching students critical thinking skills so that they can learn to make informed decisions and judgments about the information they encounter on the Internet

(Whitworth & Berson, 2003, p. 480). The use of such filtering tools can also be a problem, as many sites that would be relevant to the study of a topic, such as war and conflict, would be inaccessible to students.

Locating useful and accurate information on the Web can be a struggle for K-12 students. The abundance of things to access via the Internet can cause students to be easily side-tracked and spend a great deal of time off task. Information gathering can easily become a mindless exercise in which quantity overrides quality. This sort of information-gathering exercise does little to promote deeper thinking and understanding. Students need to be instructed in and have opportunities to practice how to critically examine and make informed choices about the information they are accessing. Critical digital literacy skills need to be carefully taught and monitored to ensure students are developing proficiency in their use. In addition to learning the skills of locating and evaluating information on the Web, students also need to learn how to select relevant pieces of that information, and synthesize and organize it in order to apply it to the learning activity and communicate it to others.

Because there is an inclination to accept the computer as an authority and view the information accessed as the truth, students need to be taught to recognize that the information on a computer represents a particular viewpoint, as does any other resource. They need to be encouraged to conscientiously use critical thinking skills to make both appropriate and ethical choices when using computer-generated information. Students need to be taught and know how to apply the skills of actively interpreting the information provided, drawing conclusions from data, seeing several points of view, distinguishing fact from opinion, and finding meaning in information, as they interact with computer technology. Sheveley and VanFossen (1999) suggest using the following questions to judge the effectiveness of Web sites:

- **Authorship:** What do we know of the creators of the site that might affect the believability of its contents?
- **Sponsorship:** What do we know of the individual(s) or group(s) who sponsored the site that might affect the believability of its contents?
- **Sources of Ideas:** What do we know about how information was obtained and verified that might affect the believability of its contents?
- **Indicators of Care:** Does the site's presentation style, tone, and format provide clues about the believability of its contents?

Critical literacy skills need to be carefully taught and monitored to ensure students are developing proficiency in their use. Children need to be instructed in and have opportunities to practice how to critically examine and make appropriate, ethical, and informed choices about the information they are accessing. They need to be taught to recognize that the information on any Web site represents a particular viewpoint and that it is important to examine several points of view on any issue. They also need to be taught how to distinguish fact from opinion.

A third limitation is that WebQuests lead students through a scaffolded inquiry experience that specifies the task, the roles and perspectives to be taken, the resources to be used, and the guides for organizing the learning with little opportunity for the students to set the direction and plan for the investigation. Heavy scaffolding can also prevent learners from participating in higher-level inquiry activities (Molebash, Dodge, Bell, Mason, & Irving, n.d.). While these initial scaffolds are very important for helping children to develop problem-solving strategies, there needs to be opportunities for releasing some of the control into the hands of the learners. Molebash (2002) notes that the support of the WebQuest can be "faded" by gradually allowing more flexibility in how and what students are to produce in the task; by gradually providing fewer URLs and expecting the learner to find more; by gradually moving the scaffolding of note taking, information organizing, writing prompts, and so forth from required to implicit; and by putting more resources in the Conclusion for learners to explore on their own later.

In order to promote higher levels of inquiry in the classroom, less specific guidance should be given to students. Web Inquiry Projects (WIPs) are one way to extend the WebQuest idea beyond structured inquiry to more open inquiry that promotes higher levels of thinking and student engagement. A WIP is defined as a facilitated learning plan for teachers to promote guided and open inquiry using online interpreted data/information that allows learners to actively pursue answers to questions that are both interesting and relevant to their required studies (Molebash, 2002). Unlike WebQuests, which provide students with a procedure and the online resources needed to complete a predefined task, WIPS place more emphasis in having students determine their own task, define their own procedures, and play a role in finding the needed online resources. More often the inquiry is sparked by the interest of the students. Guided inquiry requires learners to investigate a teacher-represented question using student-selected procedures, and open inquiry has students investigate topic-related questions that are student-formulated through student-designed procedures. The learners need to plan the problem-solving process, gather the necessary information, and decide on the steps to be taken, suggest solutions, and demonstrate their reasoning.

Numerous examples of Web Inquiry Projects for the K-12 level can be viewed at http://edweb.sdsu.edu/wip/examples.htm. In the WIP entitled "The AIDS Epidemic: Can It Be Stopped?" (http://edweb.sdsu.edu/wip/examples/aids/index.htm), for example, students are presented with the following hook:

The HIV/AIDS Epidemic is still occurring today. Currently medical research in finding a cure for AIDS have [sic] not progressed beyond prolonging HIV before it turns into AIDS. Although we don't see HIV/AIDS in the news today, it is still a problem around the world. Many people feel that they are not at risk for contracting this disease, but it is important for individuals to realize that they may be at risk. The first case of HIV/AIDS was diagnosed in the United States in the early 1980s. When will the last case be diagnosed?

In order to address this challenge, students need to determine what investigative tools to use, what types of data they will need, and how they will manipulate that data in order to predict an answer. As a part of their investigation, they also conduct detailed research on AIDS in order to increase their understanding of the issues surrounding AIDS and HIV.

In another example, North American Perspectives (http://eprentice.sdsu.edu/F034/sjohnson/teacher_template2.html), students are hooked into the inquiry through a series of questions that they are to answer, initially from their own perspective then from "behind Native American eyes." They are encouraged to think of some questions related to this topic that they might like to investigate, as well as being provided some teacher-initiated ones. There are some pre-selected resources provided, but students are encouraged to locate their own as well. Some ideas for how to re-present their learning are made available, but once again students are encouraged to come up with their own ideas too.

Each of these examples allows for a greater degree of student control over the learning experiences.

CONCLUSION

Attention to digital literacy has become an essential aspect of K-12 education in the 21st century. This chapter began by identifying important 21st-century skills including digital literacy, higher-level thinking, problem solving, and communication. The research on best practice with computers for supporting and enhancing teaching and learning was then examined. Emerging from this review was an acknowledgment of constructivism as an effective way of framing learning experiences with computer technologies. The remainder of the chapter looked at WebQuests and their extension, Web Inquiry Projects (WIPs), as approaches that have the potential to effectively address constructivist learning principles as well as the development of digital literacy, thinking, problem solving, and communication skills at the K-12 level.

What makes these approaches to technology use in K-12 schools most effective are:

a. the emphasis on student-directed learning and active student engagement;
b. the level of student control over the decision making about the learning;
c. the emphasis on real-world authentic issues and questions of interest to students, and in the case of WIPs, generated by students;
d. the focus on collaborative learning both within and beyond the classroom walls; and
e. the emphasis on learning to manage information and to work with it at a higher level of thinking and understanding.

All of these features support the call for K-12 learning experiences that attend to digital literacy and developing the thinking, problem solving, and communication skills of today's learners.

REFERENCES

Anderson, R., Becker, H., Dexter, S., Ravitz, J., Riel, M., & Ronnkvist, A. (2000). *Pedagogical motivations for student computer use that lead to student engagement.* Retrieved from http://www.crito.uci.edu/tlc/findings/spec_rpt_pedegogical/content.html#frequency

Bruner, J. (1961). The act of discovery. *Harvard Educational Review, 31,* 21-32.

Bruner, J. (1966). *Toward a theory of instruction.* Cambridge, MA: Belknap.

Doolittle, P., & Hicks, D. (2003). Constructivism as a theoretical foundation for the use of technology in social studies. *Theory and Research in Social Education, 31*(1), 72-104.

Gardiner, H. (2000). Can technology exploit our many ways of knowing? In D. Gordon (Ed.), *Digital classroom: How technology is changing the way we teach and learn.* Cambridge, MA: Harvard Education Letter.

Girod, M., & Cavanaugh, S. (2001). Technology as an agent of change in teacher practice. *Technological Horizons in Education (T.H.E.) Journal, 28*(9). Retrieved from *http://www.thejournal.com/magazine/vault/A3429B.cfm?kw=0&gw*

Githiora-Updike, W. (2000). The global schoolhouse. In D. T. Gordon (Ed.), *Digital classroom: How technology is changing the way we teach and learn* (pp. 60-66). Cambridge, MA: Harvard Education Letter.

Jonassen, D. (1996). *Computers in the classroom: Mindtools for critical thinking.* Englewood Cliffs, NJ: Prentice-Hall.

Jonassen, D., Howland, J., Moore, J., & Marra, M. (2003). *Learning to solve problems with technology.* Upper Saddle River, NJ: Merrill-Prentice-Hall.

Jonassen, D., Peck, K., & Wilson, B. (1999). *Learning with technology: A constructivist perspective.* Upper Saddle River, NJ: Prentice-Hall.

Kleiman, G. (2000). Myths and realities about technology in K-12 schools. In D. Gordon (Ed.), *Digital classroom: How technology is changing the way we teach and learn.* Cambridge, MA: Harvard Education Letter.

Laferriere, T., Bracewell, R., Breleux, A., Erickson, G., Lamon, M., & Owston, R. (2001, May 22-23). Teacher education in the networked classroom. *Proceedings of the Pan-Canadian Education Research Agenda Symposium. Teacher Education/Educator Training: Current Trends and Future Directions,* Laval, Quebec.

Levine, R. (2002). *Comparing problem-based learning and WebQuests.* Retrieved from http://www.coollessons.org/compare.htm

Mason, C., Alibrandi, M., Berson, M., Diem, R., Dralle, T., Hicks, D., Keiper, T., & Lee, J. (2000). Waking the sleeping giant: Social studies teacher educators collaborate to integrate technology into methods' courses. *Proceedings of the Society for Information Technology and Teacher Education International (SITE) Conference,* (1), 1985-1989.

McAdoo, M. (2000). The real digital divide: Quality not quantity. In C. Dede (Ed.). (1998). *Learning with technology* (pp. 143-151). Alexandria, VA: Association for Supervision and Curriculum Development Yearbook.

Means, B. (2001). Technology use in tomorrow's schools. *Educational Leadership, 58*(4), 57-61.

Milson, A. J. (2002). The Internet and inquiry learning: Integrating medium and method in a sixth grade social studies classroom. *Theory and Research in Social Education, 30*(3), 330-353.

Molebash, P. (2002). Web Inquiry Projects: Inquiring minds want to know. *Proceedings of the Fall CUE Conference.* Retrieved from http://edweb.sdsu.edu/wip/WIP_Intro.htm

Molebash, P., Dodge, B., Bell, R., Mason, C., & Irving, K. (n.d.). *Promoting student inquiry: WebQuests to Web Inquiry projects (WIPs).* Retrieved from http://edweb.sdsu.edu/wip/overview.htm

Norton, P., & Wiburg, K. (2002). *Teaching with technology: Designing opportunities to learn.* Belmont, CA: Thomson Wadsworth.

Roschelle, J. M., Pea, R. D., Hoadley, C. M., Gordin, D. N., & Means, B. (2000). Changing how and what children learn in school with computer-based technologies. *Future of Children, 10*(2), 76-101.

Shiveley, J., & Van Fossen, P. (1999). Critical thinking and the Internet: Opportunities for the social studies classroom. *The Social Studies,* (January/February), 42-46.

Staley, D. J. (2000). Technology, authentic performance, and history education. *International Journal of Social Education, 15*(1), 1-12.

Watson, K. (1999). *WebQuests in the middle school curriculum: Promoting technologi-*

cal literacy in the classroom. Retrieved from http://www.ncsu.edu/meridian/jul99/webqeust/index.html

Whitworth, S., & Berson, M. (2003). Computer technology in the social studies: An examination of the effectiveness literature (1996-2001). *Contemporary Issues in Technology and Teacher Education, 2*(4), 472-509.

Windschitl, M. (2002). Framing constructivism in practice as the negotiation of dilemmas: An analysis of the conceptual, pedagogical, cultural, and political challenges facing teachers. *Review of Educational Research, 72*(2), 131-175.

Chapter XIX
Let Them Blog:
Using Weblogs to Advance Literacy in the K-12 Classroom

David A. Huffaker
Northwestern University, USA

ABSTRACT

This chapter introduces the use of blogs as an educational technology in the K-12 classroom. It argues that blogs can be used to promote verbal, visual, and digital literacy through storytelling and collaboration, and offers several examples of how educators are already using blogs in school. This chapter also reviews issues such as online privacy and context-setting, and ends with recommendations for educators interested in implementing blogs with current curricula.

INTRODUCTION

As Internet technologies continue to bloom, understanding the behaviors of its users remain paramount for educational settings. For teachers, parents, school administrators, and policymakers, learning *what types* of activities and applications students are using on the Internet is only the surface. Understanding *how* they are using these applications can provide innovative strategies for learning environments.

Previously, many scholars have explored how Internet users communicate and present themselves online, using computer-mediated communication venues such as e-mail, chat rooms, instant messaging, newsgroups, multi-user domain (MUDs), and personal home pages to examine communication patterns, online identity construction, and even gender differences (Crystal, 2001; Döring, 2002; Greenfield & Subrahmanyam, 2003; Herring, 2000; Lee, 2003; Turkle, 1995; Witmer & Katzman, 1997).

Internet technologies continue to evolve, and it is important for scholars to examine the latest CMC arenas in comparison with past research in hopes of finding new ways to find creative learning solutions and enhance pedagogical method in educational technology. Weblogs, commonly referred to as *blogs,* represent one of the latest advances in CMC.

A blog can be simply defined as an online journal. Made up of reversed chronological entries infused with text, images, or multimedia, blogs embody a place where individual expression and online community development coexist. Not only do the authors, or *bloggers*, post thoughts and feelings on a Web page for the world to view, but blog readers can comment, creating a dialogue between the blogger and the community he inhabits. Furthermore, bloggers link to other bloggers, creating an interwoven and perhaps interdependent online community of writers and readers. Blog popularity continues to resonate throughout the media, with many scholars suggesting an evolution in individual self-expression, education and research, online journalism, and knowledge management (Alterman, 2003; Blood, 2003; Herring, Scheidt, Bonus, & Wright, 2004; Lasica, 2003; Moore, 2003; Mortenson & Walker, 2002; Oravec, 2002; Pollard, 2003b; Schroeder, 2003).

In a recent survey, Perseus Development Corporation found that among the four million published Weblogs, almost 53% are created by children and adolescents between ages 13-19 (Henning, 2003). With such a strong population of young bloggers, understanding its potential uses within a classroom remains an exciting prospect for educators and parents. Can blogs be used to enhance learning? In what ways can they be used in the classroom?

This chapter hypothesizes that blogs can be effective educational tools in the following ways: (1) they promote verbal and visual literacy through dialogue and storytelling, (2) they allow opportunities for collaborative learning, and (3) they are accessible and equitable to a variety of age groups and developmental stages in education.

In order to evaluate this hypothesis, this chapter will proceed as follows. First, it will provide a thorough explanation of what blogs are, fundamental blog features, how they are used, and the demographics of the blog population. Second, this chapter will define verbal, visual, and digital literacy, and their importance in learning. Third, it will explain how blogs foster literacy through storytelling and peer collaboration. Fourth, this chapter will describe examples where blogs are used in K-12 classrooms, with an emphasis on the previous concepts of storytelling, peer collaboration, and literacy. Finally, this chapter will provide specific recommendations for educators and school administrators interested in implementing blogs in their schools and classrooms.

WHAT IS A BLOG?

Blogs are personal journals written as a reversed chronological chain of text, images, or multimedia, which can be viewed in a Web page and are made publicly accessible on the Web (Huffaker, 2004a; Winer, 2003). As depicted in Figure 1, blogs typically contain text in the form of a "blog post," offer the ability for readers to comment or provide feedback, contain archives to past blog posts, and link to other blogs and bloggers.[1]

Figure 1. An example blog

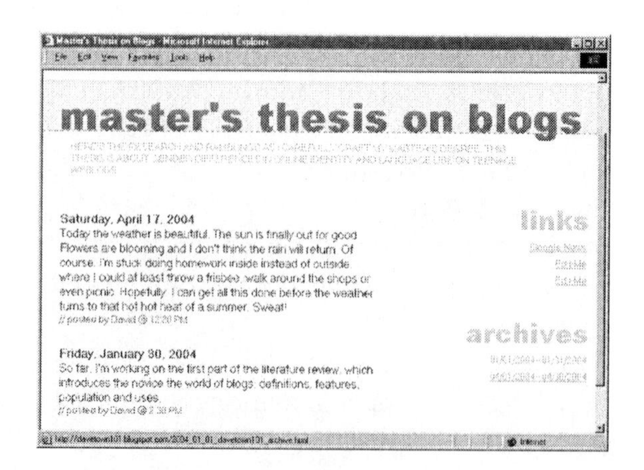

Blogs are inherently different from personal home Web pages. First, bloggers post entries through manual software, such as a Web browser, or automatic software, which is downloaded off the Internet and used to instantly publish content to the Web. Therefore, bloggers do not need to understand HTML or other Web programming languages to maintain their blogs. Second, the resulting blog page resembles a personal diary with entries sorted by time and date, a much stricter format than personal Web pages of the past.

Blog Features

David Winer, a fellow at Harvard Law School, considered one of the more visible writers and developers of Weblogs, describes some of the more important features of blogs:

- **Weblog Posts:** Weblog posts include a subject title and a body message. Posts can be relatively short in length such as one or two paragraphs, or they could be a long, thoughtful exposé, encompassing a dozen paragraphs. They can comprise a variety of media objects, including text, pictures, graphics, or multimedia, and even popular file formats such as *Microsoft Office* documents or *Adobe* PDFs (Winer, 2003). These posts receive a timestamp to denote time and date of the post. Figure 2 demonstrates the format of a typical blog post.
- **Comments:** Readers of a Weblog have an opportunity to respond to a blog post through a comment or feedback link. These comments create a thread, as many readers can comment on a single post (Winer, 2003). They contain a timestamp and are viewable to the public. Bloggers can also respond back to reader comments. Figure 2 highlights the comment section. Blog readers can click "Post a Comment" to add a new comment or "1 Comment" to read previous ones.
- **Archives:** The front page of a blog contains only a certain amount of posts, sometimes two or three and sometimes twenty. For authors who have maintained their blogs for longer periods of time, they can store past blog posts in an accessible, often searchable archive. As depicted in Figure 3, archives can be organized by month, by week, or even by number of posts.
- **Templates:** Another useful feature for Web authors is a set of presentation tools that allow pages to be built from preexisting templates. Blog authors can choose from a variety of graphical layouts, typog-

Figure 2. Blog post with comment link

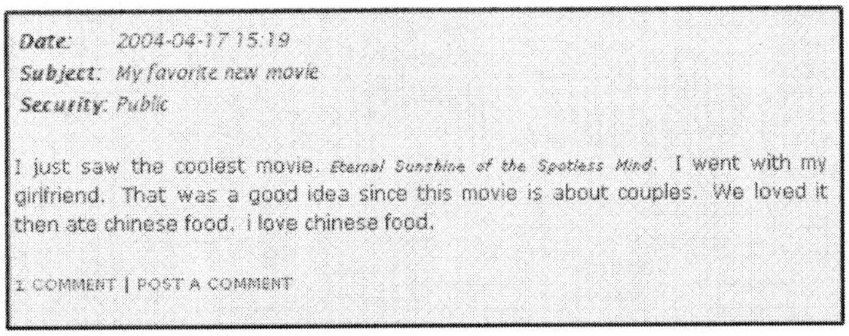

Date: 2004-04-17 15:19
Subject: My favorite new movie
Security: Public

I just saw the coolest movie. *Eternal Sunshine of the Spotless Mind*. I went with my girlfriend. That was a good idea since this movie is about couples. We loved it then ate chinese food. i love chinese food.

1 COMMENT | POST A COMMENT

Figure 3. An example of blog archives

raphy, and color schemes. This allows complete customization and a feeling of personalization for bloggers without any sophisticated technical expertise (Winer, 2003). However, some bloggers like to tinker with a Web programming language, to add third-party applications or other bells-and-whistles to their blogs, and the capability to reprogram a blog is available. Figure 4 represents the types of templates Blogger.com, one of the most popular blog-hosted sites, offers.

What is the Blogosphere?

Although some studies suggest the majority of blogs are still highly personalized venues for self-expression (Herring et al., 2004), many blogs contain links to other bloggers, creating an online community that is often referred to as the *blogosphere*. These blog communities typically share a common purpose or responsibility (Carl, 2003), for example, a group of friends spread out across the world may use blogs as a means to communicate with each other, or a support group might encourage each other's therapeutic development, or a group of amateur journalists might be spreading news on situations of political strife or violent conflict in an area that a global news agency may be unaware of.

The blogosphere also represents the *total population of bloggers*, an Internet culture that is continuing to grow. Several reports and

Figure 4. Example of blog templates

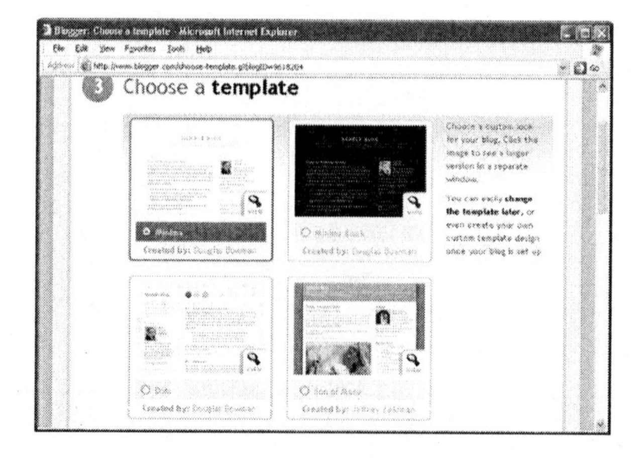

Web sites that gather information on blog statistics, which estimate between two and four million blogs, demonstrate the vastness of the blogosphere; many expect this growth to continue (Henning, 2003; Herring et al., 2004; Kumar, Novak, Raghavan, & Tomkins, 2003). Certainly, the acquisition of population blog software or services within online commercial giants such as AOL, Yahoo, and Google suggest that the world may be on the cusp of a blog saturation in Internet culture (Munarriz, 2003; Olsen, 2004; Shirpy, 2003).

Size of the Blogosphere

Blogs continue to be created and abandoned each day, so the exact population figures on the size of the blog population remains in transit. The population size is captured using software that indexes blogs or comes directly from sites that host blogs such as *LiveJournal* or *Blogger*. It must also be considered how many blogs are "active" or updated within the last three months, because many blogs become abandoned (Henning, 2003).

- **Blogcount** (http://dijest.com/bc/), a Web site dedicated to understanding how vast the Weblog community is, collects scholarly and industrial reports that discuss technical and demographic issues, from frequency of blog posts to mapping the blog community. By June 2003, **Blogcount** estimated there were 2.4 to 2.9 million active Weblogs (Greenspan, 2003).
- **BlogCensus** (http://www.blogcensus.net/) uses a software program to search and locate Weblogs and categorize them by language and authoring tool. As of April 2004, the BlogCensus[2] has indexed 1.83 million blogs and estimates at least 1.21 million of these are active.
- **LiveJournal** (http://www.livejournal.com/) publishes daily statistics direct from its servers, citing that there are over 2.9 million "livejournals," with 1.4 million active, as of April 2004 (specific data can be found at http://www.livejournal.com/stats/stats.txt).
- **Perseus Development Corporation**, which offers enterprise-level surveys regarding software and technology, recently published a white paper that estimates there are 4.12 million blogs, but at least 2.7 million have been temporarily or permanently abandoned (Henning, 2003).

Age and Gender Demographics of the Blogosphere

Most of these surveys suggest that a significant portion of the total blog population is made up of teenagers, and almost evenly partitioned between genders. The largest age distribution of bloggers typically ranges between 13 and 20, which assumes most bloggers are either in early secondary school or beginning college or university. There are some subtle discrepancies in gender within the studies listed below; however, the margins are not far enough apart to suggest severe differences in gender use of blogs.

- Perseus Development Corporation, for instance, finds blogs are dominated by the youth, with 51.5% of all blogs being developed and maintained by ages 13-19. They also find 56% of the *total* bloggers are female and 44% are male (Henning, 2003).
- A recent academic study of 203 randomly selected Weblogs revealed 54.2% male authors and 45.8% female authors, as well as 40.4% of blog authors being under age 20 (Herring et al., 2004).

- Another academic study of 358 randomly selected blogs found 52% male and 48% female bloggers, and 39% of bloggers are less than 20 years old. However, they also found there are more females than males in the 'teen' category (Herring, Kouper, Scheidt, & Wright, 2004).
- BlogCensus randomly sampled 490,000 blogs to find 39.8% male and 36.3% female, with the rest of the blogs unidentifiable in terms of gender (see http://www.blogcensus.net/weblog/).
- Finally, Jupiter Research found that blogging is split evenly among genders (see http://www.jup.com/bin/home.pl) (Greenspan, 2003).

Use of Blogs

Understanding the features of the blogs helps distinguish them from other Internet applications, and grasping the size of the blogosphere signifies the popularity of blogs in Internet culture. The next question involves the content of blogs. What are bloggers writing about? The answer not only provides a context for online community interaction, but possible application for educational technology. These can be divided into five areas: (a) personal blogs, (b) community blogs; (c) journalism blogs; (d) education and research blogs; and (e) knowledge blogs.

Personal Blogs

The most popular use of blogs are similar to personal Web sites authored by individuals, which include chronological posts as well as links to other Web sites or Weblogs (Lamshed, Berry, & Armstrong, 2002). Despite the popular notion that Weblogs lean toward external events, or remain highly interlinked with the blogosphere, the majority of Weblogs are still individualistic self-expressions written by one author (Herring et al., 2004).

Community Blogs

Virtual communities develop through the use of a blog (Lamshed et al., 2002). Examples might include a community support group, a site for parents to ask questions and exchange answers, a research community sharing resources and data, or a mirror of an offline community, like a softball team or neighborhood newsletter. Although personal blogs may dominate the blogosphere, the ability for individuals to choose their level of community participation may be another reason for blog popularity, as it allows the blog author to explore individual needs while benefiting from community interactions (Asyikin, 2003). The linkages with other Web sites, people, and ideas even form micro-communities with varying levels of involvement.

Journalism Blogs

The idea of alternative forms of journalism manifesting through Weblogs has received increasing attention in media and scholarship (Alterman, 2003; Blood, 2003; Lasica, 2003). Where is Raed? (http://dear_raed.blogspot.com), for instance, is a blog by an Iraqi that discusses what is happening in Iraq since September 2003. He discloses a candid view of the U.S. occupation, but also introduces readers to fellow Iraqi bloggers. For most, the global news agency is the only link to international affairs—having personal, subjective commentary within a foreign world provides a unique view to outsiders.

A different, but equally unique log is *J-Log*, which provides community critiques and commentary on current journalism and news. The community not only shares interesting news items, but also poses questions such as, "Is this news fair and balanced?" Perhaps *J-Log* (http://www.mallasch.com/journalism/) and individual reports such as Raed demonstrate new forms of online journalism; critiques, however, as to the viability of these news sources remain an issue, including the resources and references and even the subjectivity amidst objective journalistic philosophy.

A link to http://blogdex.net/, an MIT Laboratory experiment that captures the fastest-spreading ideas in the blog community, typically results in news headlines as the most contagious information.

Education and Research Blogs

Blogs have been heralded as opportunities to promote literacy in learning by allowing students to publish their own writing, whether it is a journal or story, or even comments on class readings (Kennedy, 2003). For more advanced students, blogs present the same opportunities: writing and thinking with Weblogs, archiving and analyzing past knowledge, and developing a social network to collaborate and critique (Mortenson & Walker, 2002).

Blogs allow educators and students to interact in the same common space and format (Wrede, 2003). Lamshed et al. (2002) find that students believe blogs are easy to use and navigate, and enthusiastic about learning features such as storing and managing information, communicating, reviewing work before posting, and "keeping on track," which refers to managing time or monitoring progress (Lamshed et al., 2002). If students are eager to adopt blogs in the classroom, teachers may have an opportunity to also keep on track with students, as communication and interactions are visible and accessible anytime-anywhere.

Several sites explore the use of blogs in education. *Weblogg-Ed* (http://www.weblogg-ed.com/), maintained by Will Richardson, collects ideas about Weblogging in school settings and facilitates dialogue between teachers. Similarly, Edublog (http://edublog.com/) is an initiative to develop and study blog-based tools for the classroom.

Knowledge Blogs

Similar to education and research, blogs provide opportunities for organizations to manage and share content, as well as communicate across the network. Dave Pollard, Chief Knowledge Officer at Ernst and Young, Inc. and popular writer on the role of blogs in the business, suggests that blogs can be used to store and codify knowledge into a virtual file cabinet. But unlike other content management tools, blogs allow authors to publish in a personal and voluntary way, creating a democratic, peer-to-peer network (Pollard, 2003b). Pollard also suggests companies can increase profitability by designing information architecture to embrace the use of Weblogs (Pollard, 2003a, 2003c).

VERBAL, VISUAL, AND DIGITAL LITERACY

Literacy has always been a focus of learning, especially considering that the foundations of education are grounded in reading and writing. In fact, reading and writing, often referred to as *verbal literacy,* serves as a benchmark for success in education. Reading and writing are not only important in language arts or humanities; they serve as prerequisites for all academic disciplines, including science and mathematics (Cassell, 2004). Verbal literacy is developed even before children enter school as parents read stories to their children, helping them to understand the relationship between words and pictures, as well as helping to develop narrative structure (Bransford, Brown, & Cocking, 1999; Huffaker, 2004b).

Scholars have recognized how communication also takes place in the form of images and symbols, what is referred to as *visual literacy* (Gee, 2003). As James Paul Glee (2003) suggests, texts are becoming increasingly multimodal, containing both text and images, and can be recognized in everything from the advertisement to the high school textbook (Gee, 2003). Our society is filled with these multimodal symbols, and their coherence is intrinsic for operating in the modern world. Therefore, reading and writing should not only include words but also images, and the development of both *verbal* and *visual* literacy is essential for success inside and outside school walls.

The use of technology represents a third type of literacy, equally ubiquitous and important as the other forms. *Digital literacy,* sometimes referred to as technological fluency, embodies the idea of using technology comfortably, as one would a natural language (Cavallo, 2000). As users of computers and other digital technology become more fluent, they learn to communicate and express themselves explicitly and eloquently using these tools; in effect, technology becomes innate. Furthermore, just as reading and writing are widely hailed as the building blocks for success in society, *digital literacy* becomes necessary for success in the technological world we inhabit.

This is what makes educational technology so exciting—it encourages and advances all three types of literacy: verbal; visual; and digital. Blogs, for example, utilize both textual reading and writing, but also involve the use of graphics in the forms of emoticons, or graphical expressions of emotions, images, and multimedia. Blogs also encourage digital literacy, which is grasped from navigating a graphical user interface, and using computers and the Internet in order to publish content.

Some CMC contexts such as e-mail and instant messaging often utilize short pieces of dialog, informal language, and even altered words from spoken language (Crystal, 2001). Educators might complain that using these forms in a classroom might reduce literacy skills, as language development is often paired with precision and formality. Blogs might combat this issue, as its medium involves longer written passages and can be contextualized by educators to promote formal language skills.

STORYTELLING AND COLLABORATION

Storytelling is a natural part of adolescent development, and children understand the fundamentals of storytelling in their first three years of life (Bransford et al., 1999). Even as a baby, a parent introduces the child to storytelling via bedtime readings or oral tales. By the time a child reaches age four, he can recall many types of stories, whether fictional or autobiographical (Bransford et al., 1999). The stories develop into more mature narrative as the child gets older, thus storytelling provides a way for children's language and reading skills to develop (Bransford et al., 1999).

Children advance literacy skills through the practice of telling stories to adults and peers alike (Ryokai, Vaucelle, & Cassell, 2003). With adults, children advance language skills because their partners have even *more* advanced language skills, and children will adapt (Ryokai et al., 2003; Vygotsky, 1980). With peers of similar age groups, children feel more comfortable to collaborate, and enjoy learning together and building on each other's knowledge (Ryokai et al., 2003). Because peer relations remain quite important to children (Bullock, 1998), effective group dynamics and learning also results in pro-social behavior through a sense of ownership, compromise, and shared responsibility (Calvert, 1999).

Storytelling does not end in childhood, but continues throughout adolescence and even adulthood. Stories help children and adults alike share experiences and feelings in an engaging and even entertaining way (Denning, 2001). From simple sandbox sketches to dinner party yarns, the importance of stories as a catalyst for conversation and dialogue is clear. Furthermore, storytelling helps create connections between people, to engage people, to captivate them. This captivation occurs on both ends—it is just as fun for the teller of the story as the listener. Storytelling fused with peer collaboration is an excellent way to improve language abilities and advance literacy for children and adolescents (Ryokai et al., 2003).

Blogs have the potential to foster both storytelling and collaboration. First, blogs serve as a natural venue for personalized self-expressions. Like diary entries, blogs take form as stories, anecdotes, or vignettes, similar to the types of oral and written stories people encounter everyday. This includes past and present activities, feelings about oneself or other people, or even hyperbolic or fictitious tales.

Secondly, blogs have technical features that offer the potential to create a dialogue between blog author and blog reader, whether it is a reader response, a question-and-answer sequence, or even general brainstorming. In some cases, blog authors discuss candid feelings and readers respond with encouraging statements, providing therapeutic connections. These comments form a chain between the author and readers, and in essence, an online community. Communities are also built as bloggers link to each other, creating a group of storytellers that provide individualistic expressions, as well as interactions with each other.

EXAMPLES OF BLOGS IN PRACTICE

Blogs are just beginning to infiltrate classrooms, as educators and school administrators consider blogs as a useful tool for communicating between teachers, schools, students, and parents, and as a way to showcase the work of students (Richardson, 2004). These practices are celebrated on the Internet through communities of educators interested in blogging and education. Will Richardson's "Weblogg-ed: Using Weblogs and RSS in Education" Web site (http://www.weblogg-ed.com/), for instance, is a useful source of information. His site focuses on best practices, offers a quick start guide for teachers, and links to other bloggers concerned with blogs in education. "Edblogger Praxis" (http://educational.blogs.com/edbloggerpraxis/) is another important Web site which unites educators who blog about their experiences or pedagogical philosophies.

This section will look at examples of blogs in practice, separating them into high school (grades 9-12), elementary and middle schools (grades 4-8), and primary school (grades K-3) in order to contextualize blog use by age and developmental stage, and to provide useful models for educators and school administrators interested in viewing how blogs work at different levels of the school system.

High School Blogs

Will Richardson is a teacher and Supervisor of Instructional Technology and Communications at Hunterdon Central Regional High School in Flemington, New Jersey. He not only initiates school-wide policies for blog use in the classroom, but has found success in using blogs for his own literature and journalism classes. In the

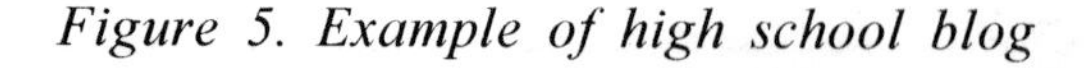

Figure 5. Example of high school blog

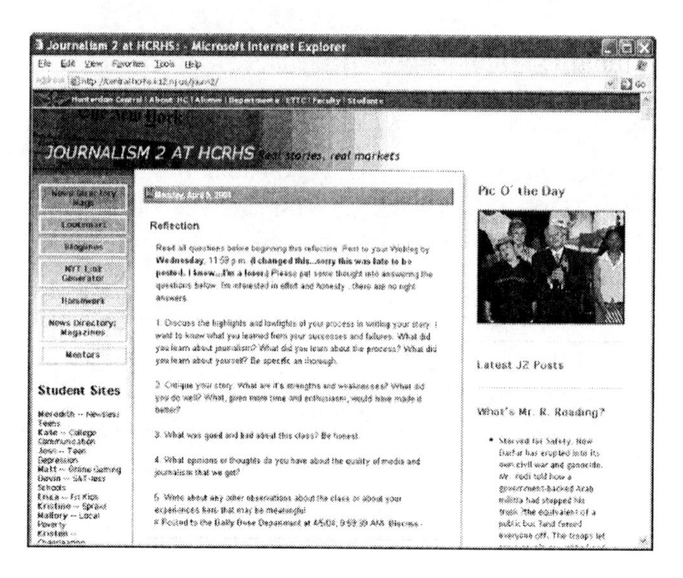

literature class, students use blogs to comment and critique on reading assignments. In the journalism class, the students collect news stories to post in their blogs; they also comment on each other's stories while serving in an editorial role. Therefore, in language arts classes, blogs can be used to generate a discussion using critical analysis, or as a collaborative tool where students comment and edit each other's work. For Richardson, blogs help students become more aware of their writing, as well as their audience (Kennedy, 2003). Figure 5 represents how blogs can be contextualized to provide assignments for students and allow sections for critique and reflection. Each student has an individual blog, which is linked to a homepage where the teacher assigns work and posts relevant news and other information.

Blogs also help create a community of practice among participating students. They can collaborate with each other and build knowledge. These types of discussions, where ideas are synthesized and news ideas created, may be intrinsic to building critical thinking skills. They may also feel that they are "part of a team," and that each individual has a responsibility to contribute in order to achieve success for the group. Again, effective group behavior involves shared ownership, compromise, and responsibility, which creates pro-social conduct (Calvert, 1999).

Blogs are accessible to the general public, and Richardson found an important side effect as students began blogging. When the class discussed a new novel, the author of the book accessed the blog and posted an impressive response to the questions students were asking (Richardson, 2004). This may have a profound affect on how students view their own work. For instance, this may give students the impression that their ideas and discussion are important in the real world, providing a sense of empowerment and self-efficacy. Literature and journalism classes and experts are not the only beneficiaries; political leaders, scientists, artists, and philosophers can also directly participate in an educational blog, making students feel their academic work remains valuable, an important consideration for motivating children to learn.

Elementary and Middle School Blogs

Blogs can be utilized in many of the same ways in other grade levels. The Institut St-Joseph in Quebec City, Canada, for instance, uses blogs among fifth and sixth graders in order to practice reading and writing. Implemented by the school principal, Mario Asselin, Institut St-Joseph bloggers use a software program to write about anything and everything that is school related. Similar to Richardson's realization, the fact that blog posts are being read by anyone in the world has an acute effect on students. They felt empowered as their blogs received comments from total strangers and even Canadian celebrities (Asselin, 2004). Similar to Will Richardson's work, Figure 6 portrays a homepage blog where projects are assigned and school- or project-related links are provided. Students can comment on homework assignments to the entire class, and still post to their own blog space.

At first, some parents and other readers complained that the language of the student blogs was too informal to be considered good writing, and even contained misspelling and grammar errors. Critics complained that blogs were teaching improper language skills, so Asselin (2004) discussed these challenges with the students. The students came up with a system where each blog post would be reviewed by students and a graphic would be posted alongside the text, which served as a stamp of approval. Suddenly, student writing improved dramatically, as no student wanted mistakes after insisting their quality of writing was excellent (Asselin, 2004).

Similarly, Anne Davis' *Newsquest* involves fourth through sixth graders who practice writing by commenting on news and events. Students felt that blogging helped them write better stories, extend their vocabulary, and even feel 'grown up' knowing their voice is on the Internet. After a period of working with blogs, these students began to collaborate with Will Richardson's high school journalists, forming the "Georgia-NJ Connection" blog. Now, older journalists and younger journalists can share ideas, comments, and critiques. Again, this

Figure 6. Example of middle school blog (http://cyberportfolio.ixmedia.com/carriere/)

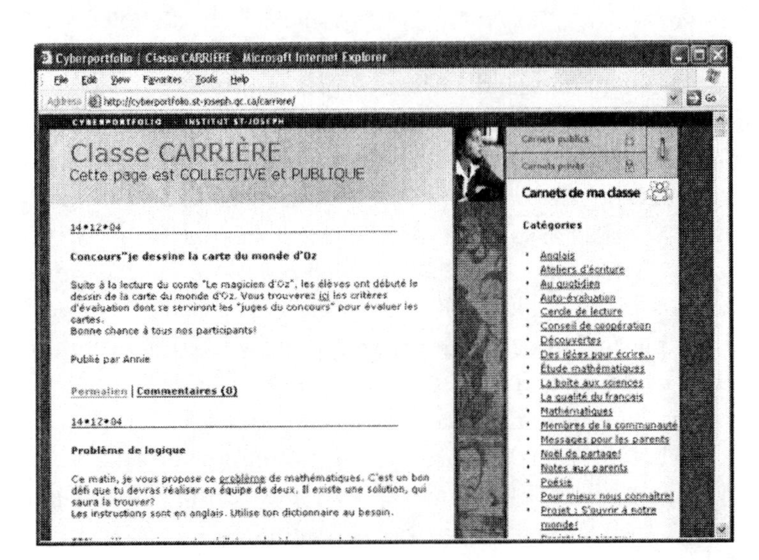

Figure 7. Example of elementary school blog (http://www.schoolblogs.com/NewsQuest/)

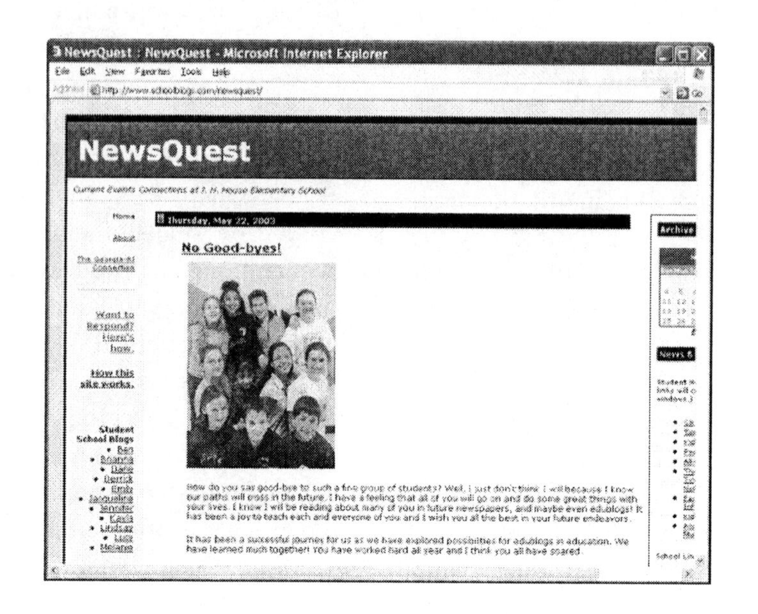

resonates with Vyogtsky's notions that children will adapt to more advanced language skills observed in adults (Ryokai et al., 2003; Vygotsky, 1980). Figure 7 demonstrates how blog information can be scaled down to reach a younger audience.

Primary School

Mrs. Dudiak, who teaches second grade at Oakdale Elementary School in Maryland, uses blogs to create writing assignments for her students. She might use a picture such as a waterfall and ask students to write a description, a story, or poetry. She also asks them to discuss favorite books, what 'types' of books and stories they like, as well as depictions of books into a "movie in our head;" students reply in the comment section of the blog. This is an excellent example of how blogs can be placed in specific contexts to meet the goals of classroom curriculum. Figure 8 exemplifies how blogs can be simplified to reach even the youngest students. Changing colors, enlarging texts, and contextualizing material to reach the needs of primary students represent easy changes for educators to make.

Lewis Elementary School in Portland, Oregon, uses a blog to showcase K-5 student work, as well as to post important information and news for parents. For instance, photo galleries of artwork are displayed alongside weekly classroom notes from each teacher. This demonstrates how blogs cannot only be effective in practicing and advancing writing and language skills, but also in communicating between teachers, students, and parents. Blogs may give parents more direct exposure to their children's school life, and thus more opportunities to have an impact on their children's learning. Little Miami Schools (http://www.littlemiamischools.com) in Morrow, Ohio, also uses a blog to communicate school goals and news, as well as spotlighting the successes of its students and teachers.

Figure 8. Example of primary school blog (http://mrsd.tblog.com/)

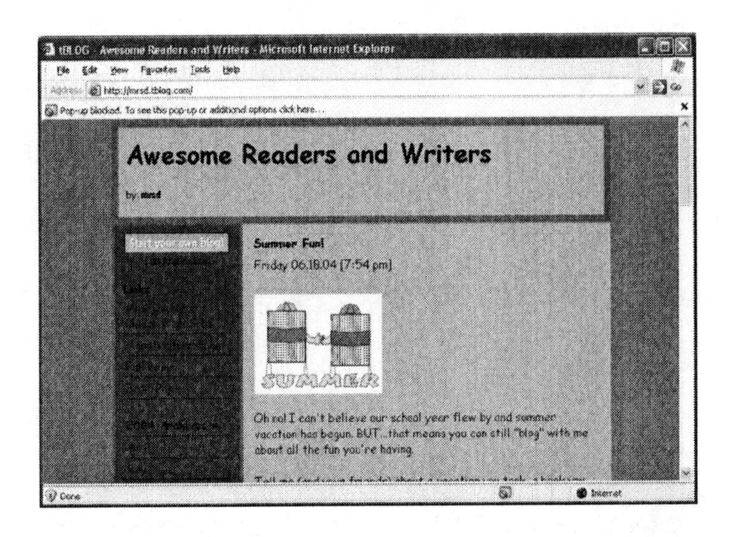

Figure 9. Example of blog as communication hub for teachers, schools, and parents (http://lewiselementary.org/)

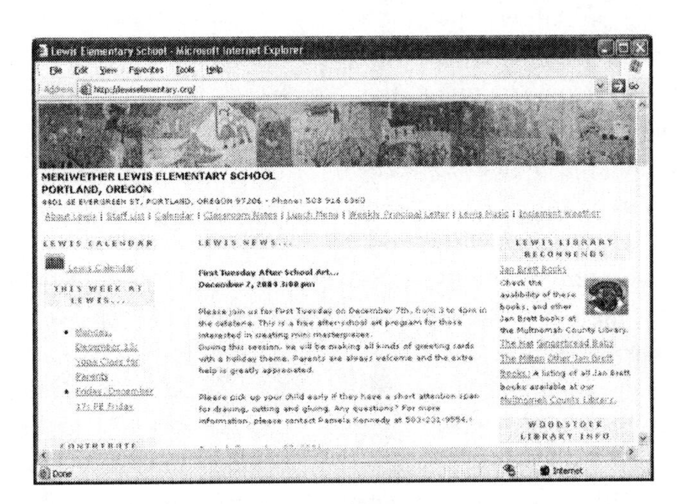

In sum, blogs can be used at a variety of age and grade levels to promote reading and writing skills, as well as to provide communication links between students, parents, teachers, and schools. Students can use blogs on their own, finding stories or news they find interesting, as demonstrated with the classes of Mr. Richardson and Mr. Asselin, or they can have specific assignments designed by the teacher, as seen in the case of Mrs. Dudiak's class. Furthermore, schools can use blogs to bridge the links between students, teachers, and parents, making demonstration of work and student progress available, as well as school news and schedules. Figure 9 demonstrates how blogs can be used as a central communication hub for teachers, schools, parents, and students to share information.

The next section will address some of the issues and considerations for educators interested in implementing blogs in their schools and classrooms.

ISSUES AND CONSIDERATIONS

School- and District-Wide Considerations

This chapter has mainly approached blog implementation from the educator perspective. However, it is important to examine the schools and districts that encompass the individual classroom. Not only do school and district administrators remain an intrinsic part to choosing educational technology initiatives, but their choices can have a positive or negative effect on how these choices become applied. For instance, schools and districts have to work within a budget when making choices for technology. Second, they have to choose which hardware and software *standards* will be equitable across the community. Finally, they have to implement technology within certain ethical, legal, and security issues.

For each of these considerations, blogs still stand as a viable solution. First, most blog software is free and open-sourced, so administrators can host blogs within local technology infrastructure. Second, blogs are delivered via the Web, so standards have already been established—blogs can be implemented anywhere a computer and Internet accessibility is present. Finally, blogs have the option of being available to the public or of being private, an important consideration that will be discussed in the next section.

In sum, the importance of school- and district-wide influence on educational technology implementation cannot be ignored. However, for the same reasons blogs can be a useful tool and practical solution for educators; administrators should also be impressed.

Issues of Online Privacy

A key concern for any educator and school system involves student privacy. Cyberstalking or sexual predation is a serious concern for adolescents using the Internet (Earley, 1999; Gilbert, 1995). Many of the blog hosts offer free Web space available to the public on the Internet; information on a blog is publicly accessible. If the content of blogs or identities of students need to remain anonymous or private to the world, blog software and hosts offer other options.

For instance, free blog-hosts such as LiveJournal offer a "Friend's Only" option, which involves a username and password for entrance. For even more security, blog software such as MoveableType, which can be implemented on local school Web servers, can remain completely in-house, that is, no one outside of school or classroom computers would even have access to the blogs.

While privacy is an important consideration for educators and school administrators, the counter-argument to keeping blogs private would be that making blogs available to the world might generate feedback that will empower bloggers and further develop communication and collaboration. For example, if a classroom is discussing a science project and a professional scientist participates in the discussion, it would add significant value to the learning experience.

Setting Context

One of the primary challenges for using blogs, as with many technologies, in the classroom is the importance of setting a context for learning to occur. While discovery and creativity does abound when adolescents are allowed freedom to explore these CMC venues, some structure needs to be in place to facilitate a learning outcome. For instance, if writing quality is a concern for a teacher, then contextualizing the language to focus on clear and succinct writing skills is a must. Similarly, students have to be encouraged to use the blog on a steady basis and focus on the classroom material.

In sum, contextualizing blogs to the learning experience will serve to produce more exciting and educational blog experiences in the classroom, experiences that parents, administrators, and students alike can observe and reflect upon.

The next section will provide some recommendations for educators interested in implementing blogs in the classroom to promote literacy and learning.

RECOMMENDATIONS

Use free blog software for easy implementation in the classroom. There are several popular Web sites that provide templates, maintenance, and hosting of blogs. Using free blog software makes it an easier decision for educators and school administrators concerned with school budgets and technology requirements. With a computer and an Internet connection, blogs can be set up in minutes. For educators who want to keep blogs inaccessible to the public, most blog hosts offer the opportunity to password-protect blogs. The most popular blog-hosts are:

- **LiveJournal:** http://www.livejournal.com
- **BlogSpot:** http://www.blogspot.com
- **Xanga:** http://www.xanga.com
- **MoveableType:** http://www.moveabletype.org
- **T-Blog:** http://www.tblog.com

For those who want to implement the blogs on local school servers, there are open source applications such as *MoveableType* and *phpBB* which can be altered and used in any expected capacity, although they have technical requirements (documentation are available on these sites).

Open-source applications such as *MoveableType* and *phpBB* allow complete customization of the interface and blog application. So educators should feel that blogs can be completely tailored to suit the needs of students and the affordances and constraints of school district policy. The examples of blogs used in this chapter are presented to demonstrate current applications and serve as a baseline for new ideas and implementation. Educators should feel complete flexibility in redesigning blog software for learning experiences.

Fuse emergent learning with curriculum-based assignments. If given the opportunity, students might begin to navigate their own learning paths, a feature that is recommended by the new science of learning as directed by the U.S. National Research Council (Huffaker & Calvert, 2003). Surely, letting students do as they want with blogs will reveal some interesting trends that could be manipulated into important learning opportunities. However, it may be a better idea to create a hybrid form where students construct their own learning, as well as assignments that teachers feel aid the curriculum. Therefore, students have some flexibility in exploring the medium of blogs, while teachers provide direction and structure in the educational practice.

Encourage collaboration. Blogs have great potential as a collaborative environment. As students read and write, they may comment and critique each other's work. They may even edit each other's work. Assignments can be created where students have to work as a team to complete the goals, such as a story "train," or chain of story parts devised by individual students, or literature critiques of different parts of a book, or even a mix of different perceptions on the same experience.

Develop a system for rating student work before it is published. One way to maintain the caliber of writing that is expected in formal writing is to create a way to 'proof' student work before it goes into the public. Most blogs contain a *preview* area before publication takes place. The teacher does not necessarily need to be the final editor—students can edit each other's work and provide the final stamp of approval before the world sees the writing. This may be more difficult at early ages, but the metaphor provides accountability for students, which the Institut St-Joseph found dramatically improved writing quality (Asselin, 2004).

Create interactions between students and the outside world. One way to make students feel their work is valuable is to demonstrate how their words are being read outside the classroom. For instance, if parents or even total strangers provide feedback on student work, children and adolescents might become more aware of their writing, as well as their audience (Richardson, 2004). Perhaps even more empowering are experts such as authors, poets, artists, scientists, engineers, or other professionals who may comment on student work or add to the discussion. These linkages are not only useful for learning and achievement, but are not difficult tasks to achieve when using blogs, especially considering their accessibility by the public sphere.

Link with other educational blogs. As exemplified in the Georgia-NJ Connection, students using blogs may connect with other classrooms despite geographic constraints, in order to expand the community of practice. Even regardless of age, students can share with other students, providing an online learning community that may become self-motivating and self-sustaining. Blogs provide an easy way to do this with built-in options for linking to other blog communities.

Get parents involved in reading and participating with blogs. As previously mentioned, getting parents to provide feedback to student work might be an excellent way for students to feel their work is valuable and even appreciated. Yet this has a two-fold benefit—parents may feel they have more exposure to, and thus more interaction with, their children's learning. Therefore, both students and parents benefit.

Furthermore, blogs can provide links between schools, classrooms, teachers, students, and parents. Parents who visit a classroom blog may encounter the latest third-grade arts and crafts, see the latest school news, or see what is on for lunch that day. This helps bridge communication between all parties responsible for educating children and adolescents.

Creating linkages between all the classrooms in the school creates an interwoven community where everyone feels connected and accessible. This community of practice may have a paramount effect on increasing the efficiency and success of learning in K-12 educational settings.

CONCLUSION

This chapter has sought to demonstrate how blogs can be a useful tool for promoting literacy in the K-12 classroom. Literacy most noticeably takes form in reading and writing, but *visual literacy* as well as *digital literacy* may be just as important for success in education and beyond school walls (Cassell, 2004; Cavallo, 2000; Gee, 2003). As an Internet technology, blogs involve the reading and writing of texts, but also the creation and interpretation of images and multimedia, as well as the need to navigate a graphical user interface, all of which serve to advance skills in verbal, visual, and digital literacy.

Storytelling, a natural part of children's language development, serves as a catalyst for promoting literacy (Bransford et al., 1999; Huffaker, 2004b). Children and adults alike use stories to express thoughts, feelings, and experiences in an engaging and entertaining way (Denning, 2001), and storytelling helps create collaborations between the teller and listener. Collaborations, whether between children and adults or among peers, are also important aspects of learning, producing ways to adapt or scaffold learning (Ryokai et al., 2003; Vygotsky, 1980). Blogs, which resemble personal journals, create a perfect environment for sharing stories and other forms of writings, and provide ways for others to participate through feedback or critique on student work.

Currently, blogs in practice highlight that blogs can be used in a broad range of age groups and developmental stages in education. Second graders and high school teenagers alike can use blogs to practice reading and writing, and help develop language skills. Blogs can also be used in a variety of disciplines, from literature and journalism to arts and sciences. Anywhere language is used to promote discussion, blogs serve as an arena where thoughts can be published and made available to everyone for additional feedback. Not only can *students* use the blogs, teachers, parents, and experts can collaborate alongside them, which may help develop a sense of empowerment and self-efficacy for children and adolescents.

Finally, this chapter has recommended several concepts for educators interested in implementing blogs in the classroom. These concepts embrace collaboration, online community building, interlinking parents with schools, student accountability, and ways to monitor student progress. Because blogs are readily available for free at a selection of blog hosts, implementation into the class does not require a high level of technical expertise or an extravagant school budget. Therefore, the benefits of advancing language skills or bridging communication between teachers, parents, and schools far outweigh the costs.

"Let *them* blog" is a simple idea—blogs provide a computer-mediated communication context where students can practice and advance different types of literacy, use storytelling to express themselves, and collaborate with peers and adults as they complete assignments. "*Letting* them blog" is even simpler—blogs are easy to use, free, and extremely accessible for anyone with a computer and Internet access. Teachers may find blogs to be one of the easiest educational technology applications to implement, and since adolescents are already using blogs outside the classroom, it may be an excellent time to connect the two.

ACKNOWLEDGMENTS

The research of this chapter was funded in part by a grant to the Children's Digital Media Center at Georgetown University (http://cdmc.georgetown.edu) by the National Science Foundation (Grant #0126014). Special thanks to Sandra Calvert for all her mentorship and support.

REFERENCES

Alterman, E. (2003). Determining the value of blogs. *Nieman Reports, 57*(3).

Asselin, M. (2004). Weblogs at the Institut St-Joseph. *Proceedings of the International Conference of Educational Multimedia*, Quebec City, Canada.

Asyikin, N. (2003). *Blogging life—An inquiry into the role of Weblogs in community building*. National University of Singapore.

Blood, R. (2003). Weblogs and journalism: Do they connect? *Nieman Reports, 57*(3).

Bransford, J. D., Brown, A. L., & Cocking, R. R. (Eds.). (1999). *How people learn: Brain, mind, experience, and school* (expanded ed.). Washington, DC: National Academy Press.

Bullock, J. R. (1998). Loneliness in young children. *ERIC Digest*. Availble from http://eric.ed.gov/ERICDocs/data/ericdocs2/content_storage_01/0000000b/80/2a/2b/28.pdf

Calvert, S. L. (1999). *Children's journeys through the information age*. Boston: McGraw-Hill College.

Carl, C. (2003). *Bloggers and their blogs: A depiction of the users and usage of Weblogs on the World Wide Web*. Washington, DC: Georgetown University.

Cassell, J. (2004). Towards a model of technology and literacy development: Story listening systems. *Journal of Applied Developmental Psychology, 25*, 75-105.

Cavallo, D. (2000). Emergent design and learning environments: Building on indigenous knowledge. *IBM Systems Journal, 39*(3&4), 768-779.

Crystal, D. (2001). *Language and the Internet*. Cambridge: Cambridge University Press.

Denning, S. (2001). *The springboard: How storytelling ignites action in knowledge-era organizations*. Boston: Butterworth Heinemann.

Döring, N. (2002). Personal home pages on the Web: A review of research. *Journal of Computer-Mediated Communication, 7*(3).

Earley, M. (1999). *1999 report on cyberstalking: A new challenge for law enforcement and industry*. Washington, DC: United States Department of Justice.

Gee, J. P. (2003). *What video games have to teach us about learning and literacy*. New York: Palgrave Macmillan.

Gilbert, P. (1995). On space, sex and stalkers. *Women and Performance, 17*.

Greenfield, P. M., & Subrahmanyam, K. (2003). Online discourse in a teen chatroom: New codes and new modes of coherence in a visual medium. *Applied Developmental Psychology, 24*, 713-738.

Greenspan, R. (2003). Blogging by the numbers. *CyberAtlas*, (June 23).

Henning, J. (2003). *The blogging iceberg: Of 4.12 million Weblogs, most little seen and quickly abandoned*. Braintree, MA: Perseus Development Corporation.

Herring, S. C. (2000). Gender differences in CMC: Findings and implications. *Computer Professionals for Social Responsibility Journal, 18*(1).

Herring, S. C., Kouper, I., Scheidt, L. A., & Wright, E. L. (2004). Women and children last: The discursive construction of Weblogs. In L. Gurak, S. Antonijevic, L. Johnson, C. Ratliff, & J. Reyman (Eds.), *Into the blogosphere: Rhetoric, community, and culture of Weblogs*. Los Alamitos, CA: IEEE Press.

Herring, S. C., Scheidt, L. A., Bonus, S., & Wright, E. (2004). Bridging the gap: A genre analysis of Weblogs. *Proceedings of the 37th Hawaii International Conference on System Sciences.*

Huffaker, D. (2004a). *Gender similarities and differences in online identity and language use among teenager bloggers.* Washington, DC: Georgetown University.

Huffaker, D. (2004b). Spinning yarns around a digital fire: Storytelling and dialogue among youth on the Internet. *Information Technology in Childhood Education Annual, 1.*

Huffaker, D. A., & Calvert, S. L. (2003). The new science of learning: Active learning, metacognition, and transfer of knowledge in e-learning applications. *Journal of Educational Computing Research, 29*(3), 325-334.

Kennedy, K. (2003). Writing with Weblogs. *Technology and Learning Magazine, 23*(February 13).

Kumar, R., Novak, J., Raghavan, P., & Tomkins, A. (2003). On the bursty evolution of blogspace. *Proceedings of the 12th Annual World Wide Web Conference,* Budapest, Hungary.

Lamshed, R., Berry, M., & Armstrong, L. (2002). *Blogs: Personal e-learning spaces.* Australia: Binary Blue.

Lasica, J. D. (2003). Blogging as a form of journalism. In R. Blood (Ed.), *We've got blog: How Weblogs are changing our culture* (pp. 163-170). Cambridge: Perseus Publishing.

Lee, C. (2003). *How does instant messaging affect interaction between the genders?* The Mercury Project for Instant Messaging Studies at Stanford University, USA.

Moore, C. (2003). Blogs refine enterprise focus. *InfoWorld,* (January 10).

Mortenson, T., & Walker, J. (2002). Blogging thoughts: Personal publication as an online research tool. In A. Morrison (Ed.), *Researching ICTs in context* (pp. 249-279). Oslo: Intermedia Report.

Munarriz, R. A. (2003). Google's big day. *Fool.com,* (September 11).

Olsen, S. (2004). Yahoo tests syndication technology for MyYahoo. *CNET News.com,* (January 12).

Oravec, J. A. (2002). Bookmarking the world: Weblog applications in education. *Journal of Adolescent & Adult Literacy, 45*(7).

Pollard, D. (2003a). *Blogs in business: Finding the right niche.* Retrieved January 30, 2004.

Pollard, D. (2003b). *Blogs in business: The Weblog as filing cabinet.* Retrieved January 27, 2004.

Pollard, D. (2003c). *A Weblog-based content architecture for business.* Retrieved January 27, 2004.

Richardson, W. (2004). Blogging and RSS—the "what's it?" and "how to" of powerful new Web tools for educators. *Information Today, Inc., 11* (January/February).

Ryokai, K., Vaucelle, C., & Cassell, J. (2003). Virtual peers as partners in storytelling and literacy learning. *Journal of Computer Assisted Learning, 19*(2), 195-208.

Schroeder, R. (2003). Blogging online learning news and research. *Journal of Asynchronous Learning Networks, 7*(2), 56-60.

Shirpy, C. (2003). AOL, Weblogs, and community. *Corante,* (July 5).

Turkle, S. (1995). *Life on the screen.* New York: Simon and Schuster.

Vygotsky, L.S. (1980). *Mind in society: The development of higher psychological processes*. Cambridge, MA: Harvard University Press.

Winer, D. (2003). *What makes a Weblog a Weblog?* Cambridge, MA: Harvard Law School.

Witmer, D. F., & Katzman, S.L. (1997). Online smiles: Does gender make a difference in the use of graphic accents? *Journal of Computer-Mediated Communication, 2*(4).

Wrede, O. (2003). Weblogs and discourse. *Proceedings of the Blogtalk Conference,* Vienna, Austria.

ENDNOTES

[1] Except where noted, all examples of blogs were created by the author in order to avoid any privacy or copyright issues.

[2] Blogcensus was developed by the National Institute for Technology and Liberal Education (NITLE), a non-profit organization funded by the Andrew Mellon Foundation.

Chapter XX
Information Problem-Solving Using the Internet

Steven C. Mills
The University Center of Southern Oklahoma, USA

ABSTRACT

Today's students must think critically and analyze and synthesize information so that they can recognize the technical, social, economic, political, and scientific problems of the information age. This chapter describes how the vast resources of the Internet can supply communication tools and information resources that facilitate the application of a robust set of instructional methodologies in the K-12 classroom to address these skills. The development of information literacy skills in today's classrooms necessitates instructional approaches that address complex sets of learning objectives and focus on rich, multidisciplinary learning. The author maintains that Internet and information technologies provide tools and resources that enable teachers to create powerful learning environments for educating students for the information age using student-centered learning approaches, interactive communication with peers and experts, and collaborative, problem-solving methodologies.

INTRODUCTION

This year middle school students at Parkview Middle School used the Internet to participate in the Constitutional Convention, explore the Great Northwest with Lewis and Clark, visit a wounded Civil War soldier in an army hospital, and assist an astronaut in preparing to make the first trip to the moon. The project, Back to America's Future, was designed to provide seventh graders with the opportunity to research historical events using the Internet, and to analyze and reflect upon the impact of these historical events in view of current social, cultural, and political issues.

The assignment was a collaboration between the classes of seventh-grade English teacher, Martha Montoya, and seventh-grade social studies teacher, Michael Huang. The teachers developed a WebQuest that allowed seventh graders in Ms. Montoya's English classes and Mr. Huang's American History classes to use computers and the Internet in developing research skills and creative writing skills. The project was designed to help students become more comfortable doing research on the Internet by identifying, analyzing, and synthesizing historically accurate information, and by creating a work product based on their research.

Using the computers in their classrooms and in the middle school media center, students worked in cooperative groups and searched history-related Web sites. Students were required to read at least five stories or articles about their chosen historical event. Student teams then wrote a diary as if they were a character in the event they were researching. The final assignment of the WebQuest was to produce a timeline and concept map to illustrate the impact of the historical event on current social, cultural, or political issues or events. Students used multimedia software to present the results of their projects to other students. Each student project was posted on the school's Web site so other teachers and students could view them.

The project went so well that the principal asked Ms. Montoya and Mr. Huang to work with teachers in other subjects in developing several multidisciplinary WebQuests. The principal wants every seventh grader to participate in at least one WebQuest project during the school year.

BACKGROUND

Today's K-12 students will become information-age workers who will be expected to process large amounts of information on the job and create the knowledge needed to solve problems or make decisions. One way for K-12 schools and classrooms to address the teaching and learning of complex skills is by deploying technology in the classroom that facilitates active, resource-rich, student-centered learning environments to help students learn to think critically, analyze and synthesize information to solve technical, social, economic, political, and scientific problems, and work productively in groups (Mills, 2006). For K-12 schools and classrooms to address these skills, instructional approaches must focus on rich, multi-disciplinary learning tasks that address complex sets of learning objectives.

Using technology to support teaching and learning makes it possible to use powerful methodologies such as cases, projects, and problems that are relevant and representative of real-world tasks. The vast resources of data and information on the Internet supply tools and resources that permit application of a broader and more powerful set of instructional methodologies in the classroom. The use of these powerful instructional methodologies, however, requires teachers to design and develop learning tasks and activities that enhance the classroom curriculum. Information and Internet technologies support instructional methodologies that encourage learning in authentic contexts, collaboration, and the use of multiple primary source materials and resources (Fulton, 1997).

Internet technologies provide access to information that would have been impossible to access just a few years ago, including virtual libraries, electronic databases, and powerful search engines. This information can be manipulated to generate knowledge for solving problems or making decisions. The Internet also permits communication and interaction that facilitates information exchanges among peers and with experts outside of the local classroom, both synchronously and asynchronously. Internet communication technologies support interaction and collaboration that allow students to share ideas, ask questions, and discuss classroom projects.

THE CASE FOR INFORMATION LITERACY

Information literacy occurs when information technology is utilized in the context of solving a problem. For example, in the case study presented at the beginning of this chapter, the students in Ms. Montoya's and Mr. Huang's seventh-grade English and history classes applied the lessons learned from multiple historical events to contemporary social, cultural, and political issues using the extensive information resources of the Internet. Many beneficial learning activities can be developed for the K-12 classroom with students collecting, compiling, comparing, and reporting different types of information to solve problems. Therefore, it is important for students to develop fluency in using information technology, because in the information age students who know how to locate and use information to build knowledge that is useful to themselves and the community will be at an advantage over those who do not (Scardamalia & Bereiter, 2002).

In a report for the Milken Family Foundation, Kathleen Fulton (1997) described "technological fluency" as the combination of information skills, communication skills, and technology skills necessary to function in a technological environment. According to Fulton, technological fluency is based on the ability to find information from a variety of sources and disciplines; to make judgments about the value, reliability, and validity of information; and to create and disseminate information using technology-mediated communication formats. Technological fluency is the difference between learning *about* technology and learning *with* technology, and is demonstrated in the classroom by applying technology skills to problem-solving learning activities.

Like Fulton's technological fluency, information literacy refers to the information-processing skills students need to complete projects that are relevant and related to almost any subject in the curriculum and at any grade level (Jukes, Dosaj, & Macdonald, 2000). Because the features of the Internet make it a unique educational resource that is useful for communications, research, and publication, it is important for students to develop information fluency in using Internet technologies, tools, and resources for learning.

The American Association for School Librarians, in partnership with the American Library Association, has developed standards for information literacy (see http://www.ala.org/ala/aasl/aaslproftools/informationpower/informationpower.htm). According to the AASL/ALA definition, information literacy calls for students to access information efficiently and effectively and evaluate and use information critically and competently. Information literacy is important for today's students because the traditional literacies of reading, writing, and mathematical reasoning are no longer sufficient for lifelong learning in a world of ever-increasing information.

Acquiring information literacy requires the active participation of students in seeking information from multiple sources rather than passively receiving and restating facts. The American Library Association (2003) lists the following teaching practices that support information literacy:

- supports diverse approaches to teaching,
- incorporates appropriate information technology and other media resources,
- includes active and collaborative activities,
- encompasses critical thinking and reflection,
- responds to multiple learning styles,

- supports student-centered learning,
- builds on students' existing knowledge, and
- links information literacy to ongoing coursework and real-life experiences appropriate to program and course level.

PROBLEM-BASED LEARNING

Cognitive learning theories influence most modern pedagogy. These place more emphasis on factors that are internal to the learner than behavioral theories, which place more emphasis on factors within the environment (Smith & Ragan, 1999). Information processing theories have been one of the most influential contributions of cognitive learning theory to instructional design practice (Smith & Ragan, 1999) and to learning and developmental psychology (Bransford, Brown, & Cocking, 2000).

Information processing theory is based on a set of hypothesized structures in the brain that work much like a computer. For learning to occur, a series of transformations of information takes place in or through these structures. Like a computer, the human brain receives information into working memory, performs operations on it to change its form and content, stores it in long-term memory, and then locates it and generates responses to it. The learning process is based on integrating or assimilating information into long-term memory in a meaningful way that includes gathering and representing information, or encoding, storing information or retention, and getting at the information when needed, or retrieval.

Transfer is an important process for acquiring a deep or meaningful understanding of learned information. Transfer of learning occurs when a learner unconsciously or deliberately applies knowledge or skills associated with one task to the completion of another task. In one sense, learning or the goal of learning is the capability to transfer knowledge and skills from one situation, context, or domain to another.

Transfer of learning can occur along a continuum of varying degrees of difficulty where transfer of certain knowledge or skills can be developed to a high level of automaticity, while transfer of other knowledge or skills requires conscious, deliberate efforts. Problem-based learning supports and promotes transfer of learning, and Internet technologies and information resources can be used to establish learning contexts that create opportunities for transfer to occur.

Problem-based learning uses realistic problems embedded in the curriculum to promote deep understandings about a subject. With problem-based learning, teachers model, coach, support, and provide resources and guidance (Barrows, 1985; Stepien, Gallagher, & Workman, 1993) to help students construct an individual understanding of the problem and then develop and present a solution.

Problem-based learning requires students to acquire and apply information literacy skills. Implementing problem-based learning in the K-12 classroom, however, involves more than simply teaching students a series of steps and then sending them to the library or letting them access the Internet. Students must acquire information literacy skills that allow them to become proficient in locating, handling, and processing information.

Internet technologies can be used to support problem-based learning in the K-12 classroom. For example, the World Wide Web offers information resources that can be organized to simulate real-world problems by providing both content and context for a problem. The World Wide Web can be explored to identify and acquire the information necessary to understand a problem. Students can use the Internet to communicate with outside experts or peers to provide context or multiple perspectives of a problem.

With problem-based learning, the teacher guides students through a problem-solving process. Students may first reason through the problem and apply knowledge they already have to the problem. This elaboration of prior knowledge helps students understand what information they need to acquire to better understand and resolve the problem. As students begin to research and acquire information about the problem and reach possible solutions, they develop the information literacy skills they need to become self-directed learners.

LEARNING STRATEGIES AND PROBLEM SOLVING

Many learning theorists believe that working memory has limited capabilities; therefore, it is necessary for learners to employ strategies to regulate their learning when performing learning tasks. Learning strategies include both cognitive strategies and metacognitive strategies. Cognitive strategies include mental activities such as acquiring, selecting, and organizing information; rehearsing material to be learned; relating new material to information in memory; and retaining and retrieving different kinds of knowledge.

Learning strategies that deal with strategic learning or "learning to learn" are called metacognitive strategies. Metacognition refers to a self-awareness of one's own cognitive processes and is often defined simply as thinking about thinking. Metacognition occurs when learners control and monitor their own learning and implement strategies that support learning and problem solving. Metacognitive strategies are techniques for allocating, monitoring, coordinating, and adjusting mental capabilities to support learning (Mayer, 2001). Metacognitive strategies involve cognitive activities such as checking, monitoring, planning, and prediction (Brown, 1987). Some common metacognitive strategies include connecting new information to former knowledge, deliberate selection of cognitive strategies, and planning, monitoring, and evaluating cognitive processes.

Student effectiveness in student-centered learning environments is highly dependent upon the ability to regulate and control one's own learning. Metacognition is an essential skill for student-centered learning because students must be able to think about their own thought processes, identify the learning strategies that work best for them, and consciously manage how they learn.

Metacognitive strategies enable students to solve problems and to transfer problem-solving skills to new situations. With problem-solving learning activities, students need to organize a series of different facts into a coherent knowledge structure. Therefore, it is important that students can study and explore their own problem-solving efforts. Reflective activities can encourage students to analyze their performance, contrast their actions to those of others, abstract the actions they used in similar situations, and compare their actions to those of novices and experts (Goodman, Soller, Linton, & Gaimari, 1998).

According to Schoenfeld (1987) metacognition occurs when students reflect upon or talk about the problem-solving process. Problem solving and related research activities can provide opportunities for developing metacognitive strategies and integrating metacognitive skill-building into the curriculum. Students can learn how to learn by developing a repertoire of cognitive or thinking processes that can be applied to solve problems. As students perform problem-solving activities, teachers can focus student attention not only on solutions, but also on how tasks are accomplished.

Two important metacognitive strategies used for problem solving are planning and self-monitoring. Teachers can use technology tools to promote planning by having students use graphical mapping or outlining programs to plan their approach for solving a problem prior to starting their research. Teachers can use Internet technologies to support self-monitoring through reflective practices during or at the end of a project with Web essays or Weblogs (or essays or activity logs not published on the Web).

When using problem-solving activities in the K-12 classroom, learning strategies can be embedded in the instructional content and procedures so that process goals as well as product goals are accomplished and evaluated. Problem-solving activities can be structured to help students recognize the metacognitive processes that improve learning. Other scaffolds teachers can employ to develop metacognitive strategies with problem-solving activities include coaching, think-aloud techniques, cases and scenarios presenting multiple perspectives, procedural guidance, questioning for critical thinking, promoting interaction and collaboration, and brainstorming for different solutions (Dabbagh, 2003). Using Internet technologies and resources, teachers can create learning environments that foster the development of metacognitive skills and make classroom learning a more engaging and meaningful process for students.

USING THE INTERNET FOR SOLVING PROBLEMS COLLABORATIVELY

Problem-based learning focuses on challenging problems or tasks that can improve higher-order thinking skills. Because problem-based learning is often used to perform complex tasks or solve ill-structured or ill-defined problems, a student working independently may not possess the knowledge, skills, or time to accomplish the task. Therefore, problem-based learning can be used to help students learn to work together collaboratively and cooperatively.

Collaborative learning refers to a variety of educational approaches that involve shared intellectual effort by peers, or peers and experts. Collaborative learning generates dialog and interaction among peers, or communities of peers and experts, for the purpose of constructing collective knowledge or shared understanding about a concept, case, or problem. The case study presented at the beginning of this chapter is an example of students in K-12 classrooms participating in learning activities that use the Internet for solving problems collaboratively.

Schrage (1990) defined collaboration as "two or more individuals with complementary skills interacting to create a shared understanding" (p. 40). With collaborative learning, peers are responsible for one another's learning as well as their own learning. Peers work in groups of two or more to search for a mutual understanding, solution, or meaning, and to create a product of their shared learning experience.

Collaborative learning activities can range from classroom discussions that may include short lectures to participation on research teams. Generally, collaborative learning activities are designed to encourage interaction among students and promote consensus-building in reaching a shared solution. Through collaboration and cooperation, peers investigate subject matter at varying levels. Collaborative learning builds student awareness of different perspectives. By justifying and defending their ideas to peers, students build deeper knowledge and understanding of a topic.

Collaborative learning is especially appropriate for complex and ill-defined problems because the construction of complex knowledge seems to be facilitated by collaborative processes (Feltovitch, Spiro, Coulson, & Feltovich, 1996). Collaborative problem solving can be supported with information technology and the Internet using synchronous and asynchronous communication tools that facilitate the exchange of ideas among distant participants. Recognition of the educational value of student collaboration has led to the use of Internet communication tools such as e-mail, chat, and threaded discussions.

Group learning can build collective knowledge based on shared problem-solving, interpersonal feedback, and social support and encouragement. Small groups can work together; Collaboration and teamwork can be facilitative and can provide scaffolding for the construction of knowledge as well as enhance student satisfaction and learning (Doran, 2001). To form and use groups effectively for collaborative learning, teachers should employ a number of techniques including incorporating team-building activities at the beginning of the year, establishing a feedback process, requiring groups to report progress to the teacher online or face-to-face, evaluating group experiences and providing evaluation or assessment information to the teacher, and employing multiple instructional strategies with group work (Doran, 2001).

Internet-based collaborative problem solving utilizes teams or groups of students within the same classroom or in classrooms in multiple locations. Therefore, forming effective groups is an important aspect of using online communication to increase student participation in learning and to develop a learning community in the classroom. Online communication tools can be used effectively for collaborative problem solving, even among students in the same classroom working in small groups, because students can access information resources, experts, or peers anywhere in the world. Small groups formed within a classroom may work best with asynchronous communication tools such as e-mail or discussion lists, while online groups from multiple classrooms may work best with synchronous communication tools such as chat.

Facilitating problem-based learning among peers by having them solve real-world problems in a collaborative learning activity is a key to developing information literacy skills. Problem solving in collaborative learning environments allows students to connect through the Internet and interact with students in other locations to work collectively on a common task. With Internet-based, collaborative, problem-solving learning activities, students meet online using synchronous and asynchronous communication tools to discuss topics and solve problems jointly, just as they might if they were in the same location.

Developing information literacy through collaborative problem solving can be facilitated in most K-12 classrooms regardless of the level of availability of information technology and access to the Internet. The problem in many K-12 classrooms is that there are not enough computers to go around or Internet access is not available. Information literacy skills can be developed only when every computer that is available is used by every student for meaningful learning activities integrated with the classroom curriculum.

Collaborative problem solving, however, can provide a way to structure learning activities when there is less-than-optimal computer access. Here are some of the possibilities for use when there are limited information technology resources in the classroom:

- In classrooms with only one or two computers, teachers can use information technology for guiding whole-class discussions and explorations.
- In classrooms with two to three students per computer, students can work on information literacy projects in groups or teams, or individual students can be cycled through for independent learning while other students work off-line.
- In classrooms where the only access to the Internet is in a computer lab where students are scheduled for a limited number of computer lab periods per week, well-designed and organized lessons that coordinate off-line and online learning experiences will allow students to maximize computer lab time.
- When no classroom or lab computers have Internet access, teachers can create Web archives on another computer and load the archived information onto classroom or lab computers.

Collaborative problem solving in the K-12 classroom can maximize the development of information literacy skills by organizing instruction according to the abilities and learning needs of students. When students work in groups, they bring different abilities and expertise to bear within the learning dynamic. Group learning can promote academic achievement by allowing students to encourage each other, ask questions of one another, require each other to justify opinions and reasoning, and reflect upon their collective knowledge (Brown & Palincsar, 1989; Cohen, 1994; Johnson & Johnson, 1994). Collaborative learning is easy and inexpensive to implement, and can promote improved behavior and attendance and positive attitudes about school (Slavin, 1987).

Having students work in groups encourages discussion and develops social skills useful for a professional working environment for which they are training. Group collaboration takes advantage of learner-learner interactions rather than learner-content interactions for learning. Small groups working on a project learn team-building skills and goal-setting strategies needed to be productive in the workplace. Collaborative learning activities can increase students' satisfaction with the learning process and can decrease the time required by the teacher for administering and structuring a course, program, or other unit of instruction.

Collaborative learning is about building learning communities in the classroom, and learning communities can provide an organizing structure and delivery system for the practice of collaborative learning. Learning communities frequently provide more time and space for collaborative learning and other more complicated educational approaches. K-12 classrooms can provide a sense of community by promoting and supporting collaborative learning.

Internet technologies make it possible to transform classrooms into virtual learning communities by connecting students with peers, experts, and resources beyond the classroom. K-12 classrooms can create collective knowledge while building information literacy skills through the discoveries of students sharing information, interests, and learning objectives via the Internet. This collective knowledge can even transcend a single classroom and a single school year through collaborative, problem-solving learning activities that allow students to build upon the knowledge and discoveries of predecessors.

TOOLS FOR COLLABORATIVE, INFORMATION PROBLEM SOLVING

Developing information literacy skills through collaborative problem solving can be good pedagogy. Teaching students the ability to recognize an information need and apply a structured, problem-solving process for finding, using, and evaluating information is an important skill for any subject in the K-12 curriculum (Mills, 2006). Developing information literacy skills should always be coordinated with the curriculum and reinforced within the classroom. Teachers can provide students with opportunities to develop information literacy within the K-12 classroom curriculum by developing high-quality classroom activities that allow students to practice and master information technology skills in the context of information problem-solving activities.

Information problem-solving tools can support problem-based learning by describing the processes people use to solve problems. Problem-solving tools can provide metacognitive scaffolds to support students in locating and using information. Information problem-solving models can provide a structure that permits learners to regulate their own learning processes. These models are generally composed of a number of steps and tasks for each step that provide a structure useful for scaffolding students in using and applying information across multiple content areas. For students to be effective learners in student-centered learning environments, they must have the ability to regulate their own learning while locating, selecting, organizing, analyzing, evaluating, and applying relevant information to the completion of learning tasks and activities.

There are several online tools and resources that are available to support collaborative information problem solving that can engage students in meaningful, challenging, and motivating inquiry and critical thinking (Mills, 2006). These problem-solving models or frameworks engage students in critical thinking and problem solving through a structured research process. Another form of problem-solving framework, WebQuests, promote collaborative problem solving through inquiry and exploration.

Frameworks for Information Problem Solving

Problem-solving consists of moving from an initial, undesired situation to a desired goal, so problem-solving is a process of planning and executing a set or series of steps to reach the goal (Moursund, 1999). There are several approaches to problem-solving. Some problem-solving strategies are domain-specific, and some strategies use general approaches for problem-solving that can be used across many domains.

Several problem-solving frameworks have been proposed that focus on information problem solving and research using the Internet including InfoSavvy, the Big6, the Research Cycle, and the Building Blocks of Research. These problem-solving frameworks focus on skills related to the appropriate use of information for problem solving and are applicable to the development of information literacy. These problem-solving frameworks can provide a foundation for creating learning activities for the K-12 curriculum that helps students develop information problem-solving skills.

InfoSavvy

InfoSavvy (Jukes et al., 2000) is a general problem-solving framework that is relevant for information literacy. Information technology and the Internet were the primary focus of the original model because appropriate use of information from the Internet required the application of information literacy skills. InfoSavvy as a problem-solving tool, however, is not necessarily limited to use with information technology or the Internet.

The InfoSavvy model attempts to address the problem of information overload by helping students learn to process information effectively. Specifically, InfoSavvy provides a framework for embedding information literacy activities and experiences at all grade levels and in all subjects. InfoSavvy advocates school curricula that focus on critical thinking and problem solving, and leads to information fluency where process skills are deeply embedded in an educational process that allows relevant content and processes to be learned simultaneously (Jukes et al., 2000).

With InfoSavvy, a problem is considered to be an information need, and teachers and students apply a five-step process, the five As of Information Literacy: Asking, Accessing, Analyzing, Applying, and Assessing information from sources like Internet sites, newsgroups, chat rooms, e-mail, and other electronic and non-electronic resources (Jukes et al., 2000). Each of the five As consists of a set of subskills built into a framework and are intended to be used to solve any information need:

1. Ask the right questions to clearly define the problem and its context in terms of questions.
2. Access the information to determine where information is and what skills are needed to locate it.
3. Analyze the information to identify missing information, manage incomplete information, separate facts from opinions, and establish the authenticity and credibility of the data.
4. Apply the information by creating products, taking actions, solving problems, or satisfying information needs in other ways.
5. Assess the information by asking questions about the process used and the information obtained, reflecting on the process, and transferring the learning to other situations.

The Big6

The Big6 by Michael B. Eisenberg and Robert E. Berkowitz (http://big6.com/) is a general problem-solving model that is used widely in thousands of K-12 schools, institutions of higher education, and corporate and adult training programs to teach information and technology skills. The Big6 information problem-solving model is applicable to almost any situation where people need to locate and use information. The Big6 is comprised of six steps and related tasks for integrating information search and application skills along with technology tools in a systematic process to find, use, apply, and evaluate information for specific needs and tasks:

1. **Task Definition:** Define the information problem and identify the information needed.
2. **Information Seeking Strategies:** Brainstorm all possible sources and select the best sources.
3. **Location and Access:** Locate sources and find information within the sources.
4. **Use of Information:** Engage in the source (read, hear, view, touch) and then extract relevant information.
5. **Synthesis:** Organize information from multiple sources and present the information.
6. **Evaluation:** Judge the process (efficiency) and judge the product (effectiveness).

The Big6 is a useful framework for integrating information literacy across the curriculum because it establishes a common terminology that fits many subject areas and provides a mechanism for finding, filtering, and using many types of information. Because the Big6 is widely used in K-12 education, many examples and lesson plans for using the Big6 across the curriculum are available on the Big6 Web site and various schools' and other related Web sites.

The Research Cycle

The Research Cycle (http://www.questioning.org) by Jamie McKenzie focuses on defining and refining essential and subsidiary questions early in the problem-solving process. There may be information that students do not know when they first plan their investigations. Furthermore, students may start gathering information without carefully mapping out the many questions they should be examining in their information search.

The Research Cycle attempts to place students in the role of information producers instead of information consumers. Students actively revise and rethink their research questions and plans throughout the process by moving repeatedly through each of the steps. Students complete several repetitions of the cycle before reporting results so that they will have developed sufficient insight into the problem (McKenzie, 2000). The Research Cycle includes the following steps:

- **Questioning:** Clarify the essential question being explored by brainstorming to form a cluster diagram of all related questions. The subsidiary questions are used to guide subsequent research efforts.
- **Planning:** Think strategically about the best ways to find relevant and reliable information that will help students construct answers to the subsidiary questions.
- **Gathering:** Collect only information that is relevant and useful.
- **Sorting and Sifting:** Systematically scan and organize data using only information that is likely to contribute to insight.
- **Synthesizing:** Arrange and rearrange information fragments until patterns begin to emerge.
- **Evaluating:** Ask if more research is needed before proceeding to the reporting stage. With complex research questions it is usually necessary to complete several repetitions of the Research Cycle. Reporting and sharing of insights is determined by the quality of information produced during the Evaluation stage.
- **Reporting:** Report findings and recommendations to an audience of simulated or actual decision makers.

The Building Blocks of Research

The Building Blocks of Research tool is an elaboration of the Big6 with emphasis on the specific research skills that build information literacy. The Building Blocks of Research tool is published on the NoodleTools Web site (http://www.noodletools.com). NoodleTools provides a set of interactive tools designed to help students and professionals with online research.

According to this tool, information literacy is considered to be a transformational process in which learners find, understand, evaluate, and use information in various forms to create knowledge for personal, social, or global purposes. Information literacy shares a basic set of thinking and problem-solving skills with other disciplines, and so students should be presented with authentic, cross-disciplinary problems that allow opportunities for observation and inference, analysis of symbols and models, comparison of perspectives, and assessment of rhetorical context. Thus, students are engaged in developing information literacy over time.

The Building Blocks of Research employ the following tasks for conducting a research activity:

- **Engage the Searcher:** Become familiar with the characteristics of print, media, and technology; navigate the library and the information world; actively listen to and communicate with peers and experts; and display interest in issues, problems, ideas.
- **Define the Search:** Identify an information need, recognize that a problem exists, distinguish between recreation and research, and specifically define an information problem.
- **Initiate the Search:** Define the focus of the investigation, develop a search plan and define search criteria, identify potential search engines and sources based on strengths and needs, develop criteria for evaluating sources, develop questions to structure and clarify the search, and develop a timeframe.
- **Locate Materials and Resources:** Understand needs and search options, and use flexible search techniques to locate information.
- **Examine, Select, Comprehend, and Assess Information:** Apply reading, listening, and viewing strategies; apply questioning skills; and assess information.
- **Record, Sort, Organize, and Interpret Information:** Use visual and linear note-taking methods to analyze and interpret information.
- **Communicate and Synthesize Information:** Communicate solutions with electronic presentations or other audio or visual formats.
- **Evaluate the Process and Product:** Evaluate the solutions with interactive assessments, collaborative conferences, portfolios, and rubrics.

Online Inquiry and Problem Solving Using WebQuests

WebQuests (http://www.webquest.org) present students with an authentic task, which is usually based on a real-world problem to solve or a project to complete. The WebQuest model was originally developed in 1995 by Bernie Dodge and Tom March at San Diego State University. WebQuests provide students with a fairly structured learning framework that encourages self-directed learning. WebQuests provide an authentic meaningful contextual learning environment, which enhances the nature of learning and thinking, problem-solving and the assimilation of knowledge.

According to Bernie Dodge, co-originator of the WebQuest concept, WebQuests are inquiry-oriented activities in which most or all of the information used by learners is acquired from the World Wide Web (Dodge, 1997). WebQuests are designed to focus on using information to create knowledge and are designed to support the development of higher-level thinking skills. Tom March, the other co-originator of WebQuests, stated that WebQuests were designed to bring together effective instructional practices into an integrated student-learning activity (March, 1998). March explained that the educational rationale for the instructional features of a WebQuest was to engage learners in real-world learning activities, develop critical thinking skills, and support teamwork and cooperation.

Christie (2002) identified several of the pedagogical principles that WebQuests employ to structure learning so that students are actively involved in the discovery and construction of knowledge. Christie maintained that WebQuests encourage students to use reasoning skills, require students to use critical thinking skills, employ a collaborative discovery process, promote the development of social skills, multiculturalism, and diversity, facilitate student reflection, and foster an interdisciplinary approach to learning. Thus, WebQuests can provide a methodology for connecting information problem-solving activities to the K-12 curriculum and can provide a metacognitive strategy for collaborative problem-solving (Mills, 2006).

The WebQuest Model

WebQuests are usually organized as a Web page with multiple sections or a set of linked Web pages. Bernie Dodge (1997) described the WebQuest model as a lesson with several components or "building blocks." These building blocks can be re-configured in many ways to accomplish a broad range of learning goals. The current WebQuest model is comprised of six building blocks or components: an introduction, a task, the process for accomplishing the task including any Web-based resources used to accomplish the task, an evaluation, a conclusion, and any credits and references. Some WebQuests include a teacher page that provides information for teachers using the WebQuest.

- **Introduction:** The Introduction page provides a real-world context for the learning activity and includes a central question of the WebQuest.
- **The Task:** The Task page describes a product that students will create to transform the information they have collected into useful knowledge.
- **The Process:** The Process page suggests steps that students follow to accomplish the learning task. The Process section should also include information resources that students need to accomplish the learning task.
- **Evaluation:** The Evaluation page describes how student performance will be evaluated, often through the use of a rubric.
- **Conclusion:** The Conclusion page describes what students have accomplished by completing this WebQuest. It may also have additional resources to encourage students to extend their learning beyond the task presented in the WebQuest.
- **Credits and References:** The Credits and References page contains a description of online and offline sources used in the WebQuest.
- **Teacher Page:** The Teacher Page includes information and explanations that will help other teachers use the WebQuest.

CONCLUSION

Educators have now come to realize that the real strength of technology in the classroom is embedded in its potential to facilitate basic changes in the way teaching and learning occurs in the classroom (Mills & Tincher, 2003). The Internet and information technologies enable teachers to implement powerful learning environments in the classroom.

The Internet is an effective teaching and learning environment because it can support media features such as text, graphics, animation, audio, video, or hyperlinks, as well as a number of a number of pedagogical methodologies (Reeves & Reeves, 1997). The Internet and information technologies have the potential to augment teaching and learning in today's classrooms in several ways:

- Creating powerful learning activities that facilitate independent and collaborative student-centered learning.

- Providing meaningful ways for students to communicate and learn from one another.
- Motivating students to work with information and content and to articulate their knowledge and understanding.
- Enabling authentic or real-world contexts to support transfer of knowledge to other contexts.
- Helping students to manage their own learning (Oliver, Omari, Herrington, & Herrington, 2000).

Many of the features of the Internet have great potential for educational and instructional use that make it more than just another medium for the delivery of instruction. The information resources and communication technologies of the Internet can support the development of information literacy skills using student-centered learning approaches, interactive communication with peers and experts, and collaborative, problem-solving methodologies. Thus, the Internet can play a critical role in creating K-12 classroom learning environments that educate students for the information age.

REFERENCES

American Library Association. (2003). *Characteristics of programs of information literacy that illustrate best practices.* Retrieved May 6, 2003, from http://www.ala.org/aasl/ip_nine.html

Barrows, H. S. (1985). *How to design a problem-based curriculum for the preclinical years.* New York: Springer.

Bransford, J. D., Brown, A. L., & Cocking, R. R. (Eds.). (2000). *How people learn: Brain, mind, experience, and school.* Washington, DC: National Academy Press.

Brown, A. (1987). Metacognition, executive control, self-regulation and other more mysterious mechanisms. In F. Weinert & R. Kluwe (Eds.), *Metacognition, motivation, and understanding* (pp. 65-116). Hillsdale, NJ: Lawrence Erlbaum.

Brown, A., & Palincsar, A. (1989). Guided, cooperative learning and individual knowledge acquisition. In L. Resnick (Ed.), *Knowledge, learning and instruction* (pp. 307-336). Hillsdale, NJ: Lawrence Erlbaum.

Christie, A.A. (2002). *What is a WebQuest?* Retrieved October 5, 2003, from http://www.west.asu.edu/achristie/675wq.html

Cohen, E. G. (1994). *Designing groupwork: Strategies for the heterogeneous classroom.* New York: Teachers College Press.

Dabbagh, N. (2003). Scaffolding: An important teacher competency in online learning. *Tech Trends, 47*(2), 39-44.

Dodge, B. (1997). *Some thoughts about WebQuests.* Retrieved October 2, 2003, from http://webquest.sdsu.edu/about_webquests.html

Doran, C.L. (2001). The effective use of learning groups in online education. *New Horizons in Adult Education, 15*(2).

Feltovitch, P. J., Spiro, R. J., Coulson, R. L., & Feltovich, J. (1996). Collaboration within and among minds: Mastering complexity, individually and in groups. In T. Koschmann (Ed.), *CSCL: Theory and practice of an emerging paradigm* (pp. 25-44). Mahwah, NJ: Lawrence Erlbaum.

Fulton, K. (1997). *The skills students need for technological fluency.* Milken Family Foundation.

Goodman, B., Soller, A., Linton, F., & Gaimari, R. (1998). Encouraging student reflection and articulation using a learning companion. *International Journal of Artificial Intelligence in Education, 9*(3-4), 237-255.

Johnson, D.W., & Johnson, R.T. (1994). *Learning together and alone. Cooperative, competitive, and individualistic learning* (4th ed.). Edina, MN: Interaction Book Company.

Jukes, I., Dosaj, A., & Macdonald, B. (2000). *NetSavvy: Building information literacy in the classroom.* Thousand Oaks, CA: Corwin Press.

March, T. (1998). *Why WebQuests? An introduction.* Retrieved October 3, 2003, from http://www.ozline.com/webquests/intro.html

Mayer, R.E. (2001). *Multimedia learning.* Cambridge, UK: Cambridge University Press.

McKenzie, J. (2000). *Beyond technology: Questioning, research, and the information literate school.* Bellingham, WA: FNO Press.

Mills, S. C. (2006). *Using the Internet for active teaching and learning.* Upper Saddle River, NJ: Merrill/Prentice Hall.

Mills, S. C., Frieden, B., & Scott, S. A. (2004). *Authentic assessment of technological fluency: Computer assessment and tutorial online.* Manuscript submitted for publication.

Moursund, D. (1999). *Project-based learning using information technology.* Eugene, OR: International Society for Technology in Education.

Oliver, R., Omari, A., Herrington, J., & Herrington, A. (2000). Database-driven activities to support Web-based learning. In R. Sims, M. O'Reilly, & S. Sawkins (Eds.), Learning to choose: Choosing to learn. *Proceedings of the 17th Annual ASCILITE Conference* (pp. 551-560). Lismore, NSW: Sounther Cross University Press.

Reeves, T. C., & Reeves, P. M. (1997). The effective dimensions of interactive learning on the WWW. In B. H. Khan (Ed.), *Web-based instruction* (pp. 59-66). Englewood Cliffs, NJ: Educational Technology.

Scardamalia, M., & Bereiter, C. (2002). Knowledge building. *Encyclopedia of education* (2nd ed.). New York: Macmillan Reference.

Schoenfeld, A. H. (1987). What's all the fuss about metacognition? In A. H. Schoenfeld (Ed.), *Cognitive science and mathematics education* (pp. 189-215). Hillsdale, NJ: Lawrence Erlbaum.

Schrage, M. (1990). *Shared minds.* New York: Random House.

Slavin, R. (1987). Cooperative learning: Can students help students learn? *Instructor,* (March), 74-78.

Smith, P. L., & Ragan, T. J. (1999). *Instructional design* (2nd ed.). Upper Saddle River, NJ: Prentice-Hall.

Stepien, W. J., Gallagher, S. A., & Workman, D. (1993). Problem-based learning for traditional and interdisciplinary classrooms. *Journal for the Education of the Gifted, 4,* 338-345.

Chapter XXI
How to Teach Using Today's Technology: Matching the Teaching Strategy to the E-Learning Approach

Moti Frank
Holon Institute of Technology, Israel

ABSTRACT

This chapter reviews the benefits and challenges of five approaches for integrating technology and teaching. Three of the models involve distance learning, while the two others utilize technology as a teaching aid. The first model is a lecture-based course also available through a Web site. The second is a fully online asynchronous course. The third is synchronous distance learning. The fourth combines a virtual laboratory and visualization with regular teaching, and the fifth fuses technology with different teaching methods. The pedagogical and operational aspects of the five approaches are discussed. The main pedagogical aspects discussed are: applying active and interactive learning principles, using multimedia, organizing the course and its lessons, and providing immediate feedback to students about their progress. By comparing the advantages and challenges the different models offer, teachers in K-12 will be able to match the appropriate model and its teaching strategy to their learning goals.

INTRODUCTION

The Scope and Aims of this Chapter

Over the past few years, as a consequence of the rapid development of technology, we are seeing the crystallization of the five most commonly used models for the integration of technology in teaching. The major aim of this chapter is to present these five models and discuss their pedagogical aspects, benefits, and disadvantages. By comparing the advantages and challenges the different models offer, teachers in K-12 will be able to match the appropriate model and its teaching strategy to their learning goals.

Three of the models involve distance learning, while the two others utilize technology as a teaching aid.

The first fundamental condition for the successful implementation of each of the five models is that the teacher and students be technologically literate. They must all possess a basic level of technological literacy in order to be able to take advantage of the benefits offered for making teaching more efficient.

We live in an era in which advanced teaching strategies depend upon technology. Teachers and students who understand and are comfortable with the concepts and workings of modern technology are better able to participate fully in technology-based learning environments. For these reasons a growing number of voices are calling for the mandatory study of technology by K-12 students worldwide. Technological literacy is the ability to use, manage, assess, and understand technology. It involves knowledge, abilities, and the application of both knowledge and abilities to real-world situations (ITEA, 2003).

Alongside presentation of the five models' pedagogical benefits and shortcomings, this chapter also discusses, for each model, ways in which teachers' and students' technological literacy can be developed so that they will be able to use these models effectively.

Five Commonly Used Models for Integrating Technology in Teaching at the K-12 Level

The first three models involve distance learning. The first model is a lecture-based course also available through a Web site. In other words, lectures are given in the traditional manner, but in parallel a Web site is built on behalf of the course for exercises and practice drills, enrichment, and in-depth study of the subject. The second model—a fully online asynchronous course—requires only a very limited number of classroom sessions. The primary teaching is conducted through the course Web site. This contrasts with the third model—synchronous distance learning. In this model, teaching resembles, in some of its features, traditional teaching, yet the teacher and his or her students are physically distant from one another.

The first model uses computer technology as a teaching aid and combines a virtual laboratory and/or visualization with regular teaching. This model is most prevalent in science teaching. The second technology as a teaching aid model fuses technology with different teaching methods, such as using the Internet as a source of information in teaching/learning, founded on inquiry-based learning, project-based learning, or the WebQuest approach.

What are Distance Learning, Asynchronous Distance Learning, and Synchronous Distance Learning?

According to Mielke (1999), distance education is a method of education in which the learner is physically distanced from both the teacher and the institution providing the instruction. Learning may be undertaken either individually or in groups. In its original form, teachers teaching distance education corresponded with students via regular mail, telephone, or fax. The students usually submitted the assignments through the mail. Using various forms of electronic media—such as radio, television, and videoconference—and advanced communication technologies increases time effectiveness, enables flexibility of location, and improves delivery of information.

Electronic delivery may mean synchronous communication, in which class members can participate simultaneously, as well as asynchronous communication, in which participants are separated by time and space.

In synchronous communication, the teaching and the learning are conducted simultaneously. In this respect, computerized synchronous communication is similar to a standard lecture given within a brick-and-mortar classroom. The difference is that there is a physical separation between teacher and students. The teacher is in the broadcasting studio, which contains the technological means to transmit voice and data (such as PowerPoint slides). Students are located in a distant learning center, which generally is technologically set up to allow communication between the teacher and students (a computer for each student or group of students, speakers, microphones, headphones, etc.). Students listen to the teacher through speakers or headphones and speak to the teacher using microphones. They see the course content on their computer monitors (or on a large, central screen). Sometimes, the communication also includes two-way video—cameras in the broadcasting studio transmit pictures of the teacher to students, and cameras in the learning center transmit pictures of the class to the teacher. In the latter case, communication includes two-way transmission of voice, data, and video. With the enhancement of the Internet and related technologies that allow high-speed Internet (for example, ADSL or cable), the use of teaching methods in which students do not meet in a distant learning center, but remain in their individual homes using their personal computers, has increased. Computerized synchronous communication courses are especially prominent in countries with large land area where it is difficult for students/pupils to come to an educational facility everyday due to the geographical constraints.

In asynchronous communication, the teacher and his or her students are separated both in time and distance. The teacher and students are not physically in the same classroom and the learning does not happen at the same time that the learning material is presented. Students learn on their own, individually, at their leisure.

In synchronous distance learning, there is no time-lag between teacher and students in spite of physical distance. Every synchronous distance learning course may have an asynchronous element that includes recording all lessons and some face-to-face meetings. Optimally, the first meeting between the teacher and his or her students in synchronous distance learning will be a face-to-face meeting. The students should be invited to the meeting in order to become acquainted with each other and with the teacher. They should also be provided with some details and information about the technologies and procedures of the program.

THE FIRST MODEL: CREATING A COURSE WEB SITE FOR FACE-TO-FACE TEACHING

This section discusses pedagogical and other aspects of K-12 learning environments that integrate traditional teaching methods and the use of a course Web site. For hundreds of years, the lecture method (sometimes called frontal teaching or face-to-face teaching) has been considered to be the primary teaching method. The lecture has its benefits and limitations. Many teachers think that the lecture is still the most efficient teaching method for delivering the basic content of a given subject matter, as the teacher has control over what is happening in the classroom. This teaching method is sometimes perceived as the most convenient and economic method for delivering ample material to a large number of students.

Participating in the lessons, asking questions, making notes, and discussing with the teacher would probably continue to be among the major aspects of the teaching in the coming years. The chief criticism about lecture as a teaching method is that students are allocated a passive role and thus their studying efficiency is low.

However, the use of innovative educational technologies is growing. In recent years, we have been witnessing considerable growth in the number of courses with Web sites meant to be tools for augmenting traditional teaching. A well-designed course Web site may serve as a complementary tool and raise the learning effectiveness through active and interactive studying. In many *learning management systems* (packaged software for building course Web sites), active and interactive learning are intrinsic, and Web site designers have many tools on hand for applying active learning principles.

An enormous amount of books, articles, papers, and chapters deal with the Internet as a learning environment in education. In much of the research, no significant differences were found between an e-teaching/learning environment and the traditional teaching/learning environment in relation to the variables that were examined. Yet, many studies did find significant differences. It seems that the real question is not whether it is possible to elicit benefits from a course Web site in a learning environment that integrates traditional teaching methods with the use of a course Web site, but under what conditions this can be achieved.

Some Benefits That May Be Derived from a Course Web Site

Creating a course Web site for K-12 students is not an easy task. It requires investment of a lot of effort and usually consumes a great deal of time. However, as with any other investment, if it is performed correctly, it should yield commensurate benefits.

Many researchers refer to organizational and operational advantages that may be extracted from a course Web site. Thus for example the argument of *accessibility anywhere at anytime for anyone* is most prevalent.

Learning management systems for building course Web sites allow course designers to choose among many options for organizing courses and applying pedagogical principles. The major organizational and operational advantages for the teacher are the ability to: continuously update the course content, learning material, syllabi, course calendars, glossaries, and assignments; communicate with registered students (billboard, whiteboard, mail, chat, discussion group, etc.); present links to relevant sites and databases; easily follow the students' progress; extract various statistical data; deliver home assignments; perform automatic checking of multiple-choice questions; and manage the course, the students, and the grades.

For students this is a comfortable means for viewing and/or downloading learning material, presentations, slides, messages, homework assignments and solutions, and past years' examination papers, and solutions. Students have access to the material at anytime, from anyplace. Certain systems enable easy submission of homework through the site. If desired, the site can be used as a communication tool between students and teaching staff.

However, the most important issue is pedagogical advantages. What is happening to the learning? What are the pedagogical advantages that may be elicited from a course Web site? How can learning be improved by using discussion groups (forums)? Animation and simulations? Multimedia? By providing feedback? And what about academic achievement in the e-learning environment compared to the traditional environment?

One inherent advantage of a course Web site is the ability to implement four dimensions of "good teaching"—applying active and interactive learning principles, using multimedia, organizing the course and its lessons, and providing immediate feedback to students about their progress. Students must be active and interactive; teachers must organize their courses and the material for the lessons in advance through "trees" that make orientation easy; the software enables easy transmission of feedback, and the Web site can assist in that it usually provides an option for using multimedia and multiple representation means such as text, charts, graphs, tables, illustrations, pictures, sketches, animations, simulations, equations, light, color, and sound. The rest of the "good teaching" dimensions (Hativa, 2000)—the ability to give (written) clear and interesting presentations and explanations, and the capacity to build a supportive learning environment—depend on the course teacher.

A well-designed course Web site should provide (automatic) immediate feedback to students as well as hints and directions on how to continue in case of mistakes. Through the Web site, the learner can be exposed to multiple realities. The teacher can place challenging inquiry tasks, present and discuss paradoxes and contradictions, and initiate reflection on the learning processes.

A course Web site should nurture development of independent learning on one hand (learning according to the pace, style, and level that is suitable for each learner), and collaborative and/or team learning on the other hand. The use of a discussion group tool (sometimes known as an asynchronous computer conferencing environment or forum) was found in research to be efficient in building shared understanding and a learning community, and enabling students to actively build knowledge through interaction with the teaching staff, colleagues, and others. For example, in certain courses the teacher can ask a team of students to create a model and run it on their computers. The discussion group tool might provide support to students having difficulties. This support may be given by colleagues or by the teaching staff.

Discussion Groups (Forums)

Forums allow participants, or members, to relate to and express their opinions about the subject at hand, to read the material submitted by other members, and to submit their response, should they wish to do so. Forums facilitate communication at different times between many participants, with similar or diverse know-how, sitting in numerous dispersed locations. Forums are particularly suited to the study of subjects based on the exchange of ideas, topics into which participants want to delve deeply, concepts that are in dispute, topics that are important to look at from a number of perspectives, and issues around which disagreements and arguments are likely to arise (such as discussions about ethical dilemmas).

The pedagogical idea at the root of using forums as a teaching/learning aid is that individual learning is enriched when there is interactive social support in the building of knowledge and development of thought processes (a concept sometimes called learning communities or shared understanding). We are not talking here about a new pedagogical idea. The well-known teaching/learning method—collaborative learning—is also based on this concept. The innovation here is that now technology is enabling this type of learning for K-12 distance learners who, though in the same class, are separated from each other both in terms of time and physical space.

Based on Fosnot's findings (1996), the forum leader should offer challenging and open questions and must, in particular, discuss anomalies, examining and explaining them. As creators of meaning, people aspire to organize and rationalize their experience. Discussion gives rise to reflection, and examining experience gives rise to learning.

Nevertheless, the use of forums as a teaching/learning platform presents teachers with a number of challenges. First, in contrast to a normal class where communication between the teacher and his or her students, and also among the students is verbal, forums mandate skills in written communication. Some students' proficiency in this type of communication will be low. Second, someone who enters the forum is likely to be confronted with an abundance of lengthy and tedious messages. Third, it has happened more than once that a forum did not take off, and then becomes an additional burden on students. In such cases, some teachers tend to require participation, and even to give marks for the quantity and quality of the messages students submit to the forum. In this situation, students may indeed participate in the forum, but only because they are required to do so and they do not invest much effort in creating pertinent messages. Fourth, the feedback that a message submitter receives is not immediate (and sometimes is never given)—an event that may frustrate many participants. Finally, fifth, some students are shy, and so may hesitate to expose themselves and their written messages before their classmates.

In light of these challenges, it is advisable for teachers to carefully check the use of forums as a teaching/learning aid. They should use it only for teaching subjects appropriate for teaching with the assistance of forums. If a teacher operates a teaching/learning forum, he or she must use different tools to spur students to participate and to ensure that the forum is a productive and nurturing site.

THE SECOND AND THIRD MODELS: SYNCHRONOUS DISTANCE LEARNING (SDL) AND ASYNCHRONOUS DELIVERY MODE

Some Benefits of SDL

The literature indicates several advantages that may be achieved by using synchronous distance learning in K-12. Thus, for example, Davey (1999) emphasized that teachers can give all students ongoing and timely performance feedback. In a course described by Frank, Kurtz, and Levin (2002a), much use was made of a tool that also permitted immediate feedback for the teacher.

According to Carr-Chellman and Duchastel (2000), the advantages of synchronous interchanges as compared to asynchronous tutorials include a more direct sense of collegial instruction and immediate resolution to the questions posed. Branon and Essex (2001) examined practices of distance educators in regard to their use of synchronous and asynchronous text-based communication tools in their online courses. Their study revealed that the reasons for using synchronous communication included: virtual office hours, creating a community, team decision making, brainstorming, and dealing with technical issues. Power et al. (1999) indicated that "...synchronous instruction allows students in multiple, rural schools to communicate in real time with each other and with their teacher, using whiteboards, chat, video, and voice communication."

According to other researchers, synchronous distance learning permits: access to students in remote places; transmissions by a single, highly qualified teacher to several remote sites simultaneously; allowing the course to be given to a relatively large number of students (an economic advantage); and the use of statistical data stored in the technological system for follow-up of students' progress.

Lister et al. (1999) found that the synchronous part of their course was of crucial importance. Learning outcomes and student retention rates in their purely asynchronous courses were often disappointing for all age groups. In the synchronous section of their distance course, they attempted to create a social construct—an interactive, face-to-face classroom. The teacher could begin each lesson by asking if there were any questions about homework, reading assignments, or group projects. To help answer students' questions, the teacher could activate the system's whiteboard and shared stored graphics, or solve analytic problems interactively by writing texts and equations on the white board, to appear on all the students' screens. He/she could also call up a question and answer tool that allowed real-time interactive quizzing and polling. Following the setup of a typical studio classroom, the teacher could then present a brief lecture on new material, sharing PowerPoint slides and multimedia material; or using synchronized Web-browsing, the teacher could lead students to Web sites with course-related content. The synchronous sessions also help to keep students abreast of course deadlines and build teams and community, allowed them to receive immediate feedback, and improved retention rates.

SDL: Challenges and Issues

Apart from the fairly complex logistics of organizing synchronous meetings, the literature also presents some issues with regard to SDL. The necessity of attending classes at specific times and a lack of face-to-face interaction with the teacher adversely affects some students. In other words, this method is not suitable for everyone in K-12.

Weak students, or those who would be shy of participating in an ordinary classroom session, are likely to be even more ill at ease when they realize that they can be heard in real time by many other students, most of whom they have never even met. "Because of the real time nature, there may be greater social pressure for conformity in participation" (Carr-Chellman & Duchastel, 2000). Monson, Wolcott, and Seiter (1999) found that "some students experienced a high degree of state-communication apprehension in synchronous distance education."

Frank et al. (2002a) tried to overcome the above-mentioned disadvantages. For instance, attempts were made to deal with the lack of face-to-face meetings between teacher and students by means of blended e-learning—a combination of face-to-face meetings and distance study delivered by technological means. The teacher was filmed throughout the course delivery, and the film was transmitted to the students' computer screens. Also, a camera was set up in the remote classrooms so that the teacher could get an overall view of the class.

Teachers' Preparation for Synchronous Teaching

Synchronous distance learning by means of advanced technology requires careful preparation by the teacher, both pedagogically and technically. Teachers should attend a workshop to familiarize themselves with the system and with distance learning. Technical teams must help the teacher set up the study program and be ready to provide technical assistance throughout the course.

At the end of the synchronous teaching program described by Frank et al. (2002a), the teachers emphasized the importance of prior training for teaching in an advanced technology environment. In their opinion, careful and detailed preparation is imperative before each lesson, and for adapting it for distance learning. This includes planning the teaching materials and, in particular, the slides to be used, all of which take up much time. One of the teachers estimates that at least five hours of preparation were required for every hour of lecturing. However, it should be noted that this preparation is a one-time occurrence, and repetitions of the course should require no extra preparation.

The teachers also felt, however, that too careful pre-course preparation resulted in a lack of spontaneity during the actual lesson. It led to less opportunity to exploit and respond to situations in the class that required modifying the planned order of instruction, or reshaping exercises for home assignments planned in advance that were sometimes not applicable to what was actually taught, or answering the questions asked during the lesson. Being overly prepared restricted humorous responses to situations arising in the course of the lesson, and which could have created a special atmosphere and made the lessons more meaningful. Most probably, more experiences with teaching using this method should help overcome these difficulties.

The teachers also remarked that the output in distance learning is relatively lower than in conventional learning, and that in the e-learning environment there are drawbacks that make it difficult, for instance, to write complicated mathematical formulae, to draw graphs, to transmit complicated messages, or to present long exercises that require several stages for solution.

The abilities required for distance teaching are: experience in teaching in general and in distance courses in particular, an ability to concentrate and coordinate with the technological system during the lesson, and familiarity with a computerized environment (Frank et al., 2002a). This list also concurs with the findings of Mortera-Gultierrez and Fernando (2000):

If teachers are to be successful distance educators, they must be capable of using at least the following types of interaction: instructor-learner; instructor-content; instructor-technology: instructor-facilitator; instructor-peers; instructor-support staff (technicians); and instructor-institution.

The teachers who taught the course described by Frank et al. (2002a) were trained to use the above-mentioned types of interaction. They also used the Garrels' (1997) model, which describes five critical elements for successful distance teaching—instructor enthusiasm, organization (i.e., preparing teaching materials in advance), strong commitment to student interaction, familiarity with the technology used, and critical support personnel.

Teacher-Students Interaction in SDL

In distance learning, teachers and students are physically remote from each other, so that interaction is quite different from teaching face to face. In the project described by Frank et al. (2002a), interaction was mediated via computer, and complications arose in that, if students wished to communicate with the teacher, they had to get permission by means of pressing the 'indicator' key. Apart from vocal interaction, the students could also write questions or comments, or send e-mail to the teacher. There are both advantages and disadvantages to written interaction. The students as well as the teachers stated in interviews that, compared to conventional teaching, teacher-student interaction in distance learning is not easy.

Nonetheless, the teachers agreed positively about the fact that, apart from the spoken and written interaction, they could also use tools for real-time feedback. For instance, they presented a multiple-choice question and quickly received the answers from all the students. Another possibility was receiving anonymous responses from students about comprehension of the study material by using the + or – keys of the distance learning system.

The teachers' reservations about interaction were essentially related to the lack of immediate contact with students—body language, facial expressions, and eye contact. These were all means that the teachers were accustomed to using in the classroom in face-to-face sessions to assess comprehension. Albeit, there was a camera in the distance classroom that provided an overall picture for the teacher, but due to low resolution, details could not be made out.

Eye contact also allows teachers to exploit their own body language. The teachers said that they found it problematic that they were unable to assign an exercise and then move about the classroom among the students to see what was going on. They were unable to use the 'Socratic' approach—a give and take of questions and answers between instructor and students.

Other reservations related to the fact that in distance teaching one cannot use one's hands to demonstrate different examples. For that matter, distance teaching rules out the use of any kind of theatrics during the lecture, though these very often help to explain and also create a positive atmosphere in the class.

SDL: Suggestions for Implementation

Based on the research presented by Frank et al. (2002a), some conclusions and suggestions may be derived for implementing the SDL approach in K-12 classes. All implementation tips in this section require careful consideration and study in order to evaluate their effectiveness, advantages, and limitations.

First, the teacher must assign students homework and return it to them after correcting (in some cases, by means of the automatic marking system). This serves two purposes: the students get practice exercises, and they interact with the teacher concerning their progress.

A teaching assistant should be selected for each DL class whose *other* job is to ensure that the classroom is ready (connected, clean, and cool/heated), the students in their places on time, and all handouts (exercises, summaries) distributed before the lesson. The teaching assistant should have consulting hours at the distance site (at least one hour for every teaching hour) for answering questions, and should be responsible for collecting and correcting exercises (and grading, if the course is for academic credit).

Third, as in standard lessons, the dilemma as to whether to distribute copies of slides before the lectures also surfaces here. It is suggested that teachers distribute 'skeleton' slides containing only part of the information, so that, on the one hand, students will not need to copy every slide, while on the other, they will be able to assimilate the exhibited data while taking notes.

Fourth, about one hour before the start of each lesson, the equipment must be tested for proper functioning in the classrooms and for complete communication with the broadcasting studio. A reliable technician must be on hand to deal with any technical problems that may occur during lessons. Division of responsibility among the personnel involved must be demarcated in advance, to prevent (as far as possible) problem situations where it is not absolutely clear who is responsible for locating and repairing them.

Fifth, it is recommended that teachers meet their students face to face at the beginning, in the middle, and at the end of the course, the mid-course meeting to be held at the distant sites.

Sixth, individual computers and headsets separate the students from each other, as well as from the teacher. A large screen, loudspeakers, and classroom microphones would eliminate the need for headsets and computer screens. Simple voting boxes (1, 2, 3, 4) can be used for multiple-choice questions and opinion polls during DL sessions. Students should be facing the camera transmitting from the classroom to the teacher so that the latter is always aware of what the students are doing at any time (talking to each other, looking at the screen, raising their hand to ask a question, etc.). Students must be able to ask questions and receive answers during the DL lecture, so that everyone can hear the question. Teachers must be able to ask questions, collectively and to individual students, and monitor classroom and inter-classroom discussions. This can only be done by means of a large screen and loudspeakers, with the classroom fully wired with microphones (as in a concert hall). There is no reason why two or even three classrooms should not be connected, with the teacher deciding whether the second classroom is visible to the first. The DL environment must be interactive.

If the students can interact with each other, without headsets, and the teacher can see and hear all the students (who are facing the camera and the classroom microphones), then it should be possible to establish a videoconference rapport as effective as that of any other teaching method used by any good teacher.

There are those who think that, as far as possible, synchronous distance learning should resemble standard classroom teaching. They insist on the advantages of this type of teaching over asynchronous distance learning. Nevertheless, the question must be raised as to whether this is what we are really aiming at. Much has been written about the disadvantages of traditional teaching, where students are passive. SDL, in opposition, together with its shortcomings, does not appear to have this built-in passivity. It easily allows teachers to reap pedagogical advantages (if they take the opportunity)—active learning, lessons and course organization, and immediate feedback between teacher and student about levels of comprehension. Students should be active, the teacher must do advance preparation for organizing the lessons and the course, and the technological systems easily allow for getting and giving feedback. The technological system enables, with relatively little difficulty, the use of multimedia—incorporating video, sound, text, animation, and simulation.

Fully Online Asynchronous Course

Benefits of Fully Online Asynchronous Course

A fully online course is inherently an asynchronous course. Such courses require a very limited number of meetings in a classroom setting. The bulk of the teaching is executed through the course Web site. The benefits and limitations of course Web sites were discussed in detail earlier. At this point some additional features will be mentioned.

In courses where there are a large number of students, asynchronous courses have an advantage over frontal teaching. In the former, each student can ask the lecturer a question when he or she needs or wants to and receive an answer (through e-mail, for instance), whereas in the latter, the teacher cannot hope to respond to each of the students. Moreover, feedback is given only to those who dare ask a question in front of so many co-students (Willis & Dickinson, 1977). Shy students are liable to hesitate in a frontal teaching or synchronous course, in which all the students will hear the question. In frontal teaching, either the time allotted for questions is very limited or students must approach the teacher during the latter's office hours (which are not always at times convenient for the student).

The main advantage for students in online asynchronous courses is the flexibility of time and place. They can study anywhere (wherever a computer can be linked to the Internet) and at any time suitable for them. The course content remains stored in the system, and each student can access any part he or she wants when he or she wants. In contrast, in synchronous courses students must show up in the learning center or be at home at the specific time of the lecture. In asynchronous courses, students submit assignments electronically through the course Web site and there is no need to present hardcopy (written) homework.

Another advantage of the asynchronous approach, as compared to the synchronous setting, is related to technology. For the latter type of course, the execution mandates unique and costly technological resources. In opposition, asynchronous course implementation is relatively simple (usually only necessitating the use of an off-the-shelf software package such as learning management systems, personal computers, and an Internet connection).

A fully online course can be given to a relatively large number of students (an economic advantage), and statistical data stored in the technological systems can be applied for follow-up on students' progress.

Other investigators have discussed other advantages that may be achieved when using asynchronous communication (comparing to synchronous communication). Branon and Essex (2001), for example, noted:

...distance educators have found asynchronous communication useful for: encouraging in-depth, more thoughtful discussion; communicating with temporally diverse students; holding ongoing discussions where archiving is required; and giving all students the opportunity to respond to a topic.

In an asynchronous course, study and learning can reach a depth unlikely to be attained in a synchronous course, since students can invest as much time and effort as required, each according to his or her learning tempo. This contrasts with synchronous and frontal teaching classes, which have a time limit and progress at a uniform pace that is not necessarily suitable for each student. When the teacher assigns questions/tasks, students in fully online courses have enough time to think before giving their answers—a luxury not always available in frontal or synchronous classes. Bhattacharya (1999) stated that her students preferred the asynchronous mode because it gave them a better opportunity to concretize their ideas before responding.

The advantages teachers have in asynchronous teaching include: the ability to continually update learning material, the capacity to send messages quickly to all students in the course, links to databases and relevant Internet sites, and the possibility to automatically check exercises and manage grades. Another benefit, which appears surprising perhaps, is that in fully online courses with a large number of students, teachers get to know the students, or at least some of them, better than that in normal frontal courses. The acquaintance is usually developed through e-mail correspondence or the course forum. Frank, Reich, and Humphries (2002b) found that in asynchronous distance learning, the teacher is able to develop a greater personal connection with students through the use of e-mail than he or she is able to develop in a conventional classroom. In a conventional classroom, a *dominant* student may monopolize the discussion (Brown, 2000), but in distance learning such a situation is unlikely to happen.

Challenges When Using a Fully Online Asynchronous Mode

Lister (1999) noted that in certain cases the problem of motivation in asynchronous courses may arise. Lacking a serious incentive, students may not make the effort needed to learn what the course teaches. The fact that the responsibility for learning is on students, who are meant to log in every so often and study, may be a problem for those with low motivation.

Freedman (1998) stressed the proximity of teachers and students in regular classes. In his opinion, some students prefer standard face-to-face classes that provide warmth, a feeling of intimacy, and help when problems crop up. Computer-aided teaching, synchronous and asynchronous, are not, according to Freedman, suited to every student: "The main aspect in the interaction between teacher and students relates to the affinity between them. The students want teachers who are able to give them human warmth and even an intimate attitude, and can help them to solve their personal problems."

According to Wolcott (1995) and Hill (1997), the interaction in distance learning is less than that of face-to-face learning. The teacher cannot see the students' reactions to the study material. He or she may miss out on facial expressions or body language, for instance. In fact, several researchers related to the difficulties arising from lack of eye contact between teacher and student, as in distance learning. Willis and Dickinson (1997), for example, wondered whether teachers can be effective if they are unable to maintain eye contact with their students, or to observe students' non-verbal behavior.

The main conclusions of research that examined the challenges with which 11- and 12-year-old students are faced when participating in a course that is based on asynchronous distance learning, indicate the importance of personal contact and direct connection between teachers and their pupils (Frank et al., 2002b). The main issue evolved in the research is the student's loneliness when learning from distance—that is, the lack of personal contact among the learners and between the teacher and the students. The authors conclude that it is important for the teacher to take into consideration and respect the varied and various human needs of the children when developing electronic learning for such young learners. Thus, it is recommended that at the beginning of an asynchronous distant course, a face-to-face meeting be held. The purpose of such a meeting is for the teacher to get to know the students and vice-versa, and for the students to get to know each other in order to reduce feelings of social isolation and alienation. Also, the authors encourage the students to work in pairs or small groups.

Finally, teachers have two more problems with fully online courses. First, teachers must invest greater effort in writing up course content (in the case that these were not prepared beforehand). And second, teachers cannot respond spontaneously to ongoing events, as teachers in standard classes can.

THE FOURTH MODEL: VIRTUAL LABORATORY, VISUALIZATION, AND MULTIMEDIA

The term multimedia refers to the combination of multiple technical resources for the purpose of presenting information represented in multiple formats via multiple sensory modalities (Schnotz & Lowe, 2003). Accordingly, multimedia resources can be considered on three different levels: the technical level (i.e., computers, networks, displays, etc.); the semiotic level, referring to the representational format (i.e., texts, pictures, sound, etc.); and the sensory level (i.e., visual or auditory modality).

Here, we will relate mainly to the sensory and semiotic levels. Many educators assume that creating learning environments that contain visual and auditory effects while using tools such as animations and videos is sufficient for promoting cognitive processing and constructing elaborate knowledge structures. However, many research studies found that the use of visual and auditory effects does not necessarily improve learning and, thus, using technology per se does not guarantee success. In order to improve learning processes, the instructor has to plan correctly the manner in which the information is presented and to refer to its sensory and semiotic aspects.

Effect of Illustrations on Learning

In a series of four laboratory experiments, Mayer (2003) investigated the conditions under which the addition of illustrations to a text, written or vocal, foster meaningful learning. It was found that students learn more deeply: from words and pictures than from words alone, when extraneous material is excluded rather than included, when printed words are placed near rather than far from corresponding pictures, and when words are presented in a conversational rather than formal style. A possible explanation for these findings is that learning is more meaningful when the information is absorbed via two channels—auditory and visual, when learners pay high attention both to words as well as to pictures; and when they integrate the verbal representations with the visual representations, and between them and prior knowledge.

Another lab experiment (Schnotz & Bannert, 2003) found that presenting graphics is not always beneficial for the acquisition of knowledge. Whereas task-appropriate graphics may support learning, task-inappropriate graphics may interfere with mental model construction. Pictures facilitate learning only if the learners have low prior knowledge and if the subject matter is visualized in a task-appropriate way. If good readers with high prior knowledge receive a text with pictures in which the subject matter is visualized in a task-inappropriate way, then these pictures may interfere with the construction of a task-appropriate mental model. The researchers behind this experiment concluded that the structure of graphics affects the structure of the mental model. In the design of instructional material including texts and pictures, the form of visualization used in the pictures should be considered very carefully.

Effect of Animations on Learning

Animation is a dynamic depiction that can be used to make change processes explicit to the learner (Schnotz & Lowe, 2003). Many educators believe that animations are superior to static illustrations as tools for learning. In order to comprehend a dynamic situation that is externally represented by a static graphic, the learner must first construct a dynamic mental model from the static information provided. In contrast, animations can offer the learner an explicit dynamic representation of the situation. On the other hand, the transitory nature of dynamic visuals may cause higher cognitive load because learners have less control of their speed of processing. Lowe (2003) and Lewalter (2003) showed that merely providing learners with the dynamic information in an explicit form does not necessarily result in better learning.

An experimental study with 60 physics students, conducted by Lewalter (2003), investigated the effects of including static or dynamic visuals in an expository text on a learning outcome. She found that either adding animations or adding static illustrations can result in better learning. However, she found no difference between animations and static illustrations with respect to knowledge acquisition about facts, and only a small non-significant difference in favor of the animation group with respect to comprehension. Kozma (2003) found that with regard to the use of representations, such as animations and video segments showing lab experiments, chemistry experts may extract more benefits than chemistry novices. Lowe (2003) found that explicit presentation of the dynamic aspects of the content in a multimedia learning environment does not necessarily have a positive impact on learning. In many cases, the use of static visuals including conventional signs for motion, such as arrows, or the use of a series of frames may be sufficient for learning.

To review, the use of advanced educational technology as such does not assure a positive effect on learning. In order to improve learning, the instructor has to thoroughly plan the use of pictures and animation according to the following principles: students learn more deeply from words and pictures than from words alone; pictures facilitate learning only if the learners have low prior knowledge and if the subject matter is visualized in a task-appropriate way; animations are more effective when the learner can control the pace and the direction, but even animations allowing a high degree of user control should incorporate considerably more support and direction if they are to function as effective tools for learning; and in science teaching, it is not sufficient to present virtual experiments. Students must participate in hands-on experiments as well.

THE FIFTH MODEL: THE WEB AS AN INFORMATION SOURCE

Information Literacy

The World Wide Web is extremely accessible and always available for retrieving information on any subject. Over the past few years the use of the Internet as a source of information for teaching/learning purposes has escalated. Teaching strategies such as inquiry-based learning, project-based learning, and WebQuests are driving its ever-increasing use.

More than ever before, K-12 students are expected to master an abundance of information skills. For example, they must know how to locate, identify, retrieve, find, process, organize, use, sort, present, analyze, evaluate, and integrate information.

One of the major tasks of the educational system in the modern era is to promote information literacy. According to Doyle (1992), information literacy is the ability to access, evaluate, and use information from a variety of sources. An information-literate person is one who: recognizes that accurate and complete information is the basis for intelligent decision making, recognizes the need for information, formulates questions based on information needs, identifies potential sources of information, develops successful search strategies, accesses sources of information including computer-based and other technologies, evaluates information, organizes information for practical application, integrates new information into an existing body of knowledge, and uses information in critical thinking and problem solving. Beyond all this, the Internet allows students to communicate with experts, co-students, and scientists all over the world.

Web Site Evaluation

Unlike a library, there is generally no control over the content available on the Internet. Therefore, evaluating information found on the Web is a necessary skill in the information age. For instance, Beck (1997) asserted that one should approach the information available on the Web with care, since anyone can publish what they want on the Web. In many cases, it is difficult to know who is the source of the information—even if the author identifies him/herself, it is not always clear what the author's qualifications and expertise are; many times the content will not have been checked prior to publication. Beck and other researchers proposed eight criteria for evaluating Web sites:

1. **Accuracy:** Is the information reliable and error-free? Is there an editor or someone who verifies/checks the information?
2. **Authority:** Is there an author? Is the page signed? Is the author qualified? Is the author expert? Who is the sponsor? Is the sponsor of the page reputable? How reputable? Is there a link to information about the author or the sponsor? If the page includes neither a signature nor indicates a sponsor, is there any other way to determine its origin?
3. **Objectivity:** Does the information show a minimum of bias? Is the page designed to sway opinion? Is there any advertising on the page?
4. **Currency:** Is the page dated? If so, when was the last update? How current are the links? Have some expired or moved?
5. **Coverage:** What topics are covered? What does this page offer that is not found elsewhere? What is its intrinsic value? How in-depth is the material?
6. **Format and Presentation:** Is the page easy to use? How are the links organized? Is the navigation through the site acceptable?
7. **Citations:** Is there a reference list?
8. **Efficiency:** Is the page always available?

Some More Limitations of Using E-Learning in K-12

Various studies identified problems, difficulties, and limitations related to computerization. The following is a partial list of the problems which educators and teachers should be aware:

- **Literacy:** Many students have difficulty finding relevant useful information by themselves.
- **Information Overload:** Some students find themselves overwhelmed by the sheer quantity of the information and cannot isolate the material they need from the mass of information.
- **Isolation and Social Estrangement:** For some students, it is hard to sit alone in front of a computer. Even when there is an ongoing forum on the Internet, the discussion will be happening via non-oral texts, without facial expressions and body language. In extreme cases, a student's ability to conduct interpersonal communication and be socially involved may be harmed to the point of feeling depressed and lonely.
- **Surfing:** As they surf the Web, students may reach commercial and irrelevant sites, and even inappropriate and harmful ones. The danger of aimless surfing—surfing without any strategy behind it—may arise.
- **Non-Linear Text:** Books are organized such that readers advance step by step according to an order. In many sites there are numerous links, with each one leading to additional links so that, sometimes, stu-

dents cannot remember where they began and where they were actually headed. Many students have difficulty with this type of reading (in opposition, many others see a welcome advantage in this type of information presentation).

- **Effects:** Students may be drawn to the site's effects, the graphics, and the design elements at the expense of scrutinizing the site's content.
- **Writing Ability:** Studying on the Web and participating in online forums requires a strong proficiency in expressing oneself in writing. This is a stumbling block for many students. At times, the wide use of the Internet is at the cost of reading books (and this apparently harms verbal and writing skills).
- **Visual Literacy:** Some students find it hard to read directly from a computer screen.
- **Addiction:** There are students who become addicted to surfing the Web. They sit for hours in front of the computer monitor and in extreme cases may lose contact with reality.
- **Motivation and Self-Drive:** Learning on the Web requires motivation and the ability to persevere. The responsibility to sit and study is the student's (in contrast to the classroom setting in which students must, at a minimum, stay until the bell rings).
- **Copying:** Sites exist on the Web offering students prepared essays on almost any topic.
- **Digital Gap:** Different populations have varying degrees of computer skills, as a result of general economic conditions that do not allow them access to computer technologies.
- **English:** Many students in non-English-speaking countries are not adept in English. This limitation shrinks their opportunities to exploit the information that exists on the Web.
- **Law Breaking:** Copying of software, hacking into Web sites, and so forth.
- **Technical Problems:** Breakdowns, "crashes", communication problems, viruses, and so forth.

DISCUSSION AND PRESENTATION OF SOME PEDAGOGICAL ASPECTS RELATED TO THE FIVE MODELS

Active Learning: The Constructivist Approach and its Implementations for Teaching

The main pedagogical basis for e-learning is active learning. Many elements of the active learning approach are derived from principles of the constructivist teaching approach. This section outlines in brief the principles of the latter approach and their application to teaching.

Constructivism is a theory concerning learning and knowledge which suggests that human beings are active learners who construct their knowledge from personal experiences and on their efforts to give meaning to these experiences. According to this approach, the learning environment should enable students to construct their knowledge through active learning and trial and error.

In the literature, three modes of constructivism are discussed: radical (Glasersfeld, 1995), contextual (Cobern, 1993), and social (Vygotsky, 1986). The focus here is on social constructivism. One of the better-known researchers who refers to social constructivism theory in education is Vygotsky (1986). He states that learners construct knowledge or understanding as a result of active learning, thinking, and doing in social contexts.

Constructivism suggests that learners learn concepts or construct meaning about ideas through their interaction with others, with their world, and through interpretations of that world by actively constructing meaning. They cannot do this by passively absorbing knowledge imparted by a teacher. Learners relate new knowledge to their previous knowledge and experience. A constructivist model of teaching has five characteristic features: active engagement, use and application of knowledge, multiple representations, use of learning communities, and authentic tasks (Krajcik, Czerniak, & Berger, 1999).

As we have seen earlier, in all five e-learning models, students constructed their own knowledge by active learning while interacting with the teaching staff, parents, experts, and other students in the classroom.

The teacher's task, according to this approach, is to tutor students and teach them how to learn. He/she is not a mere purveyor of knowledge or provider of facts, but is, rather, a mentor, facilitator, helper, and mediator for learning. The teacher must create a learning environment that will allow the student to construct his or her own knowledge by experiencing and interacting with the environment (Hill, 1997). An e-learning strategy, if designed correctly, may provide precisely such a learning environment.

Many researchers testify to the efficiency of the active learning. For Example, Hake (1998) examined 6,542 students who participated in physics courses. He found that the conceptual understanding and the problem-solving ability of students who applied interactive-engagement methods in their studies was significantly higher than students who studied in traditional methods.

The Effects of Computerized Feedback Intervention on Learning

This section discusses feedback intervention given to the student by the computer both in SDL and asynchronous online courses. "Feedback interventions are defined as actions taken by (an) external agent(s) to provide information regarding some aspect (s) of one's task performance" (Kluger & DeNisi, 1996). This definition excludes several areas of investigation: (1) natural feedback processes such as homeostasis, intrinsic feedback, or the negative-feedback-loop of a control system that operates without an external intervention; (2) task-generated feedback which is obtained without intervention; (3) personal feedback that does not relate to task performance; and (4) self-initiated feedback-seeking behavior. We concentrate here on feedback intervention given to the student by an external agent (the teacher) as regards certain aspects and outcomes of the learning process. The feedback could also be automatic—the computer, both in SDL and asynchronous online courses, returns feedback, which is prepared by the teacher in advance.

Following a literature review it seems that the question on which we should focus is not whether feedback should be given, but how should the feedback be designed in order to improve learning. Based on research findings, a short discussion about the conditions under which computerized feedback has a positive effect on learning is presented below.

The Effect of Feedback on Performance

Many organizational psychology research studies show that feedback has a positive effect on performance level. Thus, for example, according to Locke and Latham (1990), a meta-analysis of 33 investigations shows that in relation to pre-defined goals, feedback is more efficient than in a situation where goals were defined and feedback was not given or a situation in which feedback was given but no goals were defined.

The educational literature has plenty of evidence showing that well-designed feedback given by teachers has a positive effect on learning (Cronbach, 1977; Natrielo, 1987; Crooks, 1988; Black & William, 1998; William, 2002). For example, according to Cronbach (1977), "...feedback or knowledge of results [is] the strongest, most important variable controlling performance and learning...It has been shown repeatedly that there is no improvement without knowledge of results, progressive improvement with it, and deterioration after its withdrawal" (p. 404). And William (2002) summarizes, "After a year, we found significant improvements in the attainment (as measured by external tests) of students taught by teachers using formative assessment, compared with controls in the same schools."

Since this section focuses on feedback provided (automatically) by the computer, let us examine if there is a significant difference between regular teacher feedback and computerized feedback in relation to the effect on learning. Early (1988) found that immediate feedback given by the computer stimulates more confidence, leads to better self-efficacy, and improves performance compared to feedback given by the teacher, verbally or in writing. A possible explanation could be that feedback given by the teacher might detour the student's attention to "himself/herself" (i.e., the student will attempt to understand the teacher's intentions, compare him/herself to others, perceive the feedback as something that is being subjectively aimed at himself/herself personally, perceive the feedback as a threat or even as offensive in certain cases). On the other hand, feedback given by the computer focuses the attention on the task. Jackson (1988) and Kumar and Helgeson (2000) also found that immediate feedback given by a computer is more efficient than feedback provided through traditional methods.

Does feedback always have a positive effect on performance? Kluger and DeNisi (1996) argued that feedback could cause various effects on performance—in certain situations feedback improves the performance level, in others there is no significant effect, and at times there is a negative effect. That is why just providing feedback is insufficient. In order for feedback to have a positive effect, one should plan it properly. The following are a few aspects to be taken into consideration when planning to provide feedback.

Negative Feedback

Here, the term "negative feedback" refers to feedback about a mistake made by a student. According to Kluger and DeNisi (1996), feedback influences the student's pleasantness and the alertness and, therefore, the performance as well. Negative feedback could also have an unintended emotional influence. When an individual is given negative feedback, he/she evaluates the level of his/her performance in relation to the goal, and accordingly, he/she can proceed using one of four strategies: redouble the effort in order to meet the goal; decrease the goal level to one that can be achieved; reject the feedback; or give up and "run away" (physically or mentally) from the situation. Repetitive negative feedback might induce a reaction of learned helplessness.

Of course, the teacher must create a learning environment that leads the student to choose the first strategy—redouble the effort in order to achieve the goal. Practically, feedback about a mistake that directs the learner to interpret the mistake and challenges him/her toward additional thinking paths would be more efficient than laconic negative feedback, such as "you made a mistake, try again!"

Positive Feedback

Surprisingly, positive feedback does not necessarily result in better learning. Many researchers (see Kluger & DeNisi, 1996) found that praise could also harm performance. For example, feedback that is "too good" may encourage low effort by the student. A teacher, who is effusive with his/her commendation, even when there is no justification for it, might cause non-confidence (why exert oneself if the teacher praises everything anyway in order to form a positive climate in the classroom or in order to encourage students). So, in order to improve performance, positive feedback and praise should relate directly to the task.

Positive feedback, just like negative feedback, should be as detailed and informative as much as is possible. It is not always sufficient to react with a "yes" or "untrue." It is advisable to add an explanation such as: "Your answer is not correct because...," "The right answer is B since...,", "Answers A and D are wrong because...," "Answer C is wrong because...," and so forth.

Feedback Components

According to Levin and Long (1981), efficient feedback is composed of three components: definition of the required goal, provision of detailed feedback about the performance, and provision of direction to the student as to how to close the gap between his/her performance and the goal. Formative assessment is better than summative assessment given at the end of the semester when there is no longer the possibility to correct mistakes and close gaps in order to achieve learning goals.

In short, immediate feedback given by the computer could, if it is correctly designed, stimulate more confidence, lead to better self-efficacy, and improve learning compared to feedback given by the teacher, verbally or in writing. Through the investment of little effort, it is possible to design feedback provided by the computer through the course Web site so that a positive effect on learning is achieved. The feedback must: be focused and specific to the task, contain relevant and detailed information, be given immediately, direct the learner to understand his/her mistake, challenge the learner towards additional thinking paths, and point at other possible solutions. The teacher should also present the aims of the course and the learning goals.

SUMMARY

This chapter reviewed the benefits and challenges of five approaches for integrating technology and teaching. The advanced technology exists, but using technology simply because it is there does not assure effective learning. Technology must be a means, not the aim. More important are the pedagogical considerations and the ways of using the technology to extract most of the pedagogical benefits. The technology should be used to drive active learning, give immediate feedback, and present external and internal multiple representations in multimedia learning. In using discussion groups, interactive features, and inquiry-based approaches, teachers should nurture a learning environment that enables students to create their meaning, and organize and rationalize their experiences. Examining experience gives rise to learning (Fosnot, 1996). Technology should be used to serve pedagogical needs and to enable meaningful learning. While motivational issues should be taken into consideration, particularly when referring to K-12 students, the teacher should use technology for creating a learning environment that assures: "Overall, students find electronic interaction a meaningful, enjoyable experience" (LaMaster & Morley, 1999).

REFERENCES

Beck, S. (1997). *Evaluation criteria. Why it's a good idea to evaluate Web sources.* Retrieved July 10, 2004, from http://lib.nmsu.edu/instruction/evalcrit.html

Bhattacharya, M. (1999, October 24-30). A study of asynchronous and synchronous discussion on cognitive maps in a distributed learning environment. *Proceedings of the WEBNET 99 World Conference on the WWW and Internet,* Honolulu, Hawaii.

Black, P., & William, D. (1998). Inside the black box: Raising standards through classroom assessment. *Phi Delta Kappan, 80*(2), 139-144.

Branon, R. F., & Essex, C. (2001). Synchronous and asynchronous communication tools in distance education. *TechTrends, 45*(1), 36-42.

Brown, B. L. (2000). *Web-based-training.* Washington, DC: Office of Educational Research and Improvements (ERIC Document Reproduction Service No. EDO-CE-00-218).

Burge, E. J., & Roberts, J. M. (1998). *Classrooms with a difference: Facilitating learning on the information highway.* Montreal: McGraw-Hill.

Carr-Chellman, A., & Duchastel, P. (2000). The ideal online course. *British Journal of Educational Technology, 31*(3), 229-241.

Cobern, W. (1993). Contextual constructivism: The impact of culture on the learning and teaching of science. In K. G. Tobin (Ed.), *The practice of constructivism in science education.* Hillsdale, NJ: Lawrence Erlbaum.

Cronbach, L. J. (1977). *Educational psychology.* New York: Harcourt Brace Jovanovich.

Crooks, T. J. (1988). The impact of classroom evaluation practices on students. *Review of Educational Research, 58*(4), 438-481.

Davey, K. B. (1999). Distance learning demystified. *Phi Kappa Phi, 79*(1), 44-46.

Doyle, C. S. (1992). *Outcomes measures for information literacy within the national education goals of 1990: Final report.* Washington, DC: National Forum on Information Literacy (ERIC Document Reproduction Service No. ED 351 033).

Early, P. C. (1988). Computer-generated performance feedback in the magazine-subscription industry. *Organizational Behavior and Human Decision Processes, 41,* 50-64.

Fosnot, C. T. (1996). Constructivism: A psychological theory of learning. In C. T. Fosnot (Ed.), *Constructivism: Theory, perspective and practice.* New York: Columbia University, Teacher College.

Frank, M., Kurtz, G., & Levin, N. (2002a). Implications of presenting pre-university courses. Using the blended e-learning approach. *Educational Technology and Society, 5*(4), 37-147.

Frank, M., Reich, N., & Humphreys, K. (2002b). Respecting the human needs of students in the development of e-learning. *Computers & Education, 40*(1), 57-70.

Freedman, Y. (1998). *Teachers and students—mutual respect relationships.* Jerusalem: Henrietta Szold Institute for Social Sciences Research (in Hebrew, abstract in English).

Garrels. M. (1997). *Dynamic relationships: Five critical elements for teaching at a distance*. Retrieved July 10, 2003, from http://www.ihets.org/learntech/distance_ed/fdpapers/1997/garrels.html

Glasersfeld, E. V. (1995). A constructivist approach to teaching. In P. Leslie & J. Gale (Eds.), *Constructivism in education*. Hillsdale, NJ: Lawrence Erlbaum.

Hake, R. R. (1998). Interactive-engagement versus traditional methods: A six-thousand-student survey of mechanics test data for introductory physics courses. *American Journal of Physics*, *66*(1), 64-74.

Hativa, N. (2000). *Teaching for effective learning in higher education.* Dordrecht, The Netherlands: Kluwer Academic.

Hill, A. M. (1997). Reconstructionism in technology education. *International Journal of Technology and Design Education,* *7*(1-2), 121-139.

Hill, J. R. (1997). Distance learning environments via the World Wide Web. In H. K. Badrul (Ed.), *Web-based instruction.* Englewood Cliffs, NJ: Educational Technology Publications.

International Technology Education Association. (2003). *Technology for all Americans project.* Retrieved from http://www.iteawww.org/taa/taa.html

Jackson, B. (1988). A comparison between computer-based and traditional assessment tests, and their effects on pupil learning and scoring. *School Science Review, 69,* 809-815.

Kluger, A. N., & DeNisi, A. (1996). The effects of feedback interventions on performance: A historical review, a meta-analysis and preliminary feedback theory. *Psychological Bulletin, 119,* 254-284.

Kozma, R. (2003). The material features of multiple representations and their cognitive and social affordances for science understanding. *Learning and Instruction: The Journal of the European Association for Research on Learning and Instruction,* *13*(2), 205-226.

Krajcik, J., Czerniak, C., & Berger, C. (1999). *Teaching children science: A project-based approach.* New York: McGraw-Hill College.

Kumar, D. D., & Helgeson, S. L. (2000). Effect of gender on computer-based chemistry problem solving. *Electronic Journal of Science Education,* *4*(4). Retrieved July 12, 2004, from http://unr.edu/homepage/crowther/ejse/kumaretal.html

LaMaster, K. J., & Morley, L. (1999). Using *WebCT bulletin board option to extend transitional classroom walls.* (ERIC Reproduction Service No ED 440 922).

Levin, T., & Long, R. (1981). *Effective instruction.* Alexandria, VA: ASCD.

Lewalter, D. (2003). Cognitive strategies for learning from static and dynamic visuals. *Learning and Instruction: The Journal of the European Association for Research on Learning and Instruction,* *13*(2), 177-190.

Lister, B. C. et al. (1999, October 26-29). The Rensselaer 80/20 model for interactive distance learning. *Proceedings of Educause '99,* Long Beach, California. Retrieved July 12, 2004, from http://www.educause.edu/ir/library/html/edu9907/edu9907.html

Locke, E. A., & Latham, P.L. (1990). *A theory of goal setting and task performance.* Englewood Cliffs, NJ: Prentice-Hall.

Lowe, R. K. (2003). Animation and learning: Selective processing of information in dynamic graphics. *Learning and Instruction: The Journal of the European Association for Re-*

search on Learning and Instruction, 13(2), 157-176.

Mayer, R. E. (2003). The promise of multimedia learning: Using the same instructional design methods across different media. *Learning and Instruction: The Journal of the European Association for Research on Learning and Instruction, 13*(2), 125-140.

Mielke, D. (1999). *Effective teaching in distance education* (Report No. EDO-SP-1999-5). Washington, DC: Office of Educational Research and Improvement.

Monson, S. J., Wolcott, L. L., & Seiter, J. S. (1999, February 19-23). Communication apprehension in synchronous distance education. *Proceedings of the Annual Meeting of the Western States Communication Association,* Vancouver, British Columbia.

Mortera-Gutierrez, F., & Murphy, K. (2000, January 25-28). Instructor interactions in distance education environments: A case study. *Proceedings of the Annual Distance Education Conference,* Austin, Texas.

Natriello, G. (1987). *Evaluation processes in schools and classrooms*. Baltimore: Johns Hopkins University, Center for Social Organization of Schools.

Power, D. (1999). *Vista school district digital intranet: The delivery of advanced placement courses to young adult learners in rural communities.* Retrieved July 12, 2004, from http://www.tellearn.mun.ca/pubs/kssn news.html

Sandalov, A. N. et al. (1999). *The development of open models for teaching physics to schools in dispersed locations in Russia and Canada.* Retrieved July 12, 2004, from http://www.tellearn.mun.ca/pubs/russia.html

Schnotz, W., & Bannert, M. (2003). Construction and interference in learning from multiple representations. *Learning and Instruction: The Journal of the European Association for Research on Learning and Instruction, 13*(2), 141-156.

Schnotz, W., & Lowe, R. (2003). Introduction. *Learning and Instruction: The Journal of the European Association for Research on Learning and Instruction, 13*(2), 117-124.

Vygotsky, L. S. (1986). *Thought and language*. Translated by A. Kozulin. Cambridge, MA: MIT Press. (Original English translation published 1962.)

William, D. (2002, August 28-30). Notes towards a theory of formative assessment. *Proceedings of the Joint Northumbria/EARLI Assessment Conference: Learning Communities and Assessment Cultures, Connecting Research with Practice,* Newcastle, UK.

Willis, B., & Dickinson J. (1997). Distance education and the World Wide Web. In H.H. Badrul (Ed.), *Web based instruction.* Englewood Cliffs, NJ: Educational Technology Publications.

Wolcott, L. (1995). The distance teacher as reflective practitioner. *Education in Technology, 34*(3), 49-55.

Chapter XXII
Integrating Computer Literacy into Mathematics Instruction

Allan Yuen
The University of Hong Kong, Hong Kong

Patrick Wong
The Mission Covenant Church Hom Glad College, Hong Kong

ABSTRACT

The focus of computer literacy in education has evolved from teaching computer programming to integrating information and communication technology (ICT) across subjects. However, most schools in Hong Kong or elsewhere consider computer literacy as a stand-alone subject. One important question is how teachers integrate computer literacy into other subject teaching. There is no simple method, however a well-defined pedagogical strategy might help teachers and educators to better understand the issues and opportunities that the integration provides for meaningful learning. This chapter endeavors to report a case study of integrating computer literacy into mathematics instruction in a Hong Kong secondary school which focused on exploring essential conditions to the integration from the different ways students and teachers experienced the implementation of the curriculum integration. Four conditions emerged from the data analysis, namely, student performance and preference, pedagogical approach, student satisfaction, and perceived constraints.

INTRODUCTION

Nowadays, policymakers see mastery of information and communication technology (ICT) as one of the key features of a competitive modern economy (Pearson, 2001). Computer literacy as a course of study was indeed first introduced in 1965 at colleges and universities (Hess, 1994). As a result of the marketing of desktop computers to both businesses and individuals in the early 1980s (Childers, 2003), the term *computer literacy* has become very popular as a buzzword describing a new type of understanding of literacy apart from reading and writing. With the proliferation of ICT, technologies have changed the way we live as well as the mode of functioning in our society, in particular the Internet technologies have been translated into a number of strategies for teaching and learning (Jonassen, Peck, & Wilson, 1999) and brought about changes in classrooms (Garner & Gillingham, 1998; Law et al., 2000). Although the literature on computer literacy or technology literacy is extensive, the definitions of the phrase seem to reflect on the different authors' academic backgrounds, preferences, and emphases, indicating that it is a multifaceted idea (Childers, 2003). Computer literacy courses are believed to play a critical role in providing students with fundamental computer concepts and skills. Nevertheless, there is still no consensus as to how this educational goal should be achieved (Hess, 1994).

Karsten and Roth (1998) conducted a study to identify the relationships that exist among computer experience, computer self-efficacy, and computer-dependent performance in an introductory computer literacy course. Results suggest that it is the relevance, rather than quantity, of computer experience students bring to class that is most predictive of performance. Accordingly, only computer self-efficacy was found to be significantly related to computer-dependent course performance. In a regression analysis on whether completion of course prerequisite improves student performance in business communications, the results indicate that the computer literacy prerequisite has no effect on student performance (Marcal & Roberts, 2000). However, the concept of computer literacy still has merit (Childers, 2003), and different innovative pedagogical approaches for teaching computer literacy have been advocated. Bretz and Johnson (2000) found that students enrolled in a Web-based, self-paced, competency-based introductory computer literacy course demonstrated positive student perceptions and student learning outcomes. It is also argued that to ensure students:

> *...build upon a computer literacy foundation essential for much professional success and personal fulfillment, teachers need sustained hands-on guidance in learning to design and schedule classroom computer activities that both relate to the curriculum and challenge their students to learn new skills.* (Hackbarth, 2001, p. 19)

The focus of computer literacy in education has evolved from teaching computer programming to integrating ICT across subjects (Hess, 1994). Can ICT be a proper subject? Pearson (2001) argued that ICT is not a stable subject since technological innovations are redefining the knowledge and skills which students need to possess, and the status of ICT as a cross-curriculum discipline is enhanced in the curriculum for technology. Jonassen et al. (1999) indicated that ICT should be used as learning tools for students to learn with.

Given that the 174 cases collected from 28 countries in Module 2 of the Second International Information Technology in Education Study (SITES M2) represented different extents of innovativeness of using technology in learning and teaching (Kozma, 2003), it would be important to note that these innovative pedagogical practices are indeed examples of integrating ICT into the curriculum. Three types of curriculum focus were found in these ICT integration: (1) single subject focus, in which ICT is used to improve students' understanding of subject matter content and concepts; (2) cross-curricula thematic focus, in which curriculum content is offered through themes and ICT is used to facilitate the implementation of lifelong learning; and (3) school-wide focus, in which ICT-supported pedagogical practices are integrated throughout the school curriculum and ICT facilitates the realization of the school's vision on teaching and learning.

Teachers are facing compelling questions regarding the integration of technology into instructional environments (Wright, 2001). However, most secondary schools in Hong Kong or elsewhere consider computer literacy as a stand-alone subject. Students in Hong Kong are expected to acquire basic knowledge and skills in computer literacy lessons during their junior secondary education. One important question is how teachers integrate computer literacy into other subject teaching. There is no simple method, however a well-defined pedagogical strategy might help teachers and educators to better understand the issues and opportunities that the integration provides for meaningful learning. This chapter endeavors to report a case study of integrating computer literacy into mathematics instruction in a Hong Kong secondary school, which focused on exploring essential conditions to the integration from the different ways students and teachers experienced the implementation of the curriculum integration.

CURRICULUM INTEGRATION

Integration is the combining of two equal groups into a unified whole, and "curriculum integration should be a two-way street" (Berke, 2000, p. 9). Davis (1997) suggests that integration focuses on students making sense of their experience in particular ways rather than imposing a form of integration on students, a particular way of making meaning. Then, how can we integrate computer literacy into mathematics curriculum with an emphasis of students' development of meaning?

Banks and Banks (1997) proposed a four-level model to explain how integration can occur within a curriculum, namely, the contribution approach, the additive approach, the transformative approach, and the social action approach. However, Bullough (1999) argues that "no model of curriculum integration has proven to be as helpful as Alberty's" (p. 157). Alberty's model of curriculum integration, first developed in 1938, include five designs: based upon separate subjects; based upon correlation of two or more subjects; based upon the fusion of two or more subjects; based upon common problems, needs, and interests of students within a framework of problem areas; and based upon teacher-student planned activities without reference to any formal structure (Bullough, 1999).

In recent years, business organizations are facing challenges posed by increasing competition, globalization, and rapid changes in ICT (Hammer & Champy, 1993). Business curricula in universities should be redesigned to meet the needs of the business organizations and their stakeholders. There is an urgent need to move beyond individual disciplines and achieve meaningful cross-functional integration in the curriculum (Porter, 1997). Hence there has been a trend of gradually shifting from fragmentation to integration and coherence in business education (Hyslop & Parson, 1995).

Hamilton, McFarland, and Mirchandani (2000) developed a framework to link up student learning and the interaction among teachers and students. They dichotomized the pedagogy used in the integrated business curricula as either experiential learning or classroom-based learning. Experiential learning, focused on the experience outside the classroom, was primarily inductive, whereas classroom-based learning was deductive (Whetten & Clark, 1996). Each of them consisted of two different contextual dimensions, namely, social themes (include areas such as environmental protection and diversity) and closer alignment with business (emphasizing the interdisciplinary nature of business). The pedagogical and contextual dimensions form a 2x2 integration framework with various proposed approaches as shown in Table 1 (Hamilton et al., 2000, p. 106).

The strategy of integration proposed in the current case study was borrowed from the "closer alignment with business/classroom-based learning" of the curriculum integration model developed by Hamilton et al. (2000), in which the focus of the strategy was grounded on the first three approaches:

1. just-in-time introduction of skills helps students to understand the importance of the skills, which are interwoven within the context of a problem, and students are likely to develop the needed skills through the immediate reinforcement, and may appreciate the interdisciplinary nature of study through hands-on experience;
2. using multidisciplinary tools demonstrates the concept of cross-functionalization and makes the study real-life orientated, and this approach would be successful if teachers and students were capable of using the tools effectively; and
3. team teaching provides students with the benefits of several instructional styles as well as a variety of perspectives throughout their study.

This three-approach integration strategy allows the students to learn in context (Jonassen et al., 1999), learn by doing, and provides opportunities for the multidisciplinary perspective of learning from teachers in different subjects. It is believed that the integration strategy gives students opportunities in making sense of their experience in particular ways, as reinforced by Davis (1997). This strategy is similar to the transformative approach in Banks and Banks (1997). The present case study aims to explore a curriculum integration using these three approaches.

LEARNING GEOMETRY WITH ICT

The mathematics instruction involved in the current curriculum integration focused on the topic of geometry. In this case study, Geometer's Sketchpad, an ICT-based dynamic geometry environment, was taught as a multidisciplinary tool in the computer literacy lessons as well as mathematics lessons.

Geometer's Sketchpad is a construction tool to replace a pair of compasses and a ruler. Constructing a figure with the appropriate geometric primitives, the main features of the

Table 1. Organization framework for integration

	Experiential Learning	Classroom-Based Learning
Social Themes	Live social-theme-based projects Mentoring	Shared teams Reinforcement of common skills and themes Course coordination through coordinated syllabi
Closer Alignment with Business	Internship/cooperative education Live multidisciplinary projects	Just-in-time introduction of skills Multidisciplinary tools Team teaching Multidisciplinary case studies New multidisciplinary courses Guest speakers Course coordination through block scheduling

figure do not collapse under mouse dragging. This is a powerful tool for student understanding on the properties of geometric figures.

Using Sketchpad to explore geometric relationships, dragging can produce continuous transformation of a figure. In this respect, the geometric relationship to be examined is not confined to the static paper-and-pencil diagram (Laborde, 2001), and users can produce various instances of the same figure. In this way, students can generate empirical evidences of the geometric properties (Hoyles & Jones, 1998) of a figure that can hardly be produced by paper-and-pencil drawing. Through continuous transformation of a figure, users can find the invariant properties of the figure.

With the use of Sketchpad, we can design learning tasks to let students get engaged to make conjectures, construct, and explain geometric relationship (Healy & Hoyles, 2001). Dragging a figure, students can produce visual evidence to support or disprove their previous predictions on the properties of the figure. Besides deductive reasoning, the properties of a geometric object can be derived from empirical operations and fostered by organization of those empirical data. This provides a new window onto geometry for teachers and students (Noss & Hoyles, 1996).

With the measuring tools and mouse dragging, students can extract empirical evidence on the behavioral changes of the inspected geometric objects, which is similar to a scientific investigation. In the beginning, students can just collect piecemeal data from the objects to be investigated. After acquiring a better understanding of the objects, students would then have greater confidence to articulate some properties of these objects. This is the starting point of making conjectures and devising methods to prove them. The design of Geometer's Sketchpad allows students to transform the examined figures. Students may discover various geometric properties through the dynamic transformation of figures.

METHOD

This exploratory study uses a case study approach to depict the conditions that are essential to the integration of computer literacy into mathematics instruction from the different ways students and teachers experienced the implementation of the curriculum integration. This case study was conducted in a Hong Kong secondary school, which has a reputation for curriculum innovation. In the current curriculum integration, the main objective of the computer literacy lessons was to make students aware of the use of the computer as a teaching and learning tool. Through the hands-on experience in using the computer-assisted learning (CAL) software, students were able to appreciate the advantages of using CAL software (Geometer's Sketchpad). With the just-in-time introduction of computer skills, students were able to apply Sketchpad to learn geometry in mathematics lessons. The integration strategy and implementation is summarized in Table 2.

Procedure

The duration of this case study lasted for two teaching cycles. In each cycle we had one computer literacy lesson and eight mathematics lessons. Altogether the study involved 16 mathematics lessons and two computer literacy lessons. We chose three classes of Secondary 3 (Grade 9) students, a total of 115 students, to take part in this study. Since the computer lessons were carried out as split classes, there were two computer literacy teachers involved, Teacher A (female) and Teacher B (male). On the other hand, there was one mathematics teacher, Teacher C (male), who took part in the curriculum integration.

Table 2. A summary of integration strategy

Integration approach	Curriculum implementation
Just-in-time introduction of skills	Just-in-time introduction of computer skills to students in learning mathematics
Multidisciplinary tools	Sketchpad as multidisciplinary tool for computer as well as mathematics lessons
Team teaching	Two computer teachers and one mathematics teacher worked in team

The arrangement of the lessons was designed to begin with a computer lesson followed by eight mathematics lessons. The sequence repeated in the second teaching cycle, and the study ended with a test in the computer lesson in the third teaching cycle. In the first and second computer lessons, a just-in-time introduction of computer skills was provided to the students.

In the first computer literacy lesson, the computer teachers taught students the basic skills of using Sketchpad with the help of a set of notes from the Geometer's Sketchpad user manual. Students learned how to draw line segments and triangles. They also learned how to measure the lengths of line segments with the software. After they had mastered the basic skills, they were asked to complete the worksheet in connection to learning geometry. With the help of the worksheet, students were guided to form conjectures, share conjectures with their classmates, and finally test the conjectures by the software. The purpose of this worksheet was to help students to discover the Mid-Point Theorem.

Since there were only 40 minutes in a computer lesson, students were requested to bring along the completed worksheet to the next mathematics lessons. During the mathematics lessons, the mathematics teacher guided the follow-up discussion on the Mid-Point Theorem. The worksheet served as a starting point for students to discuss what they had discovered.

In the second computer lesson, students learned how to draw parallel lines and perpendicular lines with the software. Also, they learned how to measure the angles. Then they were asked to follow the instructions of the teaching notes and worksheet to construct parallelograms. Students were asked to form conjectures about the properties of parallelograms. Discussion with classmates was encouraged during the lesson. In order to ensure students had sufficient time to share and explore their conjectures with their classmates, they only needed to finish one question in the worksheet during the computer lesson. However, they would complete all the questions in the worksheet and bring it to the next mathematics lessons. Similarly, the mathematics teacher would guide the follow-up discussion based on the worksheet.

Data Collection

Altogether 115 Secondary 3 (Grade 9) students, two computer literacy teachers, and one mathematics teacher took part in teaching and learning geometry with Geometer's Sketchpad.

In order to portray the case, data were derived from two major sources—individual interviews with teachers and post-activity questionnaires for students. The student questionnaire consisted of items such as attitude towards learning Geometer's Sketchpad, competence in using the software, ability to learn with the discovery approach in geometry, and satisfaction with the curriculum integration.

Interviews with the mathematics and two computer literacy teachers were conducted to collect teachers' reflections and views on the curriculum integration. The interviews of the mathematics teacher and computer teachers were conducted separately.

RESULTS

There are a number of points worth noting from the findings of the data analysis. First of all, we focus on the analysis of the student questionnaire and description of the development of constructs, and then present four conditions emerging from the analysis of quantitative as well as qualitative data.

Student Questionnaire and Construct Development

There were 16 questions designed to collect students' attitudes and views in the student questionnaire (Appendix), in which responses were in a five-point Likert scale: 1 for "strongly disagree," 2 for "disagree," 3 for "neutral," 4 for "agree," and 5 for "strongly agree." To explore the factor structure of the questionnaire, the principal component analysis with varimax rotation (Kaiser normalization) was used. The factor analysis of the data generated three constructs: (1) Q1-5, (2) Q9-11 & Q14-16, and (3) Q8 & Q12-13, which were labeled as "competence in using Sketchpad" (Competence), "satisfaction with the curriculum integration" (Satisfaction), and "preference of learning mathematical software in computer literacy lessons" (Preference) respectively, and total variance explained was 67%. Table 3 shows the factor loading and reliability coefficients (Cronbach's alpha) of the three factors.

Apart from these three constructs (14 items), Q6 (I can use Sketchpad effectively to take part in discovery learning activities) and Q7 (I like taking discovery approach to learn mathematics) represent two aspects of discovery learning, which has been advocated as a new approach to learning (Rowe, 2004). Table 4 shows the descriptive statistics of Q6, Q7, and the averages of the three constructs. These

Table 3. Results of three factors

Measurement items	Factor 1	Factor 2	Factor 3	Cronbach's alpha
Q4	**0.888**	0.038	0.098	0.918
Q5	**0.886**	0.114	0.145	
Q1	**0.821**	0.353	-0.076	
Q2	**0.807**	0.273	0.093	
Q3	**0.779**	0.299	0.018	
Q15	0.074	**0.858**	0.102	0.841
Q14	0.305	**0.806**	-0.126	
Q16	0.331	**0.672**	0.210	
Q9	0.219	**0.649**	0.268	
Q10	0.122	**0.585**	0.386	
Q11	0.212	**0.523**	0.305	
Q12	0.100	-0.040	**0.877**	0.725
Q13	0.005	0.392	**0.712**	
Q8	0.031	0.222	**0.701**	
Eigenvalue	5.743	2.272	1.397	
Percentage of variance	41.020	16.228	9.980	

Table 4. Descriptive statistics of variables

Variables	N	Mean	S.D.
Q6 (Use Sketchpad in discovery learning)	115	2.983	1.026
Q7 (Discovery learning in mathematics)	115	3.183	1.113
Competence (Q1-5)	115	3.496	0.899
Preference (Q8 & Q12-13)	115	3.052	0.831
Satisfaction (Q9-11 & Q14-16)	115	3.277	0.736

variables will be discussed in the following sections.

Student Performance and Preference

The mean score of the Competence (3.496) is relatively higher than other constructs, indicating students slightly were inclined to agree that they were confident using Geometer's Sketchpad to draw and measure simple geometric figures in learning geometry. However, the students were not sure whether they could effectively use Sketchpad to take part in discovery learning activities (mean score of Q6 is 2.983).

It is not clear that students could use the discovery approach to learn mathematics (mean score of Q7 is 3.183). One interpretation of this would be students were adapted to the conventional method of teaching and learning. They could not change their ways of learning in such a short period of time. The mathematics teacher might have to try out this approach in more areas in order to familiarize students with the new ways of learning and thus enable them to learn with the discovery approach. In addition, there is no noticeable preference for the students to learn the CAL software in computer lessons (mean score of Preference is 3.052).

When the mathematics teacher was asked how students performed in the mathematics lessons before and after the curriculum integration, Teacher C said:

Students were more motivated and actively participated in the class discussions than before. Most of them were able to find out the geometric relationships, which they were supposed to find out by using the software. I could feel a sense of success from their faces.

Teacher A, who had one year teaching computer literacy experience, was asked about how the students preformed in her computer class. She said:

Generally the students were able to follow the steps listed in the notes to draw simple geometric figures. However, it seemed to me that they learned it as if they were just learning a new drawing tool. It was difficult for them to find the relationship between the software and learning Mathematics. When they were asked to form conjectures about the properties of geometric figures, most of them did not know what to do. So I tried to give them some examples to let them know how to think and how to form conjectures. Then some of them could make it. They were not used to learning Mathematics in this way.

Sharing a similar view of Teacher A, Teacher B made the following comment:

In fact students were very excited when they studied in the computer laboratory. They were willing to learn anything about computers but the learning experience was usually limited to the operation of the software. They seldom treated the software they learned as a tool to learn.

It seems that the mathematics teacher was positive towards the curriculum integration, whereas the two computer teachers were not certain about the benefits of the curriculum integration, in particular the effect of Sketchpad as a multidisciplinary tool. The result raises an issue whether such curriculum integration would be adequate if mathematics teachers are able to teach related CAL software in mathematics lessons. Furthermore, we need to explore whether students would prefer mathematics teachers to teach mathematics CAL software during mathematics lessons. However, the effect of the integration approaches, "multidisciplinary tools" and "team teaching," is unclear in general.

Pedagogical Approach

The responses to Q6 did not give clear evidence that students were capable of using Sketchpad in the discovery approach. However, Teacher C's observation was to some extent different from the feedback we collected from students. He explained:

In the first two lessons, I brought a computer to the classroom and asked students to come out to share their geometric conjectures. Then I divided the students into three groups. The first group agreed to the conjectures. The second group disagreed with the conjectures, and the students who were not sure about what they thought were put in the third group. The students challenged others' opinions and defended their own conjectures. The participation of the students was very good. They performed as if they were Mathematicians to invent some new Mathematics. However, when I switched back to the normal teaching method afterwards, and asked them to follow the formal proof and to do the exercise in the textbook, their enthusiasm significantly decreased. The situation repeated when I used the same teaching method to teach the properties of parallelograms.

The curriculum integration provided opportunities for pedagogical change. Students were positive to the discovery learning, however they were not so enthusiastic when the mathematics teacher switched the discovery approach back to normal. Teacher C was then asked whether he could extend the number of lessons to use the discovery approach or keep on using normal approach. He said:

The teaching schedule was very tight. Although the new teaching method was welcomed by the students, it cost us a lot of time. Besides, the abilities of our students were not so high that they needed traditional drillings on exercises to increase their competency on this topic.

This indicates that Teacher C had a dilemma in pedagogical approach. However, it is found that the discovery approach (Q6 & Q7) was clearly correlated to Competence, Preference, and Satisfaction (Table 5), in particular student satisfaction.

Obviously, all variables are significantly correlated except the correlation between Competence and Preference. The results clearly

Table 5. Correlations between variables

	Q6	Q7	Competence	Preference	Satisfaction
Q6	-				
Q7	0.587***	-			
Competence	0.401***	0.379 ***	-		
Preference	0.334***	0.280**	0.179	-	
Satisfaction	0.485***	0.538***	0.508***	0.451***	-

*p<0.05, **p<0.01, ***p<0.001

Table 6. Grouping of satisfaction

Group		Frequency	Percent
1	Low satisfaction (mean ≤ 3)	48	41.739
2	Medium satisfaction (3 < mean < 4)	45	39.130
3	High satisfaction (mean ≥ 4)	22	19.131
	Total	115	100

demonstrate that Competence, Preference, and Satisfaction are factors in connection to the changing of pedagogy to discovery approach.

Student Satisfaction

The mean score of Satisfaction is 3.277 (Table 4), indicating students were in general satisfied with the curriculum integration. In order to further analyze student satisfaction, we classified student satisfaction into three groups (Table 6).

Analysis of variance (ANOVA) was applied to compare the variables of the three groups of student satisfaction, and it is found that the F-ratios are statistically significant (Table 7).

The post hoc test for the three groups showed: (1) Group 3 was significantly higher than Group 1 in all variables, (2) Group 2 was significantly higher than Group 1 in all variables except the variable Preference, and (3) no significant difference between Group 2 and Group 3. It seems the higher satisfaction group (Group 2 and 3) was inclined to show agreement in terms of all variables, indicating the importance of student satisfaction in the implementation of the curriculum integration. Nevertheless, the challenge of fostering student satisfaction in the curriculum integration deserves further attention.

Perceived Constraints

Apart from the aforementioned conditions, we found that the constraints being faced by teachers during the implementation was another essential condition that emerged from the analysis. When Teacher C was asked how he felt about the collaboration between the computer subject panel and the mathematics subject panel, he said:

Table 7. Results of ANOVA

	Group	N	Mean	S.D.	F	Post Hoc Test
Q6	1	48	2.563	0.920	9.650 ***	
	2	45	3.133	0.869		Gp2 > Gp1
	3	22	3.591	1.182		Gp3 > Gp1
Q7	1	48	2.625	0.815	13.513 ***	
	2	45	3.467	0.991		Gp2 > Gp1
	3	22	3.818	1.368		Gp3 > Gp1
Competence	1	48	3.025	0.943	15.510 ***	
	2	45	3.720	0.671		Gp2 > Gp1
	3	22	4.064	0.710		Gp3 > Gp1
Preference	1	48	2.799	0.765	4.730 *	
	2	45	3.156	0.597		
	3	22	3.394	1.185		Gp3 > Gp1

*$p<0.05$, **$p<0.01$, ***$p<0.001$

It was very time-consuming. During last summer holiday, the whole panel had spent a lot of time discussing the lesson plans, worksheets, teaching schedule and teaching approach. Once the plan was fixed, I could not change the schedule according to the need of my students since any change of the plan would affect the teaching of Computer Literacy. It was not flexible at all.

The responses of Preference (Table 3) in the student questionnaire suggested that students were not sure whether it was appropriate to learn mathematics CAL software in computer lessons. When Teacher C was asked whether he would like to teach the mathematics CAL software by himself in mathematics lessons, he responded:

Personally, I am not familiar with this software. I learned it because I had to use it in this project. It was OK for me to use the software for demonstration purpose during the lesson but it was different if I was asked to teach the students how to use it. The required knowledge about the software was more demanding. Besides, there was no computer facility in normal classrooms. If I had to teach the students to use the software, I had to use the computer rooms. To everyone's knowledge, the slots of using the computer rooms had already been filled up by computer lessons. How could I get into it during normal school time?

When he was asked about the possibility of continuing such curriculum integration in the coming year, he commented that if it was the policy and direction of the mathematics panel to promote the CAL software and to use the new teaching method in the coming school year, he would accept it, but if it was only a pilot test to evaluate the effectiveness of the new approach, he preferred to pass the project to other colleagues.

When Teacher A was asked whether there was any problem in teaching Geometer's Sketchpad in computer literacy lessons, she said:

Technically I did not have any problem but since I did not know much about the syllabus of Secondary 3 Mathematics, I found some difficulties in relating the software with what they might actually come across in mathematics lessons. If I were the mathematics teacher, the teaching would be more effective.

Teacher A compared the difference between teaching Sketchpad and other software in computer lessons. She commented:

Since Geometer's Sketchpad is a learning tool, it is not as attractive as learning Flash. The motivation to learn was lower than learning other software like Flash or PhotoImpact.

To a certain extent, both the computer and mathematics teachers indicated difficulties arising from the rigidity of the project arrangement and teaching schedule, though the two subject panels had spent time in planning the project. The effectiveness of this curriculum integration was not clear. We summarize the constraints of the current curriculum integration project as follows: it required teachers and students to have special preparation, it required a great deal of coordination among participating teachers, it reduced teacher autonomy, and it required significant commitment by teachers. These constraints raise a number of challenges in coordinating the approach of "team teaching."

These constraints are consistent with the experience described in Hamilton et al. (2000). The most difficult problem we came across was the time spent in the coordination of the project. Also, the integration needed the commitment of teachers involved. They needed to spend extra time on project administration and preparation.

DISCUSSION

This is a case study of a curriculum integration implementation in a Hong Kong secondary school. There are obvious limitations in generalizations as well as stability of the results. This study aims to describe an attempt at integrating computer literacy into mathematics through the approaches of just-in-time learning, multidisciplinary tool, and team teaching, in the hope of stimulating discussion in issues of integrating technological literacy into other contexts of student learning.

"The costs of curriculum integration are high, real, and certain. The benefits of integration are low, vague, and difficult to measure" (Schug & Cross, 1998, p. 56). Results in the current case study indicate that students were positive in just-in-time learning of Geometer's Sketchpad and were confident of mastering the software. However, they were not certain whether they could effectively use the discovery approach to learn geometry with Geometer's Sketchpad. Analysis of teachers' interview reflected the curriculum integration could help students learn in context as well as learn by doing, and meet the needs or request of the community. However, it was not clear whether the curriculum integration provided multidisciplinary perspectives. Teachers also indicated concerns faced during the implementation of team teaching: extra time spent in the coordination and administration, the integration needed the commitment of the teachers, and there was less flexibility in pedagogical practice. The results also bring about issues and challenges in connection to the integration of computer literacy into other subject instruction in schools.

How should the computer literacy be treated in schools? Can curriculum integration help to answer the question? In this chapter we present a case of curriculum integration on computer literacy and mathematics based on three integration approaches: just-in-time introduction of skills, multidisciplinary tools, and team teaching. Though the empirical evidence of the current case study did not provide clear support to the integrated approaches of teaching computer literacy, we did observe that students could answer questions that require them to examine a concept from two subject areas; in particular the high satisfaction group demonstrated positive outcomes in connection to discovery learning and Sketchpad competence. This is a signal that "integration has been achieved" (Berke, 2000, p. 11).

Curriculum integration requires support from staff development and planning, as Schug and Cross (1998) argued, "Curriculum integration requires teachers who are or can become sufficiently expert in understanding their subjects to be able to make meaningful connections across the disciplines" (pp. 56-57). Thus, in the case study, computer literacy teachers needed to understand mathematics, whereas the mathematics teacher needed to be computer literate. Indeed, the integration of computer literacy training into subject teaching for teachers does provide "future teachers with the confidence to transfer their computer skills into their classrooms based on their own exploratory experiences" (Halpin, 1999, p. 135). It appears that staff development is an essential component of the curriculum integration. Nevertheless, enabling teachers to envision more clearly how various technologies can be orchestrated to achieve learning objectives within assessment tasks in different subjects is also significant, since we need new ways of assessment in the integration of computer literacy with other subjects.

Young (1998) proposed two major concepts in curriculum implementation, namely curriculum as fact and curriculum as practice. In the notion of curriculum as fact, knowledge is embodied in syllabi and textbooks, and external to teachers and students. The pedagogical implication is that knowledge is something to be transmitted. However, knowledge transmission is no longer adequate in the idea of curriculum as practice. Knowledge is produced by people acting collectively and becomes that which is accomplished in the collaborative work of teachers and students. The idea of curriculum as practice, in particular the notion of knowledge production through collective effort, may have implications for the implementation of curriculum integration strategy—multidisciplinary approach and team teaching.

What are the implications of this case study for the school education? First, it is important not to overlook the complexity of integrating technological literacy into other curricula. Second, the approach of just-in-time learning of technological skills is particularly helpful to students in applying technology to other learning contexts. Third, the selection of an appropriate multidisciplinary tool is extremely important in the integration strategy in technological literacy. Fourth, the arrangement of team teaching in the curriculum integration needs careful planning and support. Finally, fostering student satisfaction of the curriculum integration process is fundamental. These are preliminary lessons arising from the study, which require a great deal of future research.

Finally, computer technology is an evolving concept, which demands evolving pedagogical practices for the integration of computer literacy with other subjects in order to advance students' development of meanings in both areas. It is believed that the three integration approaches and four essential conditions emerging from the experience of the current case study would help to stimulate discussion of the development of well-defined pedagogical strategies for integrating technological literacy into other curricula. This study does not claim to be able to address any theoretical issues in curriculum integration, however it provides initial thinking of a pedagogical practice in integrating computer literacy into mathematics instruction.

REFERENCES

Banks, J. A., & Banks, C. A. M. (1997). *Multicultural education: Issues and perspectives* (3rd ed.). Boston: Allyn & Bacon.

Berke, M. K. (2000). Curriculum integration: A two-way street. *General Music Today, 14*(1), 9-12.

Brentz, R., & Johnson, L. (2000). An innovative pedagogy for teaching and evaluating computer literacy. *Information Technology and Management, 1*(4), 283-292.

Bullough, R. V. Jr. (1999). Past solutions to current problems in curriculum integration: The contribution of Harold Alberty. *Journal of Curriculum and Supervision, 14*(2), 156-170.

Childers, S. (2003). Computer literacy: Necessity or buzzword? *Information Technology and Libraries, 22*(3), 100-104.

Davis, O.L. Jr. (1997). The personal nature of curricular integration, *Journal of Curriculum and Supervision, 12,* 95-97.

Garner, R., & Gillingham, M. G. (1998). The Internet in the classroom: Is it the end of transmission-oriented pedagogy? In D. Reinking et al. (Eds.), *Handbook of literacy and technology.* Mahwah, NJ; London: Lawrence Erlbaum Associates.

Hackbarth, S. L. (2001). Changes in primary students' computer literacy as a function of classroom use and gender, *TechTrends, 45*(4), 19-27.

Halpin, R. (1999). A model of constructivist learning in practice: Computer literacy integrated into elementary mathematics and science teacher education. *Journal of Research on Technology in Education, 32*(1), 128-138.

Hamilton, D., McFarland, D., & Mirchandani, D. (2000). A decision model for integration across the business curriculum in the 21st century. *Journal of Management Education, 24*(1), 102-126.

Hammer, M., & Champy, J. (1993). *Reengineering the corporation.* New York: Harper Collins.

Healy, L., & Hoyles, C. (2001). Software tools for geometrical problem solving: Potentials and pitfalls. *International Journal of Computers for Mathematical Learning, 6,* 235-256.

Hess, C. A. (1994). Computer literacy: An evolving concept, *School Science and Mathematics, 4,* 208-214.

Hoyles, C., & Jones, K. (1998). Proof in dynamic geometry contexts. In C. Mammana & V. Villani (Eds.), *Perspectives on the teaching of geometry for the 21st century: An ICMI study* (pp. 121-128). Dordrecht: Kluwer Academic.

Hyslop, C., & Parsons, M.H. (1995). Curriculum as a path to convergence. *New Directions for Community Colleges, 91*(2), 41-49.

Jonassen, D. H., Peck, K. L., & Wilson, B. G. (1999). *Learning with technology: A constructivist perspective.* Upper Saddle River, NJ: Prentice-Hall.

Karsten, R., & Roth, R. (1998). The relationship of computer experience and computer self-efficacy to performance in introductory computer literacy courses. *Journal of Research on Computing in Education, 31*(1), 14-24.

Kozma, R. (Ed.). (2003). *Technology, innovation, and educational change: A global perspective.* Eugene, OR: International Society for Technology in Education.

Laborde, C. (2001). Integration of technology in the design of geometry tasks with Cabri-geometry. *International Journal of Computers for Mathematical Learning, 6*(3), 283-317.

Law, N., Yuen, H. K., Ki, W. W., Li, S. C., Lee, Y., & Chow, Y. (Eds.). (2000). *Changing classrooms & changing schools: A study of good practices in using ICT in Hong Kong schools.* Hong Kong: Center for Information Technology in Education, The University of Hong Kong.

Marcal, L., & Roberts, W. W. (2000). Computer literacy requirements and student performance in business communications. *Journal of Education for Business, 75*(5), 253-257.

Noss, R., & Hoyles, C. (1996). *Window on mathematical meanings: Learning cultures and computers.* Dordrecht: Kluwer Academic.

Pearson, M. (2001). ICT in the national curriculum—revised but not resolved. In C. Cullingford & P. Oliver (Eds.), *The national curriculum and its effects* (pp. 193-205). Aldershot: Ashgate.

Porter, L. W. (1997). A decade of change in the business school: From complacency to tomorrow. *Selections, 13*(2), 1-8.

Rowe, A. (2004). *Creative intelligence.* New York: Prentice-Hall.

Schug, M. C., & Cross, B. (1998). The dark side of curriculum integration in social studies. *The Social Studies, 89*(2), 54-57.

Whetten, D. A., & Clark, S. C. (1996). An integrated model for teaching management skills. *Journal of Management Education, 20*(2), 152-181.

Wright, C. (2001). Children and technology: Issues, challenges, and opportunities. *Childhood Education,* (Fall), 37-41.

Young, M. (1998). *The curriculum of the future.* London: Falmer Press.

APPENDIX: ITEMS OF THE STUDENT QUESTIONNAIRE

Q1. I can use Sketchpad to draw triangles.
Q2. I can use Sketchpad to draw parallelograms.
Q3. I can use Sketchpad to draw squares.
Q4. I can use Sketchpad to measure angles in geometric diagrams.
Q5. I can use Sketchpad to measure lengths in geometric diagrams.
Q6. I can use Sketchpad effectively to take part in discovery learning activities.
Q7. I like taking discovery approach to learn mathematics.
Q8. I think it is better to learn the use of mathematical software (Sketchpad) in computer literacy lessons than mathematics lessons.
Q9. The integration of computer literacy and mathematics instruction helps me to understand how to use information technology in learning.
Q10. The integration of computer literacy and mathematics instruction raises my interest in the subject of computer studies.
Q11. I am satisfied with the integration of computer literacy and mathematics instruction.
Q12. I think computer literacy teachers are better than mathematics teachers in teaching Sketchpad.

Q13. I think it is more effective to learn Sketchpad in computer literacy lessons than mathematics lessons.
Q14. Sketchpad is very useful for my learning of geometry.
Q15. Information technology enhances my interest in learning mathematics.
Q16. The integration of computer literacy and mathematics lessons was in very good cooperation.

Chapter XXIII
Teaching and Learning with Tablet PCs

Leo Tan Wee Hin
National Institute of Education, Singapore

R. Subramaniam
National Institute of Education, Singapore

ABSTRACT

This study explores the experiences of a secondary school in Singapore which was the first to introduce Tablet PCs to the entire cohort of its Secondary One (Grade 7) population. Except for physical education, the workshop aspects of design and technology, and the laboratory aspects of home economics, all other subjects are taught and learned using Tablet PCs. The implementation issues, including the business model which enabled 95% of the students to own their Tablet PCs, and the students' perceptions of this new technology for learning are explored. Results show that the use of Tablet PCs for education is viewed very positively by both students and teachers. Being among the very first (and very few) studies investigating the effectiveness of Tablet PCs for educational use, it is suggested that the findings from this study have some implications for other educational institutions considering the use of Tablet PCs for their students.

INTRODUCTION

The educational landscape has undergone tremendous transformations in the past decade, and is still dynamically configuring itself for its strategic positioning in the information age. Two factors that have greatly influenced this course of development are the availability of affordable personal computers and the ease of connectivity to the Internet. These are increasingly helping to transition the move from a reduction in teacher-centric pedagogy to increased learner-centric pedagogy, with implications for instructional methods to engage students in class.

That appropriate technological tools can enhance the effectiveness of classroom tasks for the teacher as well as open up new pathways for interacting with students are supported by a number of studies. Indeed, in recent times there has been a pronounced emphasis on incorporating instructional methods that encourage student engagement. This can create active learning environments for the meaningful assimilation of knowledge (Magolda, 1992). When technology-based instructional tools present lessons in interesting ways that capture student attention, possibilities for promoting further student learning are afforded (Macdonald, 2003). Some of the technological tools in this regard include electronic whiteboards (Smith, Higgins, Wall, & Miller, 2005) and PowerPoint presentations (Susskind, 2005). A more recent development is the Tablet PC, which promises to revolutionize education even further.

The Tablet PC is basically a next-generation PC with a slew of features. It is lightweight, portable, has a long battery life, possesses powerful processing capabilities, and can connect to the Internet via a wireless network. Other advantages include:

- It allows students to work from anywhere—class, library, or home—with the summative efforts having the potential to enhance cognitive gains in the learning process.
- With the use of digital ink, it becomes possible to handwrite notes, organize these in a searchable format, and integrate them with other applications software running on the Tablet PC.
- In classrooms and lecture theatres, the use of the linked audio facility in the Tablet PC allows voice to be recorded when taking down notes.
- Besides use of keyboard and digital ink, input can also be via voice commands.

Migrating to the Tablet PC regime affords schools an opportunity to do away with expensive cabling and equipment idling time in computer laboratories. Desktops PCs, with their attendant paraphernalia such as cables and network points, represent a significant investment, including recurrent expenditure.

As the Tablet PC is a very recent innovation, having been introduced only in January 2003, there is a scarcity of studies in the literature on its use for education. In fact, we are not aware of any studies in the mainstream journal literature in this regard. A few studies have been reported in conference papers, and these focused on its use for teaching computer science (Anderson et al., 2004) and mathematics (Golub, 2004). Positive feedback on their use is evident from these studies.

The objective of this chapter is to explore the experiences of one school in Singapore which was the first to introduce Tablet PCs to the entire cohort of its Secondary One (Grade 7) population. The implementation issues, including the business model that allowed 95% of the students to own their Tablet PCs, are discussed, and the students' views on the new technology are ascertained with the aid of an evaluation instrument specially developed for this study. Being among the very first studies on the use of Tablet PCs for educational use, it is suggested that aspects of this study can offer some lessons to other institutions considering the introduction of Tablet PCs for their students.

METHODOLOGY

Instrument Design

An evaluation instrument was designed to assess the effectiveness of Tablet PCs for use in an educational setting. The instrument has three subscales: *Learning Environment,* which refers to the new setting that the Tablet PC brings to the teaching and learning process; *Effectiveness of Learning,* which refers to the cognitive enhancements promoted by the Tablet PC; and *On the Tablet PC,* which refers to some of the perceptions and features of this equipment. It was felt that these three subscales were adequate to obtain some reliable information about the educational potential of the Tablet PC. The instrument aimed to solicit the views of students.

For each subscale, a draft list of eight statements was formulated based on the authors' experiences in constructing survey instruments and the use of technological tools for learning. All the statements were worded positively. Though survey pundits recommend that negatively worded statements be interspersed among the positively worded statements in a survey instrument, we have refrained from doing this, as all the subjects in this study are young students (14 years of age) who may not be able to construe the full import of negatively worded statements In fact, a study by Benson and Hocevar (1985) involving students in grades 4-6 in the U.S. indicated that it was rather difficult for young subjects to express disagreement with a negatively worded statement. Another study by Schriesheim and Hill (1981) found that even older respondents often find difficulty in this regard. Our recent study on the effectiveness of a super computer-based virtual environment for learning also attests to the need to be careful when using negatively-worded statements (Tan, Subramaniam, & Anthony, 2003). The statements were edited for clarity, ambiguity, redundancy, and item phrasing before trimming these down to five statements for each subscale. Care was taken to ensure that the readability of the statements was appropriate for the educational level of the students used in this study. For ease of administration, the scalability of the statements was set on a five-point Likert scale ranging from Strongly Agree (SA) to Strongly Disagree (SD). The corresponding numerical measures, for the purpose of statistical analyses, were 5 for SA and 1 for SD. An optional fill-in field at the end of the evaluation instrument allowed for comments to be made by the respondents. The 15-item evaluation instrument was considered adequate for the purpose of this study. A longer evaluation instrument was not desirable for two reasons: the young subjects may be amenable to respondent fatigue, and also curriculum time needs to be used for the survey administration.

As the Tablet PC is used for several subjects in the school curriculum, we have not incorporated any subject-specific statements in the evaluation instrument. Our aim in this exploratory study is to seek some general perceptions of its utility in the curriculum rather than study its effectiveness for use in a particular subject. Its utility value for specific subjects would be worthy of further study.

A teaching fellow who has extensive experience in the use of computers for learning helped to validate the survey instrument. He confirmed the face validity of the statements, and provided minor comments that were incorporated into the evaluation instrument. It was then piloted on a class of Secondary One students (N = 38) not included in the main study. Analysis of the data indicated that the overall reliability of the instrument, as reflected by Cronbach's Coefficient Alpha, was 0.86. This was more than the norm of 0.70 recommended by Nunally (1978) for good reliability of a survey form. It was also ascertained that the phrasing of the statements in the survey form was generally clear and unambiguous. The piloting also revealed the need to slightly refine two statements to better reflect the principal objectives of the survey.

Sample

The subjects in the main study were 67 students from two Secondary Two classes in Crescent Girls' School in Singapore. Their average age was 14 years. All of them have been using Tablet PCs for about a year. The medium of instruction in the school, as in all government schools in Singapore, is English. The school has a reputation for engaging in innovative teaching practices.

Procedure

The final version of the evaluation instrument was administered to two classes of Secondary Two students. Prior to the completion of the evaluation instrument, the students were briefed by the teacher about the purpose of the study and given instructions on how to complete it. The instrument took about five minutes for completion.

Data Analyses

Data were collated as one aggregate score for the purpose of extracting the following parameters: descriptive statistics, inter-item correlation, corrected item total correlation, and Cronbach's Coefficient Alpha.

Where appropriate, we have included data and information from the results of the school's internal study done in 2004 involving the entire Secondary One population (Crescent Girls' School, 2004).

IMPLEMENTATION ISSUES

The introduction of new technologies in any school setting often brings along an attendant set of problems which need to be addressed before it could be purposefully deployed for teaching and learning. This section surveys some of the important issues in this regard.

The Tablet PC initiative in Crescent Girls School is part of a pilot program called BackPack.NET which aimed to explore the feasibility of introducing Tablet PCs into the educational setting. It is a $200 million collaborative effort between the Infocom Development Authority of Singapore (IDA), Microsoft Singapore, industry partners, and schools. The initiative is supported by the Ministry of Education.

Business Model for Getting Each Student to Own a Tablet PC

Tablet PCs do not come cheap—a typical model costs about S$4,000. It is not easy to convince students (or rather, their parents) that the school bag containing textbooks, exercise books, and stationery be discarded in favor of a new technology which has not been proven but which has tremendous potential to augment learning. The right business model is therefore imperative if every student is required to own a Tablet PC.

The school held two rounds of meetings with parents to share its vision of how Tablet PCs can empower students to be better learners. The vendor, Fujitsu, came up with a novel financing scheme that included discounts and interest-free financing for the sale of the Tablet PCs. This helped to make the Tablet PC more affordable to students, who came from different socio-economic backgrounds. The foregoing initiatives helped to win the parents over—in fact, 95% of the parents wanted their children to own a Tablet PC. For the remaining 5%, their children were given the opportunity to sign out a Tablet PC when they come to school and store their work on a thumb-drive before returning the former at the end of the day.

The Tablet PC in use in the school is a Fujitsu Model T3010 weighing 1.9 kg. It has a display screen that is foldable and can be positioned in an angular manner for inputting text and other data.

Digitizing Textbooks

The use of Tablet PCs means that print-based textbooks are a thing of the past. Most of the recommended textbooks were therefore digitized and loaded onto the Tablet PC.

A vendor, Popular E-Learning, came up with a program called *Living Textbook* to automate the conversion from print media to digital media. The respective book publishers provided the pdf document of their books which Popular E-Learning converted to digital format. Typically, a 220-page book is equivalent to 128 Mb of memory size. Additional resources provided by the publisher to complement the books were also integrated by Popular E-Learning into the digital books. For example, in the case of the geography book, Earth Our Home, 414 Mb of additional resources, both textual and video clips, were provided by the publisher, Marshal Cavendish International, and integrated into the requisite sections of the e-book to provide added value to the learning experience, particularly for those with different learning styles. For all of the e-learning books, rich learning resources developed by the school's teachers have also been incorporated. Currently, textbooks in science, English, home economics and geography for Secondary One have been digitized. For Secondary Two, books on science, English, mathematics, home economics, design and technology and geography have been digitized.

Training

Even though students in Singapore are rather ICT-savvy, introduction of new technological delivery systems for teaching and learning necessitate the need for adequate training for both students and staff so that the full functionalities of the Tablet PC can be cognized for better utilization. If anything, the high level of ICT in schools in Singapore made the training somewhat easier for both students and teachers.

It is to be noted that the Lower Secondary Computer Education Programme, which provides for one hour of PC-based training each week, has already equipped students with the necessary skills to learn with computers. The training program for the students is spread over a year. Details are reflected in Table 1.

The school's IT Committee briefs students on issues such as code of conduct when learning with Tablet PCs, security of Tablet PCs, responsible use of Tablet PCs, and personal responsibility for their care. This helps to ensure that the downtime of the Tablet PCs due to improper use is kept to a minimum.

It is recognized that teachers need to be briefed adequately prior to the launch of the system. Two day IT seminars are held biannually in the school to hone the core competencies of teachers in IT. A unique feature of these seminars is the holding of concurrent sessions, which allow teachers to opt for a session that best meets their training requirements. A sharing session is an integral part of these seminars as it allows them to reflect on their practice. Training sessions on the use of specific applications software were conducted by a vendor, Heuristix Labs, and this was followed up in later sessions by the school's IT support team. Ongoing sessions include the twice-a-month sharing session for teachers using Tablet PCs—these allow them to learn from their peers and cognize best practices. Training on learning profiles are also conducted so that teachers can better understand issues related to students and thus customize appropriate learning packages for them. For example, in the case of visual learners, presenting

Table 1. Students' training program

Jul-Aug 2004	Tablet PC functionalities Microsoft Journal Microsoft One Note
Aug-Sep 2004	Living Textbooks
Sep-Oct 2004	Fun with Construction Virtual Classroom
Jan 2005	Artrage Living Textbooks Fun with Construction (enhanced)
Feb 2005	Microsoft Producer CrezSphere (Students' portal) Virtual Classroom (enhanced)
Mar 2005	Microsoft Movie Maker Microsoft Learning Gateway
Apr-May 2005	Adobe Photoshop Macromedia Flash Mindbook
May 2005	Macromedia Dreamweaver
Jul 2005	Advanced Macromedia Dreamweaver
Aug-Sep 2005	3D Maya

information in the form of mind maps, graphics and videos would facilitate better cognition of concepts.

A brainstorming session among teachers yielded ideas on how Tablet PCs can be used in the teaching and learning of the various subjects. These ideas formed the thrust of the applications that the school co-developed with IDA and Heuristix Labs. One example is the virtual classroom, an applications system that allows the teacher to wirelessly engage a class of 40 students. The Virtual Classroom allows the teacher to see whether all students have logged on, observe what each student is doing on her Tablet PC and, if need be, to shut down all their screens if he/she wants them to pay more attention to his/her oral presentation. Another example is the Fun with Construction application developed for making use of the inking technology in mathematics. The popularity of the latter application also spilled over into other subjects such as science and geography.

The school recognized that for the Tablet PC to be entrenched in teaching and learning, adequate structures need to be placed to support and encourage staff as they went about learning, experimenting and teaching with the new technology. A three-tier framework formed the basis of this structure. First, the Head of Department Instructional Programme provided teachers with the necessary pedagogical expertise and guidance. Second, an ICT Committee, comprising the Head of Department for IT and representatives of the different subject disciplines, provided coaching to staff who are new to the technology. Third, an IT support team provided help desk support and other assistance for classroom activities that need to leverage extensively on Tablet PCs.

Infrastructure

A wireless infrastructure was set up in the school by a vendor, CET Technologies, to provide Internet connectivity for the use of the Tablet PCs. Comprising 60 access points, it provides a 4 Mbps uplink and downlink connectivity within the school premises, both in built-up areas such as the school hall, library, and classrooms as well as in open areas like the Ecopond.

The school's communication portal, called CrezSphere, was set up in early 2005. A variant of the Microsoft Learning Gateway, CrezSphere allows students, staff, and parents to communicate with each other 24/7. Another advantage of CrezSphere is that it allows students to be part of a close-knit community even when they are geographically dispersed. For example, they could communicate with their teachers if they are doing a project at the science center, bird park, or zoo. The servers are hosted in the school.

Limited Battery Life

As Tablet PCs are not wired to the electricity grid, they have to be powered by batteries for operation. A typical battery, when fully charged, can provide sufficient power for 4-6 hours of operation, depending on the applications run. Interruptions were commonly encountered during the initial phase of deployment when students used the Tablet PCs intensively for a continuous period, thus causing the battery to discharge prematurely. Though this related mainly to batteries which did not have a full charge at the start of the day, realignment of the seating arrangement of the pupil is needed when she accesses the nearest power point to recharge the battery. To address this problem, simple solutions were offered—students were reminded of the need to fully charge their batteries at home before coming to school, advice on proper battery care was provided, and the option for purchasing at a discount an additional battery for back-up purposes was dangled. Also, when not accessing the wireless network, students were encouraged to switch off the Centrino card so that the resulting energy savings could translate into additional operating time for the Tablet PC.

RESULTS

Data related to the evaluation instrument are displayed in Tables 2-4.

For all the subscales, the readability statistics (Table 2) show that the statements in the evaluation instrument are well within the com-

Table 2. Readability statistics of evaluation instrument

Subscale	Flesch Reading Ease Index	Flesch-Kincaid Grade Level
Learning Environment	71.8	5.2
Effectiveness of Learning	71.8	5.2
On the Tablet PC	71.8	5.2

Table 3. Descriptive statistics and corrected item total correlations for evaluation instrument

A	*Learning Environment*	Min	Max	Mean	Std Dev	r
1.	I find that learning with Tablet PC is enjoyable.	3	5	4.48	0.58	0.55
2.	I find that learning with Tablet PC is fun.	3	5	4.48	0.58	0.62
3.	I get excited when I use Tablet PC for learning.	2	5	4.10	0.73	0.60
4.	I look forward to learning when I use Tablet PC.	2	5	4.16	0.68	0.56
5.	Using Tablet PC makes me want to study harder.	1	5	3.61	0.81	0.62
B	*Effectiveness of Learning*					
6.	Learning is generally more effective when I use Tablet PC.	2	5	3.68	0.74	0.63
7.	Using Tablet PC has increased my motivation to learn.	2	5	3.78	0.75	0.65
8.	I am more engaged in learning when I use Tablet PC.	2	5	3.87	0.78	0.71
9.	Tablet PC arouses my interest in learning.	2	5	3.97	0.80	0.77
10	I am more focused on learning when using Tablet PC.	1	5	3.25	0.85	0.67
C	*On the Tablet PC*					
11.	Tablet PC is an exciting media for use in learning.	3	5	4.42	0.63	0.66
12.	I prefer to be taught using Tablet PC rather than the traditional whiteboard/blackboard.	1	5	4.00	0.91	0.48
13.	I find that the Tablet PC is easy to use.	2	5	4.19	0.73	0.63
14.	When using Tablet PC, I do not miss my textbooks, paper, and pen.	1	5	3.74	1.01	0.65
15.	I like the portable nature of the Tablet PC.	1	5	4.30	0.88	0.54

r refers to corrected item total correlation

prehension level of the Secondary Two students used in this study.

Table 3 displays the descriptive statistics and the corrected item total correlation for the entire instrument. The means of the statements ranged from 3.25 to 4.48; standard deviations were in the range 0.58 to 0.88, except for Item 14 where it was 1.01. Cronbach's Coefficient Alpha for the entire instrument was 0.91, thus indicating high reliability of the overall instrument (Cronbach, 1951). Deletion of any one item did not change the value of Cronbach's alpha for the rest of the instrument—the only exception was Statement 9, where the alpha

Table 4. Descriptive statistics and reliability data for subscales of survey form

Subscale	N	Mean	Standard Deviation	Cronbach's Coefficient Alpha
Learning Environment	5	4.17	0.68	0.86
Effectiveness of Learning	5	3.71	0.78	0.88
On the Tablet PC	5	4.13	0.83	0.84

N refers to number of items in subscale

value was lowered by an insignificant 0.01. Eighty-six-point-seven percent of the item-item correlations were within the expected range 0.20 to 0.70 while the corrected item-total correlations were in the range 0.48 to 0.77. These further indicate a reliable overall scale.

When the data are viewed subscale-wise, it is of interest to note that the reliability of the subscales ranged from 0.84 to 0.86, which is still high, thus indicating good internal consistency (Table 4). Fraser (1989) has reported that for subscales which contain five items each, a satisfactory alpha value would be between 0.59 and 0.81. The means of the subscales were all above the neutrality point, thus attesting to the general truth of the statements.

Some of the interesting findings emerging from an examination of the results obtained from the evaluation instrument are summarized as follows:

- The use of the Tablet PC has promoted an environment conducive for learning. The means for Statements 1 and 2 are both 0.48, the highest for the instrument. Learning with Tablet PC is thus perceived by students to be enjoyable and fun. Indeed, students look forward to learning when using the Tablet PC—mean score of 4.16 for Statement 4.
- There is support to show that the effectiveness of learning has been enhanced with the introduction of Tablet PCs. The mean for this subscale is 3.71. Students find that learning has generally become more effective (mean score of 3.68 for Statement 6), that it has increased their motivation to learn (mean score of 3.78 for Statement 7), and that they have generally become more focused on learning (mean score of 3.25 for Statement 10).
- The introduction of Tablet PCs has meant the dispensation of the traditional paraphernalia of learning—books, pens, and exercise books. This has not caused significant resentment among the students (mean score of 3.74 for Statement 14). Indeed, students prefer to be taught using Tablet PCs (mean score of 4.00 for Statement 12) because they perceive it to be easy to use (mean score of 4.19 for Statement 13), portable (mean score of 4.30 for Statement 15), and an exciting media for use in learning (mean score of 4.42 for Statement 11). However, the large standard deviations for Statement 14 means that the variability of the responses (relating to whether students miss their textbooks, paper, and pen) span the entire Likert scale. Some further analysis is needed to determine whether the large standard deviation is a result of the data

Table 5. Analysis of Statement 14 in evaluation instrument

Likert Score	Number of Respondents	Percentage
5	16	23.2
4	23	33.3
3	22	31.9
2	5	7.2
1	3	4.3

Because of rounding off, the percentages do not add up to 100.0.

being skewed towards an end of the scale continuum. Table 5 presents the data in this regard.

It is clear from Table 5 that more than half of the students surveyed (56.5%) do not miss their textbooks, paper, and pen; 31.9% were neutral; and 11.5% do miss these accoutrements of learning. These data suggest that the large standard deviation for Statement 14 does not negate the essential positivity of this statement.

Some representative comments appended at the end of the survey form by students are instructive:

Tablet PC has indeed assisted in my learning as it allows me to log onto the Internet regularly to do more research on my own. (Eileen Foo, Secondary 2C3)

We can store many information and study materials in our Tablet PCs instead of carrying many books to school. (Joyce Ho, Secondary 2C3)

Tablet PC is a fun and enjoyable tool for learning although it would distract us sometimes. (Tang Yan Yue, Secondary 2G3)

I hope that we will get to continue with this enjoyable learning experience. I love my Tablet PC very much. (Lidya A, Secondary 2G3)

If it would be lighter.... (Wong Wan Mei, Secondary 2G3)

Tablet PC is fun and enjoyable to use (and it) arouses my interest to study. (Nur Hanisah Sukaman, Secondary 2G3)

It could be quite inconvenient when the Tablet PC's battery runs out and some may not have additional batteries as we tend to use [them] for quite a long time. (Ng Yali Shermain, Secondary 2C3)

They are really portable and convenient. (Wirda Syazwani, Secondary 2G3)

These comments by students further reiterate the general usefulness of Tablet PCs.

Data in Table 6 from the school's internal study suggest that the use of Tablet PCs has significantly empowered students to take charge of their own learning: 79% of the students are accessing online resources to augment their own learning more frequently and; 71% of students reported that they are accessing their digitized textbooks more often. The ready avail-

Table 6. Learning practices by students

Item	More frequently (%)	Just as frequent (%)	Less frequent (%)
Use of online materials for learning	79	18	15
Use of digital textbooks	71	22	7
Selection of own research areas	63	35	2

Table 7. Benefits of Tablet PC

Item	Agree (%)	Same (%)	Disagree (%)
Makes schoolwork more interesting	73	22	5
Allows me to be more IT savvy	72	23	5
Allows me to learn independently	69	26	5

ability of the digitized textbooks on the Tablet PC, coupled with the online resources just a click away, suggests that the blending of these learning resources in the same device does provide opportunities for students to make cross-references while learning. In addition, 63% of students said that they are motivated to select their own research areas for exploring a topic further.

Further support on the perceived advantages of Tablet PCs is provided by data in Table 7, which is also obtained from the school's internal study.

Clearly, the shift from whiteboard-centric pedagogy to the Tablet PC platform has made schoolwork more interesting for 73% of the students. The functionalities, features, and conveniences of the Tablet PC are factors in this regard. As the Tablet PC requires a suite of new skills to be learned by students via training, it is not surprising that it has made students more IT savvy (72%). For 69% of the students, it has provided them the impetus to learn independently.

One feature of the Tablet PC that has evoked tremendous interest among students is the use of inking technology. Eighty-two percent of them had used it for drawing diagrams, 78% for mathematical construction, and 75 % for note taking. This is not surprising as the inking feature of the Tablet PC is a distinctive tool that allows students to annotate notes in the course of a lesson; also this input can be digitized and subsumed into a retrievable index. Three softwares also evoked popular appeal: Internet (88%), note taking (81%), and e-communication (75%).

The permeation of ICT into work processes and the teaching environment has seen teachers making more use of ICT to enhance their teaching. The school's survey among teachers indicated several interesting findings in this regard (Crescent Girls' School, 2004):

- 50% of curriculum time per week in 2004 included the use of Tablet PC and its applications.
- 96% of the teachers found that lesson preparation has become more efficient and effective.
- 82% reported improvement in ease of sourcing for relevant online educational resources to make their lessons interesting.
- 86% of the teachers observed that the learning climate in the school for students has become more congenial after the introduction of Tablet PCs—they felt that students learned more in terms of content and that there has been a distinct improvement in the quality of learning by students.
- 77% of teachers were able to provide customized learning opportunities for their students, with 61% indicating that they were able to provide work pegged to the individual competencies of their students during curriculum time.
- 75% of the teachers felt that, with the Tablet PCs empowering students to take more control of their learning, their core duties have veered more towards facilitating and guiding students.
- The shared virtual workspace afforded by the Tablet PC has allowed opportunities for students to share their work with others in the class, in the process developing their presentation skills further. In fact, 72% of the teachers observed a definite improvement in the quality of the interaction among students in their classes.

These data suggest that the introduction of the Tablet PC has also empowered teachers to be more effective in their schoolwork.

DISCUSSION

There is a pronounced emphasis in today's society, which is highly networked and globalized, to provide opportunities for students to be imbued with the necessary suite of skills that would allow them to participate meaningfully in the networked economy (Tan & Subramaniam, 2001). A key consideration is ICT skills, which span a wide range.

In Singapore, ICT has permeated into the educational system in a pervasive manner. All schools are well equipped with computer laboratories and have high-speed access to the Internet (Tan & Subramaniam, 2000, 2003). About 30% of curriculum time is devoted to ICT-based lessons. The initiative to introduce Tablet PCs in a few selected schools in order to assess its viability in the educational setting is a conscious attempt to see how this new technology, which has been touted to offer tremendous potential for education, can enable teachers and students to be more effective in their respective roles. In due course, it is expected that an informed appraisal would help decide on their deployment in other schools. Early adoption of new technologies can confer strategic entry-level advantages from an educational standpoint.

In our opinion, the introduction of Tablet PCs with wireless support backs the digital culture of the younger generation. The impetus given to the acquisition of ICT skills in Singapore and the tendency for the younger generation to be hooked onto mobile phones, computer games, and gaming devices have spawned a generation that is at ease with new technologies. Leveraging on technologies to support learning is a natural extension to enlarge their learning horizon as well as promote the formation of learning communities. Consequently, there is a need for schools to engage students with new technologies with which they are at ease so as to foster further learning gains (Holloway & Valentine, 2003; Facer et al., 2004). It cannot simply be "action replays of earlier pedagogic encounters" (Hawkey, 2002). The consequent shift in 'technical praxis' (Roshcelle & Pea, 2002) would thus help to promote educational experiences, which are enriching for the younger generation, as well as present new pedagogical challenges for teachers.

The present independent study conducted by the authors using a validated evaluation instrument in one particular school shows that Tablet PCs can augment teaching and learning effectively. It supports and extends the findings reached by the school earlier. The internal consistency of the evaluation instrument, as reflected by the high value of Cronbach's Coefficient Alpha, attests to the overall reliability of the instrument and, thus, the conclusions reached. Students have generally warmed towards the introduction of the Tablet PC, in the process dispensing with their textbooks, exercise books, and stationery. An environment conducive for learning has been promoted by the Tablet PCs. Students perceive it as being fun and enjoyable. When learning becomes fun and enjoyable, it is no longer seen as a chore.

As the Tablet PCs are owned by the students and, with their reported perceptions of it being 'cool', the utilization factor must perforce be high since they use it both in class and at home. Thus, it is not surprising that the learning effectiveness has been enhanced—an observation also made by the school teachers. Another possible reason for the enhancement in learning effectiveness could be that the technology is well integrated into the curriculum. The introduction of Tablet PCs has seen students picking up new ICT skills through the school's comprehensive training program as well as honing these further as a result of ongoing experiences with their Tablet PCs. These experiences, besides serving immediate needs, also prepare them to be more effective participants in the future workforce of the networked economy. It is worth stressing that a recent study by McAvinia and Oliver (2002) provides support that the acquisition of such core skills needs to be encouraged at every level of the education system as it can help students to manage their own learning. Though such skills are distinct from specialist knowledge, the transferable aspect of these skills to new settings and situations does empower them in their pursuit of new knowledge in appropriate settings.

Perhaps the greatest benefit following the introduction of Tablet PCs is that it allows students to take charge of their own learning. Educationists have often reiterated the point that when students take charge of their own learning, the outcomes of the learning process can be optimized. The finding that 75% of the teachers surveyed by the school find that their roles have morphed more towards facilitating and guiding students is not without significance.

Of particular significance is the fact that the Tablet PCs have provided an environment for students to share their work with others, in the process further promoting interaction with their classmates. Further support is afforded by the observation made by 72% of the teachers in this regard. These have strains of the social constructivist approach, which maintains that learning occurs principally through a social interactive process (Pear & Crone-Todd, 2002). Another study by Zurita and Nussham (2004) on the use of handheld computers supported by a wireless network also noted that such environments promote a constructivist setting for social interaction and collaboration.

Veering from the traditional instructivist approach to the Tablet PC regime requires significant changes to be made to teaching practices, assessment frameworks, and curriculum re-organization. These are important in promoting the appropriate learning environment for students. The high level of ICT skills among students and teachers in Singapore and the good technology support available for the implementation of such initiatives are factors which have played an important role in this regard, and these may not be available in other settings. When teachers do not have the necessary background experiences and skill sets, implementation of such practices is unlikely to support student learning.

Though the findings from this study are very positive on the use of Tablet PCs in an educational setting, the small but significant percentage of responses against their use and disagreement over their effectiveness should not be construed as votes against the system, but as problems and (mis)perceptions which need to be addressed by research.

CONCLUSION

The present study conducted in a secondary school in Singapore shows that the use of Tablet PCs for education is viewed positively by both students and teachers. Desirable outcomes in the teaching and learning processes have been brought about with the introduction of Tablet PCs. It is suggested that the findings from this general study have some relevance for other educational institutions contemplating the introduction of Tablet PCs for their students.

ACKNOWLEDGMENTS

We gratefully thank Mrs. Lee Bee Yann, Principal of Crescent Girls' School, for permission to conduct this study in her school and for providing us with a copy of the school's internal report on m-learning@Crescent as well as detailed written responses to our queries. We also thank Ms. Goh Wee Suan for her assistance in administering the survey forms to the three classes of students used in this study.

REFERENCES

Anderson, R., Anderson, R., Simon, B., Wolfman, S., VanDeGrift, T., & Yasuhara, K. (2004). Experiences with a Tablet PC-based lecture presentation system in computer science course. *Proceedings of the 35th SIGCSE Technical Symposium in Computer Science Education* (pp. 55-60). Norfolk, Virginia.

Benson, J., & Hocevar, D. (1985). The impact of item phrasing on the validity of attitude scales for elementary school students. *Journal of Educational Measurement, 22*(3), 231-240.

Crescent Girls' School. (2004). *m-learning@Crescent.* Report 1.

Cronbach, L. J. (1951). Coefficient alpha and the internal structure of tests. *Psychometrika, 16,* 297-304.

Facer, K., Joiner, R., Stanton, D., Reidt, J., Hull, R., & Kirk, D. (2004). Savannah: Mobile gaming and learning? *Journal of Computer Assisted Learning, 20,* 399-409.

Fraser, B. J. (1989). *Assessing and improving classroom environment: What research says to the science and mathematics teachers* (no. 2). Perth, Australia: The Key Centre for School Science and Mathematics, Curtin University.

Golub, E. (2004). Handwritten slides on a Tablet PC in a discrete mathematics course. *Proceedings of the 35th SIGCSE Technical Symposium in Computer Science Education* (pp. 51-55). Norfolk, Virginia.

Holloway, S., & Valentine, V. (2003). *Cyberkids: Children in the information age.* London: Routledge.

Hawkey, R. (2002). The life long learning game: Season ticket or free transfer. *Computers and Education, 38*, 5-20.

Macdonald, J. (2003). Assessing online collaborative learning: Process and product. *Computers and Education, 40,* 377-391.

Magolda, M. B. W. (1992). Students; epistemologies and academic experiences: Implications for pedagogy. *The Review of Higher Education, 15*(3), 265-287.

McAvinia, C., & Oliver, M. (2002). But my subject's different: A Web-based approach to supporting disciplinary life long skills. *Computers and Education, 38,* 209-220.

Nunally, J. (1978). *Psychometric theory.* New York: McGraw-Hill.

Pear, J., & Crone-Todd, D.E. (2002). A social constructivist approach to computer-mediated instruction. *Computers and Education, 38,* 221-231.

Roschelle, J., & Pea, R. D. (2002). A walk in the WILD side: How wireless handhelds may change computer supported collaborative learning. In G. Stahl (Ed.), *Proceedings of the International Conference on Computer Support for Collaborative Learning* (pp. 51-60). Mahwah, NJ: Lawrence Erlbaum.

Schriesheim, C. A., & Hill, K.D. (1981). Controlling acquiescence response bias by item reversals: The effect on questionnaire validity. *Educational and Psychological Measurement, 41*(4), 1101-1114.

Smith, H., Higgins, S., Wall, K., & Miller, J. (2005). Interactive whiteboards: Boon or bandwagon: A critical review of the literature. *Journal of Computer Assisted Learning, 21,* 91-101.

Susskind, J. (2005). PowerPoint's power in the classroom: Enhancing students' self-efficacy and attitudes. *Computers and Education, 45,* 203-215.

Tan, W. H. L., & Subramaniam, R. (2000). Wiring up the island state. *Science, 288,* 621-623.

Tan, W. H. L., & Subramaniam, R. (2001). Science education: The paradigm shift. *Science's Next Wave,* (September).

Tan, W. H. L., & Subramaniam, R. (2003). Information and communication technology in Singapore: Lessons for developing nations on the role of government. In F. Tan (Ed.), *Advanced topics in global information management* (pp. 293-311). Hershey, PA: Idea Group Publishing.

Tan, W. H. L., Subramaniam, R., & Anthony, S. (2005). Cave automated virtual environment: A supercomputer-based multimedia system for learning science in a science centre. In S. Sharma & S. Mishra (Eds.), *Interactive multimedia in education and training* (pp. 327-349). Hershey, PA: Idea Group Publishing.

Zurita, G., & Nussbaum, M. (2004). A constructivist mobile learning environment supported by a wireless handheld network. *Journal of Computer Assisted Learning, 20,* 235-243.

Chapter XXIV
Using Technology to Create Children's Books for Students by Students

Lyn C. Howell
Milligan College, USA

ABSTRACT

This chapter describes a children's book project in which high school students used technology to create e-books for younger students. The benefits of the project for both younger and older students are discussed. Older students developed technology and writing skills; younger students developed letter writing and reading skills. The process is also detailed in the hope that others who might be interested in replicating the project in their own classroom would be able to do so.

INTRODUCTION

This project combined a high school English class writing instruction project with an elementary school unit on writing letters and served to develop both creativity and reading skills. High school students used a variety of technological resources including MSWord, PowerPoint, Paint, the Internet, and scanners to create electronic children's books. These were saved on CDs and floppy disks, and e-mailed to elementary students. Publishers report the growing popularity of electronic books (e-books), reporting double-digit growth in the sale of e-books. (Reid, 2002) When students in a class that used e-books exclusively were surveyed about the experience, 16 of the 19 students rated it "extremely positive" or "positive"; only one student rated it "negative" (Simon, 2002). This project gave students a chance to experiment with their own e-books.

Both sets of students used e-mail to communicate about the books. Technology integration enhanced this assignment for both groups of students. Older students developed their creativity and writing skills, and younger students improved their communication and reading skills.

TECHNOLOGY IN THE CLASSROOM

Arends (2004) defines some of the challenges for teaching in the 21st century as "teaching in a multicultural society, teaching for the construction of meaning, teaching for active learning, teaching with new views about abilities, and teaching and technology" (p. 9). This children's book project finds a way to meet each of those challenges.

The best use of technology is when it is an integral part of the lesson, not as a stand-alone piece or a separate, unique activity, but as a means of enhancing the learning opportunity. This is particularly true in a time when students are being tested yearly to ensure that they meet state standards. Bitner and Bitner (2002) suggest that "learning should be the impetus that drives the use of technology in the school. Its use can allow teachers and students to become partners in the learning process" (p. 95). Using technology simply to use technology separates it from the learning process. It is only when technology is a part of the context of the class that it is truly valuable. Tyack and Cuban (2000) contend that the use of computers and related technologies are potent tools for both teaching and learning. When teachers know how to integrate them appropriately into their classrooms, students benefit in a myriad of ways. Students benefit enormously from day-to-day interaction with technology when it is used to support and extend learning.

At every level, teachers are required to cover a wide variety of standards and course objectives; when technology is integrated with those requirements, it not only provides the opportunity to learn specific information, but also gives the learner tools to use to expand learning. In addition, for most students, using technology adds a level of fun and a sense of accomplishment to the experience. Goddard (2002) argues:

> *Computers in the classroom should support, not carry, the curriculum as a tool for real-world applications, inquiry composition, and communication. Integrating technology with the curriculum fosters creativity, which, in turn, can lead to classrooms where engagement is nourished and learning enhanced.* (p. 19)

But teachers often perceive difficulties with integrating technology. According to the Heller Report on Educational Technology Markets (2004), the three areas that teachers most often indicated as being a moderate to great barrier to using technology all had to do with time limitations: limited time to develop new activities that incorporate technology, limited time in the school schedule to conduct activities, and limited time to practice technology skills. With a little creativity, instead of developing new activities and trying to fit additional activities into an already full schedule, teachers can enhance current assignments by incorporating the use of technology into existing lesson plans. In addition, the teacher does not have to be an expert in all facets of technology. Students are anxious to figure out ways to use technology, and equally anxious to show others.

Giving students the opportunity to discover how to use a particular piece of software or allowing students to teach their peers is a valuable learning experience in and of itself.

Children's Book Project

One example of a learning activity that was adapted to integrate technology is a multi-age children's book project. Children's books are not just for children. Combining the simple children's book with technology can provide a wide range of learning experiences for both older and younger students while meeting a variety of learning objectives.

Boyd (2002) describes a cross-aged literacy program and its contribution to low-achieving ninth-grade students' development of perceived self-competence, autonomy, and relatedness. Sanacore (1981) describes a creative writing course in which high school seniors first become familiar with children's literature in order to identify and appreciate story devices used by professional writers. They then use the skills learned to create children's stories, practice storytelling techniques, and present them to preschool children. Sanacore indicates that the experience is a positive one for both seniors and children. My experience supports the findings from both studies.

Using some of the ideas from these programs, we gave older students a chance to author their own children's book. This was a successful project without the use of technology; adding technology not only improved the original project in several ways, but also added additional learning opportunities.

Almost everyone has a favorite children's book. These books bring back pleasant memories. They're bright and colorful; they're short and easy to read. And, they are a perfect vehicle to use to teach students about plot, characterization, sequence, and writing for a particular audience.

For years I required my tenth- and eleventh-grade Communication Skills students to create children's books. They would begin by looking through a variety of children's books, identifying the different types; some students even brought in their favorites from childhood. We would choose a few to read aloud, and students would comment on those they like or did not like and what makes the difference between the two.

While they were evaluating children's books, they were also beginning to correspond with elementary school children. Once a partner elementary school class was determined, my students began the process of writing letters back and forth with the elementary school students. High school students learned to ask good questions in order to find out what kinds of things their pen pals were interested in. Younger students learned to write letters using correct letter format.

Because they had a reason to learn the skills, students were conscious of trying to produce quality work. Siccone and Canfield (2003) point out that one cause of low motivation is a lack of meaningful goals. When students can see a purpose in the work, they are more willing to put forth the effort needed for quality work. In this case, since the younger students knew that they were writing to high school students who would actually read their letters and respond, the younger students were always careful to learn the correct format for writing letters such as the greeting and the close. They also took special care to look up the words they wanted to use in the letter to make sure that all were spelled correctly. In addition to getting a letter (for some students, the first one they ever received), the elementary school students knew that, in return for the letter they wrote, they would receive a book that was written especially for them by their high school pen pal.

High school authors were motivated to do their best work because they knew that an audience would see it. Avery and Avery (1993) found that students who have the opportunity to display their work are motivated to put forth their best efforts, and using technology gives students the ability to publish their work. Most assignments go to the teacher and back to the student. This assignment went on a class CD so that students all had access to each other's books, and it went to a classroom full of students who had expressed how eager they were to see the books. Students also realized that they could e-mail their work to family and friends.

The project also lends itself to intrinsic motivation for students: having the opportunity to make their own decisions about their work, being responsible for the outcome of their efforts is encouraging and motivating for students. Many researchers (Ames & Ames, 1991; Pardes, 1994; Slavin, 1991; Wlodkowski & Jaynes, 1990) have found that feeling a sense of responsibility for their work motivates students to be more involved in learning.

CREATING THE STORY

One of the objectives for this unit was to help students understand about writing for a specific audience. As a result, student authors not only had to adjust their language to the age group for whom they were writing, they also took care to incorporate the information from the letters in their stories. They often used the name of their pen pal as the main character in the story. A young boy who confessed that he was afraid of dogs might find himself the hero who helped a dog and was later rescued by that dog. A student new to the school might read about a fish who had to join a new school. A student whose family speaks Spanish at home might receive an autobiographical story written in Spanish about the author's experiences in coming to the United States.

After students came up with plots as a result of the letters from their elementary school pen pals, they made storyboards to help them decide how to divide their thoughts and the action of the story. They then had to decide what pictures to use to illustrate those thoughts. Before the use of technology, students used a variety of materials—from construction paper, to cardboard, to felt—in order to create their children's books. They carefully wrote out their story, added illustrations, and devised a way to hold the book together. Some used commercial binders, others used glue, and still others connected the pages with yarn or rope woven through holes in the margins of the paper. The final step was to share their books with each other through reading them aloud in class. After that, the books were boxed up and sent to their elementary school pen pals.

This was always a rewarding experience for both my high school students who made the books and the younger students who received the books. My students enjoyed being creative, learning to write for a specific audience, and feeling good about participating in a service-learning project that does something nice for someone else. The elementary students learned to write letters. Having their own special book encouraged them to read, and knowing that a high school student took the time to write it just for them was a very big deal for second- or third-grade students. Those books became their favorites; the teacher's major complaint was that students would hurry through their other work in order to have time to re-read their special books.

I always encouraged my students to include an author's biography and possibly a picture at the end of the book. These author's notes were often the highlight. "Look, Mrs. Dowdy, this girl is a cheerleader; that's what I want to do." "This guy plays in a band, how cool!" "Wow, she has an older brother just like me!" The second-grade students liked being able to make a personal connection with *their* author.

PROBLEMS WITH THE PROJECT

However, along with the positive aspects of this project, there were some negative ones. Because we were corresponding with students in another state, it seemed to take a long time for letters to travel back and forth from one school to another. It took nearly two weeks of class time to mail the letters, have the younger students write a response, and get those letters returned. This caused momentum to be lost and students to lose interest. Another problem was the expense. The cost of shipping the completed books was almost prohibitive, and, again, it took several days for the books to reach the students.

My students often spent a lot of money on the materials to make their books; they also poured a great deal of effort and creativity into the creation of their children's books. For those reasons, students frequently withheld their books from the shipment, preferring to keep their books rather than taking part in the service-learning project by sending them to the elementary school. Even professional publishers recognize that it is cheaper to publish a book electronically than the regular way. One company figures that it can sell an e-book for $6 when the book would regularly sell for $24 (Schuyler, 1998).

Since the books that were sent were irreplaceable, the elementary students were not allowed to check them out, and so they could only read their books during school hours. By using technology, we were able to overcome all of these obstacles.

OVERCOMING OBSTACLES WITH TECHNOLOGY

Our district provides free e-mail addresses for students, but even without that benefit, most high school students have their own e-mail addresses. We were able to speed up the mail dilemma by having each of my students write an e-mail that they sent to the elementary teacher's address. Their e-mails introduced themselves and asked general questions such as what games or television shows students like, what their hobbies were, or how many brothers and sisters they had. The second graders responded by typing their answers in Word; their teacher copied and pasted those answers into e-mails that she sent back. Thus, the process of exchanging letters, which had formerly taken several weeks, was completed in just over a week. Elementary students still learned correct letter format and were also able to practice using a word processing program. An added bonus was that the older students were able to read their younger pen pal's letters. In the past, it was often difficult to decipher second-grade students' handwriting. And, since they had received replies so quickly, my students were motivated to complete their books to be sent to these children.

Using technology to create the books gave students the opportunity to practice with those technologies as well as increase creativity. Schwartz (1986) believes that using computers in the English classroom can help students enjoy both reading literature and writing. The computer enables students to integrate the two skills.

Instead of using paper and pencil to create a storyboard, my students turned to PowerPoint. They set up the required number of pages (slides) and listed a main point or idea for the story on each one. Writing only a few words or a phrase on each slide enabled them, from the slide sorter view, to easily see how the story flowed. They could also easily rearrange, insert, or delete slides to make their story better.

After planning how to organize their books, students wrote their stories on the slides. Before using PowerPoint, students had to either type the story and physically cut and paste it on to the paper, or they had to carefully print the words since second-grade students usually have difficulty reading cursive writing. This was time-consuming, painstaking work. If a student made a mistake, he or she had to re-do the entire page. Students often settled for an inferior product because making changes was so difficult. With PowerPoint, students could type their story directly onto the page, enlarge the font to make it easier for second-grade students to read, and arrange the lines of print in any way they chose. If they made mistakes or thought of a more pleasing arrangement, it was easy to manipulate the elements without having to start over again. The spell and grammar check functions helped to ensure that words were spelled correctly and sentences were grammatically correct.

Once the stories were written, it was time to add color and graphics (Figure 1). Instead of trying to find pictures in magazines to illustrate their stories, students began by using the clip art provided with PowerPoint. They learned to change the size and shape, recolor the graphics to meet their needs, and even rearrange elements in the graphics. Using magazine pictures, students had to be content with pictures that were close to their vision. Using PowerPoint allowed students to choose exactly the colors and sizes they wanted. Using the Draw button at the bottom of the page, students could choose "flip horizontal" to change the picture from facing left to facing right. When they click on the picture, a "picture" bar appears with a "recolor" option. Clicking on that option opens a box with each color from the picture separated. Students can click on any color and change it to any other color they like so that each picture could be customized to fit the author's vision and needs.

When they found out that they could manipulate their illustrations, students became intent on using graphics that were precisely right for their stories, often combining elements. One example is the use of a picture on which students drew flowers with faces having the precise colors they needed and a foot about to step on one of the flowers.

They quickly moved to other forms of art as well. Some used the free graphics sites found on the Internet (remembering, of course, to cite the source in their Acknowledgements page); other students who could not find exactly what they wanted used the Paint program to create their own art. These students were able to create the exact picture they wanted in the perfect colors. Students often learned that they had talent that they did not know about. Students also volunteered to draw for others who were having difficulty. This part of the experience encouraged cooperative learning. The enthusiasm they demonstrated supports

Figure 1. Graphics from an e-book

Conrad's (1994) belief that cooperative learning is the key to increasing and enhancing student motivation. An increase in cooperative learning is an additional advantage to incorporating technology in the classroom. Ringstaff, Sandholtz, and Dwyer (1995) discovered that technology-rich classrooms encouraged students to work together more often and, as a result, become more interested in school.

Students also found pictures that they wanted to use from a variety of other sources. They scanned those pictures and inserted them into the PowerPoint slide. One student used a picture drawn by an elementary school student as the cover for his story about Buster, the dog (Figure 2).

Another option is the use of digital cameras. With digital cameras students can set up shots to illustrate their stories with pictures they take; one student used his cat as the main character and took pictures of it in a variety of situations. Students also took pictures of themselves with a digital camera and used them on the author's notes pages. These digital pictures were easy to insert into their PowerPoint slides.

Having a variety of options, limited only by their imaginations, set up an ideal atmosphere for constructivist learning for these students. Johnson and Johnson (1994) maintain:

> *Experiential learning is based upon three assumptions: that you learn best when you are personally involved in the learning experience, that knowledge has to be discovered by yourself if it is to mean anything to you or make a difference in your behavior, and that a commitment to learning is highest when you are free to set your own learning goals and actively pursue them within a given framework.* (p. 7)

This project met all three assumptions. Students became completely immersed in the project, looking for new ways to express themselves or demonstrate their creativity. They worked hard to find a way to create those aspects that they wanted in their books. If they were not sure how to achieve their goals, they asked for help from other students and experimented until they accomplished their vision. Finally, within the general framework of a children's book, students could choose to develop their plots and construct their books in any way they chose.

Figure 2. Buster's family

As Arends (2004) points out:

> *Learning from a constructivist perspective is not viewed as students passively receiving information from the teacher but instead as actively engaging in relevant experiences and having opportunities for dialogue so meaning can evolve...characterized by high levels of participation and engagement.* (p. 14)

The experience of finding ways to get the perfect graphic provided opportunities for both constructivist learning and cooperative learning.

As students became proficient at working with graphics in one form or another, they often took on the role of teacher, helping classmates to learn a new technique. Students helped each other to find the perfect example of clip art or demonstrated drawing techniques. One student became the "color expert," helping others to achieve just the right background or the perfect color for the dog's coat. Another student taught others to insert hyperlinks so that they could jump back and forth between pages. Arends (2004) contends that this type of cooperative learning gives students who have different backgrounds and different skills the opportunity to work interdependently. Each has a chance to demonstrate his or her abilities, and students learn to appreciate the skills of each other. In addition, it teaches cooperation and collaboration.

EMBELLISHMENTS

At this point, using the hardcopy technique, the book would be finished. However, using PowerPoint gave us additional options. Students added animation and sounds to their books: fish swam, car horns blared, and people who collided made a crashing sound.

Students also had the opportunity to demonstrate their dramatic side by creating an audio book. Students plugged in a microphone to the computer; in PowerPoint, under "Slide Show," they clicked on "Record Narration." They then read their book aloud, clicking on the slides so that each slide would change at the appropriate time, in sync with the narration. This gave students who received the books the option of reading the words themselves or following along, listening to the book as it was read to them. The ability to follow the words and hear them read was particularly helpful for English language learners in the elementary class.

Authors also added a page of questions at the end of the book. These questions were each hyperlinked to the page that contained the answer so that the second graders could test themselves on what they remembered about the story. The students clicked on the question which linked back to the page where the answer was underlined. Clicking on the underlined answer took them back to the review page. This activity was a game for the student and a ready-made review quiz for the teacher.

Since copies of the CD containing the books were easy to make and inexpensive, I was able to keep a copy for the classroom to use as a model for students in subsequent classes. Showing students examples of excellent work helps them to understand the characteristics of quality. Even examples of mediocre work can be advantageous. Asking students to distinguish between average and excellent work gives them insight into ways that they can improve their own work and gives them a goal to aim for. When students understand the criteria for excellence, they often produce higher quality products. Being able to envision an excellent product and seeing the difference between that and an average one can help students become more self-directed and more able to evaluate and improve their own work (McCombs, 1984).

Another advantage was increased availability. As a result of using technology, everyone could have a copy of the books. Instead of having to give up their books in order for them to be sent to the elementary school, my students could save their work on a floppy disk for themselves and send an e-mail copy to the teacher for their second-grade pen pal. Students still had the option of making a hard copy by printing their book on a color inkjet printer, and the elementary teacher could print as many copies as she wanted to for her students. But, the initial cost of materials for creating a book was reduced to the cost of a floppy disk. We were also able to put all of their books on one CD to keep a permanent record of the work the class completed, as well as sending the entire library of books to the second-grade class for their collection.

The cost of packing and mailing a large box of books and the time it took for the books to be shipped were both eliminated. All 30 books could be sent for free in a matter of minutes through e-mail. A bonus was that the elementary students now had access to all of the books, not just the one written for them. Those who had computers at home could ask the teacher to e-mail a book to their home computer so that they had access to it at any time, and the teacher did not have to worry about books being torn or lost.

The high school is just beginning to consider portfolios as part of the graduation requirement. Screen shots of the book or a copy of the floppy disk could become a part of that portfolio. If the school moves to electronic portfolios, it will be easy for a student to include the book as a part of his or her individual portfolio.

Adding the element of technology was particularly important for the boys in the classroom. Alloway, Freebody, Gilbert, and Muspratt (2002) found that boys had a strong interest in electronic and graphic forms of literate practice; they were more willing to 'do' literacy in active, public ways. This held true for the boys in my classroom. They were eager to demonstrate their technological skills. They enjoyed creating vivid pages and adding as many "bells and whistles" to their stories as they could. These students made extensive use of animation and sound to enhance their books, and were eager to demonstrate every technique they learned.

WORKING WITH A PARTNER SCHOOL

Elementary school students also benefited from the project. They learned to write letters and thank-you notes and use a word processing program. They also were motivated to read and had additional books for their classroom. It is necessary to find a partner who understands the benefits of technology and is willing to incorporate it into his or her lessons.

The students who participate can be from a school across town or across the country. My students partnered with a third-grade class 1,500 miles away. There are several ways to find partner schools. One easy way is to talk to people at conferences to find those teachers interested in a pen pal type of exchange and willing to incorporate technology into their lessons. Another way is through the Internet. Many schools have Web sites listing teachers' e-mail addresses. One way to identify a teacher who is interested in using technology is to go to www.teacherweb.com. There you can choose a state, select a school, and find the e-mail addresses of teachers who have classroom Web pages connected to the site. Teachers who create and maintain their own Web site are generally interested in technology and in helping their students to become comfortable with technology, and so are good candidates for partners in this project.

It is important to establish timelines with the partner school before introducing students to the project. If one school has a school vacation in the middle of the project, it stops the momentum for that class and frustrates the partner class since e-mails go unanswered for the duration of the time the class is not in session. If students lose interest in the project, it is very difficult to motivate them again.

CONCLUSION

Benson (2003) remarks that, with the emphasis on meeting state standards, teachers often comment that the boundaries between content area disciplines are blurring. She continues, "This is appropriate because if we want to prepare students for success in a complex world, we must see that discipline knowledge is not used in isolation. In fact, real world problems are, by their very nature, interdisciplinary" (p. 4). Teachers are sometimes hesitant to involve their students with technology because of this emphasis on meeting state standards. They feel that the time spent on technology will be time away from state requirements. In contrast to that fear, teachers are finding that incorporating technology can enable the teacher to meet a variety of standards within the same unit.

Using PowerPoint to create children's books fulfilled all of the objectives of the lesson for both age groups in addition to introducing students to tools of technology. Older students learned to write for a specific audience when they created books for younger students to enjoy. They learned to organize their thoughts, use vivid words, write description and/or dialogue, and they were able to practice creative writing. They also fulfilled the school's service-learning requirement by giving the books to the elementary school. In comparison to the paper-and-pen method, it was also quicker and less expensive, both in materials to create the books and in mailing costs, and it also provided increased access to the finished product.

Elementary students learned to use word processing, were able to send and receive messages through e-mail, and could use PowerPoint to read or listen to their books. The second-grade teacher reported that, "Students love to get on the computer; they just like to click." This project is an easy way to incorporate technology into the elementary classroom and, "Because it provides students with a different media for reading, it also encourages literacy."

There were several unexpected bonuses. One was the high school students' increased motivation. Students who had never been on time to class were in the computer lab at their computers and working before the bell rang. Students who had never said a word in class were either actively seeking information or were demonstrating a technique to other students.

> *Many teachers have observed that when given the opportunity to produce a tangible product or demonstrate something to a real audience (for example, peers, parents, younger or older students, community members), students often seem more willing to put forth effort required to do quality work.* (McTighe, 1996/1997, p. 7)

In addition, my English Language Learning students were able to participate fully since they understood the icons and could write the books in their own language, and the second-grade classroom was able to add books in their native language for its ESL students at no cost.

Finally, the second-grade students who received the books could now share them at home with their parents or siblings.

Incorporating technology into a good educational unit made it an outstanding unit for both the secondary and elementary students involved.

REFERENCES

Alloway, N., Freebody, P., Gilbert, P., & Muspratt, S. (2002). *Boys' literacy and schooling: Expanding the repertoires of practice.* Commonwealth Department of Education, Science and Training. ADD—Canberra: Australian Government Publishing Services.

Ames, R., & Ames C. (1991). Motivation and effective teaching. In L. Idol & B.F. Jones (Eds.), *Educational values and cognitive instruction: Implications for reform* (pp. 247-271). Hillsdale, NJ: Lawrence Erlbaum.

Arends, R.I., (2004). *Learning to teach* (6th ed.). Boston: McGraw-Hill.

Avery, C., & Avery, B. (1993). Eight ways to change a school: Reading promotions that work. *The Florida Reading Quarterly, 29,* 6-10.

Benson, B. P. (2003). *How to meet standards, motivate students, and still enjoy teaching.* Thousand Oaks, CA: Corwin Press.

Bitner, N., & Bitner, J. (2002). Integrating technology into the classroom: Eight keys to success. *Journal of Technology and Teacher Education, 10,* 95-100.

Boyd, F. B. (2002). Motivation to continue: Enhancing literacy learning for struggling readers and writers. *Reading and Writing Quarterly: Overcoming Learning Difficulties, 18*(3), 257-77.

Conrad, L. M. (1994). *Student motivation and cooperative learning.* (ERIC Document Reproduction Service No. ED407128).

Goddard, M. (2002). What do we do with these computers? *Journal of Research on Technology in Education, 35,* 19-26.

Heller Report on Educational Technology Markets. (2004). U.S. Department of Education's Integrated Study of Educational Technology. Retrieved March 27, 2005, from http://www.ed.gov/rschstat/eval/tech/iset/summary2003.pdf

Johnson, D.W., & Johnson, F.P. (1994). *Joining together: Group theory and group skills* (4th ed.). Englewood Cliffs, NJ: Prentice-Hall.

McCombs, B. (1984). Processes and skills underlying intrinsic motivation to learn: Toward a definition of motivational skills training intervention. *Educational Psychologist, 19,* 197-218.

McTighe, J. (1996/1997). What happens between assessments? *Educational Leadership, 54*(4), 6-12.

Pardes, J. R. (1994). Motivate every learner: How to replace motivation myths with strategies that work. *Instructor, 104*(1), 98-100.

Reid, C. (2002). Survey shows steady growth in e-book sales. *Publishers Weekly, 249*(30), 19.

Ringstaff, C., Sandholtz, J. H., & Dwyer, D. C. (1995). Trading places: When teachers utilize student expertise in technology-intensive classrooms. *Apple Education Research Reports.* Eugene, OR: International Society for Technology in Education.

Sanacore, J. (1981). *Creative writing and storytelling: A bridge from high school to preschool*. (ERIC Document Reproduction Service No. ED 225177).

Schuyler, M. (1998). Will the paper trail lead to the e-book? *Computers in Libraries, 18*(8) 40.

Siccone, F., & Canfield, J. (1993). *101 ways to develop student self-esteem and responsibility*. Boston: Allyn & Bacon.

Simon, E. J. (2002). An experiment using electronic books in the classroom. *Journal of Computers in Mathematics and Science Teaching, 21*(1), 53-66.

Slavin, R. (1991). Synthesis of research on cooperative learning. *Educational Leadership, 48*(6), 71-77.

Swartz, H.J. (1986). *The student as producer and consumer of text: Computer uses in English studies*. (ERIC Document Reproduction Service No. ED283211).

Tyack, D., & Cuban, L. (2000). Teaching by machine. *The Jossey-Bass Reader on Technology and Learning*. San Francisco: Jossey-Bass.

Wlodkowski, R., & Jaynes, J. (1990). *Eager to learn*. San Francisco: Jossey-Bass.

Chapter XXV
Electronic Portfolios and Education: A Different Way to Assess Academic Success

Stephenie M. Hewett
The Citadel, The Military College of South Carolina, USA

ABSTRACT

The use of electronic portfolios for students as an assessment tool is explored in this chapter. Portfolios have expanded from use in the arts and humanities to the field of education. Teachers, administrators, and students understand the benefits of portfolio assessment. The age of technology has improved the use of portfolio assessment by allowing the portfolio information to be transmitted and shared worldwide. No longer are portfolios limited to the single assessment of one person. Based on the current literature on electronic portfolios, the simplicity of creating electronic portfolios, the efficiency of collecting and organizing massive amounts of work, the ease of worldwide transmission of portfolio material, and the promotion of candidate-centered (student-, teacher-, professor-centered) assessment through the use of e-portfolios, the author hopes to promote the electronic portfolio as a beneficial way for the student, teacher, and professor to highlight their achievements for assessment.

INTRODUCTION

Portfolios have been used in a variety of careers including art, architecture, photography, and modeling. The portfolios are used to display a person's skills and talents. Portfolios have opened the doors to many opportunities for the person who has a professionally organized display of their finest works. Portfolios are strong representations of the identity of the person. Portfolios can help describe the person and his/her talents. Looking through a person's portfolio provides insights into the person's thinking and personality. The architect's portfolio allows the prospective builder to look at what the architect has designed and determine if the designs match the building ideas wanted by the builder. The artist can demonstrate types of art that he/she has produced to get commissions from buyers. Models get jobs with their best pictures and poses placed in a portfolio. Advertisers search for certain looks to sell their products to a target audience. The model's portfolio projects the different looks of the model so that advertisers can match their products with the appropriate model. Portfolios display the best products of the person creating the portfolio.

Education has been behind the times in the use of the portfolios. For many years, teachers have had only one way for students to show their knowledge. The one way typically used by teachers to find out what a student knows is through a test. Tests can be standardized or informal but only provide one way to show knowledge. A student must be able to read the questions and be able to write the answers to show what they know on a test. This type of assessment is a linguistic approach according to Gardner (1994), who has written many articles and books on the theory of multiple intelligences. Gardner believes that a person is born with not one, but several intelligences. The intelligences include intrapersonal, interpersonal, musical, linguistic, logical/mathematical, spatial, and artistic. Gardner states that a person has a dominant intelligence that is the best way to demonstrate a person's knowledge. For example, a student may be studying about crustaceans in biology. A typical test may not be the best way for a person to show what all they know about crustaceans. The person may be able to draw and label the crustaceans to visually show what they know or show his/her knowledge through some form of music.

The learner-centered philosophy of education recognizes the need to provide choices for students to show their knowledge. They may not be able to linguistically present their knowledge. Typical tests require the learner to show what they have learned through the linguistic intelligence. Recently, educators have begun using portfolios to allow students to show off their best works as well as show what knowledge they have gained. Educational portfolios give students choices in the way to present their knowledge. An educational portfolio is "a purposeful collection of student work that exhibits the student's efforts, progress and achievements in one or more areas" (Paulson, Paulson, & Meyer, 1991). A portfolio gives a broader picture of what a student has achieved than typical assessments. Herman and Michael (1999) argued that portfolios shift the balance from teacher-centered learning to student-centered learning. The student takes the responsibility of selecting what products best display their learning. The students also decide how to professionally present their materials.

The K-12 setting is ideal for the introduction of portfolios for assessment. In the early years, students love to show off their work to anyone who will listen! As a first-grade teacher, I endlessly sought ways to keep and display students' artwork and important writings and math assignments to show parents and students the progress that they made throughout the year. The introduction of portfolios at this level enables that collection and display of work, as well as provides a forum for students to present their portfolios to parents and other students. As the student grows and matures, he/she wants to show off his/her work, but does not know how to do it without looking childish. For elementary and middle school children, the portfolio offers the opportunity to display their work in a professional manner. High school students are able to add their creativity and computer knowledge as they develop their portfolios and even display those portfolios electronically. The opportunities for portfolios do not end in high school. Many colleges are requiring students to have writing portfolios to demonstrate their writing proficiencies.

Students are not the only people who can use portfolios to exhibit their achievements. Teachers and professors can create portfolios to highlight their educational careers. Teaching portfolios have been used over the years for teacher candidates to show their skills in planning and assessment. The portfolios are typically used to show growth in planning and professional knowledge. Professors use portfolios to collect and show their professional growth in teaching, scholarly activities, and service to the field. The quantity of materials for professors is massive and can be displayed in numerous file boxes, which are difficult to handle and maneuver to different locations.

The problem with the typical portfolio for K-12 students, teachers, and professors is the cumbersome nature of portfolios. The student portfolios can contain videos, artwork, essays, dioramas, and many other creative presentations and products. The teacher and professor portfolios can contain journals, conference programs, books, lesson plans, syllabi, and evidence of service activities. Keeping all of these materials together in a neat and organized fashion is a challenge within itself. The rise of technology has solved many of the problems associated with the collection and presentation of products. The electronic portfolio, sometimes called an e-portfolio, simplifies the collection and storage of the products. Written assignments, artwork, videos, music, and even tests can be saved to a disk and linked to a Web page to provide a simple way to store and organize the products.

Based on the current literature on electronic portfolios, the simplicity of creating electronic portfolios, the efficiency of collecting and organizing massive amounts of work, the ease of worldwide transmission of portfolio material, and the promotion of candidate-centered (student-, teacher-, professor-centered) assessment through the use of e-portfolios, the electronic portfolio offers a beneficial way for the student, teacher, and professor to highlight their achievements for assessment.

BEST PRACTICES IN ASSESSMENT

Positive effects on students' learning have occurred through the use of portfolio assessment (Santos, 1997; Sweet, 1993; Tierney, Carter, & Desai, 1991; Wolf & Siu-Runyun, 1996). As an assessment, a portfolio:

...matches assessment to teaching, has clear goals, gives a profile of learner abilities, is a tool for assessing a variety of skills, develops awareness of own learning, caters to individuals in the heterogeneous class, develops social skills, develops independent and active learning, can improve motivation for learning and thus achievement, is an efficient tool for demonstrating learning, and for student-teacher provides opportunity dialogue. (http://www.etni.org.il/ministry/portfolio/default.html)

Educators utilize portfolios to get the most effect from assessments by:

...encouraging self-directed learning, enlarging the view of what is learned, fostering learning about learning, demonstrating progress toward identified outcomes, creating an intersection for instruction and assessment, providing a way for students to value themselves as learners,

and offering opportunities for peer-supported growth. (http://www.pgcps.pg.k12.md.us/~elc/portfolio.html)

The research points to performance assessment as the most commonly used assessments.

Performance assessment is a dynamic process calling for students to be active participants, who are learning even while they are being assessed. No longer is assessment perceived as a single event...The purpose of assessment is to find out what each student is able to do, with knowledge, in context. (Wiggins, 1997, p. 20)

Performance is an umbrella term that embraces both ***alternative assessment*** *and* ***authentic assessment****. The term alternative assessment was coined to distinguish it from what it was not: traditional paper-and-pencil testing. There are even now distinctions within performance assessment, a distinction which refers to the fact that some assessments are meaningful in an academic context whereas others have meaning and value in the context of the real world, hence they are called 'authentic'.*

Performance assessment is ***a continuum of assessment formats*** *which allows teachers to observe student behavior ranging from simple responses to demonstrations to work collected over time. Performance assessments have two parts: a clearly defined task and a list of explicit criteria for assessing student performance or product.* (Rudner & Boston, n.d.)

Astin and others (2005) discuss the characteristics of best practices in assessments which include the following concepts regarding assessment.

Assessment:

- begins with education values;
- is most effective when the assessment reflects an understanding of learning as multidimensional, integrated, and revealed in performance over time;
- must have clear, explicitly stated purposes;
- looks at outcomes and experiences that lead to those outcomes;
- is ongoing and not sporadic;
- is enhanced when representatives from across the educational community are involved;
- illuminates questions that people really care about; and
- promotes change by being a part of a larger set of conditions.

As many states have turned to standardized tests to assess educational progress, the educational community recognizes the pitfalls of having a single assessment to evaluate educational progress. Suskie (2000) disputed the thought that one simple assessment could fairly assess educational progress. In the May 2000 issue of the *American Association for Higher Education Bulletin,* Suskie wrote an article for the Fair Assessment Practice Column entitled "Giving Students Equitable Opportunities to Demonstrate Learning." She stated:

An assessment score should not dictate decisions to us; we should make them based on our professional judgment as educators, after taking into consideration information from a broad variety of assessments.

The characteristics of best practice assessments match the characteristics of portfolios. The research literature on portfolios follows and demonstrates the connection of a best practice assessment and electronic portfolios.

RESEARCH LITERATURE ON ELECTRONIC PORTFOLIOS

The research shows that people generally see assessment as "something that is done to them" (Sweet, 1993). People have little knowledge in what is actually involved in the evaluation process. The authentic and performance-based measures of the portfolio clarify the assessment process. Lankes (1995) identified six different types of portfolios:

1. Developmental portfolios to document improvements and growth over an extended period of time.
2. Planning portfolios to identify weaknesses and develop an action plan to address those weaknesses.
3. Proficiency portfolios to demonstrate competencies and performances in a variety of areas.
4. Showcase portfolios to document a person's best work.
5. Skills portfolios to demonstrate proficiency of skills required to accomplish a variety of specific tasks.
6. Admissions/employment portfolios to show a person's capability to perform at an expected level in the specific setting.

Regardless of the type of portfolio assessment, it is a multi-faceted process with the following qualities:

- It is continuous and ongoing, providing both formative (i.e., ongoing) and summative (i.e., culminating) opportunities for monitoring progress toward achieving essential outcomes.
- It is multidimensional, i.e., reflecting a wide variety of artifacts and processes reflecting various aspects of the learning process(es).
- It provides for collaborative reflection, including ways for people to reflect about their own thinking processes and metacognitive introspection as they monitor their own comprehension, reflect upon their approaches to problem solving and decision making, and observe their emerging understanding of subjects and skills. (George, 1995)

The major characteristics of effective portfolio assessments are that they:

1. reflect clearly stated outcomes;
2. focus on performance-based experiences, and the acquisition of key knowledge, skills, and attitudes;
3. contain samples of work over an extended period of time;
4. contain works that represent a variety of different assessment tools; and
5. contain a variety of work samples and evaluations based on different sets of audiences. (George, 1995)

Kemp and Toperoff (1998) prepared a set of guidelines for portfolio assessment. The guidelines include a list of reasons that a person should use a portfolio to organize and display their works. Portfolios should be used to:

1. match assessment to the purpose for the collection of products;
2. clarify goals;
3. give a profile of the individual, including depth of quality of work available because of lack of time restraints, breadth of a wide range of products, and growth over time;
4. assess a variety of different skills;
5. develop awareness of one's own learning and growth;
6. cater to different learning styles and allow expression of different strengths;

7. develop active independent learning;
8. improve motivation and thus achievement;
9. demonstrate different types of learning and growth; and
10. provide opportunities for dialog with the evaluator.

Electronic portfolios are a collection of works made available on the Internet. Electronic portfolios are technology based. Electronic portfolios:

1. are used to foster active learning,
2. are used to motivate,
3. are instruments of feedback,
4. are instruments of discussion,
5. demonstrate benchmark performances,
6. are accessible, and
7. can store multiple media, are easy to upgrade, and allow cross-referencing. (Creating and using..., 2003)

Bull, Montgomery, Overton, and Kimball (2000) state that electronic portfolios promote self-evaluation and maximize the use of a variety of independent learning strategies. In addition, electronic portfolios serve as an excellent activity to enhance problem-solving skills (Barrett, 1994). The person takes responsibility for the compilation and organization of their work, therefore having a degree of control over the learning process (Campbell, Cignetti, Melenyzer, Nettles, & Wyman, 1997). The creation of the portfolio requires that the person become an assessor of his/her own products. The self-evaluation involved in selecting the most important and best representation of the person's work is a major instructional benefit. The final selection and display of the person's accomplishment demonstrates the process of learning that the person underwent to reach the present point. The key benefit of this type of assessment is that the portfolio enables the process of learning to be assessed as well as the products. The multiple sources of evaluation, in combination with the self-evaluation required in portfolio development, aids in the recognition of the person's strengths and weaknesses (Barrett, 2000).

The purposes, qualities, and benefits of electronic portfolios have been well documented in the research literature. Electronic portfolios have been portrayed in the literature as a best practice for authentic performance assessments. As a best practice, it is important that it is easy to create, or it will become an underutilized assessment tool.

CREATING ELECTRONIC PORTFOLIOS

Before beginning an electronic portfolio, one must understand and have working knowledge of a portfolio. In order to create an effective portfolio, one must first identify the goal of the portfolio. If the goal is to assess growth, then specific goals and criteria should be set so that everyone is clear as to what they are trying to attain. The goals will guide the selection of the inclusion of work in the portfolio. The required portfolio contents should also be identified. For a student, the portfolio may have examples of writings, auditory presentations, drawings, and reflections. Teachers may include sample lesson plans, pictures of classes, videos of teaching, examples of assessments, and reflections of teaching experiences. Professors would include evidence of teaching excellence, scholarly activities, and service to the field. Everyone must understand the purpose of collecting the work so that their selection process would meet the assessment goal.

Barrett (2002) identified five steps inherent in the development of effective electronic portfolios:

1. **Selection:** The development of criteria for choosing items to include in the portfolio based on established learning objectives.
2. **Collection:** The gathering of items based on the portfolio's purpose, audience, and future use.
3. **Reflection:** Statements about the significance of each item and of the collection as a whole.
4. **Direction:** A review of the reflections that looks ahead and sets future goals.
5. **Connection:** The creation of hypertext links and publication, providing the opportunity for feedback. The power of a digital portfolio is that it allows different access to different artifacts. The user can modify the contents of the digital portfolio to meet specific goals. As a student progresses from a working portfolio to a display or assessment portfolio, he or she can emphasize different portions of the content by creating pertinent hyperlinks. For example, a student can link a piece of work to a statement describing a particular curriculum standard and to an explanation of why the piece of work meets that standard. That reflection on the work turns the item into evidence that the standard has been met. (Barrett, 2002)

Once the goal of the portfolio is established along with the standards stating the criteria for success, the organization and planning of the portfolio must occur. The major task in the planning and organization of portfolios is deciding on what types of authentic products need to be included and how the works will be organized. Mandatory performance measures including specific evidence of knowledge, skills, and activities should be identified. A timeframe of works should also be developed. The collection of the samples over a period of time shapes the depth of the quality of the sample collection. The ability to collect work over a period of time makes portfolio assessment a needed alternative in a world where people recognize that knowledge can be shown in a variety of ways.

The next step is the actual collection of the work samples. The selection process is of as much benefit to the portfolio creator as the preparation of the works themselves. The collection of the works is dependent on the type of portfolio being established. The collection and selection of the products for the portfolio offer the creator a chance to look back and reflect on his/her own learning. As the person selects the work samples to be included in the portfolio reflecting on individual growth, he/she completes a self-assessment that identifies strengths and weaknesses. The weaknesses then become improvement goals. As the collection continues, the person strives to strengthen the weak areas and attain the improvement goals. This process deepens the person's understanding of his/her own growth and learning process. The reflections of the person's achievements and process to attain those achievements lead to the major self-evaluations, creating further growth.

The reflections become a crucial part of the work collection in the portfolio. The reflections may include learning logs, reflective journals, experience logs, and descriptions of the thinking processes employed in the development of the work samples. As the reflections continue while the portfolio is developed, the person's overall assessment of his/her progress is developed. People gradually recognize where they began and the progress that they have made. In any portfolio development, the recognition of growth and the overall self-assessment are major benefits of portfolio assessment.

As the development of a portfolio continues, problems also can emerge. The enormous amount of paperwork that is collected as work samples within the portfolio creates organization and storage issues. Large numbers of portfolios cannot be easily handled and maintained. Paperwork can easily be misplaced or filed incorrectly. Multimedia presentations, videos, and auditory tapes require specialized equipment to view and hear. With the increase in the use of technology, it became evident that a portfolio that could be electronically developed and maintained would solve many of the storage and maintenance issues.

Electronic portfolios have the capability of storing a large amount of works. Pictures, artwork, writing samples, videos, and auditory samples can be easily linked to the Web page. Work can be scanned that normally would not be computer based. In creating an e-portfolio, one must have knowledge of technology and the appropriate terms. The use of computers has necessitated a whole new language. Literacy of technology is essential to insure that the intended message is communicated when speaking to the expert or the novice. Key technology terms that need to be defined before the discussion of e-portfolios continues are:

- **Web Page:** A starting point for the electronic portfolio; a page designed to lead the viewer through the different products.
- **Hot Spot:** A blue underlined word or phrase that when clicked on links/opens the product the person is presenting.
- **Links:** The products that are part of the portfolio and must be opened by clicking on hot spots.

The electronic portfolio utilizes a Web page to link to the sample products. Most Internet service providers have free space that can be used for a personal Web page. There are many programs available to assist in Web page design and development. Netscape Communicator offers a free and easy program to develop a Web page. Creation of a Web page does not require a high level of computer skill and/or literacy; however, the level of technology skill and knowledge will increase as the person creates the e-portfolio. A trial-and-error creation process will result in the most professional e-portfolio design. The electronic portfolios allow the work samples to be posted on the Internet, where it can be easily accessed. The Web page is the start of the electronic portfolio. Hot spots are created with hyperlinks so that a hot spot is clicked on and it links the viewer with the work sample. Artwork and pictures can be scanned. Video and audio clips can be linked to the page. Electronic portfolios can be more comprehensive with the inclusion of multimedia presentations. The e-portfolio can be as basic as a Web page with links to Microsoft Word documents, to a more sophisticated page with multimedia presentations.

COLLECTION AND ORGANIZATION OF WORK SAMPLES

When collecting work samples, the portfolio should include examples with:

1. evidence of reflection and productive thinking processes;
2. growth and development in relation to the purpose;
3. understanding and application of key processes;
4. completeness, correctness, and appropriateness of work samples and reflections; and
5. a variety of formats to demonstrate performance and growth. (http://www.pgcps.pg.k12.md.us/~elc/portfolio5.html)

DeFina (1992, pp. 13-16) lists the following assumptions about portfolio assessment:

- "Portfolios are systematic, purposeful, and meaningful collections of students' works in one or more subject areas.
- Students of any age or grade level can learn not only to select pieces to be placed into their portfolios but can also learn to establish criteria for their selections.
- Portfolio collections may include input by teachers, parents, peers, and school administrators.
- In all cases, portfolios should reflect the actual day-to-day learning activities of students.
- Portfolios should be ongoing so that they show the students' efforts, progress, and achievements over a period of time.
- Portfolios may contain several compartments, or subfolders.
- Selected works in portfolios may be in a variety of media and may be multidimensional."

The actual collection of the work samples is not thought provoking. All work samples can be saved on a flash drive or other storage device to be accessed at any time. To link to the Web page, all documents should be saved on the same storage device (CD, flash drive, zip disk, or DVD). Saving documents on different devices is inefficient and limits the ease of linking to the documents from the Web page. Although collecting the work samples does not require much thought, the organization of the Web page and e-portfolio is extremely thought provoking. The selection of the products to be included in the portfolio is one of the best learning tools that an assessment can offer. While selecting the products, students not only review what they have done over a period of time, but can see the progress and growth that has been made. The review of the material is actually a wonderful study guide to help students recall what was studied and to review major concepts. The selection of material is the self-assessment component that is missing in most assessment devices. The self-assessment creates opportunities for the person to reflect on what he/she learned and the mistakes that have been made. People typically learn from their mistakes, making the process of creating an electronic portfolio an important learning process.

A frequently missed learning tool in creating an electronic portfolio is the increase in technology skills that go along with development of the Web page and electronic portfolio components. Presenting the material in the most professional manner is more difficult with an electronic portfolio than with a regular portfolio. An electronic portfolio offers so many different options. Creative fonts and word art, along with clip art and animations, can be added to the electronic portfolio to enhance its appearance. Instead of boxes of work samples, the electronic portfolio is a blank screen with the opportunity for creative integration of technology. The electronic portfolio is eye catching and provides an additional insight into the technology skills of the creator. For the typical young person, the technology skills are a given. According to *USA Today* (2003), 83% of college students regularly use information technology. With basic computer knowledge, the person can easily create and transmit his/her professional portfolio across the world.

TRANSMISSION OF THE PORTFOLIO

Electronic portfolios can be easily transmitted worldwide on the Internet. Once the Web page has been developed, the Internet service provider may offer a free Web page to upload the portfolio to the Internet. The Web page, all pictures, scanned items, Microsoft Word documents, video clips, and audio clips must be uploaded. For ease of uploading, save all items in one folder marked "portfolio." The opportunity to publish the portfolio on the Internet has its advantages. The Internet address of the Web page can be e-mailed and distributed to interested people worldwide. The assessment process can be enhanced with electronic portfolios by inviting educators from around the world to look at the electronic portfolio and assess it. The diversity of assessments gives a fair indication of the overall quality of the work.

If a person does not decide to make the electronic portfolio available to all on the Internet, he/she can save the portfolio on a CD or disk, make copies, and share with other interested people. When exploring the possibilities of information sharing through technology, it is noteworthy that electronic portfolios:

- increase opportunities for peer review,
- provide flexibility in the overall assessment process,
- serve as excellent introductions of the professionals who create the portfolios,
- encourage feedback from people outside of the education profession,
- eliminate barriers to parent participation in the schools, and
- ensure fair assessments based on a variety of assessors.

The purpose of the electronic portfolio dictates whether it will be saved on a disk or transmitted worldwide. For student assessments, saving on a disk to provide a type of scrapbook of learning does not require transmission of the portfolio on the Internet. In order for students to have peer review and share information with other students around the world would require transmission through the Internet. Teacher electronic portfolios can be used for awards and employment opportunities. The transmission of the documents through the Internet is beneficial for both the teacher and the future employer and award committees. An e-mail with a link to the electronic portfolio is a quick and easy way to share the professional documents with the awards committee members and future employers. The electronic portfolio also offers a unique way for professors to present their documentation for employment and promotion. The ease of transmission without the massive paperwork makes the electronic portfolio the best way to present professional documents.

Using electronic portfolios can change the way that people think about assessments. Instead of the typical dread that a person feels when undergoing an evaluation, the electronic portfolio offers an opportunity to "show off" his/her works. He/she has ownership in the collection and selection of works to be included, as well as the option of transmitting the portfolio to whomever he/she chooses.

CANDIDATE-CENTERED ASSESSMENTS

The Citadel School of Education has adopted a learner-centered philosophy of teaching. As written by Reilly (2000):

Learner-centered education is defined by McCombs and Whisler (1997, p. 9) as: the perspective that couples a focus on individual learners (their heredity, experiences, perspectives, backgrounds, talents, interests, capacities, and needs) with a focus on learning (the best available knowledge about learning and how it occurs and about teaching practices that are most effective in promoting the highest levels of motivation, learning, and achievement for all learners). This dual focus, then, informs and drives educational decision-making. Learner-centered education in this perspective embodies the learner and learning in the programs, policies and teaching that support effective learning for all students.

Administrators are responsible for developing, maintaining, and enhancing a school environment that enhances effective learning. They are also responsible for assuring teachers are knowledgeable about their students and how learning best occurs. Teachers are responsible for having classrooms that promote effective learning for all, as well as being familiar with the instructional techniques that promote effective learning for all. School counselors are concerned with improving both the conditions for learning (parent education, classroom environment, teacher attitude), as well as assisting each learner develop his/her fullest potential. The following five premises support these assertions:

1. Learners have distinctive perspectives or frames of reference, contributed to by their history, the environment, their interests and goals, their beliefs, their ways of thinking, and the like. These must be attended to and respected if learners are to become more actively involved in the learning process and to ultimately become independent thinkers.
2. Learners have unique differences, including emotional states of mind, learning rates, and learning styles, stages of development, abilities, talents, feelings of efficacy, and other needs. These must be taken into account if all learners are to learn more effectively and efficiently.
3. Learning is a process that occurs best when what is being learned is relevant and meaningful to the learner, and when the learner is actively engaged in creating his or her own knowledge and understanding by connecting what is being learned with prior knowledge and experience.
4. Learning occurs best in an environment that contains positive interpersonal relationships and interactions, and in which the learner feels appreciated, acknowledged, respected, and validated.
5. Learning is seen as a fundamentally natural process; learners are viewed as naturally curious and are basically interested in learning about and mastering their world. (Reilly, 2002)

With those premises in place, the electronic portfolio becomes the assessment tool of choice. Learners are encouraged to become actively involved in the learning process as they collect, select, design, and create their electronic portfolios. Their different perspectives of what they have learned are incorporated in the e-portfolio, as these include works that address a history of their learning, their beliefs, and their way of thinking. The electronic portfolio gives learners a chance to embrace their unique differences, including their abilities and talents. As learners create electronic portfolios, they maximize their learning potential. The products of the portfolio are relevant and meaningful to the learner, and actively engage him/her in creating and understanding the knowledge gained and experiences that lead to the knowledge gain. The portfolio serves as a tool in which the learner begins to feel appreciated, acknowledged, respected, and validated through sharing his/her portfolio with others. Learners' curiosities are challenged as they learn the technology skills required to produce an electronic portfolio. Learners begin to see the assessment process as meaningful and fun!

E-PORTFOLIO RESOURCES

The Internet offers a wide range of resources in the development, implementation, and assessment of electronic portfolios. Interesting Internet sites that are useful in the creation of electronic portfolios include:

- http://www.essdack.org/port/ — Tammy Worcester (2005) from Soderstrom Elementary School in Lindsberg, Kansas, presents information on why to use electronic portfolios, what to include in electronic portfolios, an assessment of electronic portfolios, and how to create electronic portfolios. She also includes examples and resources for electronic portfolio preparation.
- http://electronicportfolios.com/ — Dr. Helen Barrett is an internationally known expert on electronic portfolio development of all ages and has sponsored *electronicportfolios.org*. The site includes listservs, blogs, wikis, resources, and special topics on electronic portfolios.
- http://www.educationworld.com/a_tech/tech/tech111.shtml — Education World at Ashland University has a site entitled "Electronic Portfolios in the K-12 Classroom" (2005), which informs about what electronic portfolios are and how they can help the teacher and benefit the student including guidelines for developing personal portfolio.
- http://www.uvm.edu/~jmorris/portresources.html — "Electronic Portfolio Resources" provides examples of online portfolios, selecting electronic portfolio programs, and links to electronic software resources and electronic articles.
- http://eduscapes.com/tap/topic82.htm — "Electronic Portfolios: Students, Teachers and Life Long Learners" provides information on electronic portfolios, including what a digital or electronic portfolio is, how to develop an electronic portfolio, and how to integrate text, photos, diagrams, audio, video, and other multimedia into the electronic portfolio, along with examples of electronic portfolios.

Many people use rubrics, a criteria-rating scale, giving the teachers a tool that allows them to track student performance to assess the quality of the work. Rubrics describe the expectations of the portfolio and the rating system. The Pearson/Prentice-Hall Publishing Company includes a rubric for electronic portfolios (Rubrics for electronic portfolios, 2005). Not only does the rubric evaluate the content choice, organization, and personal reflections, but the rubric also assesses the creative use of technology. The rubric encourages the use of varied technology. There are other excellent examples of rubrics on the World Wide Web, but this template provides the flexibility for evaluators to develop the scoring to their needs and purposes.

The rubric template shows evaluation of the following elements:

1. the creative use of technology,
2. the content choice,
3. the organization and mechanics, and
4. the personal reflections.

Rubrics guide the students as they create their portfolios. When creating e-portfolio directions and rubrics, explore the World Wide Web to determine if there are existing directions and rubrics which will be useful in the planning for portfolio creation.

CONCLUSION

As computer technology progresses, so will the uses and benefits of electronic portfolios. As students and teachers continue to refine electronic portfolios, the assessment benefits will also continue to emerge. The research literature supports the use and benefits of electronic portfolios; and based on the ease of creating electronic portfolios, the efficiency of collecting and organizing massive amounts of work, the possibilities of worldwide transmission of portfolio material, and the promotion of candidate-centered (student-, teacher-, professor-centered) assessment through the use of e-portfolios, the electronic portfolio is becoming the most effective and efficient way to showcase and assess K-12 students', college students', teachers', and professors' academic growth and progress.

REFERENCES

Astin, A. W., Banta, T. W., Cross, K. P., El-Khawas, E., Ewell, P. T., Hutchings, P., et al. (2005). *9 principles of good practice for assessing learning.* [online]. American Association for Higher Education. Available at http://www.aahe.org/assessment/principl.htm

Barrett, H. C. (1994). Technology-supported portfolio assessment. *The Computing Teacher, 21*(6), 9-12.

Barrett, H. (2000). Electronic teaching portfolios: Multimedia skills + portfolio development = powerful professional development. *Proceedings of the Society for Information Technology and Teacher Education* (pp. 1111-1115). Charlottesville, VA: Association for the Advancement of Computing in Education.

Barrett, H.C. (2002). Retrieved April 1, 2005, from http://electronicportfolios.com

Bull, K. S., Montgomery, D., Overton, R., & Kimball, S. (2000). *Developing teaching portfolio quality university instruction online: A teaching effectiveness training program.* Retrieved February 9, 2002, from http://home.okstate.edu/homepsages.nsaf/toc

Campbell, D. M., Cignetti, P. B., Melenyzer, B. J., Nettles, D. H., & Wyman, R. M. (1997). *How to develop a professional portfolio: A manual for teachers.* Boston: Allyn and Bacon.

Chriest, A., & Maher, J. (Eds.). (2005a). *Why use a portfolio?* Retrieved March 31, 2005, from http://www.pgcps.pg.k12.md.us/~elc/portfolio.html

Chriest, A., & Maher, J. (Eds.). (2005). *How can portfolios be evaluated?* Retrieved March 31, 2005, From http://www.pgcps.pg.k12.md.us/~elc/portfolio5.html

Corbett-Perez, S., & Dorman, S. M. (1999). Technology briefs. *Journal of School Health, 6*(69), 247.

Creating and using portfolios on the alphabet superhighways. (2003). Retrieved October 15, 2003, from http://www.ash.udel.edu/ash/teacher/portfolio.html

DeFina, A. (1992). *Portfolio assessment: Getting started.* New York: Scholastic Professional Books.

Electronic portfolios in the K-12 classroom. (2005) Retrieved April 1, 2005, from http://www.educationworld.com/atech/tech/tech111.shtml

Gardner, H., & Boix-Mansilla, V. (1994). Teaching for understanding: Within and across disciplines. *Educational Leadership, 51,* 14-18.

George, P. S. (1995). *What is portfolio assessment really and how can I use it in my classroom?* Gainesville, FL: Teacher Education Resources.

Herman, L. P., & Morrell, M. (1999). Educational progressions: Electronic portfolios in a virtual classroom. *T.H.E. Journal, 26*(11), 86.

Kemp, J., & Toperoff, D. (1998). *Guidelines for portfolio assessment in teaching English.* Retrieved March 31, 2003, from http://www.etni.org.il/ministry/portfolio/default.html.

Lamb, A. (2002). *Electronic portfolios: Students, teachers, and life-long learners.* Retrieved March 30, 2005, from http://www.eduscapes.com/tap/topic82.htm

Lankes, A. M. (1995). *Electronic portfolios: A new idea in assessment.* ERIC Digest, ED390377.

Marklein, M. B. (2003). Students aren't using info technology responsibly. *USA Today,* (November 9).

McCombs, B. L., & Whisler, J. S. (1997). *The learner centered classroom and school: Strategies for enhancing student motivation and achievement.* San Francisco: Jossey-Bass.

Morris, J. (2005). *Portfolio resources.* Retrieved March 28, 2005, from http://www.uvm.edu/~jmorris/portresources.html

Paulson, L. F., Paulson, P. R., & Meyer, C. (1991). What makes a portfolio a portfolio? *Educational Leadership, 49*(5), 60-63.

Reilly, D. H. (2000). The learner-centered high school: Prescription for adolescents' success. *Education, 121*(2), 219.

Rubrics for electronic portfolios. (2005). Retrieved April 12, 2005, from http://www.phschool.com/professional_ development/assessment/rub_electronic_portfolio.html

Rudner, L., & Boston, C. (n.d.). *A long overview on alternative assessment.* Halifax, Nova Scotia: Norwood Publishing Company.

Santos, M. G. (1997). Portfolio assessment and the role of learner reflection. *Forum, 35*(2), 10-16.

Suskie, L. (2000). Fair assessment practices: Giving students equitable opportunities to demonstrate learning. *American Association of Higher Education Bulletin,* (May).

Sweet, D. (1993). Student portfolios: Classroom uses. *Office of Educational Research: Consumer Guide, 8.*

Tierney, R. J., Carter, M. A., & Desai, L. E. (1991). *Portfolio assessment in the reading-writing classroom.* Halifax, Nova Scotia: Norwood Publishing Company.

Wiggins, G. (1996). Practicing what we preach in designing authentic assessments. *Educational Leadership, 55*(1), 18-25.

Wolf, K., & Siu-Runyan, Y. (1996). Portfolio purposes and possibilities. *Journal of Adolescent & Adolescent Literacy, 40*(1), 30-36.

Worcester, T. (2005). *Electronic portfolios.* Retrieved April 13, 2005, from http://www.essdack.org/Port

Section III

Issues Related to Teacher Education and School-Based Matters

Chapter XXVI
Teachers and Technology: Engaging Pedagogy and Practice

Karen Cadiero-Kaplan
San Diego State University, USA

ABSTRACT

This chapter focuses on the pedagogy necessary in critically considering technology development for K-12 teachers and their students'. Three key questions frame this analysis: First, what literacies are necessary in the learning and use of technology? Second, what methods or processes are most effective in developing and implementing such technological literacy? Third, how do teachers best develop skills in using computers which ultimately ensure the development of skills and knowledge for students in classrooms? The chapter will illustrate, through the author's work in professional development settings, pedagogical techniques and strategies that have been implemented successfully in building capacity among new and experienced teachers in using technology for lesson planning, teaching enhancement, and portfolio development. Finally, Pailliotet and Mosenthal's (2000) four "I's" of media literacy—identity, intermediality, issues, and innovations—are utilized to analyze the case studies and provide a framework for implementing student-centered processes for technology use and literacy development.

INTRODUCTION

Presently many initiatives encourage both classroom teachers and faculty in higher education to acquire technology skills. The goal is to better prepare teachers to utilize technology in K-12 classroom settings. The focus of development is to have teachers go beyond the 'learning of technology' to 'integrating technology' into classroom practice. In the U.S. many institutions of higher education have been supported in this effort through the Preparing

Tomorrow's Teachers to Use Technology grants (U.S. Department of Education, 1999) and the Technology Innovation Challenge Grant Program (U.S. Department of Education, 1995). These projects have a common goal: to develop capacity for the use and integration of computer and media technologies by teacher education faculty, new teachers, and experienced K-12 teachers. Such initiatives further support universities and school districts in acquiring up-to-date hardware and software. This includes funding for computer labs, smart boards, smart classrooms, digital cameras, laptop computers, and many other related software and hardware tools.

The focus of these initiatives and resources on developing teachers' technology skills and knowledge provides an opportunity to analyze assumptions made regarding pedagogy and practice. Too often the elements that are taken for granted in professional development practices for technology are literacy and pedagogy. In order to consider these elements, I will address the following questions:

- What literacies are necessary in the learning and use of technology?
- What methods or processes are most effective in developing and implementing such technological literacy?
- How do teachers best develop skills in using computers, virtual environments that ultimately ensure the development of skills and knowledge for students in classrooms?

Most important to consider in responding to these questions is the changing role of the teacher. In many instances professors and teachers who are proficient in traditional teaching practices "are being challenged by the introduction of constructivist-based pedagogy, where the teacher's role is redefined from being that of a "sage on the stage" to a "guide on the side" (Ferneding, 2003, p. 89). This movement away from traditional pedagogy does not cause much argument from those advocating a student-centered approach for technology development. Such constructivist methods advocate the need for scaffolding new information, while accessing prior knowledge so students create knowledge rather than only being vessels to receive knowledge (Bruner, 1996). I support constructivist pedagogy and further advocate a more transformative pedagogy. A transformative pedagogy is one where students contribute to knowledge meaning and making, with teachers open to learning from students (Freire, 1993). This process, known as a co-construction of knowledge, is a key element missing from both teacher-directed and constructivist approaches for skills development. To this end, Ferneding (2003) states:

> *Within the context of integrating computer technology with print and oral traditions, teachers need to take a more active role in synthesizing the complex interactions of these various modes and mediums of communication, thus standing on the sidelines may not be an appropriate place for teachers.* (p. 89)

What is problematic with the "guide on the side" is that it places teachers in the role of a technician—one who only 'facilitates' as needed rather than making important cognitive and literate connections that are inherent in student-centered and more critical pedagogical approaches of teaching.

This chapter will focus on the processes necessary in considering technology development for K-12 teachers. Processes here are those that are critically student centered and are applicable to staff development and higher education settings as well as K-12 classrooms. This analysis begins first by defining literacy as it relates to technology and teaching. Second, based on work with pre-service and in-service teachers, processes of engagement will be shared which illustrate pedagogical techniques and strategies that have been implemented successfully in building capacity among new and experienced teachers in using technology for lesson planning, teaching enhancement, and portfolio development. These critical student-centered approaches include project-based learning, and teacher as learner and mentor models. The crucial point to consider is how capacity is built and continually supported to help foster development and extension to new and experienced K-12 teachers, while at the same time recognizing teachers as professionals who are part of the decision-making process. It is hoped that through this modeling, teachers will implement the same techniques into their classroom teaching. This learner-centered approach runs counter to technology-infusion models where university educators and school district technology trainers utilize a top-down model where "teachers [are] rarely consulted, though it [is] mainly their job to make it work in the classroom" (Tyack & Cuban, 2003, p. 248). This is not to imply that participants are not engaged in collaborative work as part of the learning process, but to recognize that the approach to initially 'develop' skills lacks a clear articulation of literacy in relation to technology and related processes of knowledge development.

LITERACY AND TECHNOLOGY IN SCHOOLS

For media literacy advocates, one of the great battles of course is trying to convince parents, administrators, and other teachers that film, television, advertising, and other media not only are 'culturally significant', but also that the messages they contain and convey should in fact be regarded as information, not simply entertainment. (Considine, 2000, p. 303)

We live in what is often referred to as the digital age, where schools and society support many forms of literacy, including computer, visual, media, digital, and information where "electronic forms of literacy are superimposed over oral and alphabetic forms and can be called multiliteracies" (Pailliotet & Mosenthal, 2000, p. xvi). Given this context, it is important to consider what counts as 'literacy' in schools and society along educational, social, and political lines. Such concerns have been the fuel for educational and political debates since the 1700s (Cadiero-Kaplan, 2004). The question "What is literacy?" is particularly important when considering the inclusion of media literacies (e.g., online discussions, e-mail, Web surfing, music download, etc.) alongside the cultural and linguistic diversity of the student population in most public schools today. An analysis of demographic characteristics of California K-12 students by the California Labor Department indicates that the number of English language learners in California has increased 100% since 1985 (about 1.5 million are children). It is estimated that the number of children who are English learners will increase to *3.0 million by 2010* (State of California, 2002). Thus, the combination of expanded notions of literacy along with a growing population of students whose family literacies encompass other languages, dialects, and cultures is an area that can add to a richer and deeper understanding of what it means to be literate in our media-driven and technologically laden society.

In stark contrast to our expanding notions of literacy both technically and culturally, processes and ideologies of schooled literacy are becoming more narrowly defined. This is occurring through articulated policies advocating standardized measures and tests as directed through No Child Left Behind (NCLB) and greater demands for teaching institutions to produce teachers that meet federal as well as state guidelines in the areas of reading, math, and science. So while many teaching processes are focused on basic skills of reading and writing, technology is also at the forefront of educational requirements and standards. At the same time, standardized views of 'normative' literacy practice (i.e., grade-level reading tests) do not represent the literacy that is part of all communities, where literacy is shaped more holistically as part of one's cultural, social, or native language environment. Further, the success or failure of many children and school literacy programs is determined by such definitions of 'normative' or standards-based practice (Powell, 1999) that take a one-size-fits-all approach. Therefore, being literate in today's society requires a re-definition of our views of what constitutes literacy processes, texts, and instruction (Pailliotet & Mosenthal, 2000).

Today's children and youth are more likely to use a variety of literacy strategies to process information and aid them "in learning to use the tools of meaning-making and reality construction in order to adapt to the world in which they find themselves" (Bruner, 1996, p. 20). This is supported by recent research, which determined "that children's digital literacies were emerging in ways that reflected their local circumstances, where children's home computing practices were strongly influenced by their technological, social, and school environments" (Ba, Tally, & Tsikalas, 2002, p. 32). As such, literacy—rather than being a subject matter to be taught—goes beyond basic skills to inform and transform subject matter (Considine, 2000; Mayer, 2000).

Literacy in schools is usually thought of as the ability to read and write, and is related directly to a notion of print literacy. Such a definition of literacy limits not only the materials that are utilized in school, but the processes of engagement and the final products that students produce to demonstrate knowledge. Left out of most teacher development and university teacher education curricula is the recognition of digital literacy. According to Gilster (2000), "Digital literacy is the ability to understand and use information in multiple formats from a wide range of sources when it is presented via computers" which includes the "ability to read with meaning and to understand" (p. 215). As such:

> *[N]ew electronic literacies [emphasis added] include computer literacy (comfort and fluency in keyboarding and computer use), information literacy (the ability to find and critically evaluate online information), multimedia literacy (the ability to produce and interpret complex documents comprising texts, images, and sounds), and computer-mediated communication literacy, that is knowledge of the pragmatics of individual and group online interaction.* (Warschauer, 2002, pp. 2-3)

Therefore children, adolescents, and teachers alike must not only acquire the skills of locating information on computers via the Internet, databases, and other search tools and have the ability to utilize various hardwares' for instruction, but additionally all must know how to apply and utilize these processes in classroom contexts.

The above supports Bailey's views of technology integration for English language learners (ELLs). Bailey (1996) argues that teachers need to examine both their own and their students' values, beliefs, and expectations for technology use as part of larger societal factors of language, race, class, gender, and ethnicity.

The point is that we do our work as educators within societal institutions that have perceived values and goals that may not always match those of the students we teach, whether they are K-12 students or teachers (Darder, 2002). In my own experiences as both a high school teacher of ethnically diverse students and as a university professor, I never introduce technology or any content without first giving students the space to share their own experiences, knowledge, and beliefs regarding the curriculum teaching processes, in this case teaching English as a Second Language. Through the following case studies and analysis, I will illustrate how these concepts have been implemented with teachers from an urban elementary school with little familiarity with technology use as part of their classroom practice. The second case study is derived from my work with bilingual students in a pre-service teacher education program.

PEDAGOGY OF TEACHING FOR TECHNOLOGY DEVELOPMENT

It has been reported that teacher preparation programs are not preparing teachers to teach effectively with technology (Moursund & Bielefeldt, 1999; Smerdon et al., 2000). Results of a survey given by the International Society for Technology in Education commissioned by the Milken Exchange on Education Technology found that "teacher educators do not model technology use in part because they lack the skills to do so, and they lack the necessary hardware and software" (Matthew, Callaway, Letendre, Kimbell-Lopez, & Stephens, 2002, p. 45). What these findings do not take into consideration is that more than skills, hardware and modeling are necessary in work with K-12 teachers in utilizing and integrating technology. These results take a deficit view of the problem—looking at the learner and their environment as opposed to examining the 'problem' from a critical pedagogical perspective. Tyack and Cuban (2003) state that the noted inability of teachers to embrace new technology "depends in good part on the ability of technologically minded reforms to understand the realities of the classroom" (p. 248) and the curriculum processes and technologies teachers utilize. That is, rather than identifying the problem as being a personal deficit, where individuals need to 'be developed', or as a resources deficit, such as a 'lack of equipment', we, as technology educators, need to examine issues of pedagogy. It is important to examine the knowledge and skills teacher educators and teachers bring to their content, and "enlist teachers as collaborators rather than obstacles to progress" (p. 248).

During the 1998-99 school year, I was involved in a school-based technology development initiative at a local urban elementary school that served predominately Latino and African-American children in grades K-6. The goal of the project was to work with novice computer-using teachers to integrate technology into their classrooms and curriculum. There were 27 teachers at this elementary school; only three had expertise in utilizing computers in classrooms. These three teachers were also experts in utilizing video editing hardware and software, and two had taught in school computer labs. The majority of teachers at the school had decided to forgo a large computer lab in order to have their classrooms equipped with two new iMac computers with connectivity to the Internet and television screens. However, during the project they received funding for a microcomputer lab with 30 iMac computers. My role was to assist in the professional development of these novice computer-using teachers. Since the district had placed the equipment and software in the school, they indicated to me that they would take responsibility for training and saw my role as being a 'coach'. The district trainers scheduled three days of training before the beginning of the school year; in the three days these teachers received intense instruction in the use of the desktop computers and software that included ClarisWorks word and database, Hyper Studio multimedia software, scanners, and related equipment and software, including digital cameras, recorders, and editing tools. The teachers also received training in the use of an e-mail program that was unique for Macintosh computers, a program that none of the staff had utilized prior to this training (Cadiero-Kaplan, 1999). Needless to say the teachers were overwhelmed with information, they were never asked what they already knew and/or were familiar with. The fact is most of these teachers had knowledge of PC environments and were familiar with Microsoft Word and Internet Explorer and Web-based e-mail software. The point of sharing this story is that at no time during the three-day training were the teachers engaged in discussing their experiences, knowledge, or skills; nor were they asked how they envisioned the computer's role in their classroom or teaching, or what grade levels or curriculum areas they were expert in. Thus, the approach taken in "teaching" or "training" these teachers was more teacher directed rather than student centered. This process, if taken as a 'model' for teachers to take back to their classrooms, goes counter to notions of student-centered learning and literacy development that the literature above articulated. It also did not reflect the inherent belief in student-centered approaches that these teachers practiced in their teaching on a daily basis.

Following the district training we had a staff meeting to debrief the experience and to complete a needs assessment survey (Appendix 1). At this initial meeting teachers were ready to turn the computers over to a lab; most felt inadequate, overwhelmed, and frustrated. Their initial excitement about the potential for the computers waned, comments included:

Well, I guess I don't know as much as I thought!

It seems we are going to be spending so much time learning how to use these computers that we won't have time to use them with the students.

We have so much to learn!

During this meeting I encouraged teachers to give the process some time. I let them know that the only software they really had to learn immediately was the e-mail program that would be utilized for all school and district e-mail correspondence. In addition, the teachers at the school with expert knowledge in computers assured their fellow teachers that they, along with me, would assist in setting up the computers and related software in individual classrooms. It was also important that every teacher complete the needs assessment survey to assist in developing a focus for the school technology plan, with the goal of developing staff training based on what the teachers already knew and what they wanted to learn. Additionally, we continued to assure teachers that the grant project pedagogically engaged a learner-centered model, one that would build upon their skills and areas of expertise in their K-6 curriculum content. It was our hope that teachers would then apply the model of development to their own classroom teaching.

Once we had initiated the discussion and addressed teacher concerns, the project took off. We held mini-workshops during staff development days that focused on the software and hardware teachers wanted to learn. Initially, the three technology teachers and myself led technology workshops. Eventually we asked teachers who we noticed using the computers in innovative ways to lead workshops on technology integration. Those who showed an interest in digital video requested sessions with school and district experts. By the end of the school year, teachers were utilizing technology and computers in a wide variety of ways and at varying skill levels. For example, after school we noticed teachers asking fourth- and fifth-grade students, who assisted our team in setting up the school network, to install software and assist in teaching teachers and several students to scan photos for classroom projects. One third-grade teacher utilized the e-mail program to set up an e-mail exchange with a class in another school. Along with traditional classroom jobs of line leader and sports equipment monitor was electronic messenger. The electronic messenger would check the e-mail in the morning, read the e-mail from their partner class, and would be responsible for sending a reply. A second-grade teacher set up a computer station for reader response. She had a short text for students to read, and they would write a response on the computer and include a clipart image. While these activities may appear to be far from complex, they were the first steps teachers took in utilizing the computers and software they were learning; by having them then share the process and student products through mini-workshops, other ideas were generated both within and across grade levels.

In the end, there was a safe climate that allowed teachers to engage technology without fear of failure. The teachers recognized that the technology leaders at the school considered their knowledge and thinking first, which are the key elements necessary to effect change (Bitner & Bitner, 2002). This project served as a model to several of the district technology trainers as well. In the end, the school site technology team began to work together with district personnel to utilize more needs assessments and teacher-led sessions in technology use and integration.

After completing this project I learned that the model of faculty development we had created at this elementary school followed a theoretical framework postulated by Pailliotet and Mosenthal (2000). In the section that follows, I will outline this framework and use it to analyze the project above and the case that will follow. Through the articulation of the framework, I will illustrate how it has been implemented and include specific applications to K-12 classroom settings.

DIMENSIONS OF MEDIA LITERACY: A CRITICAL THINKING PROCESS

Pailliotet and Mosenthal (2000) identify what they call the four "I's" of media literacy—identity, intermediality, issues, and innovations. They define the four "I's" as concepts that address the role of the learner and the context in which media is addressed. The authors define *media literacy* as the ability to access, analyze, evaluate, and communicate messages in a variety of forms. *Identity* refers to the impact media has on a person's construction of self and their personal behaviors. *Intermediality* assumes that "texts are laden with shifting meanings and reflect cultural ideologies" (p. xxvii). *Issues* looks at the social contexts in which media is constructed and implemented, including the use of assessment, standards, and social and institutional conditions. Finally, *innovations* are concerned with the changing conditions and conceptions that media literacy requires of individuals.

As a critical educator learning of this model, I realized that these were the factors I considered when designing and implementing education processes for engaging learners in developing technology skills. It mirrored my view that technology is a tool that is valuable as part of our learning processes and goals, but more importantly, I was encouraged to examine from a learning standpoint the four I's as part of my own teaching. I see now how these concepts were an implicit part of my work. In the following sections I will define the terms and analyze the teaching processes in which I have engaged. I will provide, through the case descriptions and analysis, a key framework for implementation of the four "I's" in educational settings.

Identity: The impact media has on a person's construction of self and their personal behaviors.

As was illustrated above in my work with the elementary school teachers, I first asked the teachers questions about their own computer knowledge and skills. That is, I began the training by first allowing the teachers' (learners') voices to be the starting point for discussion. This resulted in the teachers seeing that they had a valuable part in the development of their technology skills and use in their classrooms. Buckingham (1998) identifies four areas that teachers can utilize to establish a learner-centered dialogue around issues of media literacy:

1. Initiate a dialogue to consider the impact popular culture has on contributing to a sense of belonging for them and their students.
2. Ask teachers to consider the extent to which young people are 'active' or 'passive' in their relationships with media. How do they determine passivity and activity? Finally, to what degree are students and adults able to reflect on these media laden relationships?
3. What is it about media technologies that appeal to the interests of young people? Adults? Teachers? Children?
4. How do these processes of engagement impact or contribute to literacy events inside and outside the classroom?

Questions such as these give both the instructional leader and participants the opportunities to share not only varying definitions of media technologies, but establish clearly the varying levels of relationships that are mediated and interpreted by participants. It is through such dialogue that teachers and educators can see that their relationship and identity with media technologies may vary drastically from their students.

In the case of the elementary school above, "identity" was taken up in two ways, first by the technology trainer who took a teacher-directed approach by beginning with his beliefs of "what teachers needed to learn." The focus was on the curriculum to be covered. This scenario occurs daily as well in K-12 classrooms where teachers know what concepts they need to address in a given class period. Many may take the efficient route of following the curriculum guide and teaching from the text to the students. However, the difference that occurred for the teachers with whom I worked was significant when I began with their knowledge base, which is how and why they utilized technology in their lives. I extrapolated from their reality the connections with what they would need to know and learn, thus valuing their knowledge and experience. I also utilized this knowledge to inform the process for further development of technology skills and implementation in their own teaching.

Another example of this process is taken from my work with high school students and pre-service teachers in my college classes. I am always impressed at how fluent many of my students are in downloading music, creating personal CD-ROMs, and downloading and utilizing various forms of visual media. These are skills that I am not strongly proficient in and do not readily identify as an active process in my teaching. Conversely, I am proficient at creating graphs, tables, and various abstract images using drawing programs, a skill that most of my students have not engaged in or found relevant to their lives. The point here is that we begin to see that we all identify with media technologies, but as they relate to our own needs, interests, and values.

Case in Point

Before exploring the concepts of intermediality, issues, and innovations from Pailliotet and Mosenthal's model, I will give an overview of the class I teach for pre-service teachers. This case will set the context for the illustrations and analysis of these important concepts.

I teach a class for pre-service bilingual teachers who spend nine months of their year-long credential program in Querétaro, Mexico. When they return to San Diego, they are required to enroll in my methods class for teaching English as Second Language (ESL), which also has to meet some of the needs for their technology requirement. In approaching this class I have to first reflect on my own positionality. That is, first I am not fluently bilingual in Spanish, and second I have not met these students, who by the time they come to me have spent nine months living in another country and building community. Hence, I am new to a community of learners who have a shared common living and working experience. I then have to consider the students, most of whom were born in the United States and have Spanish either as a first language or acquired it as a second language. I also have to consider that only weeks before enrolling in my class in California, they experienced teaching in rural schools in Mexico where computer technology in schools was non-existent and, in fact, students had to teach utilizing mostly teacher-created materials and community resources. The class these students take with me is taught in what the university calls their "experimental classroom." This classroom has wireless laptops, both DOS and Macintosh operating systems. In addition the classroom has the latest in classroom devices including a smart board, two projection units, and the ability to link laptops to the projection podium, which has a wireless keyboard and remote mouse.

When students first enter this class, they are overwhelmed by the amount of technology, and one of the first concerns students voice is that they not only have never seen such access in their own learning experiences, but are aware that most of the schools where they have worked and will teach in most likely will not have even half of the technology hardware and software that we will utilize in this classroom. In response to this I explain that the purpose of using this room is not the expectation that they will implement all the technology in their teaching, but to give them exposure to the wealth and breadth of what can be done when it is present. Also, to indicate that if they see how the technology can be applied to the content, that of teaching ESL, then when they are employed in schools and funding for technology comes about, they will have some knowledge of what can be purchased and how it can be used.

Intermediality: Assumes that literacy and texts have varying contextual meanings that reflect cultural ideologies.

Pailliotet and Mosenthal (2000) define text as "laden with shifting meanings that are reflective of cultural ideologies" and further regard all "texts as constructions involving active, varied transactions of meaning making" (p. xxvii). This concept is closely related to the experience of the pre-service teacher with whom I work. I have found that in the past three years, the amount of knowledge these students bring about to computer use has dramatically increased. The first year I used this classroom with the Mexico program, only half of the students had experience using computers other than e-mail and Internet. Now almost all of the students have access to, or have, personal computers, and have used them for a variety of personal tasks including digitizing music and videos, digital photography, and for some, creating personal Web pages.

As a result, now more than before, these students bring with them a wide variety of literacies, both linguistic and media laden, and their literacies to this end far exceed my own. So, to ensure I approach my teaching to both value the skills, literacies, and knowledge they bring with them and further develop and nurture their growth, I teach my class utilizing both the methods they will need to develop English for non-native speakers in elementary settings, that is making learning contextual and real. To begin, I present the course content via the smart board and projection units utilizing PowerPoint, the Internet, and video as appropriate. Students utilize the laptops to access the course Web page that has templates for lesson planning and links to resources they will utilize for lesson planning activities and for research to be done in collaboration with peers. It is important to point out that in teaching this class I also use audiotapes, physical movement, butcher paper, and a variety of art materials that are also necessary tools for developing English literacy. Through the use of a variety of tools, students see that the technology is not the center of the instruction, which is imperative if I am to prepare them for the reality of the classrooms they will utilize, and more importantly to value all the literacies that they as well as their future students will bring.

Based on four class sessions where I have shared processes, both technological and other, students, in groups of 2-3, are required to develop two lesson plans for teaching English. For one lesson they need to incorporate PowerPoint and at least one other media technology. The students then lead the class to model the activity they have developed for a K-6 lesson. As a result, the students always teach me more about the capabilities of the software and hardware that I have only shared at a minimal level. For example, while I only model once the use of the smart board with PowerPoint and remote mouse and pen function,[1] the students apply this knowledge in a variety of ways to their ESL lessons. I believe this is a powerful model, for when these students enter their own classrooms, they may have children that bring more technology expertise and readiness to learn than they. For example, last semester one group had the class sit around the Smart board for a shared story read-aloud, and with students sitting on the floor one student read a picture book to the class and while reading and showing the pictures from the book, another student, using the remote mouse, flashed keywords and pictures for the parts of the story the students were to read aloud in unison. In another instance, a group had typed sentences with punctuation errors and asked students to come up to the smart board and rewrite the sentences, converted the handwriting to text, and then further identified spelling errors. They then used the remote keyboard to make corrections students called out. In addition to these activities, students have utilized the software and hardware to teach poetry, music, and dance. In the majority of instances, students also incorporated hands-on art, listening, speaking, and writing activities, utilizing the technology as only one of the elements of their lesson.

The outcome is, students begin to recognize the literacy functions enhanced by technology while at the same time going beyond the skill and modeling I have provided. Students tell me that they enjoy the class because the technology is not overwhelming and they have less fear due in part to my modeling not the bells and whistles, but the practicality of the software and hardware. As a result, these activities and dialogue add to the content and methods I am teaching. These processes also provide students with opportunities to learn from their peers; since each group works together, those who are more knowledgeable teach and model for others and each share their strengths, then each group in turn models different uses of the same equipment and software, thus providing a much stronger model, building capacity among the group, and furthering my own abilities, applications, and media literacy.

Students' comments regarding the use of the experimental classroom at the end of this class include:

LOVED all the access to technology we had, it was a great introduction to future classrooms (hopefully!). I would strongly recommend using that facility again if you can; it was really a great class.

I am not a computer person and when we came into the classroom the first day I was afraid, but you taught us in a way that I was not afraid. I learned a lot in this class about teaching English and using computers.

I really liked having the use of the tech room because it gave me some exposure to resources that are out there like the touch board computer. It was also helpful because, even though I don't think I'll have those resources in my classroom in the next year or two, I'm sure that fully wired rooms with laptops for each student are what is coming our way in the not so far off future.

It gave us a chance to practice using PowerPoint for lessons and made us think about how to incorporate technology in the classroom. I thought the room was great.

In the end, *intermediality* was enacted when students had to draw from my modeling and experience to reflect, practice, consider, and then implement lesson plan activities that brought in technologies they had not previously considered. Also, in most cases the fear or uncertainty students had about using such a technology-rich environment was reduced since they were able to engage with it both critically and without stringent requirements for technique; they had to incorporate the hardware and software as it related to the content they had to teach. Thus, they were learning through experimenting and sharing. As Semali and Pailliotet (1999) point out: "Intermediality puts theory into praxis, the process of dialogue, action, and reflection." Praxis here then requires that "students and teachers transform their newfound critical understandings into agency, positive acts and effect in themselves and others" (Semali & Pailliotet, 1999, p. 8). This was seen in my classroom when students at the end of the session had the opportunity to share their lesson plans and activities with all the electronic media attached through a Web-based discussion board. In this way all of the students had access to the varying forms, models, and activities they created and viewed throughout the course.

Issues: The social contexts in which media is constructed and implemented, including the use of assessment standards as they interact with social and institutional conditions.

Issues reflect the ever-changing roles, understandings, and contexts of media literacies within socio-political and institutional contexts (Pailliotet & Mosenthal, 2000; Tyack & Cuban,

2003). Within this context, educators need to consider issues of assessment and performance standards, technology applications, and the "social and institutional conditions of schools that foster or prevent change" (Pailliotet & Mosenthal, 2000, p. xxviii). Such issues were at the forefront of the work I was involved in with the elementary school staff development for technology. The impetus for the project came from an institutional need to meet technology standards and to ensure that teachers were implementing technology. Further, funds for technology in-service and development that were available for that project were part of a growing trend and requirement to meet society's need to have more citizens with the ability to be contributors to the workforce. As Levy and Murnane (2004) point out, "Technological changes and outsourcing have left American workers who lack strong skills unable to earn a decent living" (p. 728). The skills they refer to here are technology skills, which are important. However, as part of examining issues, teachers must be critically aware to see if these skills to be developed will place their students at an advantage. This requires asking if students are being prepared for high-level or low-level careers. Therefore, it is imperative to consider not just the skills, but the motivation and rationale for skill development and the buy-in of those who are to be taught since "externally mandated change can never substitute for internally motivated change" (Pailliotet & Mosenthal, 2000, p. xxviii). Again, my work with the elementary school is a prime example of what began as 'externally' motivated was rejected, but once there was internal motivation, then there was more engagement at greater depth and interest. In the end, it is the "skill and attitude of the teacher that determines the effectiveness of technology integration into the curriculum" (Bitner & Bitner, 2002, p. 95).

Innovations: The changing conditions and conceptions that media literacy requires of individuals.

This last concept takes us to the future—that is, as teachers and educators who take up constructivist and critical pedagogies, we are paving the way for our students to be the innovators. It is only when we become part of a learning community, not only with other teachers and experts, but also with our students and the teachers with whom we work, that we become innovators. Mosenthal (2000) has illustrated, and I have found in my own teaching, that "learners update their knowledge in an applied domain that is represented both linguistically and visually" (p. 358). This occurs when students are engaged in learning situations that provide a deep context that is more than student centered and goes beyond the teaching of 'basic skills' or 'literacy', but is rich in concrete experiences. In such learning environments educators recognize the multiple ways students learn and engage in tasks that are derived from the literary experiences and knowledge students bring with them. This requires being aware that these experiences may not be the same as those of the teacher. At the same time, the history and knowledge our learners bring may be rich in ways not immediately recognizable; their experiences and literacies could be based on a native language other than English, or a different culture, ethnicity, class, social cultural context, or learning experience that we have yet to encounter. When these factors are taken into consideration in the teaching of technology, an empowering and motivating learning environment is in place. The case of the preservice teachers illustrates this concept, since my own lived and learning experiences are inherently different from my students, who come from a variety of geographical areas and many possess a native language and ethnicity different from my own.

Through these examples and discussion, I have attempted to illustrate the power of expanding notions of literacy and valuing students' voices and experience for developing capacity and skills as future teachers in a technological age. The activities students engaged in throughout this four-week session were student centered and required me to do more than facilitate, but to work with students in small groups as they negotiated both the content of the course and the skills they needed to integrate technology into their lessons. In addition, we had real discussions about the realities of school and society, issues of access and equality for both our language minority students in mainstream culture, and issues related to access to technology in the schools where they worked. Creating the space for these critical discussions was important because without dialogue these students would have *felt* the digital divide, but would have lacked strategies to develop greater access for themselves and their future students. At the end of the class, I share Web sites and resources for teacher mini-grants and K-12 technology initiatives that they as future teachers can pursue. Thus, literacy in this classroom does fulfill Bruner's (1996) definition of media literacy where students are learning and applying the "tools of meaning-making and reality construction in order to adapt to the world in which they find themselves" (p. 20).

NEXT STEPS OF ENGAGEMENT

At all levels of education, from K-12 to higher education, when teachers begin to engage students in technology learning that is critically student centered, it has an impact on how technology is considered within our culture. In a culture impacted upon daily by media technologies, it is imperative to develop attitudes that question the role of technology. Especially within the media, young people in our society develop their media literacy before even entering school. Freire and Macedo (1987), when addressing literacy development, speak to the need of *reading the world to read the word*. This process is true not only when speaking of print literacy, but of any literacy that involves cognitive and conceptual development. To this end, I strongly believe, and have experienced, that innovation occurs when starting with the learner's reading of the world and linking it to the word. Whether the words are related linguistically to nouns, verbs, and adjectives, or technologically to bits, bytes, and modems, all students come with literacy, knowledge, and experience in the technology-rich world that we share. This view recognizes that we do not experience the world in quite the same way, so it becomes the responsibility of the teacher to do more than teach and test skills. It requires us to make connections and bridge knowledges and literacies in our classrooms, thus leading to a transformative and inspiring pedagogy—a pedagogy that calls for "intercultural learning that is responsive to the economic, scientific, environmental, and cultural realities of today's world" (Cummins & Sayers, 1997, p. 9). In order to leave no child behind, we must engage all learners, both teachers and students alike, and forge new pathways of learning together into the future.

REFERENCES

Ba, H., Tally, W., & Tsikalas, K. (2002). Investigating children's emerging digital literacies. *Journal of Technology, Learning, and Assessment, 1*(4). Retrieved from http://www.jtla.org

Bailey, J. (1996). Teaching about technology in the foreign language class. *Foreign Language Annals, 29*(1), 82-90.

Bitner, N., & Bitner, J. (2002). Integrating technology into the classroom: Eight keys to success. *Journal of Technology and Teacher Education, 10*(1), 95-100.

Bruner, J. (1996). *The culture of education.* Cambridge, MA: Harvard University Press.

Buckingham, D. (1998). Introduction: Fantasies of empowerment? Radical pedagogy and popular culture. In D. Buckingham (Ed.), *Teaching popular culture: Beyond radical pedagogy* (pp. 1-17). New York: Routledge.

Cadiero-Kaplan, K. (1999). Collaborative technology development: A staff development model for integrating computers into school curriculum. *Proceedings of the Annual Conference of the Association for the Advancement of Computing in Education, Society for Information Technology (SITE),* San Antonio, Texas.

Cadiero-Kaplan, K. (2004). *The literacy curriculum & bilingual education: A critical examination.* New York: Peter Lang.

Castells, M. (1998/2000*). End of millennium* (2nd ed.). Malden, MA: Blackwell.

Considine, D. (2000). Media literacy as evolution and revolution. In A. W. Pailliotet & P. B. Mosenthal (Eds.), *Reconceptaulizing literacy in the media age* (pp. 299-327). Stamford, CT: Jai Press.

Cummins, J., & Sayers, D. (1995). *Brave new schools: Challenging cultural illiteracy through global learning networks.* New York: St. Martin's Press.

Darder, A. (2002). *Reinventing Paulo Freire: A pedagogy of love.* Boulder, CO: Westview Press.

Ferneding, K. (2003). *Questioning technology: Electronic technologies and educational reform.* New York: Peter Lang.

Freire, P. (1993). *Pedagogy of the oppressed.* New York: Continuum Publishing.

Freire, P., & Macedo, D. (1987). *Literacy: Reading the word and the world.* Westport, CT: Bergin & Garvey.

Gilster, P. (2000). Digital literacy. *Technology and Learning* (pp. 215-228). San Francisco: Jossey-Bass.

Levy, F., & Murnane, R.J. (2004). A role for technology in professional development? Lessons from IBM. *Phi Delta Kappan, 85*(10), 728-34.

Matthew, K., Callaway, R., Letendre, C., Kimbell-Lopez, K., & Stephens, E. (2002). Adoption of information communication technology by teacher educators: One-on-one coaching. *Journal of Information Technology for Teacher Education, 11*(1), 45-62.

Mayer, R.E. (2000). The challenge of multimedia literacy. In A. W. Pailliotet & P. B. Mosenthal (Eds.), *Reconceptaulizing literacy in the media age* (pp. 363-376). Stamford, CT: Jai Press.

Mosenthal, P. B. (2000). Assessing knowledge restructuring in visually rich, procedural domains: The case of garbage-disposal repair writ/sketched large. In A. W. Pailliotet & P. B. Mosenthal (Eds.), *Reconceptaulizing literacy in the media age* (pp. 357-362). Stamford, CT: Jai Press.

Moursund, D., & Bielefeldt, T. (1999). Will new teachers be prepared to teach in a digital age? *Research study by the International Society for Technology in Education, commissioned by the Milken Exchange on Educational Technology.* Retrieved from http://www.mff.org/pubs/ME154.pdf

Pailliotet, A.W., & Mosenthal, P.B. (Eds.). (2000). *Reconceptaulizing literacy in the media age*. Stamford, CT: Jai Press.

Semali, L., & Pailliotet, A.W. (1999). *Intermediality: Teaching critical medial literacy*. Boulder, CO: Westview.

Shetzer, H., & Warschauer, M. (2000). An electronic literacy approach to network-based language teaching. In M. Warschauer & R. Kern (Eds.), *Network-based language teaching: Concepts and practice* (pp. 171-185). New York: Cambridge University Press.

Smerdon, B., Cronen, S., Lanahan, L., Anderson, J., Iannotti, N., & Angeles, J. (2000, September). *Teachers' tools for the 21st century: A report on teachers' use of technology* (Report No. NCES 2000-102). Washington, DC: U.S. Department of Education, National Center for Education Statistics.

State of California. (2002, May). *Department of Finance, E-1 city/county population estimates, with annual percent change, January 1, 2001 and 2002.* Sacramento, CA.

Tyack, D., & Cuban, L. (2003). Teaching by machine. *Technology and learning* (pp. 247-254). New York: Rowman & Littlefield.

U.S. Department of Education. (1995). *Technology Innovation Challenge Grant Program.* Retrieved from http://www.ed.gov/programs/techinnov/index.html

U.S. Department of Education. (1999). *Preparing Tomorrow's Teachers to Use Technology.* Retrieved from http://www.ed.gov/programs/teachtech/index.html

Warschauer, M. (2002). A developmental perspective on technology in language education. *TESOL Quarterly, 36*(3), 453-475. Retrieved November 2, 2004, from http://www.gse.uci.edu/markw/developmental.html

ENDNOTE

[1] The ability to write or print and convert the print to computerized text.

APPENDIX 1: COMPUTER TECHNOLOGY SURVEY

COMPUTER TECHNOLOGY SURVEY

Name:______________________ Room # ___________ Grade:__________

Please complete the following survey and return by September 2nd

1. Do you own a computer? □Y □N
   Brand of computer ______________________________

2. What do you primarily use your home computer for? ______________________________
   ____

3. What software programs are you proficient in using? ______________________________
   ____

4. What software programs are you familiar with? ______________________________
   ____

5. What software programs would you like to learn to use? ______________________________
   ____

6. What is your opinion about computer use in the classroom? How do you currently utilize computers in your curriculum ______________________________
   ____
   ____

7. If you do not utilize computers in the curriculum, how would you envision incorporating this technology if it were available? ______________________________
   ____
   ____

8. Do you have an e-mail account? □Y □N
   E-mail address ______________________________

9. Do you have an Internet service provider at home? □Y □N
   E-mail address ______________________________

10. What Internet software do you most commonly use? □Netscape □Explorer
    Other ____

11. Circle the answer that is most true as to how often you use the Internet?
    □ Never
    □ Once a week
    □ Daily
    □ Other

12. Have you utilized CD-ROM or Online Encyclopedia programs in your classroom in the past year? □Y □N
    If Yes, which programs or Web sites have you utilized? ____________________

13. What is your favorite or most exciting curriculum area to teach? ______________
    ____

14. Do you think it is possible to incorporate technology into the curriculum area?
    □Y □N
    If Yes, how would you like to incorporate technology? Software? Activity? Etc. _____
    ____

15. What information would you like to see included on the Web site for Cesar Chavez Elementary School? ____________________________
    ____

16. How have you utilized computer technology to assist with your lesson planning, curriculum development, or classroom management? ____________________
    ____

17. What suggestions do you have for future computer technology training? __________
    ____

Exploring the World Using Technology

Chapter XXVII
Literacy in K-12 Teacher Education:
The Case Study of a Multimedia Resource

Kristina Love
The University of Melbourne, Australia

ABSTRACT

Midway through the first decade of the new millennium, teachers are still facing considerable challenges in dealing with the complex forms of literacy that are increasingly required for success across the K-12 curriculum in Australia. Three critical areas in particular need to be addressed in teacher education in this regard: teachers' knowledge about text structures and about how language functions as a resource in the construction of a range of spoken, written, and multi-modal genres; teachers' understanding of language and text as critical socio-cultural practices and how these practices build disciplinary knowledge across the K-12 curriculum; and teachers' capacity to choose models of pedagogy that allow learners to master new literacy practices, transform meanings across contexts, and reflect substantively on learning through language. In this chapter, I will outline how a video-based interactive CD-ROM entitled BUILT (Building Understandings in Literacy and Teaching) was developed for use in teacher education to address these concerns. I will conclude by signalling some of the challenges that remain for teacher educators training novice teachers to scaffold, through ICT, their K-12 students into an important range of literacies.

LITERACIES IN SCHOOL EDUCATION IN AUSTRALIA

For the last decade or so, state and national K-12 school curriculum documents in Australia have begun to reflect a concern with literacy and its role in learning across the curriculum, though the depth and embodiment of this concern has been variable from state to state and from Key Learning Area (KLA) to KLA. This concern has been foregrounded in English (as a mother tongue) and Literacy curriculum and policy material, which is generally underpinned by a view of language (in all of its modes) as a social resource, as much as a set of cognitive skills. As such, areas of the official curriculum are beginning to reflect a number of key theoretical assumptions prevalent in the research literature, in summary:

- that literacy consists of a complex set of social practices, rather than being a unitary psychological concept;
- that language and literacy practices vary according to social contexts and therefore need to be studied as they occur in those different contexts; and
- that language and literacy constitute powerful semiotic systems for the construction of meanings. (Hammond, 2001)

Such a socially oriented view of literacy, it has been argued, is particularly important in a world where globalization of communication and proliferation of multimodal texts is increasing daily (Cope & Kalantzis, 2000).

Teachers across the K-12 curriculum are struggling to come to terms with these more socially oriented versions of literacy, with more traditional approaches to literacy being further reconfigured through new information and communication technologies. At the very least, traditional notions of literacy as reading, writing, speaking, and listening need to incorporate the processes of viewing, navigating, and composing in a multimodal environment.

Reading can no longer be simply viewed as a process of decoding symbols on a page, but must be seen as an increasingly complex process which includes:

- understanding a range of semiotic systems, in both their individual and blended forms;
- making inferences around these based on available cultural knowledge;
- testing understandings of texts in pragmatic contexts; and
- engaging in critical analysis of how texts deploy various semiotic systems to position readers.

Thus, while traditional 'basic skills' such as word recognition, spelling, and comprehension are still seen as necessary for successful individual literacy development, they are no longer sufficient for young people to meet the increasingly complex demands of communication in the 21st century. Freebody and Luke's (1990) model of competent readers as those who are able to draw on four resources simultaneously is now central to many literacy programs, those resources being: encoding and decoding resources, semantic or meaning-making resources, pragmatic or text-using resources, and critical or text-analysing resources. Helping students exploit these resources, or develop practices in using them across K-12 contexts, is now seen as central to the development of a generation of literate individuals, able to function effectively in a modern technological society.

The development of readers in these four practices is particularly important in a global context where more traditional forms of spoken and written language are increasingly embedded in larger multi-modal texts, many of these in digital electronic format. Students across the K-12 curriculum are engaging with various forms of meaning-making, many of which combine modes such as the linguistic, visual, auditory, gestural, and spatial (The New London Group, 2000, p. 6). In the hypertextual environment of the World Wide Web for example, to access the information so often required for school research, students must read information that is coded in a variety of often blended formats, including, besides the verbal, animation, symbols, photos, movie clips, and graphics. These multi-embedded texts require young navigators to draw on "a range of knowledges about traditional and newly blended genres or representational conventions, cultural and symbolic codes, as well as linguistically coded and software driven meanings" (Luke, 2000, p. 73). The processes of reading are further complicated by the fact that each mode in a multimodal text carries its own code, while simultaneously blending to produce a meaning that is often greater than the sum of its parts (Kress, 2003). Reading effectively is thus seen as a process of drawing not only on a knowledge of each code and its relationship to other codes in a multi-modal text, but on a knowledge of the socio-cultural contexts in which these codes are embedded and of how they combine ideologically to position the reader.

Writing, as well as reading, is also a more complex process in this new educational context, where it is seen as much more than simply reproducing valued texts in the standard form of a national language. Increasingly more, social semiotic approaches to language are informing a view of writing as the staged production of texts which achieve distinctive social purposes in particular contexts of situation and culture (Derewianka, 1990, 1998). Effective writers are seen as able to make appropriate language choices informed by an understanding of how register variables operate in appropriate contexts. They are also able to produce multimodal texts which combine elements of the written code, in the appropriate register, with appropriately selected elements of other visual codes (Unsworth, 2001). While there is still relatively little support for K-12 teachers interested in such a view of writing as multi-modal production, important work is underway to map the structures of multimodal texts commonly required to be read and produced in education (Cope & Kalantzis, 2000; Kress, 2003).

In the meantime, work drawing on theories of genre has been influential in the development of a writing pedagogy that addresses the demands of language and literacy (see most notably Christie, 1999; Derewianka, 1990, 1999); such models having variously successful uptake in K-12 curricula in different states of Australia (Hammond & Macken-Horarik, 2001). Genre-based writing pedagogies are framed within a critical socio-cultural context such that learners are taught to evaluate their growing mastery of a written genre in relation to the historical, social, cultural, political, ideological, and value-centred relations of particular systems of knowledge and social practice. Such genre-based writing pedagogies are also underpinned by updated theorisations about the nature of scaffolding in education (see most notably Hammond, 2001; Gibbons, 2002), and employ open-ended and flexible functional grammars that are able to describe language differences (cultural, technical, and disciplinary specific) and multi-modal channels of meaning. Most importantly, genre, as a model of semiotic choices staged to achieve distinctive social purposes, "is essential in all attempts to understand text, whatever its modal constitution" (Kress, 2003, p. 107).

Just as traditional notions of reading and writing have been reconfigured as aspects of literacy in the new communicational landscape, so too have notions of speaking and listening.

Focusing for the moment solely on the discourses of schooling, research into the structures of spoken texts commonly occurring in classrooms has run parallel with research into the socially motivated structures of written texts. Christie (2002), for example, identifies spoken curriculum genres as diverse as morning news sequences in early primary classrooms, exploratory task-oriented language in middle school science, and text response discussions in secondary English. In each of these curriculum genres, spoken language is used by the teacher and students in patterned ways that variously support student learning across different discipline areas. Likewise, examinations of how students are apprenticed through spoken language into the more abstract discourses of science (Lemke, 1990) and maths (Veel, 1999) point to the role of well-structured talk in scaffolding the literacies of these disciplines. Such insights from genre theory into the patterned choices that speakers and listeners make in interaction sit well with insights from educational sociologists such as Bernstein (1996) who illustrate how educational disadvantage can often result for students who cannot adopt the appropriate spoken register in a given context. From this research emerges a clear proposition that teachers and their students would considerably benefit from a knowledge of how various registers of spoken language can be used most effectively in different contexts for different purposes. Such knowledge can be used to support student learning, and to make explicit the forms of reasoning that are valued in particular school disciplines.

In short, the reconfiguration of traditional notions of literacy as reading, writing, speaking, and listening in the new communicational landscape of the 21st century has resulted in a view of new literacies as purposefully staged social interaction, as inherently multi-modal, and as structured in ways that can be explicitly modelled and critically deconstructed. This view of the new literacies has profound implications for teacher education, where a new generation of teachers is simultaneously grappling with new technologies as tools for their own learning and the learning of their school students.

THE NEW LITERACIES, TECHNOLOGY, AND TEACHER EDUCATION

Working with notions of new literacies as critical socio-cultural practice, K-12 teacher education programs in Australia are increasingly seeking to give novice teachers experience in working with school texts whose forms and functions vary with increasingly more diverse purposes and technologies. Recognising that all teachers need to engage critically with the multiplicity of communication channels which their own students will be using, most teacher education programs at the same time require novice teachers to work at an advanced level with new technologies. In my own teacher training context at the University of Melbourne, while we are accounting for the burgeoning variety of text forms associated with information and multimedia technologies, we are also supporting student teachers into recognising the discipline-specific forms of text reception and production required across the K-12 curriculum.

We have had considerably success with those student teachers training as generalist primary school teachers or specialist secondary English/literacy teachers, than we have had with secondary subject specialists. These language-oriented student teachers have a more sustained opportunity to come to terms with what is involved in scaffolding K-12 students into producing the genres demanded in schooling and in supporting their students to read more effectively by exploiting the four resources identified by Freebody and Luke (1991) mentioned above. Student teachers training in other disciplinary areas—for example, in secondary maths, science, history, geography, commerce, music, or art—have a much smaller component (18 hours in total) of their one-year pre-service course dedicated to the subject Language in Education. In the remainder of this chapter, I will outline the challenges that faced me as the coordinator of that subject, responsible for supporting such content area teachers into recognising the specialised literacy demands of their discipline areas. The aim of my program was to help student teachers (henceforth novice teachers) understand that, through a little knowledge of how language works to make different meanings in various contexts and modes, they can support their students into developing the literacy and multi-literacy skills needed in their specialisation, rather than seeing these (erroneously) as having been established previously in the K-6 program, or as the sole responsibility of the English/Language Arts/ Literacy teacher.

I have identified three critical issues in particular that needed to be addressed in a program supporting novice teachers in their roles as teachers of literacy across the curriculum:

- teachers' knowledge about text structures and about how language functions as a resource in the construction of a range of spoken, written, and multi-modal genres;
- teachers' understanding of language and text as critical socio-cultural practices and how these practices build disciplinary knowledge across the K-12 curriculum; and
- teachers' capacity to use models of pedagogy that allow learners to master new literacy practices, transform this knowledge across various contexts, and reflect substantively on learning through language.

Each of these issues will be addressed under separate headings below, as they underpin the design of a multimedia resource, a CD-ROM entitled BUILT (Building Understandings in Literacy and Teaching), used in my program at the University of Melbourne. First, however, I will describe the context in which this resource was developed.

BUILT: A K-12 MULTIMEDIA LITERACY RESOURCE IN TEACHER EDUCATION

For more than a decade now, the subject Language in Education has been a compulsory component for all students in the Diploma of Education and Bachelor's of Teaching courses at the University of Melbourne, regardless of which discipline they were preparing to teach. The development of this subject followed from a Project of National Significance on the Preservice Preparation of Teachers for Teaching English Literacy, which made an important recommendation that "as a compulsory component of their preservice education, all teachers should receive a substantial preparation in knowledge about language and literacy and the pedagogical principles for their teaching" (Christie et al., 1991, p. 98).

Finding ways to fully engage novice teachers in building their understandings about language and literacy across the curriculum proved difficult, and various combinations of workshop/lecture approaches, essentially based on reading packs with additional video resources, could not create the motivating and active learning environment sought by the course designers. A particular challenge was to convince non-language-background graduates that language in its various modes plays a central role in learning, and to make vivid for them the classroom contexts in which their developing knowledge about language and its relationship to learning could be applied. This challenge was compounded by the fact that only 18 contact hours were available for a subject in which nearly 1,000 students are enrolled annually. The course designers thus applied for and received a small internal grant to develop a CD-ROM that could provide innovative video-based and interactive materials displaying authentic learning of and through language and literacy.

BUILT was thus developed in an environment of mass education, of time-restricted face-to-face teaching, and of diminishing resources in a period of economic restraint, an environment currently typical of the tertiary sector as a whole (Sheely, Veness, & Rankine, 2001). Under such circumstances, it would have been tempting for the multimedia designers simply to create an "electronic textbook" (Herrington & Standen, 2000, p. 197) that would be conceptually simple to develop since it would rely on linear transmission of knowledge, with the user the passive recipient of instruction (Reeves & Okey, 1996). The designers eschewed that approach, deliberately opting for a design that allowed users to learn by "actively making sense of new knowledge—making meaning from it and mapping it into their existing knowledge schema" (Gipps, 1994, p. 22).

Developed across four units, BUILT covers the central issues of: semiotic systems and language as a key meaning-making system; spoken language in the home, community and classroom; written language, with a focus on the text types commonly required in schooling; and reading as a combination of decoding, semantic, pragmatic, and critical practice. Each of the four units is further divided into two topics. Topic A provides the principles and the bigger socio-cultural picture related to an aspect of language or literacy. Topic B provides a pedagogic focus, illustrating, through QuickTime video (QTV) clips, how these principles are put into practice in authentic, unscripted classroom interactions as teachers scaffold their students into learning through language in exemplary ways. A model of a five-stage learning/teaching cycle of Engagement, Building Knowledge, Transformation, Presentation, and Reflection, based on neo-Vygotskian notions of language and learning (e.g., Hammond, 2001; Maybin, Mercer, & Stierer, 1992) provides a framework for novice teachers to reflect on the many classroom interactions in the B topics. Table 1 summarises the structure of BUILT.

Despite spotlighting each of the modes separately in its four units, BUILT's model of literacy involves the integrated development of reading (including viewing and browsing), writing (including composing with visual text), speaking, listening, and critical thinking. Within this

Table 1. The structure of BUILT

Unit 1. Language and Learning	
Topic A: Text and context	**Topic B**: Scaffolding learning
Unit 2. Oral Language and Learning	
Topic A: Oral texts in context	**Topic B**: Scaffolding learning through talk
Unit 3. Written Language: Writing	
Topic A: Written texts in context	**Topic B**: Scaffolding students' writing
Unit 4. Written Language: Reading	
Topic A: Reading in context	**Topic B**: Scaffolding students' reading

dynamic model of literacy as social practice, BUILT focuses predominantly on verbal language as a key semiotic system in education, using functional models that can be applied to all forms of representation and communication. Across each of the four units of BUILT, cultural knowledge is seen as enabling speakers, readers, and writers to recognise, use, and critique language appropriate to particular cultural contexts and social settings.

BUILT's own learning environment is designed around a social constructivist pedagogy that promotes active learning rather than the passive consumption of mechanistically organised content. In learning about language in the social interaction of various classroom communities through "legitimate peripheral participation" (Brown, Collins, & Duguid, 1989, p. 40), novice teachers can enter the culture of teaching practice virtually. Authentic video footage is presented in the form of QTVs of classroom interactions across K-12 contexts, where school learners are seen to be supported through the effective use of language in its oral, written, and visual modes. For novice teachers, key concepts about language, literacy, and pedagogy are thus grounded in the virtual experience of actual classroom interactions. Users can freeze and systematically reflect on the language of classroom interactions in ways that are not possible in the real context. A notepad facility allows users to record reflections on these frozen moments and respond to key questions, both of which can be saved to floppy disk, a hard drive, or a server, while still using the CD-ROM. While users as linguistic novices are scaffolded into progressively more sophisticated understandings of texts as multimodal socio-cultural products, analysable using an appropriate meta-linguistic toolkit, they are encouraged to use the electronic notepad to reflect on their emerging understandings and share their reflections with others. QTV clips of teacher interviews provide further insights into teachers' theorising, planning, teaching, and assessment decisions, and how these impact on their students' language and literacy development.

A number of other features of the design of BUILT will be briefly mentioned here as they illustrate the situated and scaffolded nature of the learning about literacy in a teacher education context (for a more detailed description of the design features of BUILT, see Love, 2002, 2003; Love & Shrimpton, 2002). Panoramic QuickTime Virtual Reality screens with numerous hotspots allow users to explore the multi-modal texts produced (written, drawn, or otherwise constructed) and consumed (read or viewed) in a number of classroom environments, and to consider the quality and range of resources for scaffolding students into various literacy practices. Tools such as a Glossary button (back-referenced to relevant screens) allow users to check their understanding of technical (often linguistic) terminology at point of need, and a Bibliography button allows them to access a list of selected references for further individual investigation. In structuring user pathways, a "hierarchical" rather than "linear" or "referential" model of hypermedia linking (Oliver & Herrington, 1995) is built into the design, enabling users to enter the CD-ROM at any unit or topic, depending on their teaching area, their teaching level, and their existing knowledge about language. Once in a particular screen, a 'Go To' facility allows users to move rapidly to any other unit, topic, and screen, providing maximum opportunity to make connections between concepts. Novice teachers needed to be able to learn new and often complex information about language while simultaneously reflecting on how they could scaffold their own students into learning through various language modes. Some of these features are illustrated in Figure 1.

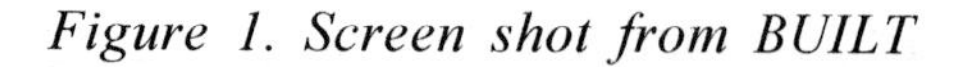

Figure 1. Screen shot from BUILT

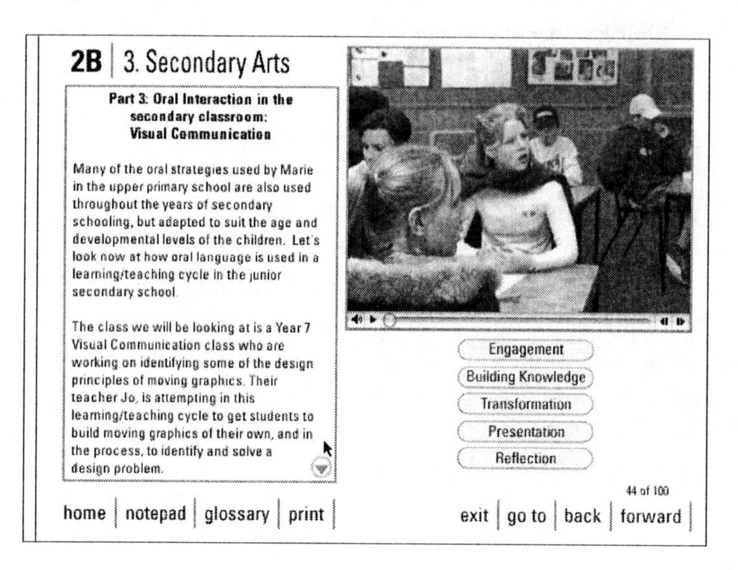

BUILT was also designed to be used in a variety of different learning, teaching, and assessment contexts. In workshop groups, novice teachers observe and discuss the QTVs of classroom interactions and teacher interviews, as these are selected by teacher educators who use data projection trolleys in specially designed collaborative teaching spaces. A regular outcome of such collaborative workshops is that novice teachers from different disciplinary backgrounds across the K-12 curriculum could identify key features of language, as these are used for learning in various modes and contexts. Novice teachers also use BUILT independently as they select QTVs relevant to their particular discipline areas and needs, and as they rehearse specific features of their own language and literacy use before their practice teaching rounds. The majority of novice teachers used BUILT on their home computers, with about one quarter using it in faculty computer labs. A third context in which novice teachers are encouraged to use BUILT is with a peer, as a means of reviewing the pedagogical decisions they made during or after their practice teaching rounds. Such paired interactions regularly take place in computer labs at the university or at the teaching practice school. While it thus provides '"authentic learning contexts," BUILT also provides "authentic assessment contexts" (Lebow & Wager, 1994), allowing novice teachers to draw cumulatively on Notepad reflections as they plan how they will address the literacy demands of the curriculum they will design in upcoming teaching rounds.

Thus, through its hypertextual structure, its interactive tools, its authentic contexts of representation, rehearsal, reflection, and assessment, BUILT was designed to scaffold novice teachers into recognizing and working with the literacy demands across the K-12 curriculum. In "situated" contexts (Brown et al., 1989; Lave & Wenger, 1991) and using authentic texts (Herrington & Oliver, 1997), novice teachers could be apprenticed into becoming "insiders" (Kramsch, 1998) in the discourse community of reflective professional educators (Schon, 1983). In learning about the nature of language as a social semiotic, while observing and reflecting on its use in specific disciplinary contexts, novice teachers using BUILT could develop insights in a virtual environment that could be rehearsed in the real environment of their upcoming teaching practice in various K-12 contexts. In particular, the potential was there for such novice teachers to learn about language as a social semiotic and about texts as multi-modal social products; to understand language and text as critical socio-cultural practices, and how these practices build disciplinary knowledge across the K-12 curriculum; and to note how an active, engaging, reflective pedagogy was modelled through multimedia, allowing learners to master new literacy practices, transform new knowledge across contexts, and reflect substantively on learning through language. Each of these issues will be dealt with in more detail under separate headings below.

KNOWLEDGE ABOUT LANGUAGE FUNCTIONS AND TEXT STRUCTURES

In Unit 1 of BUILT, various semiotic systems are illustrated, including those which make their social meanings through grapho-phonic, musical, mathematical, visual, and oral codes. However, while illustrations of the other semiotic codes (in both their independent and blended forms) are offered, BUILT focuses on verbal language as a key semiotic system in the area of literacy in education. Language is thus examined in BUILT primarily in its verbal modes, with attention to the visual, audio, and hypertextual codes as required, but using a model of language structure and function that can be applied to other modes besides the verbal (e.g., Unsworth, 2001).

A central premise of BUILT is that teachers across the curriculum and across the K-12 years need a language to describe the forms of meaning through which their discipline builds its content. They need a language for talking about language in each of its modes—a meta-language. The model of language underpinning BUILT is a functionally oriented one, concerned with how meanings are made in real social and cultural contexts where language serves a variety of purposes (Halliday, 1973, 1994). This functional grammar is more open-ended and flexible than traditional or formal grammars, whose use is restricted to the prescription of standard written national languages. It is premised on the assumption that language is a system of choices, made in ways that fulfil particular social functions in particular contexts. As such, it is a view of language particularly suited to education, where the literate practices that are valued in the K-12 curriculum can be developed through attending to language choices that achieve distinctive purposes in different subject areas. Accessible outlines of this functional grammar have been developed in book form (Derewianka, 1998; Droga & Humphrey, 2003) and are being used increasingly by primary school teachers in Australia. They are used considerably less by secondary school teachers which is particularly unfortunate, given the value of an educational meta-language for identifying and explaining differences between texts as they operate in different contexts in the modern technologised world.

In Unit 1 of BUILT, novice teachers are given a taste of how certain language choices operate to construct different bodies of knowledge (field), different relationships (tenor), and different channels of communication (modes) in different situations. For example, they can view an authentic QTV clip of a mother teaching her daughter to bake a cake, side by side with the transcript of that interaction (see Figure 2). This transcript is animated such that certain language choices which operate primarily to construct the intimacy of the relationship (e.g., modality, interrogative mood, intimate

Figure 2. Screen shot from Unit 1A of BUILT illustrating genre and register

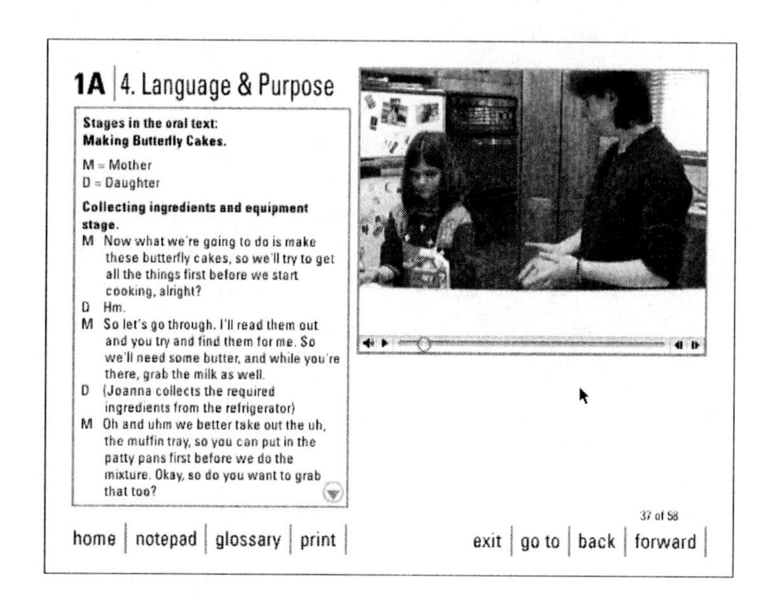

terms of address) are progressively highlighted in red; those which operate primarily to construct the technicality of the field of cooking (e.g., kitchen-specific terms such as 'patty pans') are in blue; and those which operate primarily to construct the mode as spoken face to face (e.g., ellipsis, use of exophoric reference) are progressively highlighted in green. A similar principle is used in the description of the language choices made in a formal telephone conversation and a personal e-mail, providing novice teachers with opportunities to apply, rehearse, and reflect on their emerging understandings of the meta-language and the concepts of field, tenor, and mode with other multimodal texts provided in the CD-ROM. In this way, novice teachers are engaged interactively with authentic texts in situated learning about the structures and functions of language as it operates to construct various social meanings.

Users of BUILT also learn deductively and interactively about the notion of genre as a conventional way of using language to get things done in a culture (Christie et al., 1991). The cooking interaction is presented as an example of a spoken instructional genre which is staged into a number of clear and predictable phases in order that the child masters the skill of baking a cake competently and without injury. The transcript of the formal telephone conversation is presented as an example of an oral transactional genre, visually and interactively represented as a series of stages moving from greeting, health enquiry, negotiation of appointment time, and farewell. The e-mail is represented visually and interactively as a personal recount genre, staged to progressively record a chronologically organised series of personal events while building a close relationship with the reader. In developing a conscious awareness of the generic structures of these everyday spoken and written texts, and the language used to achieve their social purposes, novice teachers are being provided with a meta-language that can be applied to their examination of the generic structures and language choices of more specialised texts used in the K-12 curriculum.

Genre and Writing

In the last two decades, substantial research has been undertaken, much of it in Australia, into the structures and associated language features of the prototypical genres required to be written and read throughout the years of schooling. Martin and Rothery (1980, 1981) initially identified the text types or genres regularly required in the primary school. They found, for example, that personal recounts were a popular written genre in early primary school, structured as they were into an introductory stage which oriented the reader to the context, followed by a series of chronologically ordered events. This quite predictable structure functioned to help the young learner recall a personal experience, the writer regularly choosing grammatical features such as the first person singular to foreground their personal involvement in the events and temporal connectors such as 'and then' to sequence the unfolding events.

Subsequent research identified the structures and grammatical features of the increasingly more sophisticated written genres required across a range of subjects in the secondary school. Building on the work of Halliday and Martin (1993), Veel (1997) for example noted the increasing complexity of the genres that are regularly required in school science, with students being expected to move from mastering the relatively simple structure of Procedures and Procedural Recounts in junior science to demonstrating control of more structurally complex genres such as Consequential Explanations, Taxonomic Reports, and Expositions in the upper school. Such research highlights the need for K-12 teachers to have some knowledge of the structures of the texts they require their students to read and write, and challenge the view that literacy is the preserve of the Language Arts/Literacy teacher.

The concept of genre has proved extremely useful for education and has considerable potential in the description of texts which involve modes other than the linguistic (see Kress, 2003, for some particularly powerful arguments). It is thus used extensively in BUILT as a heuristic for helping teachers identify the patterned ways in which language, across its various modes, is used to achieve educational outcomes. In Unit 3 of BUILT, the focus is on six of the prototypical genres that are required to be written across the K-12 curriculum: procedures, recounts, narratives, reports, explanations, and arguments. As a multimedia resource, BUILT can exploit more interactive means for novice (and indeed experienced) teachers to learn about the structures and grammatical features of these genres than are available in print-based textbooks (e.g., Derewianka, 1991). The way in which this is done in BUILT will be illustrated with reference to a sequence of screens in Unit 3A, concerned with the generic structures and key grammatical features of Information Reports.

Users of BUILT are first presented with authentic samples of a range of Information Reports read and written by students in Grades 2 and 5 of the primary school, and in the secondary school, in Grade 7 Science and in Grade 11 International Studies. They are prompted (via Notepad and drag-and-drop activities) to deductively build insights into how each of these Information Reports are staged, as they firstly classify the phenomenon being described (whether it be dolphins, a particular country, an aspect of the plant world, or a concept such as third-world debt), and secondly as they describe various aspects of that phenomenon in subsequent paragraphs. In using the Notepad to record exploratory and

Figure 3. Screen shot of report outline from BUILT

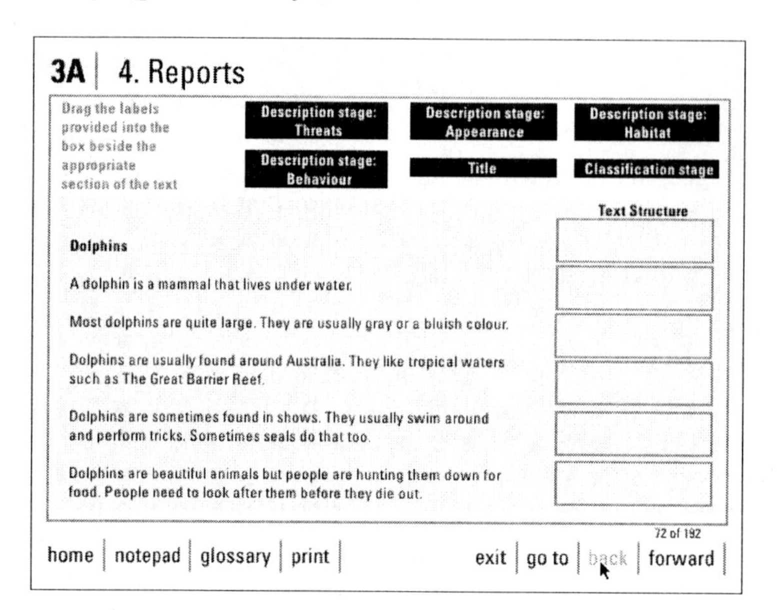

immediate insights about text structure and grammatical features, novice teachers become personally engaged in the task of learning about the structure of Information Reports. Engagement is represented as the first stage in the model of the learning/teaching cycle used in BUILT and is also a central design feature of the architecture of the CD-ROM, as will be discussed further below.

Through subsequent screens that combine verbal, pictorial, and video information, animation, and hypertext, users are then provided with overt instruction (the Building Knowledge stage of the learning/teaching cycle) about the generic structure and grammatical features of Information Reports as these are written and read by students across the K-12 curriculum. Users are subsequently given opportunities to rehearse their emerging understandings in a series of screens featuring drag-and-drop, Glossary, and Notepad activities (the Transformation stage of the learning/teaching cycle). Figure 3 is an illustration of one screen which allows users to play with a drag-and-drop facility to consolidate their understandings about the generic structure of Information Reports.

While users learn about generic structure, they also learn about a key number of grammatical features. For example, four linguistic resources are particularly significant in the effective reading and writing of Information Reports: the use of subject-specific technical terminology to build taxonomies, the use of the present tense to create a sense of the phenomenon as an enduring one, the use of *being* and *having* verbs to classify the phenomenon and its attributes, and the use of nominalisation to compress information and reflect the abstractions inherent in the study of the phenomenon. BUILT is designed for users with no prior knowledge of linguistics, so these grammatical features are explained in a highly scaffolded way, exploiting the animation and interactive features of the multimedia. Through the hyperlinked Index, the Go To facility, and the glossary functions, users are also provided with extensive opportunities to review and connect information at point of need.

Finally, novice teachers are encouraged to reflect on the applications of their understandings about the generic structures and grammatical features of Information Reports to their own subject areas. Reflection is also the final stage in the model of the learning/teaching cycle proposed in the CD-ROM; also, in modelling the learning/teaching process with novice teachers, the designers intended to provide an opportunity for situated learning of the sort that could be applied in users' own teaching. Thus, novice teachers experience first hand the value of learner reflection in its various modes, whether this be face to face in group discussion about their own learning and its applications, or the use of electronic reflective resources such as the Notepad in BUILT.

In these ways, novice teachers using BUILT are supported in developing a confidence with the generic structures and linguistic features, not only of Information Reports, but of all the prototypical school genres as they are used across the K-12 curriculum. Most come to recognize the socially motivated structures of these genres, and that a knowledge of these structures and grammatical features can be powerful for themselves as, for example, science, history, or music teachers interested in supporting their students into the writing demands of their discipline areas.

Genre and Reading

Such a knowledge of text structures and linguistic features enables content area teachers to better support their students in unpacking the reading, as well as the writing demands inherent in their discipline areas. In their roles as text participants (Freebody & Luke, 1991), drawing on semantic or meaning-making resources, readers bring prior knowledge of a text's purpose, structure, and grammatical features to the meaning-making process. BUILT demonstrates how content area teachers can assist those students who do not have that prior knowledge by explicitly providing the schematic structures that aid in comprehending texts. A series of interactive screens in Unit 4 offers a number of strategies for helping readers develop the required semantic practice, including building technical field knowledge prior to reading, or providing graphic aids to support understanding of text structure. Strategies to support the reader's comprehension account for the visual as well as the verbal components of the text, with students being overtly taught how to read the diagrams in association with the remaining text. Throughout Unit 4, users learn, again deductively and interactively, how to evaluate the effectiveness with which the visual, verbal, and hyptertextual elements combine to make meaning in materials which are regularly used across the K-12 curriculum (see Figure 4). They also see QTV clips of teachers making explicit to students the relationship between the visual, verbal, and hypertextual elements of a Web page as they search for information electronically.

Novice teachers are thus overtly instructed into attending to the individual and blended codes of multimodal texts in order to make explicit to their students the separate and integrated operation of the visual, verbal, and hypertextual codes of the texts they read.

Genre and Oral Interaction

Building on the introduction to genre of Unit 1, Unit 2 of BUILT offers novice teachers QTVs of a range of spoken genres used in schooling, from those where oral language is used in exploratory, small-group interactions to those more formal oral presentations by students and teachers. Users can freeze these interactions and, by referring to the transcripts which accompany many of the QTVs, examine the structure and language choices used in differ-

Figure 4. Screen shot outlining components of hypertext from BUILT

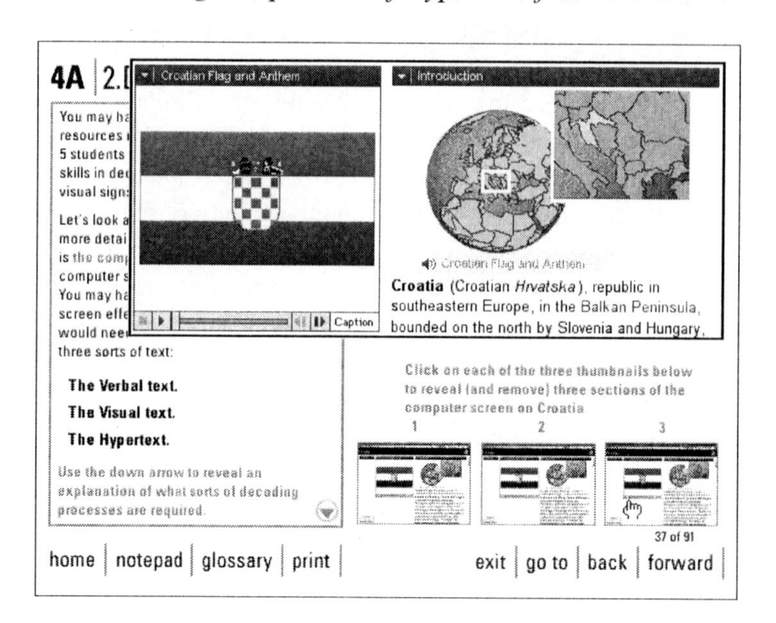

ent K-12 contexts. As they examine the transcripts of these interactions, novice teachers are prompted through sets of guided questions and Notepad activities to attend to how the teachers in each case scaffold the students through talk into the more literate forms of the language that they will be required to use in their subsequent writing. Such a view of language is potentially helpful for novice teachers grappling with the different language experiences and different spoken registers used by speakers of various linguistic and cultural backgrounds. Figure 1 above is a screen depicting how different forms of talk are used at different stages of a learning/teaching cycle in a Year 7 Visual Communication class.

In summary, through the situated nature of the practice they experience in using authentic texts, through the overt instruction regarding generic structure and key linguistic features, and the opportunities for rehearsal, transformation, and reflection, users of BUILT are well positioned to begin to support their students into the spoken, reading, and writing demands across the K-12 curriculum. However, without what Cope and Kalantzis (2000, p. 33) identify as a theory of "critical framing," any pedagogy which purports to support the multi-literacies that are essential for effective communication in a modern globalised and technologised world is limited. A theory of genre and its associated meta-language used in education must also be capable of "supporting a sophisticated critical analysis of language and other semiotic systems, yet at the same time not make unrealistic demands on teacher and learner knowledge" (Cope & Kalantzis, 2000, p. 24). We will now look at how this requirement is met in the design and implementation of BUILT.

LANGUAGE AND TEXT AS CRITICAL SOCIO-CULTURAL PRACTICES

Many novice teachers do not have a consistent critical meta-language for talking about the structures and processes whereby texts construct their meanings and social positions. Indeed, despite the prevailing rhetoric of critical literacy in Australia, many experienced teachers still do not have the critical language awareness required to identify how people and social practices are represented in texts across the school curriculum (Morgan, 1996; Wyatt-Smith, 2000). As a result, "students are sold short of the knowledge that they need to live in today's world, saturated as it is with ideologically loaded texts" (Wyatt-Smith, 2000, p. 78).

A Critical Framing

The position adopted in BUILT is that all texts, those used both in and outside of school, are products of particular contexts and ideologies, and that a capacity for "critical framing" (Cope & Kalantzis, 2000) of such texts is essential in the development of teachers in the 21st century. The approach to critical practice used in BUILT is guided by Kress's (1985) three questions to ask of any text:

1. Why is this topic being written about?
2. How is this topic being written about?
3. What other ways are there of writing about this topic?

BUILT illustrates how these questions can be asked of a number of texts used in a range of classrooms, thus modelling strategies for reading as critical practice.

These strategies can be briefly illustrated with reference to one such text, entitled "The Scottish Soldiers at Lucknow" taken from *A Nursery History of England* by Elizabeth O'Neil, published in the early 1950s (n.d.) by Thomas Nelson and Son. As a historically distant text, this extract allows novice teachers to examine how, even in a children's history book, the ideologically laden language constructs a view of English colonization of India as benign and of Indian soldiers as malevolent. The drawings accompanying the verbal text can also be analysed as they reinforce the ideological positioning carried in the verbal mode. Novice teachers first record, in their electronic Notepads, responses to Kress's three questions regarding this text in its blended visual and verbal format. They then discuss these individual responses in small face-to-face groups, where the ideologically transparent nature of this text is consolidated. They noted that while the English point of view was made highly visible, the Indian point of view was silenced and discussed some of the resources, both verbal and multi-modal, that the writer and illustrator used to construct the text's 'view of the world'. This workshop process provided a model of critical practice that could be applied across the K-12 curriculum, as teachers support students to examine how "ideologies find expression in and through ways of combining discourse, genre, register and textual features" (Wyatt-Smith, 2000, p. 77). Such a model of critical framing is particularly important as teachers and students work not only with print-based text books, but increasingly more with commercially produced multimedia and Internet resources which have various degrees of credibility.

In the remainder of Unit 4 of BUILT, this critical framing is then applied to the classroom context. Users first observe a Grade 5 teacher working with her students to examine how the visual and verbal elements of travel adverts are designed to persuade people to visit a particular tourist spot. The teacher engages her students by getting them to identify what features of their own community they would tell people about and which features they would leave out. Subsequently, in examining travel adverts, she gets students to examine the design of the visual and verbal information that has been included and to imagine what visual and verbal information may have been omitted. She thus systematically trains her students from a young age in examining the gaps and silences of texts as ideological constructs. With such early exposure to the idea that no text is innocent, and with the early development of a critical metalanguage, these young students are well positioned to take a more critical stance to the materials that they will read later on in schooling. Further on in Unit 4 of BUILT, for example, we see QTV of students in a Year 11 International Studies class researching the Web for information about third-world debt. One student expresses uncertainty about the validity of the information he has discovered on a site that may not be the official site of the IMF (the International Monetary Fund). Again, the teacher has scaffolded her students into being able to critically evaluate the content of the Web-based information by identifying the stance and credentials of the author, the potential purposes of the text, the language and visual features that are used to convey that purpose, and the gaps and silences of the text.

A Critical Meta-Language

By asking the kinds of questions which Kress (1985, 2003) suggests should be asked of any text, both the Grade 5 and Year 11 teachers provide their students with a useful critical meta-language that is supplemented in BUILT through further examination of how specific linguistic choices operate to position readers/listeners ideologically. For example, the grammatical process of nominalisation is demonstrated in Unit 3A, Screen 111 of BUILT as one which often serves ideological purposes. Nominalisation is the transformation of a grammatical item that is not a noun, into a noun or a nominal group. An example that is presented in animated form in BUILT is:

Large international companies log trees, and the native animals that live in those trees are dying and becoming extinct.

Tree logging causes **the extinction** of many species of native animals.

In transforming the verb *log* into the nominalisation *tree logging*, the agency of the large international companies is elided and questions need to be asked about who is responsible. Nominalisation can thus be seen as a means for a speaker or writer to elide agency in a text where naming the agent may have undesirable ideological consequences. Identifying nominalisation and reinstating agency is a means of further identifying the gaps and silences in a text. A knowledge of language as a functional system of meaning choices and of texts as socio-cultural genres thus provides a confident base on which novice teachers and their students can conduct a critical examination of texts as the embodiment of various ideologies.

Disciplinary Specific Literacies

A critical framing of literacy practices, particularly when accompanied by a critical metalanguage of the sort identified above, also allows novice teachers to see that disciplinary-specific ways of reasoning are themselves ideologically laden. The disciplines of Science and those of Art, for example, see the world differently, those differences being construed largely through language. Linguistically oriented research in the area of mathematics (Veel, 1999), history (Coffin, 1997), and geography (van Leeuwen & Humphrey, 1996), for example, point to the profound ways in which the spoken and written discourses of each of these disciplines both shape and reflect the forms of reasoning that are valued in each.

Through BUILT, novice teachers are provided with a means of identifying some of the discipline-specific literacies they may not be aware of. In Unit 2B (Screen 81), for example, we see Year 8 Science students developing a growing confidence in their use of the terminology and forms of classificatory reasoning required of biologists, as their teacher scaffolds them into using these concepts in their oral presentations. We can compare this with the spoken and written discourses of the Art class, where a more aesthetically oriented form of reasoning is valued. Likewise, through the interactive tasks in Unit 3 of BUILT, novice Geography teachers recognise what they need to teach about the specific language features of Explanations, including the use of temporal and causal conjunctions, in order for their students to describe the Water Cycle effectively, while History teachers recognise how to teach the specific language features of Historical Recounts, including the effective use of circumstances of time and place.

Such explicit teaching about the generic and linguistic features of texts required to be written and read across the K-12 curriculum is particularly important as students move into increasingly more complex and specialised discourses while progressing through the years of schooling. For example, Veel and Coffin (1996) have shown that, as students move from the genres which chronicle history (e.g., Biographical Recount, Historical Recount) to those which explain and interpret history (e.g., Consequential Explanation, Analytical Exposition), the presence of people as participants is effaced, event sequences are nominalised, and in a process of further abstraction, event sequences are related to other abstract entities. Explicit pedagogical support has been shown to be effective in helping K-12 students deal with these increased demands (Unsworth, 2000), especially for those students with little prior orientation to these linguistic or generic forms.

Support of the sort modelled in BUILT at the same time offers a means for novice teachers to critically frame their own disciplinary discourses, themselves forms of language which operate to privilege certain ways of reasoning about the world. Through such a critical framing of their literacy practices, novice teachers and their students across the K-12 curriculum "can gain the necessary personal and theoretical distance from what they have learned; constructively critique it; account for its cultural location; creatively extend and apply it; and eventually innovate on their own, within old communities and in new ones" (Cope & Kalantzis, 2000, p. 34). However, without a model of pedagogy, novice teachers' knowledge of genre as 'staged goal-oriented social processes' and of critical framing as central to an understanding of language as a social semiotic is not sufficient. Novice teachers still need a model of pedagogy within which these insights can be realised in the classroom.

A PEDAGOGY FOR IMPLEMENTING NEW LITERACY PRACTICES

The model of pedagogy proposed in BUILT is based on carefully theorised notions of scaffolding. Despite some obvious limitations, the metaphor of scaffolding is a valuable one for describing the temporary supporting structures provided by teachers of any sort as they assist learners to develop new understandings and skills. Importantly, the metaphor captures both the centrality of the teacher's withdrawal of support as learners develop control over specific understandings and skills, and the erection of further scaffolding to support the development of a new set of specific understandings and skills. The concept of scaffolding was originally used by Wood, Bruner, and Ross (1976) to portray the temporary, but essential nature of parental support in the language development of young children. This concept has proved to be attractive in socio-cultural models of learning in general (Mercer, 1994) and of language learning in particular (Halliday, 1973, 1994). Based on a profoundly "intersubjective" (Bruner, 1986) view of learning, it has extensive implications for K-12 teachers, as they assume responsibility for organising the social processes in their classrooms to maximise learning.

In the area of language and learning, three key factors have emerged as distinguishing scaffolding from other forms of teaching (Maybin et al., 1992; Hammond, 2001). Firstly, the task, skill, or understanding being scaffolded is a specific learning activity with finite goals. Secondly, the teacher must determine what skill or understandings learners currently have, identifying the learner's Zone of Proximal Development (ZPD Vygotsky, 1978, p. 86) in order to help them build on existing knowledge in constructive ways. Thirdly, the teacher structures the learning activity such that her/his own expertise can be gradually and judiciously withdrawn as the learner or apprentice can complete the task independently. In the area of multimedia development, very similar characteristics of scaffolding have been identified (Herrington & Oliver, 1997; McLoughlin, Winnips, & Oliver, 2000; Winnips, Collis, & Moonen, 2000), with terms such as *coach* often replacing that of *expert* or *teacher*.

Based on these key features of scaffolding, a five-stage 'learning/teaching cycle' is represented throughout BUILT which allows novice or apprentice teachers to more systematically examine the skilled language and literacy practices of experienced teachers modelled in the video clips. This five-stage cycle of Engagement, Building Knowledge, Transformation, Presentation, and Reflection is graphically represented in the form of a series of pentagons, where each pentagon represents a single learning/teaching cycle which combines with others to form a cumulative, or "spiral" curriculum (Bruner, 1986).

- **Engagement** is the stage where the teacher identifies a gap in students' understandings, plans strategies to bridge the gap, and engages with learners' prior understandings.
- **Building Knowledge** is the stage where the teacher provides learners with new information, so that together, teacher and students work towards building "common knowledge" (Edwards & Mercer, 1987) and shared understandings of the purpose of the set tasks.
- **Transformation** is the stage where learners are provided with opportunities to build their own insights about the new information, the teacher helping as needed and controlling possibilities of error, but judiciously withdrawing help as appropriate.
- **Presentation** is the stage where the teacher provides opportunities for learners to complete the task independently and to demonstrate, either formally or informally, their understandings.
- **Reflection** is the stage where the teacher and learners reflect on information, insights, and understandings, and together identify a new gap.

Through the QuickTime videos, the QTVRs, and the interactive tasks with authentic texts, this learning/teaching cycle is illustrated in a wide variety of K-12 classroom contexts. QTVs of interviews with teachers consciously working with the notion of scaffolding further deepens novice teachers' understanding of the importance of a carefully theorised model of pedagogy. Users could choose to focus on material related to their own teaching area, though were encouraged (both through the design of the CD-ROM and through the assessment tasks for the subject Language in Education) to explore other areas of the curriculum as antecedents, precedents, or co-occurring with their own areas. In this way, novice teachers could recognise how learning/teaching cycles accumulate over the K-12 years and across discipline areas into a "spiralling curriculum" (Bruner, 1986) through which learners are socialised into the knowledge, skills, and values of our society. With this model of learning/teaching cycles, novice teachers could plan how to scaffold students into the genres (spoken, written, and multi-modal) valued in their discipline areas within a critical framing.

Scaffolding as Multimedia Architecture

As well as providing a framework for the instructional content, the metaphor of scaffolding also underpins the instructional design of BUILT as a learning tool that attempts to provide authentic opportunities for cognitive apprenticeship in situated learning. The progress of novice teachers into a community of reflective practitioners is assisted as a result of both overt instruction at key stages of the learning/teaching cycle inherent in the design of the CD-ROM, and as a result of the situated practice made available to them in the virtual classrooms presented in the QTVs. BUILT thus provides a multimedia "scaffolding" (McLoughlin et al., 2000; Winnips et al., 2000), a form of intermediate support in their development as professionals who can recognize the role of language in learning and the importance of literacy across the K-12 curriculum. The five-stage learning/teaching cycle advocated in BUILT's instructional content was also built into the design of the CD-ROM such that users are themselves recursively engaged in issues of language, literacy, and learning; helped to build knowledge in this new area; guided into transforming that new knowledge into understanding; provided with various means of presenting that new understanding; and provided with the means of reflecting on that new understanding. Thus, while they are learning about the principles of scaffolding children into learning through language and literacy, they are themselves scaffolded into new professional understandings as they move through the sequences of learning/teaching cycles which underpin the design of the CD-ROM.

Various multimedia features are exploited at different stages of the learning teaching cycle. In engagement phases, the Notepad facility is regularly available for novice teachers to articulate their existing but tacit linguistic, disciplinary, or cultural knowledge. Making such tacit knowledge explicit in the engagement phases allows their teacher educators to plan appropriate strategies for moving teachers (as apprentices) into a new ZPD. In the Building Knowledge phases of the CD-ROM, different modalities are exploited to provide more overt instruction to apprentice teachers, such modalities regularly including QuickTime video clips of authentic classroom interaction. In the various Transformation phases designed into the architecture of BUILT, novice teachers are provided with further examples of authentic texts to work on interactively, either through drag and drop, self-correcting exercises, or through notepad activities that can be shared in pairs or workshops. Supported by point-of-need resources such as hyperlinked explanations, a Glossary, and a Bibliography, they can rehearse newfound knowledge in contexts designed to control the possibilities of user error and are thus scaffolded into increasingly more independent understandings of language as an ideologically motivated social semiotic and of genres as identifiably staged social products. In the Presentation phases novice teachers are provided with the multimedia opportunities to demonstrate their independent understandings of new concepts and skills, either formally (e.g., as they copy a sample text from the screen into their notepads, where they can identify its key structural and linguistic features) or informally (e.g., where they can discuss which strategies demonstrated in QTV clips they would use in their own teaching areas). A selection of these situated tasks constitutes the assessment tasks for the subject Language in Education.

Finally in the Reflection phases throughout BUILT, apprentice teachers are able to evaluate the effectiveness of the teaching, the relevance of their learning, the problems or challenges, and their own teaching goals (Schon, 1983). In becoming aware of these processes in their own learning, they will hopefully be more attentive to their significance in their classroom teaching. Regular, smaller reflective tasks are thus built into the design of the CD-ROM, typically at the end of each learning/teaching cycle, and typically asking users to respond to broad prompt questions in notepads or face-to-face discussion groups. However, larger reflective tasks also occur at the end of each Unit, requiring users to review all notepad entries for that Unit, and feeding into workshop or assessment reflections. By incorporating the five stages of the learning/teaching cycle into its fundamental design, BUILT as a multimedia resource meets the conditions of true scaffolding in situated learning (Herrington & Oliver, 1995; see Love, 2002, for further details). Novice teachers can potentially be scaffolded into disciplinary understandings about the structures and functions of language in its various modes as these impact on learning, and into 'virtual practice' regarding how to apply this knowledge in their own teaching context.

FUTURE CHALLENGES

BUILT has attempted to provide novice teachers with a model of literacy aimed at teaching their K-12 students how to read and produce the texts typical of the current information age. The critical framing, the functional meta-linguistic tools, the view of genre as purposefully structured text, and a pedagogical model based on theorised notions of scaffolding have been incorporated into an interactive multimedia format designed to encourage novice teachers to apply their insights to their own emerging practice. The question remains as to how much such a resource can impact literacy practices in K-12 classrooms.

A series of formal evaluations conducted over the three-year period of BUILT's implementation in the Language in Education course suggests that the multimedia resource has indeed achieved some of its goals during the pre-service period (see Love & Shrimpton, 2002, for further details of this evaluation). Most significantly, the vast majority of novice teachers recognised the value of the CD-ROM in providing opportunities for them to work interactively with authentic examples of language and literacy used in teaching, within a strongly theorised framework. The quality and authenticity of authentic texts and classroom video footage were seen as crucial factors in the effectiveness of the resource, as were the QTVs of interviews with teachers. Interactive features such as the animations, the transcripts (used in conjunction with video footage), the drag-and-drop and the roll-over facilities, the Notepad, and the Glossary were all highly valued as various means of helping novice teachers rehearse what for many were unfamiliar features of language. University teaching staff working with BUILT in workshops also reported generally favourably on its effectiveness as a multimedia resource in the subject Language in Education. The biggest difficulties in implementation centred around technical issues such as the use of projection equipment in workshop rooms (see Love & Shrimpton, 2002, for further details).

However, student and staff feedback about the implementation of BUILT in the context of an 18-hour program has demonstrated variable capacity in terms of the three critical areas addressed in this chapter. In terms of the first area, novice teachers using BUILT in the subject Language in Education reported being generally well scaffolded into a knowledge of the structures of the key written genres required across the K-12 curriculum, a perception that was further evidenced in their performance on assessment tasks. They were however less able to demonstrate a knowledge about how language functions as a resource in the construction of these texts. In their assessment tasks in particular, many demonstrated their confusion with concepts such as nominalisation, and expressed frustration in their end-of-semester feedbacks at not having had sufficient time in such a short course to develop the required meta-language.

In terms of the second concern mentioned in the opening of this chapter, students' performance in a final assessment task (where they had to account for the literacy demands of a unit they were planning to teach) also demonstrated less confident knowledge of how language operates as critical socio-cultural practices and how these practices build disciplinary knowledge across the K-12 curriculum. Student feedback and performance on assessment tasks likewise indicated that the limited time available for the course (18 hours across only one semester) is not sufficient to develop the more sophisticated understanding of language required for these insights in graduates with no prior meta-linguistic study.

In terms of its capacity to support novice teachers in choosing a model of pedagogy, BUILT appeared to be more successful, with assessments before and after their practice teaching rounds demonstrating novice teachers' in-principle endorsement of the five-stage learning/teaching cycle as relevant to their own teaching. Particularly heartening here was novice teachers' awareness that different choices of spoken and written language could be beneficially used at different stages of this learning/teaching cycle to support not only disciplinary knowledge, but also disciplinary specific literacy.

It has been much harder to evaluate the impact of BUILT—as a constructivistly designed, interactive multimedia resource—on novice teachers' own teaching with and about literacy in actual K-12 classrooms. 'Input' does not always translate into 'uptake', as was evident in many of the novice teachers' questionnaire responses, interview comments, and assignment tasks. However, the comment below was typical of many in student questionnaire responses.

One of the strengths of BUILT was that it allowed us to see language in action, rather than focusing on written texts only...I got many ideas from the CD-ROM and actually used these during my school experience. (Sam)

Clearly, there is a need to follow students like Sam, who is a novice English and History teacher, as he progresses through the first years of his teaching career, in order to identify the extent to which they incorporate fundamental principles (not just 'ideas') from BUILT into their K-12 teaching. The enactment of principles of critical framing, of the value of metalinguistic tools, of genre as purposefully structured text, and of scaffolding as central to a teacher's role will be central to the work of novice Science, Math, Commerce, Art, Health, and other teachers across the K-12 curriculum. Because they will be teaching widely divergent groups of students in very different educational contexts with only 18 hours of pre-service preparation, it will not be easy to evaluate the impact of BUILT, as an isolated resource, on K-12 students' performance. But such evaluation is necessary if teacher educators want to provide their novice teachers with manageable models of literacy which support their students in a new communicational world where "there are now choices about how and what is to be represented: in what mode, in what genre, in what ensembles of modes and genres and on what occasions" (Kress, 2003, p. 117).

REFERENCES

Baird, J., & Love, K. (2003). Teaching purposes, learning goals and multimedia production in teacher education. *Journal of Educational Multimedia and Hypermedia, 13*(3), 243-265.

Bernstein, B. (1996). *Pedagogy, symbolic control, and identity: Theory, research, critique.* London; Washington, DC: Taylor & Francis.

Brown, J. S., Collins, A., & Duguid, P. (1989) Situated cognition and the culture of learning. *Educational Researcher,* (18), 32-42.

Bruner, J. (1986) *Actual minds, possible worlds.* Cambridge, MA: Harvard University Press.

Christie, F. (Ed.). (1999). *Pedagogy and the shaping of consciousness. Linguistic and social processes.* London; New York: Cassell.

Christie, F. (2002). *Classroom discourse analysis.* New York: Continuum Press.

Christie, F., & Martin, J.R. (Eds.). (1997). *Genre and institutions. Social processes in the workplace and the school.* London; Washington, DC: Cassell.

Christie, F., Devlin, B., Freebody, P., Luke, A., Martin, J., Threadgold, T., & Walton, C. (1991). *Teaching English literacy: A project of national significance on the preservice preparation of teachers for teaching English literacy.* Centre for Studies of Language in Education, Darwin University.

Coffin, C. (1997). Constructing and giving value to the past: An investigation into secondary school history. In F. Christie & J. R. Martin (Eds.), *Genre and institutions. Social processes in the workplace and the school.* London; Washington, DC: Cassell.

Cope, B., & Kalantzis, M. (Eds.). (2000). *Multiliteracies: Literacy learning and the design of social futures.* Melbourne: Macmillan.

Derewianka, B. (1990). *Exploring how texts work.* Newton, NSW: Primary English Teachers Association.

Derewianka, B. (1998). *A grammar companion.* Newton, NSW: Primary English Teachers Association.

Droga, L., & Humphrey, L. (2003). *Grammar and meaning: An introduction for primary teachers.* Berry, NSW: Target Texts.

Edwards, D., & Mercer, N. (1987). *Common knowledge: The development of understanding in the classroom.* London: Methuen.

Freebody, P., & Luke, A. (1990). 'Literacies' programs: Debates and demands in cultural context. *Prospect, 5*(3), 7-16.

Gibbons, P. (2002). *Scaffolding language and scaffolding learning: Teaching second language learners in the mainstream classroom.* Portsmouth, NH: Heinemann.

Gipps, C. (1994). *Beyond testing—Towards a theory of educational assessment.* London: Falmer Press.

Halliday, M. A. K (1973). *Explorations in the functions of language.* London: Edward Arnold.

Halliday, M. A. K (1994). *An introduction to functional grammar.* London: Edward Arnold.

Halliday, M. A. K., & Martin, J.R. (1993). *Writing science: Literacy and discursive power.* Pittsburgh, PA: University of Pittsburgh Press.

Hammond, J. (Ed.). (2001). *Scaffolding teaching and learning in language and literacy education.* Newton, NSW: Primary English Teachers Association.

Hammond, J., & Macken-Horarik, M. (2001). Teachers' voices, teachers' practices: Insiders' perspectives on literacy education. *Australian Journal of Language and Literacy, 12*(2), 112-32.

Herrington, J., & Oliver, R. (1995). Critical characteristics of situated learning: Implications for the instructional design of multimedia. In J. Pearce & A. Ellis (Eds.), *Learning with technology* (pp. 235-262). Parkville, Victoria: University of Melbourne.

Herrington, J., & Oliver, R. (1997). Multimedia, magic and the way students respond to a situated learning environment. *Australian Journal of Educational Technology, 13*(2), 127-143.

Herrington, J., & Standen, P. (2000). Moving from an instructivist to a constructivist multimedia learning environment. *Journal of Educational Multimedia and Hypermedia, 9*(3), 195-205.

Kramsch, C. (1998). *Language and culture.* Oxford: Oxford University Press.

Kress, G. (2003). *Literacy in the new media age.* London: Routledge.

Lave, J., & Wengler, E. (1991). *Situated learning: Legitimate peripheral participation.* Cambridge: Cambridge University Press.

Lebow, D., & Wager, W. W. (1994). Authentic activity as a model for appropriate learning activity: Implications for emerging instructional technologies. *Canadian Journal of Educational Communication, 23*(3), 231-244.

Love, K. (2002). Scaffolding as a metaphor in disciplinary content and in multimedia architecture: A CD-ROM on Building Understandings in Literacy and Teaching. *Australian Journal of Educational Technology, 18*(3), 377-393. Retrieved from http://www.ascilite.org.au/ajet/ajet18/res/love.html

Love, K. (2003). Mediating generational shift in secondary English teaching in Australia: The case study of BUILT. *L1:Educational Studies in Language and Literature, 3*(1-2), 21-51.

Love, K., Pigdon, K., & Baker, G. (2002). *Building Understandings in Literacy and Teaching* (2nd ed.) (CD-ROM). Melbourne: The University of Melbourne. (A demonstration can be retrieved from http://www.edfac.unimelb.edua.au/LLAE/)

Love, K., & Shrimpton, B. (2002, December). Can one size fit all? The case study of a CD-ROM in teacher education. In A. Williamson, C. Gunn, A. Young, & T. Clear (Eds.), *Winds of Change in the Sea of Learning: Proceedings of the 19th ASCILITE Conference,* Auckland (pp. 389-399).

Martin, J. R., & Rothery, J. (1981). *Writing project report number 1: Working papers in linguistics.* Department of Linguistics, University of Sydney, Australia.

Martin, J. R., & Rothery, J. (1982). *Writing project report number 2: Working papers in linguistics.* Department of Linguistics, University of Sydney, Australia.

Maybin, J., Mercer, N., & Stierer, B. (1992). Scaffolding learning in the classroom. In K. Norman (Ed.), *Thinking voices: The work of the National Curriculum Project.* London: Hodder and Stoughton for the National Curriculum Council.

McLoughlin, C., Winnips, J. C., & Oliver, R. (2000). Supporting constructivist learning through learner support online. *EDMEDIA,* (June).

Mercer, N. (1994). Neo-Vygotskian Theory and classroom education. In B. Stierer & J. Maybin (Eds.), *Language, literacy and learning in educational practice*. Clevedon, UK: Multilingual Matters.

Oliver, R., & Herrington, J. (1995). Developing effective hypermedia instructional materials. *Australian Journal of Educational Technology, 11*(2), 8-21.

Reeves, T. C., & Okey, J. R. (1996). Alternative assessment for constructivist learning environments. In B.G. Wilson (Ed.), *Constructivist learning environments: Case studies in instructional design* (pp. 191-202). Englewood Cliffs, NJ: Educational Technology.

Schon, D. (1983). *The reflective practitioner: How professionals think in action.* New York: Basic Books.

Sheeley, S., Veness, D., & Rankine, L. (2001). Building the Web interactive study environment. *Australian Journal of Educational Technology, 17*(1), 80-95.

Unsworth, L. (2001). *Teaching multiliteracies across the curriculum.* Milton Keynes, UK: Open University Press.

van Leeuwen, T., & Humphrey, S. (1996). On learning to look through a geographer's eyes. In R. Hasan & G. Williams (Eds.), *Literacy in society* (pp. 29-49) London; New York: Longman (Applied Linguistics and Language Studies, Series Ed. C.N. Candlin).

Veel, R. (1997). Learning how to mean—scientifically speaking: Apprenticeship into scientific discourse in the secondary school. In F. Christie, & J. R Martin (Eds.), *Genre and institutions: Social processes in the workplace and the school.* (pp. 161-195). London: Cassell.

Veel, R. (1999). Language, knowledge and authority in school mathematics. In F. Christie (Ed.), *Pedagogy and the shaping of consciousness: Linguistic and social processes* (pp. 185-216). London: Cassell.

Veel, R., & Coffin, C. (1996). Learning to think like an historian: The language of secondary school History. In R. Hasan & G. Williams (Eds.), *Literacy in society* (pp. 185-216). London: Longman.

Vygotsky, L. (1978). *Mind in society: The development of higher psychological processes.* Cambridge, MA: Harvard University Press.

Winnips, K., Collis, B., & Moonen, J. (2000). Implementing a 'scaffolding by design' model in a WWW-based course considering cost and benefits. *Proceedings of the Ed Media Conference.*

Wyatt-Smith, C. (2000). The English/literacy interface in senior school: Debates in Queensland. *English in Australia, 127-128* (May), 71-79.

Wood, D., Bruner, J., & Ross, G. (1976). The role of tutoring in problem solving. *Journal of Child Psychology and Psychiatry, 17.*

Chapter XXVIII
Demystifying Constructivism: The Role for the Teacher in New Technology Exploiting Learning Situations

Paul Adams
Newman College of Higher Education, UK

ABSTRACT

This chapter introduces constructivism as a pedagogical construct from which educational professionals might begin to analyse new technology exploiting learning-teaching interactions. Following a brief history of constructivism as both epistemology and pedagogy it presents an overview of published literature through an analysis of the characteristics of constructivist learning and learning environments and the characteristics of constructivist teachers. Finally, seven principles by which teachers might begin to analyse practice are proposed and discussed via the deconstruction of three fictional, new technology exploiting, learning-teaching vignettes. In this way it is hoped that educators in a variety of contexts will be able to engage in reflection concerning the theory and practice of constructivist pedagogy as related to personally held professional positions.

INTRODUCTION

For many years, analysis of what goes on in the classrooms and tutorials of our schools and colleges has been concerned with *teachers* and what *they do*; the agenda seemed totally fixed on describing their *work*. Whilst it is not possible to disaggregate the teacher form the learning-teaching environment, it seems curious that the focus was so much slanted in this direction. What is noticeable now is the way that teacher impact on *learning* is being scrutinised as the vital component to understanding how learning occurs and can be improved. Having said this, the picture is not entirely positive. Unfortunately, all too often what teachers do is separated out from what teachers think and believe; technical competence sometimes seems to be uppermost in the minds of policymakers. This is somewhat understandable: observable behaviours do help form an impression of any situation, and in this respect teaching is no exception. There is no doubt that being a positive role model is an important feature of professional requirements. However, how one presents oneself must, to some degree, reflect underpinning ideas, concepts, and values. We all have our views, and these *do* influence the way we act and think.

If contemporary attempts to analyse and improve learning are to continue, then it would seem pertinent to spend time considering the foundations upon which learning occurs; this is what this chapter is about. What is presented here concerns ways of thinking about knowledge and pedagogy, how these influence the construction of teacher identity, and the ways in which such constructions are both influenced by and in turn influence learning-teaching interactions. What I propose is a realignment of the learning-teaching interaction along constructivist principles. At the heart of such moves is a need to paradigmatically alter the very concept of teacher, pupil, and learning. A consideration of the use of new technologies through the analysis of fictional vignettes demonstrates how teacher-knowledge discourse, teacher-process interaction, and teacher-pupil relationships are crucial in understanding the place and location for technological application and how such tools cannot themselves alone be viewed as providing structure for a constructivist learning environment. To start then, consider these three fictional cameos.

CASE STUDY ONE

Michael (aged 14) is following a distance-learning information and communications technology (ICT) course. He still attends his local community college[1] for other subjects, but has been placed in this course, as it offers an easy way for the college to meet the requirements whilst still maintaining control over the quality of the work that is produced. The course uses a Virtual Learning Environment (VLE) to deliver the substantive course content. This is supported by group tutorials delivered by the college's own ICT teachers where students come together to discuss the information presented to them, share thoughts and ideas, and highlight any confusion they might have. The rest of the time is spent engaging with the various activities and readings presented by the VLE tutors. In this way, students are responsible for negotiating their own timetable. Even though they are required to work to externally set targets and timescales, essentially they are responsible for charting their own learning progress. The course is part coursework and part examination. Both have been pre-written to meet the requirements of the responsible examining board, requirements that have been established by individuals in the examining organisation.

CASE STUDY TWO

Sarah (17) is watching a DVD she has been given by her tutor to help her explore an aspect of her A2[2] maths course that she does not understand. She has been asked to view the material, make notes about what she has seen, write down anything she feels she does not understand, and come to the next individual tutorial armed with these responses. To aid her note taking, Sarah has been provided with a detailed sheet that explains the structure of the DVD, the topics it focuses on, how to operate the disc so that viewing is supported, and possible ways of recording her thoughts. However, it is entirely up to her as to how she uses the DVD and associated notes. Sarah is able to watch the DVD as many times as she likes before the next meeting. She is also able to review previously observed information, as well as try out some of the simple tasks the programme recommends. The responses she has made and the areas of work she feels she needs to address then set the agenda for the meeting. At this meeting, the teacher listens to Sarah's thoughts and ideas, discusses issues and problems with her, provides support and direct teaching, and facilitates Sarah's own decision making.

CASE STUDY THREE

Jamul and Karen are working on a globalisation topic within their Citizenship Studies year-7 programme.[3] As part of this work, they have been e-mailing certain groups: a multi-national company, an ethical bank, small producers in the developed world, and other school/college students in a developing nation. They have been using a normal e-mail system, and have begun to formulate their own ideas and thoughts about globalisation and the effects the world market can and does have on countries and individuals. The class teacher sets the agenda they worked to, and the students have been given certain tasks they must achieve within the given timeframe. These tasks are both procedural and knowledge based. For example, one of the tasks is to successfully send e-mail and receive a response (procedural), whilst another is for them to identify the main benefits the global market provides to the various groups (knowledge-based). Although ultimately they decide what they write, they have to follow strict protocol to ensure that certain objectives are met and pre-identified issues are raised and discussed.

BACKGROUND

What is knowledge? What is it to know something? When do beliefs become knowledge and how can we be sure that such a shift has, or indeed can, take place? In short, what is the nature and possibility of knowledge? For some, such questions may seem interesting but ultimately superficial. For others, it may seem a total waste of time and effort to even begin to discuss such questions when thinking about learning and new technologies. I would disagree with both views; surely if we, as educators, wish to enhance the cognitive structure of learners in line with some form of information, then the nature of that information—that is, knowledge—must be identified and agreed upon. For example, prior to teaching a child to add up, there must be some sort of agreement as to what the process of adding up looks like, what the outcome of adding up looks like, and why it is important that the child learns such a skill. So, if it is important that we consider what we mean by knowledge, it becomes essential to, if not answer, then at least consider some of the aforementioned questions and quite probably more. Such questions have themselves been the subject of books, doctoral theses, and numerous articles and discussions since the dawn of recorded history (and probably before). I do not propose to delve too deeply into epistemological wrangling; there is not the time and I am not a philosopher. Rather, what I propose to undertake is a consideration of knowledge in two forms: Constructivist and Objectivist.

In order to illuminate the essential differences between these two viewpoints, consider the theory of evolution. Until the middle of the nineteenth century, it was generally accepted in western thought that God created the world in seven days and that 'man' was made in his image. The idea that any other process could have existed was, if not heretical, certainly non-mainstream. With the publication of "The Origin of the Species," Charles Darwin proposed a radical departure from accepted academic and populist thought. His ideas were not immediately accepted, and one wonders what his fate would have been had he proposed them centuries before. However, what was at the time radical and groundbreaking is now considered to be déclassé. This notwithstanding, due to religious or moral reasons, some groups and individuals still hold that such ideas are incorrect. However, they are now considered to hold extremist or out-of-date views that just do not concur with accepted wisdom. I make no judgement here about the veracity of either viewpoint, rather I use this debate to illuminate what should be obvious: that knowledge in whatever form is ever shifting, temporal, and more importantly cannot be disaggregated from beliefs and values. Those who hold creationist views of the assent of humans do so within a belief system. Similarly, those who firmly believe in Darwinian evolution in an extreme form that promotes cultural and/or racial superiority do so within a value set. We may or may not agree with a person's view; indeed we may find it abhorrent. However, it is important we acknowledge that such interpretations are not only part, but also manifest of a set of ideals and values.

In a similar way, the views a teacher holds about knowledge and the processes of arriving at what s/he knows, must ultimately play a large part in the formation of a pedagogical position. For example, if the belief is that learning results from the passive absorption of set truths, then surely this guides the structuring of the learning-teaching environment towards transmission and rote learning. Thus, it is crucial that we highlight the possible origins of such thoughts. Additionally, even though it is essentially individuals who hold beliefs and understanding, it would seem disingenuous to deny the role played by social structures, discourses, and practices in developing such positions. The place for family, cultural identity, social class, gender, and language cannot be overstated in any discussion about the origins and form of knowledge (Bartlett, Burton, & Peim, 2001, p. 82).

Although we now seem to have evidence that the world is somewhat older than a few thousand years and that humankind in its homo-sapiens form has only populated the planet for an extremely short period of time (relatively speaking), we have no real proof that our ideas are anything more than the product of well-informed conjectures and debate. Such arguments are of course a product of certain dispositions. Statements such as "my research proves..." or "there is proof that x leads to y" are manifest of a certain knowledge and research paradigm. However, such "scientifc views"—although often presented as epistemological "fact"—are actually nothing more than the result of investigations within a particular mindset that sees knowledge as hard, real, and capable of being transmitted and acquired in a tangible form (Cohen, Manion, & Morrison, 2000). However, it should be remembered that at present we do not know everything, and that as we investigate further, new forms of knowledge emerge. But, should we as educators worry about whether or not knowledge is 'real and hard' or not? My answer: yes! We should care because pedagogic principles are to a large degree shaped by our epistemological stance (Hein, 1991).

What this leads to are questions about the very nature of knowledge itself: Is knowledge objective and verifiable? Can we say with authority that what we observe is a true, objective result? It is probably the case that those engaged in the physical sciences will answer yes to both of these questions; after all, why else set up experiments with a control and all other attendant issues? Importantly, we must recognise that many now desire to see the social sciences, and in particular education, develop a similar, evidence-based stance (i.e., the gathering of empirical, often statistical data to demonstrate success or otherwise). For such individuals and groups, knowledge exists as an objective, separate entity, verifiable through positivist empirical research. In this vein, truth is judged as *correspondence* between the *research account* and what *is,* independent of the researcher (Pring, 2000). The epistemology is one of knowledge as awareness of objects (a behaviour, a social event, for example) that exists independently of any subject and which has intrinsic meaning: knowledge, as correspondence to a real-world reality, can only be thought of as true if it correctly reflects that independent world (Murphy, 1997). In this vein, knowledge is stable because the essential properties of objects are knowable and relatively unchanging; the purpose of the mind is to mirror reality, and meaning is imposed by the structure of the real world, not the knowledge holder (Jonassen, 1991, p. 28).

For others, the knowledge we posses that makes up our reality is in fact nothing more than a personal construction of the mind (Bodner, 1986). Interaction between the researcher and the researched creates findings with truth defined as *consensus* between informed and sophisticated constructors (Pring, 2000). In this light, knowledge is seen as an indication of how the world might be. Given the vagaries of present wisdom, information is seen not as truth, but rather as prediction (Postlethwaite, 1993); various theories and ideas are accepted not because of their rightness, but because of their superior ability to predict what will happen given a series of predetermined events (Hanley, 1994; Adams, 2003).

To the constructivist, concepts, models, theories, and so on are viable if they prove adequate in the contexts in which they were created (Von Glasersfeld, 1995, p. 7).

Let us spend a bit more time on this: consider the theory of gravity. We cannot see, hear, taste, or smell gravity; we can only experience a force that acts to pull us towards the ground, a force that has been named 'gravity'. What contemporary society accepts is that Newton and subsequent scientists are, given our particular and contemporary erudition, correct in their explanations for these experiences. An objectivist viewpoint would argue that the theory of gravity is scientific *fact*, verifiable by empirical testing: gravity *exists*; it *is* the force of attraction one body exerts on another and was *discovered* by Newton. Conversely, constructivists would posit that the theory of gravity is presently *the best way to describe* what will happen when two objects come into close proximity. In their view, it is the best *prediction*. This position contends that Newton did not discover gravity, rather what he did was describe a series of experienced events; he *constructed* the knowledge we now share. Whilst objectivists argue that knowledge is something we *acquire,* constructivists contend that knowledge is something we *produce*; the objects in an area of enquiry are not there to be discovered, but are invented or constructed (Mautner, 1996). Constructivist epistemology holds that knowledge reflects an ordering and organisation of a world constituted by our experiences (Von Glasersfeld, 1984, p. 24).

Now some might argue that the above representation does an injustice to objectivist interpretations. They might say that no rational scientist would describe contemporary views on gravity (or any other scientific theory) as immutable fact and that what I have presented here is an extreme form of the objectivist viewpoint. This might well be true, but I would counter that often science, if not presented as fact, is often viewed in such a way. How many people accept a doctor's decision unquestioningly or use the term scientifically proven to support an argument?

A BRIEF HISTORY OF CONSTRUCTIVISM

The above discussion outlines reasons for the importance of epistemological enquiry for education and one of the key debates central to such enquiry. However, as with most things it is beneficial to locate thoughts and ideas within a historical framework. Finding such an account of the emergence of constructivism is not particularly easy; most start with either John Dewey or Jean Piaget. However, one of the foremost writers in the field, Ernst von Glasersfeld (1995), provides an illuminating history of how, in his view, the birth of constructivist thinking can be traced back to around the 6th century BC. The path such developments seem to follow originates in the work of sceptics such as Xenophenes and Protagoras who came to the realisation that knowledge of our experience cannot exist, for even if we could say what is true, we could not know it was so; we can have no real knowledge of the world, as we are unable to determine the relationship between our experience and what there is *before* we experience it. Philosophers such as Locke, Berkley, and Hume kept this tradition alive through their views, respectively, on the source of ideas being a reflection upon mental operations, indispensable concepts (time, number, etc.) as mental constructs, and the act of association forming relational concepts. Von Glasersfeld (1995) concludes this history of ideas with Kant's analysis of the rational domain confirming the inaccessibility of anything posited beyond the reach of experience (p. 49); the world we understand is only complete due to rational heuristic fictions.

This epistemological time-trip has, I feel, served the purpose for which it was intended: to demonstrate that knowledge can be said to exist as a mind construct, subjective in character and representative of personal reality. What this notes is the journey that thinkers and philosophers have undertaken in order to attempt to clarify the relationship between the known and the knower. It should be noted, however, that Von Glasersfeld's work was written to promote a particular version of constructivism to which I shall later refer. What is interesting, and possibly more pertinent to readers of this chapter, indeed this book, is how such ideas have been translated into pedagogy.

Once again this is not an easy path to tread. Whilst it is accepted that writers such as Dewey, Piaget, Vygotsky, and Bruner have all played a major role in the application of constructivist principles to education, it should be noted that application did not always stem from such epistemological examinations as those presented above. Certainly, the latter three drew their theses from cognitive psychological research into the ways in which children learn and develop. Dewey, however, utilised principles that could be said to be constructivist in approach as a rejection of his own static and authoritarian education, rejections which led to the establishment of the 'laboratory school' at the University of Chicago in 1896. Here, education focused on:

- The environment and the practical nature of learning.
- The quality of teacher-pupil interactions.
- The child as the starting point for learning.

- The importance of the social experience and impulses as a starting point for learning.
- The belief that education should not focus on some dimly perceived view of the future, but rather on the here and now of the pupil's existence.

However, it should also be noted that as far back as the 18th century, Pestalozzi was maintaining that education should rest on children's natural development and sensory influences gleaned from experiences within their home and family life.

What is now widely accepted is that constructivist pedagogy grew out of a rejection of behavioural approaches to learning and instruction. Behavioural psychology is premised on the view that as we can never truly understand a person's thought processes, we should not concern ourselves with them, rather we should use observable behaviour as an indication of a person's learning progress. Thus, Behaviourists use stimuli to describe, predict, and control specific actions. In their view, learning occurs as a result of adaptation: a process of making associations that leads to alterations in displayed behaviour. Although there are a multitude of Behaviourist approaches, they all operate from the same fundamental principles. They all:

- *Locate positivist, empirical research as key.*
- *Follow environmentalist principles*: The belief that human nature is learnt through the environment and interaction with it.
- *Adhere to egalitarian principles*: The assumption that all individuals have a common identity; differences are merely attributable to the reactions we display.
- Stem from the belief that humans are analogous to machines that can be programmed and re-programmed *(reductionism)*.

In an educational sense, behaviourism considers the learner to be a 'tabula rasa' or blank slate to be filled by teaching that adheres to common principles and procedures; to improve learning teachers must improve connections between stimuli and response. It utilises a transmission approach borne out of a desire to inculcate the learner into ways of understanding a rational and objective world. Its epistemological stance is Objectivist in nature.

The dissatisfaction with Behaviourism as a pedagogical underpinning led some to contest the assumptions upon which it is based. It was not enough that the learner be considered an empty vessel into which knowledge is poured; rather, the idea that learners actively construct meaning within the learning context began to gain currency. The most notable initial protagonist in the cognitive movement was undoubtedly Jean Piaget (1896-1980). Whilst conducting intelligence tests for Alfred Binet, he began to notice that children of certain ages always made the same mistakes when answering questions. Through empirical research Piaget began to question the assumptions of the rationalists and empiricists and formulated his own standpoint concerning the development of epistemological positioning: as children are essentially little scientists constantly asking questions and posing explanations, Piaget concluded that the active construction of knowledge rather than mere acquisition must be occurring. As his work was with young children, his findings quickly became attributable to the educational context. From this emerged constructivist pedagogy.

Whilst Piaget was working, a young Russian scientist by the name of Lev Vygotsky was conducting research in Stalin's Russia. As a critical fan of Piaget's work (Sutton, 2000), he became fascinated with the idea of the construction of knowledge in children. The work he undertook led to interesting insights into the role for language and culture on the construction of knowledge in the learner (Sutton, 2000). His untimely death whilst still in his 30s and the closed and secret society within which he lived meant that his work did not come to international prominence until the 1970s, whereupon it became a worthy addition to the constructivist field.

As Piaget influenced Vygotsky, so Vygotsky influenced Jerome Bruner. Until the 1970s Bruner's work centred mainly around interpersonal influences on human cognition, particularly that of young children. As Vygotsky's work came to prominence, he became increasingly critical with this direction and began to focus his work on the social and political context for learning. From this he postulated his theory of Instrumental Conceptualism, whereby learning consists of information acquisition, manipulation and transformation, and testing. Once again, the overriding feature for his work was the belief that learning does not simply occur *to* individuals, but something they themselves make happen (Fontana, 1995).

CONTEMPORARY CONSTRUCTIVISM

At the heart of the constructivist epistemology/ pedagogy debate lie beliefs that:

1. Learning is an active process.
2. Knowledge is constructed rather than an innate attribute or passively absorbed.
3. Knowledge is invented, not discovered.
4. Learning is essentially a process of making sense of the world.
5. Effective learning requires meaningful, open-ended, challenging problems for the learner to solve.

Although this signals a shift towards more active and experiential learning-teaching episodes, epistemological underpinnings vary between the different constructivist camps. Although the four main protagonists cited above form the backbone of present-day constructivist epistemology, it should be noted that constructivism is not one coherent set of proposals or features. Indeed, if one considers the work of the above, it becomes clear that Piaget's work was dominated by a desire to understand the role for action both on and in the physical world, whereas Vygotsky viewed understanding as social in origin. The work of Bruner followed in Vygotsky's footsteps to some degree, whereas Von Glasersfeld's idea of Radical Constructivism stems from a different premise yet again. Essentially, as an alternative to behavioural or objectivist modes of delivery, constructivism asks a range of questions that seek to understand how we arrive at the knowledge constructs we have and what these knowledge constructs mean for understanding influences on our thought processes. The fluid nature of learning that for most constructivists now predominates necessitates that teachers work from the view that different learners will understand (construct) knowledge in different ways, and that these differences stem from the alternate ways that individuals acquire, select, interpret, and organise information.

As an epistemology, constructivism considers two aspects: the relationships between the learner and reality, and the learner and knowledge (Kanuka & Anderson, 1999). For some, reality is an objective truth, whilst for others it exists merely as subjective speculation. This is not to suppose solipsism however; there is no dispute that the *physical* world exists, rather the disagreement centres over how an awareness of reality is constructed. The first position posits that there is one external interpretation of reality and that individuals need to construct a valid representation that comes as close to true understanding as is possible. The second position signals a belief that there is no shared reality (Suchman, 1987); individuals construct an interpretation that is bounded by personal, cultural, and contextual factors. In other words, reality is nothing more than speculation. Additionally, at the heart of the constructivist debate reside discussions concerning the location for the learner's construction of knowledge. Essentially, the discussion here centres on how individuals come to construct knowledge about reality: do they do so individually or socially? Thus, one theorist might decree that knowledge construction is born out of individual interpretation and thought, whilst for another the knowledge a learner develops is a result of the social processes through which they move. From these two positions, it is possible to identify a typology for the epistemological basis of constructivism (Kanuka & Anderson, 1999). The typology identifies four interpretations. The first two speculate that knowledge is individually created, whilst the third and fourth identify knowledge creation as stemming from social interaction.

Cognitive Constructivism

Here knowledge construction occurs as a result of dissonance between previously constructed cognitive structures and new information. Through a process of assimilation (including new information in previously held knowledge forms) and accommodation (the creation of new mental structures or schemas), an individual constructs new ways of thinking that aspire to understand an objective, external reality.

Radical Constructivism

This position differs from cognitive constructivism in that it hypothesises that reality is speculation—that is, nothing more than individual opinion. As no two people can ever experience the same situation in the same way, the individualistic nature of the interpretation creates individual positions. Thus, knowledge is tied not to an external reality, but to whatever the knower conceives it to be (Jonassen, 1990). According to Von Glasersfeld (1995), there is no objective reality independent of our thoughts.

Situated Constructivism

In a similar way to radical constructivism, this position locates knowledge as an indicator of unique perspectives; there is no one reality. Where this position differs from both radical and cognitive constructivism is in its view that knowledge is constructed as a result of social processes. Social patterns are observed, and these in turn lead an individual to create a personally held and unique construction of a reality interpretation.

Social Constructivism

This view is the most widely held and explored. In support of situated constructivism, it identifies knowledge as being the product of social interaction, interpretation, and understanding, with language as the mediating key (Vygotsky, 1962); the creation of knowledge cannot be separated from the social environment in which it is formed. Although an individual creates knowledge through the aforementioned social interaction, the process of communication is itself the reality due to shared meanings and common understandings. Although this position does not advocate that an external reality exists in the same way as cognitive constructivism does, it promulgates the view that shared meaning and understanding can ensue due to the negotiations mediated by language and communication; knowledge is individually constructed, but socially mediated and shared. Consensus between different subjects is held to be the ultimate criteria upon which to judge the veracity or otherwise of knowledge. Social constructivism has been likened to the middle ground (Heylighen, 1993). In identifying the need for social consensus as the grounds for agreeing "on truth" or "reality", it avoids the individual relativism of radical constructivism and the need for personal dissonance that describes the cognitive constructivist perspective.

TOWARDS A PEDAGOGICAL POSITION

I stated earlier that constructivism has been both conceived of as an epistemological position and a pedagogical one. So far the discussion has centred on constructivism as a way of thinking about knowledge. As stated earlier, such views will form part of the construction of a variety of belief systems, which in turn will influence individual thoughts and actions. With this in mind, it should be clear that to merely locate constructivism as an epistemology is not entirely helpful in the learning-teaching debate. If educators and indeed learners are to understand and develop pedagogy, then such pedagogy must be articulated. Such deliberations have been at the heart of the constructivist pedagogical debate for many years, have thrown up many ideas and variants and are usually directed in either or both of two directions: the characteristics of constructivist learning and learning environments, and the characteristics of constructivist teachers.

The Characteristics of Constructivist Learning and Learning Environments

When constructing any sort of learning and teaching environment, the relationship between the knowledge to be learnt, the pupil, and the teacher is of vital importance. The differing constructivist positions emphasise different aspects, there is commonality in their position (Ernest, 1995). For all, the nature of the learning environment is one of experimentation and dialogue, where knowledge is seen within the context of problems to be discussed and solved. The aim is to become aware of the realities of others and their relationship with one's own. What a constructivist environment posits is an appreciation that learning is a process of active, individual knowledge construction (Woolfolk, 1993) within social forms and processes. What learners require therefore is language that enables negotiated, social interaction within prevailing personal-social constructs. Thus, instead of seeing learning as a process of acquisition, assimilation, and application, a more appropriate view is interpretation, synthesis, and evaluation. As the knowledge we possess is dependent upon socially defined ideas, one must first learn to decode attendant language (Goodman & Goodman, 1990; Kanuka & Anderson, 1999). Thus, students who fail to 'understand facts' can actually be said to have failed to accurately synthesise information in order to relay a socially acceptable interpretation (Cognition & Technology Group, 1991); problems reside in personal interpretations that have less accurate predictive validity within the mediated social environment. In the Constructivist sense, learning becomes the development of personal meaning that is more able to predict socially agreeable interpretations. In order that it might effectively occur, students must be enabled to access those elements of learning that support the development of personal interpretation (Hein, 1991). Through an appreciation of thought processes, cognitive conflict, and predictive ability, learning ceases to be the acceptance of fact and associated problems of wrongness, and becomes personal interpretation, question creation, and appreciation of validity as defined by socially recognisable and appropriate forms. Thus, student discourse becomes shared and transmutable, rather than owned and unchanging. Essentially, learning ceases being bulimic (accommodation and regurgitation) and becomes satiation (internal modification through social negotiation). As knowledge is an indication of how the world might be, a variety of theoretical possibilities are acceptable, not because of their rightness, but because of their ability to predict. Truth discovery is not important, the construction of viable explanations of experiences is.

From the above, it is possible to identify two underpinning principles for a constructivist, learning environment (Hein, 1991, p. 1):

- A *focus on the learner* in thinking about learning and not on the subject matter to be taught.
- Recognition that there is no knowledge *independent of the meaning attributed to experience* (constructed) by the learner, or community of learners.

From this, Hein (1991: 2-3) highlights nine principles for constructivist oriented learning:

1. Learning is an active process; that is, learners need to do something rather than just listen and observe.
2. People learn to learn as they learn; pupils will construct meaning and systems of meaning, for example sorting a set of cards leads to learning about the relationship between concepts as well as what a group is.
3. Constructing meaning is mental; although all of the senses are vital for effective learning to take place, ultimately learning happens in the mind.
5. Learning involves language; the language we use influences our learning and the way we use language; for example, egocentric speech (talking through actions as they are undertaken) will guide and assist in the construction of knowledge.
5. Learning is a social activity; connections with others enrich and aid our understanding.
6. Learning is contextual; we learn in relation to what we already know, we build on existing experiences and knowledge.
7. One needs knowledge to learn; the previous knowledge structures we have direct and assist in the construction of new thoughts and ideas.
8. It takes time to learn; repeated exposure and thought give rise to further constructions of meaning.
9. Motivation is the key; this should not be taken to only mean feelings of desire, rather it should also be taken to encompass those aspects of learning that include 'the reasons why' and thereby give us a reason to learn.

What this provides is a framework for constructivist, learning environments. Taking this further, Wheatley (1991) identifies a three-component, problem-oriented approach whereby tasks that are problematic for the students are selected, small-group work is utilised as a means to stimulate dialogue, and debate provides a culmination where the whole class shares their thoughts and ideas. In a similar way, Saunders (1992) discusses a four-step approach to science education: the creation of investigative situations; the utilisation of active cognitive involvement (for example, egocentric speech, data interpretation, or hypothesising); the use of small-group work to stimulate higher cognitive activity and discussion; and higher-level assessment that is fit-for-purpose. This is very similar to Yager's (1991) strategy, somewhat more explicit about the various stages. Here, the teaching cycle is proposed as:

- The use of starter activities that ask questions, pose possible solutions, and identify different perceptual positions.
- The use of focussed "play" and discussion that seeks to experiment, observe, analyse, and solve problems.
- The communication of thoughts and ideas to peers in order to locate new knowledge within existing structures and review other solutions.
- The use of the above three parts to take action on a specified problem. The action itself will generate new problems and possible areas for action, and so the cycle begins again.

What all these views share is the use of:

- Problems or situations to stimulate debate within an area students find interesting and exciting.
- Open-ended questioning to generate further exploration.
- Cognitive conflict through the inclusion of situations and ideas that do not conform to existing ways of thinking.
- Group work to facilitate communication and dialogue.
- Assessment that is fit-for-purpose, real, and informative, and that uncovers not only outcomes, but the processes gone through to arrive at new knowledge constructs as well. The paradigmatic emphasis is not on arrival, but on the journey.

It appears that the constructivist-learning environment is a place where experimentation, discussion, and debate exist. What this provides is an alternative to the objectivist teaching-learning approach (Tam, 2000). In order that learning might effectively occur, the student must be able to access those elements of learning that support development towards personal interpretation (Hein, 1991); constructivist teaching facilitates the development of personal meaning.

The Characteristics of Constructivist Teachers

For objectivists, the main aim of the teacher is to communicate and instil a set of predetermined and agreed facts (Reeves, 1992). In this way, the learning-teaching situation ensures that learners acquire required information. More knowledgeable others (in this case teachers) interpret the information to be transmitted, and decide on the best way to communicate and impart this so that learners can best replicate the acquired and required structure and content. The logical conclusion of such a view is that classes should be dominated by teacher exposition, agreed texts, and methods of instruction that best assist students in negotiating summative assessments designed to unpack and identify what has been learned and understood. This should not presuppose that pupil involvement and discussion do not take place, but ultimately the purpose and direction of this peer interaction has been preset. Rather than using debate and discussion as a means to elucidate and unpack personal ideas and theories, such activities become a means whereby teachers highlight and correct misunderstandings and inconsequential knowledge.

As a counterpoint, the constructivist-oriented teacher is no all-knowing oracle, rather he or she is an organiser and but one source of information (Hanley, 1994; Crowther, 1997). His or her role is as facilitator (Copley, 1992), working to provide students with opportunities and incentives to build up knowledge and understanding (Von Glasersfeld, 1996). Quintessentially, the teacher becomes learner: the teacher provides support and guidance whilst diagnosing student interpretation to inform and direct further action (Driver, Aasoko, Leach, Mortimer, & Scott, 1994). What this necessitates is a fundamental re-conceptualisation of teacher identity towards one centred on *students* and their *learning,* rather than subject and attendant knowledge forms. Essentially what is required is a paradigm shift: the abandonment of the familiar to embrace the new (Brooks & Brooks, 1993). Such a position requires a change in emphasis from oracle to but one source in many, from director to co-actor, from outcomes driven to process oriented, from questioner to listener. In a practical sense this re-conceptualisation focuses thinking on activities that provide real-world, case-based learning to enable authentic, context-oriented, reflective practice within a collaborative and social environment (Jonassen, 1994; Rice & Wilson, 1999). Most contentiously, the constructivist environment advocates the transference of power to set the learning agenda to the learner. What must be remembered however is that constructivism does not remove the need for the teacher, rather it re-directs teacher activity towards the provision of a safe environment whereby student knowledge construction and social mediation are paramount. Such orientations require teachers to understand the requirements and stages through which students travel on their journey towards personal understanding. The process of scaffolding learners' journeys from unconscious incompetence to unconscious competence and beyond is a key teacher requisite (Vygotsky, 1978; Omrod, 1995). On a practical level, teachers should:

- Be accessible,
- Invite student decision making,
- Encourage 'what-if' questions,
- Encourage students to use their own methods,
- Promote discussion and communication,
- Use an element of surprise in lessons, and
- Demonstrate that they enjoy the work and make activities enjoyable as well. (Wheatley, 1991)

Not only does constructivism challenge views about the nature of knowledge, it also necessitates a revisiting of the learner-teacher relationship and a reappraisal of the construction of teacher/pupil identity. More importantly, it fundamentally re-describes the whole nature of learning itself. When one considers this, one realises that it is more than just pedagogy; its underpinnings challenge some of the very bedrocks upon which contemporary educational policy has been built.

CONSTRUCTIVISM AND NEW TECHNOLOGIES

One area where constructivist-learning theory has been widely discussed and debated concerns its relationship with new technologies. In many ways, such advances have transformed education: consider the rise of the World Wide Web and how this has altered the ability of individuals to glean information within seconds following the touch of a few keyboard keys. One can see that such developments are far reaching both in terms of application and implication. The increase in access to new knowledge forms highlights the pressing need for education to understand itself as both transformed and transforming. Ultimately, individuals must learn to manage information rather than regurgitate it (Mann, 1994). However, even in a technology-rich environment, it is imperative that new technologies be seen for what they are: resources. The goal is not the technology itself; this is merely an aid to the learning process (Strommen & Lincoln, 1992). However, it is becoming increasingly obvious that new technologies of all sorts are adding to the learning experience (Salomon, 1991), and that where such innovations are appropriately and sensitively employed, changes to the learning-teaching environment have often ensued (Collins, 1991).

Constructivism, although not originally described with such developments in mind, seems to have found itself a welcome bedfellow in new technological advances and their applications to learning and teaching (Tam, 2000). The fact that constructivist-learning environments encourage play and experimentation is synonymous with that which new technologies have to offer. The manipulation of ideas, data, information, and thoughts, and the ease with which conversation can be facilitated and extended through the written and spoken word, is something that technological advances not only advocate but make happen. In fact, their very existence stems from human desires to extend the frontiers of knowledge, part of which necessitates communicating and understanding with new ideas and cultural forms. The individual-social duality of technological use, coupled with its encouragement of experimentation, reflection, dialogue, and debate, seems to naturally orient its underlying pedagogical basis towards that of constructivism.

However, as was noted earlier, pedagogical processes are value laden and context driven; the values and ideals that the teacher, institution, or exam board holds and expresses are guaranteed to be a major source of educational and procedural direction. It is not the tools themselves that define or even redefine the learning environment; it is the collective wisdom and vision of the community and the shape given to such beliefs (Riel, 1990). Similarly, just because technological applications are utilised does not automatically mean that constructivist principles are followed. For example, there exist many computer packages that merely require students to fill in blank spaces or match ciphers on the screen with associated pictorial representations. Indeed, many aspects of VLEs can be the complete antithesis of constructivist learning. Nowhere is this more acute than in some of their assessment tools where cloze procedures and multiple-choice activities do nothing more than examine an individual's ability to attach previously transmitted knowledge to a predetermined question set, with no investigation of attendant processes on the road to the identified solution.

If new technologies are to keep in step with constructivist-learning philosophies, it is important that the principles of constructivist learning as related to such developments be articulated and debated. The above journey painted a picture of constructivist-pedagogy as being deeply rooted in discussions about the nature of knowledge, the psychology of learning, the role for socio-cultural aspects, the construction of teacher and pupil identity, and the nature of their relationship. Furthermore, the place for dialogue as a precursor to and antecedent of socio-constructivist learning that engenders socially mediated, predictive abilities requires that the application of new technologies actively facilitates the utilisation of intercommunication. With this in mind, I offer the following as underpinning principles against which the use of new technology can be examined, with some associated, practical thoughts.

Principle 1

Constructivist learning-teaching environments create a framework within which students develop skills and understanding to extend and develop their prior knowledge. Learning is essentially a process of sense making.

It is imperative that all teaching initially identifies prevailing student-held ideas and theories. To start designing learning-teaching episodes from the point of view of that to be imparted misses the constructivist view that 'new' information can only become meaningful in relation to that already constructed and held.

Principle 2

Learning-teaching interactions should reflect real-world situations and involve authentic problem solving. Reflection and meta-cognition are essential aspects of constructing knowledge and meaning.

In designing learning opportunities the question needs to be asked: How is this meaningful for my students given their life-world? The requirement to reflect on that, which has been personally constructed within the social world, can only carry meaning if it can be related to personal reference points. It is crucial that learning enables students to solve meaningful problems.

Principle 3

Knowledge and beliefs are formed within the learner. Active exploration of personal experiences should be encouraged so that meaning might be imbued upon such experiences.

Whenever students are asked to construct understanding, opportunities must be provided that require the deconstruction of such views from within personally held constructs. Thus, rather than asking "What do you think and why?", learners must be encouraged to explain what they think, why, and how this is itself expressive of alterations in personally constructed understanding.

Principle 4

Learning is a social activity. Shared enquiry, social negotiation, mediation, and interaction are crucial; the learner must engage in a dialogue both with and within the learning environment.

To engage with and in constructivist learning environments requires talk. Students must be provided with opportunities to share thoughts in safe and constructive ways. Merely grouping students is not enough; they need guidance in how to discuss so that the social mediation of personally constructed knowledge forms can be undertaken. The role for the teacher is to provide the technical means and language so that discussion might be meaningful and beneficial.

Principle 5

Student self-regulation and motivation are crucial. In order that learning is meaningfully attached to prior knowledge, students need to feel a sense of ownership over the material, learning process, and outcomes.

Students need to feel involved in the production and direction of learning opportunities. Asking pupils what they wish to consider and how they wish to go about investigating and presenting their work not only helps to meet individual learner needs and preferences, but also engenders feelings of importance and worthiness.

Principle 6

The teacher's role is to act as a guide and facilitator. Teachers exist not to 'instruct', but to draw out knowledge and understanding as students create and socially mediate personal perspectives into the social domain. They are but one source of information.

It is often said that if one does not know the answer, one should say so. As a starting point this is probably as good a place as any. However, underneath this lies a deeper issue: How can any one person be the sole keeper of all that there is to know? Furthermore, if learning, as constructivist principles posit, is the creation of knowledge forms for social mediation and consumption, then no teacher can ever be an oracle. In practical terms this means a recognition that one's own knowledge forms are nothing more than that: one's own views, tried and tested in the social domain for "accepted" predictive ability.

Principle 7

Assessment should form the cornerstone of a negotiated, process-oriented learning-teaching interaction; formative assessment is therefore crucial and should be used to direct both teaching and learning. Such assessment should be self, peer, and teacher initiated and oriented, and undertaken with a view that outcomes are varied and often unpredictable.

Assessing student learning should be considered akin to archaeology: unearthing that which so far has remained fully or partially hidden so that others might ponder its complexities, meanings, and implications. What assessment should not seek to do is merely attribute "right" or "wrong" to that which has been presented. Considering the outcomes of assessments from a constructivist viewpoint reorients the outlook; rather than illuminating misunderstanding or incorrectness, assessment should provide the basis for a consideration of why answers are considered to be incorrect. In other words, how and why does this position taken by this student not successfully mediate into the social domain?

USING THE PRINCIPLES TO ANALYSE THE VIGNETTES

It would appear that connections between constructivism and the use of new technologies within the learning-teaching environment are readily and easily made. Not only can such environments actively utilise new technologies, the principles of constructivism might also direct the creation of technologically exploiting, learning-teaching environments. With this in mind, it would seem useful to return to the three case studies presented at the beginning of this chapter and analyse them using those principles.

The environment with which Michael is engaged is something that increasing numbers of students encounter. VLEs are progressively being used as a means to support learning and teaching. They are multi-faceted and employ a variety of facilities such as Web pages to impart information, assessment packages to unearth achievement, synchronous (e.g., chat rooms) and asynchronous (e.g., e-mail and Web boards) communication tools, and storage facilities such as diaries and virtual lockers within which to post and store work. In Michael's case, the community college is using such an environment so it might deliver course material in a cost-effective way whilst still maintaining control over the quality of some of the learning-teaching interaction. A team of tutors who are not necessarily connected with the college or the students themselves write the VLE materials. Their work is reinforced by regular tutorials with teachers from Michael's college. These tutorials are designed to support individual learning within a social environment.

It could be argued that one of the main strengths of a VLE is that as students can gain access from any computer at any time via a user name and password they are able to work when and where they wish. The only requirement is access to a computer and data line. Michael is following a course that has been written by an exam board; tutors have to construct the VLE to meet these requirements. It would appear that both environments (online and group tutorial) permit connections with and extensions of prior knowledge. Additionally, the nature of both could utilise problem-solving techniques within the real-world scenario of the learner. The group tutorials offer the opportunity for students to share thoughts and ideas, make connections, and begin to mediate their thoughts into the public domain. This aspect can also be enhanced by the use of online communication tools; synchronous chat offers similar opportunities to those prevailing in the face-to-face group tutorials, whilst asynchronous messaging systems offer students opportunities to present carefully thought-through personal constructs. In both ways, the social aspect of learning is actively supported. Apart from timetabled group tutorials, students are responsible for organising learning opportunities including time and place. The process seems entirely driven by personal needs and desires; motivation is intrinsic in nature. Both the VLE tutors and the ICT teachers at Michael's college seem to act as facilitators of the learning environment and attendant processes. As the work is presented on a Web-based platform, opportunities exist for the integration of knowledge forms from sources other than the VLE, such as the World Wide Web.

What has been presented above is a theoretical exposition of the virtues of Michael's VLE against the seven previously articulated principles for constructivist learning. However, the material presented and the nature of the group tutorials will play a major part in deciding whether this has constructivist overtones or not. This demonstrates the very problem with attempting to apply constructivist principles to a learning-teaching environment that exploits new technologies. Whilst tools and applications might be perfectly suited to such orientations, essentially it is the nature of the human interface both with the VLE and other learners/teachers that determines the nature of the environment. It would be just as easy to go through the above discussion and describe the complete antithesis of a constructivist environment due to static and content-laden material, activities that require mere regurgitation of fact, Web discussions designed to check for content understanding rather than procedural engagement, and assessment tools used as a means to identify those who have understood described truths and those who have not. Furthermore, group tutorials might offer students nothing more than opportunities to check-in and receive feedback about assignments and progress through the VLE. What is also noteworthy is that the course itself has been established by a group of individuals with no prior knowledge of the students. It is a course that might well be followed by thousands of students throughout the country with set outcomes that aim to establish whether certain skills have been completed satisfactorily and whether certain gleaned information can be re-organised and regurgitated. This is most definitely not akin to a constructivist ideology.

I highlight these issues to reiterate a previous point: although constructivism necessitates a re-conceptualisation of teacher identity, it does not remove the need for the more experienced other to scaffold and support the work of the learner (Vygotsky, 1978). Such is the nature of the constructivist-learning environment that the role for the teacher whilst ceasing to be 'traditional' is nonetheless essential. The same points can be made about cameo 3—the use of e-mail to contact others around the globe to glean their perspective seems at face value to square nicely with constructivist principles. Not only does it utilise dialogue, it also necessitates the construction of personal knowledge forms that have to be mediated into the social arena. Given the contested nature of globalisation (Delanty, 2000), one could also assume that the whole project is premised on the desire to see learners discover, construct, share, and agree. However, what is evident here is the presetting of the learning agenda by the teacher. Objectives have been written to ensure that learning occurs and strict protocols have to be followed to ensure that predefined issues are raised and discussed; essentially the agenda is predetermined. Once again, the role played by the teacher is crucial in prescribing the orientation of the learning environment, one that seems far removed from the principles of constructivism.

The identification of the importance of the teacher in orienting the learning environment is nicely demonstrated in cameo 2. Here, Sarah is experiencing problems with her maths. Rather than present her with a series of instructions to be remembered and operated, the teacher uses the DVD to assist Sarah in structuring her own learning. What Sarah is required to do is analyse the information presented to her and make sense of that which she views: this can only be undertaken within the constraints of her understanding. Through the active exploration of her experiences, Sarah is encouraged to identify the problems and successes she encounters. As with the other two cameos, the learner is tasked with organising the time, location, and use of the new technology, but the outcomes have not been established; indeed, the process of discovery that Sarah undertakes itself formulates the outcomes. The knowledge she constructs and adapts forms the basis for subsequent tutorials at which she takes the lead. The tutor's role is to listen and support the development of Sarah's personal ways of knowing and decision making, thus permitting Sarah to mediate her personal cognitive development into the social domain.

IMPLICATIONS FROM THE CASE STUDIES

Essentially what is different between this case study and cameos 1 and 3 is the role of the teacher. In the first the role was deliberately ambiguous, whilst in the third it was more traditional. Sarah's vignette, however, highlighted a shifting emphasis for the tutor's role; the paradigm had altered. The professional was not in place to direct and instruct, but rather to facilitate and guide. All three situations carried with them elements of constructivist ideology, but it was only when the role for the teacher was specified that the nature of the situation became clear. Certainly, it is probably safe to assume that case study three, although carrying elements of constructivism, is not so inclined. Whilst it might have initially appeared so, the predetermined nature of the work—coupled with the pre-directed nature of the objectives and processes—meant that student control was never really evident. What this sketch described was a series of active learning opportunities and nothing more. Although the students have control over the form of the e-mail messages, and the thoughts and ideas they have about the topic, essentially they are being directed to a series of teacher-decided outcomes. In this situation the use of new technologies was interesting and probably gave the students a richer learning experience than they might otherwise have had; certainly they had opportunities that mere bookwork could never have provided. They were also immersed in a truly social environment: the need to work as a pair to determine the types of questions to ask, the format in which to locate these, and the increase in the immediacy of response would most certainly have added to the enjoyment these students would have experienced. However, what was lacking was any real sense of ownership of the project. Not only had the original theme been set for them, but the necessity to engage with certain tasks and procedures that actually might not have been relevant to the work with which they engaged meant that for Jamul and Karen, learning became nothing more than hoop jumping.

Scenario 1 presented a different set of considerations. Notwithstanding issues surrounding the closed nature of many VLE assessment tools, this cameo demonstrates many of the features of a constructivist-learning environment. Although we are not sure about the nature of the material being presented and the format and structure of the group tutorials, it is quite possible to conceive of this as a learning-teaching environment full of promise in constructivist terms. However, the very information that is missing is that which we require if we are to make an informed judgment. The role for the teacher in the constructivist classroom is thus accentuated and emphasised. Not only this, but the way in which the very material is conceptualised and presented as a reflection of teacher-orientation influences decisions about the nature of the environment.

SUMMARY

The role for constructivist thinking in education cannot be overstated. Its underlying epistemology presents a basis upon which to begin to comprehend why knowledge—even when described as immutable, scientific fact—can be so transitory and temporal in nature. Such questioning is the hallmark of an enquiring mind, something that most educators wish to engender in their students. However, the very structure of professional belief can itself present a barrier to the recognition that other methods and ideas are acceptable. This however becomes less problematic when considering the works of significant others such as professors and scientists, as here breakthroughs in knowledge come either as a result of deep reflective practice far beyond our ability or as the result of an 'agreed' scientific process. Yet, what is conveniently forgotten in both situations is that the knowledge forms upon which these ideas or experiments are built are in fact nothing more than models conceived of and described by other human beings. There is no realm of ideas that exists mind-independently of us.

Following this argument into the pedagogic realm presents an interesting set of issues to overcome. If knowledge is nothing but construction, then how does one avoid relativism? Similarly, if teaching is about assisting students to construct their own knowledge forms, then how do we conceive of curriculum? The answer to both of these lies in dialogue, communication, and mediation. Although thoughts and ideas are ultimately the property of the holder, their validity is determined by social mediation and agreement, the tool for which is language. Through the articulation of ideas and thoughts, the knower not only enters into a cultural dialogue, but is also part of the construction of that very culture. The student who reveals new understanding about a previously perplexing topic not only uses the cultural tool of language but also, through sharing and describing thoughts and ideas, creates the very culture within which they discuss. The articulation and transformation of personal ideas and values through agreed linguistic and physical actions creates the required forum for mediation. In this way, curriculum is developed: dialogue between learner and facilitator, presented within a cultural realm, creates a tapestry of ideas and thoughts which sits within the gallery of prevailing models and ways of thinking, but which is demonstrative of new ideas and ways of explaining. The knowledge displayed is not discovered, rather it is created, woven into the personal fabric and displayed for social consumption and critique.

Within contemporary society the methods by which individuals may create and structure knowledge and understanding and display this for social review are ever expanding. The advent of digital forms of communication means that the social realm within which to deposit ideas and thoughts is ever increasing. With this in mind, learning-teaching situations need to utilise new technological enterprise. Not only does this increase the group of people for whom the knowledge is accessible, it also enables thought and application to be posted and discussed in a variety of forms and structures. In a similar way, the learner-teacher interface can be redesigned to make use of new technologies. However, such redesigns must shift to accommodate certain principles if the educational process is to be considered constructivist in nature. At the heart of this however are constructions of learner and teacher identity. Although the move from objectivist principles requires a paradigmatic shift, it is this shift that is fundamental to the creation of constructivist learning-teaching environments. As was seen in the three case studies, role definition and the nature and purpose of the learning-teaching interaction are what define technologically oriented, constructivist learning-teaching environments. Epistemological, pedagogical, and ideological positions are what drive and determine whether new technologies are used in a constructivist sense, not the tools themselves.

REFERENCES

Adams, P. (2003). Thinking skills and constructivism. *Teaching Thinking, 10* (Spring), 50-54.

Bartlett, S., Burton, D., & Peim, N. (2001). *Introduction to education studies*. London: Paul Chapman.

Bodner, G. M. (1986). Constructivism: A theory of knowledge. *Journal of Chemical Education, 63,* 873-878.

Brooks, J. G., & Brooks, M.G. (1993). *In search of understanding: The case for constructivist classrooms*. Alexandria, VA: American Society for Curriculum Development.

Cognition and Technology Group at Vanderbilt. (1991). Some thoughts about constructivism and instructional design. In T.M. Duffy & D.H. Jonassen (Eds.), *Constructivism and the technology of instruction: A conversation* (pp. 115-119). Hillsdale, NJ: Lawrence Erlbaum.

Cohen, L., Manion, L., & Morrison, K. (2000). *Research methods in education* (5th ed.). London: Routledge Falmer.

Collins, A. (1991). The role of computer technology in restructuring schools. *Phi Delta Kappa, 73,* 28-36.

Copley, J. (1992). The integration of teacher education and technology: A constructivist model. In D. Carey, R. Carey, D. Willis, & J. Willis (Eds.), *Technology and teacher education* (p. 681). Charlottesville, VA: AACE.

Crowther, D. T. (1997). Editorial. *Electronic Journal of Science Education, 2*(2). Retrieved January 17, 2004, from http://unr.edu/homepage/jcannon/ejse/ejsev2n2ed.html

Delanty, G. (2000). *Citizenship in a global age*. Buckingham: Open University Press.

Driver, R., Aasoko, H., Leach, J., Mortimer, E., & Scott, P. (1994). Constructing scientific knowledge in the classroom. *Educational Researcher, 23*(7), 5-12.

Ernest, P. (1995). The one and the many. In L. Steffe & J. Gale (Eds.), *Constructivism in education* (pp. 459-486). Hillsdale, NJ: Lawrence Erlbaum.

Fontana, D. (1995). *Psychology for teachers*. Basingstoke: Macmillan Press.

Goodman, Y.M., & Goodman, K.S. (1990). Vygotsky in a whole language perspective. In L.C. Moll (Ed.), *Vygotsky and education, instructional implications and applications of sociohistorical psychology* (pp. 223-250). Cambridge: Cambridge University Press.

Hanley, S. (1994). *On constructivism*. Retrieved February 12, 2004, from http://www.towson.edu/csme/mctp/Essays/Constructivism.txt

Hein, G. E. (1991, October 15-22). Constructivist Learning Theory, the museum and the needs of people. *Proceedings of the CECA Conference*, Jerusalem, Israel.

Heylighen, F. (1993). *Epistemology, introduction*. Retrieved March 2, 2004, from http://pespmc.vub.ac.be/EPISTEMI.html

Jonassen, D. H. (1990). Thinking technology: Toward a constructivist view of instructional design. *Educational Technology, 30*(9), 32-34.

Jonassen, D. H. (1991). Evaluating constructivist learning. *Educational Technology, 36*(9), 28-33.

Jonassen, D. H. (1994). Thinking technology. *Educational Technology, 34*(4), 34-37.

Kanuka, H., & Anderson, T. (1999). Using constructivism in technology-mediated learning: Constructing order out of the chaos in the literature. *Radical Pedagogy, 1*(2). Retrieved from http://www.icaap.org/iuicode?2.1.2.3

Mann, C. (1994). New technologies and gifted education. *Roeper Review, 16,* 172-176.

Mautner, T. (Ed.). (1996). *The Penguin dictionary of philosophy*. London: Penguin.

Murphy, E. (1997). *Constructivism from philosophy to practice*. Retrieved April 9, 2004, from http://www.stemnet.nf.ca/~elmurphy/emurphy/cle.html

Omrod, J. (1995). *Educational psychology: Principles and applications*. Englewood Cliffs, NJ: Prentice-Hall.

Postlethwaite, K. (1993). *Differentiated science teaching*. Philadelphia: Open University Press.

Pring, R. (2000). *Philosophy of educational research*. London: Continuum.

Reeves, T. (1992). Effective dimensions of interactive learning systems. *Proceedings of the Information Technology for Training and Education Conference* (ITTE '92).

Rice, M. L., & Wilson, E. K. (1999). *How technology aids constructivism in the social studies classroom*. Retrieved February 2, 2003, from http://global.umi.com/pqdweb

Riel, M. (1990). Building a new foundation for global communities. *The Writing Notebook,* (January/ February), 35-37.

Salomon, G. (1991). From theory to practice: The international science classroom—A technology-intensive, exploratory, team-based and interdisciplinary high school project. *Educational Technology, 31*(3), 41-44.

Saunders, W. (1992). The constructivist perspective: Implications and teaching strategies for science. *School Science and Mathematics, 92*(3), 136-141.

Strommen, E. F., & Lincoln, B. (1992). Constructivism, technology and the future of classroom learning. *Education and Urban Society, 24,* 466-476.

Suchman, L. A. (1987). *Plans and situated actions*. New York: Cambridge University Press.

Sutton, A. (2000). Would the real Vygotsky please stand up? *Teaching Thinking, 1*(Spring), 52-55.

Tam, M. (2000). Constructivism, instructional design, and technology: Implications for transforming distance learning. *Educational Technology and Society, 3*(2), 50-60.

Von Glasersfeld, E. (1984). An introduction to radical constructivism. In P. Watzlawick (Ed.), *The invented reality: How do we know what we believe we know?* (pp. 17-40). New York: W.W. Norton & Company.

Von Glasersfeld, E. (1995). *Radical constructivism: A way of knowing and learning*. London: Falmer Press.

Von Glasersfeld, E. (1996). Introduction: Aspects of constructivism. In C. T. Fosnot (Ed.), *Constructivism: Theory, perspective and practice* (pp. 3-7). New York: Teachers College Press.

Vygotsky, L. S. (1962). *Thought and language*. Cambridge, MA: MIT Press.

Vygotsky, L. S. (1978). *Mind in society: The development of higher psychological processes*. Cambridge, MA: Harvard University Press.

Wheatley, G. H. (1991). Constructivist perspectives on science and mathematics learning. *Science Education, 75*(1), 9-21.

Woolfolk, A. E. (1993). *Educational psychology*. Boston: Allyn and Bacon.

Yager, R. (1991). The constructivist learning model, towards real reform in science education. *The Science Teacher, 58*(6), 52-57.

ENDNOTES

1 In England, community colleges exist in certain areas to provide education for pupils of statutory school age between 11 and 16/18. They are state funded and seek to work closely with the local community in which they reside, although there is no set model or format. Similar, although not identical, arrangements exist in the rest of the UK (Wales, Scotland, and Northern Ireland).

2 A2 denotes a form of pre-higher-education qualification usually undertaken by students aged between 16 and 18. The qualification can be studied in schools, community colleges, and colleges of further education in England and Wales. Scotland and Northern Ireland have slightly different arrangements.

3 All pupils follow the National Curriculum Year 7 programme between the ages of 11 and 12 in state-maintained schools in England. Part of these orders relate to requirements for the study of citizenship.

Chapter XXIX
K-12 Educators as Instructional Designers

Kendall Hartley
University of Nevada, Las Vegas, USA

ABSTRACT

This chapter will describe the realities of K-12 classroom practice and how this compares to common tenets in the field of instructional design. Specifically, the chapter will describe how trends towards the use of information and communication technologies in classrooms might be made more advantageous through the use of an instructional design approach. The chapter will include an introduction to the field and history of instructional design, an overview of current teacher preparation as it relates to designing instruction, and how a systematic instructional design perspective might differ. The chapter will conclude with a description of how changes in teacher preparation and access to the appropriate tools could facilitate increases in student achievement.

INTRODUCTION

Ask a K-12 teacher to identify his or her profession and one of the responses you will likely *not* hear is instructional designer. In reality, instructional designer is likely a better description of the emerging responsibilities of the 21st-century educator. While educators and instructional designers agree on many principles and share common goals, they have long enjoyed a separate existence (Rose, 2002).

A growing need exists to introduce educators to the field and principles of instructional design within the emerging context of the technology-infused curriculum. The convergence of the following trends point to this need: (a) better access to powerful learning application development tools, (b) greater dependence upon

Web-based learning, (c) a need to promote the technological literacy of students and teachers, and (d) a greater emphasis on accountability.

The first part of this chapter will begin with a review of the current landscape of K-12 education followed by an introduction to the field and history of instructional design. The second part of the chapter will: (a) provide an overview of current teacher preparation (and to some degree, current practice); (b) contrast this with how a systemic instructional design perspective might guide practice; and (c) provide a description of how this progression could be achieved in terms of teacher preparation and access to the appropriate tools.

CHANGING LANDSCAPE

Technology in Schools

The advances in educational technology over the past 15 years are tremendous. From a technological standpoint, the processing power available today, for relatively low cost, represents the unthinkable only a few years ago. In terms of software, the tools available are much more user friendly and significantly more reliable. From a communications standpoint, few people recognized the dramatic significance of high-speed Internet connections in terms of access to educational opportunities. Science classrooms have benefited enormously from powerful simulations and increased access to real scientific databases (e.g., Centers for Disease Control). However, these advances have had little impact on many of today's classrooms. In observations of two technology-rich Silicon Valley high schools, Peck, Cuban, and Kirkpatrick (2002) reported that:

Teacher use of technology during our random observations was the exception rather than the rule. Of the 35 teachers we saw on random days, 23 in social studies, science, English, math, and foreign language had a familiar teaching repertoire—lecture, review of homework, recitation, and whole-group instruction—that eschewed any use of electronic technology. (p. 49)

Of course identifying *non-users* implies that there are a number of *users*. In fact, in a significant number of classrooms, the technology *is* changing the way many teachers prepare for, plan, and deliver instruction. In a study of three teachers at a school where each student had a laptop, Windschitl and Sahl (2002) found that the technology could serve as a *catalyst* for changing professional practice. That catalyst is becoming more and more prevalent in classrooms. This can be seen in the growth of the Internet-connected classroom (Kleiner & Lewis, 2003) and the popularity of either equipping or requiring students to have notebook or handheld computers (e.g., Windschitl & Stall, 2002; Lowther, Ross, & Morrison, 2003; Edwards, 2003; Kleiner & Lewis, 2003).

Teachers as Instructional Materials Designers and Adopters

Advances in technology have brought powerful tools to the desktops of educators. Once only the purview of large publishing companies, the development and dissemination of quality instructional materials now is well within the technological reach of anybody with a modestly equipped personal computer. Teachers are also receiving professional development opportunities related to technology more frequently. Of the U.S. schools with Internet connections (99% in 2002), 87% reported that they had offered professional development activities to their teachers in the past 12 months (Kleiner & Lewis, 2003). A substantial amount of this professional development involved guidance in developing original electronic instructional materials (e.g., WebQuests).

However, as large publishers and professional instructional designers can attest, the development of effective instructional materials requires more than the technological tools. Effective materials development is often done by teams of designers, educators, subject matter experts, and technical consultants. The development of a textbook involves the work and expertise of a variety of individuals—not simply the author.

Some of the resources that were provided in the past by these development teams are quickly becoming accessible to anyone with access to the Internet. For example, the need for a subject-matter expert can be significantly reduced with the informational resources becoming available online. These resources include not only the typical Web site, but also the increasingly substantial collection of online resources available through entities such as the local university library and the U.S. Library of Congress. Organizations such as the U.S. National Aeronautics and Space Administration and the U.S. National Oceanic and Atmospheric Administration provide science teachers with extensive data access and analysis tools. Thus, all the resources necessary for designing high-quality instructional materials are within reach of most teachers.

However, not all teachers are diving into the designing role. Some are more interested in finding already-developed materials that meet their curricular needs. Regardless of whether they design their own or seek already-developed materials, this represents a significant shift from the days when someone other than the classroom teacher (e.g., curriculum coordinator or the textbook adoption committee) selected the central instructional materials for a class (i.e., the textbook). This shift puts the classroom teacher in a position where they need to be able to evaluate a wide range of materials. One reaction to this shift can be seen in the increased emphasis upon teacher (and student) evaluation skills as each group becomes more dependent upon the Web for information (American Library Association, 1998).

Regardless of whether educators become designers or merely adopters, the shift in need is evident. Educators must be able to choose between a variety of products from a wide assortment of producers to meet their instructional needs. While curriculum materials development and selection in the past was done by others, it is now typically done by the classroom teacher. The field of instructional design has developed an extensive array of guidelines and models to assist in these efforts.

THE FIELD OF INSTRUCTIONAL DESIGN

Historical Foundations

A sudden entrance into World War II presented many challenges to the United States. One rather daunting challenge was: How do you best train large numbers of individuals very quickly on a wide variety of tasks? To address this and other training needs, the military looked to educators and psychologists such as Robert Gagne and Leslie Briggs (Reiser, 2001; Rose 2002). Many of the tenets of instructional design can be traced back to the work of Gagne, Briggs, and others in these efforts.

The early work of instructional designers depended largely upon the dominant psychological paradigm of the time, behaviorism. Although the field has since adapted to more contemporary models of psychology, its early foundations in behaviorism often result in it being misclassified as inflexible and of limited applicability. Early dependence upon behavioral psychology was reduced in part as cognitive psychology and constructivist theories began to make important inroads into our understanding of human learning (Gardner, 1985; Bruner, 1966). In addition to behaviorism, the field of systems engineering figured prominently in the development of instructional design principles (Molenda, 1997). Unlike behaviorism, systems engineering continues to be a driving force in instructional design today.

One of the important ideas to emerge from early work in this area was Gagne's (1985) identification of the *conditions of learning* and *events of instruction*. The former was concerned with identifying the important conditions necessary for learning to occur. This included internal conditions (e.g., background knowledge) and external conditions (e.g., presentation method). Conditions of learning served as a foundation for future instructional design models. It also lends support to the notion that a wide variety of learning outcomes can be addressed by considering a few general principles (Ragan & Smith, 2004). The events of instruction can be viewed as a generalized strategy or sequence that could be utilized as a template for addressing an instructional need.

Another important idea, programmed instruction, emerged from the writings of B. F. Skinner. Programmed instruction is described by Reiser (2001, p. 59): "Data regarding the effectiveness of the materials were collected, instructional weaknesses were identified, and the materials were revised accordingly." The need to have programmed instruction that was highly effective resulted in designers utilizing a cycle of analysis, design, testing, and revision (Molenda, 1997). This cycle is a clear precursor to the more fully developed instructional design models in use today. A final idea that is critical to the historical development of this field is the emergence of Mager's concept of behavioral objectives and Glaser's notion of criterion-referenced testing (Rose, 2002; Reiser 2001).

The history of the field progressed along two parallel but distinct paths. One path reflects the changes and growth in the area of educational psychology. The other path represents changes in model and principle development. The distinction is important because it helps to clarify some important differences among instructional designers. A designer may hold a behavioral view of learning yet utilize a more contemporary model of instructional design. In other words, using an instructional design model does not dictate a certain psychological perspective. Regardless of the psychological perspective that is taken when one utilizes an instructional design model, the systems approach is still a prominent (if not paramount) characteristic of the field.

It is also important to acknowledge the tremendous impact that the computer has had on the field of instructional design. As one reviews the history and principles of the field, it should be clear that advances in technology have occurred simultaneously and in many ways served as a driving force.

Current Definitions

What has emerged from this progression is a variety of instructional design models that have several components in common. These common components include problem analysis, learner analysis, design, development, implementation, and evaluation (Reiser, 2001; Smith & Ragan, 1999).

Smith and Ragan (1999, p. 2) define instructional design as "the systematic and reflective process of translating principles of learning and instruction into plans for instructional materials, activities, information resources, and evaluation." Thus, instructional design is a process.

The exact mechanisms and sequence of the process are described by the various models noted above. This deviates little from what we currently expect of K-12 educators. However, this definition, while accurate, does not tell the whole story. The instructional designer (or more generally, the instructional design team) is tasked with a much more in-depth analysis of each of the steps noted above than we would expect for the typical classroom lesson.

Instructional designers generally share a number of core beliefs regarding instruction (Smith & Ragan, 1999). These beliefs include:

1. Instructional design is learner-centered.
2. A systems view of instructional development is critical.
3. Learning is more efficient when objectives, instruction, and assessments are congruent.
4. Evaluation (including, but not limited to student assessment) and subsequent revision of the instruction is a critical component of the instructional design process.

Each of these beliefs will be addressed in more detail in the following sections. These beliefs will be described from the perspective of the instructional designer. Later, a comparison between this perspective and the realities of the K-12 classroom will be made.

Learner Centered

Designing instruction with the learner in mind might seem obvious, yet somehow they do often get lost in the process. In secondary and postsecondary institutions, the teacher-centered classroom has long been the norm in spite of regular and repeated efforts to remove the "sage on the stage" (King, 1993). Being learner centered is not only concerned with the design of the classroom and the methods of discourse within the classroom, but also the conception and design of instructional activities. This view of "learner-centered" is consistent with Norman's (1988) conception of user-centered design. Successful design of instruction (and phones, software, doors, etc.) is contingent upon an accurate and complete picture of the learner.

In instructional design terms, this translates into an extensive analysis of the learner to include characteristics that might influence the success of an activity or lesson. This might include background knowledge, general aptitude, and language. In the southwestern United States, it is not uncommon to have schools where English is the first language of a minority of the students. Few would question the importance of the learner's native language; unfortunately, the adaptations necessary to accommodate the needs of learners who speak languages other than English are difficult if not impossible in the K-12 classroom.

Learner-centered contexts can also translate into instructional materials that reflect the community and cultures of the individuals. Science educators have long advocated an inquiry model of instruction, which depends upon the learner being actively engaged in authentic and relevant problem-solving activities (National Research Council, 1996). A more personalized instructional experience has strong theoretical and research support (Brown, Collins, & Duguid, 1989; Moreno & Mayer, 2004; Park & Lee, 2004; Ysseldyke, Spicuzza, Kosciolek, & Boys, 2003). Instructional designers will attempt to include this type of information in their analysis of the learners and utilize it in the subsequent development of instruction.

Being learner centered is recognition that many things can impact a student's opportunity for success. Maslow's hierarchy of needs is ever-present and thus analyzing the learner may extend beyond what is generally known about students (Reigeluth & Beatty, 2003). For example, it is estimated that approximately 1.35 million youths in the U.S. are homeless (National Coalition for the Homeless, 2002). There are tremendous opportunities afforded via computer-based technologies to customize learning to student interests (Reigeluth, & Beatty, 2003) and instructional needs (Ysseldyke et al., 2003).

While instructional design does place a great emphasis on the individual, it is not synonymous with individualized instruction (Dick & Carey, 1996). Instructional design models do not dictate learning strategies. The designer's psychological and epistemological beliefs will strongly influence the product of any design and development activity. This is probably most evident when the designer chooses instructional activities.

Systematic

An indication of the importance of systems thinking is reflected in the fact that the terms *instructional design* and *instructional systems design* (ISD) are often used interchangeably. Most instructional design models claim to utilize a *systems approach*. Mood (as cited in Moldenda, 1997, p. 42) describes a systems approach by stating: "It is simply the idea of viewing a problem or situation in its entirety with all its ramifications, with all its interior interactions, with all its exterior connections and with full cognizance of its place in its context."

The most notable instantiation of this ideal is Dick and Carey's (1996) *Systems Approach Model for Designing Instruction*. This approach is viewed as an effective means for addressing "the design, development, implementation and evaluation of instruction" (Dick & Carey, 1996, p. 4). This is acknowledgment of the incredible complexity and inter-dependencies of the teaching and learning enterprise. Similarly, environmentalists understand that any changes in the ecosystem can have a tremendous impact on all the inhabitants of that ecosystem (Wilson, 1992). The introduction of a pesticide to a small plot of land seems like a minor modification for which the benefits far outweigh any potential adverse impact. The contamination of the community water supply may or may not have been considered a potentially adverse impact when the use of pesticides became the norm.

This might exemplify the greatest difficulty of a systems view. One must anticipate all the consequences of one change in the system to all other components of the system (Weinberg, 1975). When the number of interrelated components are few (e.g., gravity between two objects—Earth and Moon), the consequences are easier to anticipate. Unfortunately, it is rare to find instances where the numbers of components are few. A solid understanding of the gravitational pull between the Earth and Moon is of little value if one ignores the pull of the Sun, Mars, Venus, Mercury, Saturn, Jupiter, Neptune, Pluto, and the other objects in the universe.

The complexity of instructional design is equally daunting when one considers the plethora of factors that exist in the classroom. In a description of the application of systems theory to adult education, Kazemek and Kazemek (1992) offer the following example:

Thus, from a general systems perspective, a young, unmarried, unemployed woman with two children cannot be seen as simply a 'single parent on welfare'. Rather we must see her as an intricately complex psychological, spiritual, and biological system who is connected to, and impacted by, a variety of other social systems, for example, her extended family, supportive women friends, particular neighborhood or community, church, welfare, and numerous other systems. The important point is that all of these different systems are integral and interrelated parts of the young woman's life. (Systems Theory section, para. 2)

This complexity should not dissuade developers in the pursuit of an appropriate design. When engineers (and astronomers) are faced with overwhelmingly complex tasks, they do two things—simplify and utilize computational tools (Weinberg, 1975). Assumptions and/or approximations can be made in any situation that can reduce the complexity of the problem while still producing an adequate solution. Teachers frequently do this by choosing a midrange target population for assignments and lectures. If the assignment is too hard, some will find the task impossible. If the assignment is too easy, some students will gain little. The key is to identify those attributes that are most critical to the student's success and ignore (or shoot for the middle) those that are less critical.

Experimental educational psychologists have long recognized the importance of identifying these critical attributes. They recognize that if they do not account for important variables such as reading comprehension ability and background knowledge, modest interventions will not reach statistical significance. Those two variables wield a tremendous influence on achievement in any learning activity.

Congruence

Consistency between learning objectives, instructional activities, and learner assessment is an important ideal in instructional design. This principle is an outgrowth of the systematic nature of the process. Each component is interrelated, and changes to one necessitate an evaluation of how it might change others. This congruence is apparent in several instructional design models that place the development of student assessments ahead of the development of instructional activities. In other words, instructional activities cannot be designed unless one has a clear idea of what the student should know and be able to do once the activity has been completed. Unfortunately, student assessment at the classroom level is often an afterthought (Young, Reiser, & Dick, 1998; Cizek, Fitzgerald, & Rachor, 1995, 1996).

Congruence should also extend beyond the classroom. This includes how instructional activities in one course mirror or complement those in another. Comparisons between two physics classrooms in the same school will reveal significant differences. While some differences are to be expected, difficulties arise when students who have taken the same course from different instructors engage in dramatically different activities and assessments. Even more challenging for the science curriculum is the overlap and gaps that occur by using the arbitrary distinctions physics, chemistry, and biology. Some have looked towards a *spiral curriculum* in science to generate a more congruent curriculum (Bruner, 1960). In addition to classroom and school-level concerns, the curriculum must be congruent with district and state standards and assessments.

Congruence is an ideal that is difficult to achieve. Instructional designers from all areas can attest to the ease with which even the simplest project can become grossly incongruent. Given limitations of time and money, assessments very often bear little resemblance to the stated learning goals. The regularity with which technical workshops are evaluated based upon satisfaction surveys is but one indicator of the difficulty of achieving congruence in even the most encapsulated learning experiences. Thus, congruence in the K-12 environment will be a difficult but worthwhile endeavor.

Student Assessment and Instructional Evaluation

The distinction between student *assessment* and the *evaluation* of instruction is vital to the improvement of instruction (Smith & Ragan, 1999). Assessment refers to the activities related to determining whether individuals achieved instructional objectives. Evaluation refers to the activities related to whether the instruction achieved the desired learning goals.

Student assessments are greatly emphasized in instructional design models and methods. These assessments constitute a broad range of activities, which include informal teacher questioning, course assignments, student portfolios, and criterion referenced tests. It is important that the assessments are valid (e.g., measure what they claim to measure), reliable (e.g., provide consistent results), and congruent with the course objectives and instructional activities. Student assessments are also viewed as learning experiences that affect subsequent instruction.

In addition to the primary function noted above, student assessments also provide information of value to the evaluation of instruction. Other information can be used to inform the evaluation of the instruction. For example, student performance on subsequent and related lessons might inform the evaluation. In spite of doing very well on the prior lessons assessment, instructors might find that the students are ill prepared for the next.

The evaluation of instruction and subsequent changes to other interrelated components is emphasized throughout the instructional design literature. This evaluation often includes some level of formative and summative evaluation. Summative refers to the evaluative tasks that take place prior to (and in some cases during) the delivery of instruction (Dick & Carey, 1996). Having individual students attempt a dry run at a laboratory experiment prior to the entire class completing the activity is one example of how teachers engage in formative evaluation. Observations during activities also would constitute formative evaluation. However, for these evaluations to be of value, they must be used to improve other components (e.g., subsequent instruction and/or assessments).

Summative evaluations constitute a more comprehensive look at the instruction. With respect to K-12 teachers and the movement towards meeting content area standards, summative assessment should include the evaluation of how students are performing on state and national assessments. This should illuminate gaps in student understandings. For this to be feasible, the data needs to be available to teachers in a meaningful format. In other words, knowing that many eighth-grade students in a large district are performing poorly on the mathematics assessment is of little value to individual teachers without much greater specificity.

Instructional Design Summary

The field of instructional design began with some concentrated attempts to enhance the efficiency and effectiveness of instruction. The early work drew heavily upon behavioral psychology and systems theory. Contemporary views of instructional design are more dependent upon ideas emanating from cognitive, rather than behavioral psychology. While initially focused on military training, the emerging field soon found outlets in government and corporate training environments. Attempts at introducing the concepts into K-12 and postsecondary education environments have been less successful (Earle, 1994). The reasons for the limited impact on traditional educational settings are likely a result of the historical foundations as well as the culture of the field (Rose, 2002). Instructional designers vary in models used and psychological perspectives taken but share some common emphases in their approach to instructional development. These emphases include focusing on the learner, taking a systematic approach, insistence upon congruence, and evaluation.

K-12 EDUCATORS DEVELOPING INSTRUCTION

Designing Instruction vs. Lesson Planning

The preceding description of instructional design core principles and general models constitute an accepted body of knowledge that strongly influences military, government, and corporate training programs. That influence does not extend to K-12 education. A number of studies have indicated that much of what is considered important in instructional design is given less consideration in teachers' lesson planning. In a survey and subsequent interviews of nine outstanding teachers, Young et al. (1998) found few parallels between the teacher's planning and a systems approach to instructional design. For example, the majority of the teachers interviewed demonstrated little concern for the explicit identification of objectives. In addition, student assessment was given little consideration in the instructional planning process. A study of K-12 teachers' familiarity with concepts such as systems thinking and needs assessment revealed that few teachers have understanding of instructional design concepts (Kennedy, 1994). The emphasis placed upon assessment is also less apparent in teacher planning. Evaluations of teacher practices by Hall, Knudsen, and Greeno (1995) and Cizek et al. (1995, 1996) revealed that assessment strategies are highly variable, unpredictable, and generally offer little information about student understanding. Teachers generally have a stronger grounding in learning theory (Kennedy, 1994).

The results of the studies above should not be viewed as a criticism of teachers. The results in many ways reflect the general methods of teacher preparation. Teachers receive instruction in lesson planning, but this bears little resemblance to the process described by instructional design professionals. This distinction will be discussed in more detail in the following section.

The perceptions of the field of instructional design also play a role in the modest degree of influence. In her analysis of the boundaries between the two areas, Rose (2002, p. 16, emphasis in original) suggests that:

Indeed, a sense of inherent superiority over the approaches that prevail in K-12 and post-secondary education is a key element of the ideology of instructional design, for one of the assumptions which frames the work and discourse of instructional designers is the idea that their field is not simply about developing instruction but, more importantly, about improving instruction.

The perception that teachers are doing something wrong and instructional designers have the answer is both unfortunate and unrealistic. In reality, instructional designers have much to learn from educators. For example, instructional designers often have the luxury of designing instructional materials for very captive audiences. Motivation is often built-in when your employer is asking you to upgrade your skills. In addition, instructional designers are often given well-defined tasks. K-12 educators have a myriad of stakeholders and considerations that frequently support contrasting views.

Teacher Preparation and Instructional Planning

Instructional planning plays a prominent role in any teacher preparation program. The view of instructional planning as it is commonly presented in teacher preparation programs can be distinguished from a systems approach to designing instruction when identifying the key components. Teachers' instructional planning often comprises the following considerations:

- Learner characteristics
- Lesson objectives
- Instructional strategies
- Sequencing of instruction
- Student assessment

Less apparent is:

- Systematic look at each of these considerations
- Instructional evaluation
- Adaptive instruction (e.g., instruction that is based upon prior assessments)
- Multilevel and across-discipline evaluations

Why this is true can be gleaned from a review of textbooks commonly used in teacher preparation courses (Callahan, Clark, & Kellough, 1998; Jacobsen, Eggen, & Kauchak, 2002; Jarolimek, Foster, & Kellough, 2001; Sparks-Langer, Pasch, Starko, Moody, & Gardner, 2000). Most will present one or two chapters on Lesson Planning. Given that lesson planning includes complex topics such as task analysis, learner analysis, determining learning goals, writing lesson objectives, instructional strategies (to include grouping strategies and teacher-directed vs. individual), and student assessment, it is not difficult to see why the depth of understanding for each of these considerable topics is limited. In addition, some important topics are not addressed. For example, the impact of each of these components upon the other is given limited attention. The ideal of adapting instruction based upon student needs and interests is often described, but little is provided in terms of practical implementation strategies.

Adapting instruction for students with special needs is often a key component of a teacher preparation program, although it often occurs in a separate course. The contrasts between teacher preparation for the regular education teacher and the special education teacher are quite telling. Special education programs are much more likely to emphasize some of the principles espoused by instructional designers. For example, the development of an individualized education program (IEP) for each student can have a tremendous impact on the educator's ability to adapt instruction to meet the student's needs. In addition, the IEP must be based upon *current* student assessments. Thus, the need exists for regular assessments that are directly related to curricular goals. Preparing special education teachers thus requires a more thorough grounding in instructional planning and meaningful assessments.

Other differences between the view of instructional designers and the view presented in teacher preparation courses can be traced to origins of both fields. While educators may start with a broad worldview of education, instructional designers will often view instructional challenges at the most basic level and then work up. For the instructional designer, an important early step in the process is an extensive evaluation of the objective—including a determination of the prerequisite skills. This sets the stage for a systematic approach to development. The more holistic view of the educator will pay less attention to the details and attempt to identify instructional strategies that are *likely* to result in objectives being met. At the curricular level, the opportunity for teachers to participate in multilevel and cross-curricular planning is limited, and thus it is not surprising that these types of connections are rare.

Implications for Teacher Preparation

As tools and resources available to educators improve, the opportunity for the development and dissemination of higher quality instructional materials becomes practical. This represents a significant shift in how we view the teaching profession. Currently, the profession emphasizes the management and delivery of educational experiences through instructional lessons and units. In accordance with this view, teacher preparation courses typically include the development of lessons and units as methods for demonstrating their understanding of the course content. For example, a typical science methods course may require students to develop a lesson that demonstrates strategies for engaging students in inquiry-based learning. The typical diversity course will require the student to develop a unit or lesson that demonstrates strategies for integrating resources regarding diverse populations in their own content area or level. A similar assignment is common in all the required education courses such as introductory educational psychology, educational technology, and special education. The emphasis is likely to be the instructional strategies at the expense of other important components such as student assessment and task analysis. Many of these courses are taught in different departments by people less involved at the program level. The view that the courses and topics are all tightly interrelated is not generally well established.

This approach can be contrasted with some of the accepted tenets of the field of instructional design described above (systematic, congruence, and assessment/evaluation). While some would argue these principles are also a part of instruction regarding the development of lesson and unit plans, most would agree that the emphasis is quite different. For example, the description of an instructional design model is virtually non-existent in undergraduate teacher preparation courses.

Courses that simply address the topic of instructional design are normally not encountered until students begin graduate coursework. Graduate programs may require a course in instructional design, but that is often limited to only those programs with an emphasis in educational technology and/or training.

Most K-12 teachers taking instructional design courses see great value in the tools that are introduced. They often comment that they would have been important supplements to their teacher preparation program. This instruction could be provided as support for meeting the needs resulting from increased state and federal accountability expectations. Subsequent coursework could use this foundation to frame and address the multitude of goals and objectives that are introduced. The increased emphasis on a systematic approach can be used in all of the courses to encourage connections with other courses in the teacher preparation program.

One should keep in mind that introducing teachers to the principles of instructional design as a 'value-free' enterprise may in fact encourage the separation that currently exists (Rose, 2002). Thus, "...base program development and delivery upon the understanding that instructional design is not a finite process or set of skills, but a culture" (p. 20). The view of this culture is one that might be especially helpful in the current K-12 environment that has greater access to technology and thus the tools to develop and/or organize their own instructional materials. In addition, the increased emphasis on accountability can be addressed with perspectives common to this instructional design culture.

NECESSARY RESOURCES

As much as a systems approach is helpful in viewing instructional planning, the same approach is important in viewing the realities and potential for today's K-12 classroom. The realities of the classroom make many of the prior suggestions difficult to implement. Adaptive instruction to a high school science teacher who has 150 students may be much more of an ideal than reachable target. The fact that the number of students with IEPs is growing is a testament to the fact that it can be done, but also that it is resource intensive.

Given the unlikelihood of a massive increase in resources to education, one has to look elsewhere for meeting these needs. An increase in the efficiency with which common teacher and administrator tasks are completed could go a long way in progressing towards some of the ideals described. Little has been said thus far about technology, but without the availability of some computer-based solutions to some very significant challenges, little will change. These solutions need to be developed (or improved) in the areas of design, assessment, and interoperability/communications.

Design Resources

Tools to facilitate the instructional design process could be of incredible value. While authoring tools are widely available, less emphasis has been placed on tools that facilitate a process that encourages a more systematic approach (van Merrienboer & Martens, 2003). A number of these systems exist. Most are used almost exclusively in areas other than K-12 schools. Examples include the Worldwide Instructional Design System (WIDS, 2004), which guides educators through a multi-step development process that includes progressively finer views of occupations, programs, courses, units, lessons, and objectives. Similarly, the ADAPT^IT program (de Croock, Paas, Schlanbusch, & van Merrienboer, 2002) supports the development of competency-based training programs. This program places an emphasis upon providing evaluative feedback to the designer based upon learner outcome and reaction measures.

One tool that does have a more K-12 focus is an online service entitled Taskstream (2004). Taskstream utilizes an approach that has more in common with lesson planning than instructional design. However, its utilization of standards and an assessment development tool represent an advance over the lesson planning described earlier.

Assessment Resources

Tools that assist teachers in the development of student assessments are quite prevalent. Some facilitate the development of multiple-choice and short-answer questions. Examples include the quiz builders that are incorporated into tools such as WebCT and online quiz developers such as Quia.com. While these represent advances over traditional test development and delivery mechanisms, they offer little in terms of assisting educators in taking a more systematic approach to their instruction.

Educators need assessment tools that:

1. are tied to existing classroom, school, and district student information systems;
2. can connect assessments with applicable standards; and

3. facilitate the storage and retrieval of assessments by a wide range of audiences for a variety of purposes (e.g., student feedback, instructional and program evaluation).

This is a significant challenge that although technically viable is not currently available. Without these types of systems in place, educational institutions will find it increasingly difficult to meet more recent accreditation requirements that emphasize the demonstration of competence by the students served.

Interconnected-Communication Resources

Possibly the most sought after and difficult to develop resources are those that enable all of these components (design/development and assessment tools) to share information within and beyond current boundaries. For example, a design tool should be able to share information with an assessment tool and vice versa. As noted earlier, the assessments can provide key information for the evaluation of instruction. In addition, the assessments need to have extensive communication capabilities. This includes the obvious such as feedback to the learner, but also the data that should be shared with other interested parties such as parents. Educators teaching the same content could benefit from access to test items developed by other teachers.

One of the more critical attributes seems also be the most elusive. This is the connections with existing student information systems. Some existing tools do provide a modest degree of interconnectedness. K12planet works with a number of student information systems to make applicable information available to students, parents, teachers, and administrators. Tools originally developed to facilitate distance education courses have begun to incorporate many of these needs into one package. While they were originally developed to provide a shared space for teachers and students, they have progressed into powerful tools that facilitate a significant degree of communication, assessment, and design.

The following scenarios exemplify the kinds of activities that could be possible if the interconnected-communications ideal were met:

1. A student who did poorly on an assessment earlier in the day is also participating in an after-school program. The facilitator of the after-school program has access to this information on their roster and the capacity to retrieve appropriate remedial exercises (Reigeluth, & Beatty, 2003).
2. The middle school textbook adoption committee can retrieve aggregated student achievement information by school and classroom. An analysis of the materials and activities used by the different classrooms can inform the selection or retention of an effective curriculum.
3. The first-year elementary school teacher can have immediate access to the plethora of activities and assessments that have already been developed for his/her grade level in the same school and district. This includes a clear description of the outcomes expected from the students coming into his or her class.
4. The veteran high school science teacher in the process of planning for the upcoming year can review student achievement data on the state standards exams to identify areas of strength and weaknesses.

Proponents of the open-source (i.e., non-proprietary, community-developed) software movement would argue that depending upon proprietary software to meet the interconnected-communication needs is unwise. By purchasing independent systems that may or may not claim to be able to communicate with other systems, the entire system becomes less flexible and too dependent upon communication among a select number of developers (who frequently work for competing companies). An open-source model could benefit from a community of developers who face similar challenges and hold similar goals.

CONCLUSION

The field and principles of instructional design combined with improved technologies could play a valuable role in meeting the changing expectations of the K-12 classroom. The changing expectations include: (a) better access to powerful learning application development tools, (b) greater dependence upon Web-based learning, (c) a need to promote the technological literacy of students and teachers, and (d) a greater emphasis on accountability. This chapter presented an overview of the field of instructional design, with an emphasis on how its principles could address these changing realities. Among the most relevant principles are that: (a) instructional design is learner-centered; (b) a systems view of instructional development is critical; (c) learning is more efficient when objectives, instruction, and assessments are congruent; and (d) evaluation (including, but not limited to student assessment) and subsequent revision of the instruction is a critical component of the instructional design process.

Implications for teacher education programs and relevant support technologies were described. Teacher education programs can inadvertently present a fragmented view of the role of the K-12 teacher. This role might be best represented by the lesson plan that should include activities that reflect diversity, meet special needs, incorporate technology, and prepare students to demonstrate proficiency. These plans often rely heavily upon learning theories and engaging activities, but place less emphasis upon specific objectives, student assessment, and lesson evaluation. This should not be surprising given the preparation provided and the tools available to the K-12 teacher. K-12 institutions would benefit greatly from user-friendly design, assessment, and interconnected-communication tools. While the challenges faced are substantial, they are surmountable. Possibly the greatest unmet need in addressing these challenges is not the lack of knowledge or desire. The greatest unmet need is the availability of appropriate software tools that can be brought to bear. The technology necessary is not incredibly sophisticated—centralized databases with Web-based interfaces. The level of complexity necessary for the databases is significant; however, given the number of institutions that could benefit, the effort will be worthwhile.

The challenges faced by the K-12 teacher in the 21st century are unlike those faced in past decades. The expectations are changing while the preparation and tools available remain the same. A systems approach would acknowledge that the changing expectations necessitate changes to the entire enterprise.

REFERENCES

American Library Association. (1998). *Information power: Building partnerships for learning.* Chicago: ALA.

Brown, J. S., Collins, A., & Duguid, P. (1989). Situated cognition and the culture of learning. *Educational Researcher, 18*(1), 32-43.

Bruner, J. (1960). *The process of education.* Cambridge, MA: Harvard University Press.

Bruner, J. (1966). *Toward a theory of instruction.* Cambridge, MA: Harvard University Press.

Callahan, J. F., Clark, L. H., & Kellough, R. D. (1998). *Teaching in the middle and secondary schools* (6th ed.). Upper Saddle River, NJ: Merrill.

Cizek, G. J., Fitzgerald, S. M., & Rachor, R. A. (1995/1996). Teachers' assessment practices: Preparation, isolation and the kitchen sink. *Educational Assessment, 3*(2), 159-180.

de Croock, M. B. M., Paas, F., Schlanbusch, H., & van Merrienboer, J. J. G. (2002). ADAPT[IT]: Tools for training design and evaluation. *Educational Technology Research and Development, 50*(4) 47-58.

Dick, W., & Carey, L. (1996). *The systematic design of instruction* (4th ed.). New York: Harper Collins.

Earle, R. S. (1994). Introduction to special issue: Instructional design and the classroom teacher. *Educational Technology, 34*(3), 4-5.

Edwards, M.A. (2003). The lap of learning. *School Administrator, 60*(4), 6-8.

Gagne, R. (1985). *The conditions of learning* (4th ed.). New York: Holt, Rinehart & Winston.

Gardner, H. (1985). *The mind's new science: A history of the cognitive revolution.* New York: Basic Books.

Hall, R. P., Knudsen, J., & Greeno, J.G. (1995). A case study of systemic aspects of assessment technologies. *Educational Assessment, 3*(4), 315-61.

Jacobsen, D.A., Eggen, P., & Kauchak, D. (2002). *Methods for teaching: Promoting student learning* (6th ed.). Upper Saddle River, NJ: Merrill.

Jarolimek, J., Foster, C. D., & Kellough, R. D. (2001). *Teaching and learning in the elementary school* (7th ed.). Upper Saddle River, NJ: Merrill-Prentice-Hall.

Kazemek, C., & Kazemek, F. (1992). Systems theory: A way of looking at adult literacy education. *Convergence, 25*(3), 5-15.

Kennedy, M. F. (1994). Instructional design or personal heuristics in classroom instructional planning. *Educational Technology, 34*(3), 17-25.

King, A. (1993). From sage on the stage to guide on the side. *College Teaching, 41*, 30-36.

Kleiner, A., & Lewis, L. (2003). *Internet access in U.S. public schools and classrooms: 1994-2002* (Project Officer: Bernard Greene). Washington, DC: National Center for Education Statistics (NCES 2004-011).

Lowther, D. L., Ross, S. M., & Morrison, G. M. (2003). When each one has one: The influences on teaching strategies and student achievement of using laptops in the classroom. *Educational Technology Research and Development, 51(*3) 23-44.

Molenda, M. (1997). Historical and philosophical foundations of instructional design: A North American view. In R. B. Tennyson, F. Schott, N. Seel, & S. Dijkstra (Eds.), *Instructional design: International perspectives* (vol. 1). Mahwah, NJ: Lawrence Erlbaum.

Moreno, R., & Mayer, R. E. (2004). *Personalized messages that promote science learning in virtual environments. Journal of Educational Psychology, 96,* 165-172.

National Coalition for the Homeless. (2002). *How many people experience homelessness?*

National Coalition for the Homeless fact sheet. Retrieved from http://www.nationalhomeless.org/howmany.pdf

National Research Council. (1996). *National Science Education standards.* Washington, DC: National Academy Press.

Norman, D. A. (1988). *The design of everyday things.* New York: Basic Books.

Park, O., & Lee, J. (2004). Adaptive instructional systems. In D. H. Jonassen (Ed.), *Handbook of research on educational communications and technology* (2nd ed.). Mahwah, NJ: Lawrence Erlbaum.

Peck, C., Cuban, L., & Kirkpatrick, H. (2002). High-tech's high hopes meet student realities. *Education Digest, 67*(8), 47-55.

Ragan, T. J., & Smith, P. L. (2004). Conditions theory and models for designing instruction. In D. H. Jonassen (Ed.), *Handbook of research on educational communications and technology* (2nd ed.). Mahwah, NJ: Lawrence Erlbaum.

Reigeluth, C. M., & Beatty, B. J. (2003). Why children are left behind and what we can do about it. *Educational Technology, 43*(5), 24-32.

Reiser, R. A. (2001). A history of instructional design and technology: Part II: A history of instructional design. *Educational Technology Research & Development, 49*(2) 57-67.

Rose, E. (2002). Boundary talk: A cultural study of the relationship between instructional design and education. *Educational Technology, 42*(6), 14-22.

Smith, P. L., & Ragan, T. J. (1999). *Instructional design* (2nd ed.). Upper Saddle River, NJ: Prentice-Hall.

Sparks-Langer, G. M., Pasch, M., Starko, A. J., Moody, C. D., & Gardner, T. G. (2000). *Teaching as decision making: Successful practices for the secondary teacher.* Upper Saddle River, NJ: Merrill-Prentice-Hall.

Taskstream. (2004). Retrieved from http://www.taskstream.com

van Merrienboer, J.J.G., & Martens, R. (2003). Computer-based tools for instructional design: An introduction to the special issue. *Educational Technology Research & Development, 50*(4), 59.

Weinberg, G. M. (1975). *An introduction to general systems thinking.* New York: Wiley-Interscience.

Wilson, E. O. (1992). *The diversity of life.* Cambridge, MA: Harvard University Press.

Windschitl, M., & Sahl, K. (2002). Tracing teachers' use of technology in a laptop computer school: The interplay of teacher beliefs, social dynamics, and institutional culture. *American Educational Research Journal, 39*(1), 165-205.

Worldwide Instructional Design System. (2004). Retrieved from http://www.wids.org

Young, A. C., Reiser, R. A., & Dick, W. (1998). Do "superior" teachers employ systematic instructional planning procedures? A descriptive study. *Educational Technology Research & Development, 15*(2), 65-78.

Ysseldyke, J., Spicuzza, R., Kosciolek, S., & Boys, C. (2003). Effects of a learning information system on mathematics achievement and classroom structure. *Journal of Educational Research, 96*(3), 163-174.

Chapter XXX
Creating a Virtual Literacy Community between High School and University Students

Tamara L. Jetton
James Madison University, USA

Cathy Soenksen
Harrisonburg High School, USA

ABSTRACT

The authors of this chapter describe a project in which a university education professor and a high school English teacher redesigned the curricula of their classrooms, so their students could participate in a literacy project that focused on computer-mediated discussions of literature. The goal of the project was to develop both the technological literacies of these students and the more traditional literacies in the form of reading and writing skills. The Book Buddy Project afforded the authors the opportunity to create a virtual literacy community in which high school and university students incorporated the traditional literacies of reading and writing within a virtual environment that facilitated communication, collaboration, and learning with text.

INTRODUCTION

Technology is an integral part of people's everyday lives as they engage in computer online chat rooms, send messages to family and friends, use their digital cell phones to conduct business and personal communication, and capture their most precious moments on digital pictures and movies. Students as early as elementary school use computer technology to create stories and draw about particular events in their lives. As students progress into middle school, they further develop their knowledge of the computer through computer literacy classes that teach them to engage in multimedia and hypertext environments. By high school, students take elective courses that focus on more sophisticated technologies that include creating digital video streams and computer programming.

Technology enables K-12 students to communicate, collaborate, and learn with people around the world as they seek to make meaning. These students use the computer as a tool for communication through computer-mediated discussions in which students engage in online discussions concerning school subject matter topics, books they are reading, and social talk. In order to facilitate this kind of communication, teachers have had to rethink their curriculum so that students are no longer engaged with one another within the limits of the classroom setting (Jetton, 2003-2004). Students can now participate in discussions that extend the boundaries of the K-12 classroom to virtually anywhere in the world. As a result of the proliferation of technology, teachers and university professors are revising their theories about the ways in which students think and learn, and are designing new course curricula that encompass these new uses of technology (Kim & Kamil, 2004).

The purpose of this chapter is to describe a project in which a university education professor and a high school English teacher redesigned the curricula of their classrooms so their students could participate in a literacy project that focused on computer-mediated discussions of literature. The goal of the project was to develop both the technological literacies of these students and the more traditional literacies in the form of reading and writing skills.

COMPUTERS AND LITERACY LEARNING

Literacy has many meanings, even within the field of education. Literacy can refer to the processes of becoming a literate citizen that might include mathematics, language, and science. With the continued proliferation of technology, literacy has now been expanded to include the development of computer-related skills such as word processing, World Wide Web searches, computer-mediated discussion strategies, multimedia presentations, and a host of other valuable skills.

Technology has also begun to change the way in which we examine the traditional literacies of reading and writing. Technology provides unique ways in which students can learn to read, collaborate through writing online, and respond to literature with others. For example, the advent of computer-assisted instruction (CAI) has enabled students to increase their comprehension of text (Boyd, 2000; Reinking, 1998; Weller, Carpenter, & Holmes, 1998). In these studies, students were provided with aids as they read the text. These aids included guided reading, vocabulary definitions, context cues, simplified texts, and additional background information.

Other studies have found that students need high levels of guidance as they engage in computer-assisted instruction that might include prescribed suggestions as to the strategies that would be the most effective to use (Gillingham, Garner, Guthrie, & Sawyer, 1989; Reinking & Rickman, 1990). In these studies, students received focused guidance such as vocabulary assistance and prompts that instructed them to reread, use their background knowledge, and locate important information in the text.

When CAI is used to teach writing skills, researchers have seen increases in motivation and task engagement, and improvements in writing quality (Daiute, 1983; Kamil, Intrator, & Kim, 2000; McMillan & Honey, 1993; Palumbo & Prater, 1992; Rosenbluth & Reed, 1992). However, these results seem to be mediated by the proficiency of students' writing skills, the quality of the instructional guidance they receive, and the students' grade level. That is, research has shown that remedial students who receive CAI about writing significantly increased the quality of their essays. Likewise, researchers have suggested that instructional support in the form of specific prompts to meet individual needs regarding specific writing tasks can improve writing performance (Bonk & Reynolds, 1992). In addition, students with lower writing ability appear to need longer interactions with CAI to achieve increases in their writing skills (Kim & Kamil, 2004).

COMPUTER-MEDIATED COMMUNICATION AND LITERACY

Technology has transformed the traditional literacies of reading and writing in the ways that students collaborate and interact with one another about the texts that they read. By sharing opinions and information about what they are reading to one another via the computer, students are finding interesting and unique ways to communicate. For example, Beach and Lundell (1998) found that online communication encourages those students who typically shy away from face-to-face interactions to interact freely though e-mail and online chats. They also noted that the online collaborations facilitate the development of social skills when students learn to infer social meaning, respond in socially appropriate ways, and write clearly to communicate to an audience.

Technology provides a forum for students to write to audiences that exist beyond the boundaries of their classrooms. By e-mailing governmental agencies and requesting information online from health organizations, museums, and various societies, students are able to communicate and collaborate with an authentic audience. Moore and Karabenick (1992) showed that when students had a clear purpose and audience, they were motivated to write lengthier essays and convey their ideas more clearly and effectively.

Researchers are also discovering that communication via technology results in different reading and writing skills than those found with the traditional literacies of reading and writing. For example, students are inventing new symbols for communicating online, integrating media files, creating links to Web sites, and sharing digital pictures as they interact with others (Merchant, 2001). Despite the proliferation of these new literacy skills as students engage in computer-mediated communication, educators may not be recognizing or valuing these skills in the schools.

Computer-Mediated Discussion

Some researchers have examined how students engage in computer-mediated communication through discussions on e-mail, discussion boards, and online chats. These researchers have referred to this particular form of discussion as computer-mediated discussion (CMD) (Bonk, 2003-2004; Fauske & Wade, 2003-2004; Jetton, 2003-2004; Schallert, Reed, & the D-Team, 2003-2004). The research on CMD is still in its infancy. The lack of studies in this area may be attributable to the time it has taken many school districts to purchase enough computers so that they are readily available to students in classrooms. Many schools in the surrounding school districts where I live in rural Virginia still only have computer labs where teachers must reserve time for their students to engage in communication with this technology. Furthermore, the school districts in my area have been slow to make available the online capabilities at school for students to engage in CMD. This is largely due to issues of privacy and possible abuses by students.

Despite the lack of studies in this area, research has focused on the advantages of CMD for students and their teachers in classrooms in three important ways: *communication, collaboration,* and the *learning environment.*

Communication

First, CMD provides other ways in which students can communicate with one another. This communication can take the form of online "pen pals," where students chat informally with one another about events in their lives and their common interests. Communication can also be more structured in that students are conversing online about their course content such as the information they learn and read (Jetton, 2003-2004). Through communication, students share knowledge, beliefs, and attitudes about an array of topics. In fact, students have reported that they find it less inhibitive to express themselves through CMD than through face-to-face discussions (Beach & Lundell, 1998). Thus, they tend to take more risks, enhance their roles and status in the electronic community, and increase the socio-emotional content of their responses (Cooper & Selfe, 1990; Kiesler, Siegel, & McGuire, 1984; Ku, 1996; Rice & Love, 1987). Beach and Lundell (1998) reported that the early adolescents in their study became more confident in their ability to express opinions and disagreements because they did not have to engage in the direct confrontation that is present in face-to-face interactions. Matusov (1996) states, "At the bottom of any agreement, there is a momentary disagreement that promotes communication..." (p. 29).

CMD provides avenues for communication because students must engage in writing during this process. Daly and Miller (1975) found that students with writing apprehension typically avoided situations involving writing, and they dreaded writing when it was placed in a public forum. Several researchers have found that computer-mediated communication benefits students who exhibit writing apprehension (Hiltz & Turoff, 1978; Mabrito, 1992; Wellman, 1997). By participating in CMD, students read others' writing and respond through their own writing, so they engage in the very process that they typically avoid (Mabrito, 2000).

Another communication advantage is that the online, written electronic environment of CMD enables students and their teachers to read and review written artifacts that are stored in memory so they can reference particular comments (Jetton, 2003-2004; Tiene, 2000). In contrast to face-to-face discussions, where responses are temporary and fleeting, electronic messages enable students and their teachers to look back and analyze certain responses that pertain to particular topics or themes. Students have opportunities to reread posted messages and construct more effective responses in light of others' contributions. In a survey of the advantages and disadvantages of online discussions versus face-to-face discussions, Tiene (2000) found that survey respondents were in strong agreement about the advantages of having a written record of the online discussions, and many of them noted that they did examine the written record of responses before posting their own ideas. These written artifacts also enable the teachers to read students' responses and, in turn, scaffold instruction that might show students how to respond more effectively when writing online.

Collaboration

Unlike the isolated reading and writing assignments that are typically assigned in schools, CMD is a communication forum that encourages social collaboration among students (Jetton, 2003-2004). Unlike face-to-face interactions in which participants can read both the verbal and nonverbal messages, CMD results in a much different student interaction. In this medium, students must rely on the written message for both the message meaning and social reasons behind it. By reading the written messages in CMD, students construct social impressions and develop assumptions about the other students with whom they are collaborating. Over time, these assumptions are tested (Walther, 1992, 1996).

Students find unique advantages to CMD that are not always found with face-to-face collaboration. Students have reported that they do not have to endure the interruptions that occur frequently during face-to-face discussions (Beach & Lundell, 1998), so they also feel safe in posting longer messages. They also find CMD to be a good medium for expressing opinions and ideas, especially diverse or controversial ones. Students feel comfortable disagreeing with others. Ferdig and Roehler (2003-2004), who describe these types of collaboration as interactivity, believe interactivity can positively impact learning because participants have opportunities to provide feedback, support, and guidance to others, make connections to others' ideas, and offer diverse viewpoints.

Learning Environment

Students benefit from the online environment of CMD in terms of processing information, increasing their knowledge, and engaging in reflective thinking (Thomas, 2002). The learning process occurs when students construct ideas, convey these ideas through writing, and reshape their ideas in response to elaboration and critique from others (Rowntree, 1995). Computer-mediated environments increase learning beyond declarative knowledge to more sophisticated knowledge structures that involve evaluation, critical analysis, and self-reflection (Thomas, 2002). The process of reflection involves defining the problem, analyzing the means to the end, and generalizing. Students who engage in this kind of critical reflection through CMD are enhancing their own learning processes and the learning processes of others.

Hara, Bonk, and Angeli (2000) examined the depth of processing and cognitive and metacognitive thinking represented by online responses during asynchronous discussions. They found that although students posted only the few messages that were required, their messages showed depth of processing in which they were using high-level cognitive and metacognitive strategies to achieve deep reflection and self-awareness. For example, the researchers found evidence in the online responses that the students were using the cognitive strategies of inference and judgment.

Computer-Mediated Discussions of Literature

To date, few studies exist that examine communication, collaboration, and learning through computer-mediated discussions of literature. Fischer (1998) and Gillespie (1998) examined how college students engaged in CMD during literature classes and found that students developed a deeper sense of purpose and audience as they engaged in CMD. As they collaborated and received responses, the students found incentive to write about the literature they were reading and became more aware of an audience. So often in literature classes, students are required to write about the literature through critical analysis and research reports that are not read by anyone but the teacher, and the only responses the students receive are the teacher's. Through CMD, students have an authentic forum for writing their thoughts and feelings about the characters, events, and themes of the stories they are reading. Fischer (1998) and Gillespie (1998) also found that when students engaged in CMD about literature, they provided textual evidence to support their responses, and they used prior experiences to connect with the literature. They also began to change each others' thinking. By assuming the roles of particular characters in the stories, students had opportunities to experiment with voice and interact with the voices of other characters in the story.

Beach and Lundell (1998) examined how 12 seventh graders engaged in CMD about literature. These seventh graders were white, middle-class, suburban students in a junior high school. Beach and Lundell focused on the influence of social factors as students participated in CMD. Students' purposes for reading the texts were to help others gain information and to share ideas about the text with others in order to build social bonds. Students also felt comfortable offering opposing and controversial ideas. As these students engaged in CMD, they assumed anonymous roles, and they exhibited introspective reflection about their personal thoughts, beliefs, and feelings. The researchers also found gender differences in that the girls wrote longer, more elaborate messages that reflected task continuative practices such as asking questions, repeating, validating, and extending others' responses.

Aside from these studies that have explored the uses of CMD as students engage in literature discussions, little research has been conducted in this area; it was difficult to find studies that explored CMD with public school students, particularly high school students. As stated previously, it has taken time for schools to purchase computers and integrate them seamlessly into the classroom environment. Furthermore, many educators may still not value the computer literacy skills of students as they participate in online discussion environments. We believe that much more research needs to be conducted with public school participants to determine the value and limitations of such discussions when students engage in literature. In the study detailed below, we examined how our students participated in discussions about literature through the use of CMD.

HIGH SCHOOL BOOK BUDDY PROJECT

The first author, Tamara, is a literacy professor in the Secondary Education Program at a Mid-Atlantic university. During the past three years, she has had the opportunity to participate in a partnership with the local school district in the town where the university is located. By examining the demographics of the local town's schools, we find an increasing diversity in ethnicity and language. The schools have seen an increase in ethnic minorities from 27% to 34% (Virginia State Department of Education, 1999, 2000, 2001). The Hispanic/Latino population has increased from 12% to 18%, a 63% growth. Additionally, according to the local city school statistics, the 875 students speak 36 different languages and are from 43 different countries (Mellott, 2002).

Most of Tamara's work has been with the high school in this district. She has helped the principal and teachers provide effective reading strategies to the diverse students who attend this school. Through these partnership activities, Tamara met the second author, Cathy, who is an English teacher at this high school. Cathy's primary teaching responsibilities entail providing English instruction to ninth-grade students in a program known as Project Achieve. Project Achieve was founded to assist a group of ninth-grade students who were designated at risk because of a number of factors, including their low test scores on the eighth-grade state criterion-referenced test. Many of these students were English language learners (ELLs) and special education students who had been placed in inclusive classrooms. These Project Achieve students received English, mathematics, science, and social studies instruction from teachers who were trained to meet their specific needs. Cathy's primary job as their English teacher was to increase their reading and writing achievement.

The students in Project Achieve are indeed diverse. During the year of our study, Cathy taught 10 African-American, 1 Bosnian, 12 Latino, 2 Middle Eastern, and 2 Russian students. Seventeen of these students were identified as English language learners. Fifteen students faced specific learning difficulties and were, therefore, provided special education services. Many of her students read far below the grade level of their ninth-grade peers. At the beginning of the school year, these students' scores ranged from a 1.8 to a 6.5 on the Star Reading (2002) test. Despite this range of scores, the average scores were between 3.0 and 4.0.

The Project Achieve students needed specific reading and writing strategies to facilitate their literacy growth, and they needed sufficient time to practice these strategies during authentic reading and writing activities. Because these students did not have the availability of technology in the home, and they had not engaged in computer-related activities in any significant way in school, they also needed computer-related literacy skills. Due to this need, Cathy proposed a Book Buddy Project, which she and Tamara subsequently planned and implemented. The theory and practice behind "book buddies" began as a program to facilitate the literacy growth of early readers who struggled with text (Johnston, Invernizzi, & Juel, 1998). The project involved pairing struggling readers with older, more advanced students. The more advanced students selected a book and read it to their "buddy" for approximately 30 minutes each week. The book buddies took turns asking and answering questions about the content of the book (Block & Delamura, 2001).

As Cathy began her search for the more advanced students, she thought of Tamara's university students. At the time, Tamara was teaching a literacy course in the special education program. The 31 students in this course were undergraduate preservice teachers in their junior year of college. One of the major goals of the course was to examine the special and diverse literacy needs of struggling readers and writers in the public schools. In addition, these preservice teachers were learning how to use technology to facilitate literacy instruction.

We decided to use technology as the tool for communication, collaboration, and learning during the Book Buddy Project. Cathy's students would be able to practice their writing skills through their communication on the computer, and they would be able to collaborate with an older, more capable student about the books they were reading. This collaboration would, in turn, give her students practice in those strategies critical to understanding text. Thus, technology became a way to bring these two very diverse classrooms together in a virtual space where the students could broaden their notions of literacy and learn beyond the bounds of the high school and university classrooms.

The Book Buddy Project was designed and implemented solely through technology. We provide a guideline of the procedures for implementing a Book Buddy Project in Appendix A. As the semester began, Tamara downloaded her course roster from E-Campus, a Web-based system that contains course rosters, schedules, and grading tools. At the same time, Cathy created a table of her students' names and reading scores in Microsoft Word. Both of us attached our class rosters via e-mail. As soon as both rosters were exchanged, the university students selected a book buddy from the high school roster, and received the buddy's reading score. This reading score enabled the university students to select books on their students' particular reading level.

Books were selected by using the technology of the high school media center. This particular high school had a Web-based system known as OPAC that catalogued books according to authors, subjects, keyword, interest, and reading level. The university students accessed the Web-based system from their computers at the university, and they began their search for books to use during the project. They narrowed their searches by specifying particular reading levels, interesting topics, and adolescent fiction or nonfiction. Using the technology of OPAC, the university students constructed a list of five books they thought might be interesting to them and their book buddy.

After selecting a number of books, the university students were ready to contact their high school book buddies via e-mail. We asked them to complete three tasks in their first e-mail. First, they were to introduce themselves in a personal way to their high school book buddies. Second, they suggested the five books that they had found on OPAC as possible books to read and asked their buddy to choose one. Third, they attached a digital picture of themselves. We included the digital pictures as another way that the students could use technology to communicate and collaborate with their buddies. The pictures provided a human link between the book buddies, so they could see the person to whom they are writing and responding.

Unfortunately, the e-mail technology of both schools provided some early glitches in the project. Since the high school students were ninth graders, they had not yet received e-mail addresses in the high school. The Book Buddy Project was delayed for three weeks while these students were given e-mails and placed on the high school computer network. When the high school students finally e-mailed the university students, many of their e-mails were dumped into the university students' junk mail folders due to the firewall imposed by the Web-based university e-mail system. Frustration set in for a couple weeks as students searched the systems for each others' e-mails. Eventually, the majority of these problems diminished.

After the initial introductions and the book buddy pairs chose a book to read, we gave the book buddies specific tasks for responding to each other through CMD. We required each buddy to respond at least once per week to his or her buddy for approximately eight weeks. Due to absences and other unscheduled interruptions, some students e-mailed each other for a longer duration. All book buddies were given a handout that detailed ways they could respond to the books they were reading. We requested that they respond to the book in the following ways:

1. Choose a word, phrase, sentence, or passage from the story that you believe is important or interesting, and explain why you chose it.
2. Relate the story to your own life experience.

3. Ask questions about words, sentences, characters, and ideas in the story.
4. Respond positively or negatively to the events in the story.
5. Explain why you think that the events in the story should have happened differently.

This handout served as a guide for practicing good reading strategies such as activating prior knowledge, questioning, and clarifying information. The high school students also completed a Reading Response Log (see Appendix A) in which they set goals for their reading and identified important literary elements within their books.

DATA ANALYSIS

All book buddy messages posted on e-mail were downloaded into a computer file and printed. The book buddy discussion responses were analyzed qualitatively using the constant comparative method (Glaser & Strauss, 1967). We employed this inductive analysis in order to determine the patterns, themes, and categories that emerged from the data (Patton, 1990). In particular, we examined the data for themes of communication, collaboration, and the learning environment. The content of all CMD and interview responses was read, examined, and open-coded in order to produce an initial code list. The responses were then reread and categories refined until all data had reached theoretical saturation.

BOOK BUDDY COMMUNICATION

We examined the students' responses and found that, while the students communicated according to the purposes and tasks that we set for the project, they also set their own purposes for communication. Although the assigned task was to respond to the books they were reading, students' correspondence often centered around their interests in their respective buddies. Many of the high school students were highly interested in the personal and professional lives of their university book buddies, as evidenced by the many questions they asked them about life on campus, their hometowns, and their interests or hobbies. One high school student wrote, "I hope you had fun during your fall break. So, where did you go?" University students also connected personally with their buddies by writing such responses as "Happy Birthday! My birthday is in December. I hope you have a great weekend. Talk to you soon."

The high school students enjoyed communicating about the exciting events in their lives. One student wrote about her Thanksgiving trip, "What are you going to do for thanksgiving (sic)? I am going to Philadelphia to see Allen Iverson. My uncle lives like 3 blocks away from him." Another student wrote, "Hey ____, I had a great spring break. I went to Newville Pa this weekend to watch wrestling. It was awesome. It was my first time goin to Pa."

Even though students engaged in brief personal communication, both groups of students remained focused on discussing the books they were sharing. They wrote about their favorite characters, the parts of the books that they liked and disliked, predictions, and how the book connected to their own lives. Many of the students also focused their communication on the parts of the story that surprised them. One high school student wrote this e-mail about his favorite character. "I really like Bonnie's character. I think that it is so nice that she talks to the kids and tried to be their friend." Through this e-mail response, the student is not merely stating his interest in the character, but also using the story to provide a reason for his opinion.

One university student wrote about her favorite part of Deuker's *Heart of a Champion* (1994) by stating:

My favorite part has been reading about how Seth has been able to find ways to practice baseball even when he is by himself. Wasn't that a cool idea to throw the ball against the part of the house where the building meets the ground so that he wouldn't know if it was going to bounce as a pop-up, line drive, or grounder?

In this example, the student is also using the story to reinforce her interest in a particular event in the story, but she is also trying to continue the thread of conversation with her book buddy by closing her response with a question.

Students expressed their interest by noting the surprises found in their stories. One high school student was surprised by the outcome of the story: "About the book, I really thought Eddie was the killer throughout the full book because it was weird how he knew everything about her and she didn't know that much about him." Another student wrote: "I was shocked when Mr. Pike got hurt. I did not think that Ric's ex girlfriend would do something like that." Other students were surprised by the characters' relationships in the book. For example, one high school student wrote: "Sometimes I am surprised that Neil and Randy & David and Terry are such good friends. They seem like complete opposites to me!" Through this response, the high school student is analyzing the characters and drawing conclusions about them through his writing.

The communication of these computer-mediated discussions was rather typical of e-mail correspondence in that the high school and university students used colloquial computer jargon. One high school students wrote: "It must b cool 2 live close by the beach. Where ru from?" Several students used computer communication to close their responses by writing, "g2g" (got to go). Some high school and university students also eliminated the necessity for many grammatical conventions such as capitalized letters and apostrophes.

Several book buddy partners found that they shared another common language besides English, and they began corresponding in such languages as Spanish and Russian. One high school student wrote in Spanish, and also provided an English translation.

Hola como estas? Espero que al leer este mensage te encuentres muy bien. Bueno espero que ya encontrastes el libro de The Parrot in the Oven. _______ y yo ya leimios hasta la pagina 80 (we read together). Y pues por lo que hemos leido, el libro se trata de un familia que se mira a ellos mismo como si no costaran nada y hay un jovensito que le gustaria seguir estudiando, pero sus padres no tienen sufisiente dinero pa' pagar sus clases, y bueno como el tiene esas ganas de estudiar, el trabaga en unos fields y tambien le quiere ensenarle a su papa que el puede salir adelante. Bueno me tengo que ir cuidate mucho.
Love, _____

Translation:

Hello, how are you.... I hope you've found the book. ______ and I have read to p80. From what we've read, the book is about a family that thinks of itself that they're not worth much. There's a boy who wants to continue studying, but his parents don't have enough money to pay for classes. Since he wants to study, he works in some fields and also because he wants to show his dad that he can make something of himself. I have to go....

Despite the more positive evidence of communication that occurred during the project, one of the major problems with communication concerned the high school students' lack of responses to their book buddies. Several high school students responded infrequently or not at all. Since CMD was the only form of communication available, and there was no possibility of face-to-face communication until the end of the semester, a vital part of the success of this project depended on the e-mail communication between the participants. Some of the university students sent three e-mails before receiving a reply from their buddies. As a result, the university students often expressed frustration to us and their book buddies that their responses were not read. One way that we thought about eliminating this problem in the future is to create newsgroups comprising two university and one high school student. By establishing newsgroups of three collaborators, a book buddy would be more likely to receive a response.

BOOK BUDDY COLLABORATION

Collaboration involves the use of communication, in this case CMD, to connect or share with someone either through the content that is discussed or through the social interactions that occur. In this study we saw evidence of collaboration in several significant ways. First, the book buddies explicitly requested collaboration through their closing comments. They wrote "E-mail when u get a chance," and "...let me know what you think about it [the book] so far." Some high school students were quite explicit in their requests for collaboration, "WRITE BACK AS SOON AS POSSIBLE!!!"

A second form of collaboration occurred as they monitored each others' progress through the literature. The book buddy partners were constantly monitoring their reading progress by setting particular dates for completing sections of the book, keeping track of the pages they had read and the characters they had met in the book. They also monitored their partners' e-mail schedule by determining the best time to e-mail. One university student wrote: "When do you check your e-mail? On Thursday or Friday? The morning or the middle of the day? I'll be sure to read the chapter and get back to you by one of those times." Another student wrote: "I'm almost done with the book. I'm on page 113. I might be a little ahead. So, I might not give out a lot about the book." In many cases, the students were very conscious of not "giving away" the contents of the book until their partners had read far enough into the book.

When the high school students read more than the university students, they often wrote to hurry them along. One student wrote, "Anyways, I'm halfway through the book and hoping you will be catching up." Other students monitored progress by noting the pages they had read and summarizing the content. For example, a high school student wrote: "I am almost at the end of the book. I am on page 176 where they meet a girl named Stephanie...Also, where their dad is missing. Where do you think he's at?"

This example illustrates another way in which many of the students collaborated. They asked each other questions to improve the likelihood of a response. Beach and Lundell (1998) refer to these responses as task continuative practices because the students are writing to receive a response, so the thread of the discussion remains unbroken. A high school student wrote:

I would not know what to do because you might have to go out and fight in the war. How would u feel if u had to make a promise to you father? I don't know if I would be able to promise my father. Would u be able to? I think that the white and black boy will become friends. The white boy is going to teach the black boy how to read. What do you think is going to happen next? What do you think about the book so far? Who's your favorite character?"

In this e-mail, the student asks several questions related to the character's dilemma and whether his partner would react in the same way as the character. The student also makes a prediction and asks his partner to make one about the next part of the story. Finally, he asks some more general questions at the end that are focused on his partners' interest in the story and characters. This type of response encouraged collaboration because his partner had many opportunities to continue the thread of conversation concerning any number of topics. His partner responded, "I do think that he will end up fighting for the North. What do you think will happen on his trip to Richmond?" Again, this university student responded by predicting the story events, but she also ended her response with another prediction question that allowed her high school partner to collaborate again without breaking the thread of conversation.

The book buddy partners collaborated by validating their partners' responses. A high school student wrote about the conclusion of her book: "I thought that the ending of the book would have given us more information... What do you think of the book? What would u like to have changed? I would like to see Bonnie and that guy get married." Her university book buddy responded: "I like the ending, but I agree with you. They should have given more information both about the hospital and about what happens between Bonnie and that guy."

High school students used collaboration with their book buddy as an opportunity to clarify parts of the books that were confusing. One high school student wrote: "Yes, I do remember about what Mr. Wagner did to Rebecca. It was very bad for a teacher to do that to a student. I kind of got lost. I don't remember reading if Rebecca's parents ever made it to Westphila (I think that's how you spell it)." Another student conveyed his confusion by stating: "I'm on page 28. I understand the beginning, but I'm kind of confused. I also wonder why the title is called *Storm*." These responses increased the collaboration among book buddy partners because the partner was now charged with the task of clarifying the text for his buddy. Some university students offered particular strategies to help the high school students overcome some of the difficulties in making meaning from the text. For example, one university student encouraged her book buddy to use strategies such as prediction and determining importance: "Making predictions is always a fun way to see if you and the author are thinking along the same lines...pick out a sentence or idea from the book that you think is important...and tell me about it in your next e-mail."

BOOK BUDDY LEARNING ENVIRONMENT

This project was a unique and rewarding method for seeing into the minds of the students and discovering the ways in which they were processing the text. We saw evidence of several strategies that are important to text comprehension—character analysis, look backs, and summaries. Some of the richest strategies were evidenced in the ways book buddies were able to analyze the characters in the literature. In analyzing these characters, students responded in a variety of ways. Some students analyzed the character by writing about how that character was one of their favorites. Other students analyzed the character's motivations in order to predict what that character might do next. Students also provided character analyses by writing about the characters' emotions and how they would act if they were the characters. In the following example, a university student analyzed the character of Mia in Cabot's *Princess in the Spotlight* (2002) by discussing the role that Mia plays within her family and relating how she might feel in Mia's position:

Mia seems like a very responsible girl...seems like she runs the house for her mom. What do you think about her mom and the algebra teacher having a baby? I think that would be a weird experience to have to go through. Mia seems to be taking it well though...she is very concerned about her mom having a healthy pregnancy. She has a very busy, crazy life.

Her high school book buddy writes back:

I would be excited yet scared 2 do an interview on national t.v. If my mom was [sic] having a baby by my algebra teacher, I would be freaked out. People would think that I am the teacher's pet. Mia is very mature 4 her age.

In this response, the students are analyzing the character by placing themselves in the role of the princess and thinking about the emotions that they would feel.

We also found that students were learning about the text by expressing their viewpoints and feelings about particular sections of the stories. One high school student referred to a part of that book that caught his attention:

I just finished Chapter 15, and I thought the way they described Neil's boy when he fell was pretty gruesome. When he fell, I was screaming 'no!' inside my head. They were so close to escaping! Now that Neil was injured, I know he would hold the others back and their chances of finding a way out would be smaller. I almost feel as if it was not his body that gave up but his mind.

In this example, the student appears to think aloud about the ideas running through his mind as he read this section of the story. Thus, we were able to see how he was making meaning with the text.

Students also conveyed their understandings of the text by summarizing the part of the story that they had just read. This high school student gave a detailed summary of the beginning of a biography about Grant Hill.

I'm on page 26, and so far it's only saying stuff that's been going on in his life. Like his mother was sort of fearful that people would be scared of her because she wouldn't let anybody mess with her or her son and his school grades. And his dad barely spent time with him because he was a professional football player, and since he had so many games, he couldn't have any time with his son. But now he can since he retired. He goes to every basketball game that Grant had. Only he isn't much of a cheering kind of father so he didn't cheer as much.

Summaries such as this one really helped us determine the learning that took place as these students engaged in text. These summaries also aided the university students in discovering how these struggling readers comprehended the text.

CONCLUSION

The Book Buddy Project afforded us the opportunity to create a virtual literacy community in which high school and university students incorporated the traditional literacies of reading and writing within a virtual environment that facilitated communication, collaboration, and learning with text. By communicating through writing, these students had opportunities to increase their writing and communication skills. This was demonstrated in the ways they expressed their interests in various characters and parts of the books, as well as their descriptions of surprising events in the stories. They also used writing to express emotions about the characters and the events within the texts.

This project gave the students the opportunity to collaborate with other readers beyond the physical space of their respective classrooms. As we analyzed the themes of collaboration, we were excited by the ways that students connected with their book buddy partners as they monitored each others' progress and asked questions to generate responses. We were also encouraged by their willingness to validate each other's reactions to the text and, more importantly, by the kind of learning that we saw in the students' responses. In typical classroom discussions, high school students are often relegated to one-sentence answers to teacher questions about the text (Jetton & Alexander, 1997). Such was not the case in this study. Instead, the high school students often wrote elaborate descriptions and analyses of the characters in their stories. They examined particular parts of the story that were sad, gruesome, or scary, and they interpreted these events in light of their own emotions and experiences.

This study offers validation for the use of CMD in K-12 education. It also reinforces the potential for CMD to create environments that transcend the traditional borders of the K-12 classroom by developing virtual environments where high school students can communicate and collaborate with authentic audiences. Despite its limitations, CMD is an excellent tool for deriving meaning from text because it allows students to examine literary characters and events, and share their conclusions, predictions, and reactions with a virtual community of other students who are engaged in the same pursuit of meaning.

REFERENCES

Beach, R., & Lundell, D. (1998). Early adolescents' use of computer-mediated communication in writing and reading. In D. Renking, M.C. McKenna, L. Labbo, & R. Kieffer (Eds.), *Handbook of literacy and technology: Transformations in a post-typographic world* (pp. 93-112). Mahwah, NJ: Lawrence Erlbaum.

Block, C. C., & Delamura, R. J. (2001). Better book buddies. *The Reading Teacher, 54*(4), 364-370.

Bonk, C. (2003-2004). I should have known this was coming: Computer-mediated discussions in teacher education. *Journal of Research on Technology in Education, 36*(2), 95-102.

Bonk, C. J., & Reynolds, T. H. (1992). Early adolescent composing within a generative-evaluative computerized prompting framework: Computer use in the improvement of writing. *Computers in Human Behavior, 8*(Special Issue), 39-62.

Boyd, R. (2000, Winter). Computer based reading instruction: The effectiveness of computer based reading instruction in positively influencing student scores in science. *CSTA Journal,* 17-23.

Cabot, M. (2002). *Princess in the spotlight.* New York: HarperTrophy.

Cooper, M.M., & Selfe, C.L. (1990). Computer conferences and learning: Authority, resistance, and internally persuasive discourse. *College English, 52,* 847-869.

Daiute, C. (1983). *Writing and computers.* Reading, MA: Addison-Wesley.

Daly, J. A., & Miller, M. D. (1975). The empirical development of an instrument to measure writing apprehension. *Journal of Research in the Teaching of English, 9,* 242-249.

Deuker, C. (1994). *Heart of a champion.* New York: HarperTrophy.

Fauske, J., & Wade, S.E. (2003-2004). Research to practice online: Conditions that foster democracy, community, and critical thinking in computer-mediated discussions. *Journal of Research on Technology in Education, 36*(2), 137-154.

Ferdig, R. E. & Roehler, L. R. (2003-04). Student uptake in electronic discussions: Examining online discourse in literacy preservice classrooms. *Journal of Research on Technology in Education, 36*(2), 119-136.

Fischer, K.M. (1998). Pig tales: Literature inside the pen of electronic writing. In D. Reiss, D. Selfe, & A. Young (Eds.), *Electronic communication across the curriculum* (pp. 207-220). Urbana, IL: National Council of Teacher of English.

Gillespie, P. (1998). E-journals: Writing to learn in the literature classrooms. In D. Reiss, D. Selfe, & A. Young (Eds.), *Electronic communication across the curriculum* (pp. 221-230). Urbana, IL: National Council of Teacher of English.

Gillingham, M. G., Garner, R., Guthrie, J. T., & Sawyer, R. (1989). Children's control of computer-based reading assistance in answering synthesis questions. *Computers in Human Behavior, 5,* 61-75.

Glaser, B. G., & Strauss, L. L. (1967). *The discovery of grounded theory: Strategies for qualitative research.* Chicago: Aldine.

Hara, N., Bonk, C. J., & Angeli, C. (2000). Content analysis of online discussion in an applied educational psychology course. *Instructional Science, 28,* 115-152.

Hiltz, S. R., & Turoff, M. (1978). *The network nation: Human communication via computer.* Reading, MA: Addison-Wesley.

Jetton, T. L. (2003-2004). Using computer-mediated discussion to facilitate preservice teachers' understanding of literacy assessment and instruction. *Journal of Research on Technology in Education, 36*(2), 171-189.

Jetton, T. L., & Alexander, P. A. (1997). Instructional importance: What teachers value and what students learn. *Reading Research Quarterly, 32,* 290-308.

Johnston, F. R., Invernizzi, M., & Juel, C. (1998). *Book buddies: Guidelines for volunteer tutors of emergent and early readers.* New York: Guilford Press.

Kamil, M. L., Intrator, S., & Kim, H. S. (2000). Effects of other technologies on literacy and literacy learning. In M. Kamil, P. Mosenthal, P. D. Pearson, & R. Barr, (Eds.), *Handbook of reading research* (vol. 3, pp. 773-788). Mahwah, NJ: Lawrence Erlbaum.

Kiesler, S., Siegel, J., & McGuire, T. W. (1984). Social psychological aspects of computer-mediated communication. *American Psychologist, 39,* 1123-1134.

Kim, H. S., & Kamil, M. L. (2004). Adolescents, computer, technology, and literacy. In T.L. Jetton & J.A. Dole (Eds.), *Adolescent literacy research and practice* (pp. 351-368). New York: Guilford Press.

Ku, L. (1996). Social and nonsocial uses of electronic messaging systems in organizations. *Journal of Business Communication, 33,* 297-326.

Mabrito, M. (2000). Computer conversations and writing apprehension. *Business Communication Quarterly, 63,* 39-49.

Matusov, E. (1996). Intersubjectivity without agreement. *Mind, Culture, and Activity, 3,* 25-45.

McMillan, K., & Honey, M. (1993). *Year one of Project Pulse: Pupils using laptops in science and English: A final report* (Technical Report No. 26). New York: Center for Technology in Education.

Mellott, J. (2002, April 12). City ranks second in Virginia for ESL percentage. *The Daily News Record,* 13-14.

Merchant, G. (2001). Teenagers in cyberspace: An investigation of language use and language change in Internet chatrooms. *Journal of Research in Reading, 24*(Special Issue: Literacy, Home and Community), 293-306.

Moore, M. A., & Karabenick, S. A. (1992). The effects of computer communications on the reading and writing performance of fifth-grade students. *Computers in Human Behavior, 8*(Special Issue: Computer Use in the Improvement of Writing), 27-38.

Palumbo, D. B., & Prater, D. L. (1992). A comparison of computer-based prewriting strategies for basic ninth grade writers. *Computers in Human Behavior, 8*(Special Issue: Computer Use in the Improvement of Writing), 63-70.

Patton, M. (1990). *Qualitative evaluation and research methods.* Newberry Park: Sage.

Reinking, D. (1998). Computer-mediated text and comprehension differences: The role of reading time, reader preference, and estimation of learning. *Reading Research Quarterly, 23,* 484-498.

Reinking, D., & Rickman, S.S. (1990). The effects of computer-mediated texts on the vocabulary learning and comprehension of intermediate-grade readers. *Journal of Reading Behavior, 33,* 395-411.

Rice, R. E., & Love, G. (1987). Electronic emotion: Socio-emotional content in a computer-mediated communication network. *Communication Research, 14,* 85-105.

Rosenbluth, G. S., & Reed, W. M. (1992). The effects of writing-process-based instruction and word processing on remedial and accelerated 11th graders. *Computers in Human Behavior, 8* (Special Issue: Computer Use in the Improvement of Writing), 71-95.

Rowntree, D. (1995). Teaching and learning online: A correspondence education for the 21st century? *British Journal of Educational Technology, 26,* 205-215.

Schallert, D. L., Reed, J. H., & the D-Team. (2003-2004). Intellectual, motivational, textual, and cultural considerations in teaching and learning with computer-mediated discussion. *Journal of Research on Technology in Education, 36*(2), 103-118.

Star Reading. (2002). New York: Perfection Learning.

Thomas, M. J. W. (2002). Learning within incoherent structures: The space of online discussion forums. *Journal of Computer Assisted Learning, 18,* 351-366.

Tiene, D. (2000). Online discussions: A survey of advantages and disadvantages compared to face-to-face discussions. *Journal of Educational Multimedia and Hypermedia, 9,* 371-384.

Virginia Department of Education. (1999). Fall 1999 membership by school division. Retrieved April 11, 2002, from http://www.pen.k12.va.us/VDOE/dbpubs/Fall_Membership/

Virginia Department of Education. (2000). Fall 2000 membership by school division. Retrieved April 11, 2002, from http://www.pen.k12.va.us/VDOE/dbpubs/Fall_Membership/

Virginia Department of Education. (2001). Fall 2001 membership by school division. Retrieved April 11, 2002, from http://www.pen.k12.va.us/VDOE/dbpubs/Fall_Membership/

Walther, J. (1992). Interpersonal effects in computer-mediated interaction: A relational perspective. *Communication Research, 19,* 52-90.

Walther, J. (1996). Computer-mediated communication: Impersonal, interpersonal, and hyperpersonal interaction. *Communication Research, 23,* 3-43.

Weller, L. D., Carpenter, S., & Holmes, C. T. (1998). Achievement gains of low-achieving students using computer-assisted vs. regular instruction. *Psychological Reports, 83*(4), 1440-1441.

Wellman, B. (1997). An electronic group is virtually a social network. In S. Kiesler (Ed.), *Culture of the Internet,* Mahwah, NJ: Lawrence Erlbaum.

APPENDIX A

Guidelines for the Book Buddy Project

1. Select K-12 students who struggle with reading, writing, and computer literacy skills.
2. Select university preservice teachers who are currently participating in a literacy course that focuses on struggling readers and writers.
3. The participating public school teacher and university instructor share class rosters via e-mail.
4. University preservice teachers are given a roster of public school students, and they sign up for a book buddy on this roster.
5. University preservice teachers obtain a digital picture of themselves.
6. University preservice teachers select five books based on their book buddy's reading level, as determined by a reading assessment administered at the public school. Other variables to consider in selecting books include topic interests of adolescents and the genre of the text. Books can be selected using technology available (see reference to OPAC in the chapter).
7. Preservice teachers write a one or two-sentence synopsis for each of the five books chosen.
8. University preservice teachers establish initial contact with their book buddies via e-mail and accomplish two tasks:
   a. Introduce themselves
   b. Attach a digital picture of themselves
9. Provide the book buddy a list of five books, along with a synopsis of each book, and ask the book buddy to select a book that they will read together.
10. The book buddies are given specific tasks for responding to each other through CMD:
    a. Choose a word, phrase, sentence, or passage from the story that you believe is important or interesting, and explain why you chose it.
    b. Relate the story to your own life experience.
    c. Ask questions about words, sentences, characters, and ideas in the story.
    d. Respond positively or negatively to the events in the story.
    e. Explain why you think that the events in the story should have happened differently.
11. Book buddies respond to each other at least once per week for approximately eight weeks.

APPENDIX B

Reading Response Log

Name______________________________
Date Due____________________________

Reading Response Log
Title____
Author____
Why did you select this book?____
____
How many pages? ________________ How many chapters?________________
How many pages will you read each day?________ Goal date for finishing?__________
Name of Book Buddy____
E-mail address of Book Buddy____
Have you and your Book Buddy agreed upon the goal date for finishing?____________

Complete one section of the reading response log for each day of reading. You do not have to complete the responses in any particular order. Use your responses to help you communicate with your book buddy!

---

* Date ___________ I have read from page ________ to page ___________.

Write 5 complete sentences to identify and describe the **main character**. Consider such information as name, age, gender, family, interests, personality, friends, joys, conflicts, etc. Does this character remind you of yourself or anyone else you know?

____
____
____
____
____
____
____
____
____.

---

* Date ___________ I have read from page ________ to page ___________.

Write two complete sentences to **identify the main conflict** of the story and another 2 sentences to **predict how that conflict will be resolved**. Have you ever been in a similar conflict or know anyone who has?

____
____
____
____
____
____
____
____.

---

* Date ___________ I have read from page ________ to page ___________.

In three complete sentences, describe the **setting** of the book you're reading. Be as descriptive as you can!

____
____
____
____
____
____
____
____.

---

* Date ___________ I have read from page ________ to page ___________.

Find examples of **figurative language** (similes, metaphors, personification, …) in your reading and record them below:

1) Example:____
   Type:__________________________
2) Example:____
   Type:__________________________
3) Example:____
   Type:__________________________

* Date___________ I have read from page ________ to page ___________.

Identify and **define** at least five **words** you have learned or become more familiar with through your reading:

1)________________________________
Definition or synonym:____
2)________________________________
Definition or synonym:____
3)________________________________
Definition or synonym:____
4)________________________________
Definition or synonym:____
5)________________________________
Definition or synonym:____

Date____________ I have read from page ______ to page _______.

Write a short paragraph about your reading so far. You may summarize the story, question a part you don't understand, predict what might happen next, or comment on your opinion of the book so far.

____
____
____
____
____
____
____
____
____.

Chapter XXXI
Online Learning Communities:
Enhancing Learning in the K–12 Setting

Chris Brook
Edith Cowan University, Australia

Ron Oliver
Edith Cowan University, Australia

ABSTRACT

This chapter reports a design framework intended to support and guide teachers in advancing K-12 literacy through principles of collaborative learning and the development of online learning communities. The study was guided by an investigation of contemporary literature focused on the community construct, online learning community development, the collaborative construction of knowledge, and the practices of experienced professionals working in the field. The intended outcome is a design framework that may be useful in guiding instructors in the advancement of K-12 literacy skills through the development of online learning communities.

INTRODUCTION

A report issued by Alberta Learning claimed that "Current technological advances are making it possible to offer online learning programs that are equal to or even superior to regular classroom instruction" (Alberta, 1999). Such programs may be seen to meet the needs of:

- Students preferring to work independently.
- Parents disagreeing with the social values taught in public schools.
- Students who need to study outside the regular school hours.
- Students who experience health challenges which restrict their access to regular schools.

- Schools unable to offer a full program of study. (p. 3)

While the potential benefits that online technologies afford K-12 education were highlighted, the report acknowledged that online technologies are not a panacea. It was suggested that care is needed in the design of learning activities to account for both the social and cognitive impact of technologies, if the full potential of online technologies to support learning is to be achieved (Alberta, 1999, p. 16).

Much has been learned in higher education settings since the time that Alberta Learning first published the report, reflecting both social and cognitive considerations in approaches to online learning that may be beneficial to the advancement of literacy in the K-12 setting.

Currently, there is growing support for the supposition that the social phenomenon of community may be put to good use in the support of online learning (Bonk & Wisher, 2000; Hiltz, 1998; Palloff & Pratt, 1999; Rovai, 2002). This suggestion is well supported by theories of learning that highlight the role of social interaction in the construction of knowledge (Bruner, 2001; Dewey, 1929; Kafai & Resnick, 1996; Vygotsky, 1978) and those that propose that knowledge is constructed within the social milieu (Cunningham, 1996). Further support may be found in contemporary literature that reports the benefits of collaborative learning settings spanning the academic, social, and psychological domain (Panitz, 1997). It has been suggested that collaborative learning settings promote increased motivation (Slavin, 1990), learning achievement (Johnson, 1991; Maxwell, 1998), and perception of skill development including satisfaction (Benbunan-Fich, 1997). Additionally, social factors such as a sense of connectedness have been shown to influence student success and satisfaction in online learning (Barab, Thomas, & Merrill, 2001).

Some researchers believe the development of learning communities should be a primary goal of online instructors (Hiltz, 1998; Palloff & Pratt, 1999). However, there is little empirical evidence to guide instructors in the development process (Bonk & Wisher, 2000; Palloff & Pratt, 1999; Paulsen, 1995). Currently, design principles tend to be process oriented (Brook & Oliver, 2003) and based on anecdotal evidence gleaned from the experience of professionals working in the field with a notable absence of empirical studies (Bonk & Wisher, 2000; Palloff & Pratt, 1999).

This chapter describes an investigation of the development of online learning communities. It proposes to establish a common understanding of the term *community* and identify the chain of events that lead to community development and the collaborative construction of knowledge, proposing a model that describes this sequence. Guidance was taken from contemporary literature, the practices of experienced professionals working in the field, and the experiences of students.

Although much of the research referred to was conducted in higher education programs, it provides a strong foundation for the K-12 setting.

UNDERSTANDING COMMUNITY

While the use of the term *community* is becoming increasingly common in education circles, it is important to acknowledge that a definitive definition of the term remains elusive (Puddifoot, 1996) with numerous definitions identified (Hillery, 1964). Due to the many disciplines that study the social phenomenon of community, presenting a range of understandings, the identification of a single definition is unlikely (Goth, 1992).

Notwithstanding continued debate, several features of community have general acceptance. Communities provide systems and processes for meeting the basic human needs for survival, nurturance, socialization, and support; cosmological or ideological perspectives; and a cohesive context from which a sense of identity, belonging, meaning, and purpose can develop (Redfield, 1960). The community experience is central to the lives of all individuals, and it is generally acknowledged that "if the sense of living in, belonging to, and having some commitment to, a particular community is threatened then the prospect of living rewarding lives is diminished" (Puddifoot, 1996, p. 327). The community experience is context specific and may vary between members (Sonn, Bishop, & Drew, 1999). Communities take many forms including those based in religion, politics, and neighbourhoods (Goth, 1992; Sarason, 1974). Of the various forms of community, a learning community is characterized by a willingness of members to share resources, and accept and encourage new membership, regular communication, systematic problem solving, and a preparedness to share success (Moore & Brooks, 2000). These characteristics clearly represent factors that may be put to good use in the support of learning, as does the social phenomenon where the sum of the parts of a community is in some way greater than the whole (Hawley, 1950). However, how these characteristics might be purposefully developed in online settings remains unclear.

Communities exist in both a geographic and relational sense (Gusfield, 1975; Worsley, 1991), with modern societies tending to develop more relational communities (Durkheim, 1964; Royal & Rossi, 1996) or communities of the mind (Tönnies, 1955). It is the relational or community of the mind that forms in the online setting (Obst, Zinkiewicz, & Smith, 2002; Surratt, 1998).

Potential Negative Aspects of Community

While these characteristics suggest a positivistic view of community, it is worth noting that the social phenomenon of community may exert negative influences on members. These include the need for members to conform, and the subsequent loss of individuality (Wiesenfeld, 1996) and the potential to hoard knowledge and restrict innovation (Wenger, McDermott, & Snyder, 2002). Also noteworthy is the potential for community structures to exert pressure on some individuals to engage in nonconforming rather than conforming behaviours, resulting in dissidents and the formation of sub-communities (Carol, 1997). It has also been suggested that, due to the socially impoverished nature of the Internet (Stoll, 1995) that encourages weak social network ties (Constant, Sproul, & Kieser, 1996; Granovetter, 1973)—void of the social support achieved through strong ties (Wellman & Wortley, 1990)—extensive participation in online communication may diminish psychological well-being (Kruat et al., 1998). Although undesirable, these characteristics cannot be ignored when the social phenomenon of community is employed to enhance the learning experience, as they represent factors that are likely to diminish rather than enhance the learning experience.

Identifying general agreement on key features of community is a useful exercise in ensuring commonality of meaning and establishing characteristics of the desired product, but does little to further insight into how community may be purposefully developed. To achieve this requires further investigation of the community construct, how this construct may be understood and measured, and the chain of events that are likely to lead to its development.

The Sense of Community

The community construct is widely accepted as a sense rather than a tangible entity (Wiesenfeld, 1996). This sense may have many referents ranging from sporting groups to neighbourhoods, and simultaneous membership to multiple communities is possible and indeed likely, although not all will give a positive sense of community (Sarason, 1974). Sense of community is based on an attachment relationship, and this relationship is not based on the interactions with any one member of the community, but instead with any member (Hill, 1996). Sense of community has been defined as "a sense that members have a belonging, members matter to one another and to the group and a shared faith that members' needs will be met through their commitment to be together" (McMillan & Chavis, 1986, p. 9). While this is not accepted as the definitive definition of community, it is accepted as a *good fit* (Sarason, 1974) and has been adopted for the purpose of this study.

McMillan and Chavis (1986) proposed that sense of community might be represented as a four-dimensional model comprising the elements of membership, influence, fulfilment of needs, and shared emotional connection, with each of the elements characterized by key attributes. Table 1 presents the four elements and their attributes.

These elements and their attributes may prove useful in guiding the development of online learning communities, keeping in mind the varying presence of each element in any given community and that shared emotional connection is considered the definitive element of true community (McMillan, 1996). Promoting these elements through a common symbol system (McMillan & Chavis, 1986; Palloff & Pratt, 1999), establishing a common purpose (Hawley, 1950), facilitating frequent and easy meetings (Worsley, 1991), and developing a sense of place (Lorion & Newbrough, 1996; Puddifoot, 1996) is likely to support community development. In addition, it has been suggested that sense of community be considered an economy where self-disclosure is the commodity for trade. In this environment trade must be perceived as fair (McMillan, 1996) and *safe*

Table 1. Elements and attributes of sense of community

Element	Attribute
Membership	Boundaries that separate *us* from *them* Emotional safety A sense of belonging and identification A common symbol system
Influence	Individual members matter to the group The group matters to the individual Making a difference to the group Individual member influences the group The group influences the individual member
Fulfillment of needs	Benefits and rewards Members meeting their own needs Members meeting the needs of others Reinforcement and fulfilment of needs
Shared emotional connection	Identifying with a shared event, history, time, place, or experience Regular and meaningful contact Closure to events Personal investment Honour Spiritual connection

(McLellan, 1998), free from shame, where individuals may trade freely. Guidance for developing this *safe environment* may be found in the literature that suggests encouraging low-risk trade, where individuals identify similarities, provide positive support, and share information (McMillan, 1996). Once this has been established, it is possible to progress to activities that require identifying differences, including strengths, weaknesses, and needs. It is not until this has been achieved that members can begin to trade freely and the community economy is established (McMillan, 1996). The sequential five-stage model developed by Salmon (2000) that includes access and motivation, online socialization, information exchange, knowledge construction, and development supports this supposition.

THE ROLE OF COMMUNITIES IN THE LEARNING PROCESS

There exists debate among theorists as to the role that communities play in the learning process. The cognitive theories of learning, which have become influential in educational environments in relatively recent times, investigate the internal process that takes place to facilitate learning. In general terms, cognitive psychologists see the learner as an active participant in the learning process, actively constructing new knowledge based on current and past experiences (Kafai & Resnick, 1996). This field of thought is known as *constructivism*. Perhaps the most widely regarded theorist in this area is Jean Piaget who proposed that mental growth is governed by continual activity aimed at balancing the intrusions of the social and physical environment with the organism's need to preserve its structural systems (Elkind, 1967). Piaget proposed that intellectual growth is a result of four contributing factors: maturation, physical experience, social experience, and equilibration (Elkind, 1967).

A second perspective of learning within the cognitive domain is the socio-cultural theory that works in contrast to the constructivist view. Where the constructivist perspective focuses on individual cognitive processes in the construction of knowledge, the socio-cultural perspective emphasizes the role of social interactions and cultural organized activities in influencing cognitive development (Cobb, 1994). Two influential theorists who advocate the importance of social interaction in the construction of knowledge are Vygotsky and Dewey (Glassman, 2001). While Vygotsky emphasizes the importance of social history, Dewey stresses the importance of individual history. Vygotsky (1978) places a heavy emphasis on the role of culture and social history in education, suggesting that the process of education works from the outside in. Dewey, with a heavy emphasis on the importance of the social history of the individual, sees the process as coming from the inside out (Glassman, 2001). Notwithstanding this philosophical difference, both theorists stress the importance of social interaction in the learning process. These social interactions are promoted in learning communities.

Cobb (1994) argues that the apparently apposing cognitive and socio-cultural theories are reality, complimentary. The socio-cultural perspective suggests the conditions for the possibility of learning, while constructivist perspective outlines what students learn and the process by which they learn. This suggestion is reflected in a third field of thought known as *constructionism*. Constructionism includes the theories espoused by Piaget, but goes beyond these to include the notion that the process of learning takes place when the learner is engaged with the construction of something external. This leads to a cycle of internalizing what is outside and then externalizing what is inside and so on (Papert, 1990). Constructionism is seen as offering an important bridge between cognitive and socio-cultural perspectives on cognitive development, by arguing that individual development cycles are enhanced by shared constructive activity in the social environment. Furthermore, social settings are enhanced by the cognitive development of the individual. The constructionist view is that shared constructions and social relations are key to individual development (Kafai & Resnick, 1996). Importantly, it is suggested that settings marked by fractured and limited social activity and less cohesive social relations may present troubling development barriers (Kafai & Resnick, 1996).

Theories of learning that emphasize the role of social engagement in the learning process and member behaviours that characterize a positive sense of community suggest that the individual's learning experience will be enhanced in learning communities. The advent of online technologies provides a mechanism through which instructors might promote social interaction and the development of a sense of community among K-12 students, affording clear advantage.

The Construction of Knowledge in Communities

Similar to community, the term *knowledge* is commonly used yet surprisingly complex. At times the term is used to refer to tangible objects that can be captured, codified, and stored—known as *structured knowledge*. At other times the term is used to refer to the human element of knowledge that cannot be articulated, codified, captured, or stored—known as less-structured knowledge (Hildreth & Kimbe, 2002). Other terms used include *formal* and *informal knowledge* that are used to separate knowledge that can be bound in books and shared from the knowledge that is used to create that which is bound in books and shared (Conklin, 1996). Hildreth and Kimbe (2002) suggested that of the many terms that are used to describe knowledge, the most controversial distinction of all is made between *tacit knowledge* (that which cannot be told) (Polanyi, 1967) and *explicit knowledge* (that which is easily expressed) (Nonaka, 1991). Despite the varying views and the continued debate, there appears to be general agreement that at some level, knowledge can be viewed as comprising both external and human elements.

While stressing the importance of the social construction of knowledge (Dewey, 1929; Von Krogh, 1998; Vygotsk yy, 1978), Hildreth and Kimbe (2002) maintain that the tacit and explicit elements of knowledge are interwoven. Attempts to advance the construction of knowledge must focus on both these elements of knowledge moving away from capturing to sharing knowledge (Hildreth & Kimbe, 2002) in accordance with constructivist philosophies (Von Krogh, 1998). Researchers argue that this sharing of knowledge is promoted in both communities of practice (Wenger, 1998) and learning communities (Moore & Brooks, 2001).

It has been suggested that the social construction of knowledge in the online environment progresses through five sequential phases (Gunawardena, Lowe, & Anderson, 1997). Table 2 lists those phases.

Statements of opinion and observation, along with corroborating examples provided by one or more participants, characterize phase one. Phase two is characterized by identifying and stating areas of disagreement and perhaps escalating conflict through reference to research or experience. Exploration of meaning and the identifying of areas of agreement characterize phase three, and phase four is characterized by testing the proposed synthesis against 'received

Table 2. Interactive analysis model for examining social construction of knowledge in computer conferencing

Five-Phase Interactive Analysis Model
1. The sharing and comparing of information
2. The discovery of exploration of dissonance or inconsistency among ideas, concepts, or statements
3. The negotiation of meaning
4. The testing and modification of proposed synthesis or co-construction
5. Agreement statements and the application of newly constructed meaning

fact' as shared by the participants and/or their culture. Metacognitive statements by the participants illustrating their understanding that their new knowledge or ways of thinking have changed characterize phase five (Gunawardena et al., 1997). The latter stages of the model require high levels of bi-directional influence between the individual and the group, an identifying characteristic of strong communities (McMillan, 1996). Of particular interest is how student interactions may be purposefully progressed through these phases to promote the collaborative construction of knowledge and the formation of a strong community.

FACTORS THAT INFLUENCE COMMUNITY DEVELOPMENT IN ONLINE SETTINGS

It appears that the decision to pursue or ignore membership in a community rests with the *will* of the individual. *Will* has been categorized as either *rational* or *natural will* (Tönnies, 1955). While *natural will* refers to more personal characteristics and traits such as character, intellect, and attitude, *rational will* refers to a rational decision-making process. An individual's *natural will* suggests a positive or negative predisposition to community orientation, while *rational will* suggests a more pragmatic view of community membership, a view heavily influenced by purpose and perceived benefits associated with membership. It has been demonstrated that individuals may exercise *rational will* to seek community membership even when antipathy is the norm (Tönnies, 1955). This suggests it is possible for online instructors to employ forms of engagement and activity that may influence an individual's *rational will* to seek community membership in the event that *natural will* is predisposed to ignore the possibility. For this to take place, there is a need for consistency between the underlying philosophy of learning and the structure of the learning setting including tasks and activities (Bonk & Cunningham, 1998) and purposeful action that moves students from a feeling of *outsider* to *insider* (Wegerif, 1998). Simply employing the software and hoping these conditions will develop is unlikely to be effective, as has been identified by Hiltz (1997) who asserts:

The development of a collaborative learning environment is not simply a matter of employing the software to facilitate a communication place and informing the students of its availability and telling them to use it at will. This will result in students not using the communication opportunity at all or dropping out of communication after a very short time. (p. 2)

Factors that may influence community development include policies (Cho & Berge, 2002), the discipline and educational level of the course (Hiltz, 1994; Palloff & Pratt, 1999), the instructor (e.g., Collins & Berge, 1996), and the students (Hiltz, 1994). At a process level, influencing factors include the purpose community serves in the lives of its members (Hawley, 1950; Palloff & Pratt, 1999), support for communication (Collins & Berge, 1996; Hill & Raven, 2000), the nature of meetings (Moore & Brooks, 2001), and the *gathering place* (e.g., Von Krogh, Ichijo, & Nonaka, 2000).

Influencing factors of this nature suggest a chain of events that may be expressed by adapting the three 'P' model of presage, process, and product (Biggs, 1989). The Biggs (1989) model describes the process of student learning and may be used to inform approaches to teaching. As described by the model, presage factors at both the student and teacher level interact to produce an approach to learning. Process factors describe the approaches students adopt to process academic tasks and the product reflects the learning outcome (Biggs, 1989). Community development may be described in a similar manner, beginning with presage factors—including the system, learning context, and students—that interact to produce an approach to community development. Progressing on, process factors describe how students process community development strategies, facilitating and concluding with, among other products, a sense of community as an outcome (Brook & Oliver, 2003) (see Figure 1).

The framework presents an integrated system representing factors that exist prior to the process of community development, the approaches supporting community development, the process of community development, and a myriad of outcomes including sense of community. The Learning Community Development Model provides a framework to identify presage factors that are likely to present barriers to conditions supportive of community development in K-12 online settings and a method for guiding the selection of processes to overcome those barriers.

THE PRESAGE COMPONENT OF THE MODEL

Presage factors are presented in three categories of system, learning context, and student characteristics. System factors refer to factors at the institutional level and include online policies and support, learning management systems, and grading policies. The learning context is broken into three sections referring to factors at the instructor level including experience, education philosophy, and skill set; factors at the course level including academic level, subject orientation, and discipline; and group factors including cohort size. Those at the student level refer, among other factors, to educational level, learning style, and willingness to engage in collaborative activity.

System Factors

In a similar manner to higher education institutions, the system that governs the learning environment in the K-12 setting may present troubling barriers to community development in online settings.

Contemporary literature suggests that providing teachers and students appropriate access to technology and appropriate support promotes community development (Berge,

Figure 1. The Learning Community Development Model (Brook & Oliver, 2003)

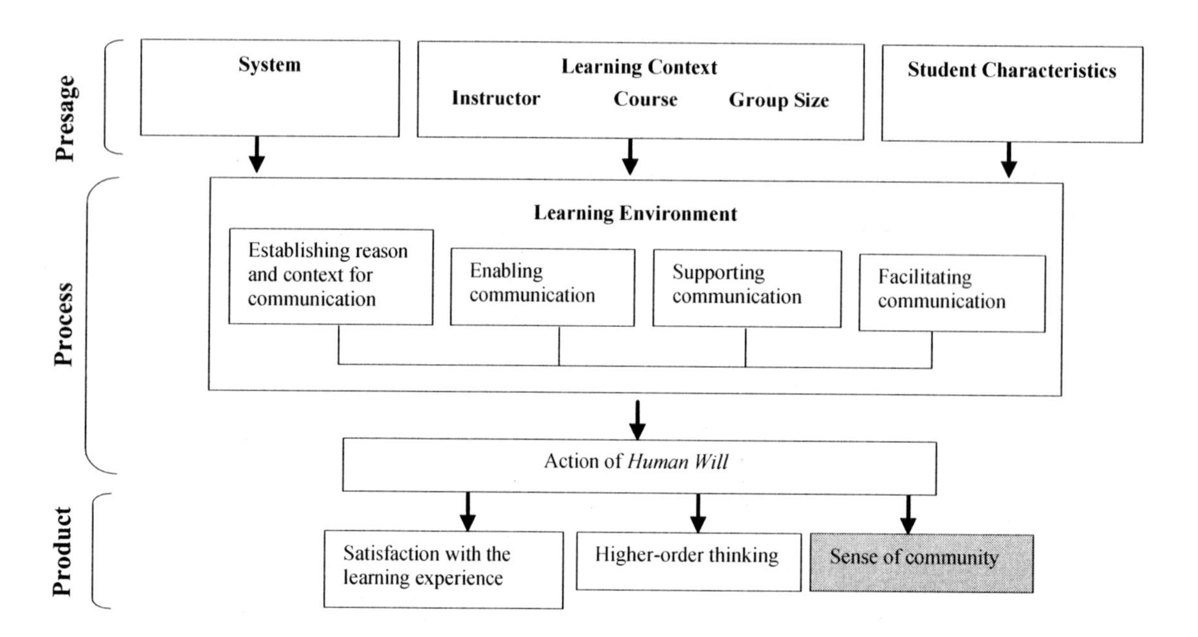

Muilenburg, & Haneghan, 2002; Collins & Berge, 1996; Salmon, 2000). Online learning environments where poor access and support are the norm tend to be characterized by low levels of participation, an identified requirement of community development (Brook & Oliver, 2002). Avoiding competition is also an important consideration, as settings characterized by high levels of competition for limited resources tend to be dominated by able members who exclude the weak (Hawley, 1950) eroding participants' sense of trust and fulfilment of needs-critical conditions in community development.

Other factors requiring consideration include ensuring that students are made aware of the processes and procedures for attaining course access, utilizing a robust technical solution that will provide reliable access, and minimizing restrictions placed on the use of computer-mediated communication (CMC) tools.

Instructor Factors

As with all educational settings, the role of teacher requires careful consideration prior to engaging in the development of online learning communities in the K-12 setting.

It has been suggested that the role of the instructor is pivotal in the development of online learning communities (Collins & Berge, 1996). The manner in which this role is approached depends on the characteristics and beliefs of the instructor (Lounsbury & DeNeui, 1996) including educational philosophies (e.g., Paulsen, 1995), perceptions of self as either *connected* or *separate* (Gilligan, 1982), and perceptions of their role. Other considerations include the instructor's online experience, the nature of the social environment they develop, and the manner in which they manage the learning setting (e.g., Collins & Berge, 1996).

Specific consideration in the role of online instructor is required in the area of the teacher's skill set and suitability for the role. Some pertinent questions to be asked of the teacher reflect their capacity to operate with the technical setting, including the resolution of technical problems; their capacity to translate appropriate pedagogic theory to practice and develop a positive social setting; and the manner in which they intend to participate in online meetings.

Course Factors

While some course-related factors such as the educational level remain beyond the control of teachers, others require careful consideration.

Factors at the course level include the educational level and discipline of study (Hiltz, 1994; Palloff & Pratt, 1999). Courses designed for the undergraduate level tend to require more structure and higher levels of instructor participation than those designed for the postgraduate level (Hiltz, 1994). In addition, some disciplines of study are more conducive to both the online setting and strategies that promote community development (Hiltz, 1994). Outcomes and objectives in course design are also influencing factors, as are the course syllabus and structure (Palloff & Pratt, 1999).

In the preparation of the course for online delivery, careful consideration is required in the planning of the course design including intended outcomes resulting from participation, the course structure, and the regularity and nature of assessment items.

Group Factors

Group factors refer to the nature of the cohort size. The characteristics of individual members are referred to as student characteristics and are explored as the final presage factor.

The nature of the cohort, including the number of participants, may influence community development strategies. In asynchronous settings group size is recommended to be no larger than 25, while 10 is suggested for the synchronous setting (Palloff & Pratt, 1999). It has also been suggested that excessively small cohorts may lack the critical mass often required to promote a healthy group experience (Allen, 2004).

In the K-12 setting the ideal cohort size supportive of community development might need to be reconsidered in light of the individual student characteristics.

Student Characteristics

While formulating forms of engagement and activity to promote community development and encourage participation may fall to the instructor (Hiltz, 1998; Palloff & Pratt, 1999), the student is not (and cannot be) a passive observer (Hiltz, 1994).

The characteristics of individual students in the K-12 setting are likely to be distinct from those of their higher education counterparts and in some ways more influential in community development.

The characteristics of participating students are likely to impact on both participation in the learning experience and the development of sense of community (Lounsbury & DeNeui, 1996). Influencing factors include the level of education and online experience (Hiltz, 1997), perceptions of self as either *connected* to or *separate* from others (Gilligan, 1982), and approaches to communication based on either a need for connection or status (Gougeon, 2002). Patterns of socialization, which tend to be gender based (Belenky, Clinchy, Golberger, & Tarule, 1986; Tannen, 1990, 1994, 1995) are also likely to impact on community development. It is suggested that students adopting the socialized female role are more likely to seek membership in learning communities than their socialized male counterparts. Culture, which governs underlying beliefs, values, and how we communicate and act among people, is also likely to impact on the development of sense of community (Triandis, 1996).

The role and responsibility of the student is to be an active participant in both the learning and community experience (Palloff & Pratt, 1999). These behaviours are often associated with adult learners, and students in the K-12 settings are likely to require a great deal of support to ensure that they engage actively in both the learning and community experience.

Summary of the Presage Phase of the Learning Community Development Model

The preceding factors, which may present troubling barriers to community development in online settings, were identified at the presage phase of the Learning Community Development Model. Table 3 lists the identified presage factors.

THE PROCESS COMPONENT OF THE MODEL

In the event that presage factors present barriers to conditions supportive of community development in the K-12 setting, there are some instructional strategies that teachers might employ to limit the negative impact presented by these barriers.

While presage factors exist prior to any direct attempt by the teacher to establish community, the process component of the Model describes the forms of engagement and activity purposefully employed by the instructor to facilitate community development. Process strategies are presented in four categories of establishing a reason and context for communication, enabling communication, supporting communication, and moderating communication, with each of these categories comprising elements outlined in more detail in the following paragraphs.

Establishing a Reason and Context

When establishing a reason and context to attract community membership, it is essential to ensure that prospective members have timely access to the setting. In the event that students

Table 3. Presage factors influencing community development in online settings

System	Learning Context			Student
	Instructor	Course	Group	
• Course location • Access procedures • Grading policy • Robust system • Communication systems • Unrestricted CMC tools • Assessment policies • Workload allocation	• Teaching experience • Educational beliefs • Perceptions of self • Management style • Technical skills • Moderating skills • Pedagogic beliefs	• Orientation • Academic level • Subject material	• Cohort size	• Education level • Experience • Learning style • Patterns of socialization • Cultural identity • Access to technology • Goals • Motivation • Personality traits • Perceptions of self

Table 4. Reasons and contexts to promote community development

Problem solving or task completion	Product development	Knowledge sharing—construction	Social activity
• Present a disorienting dilemma • Present complex problems/tasks • Present authentic tasks • Present controversial issues • Present an onerous workload • Grade participation	• Develop an artefact • Develop a plan • Develop a report	• Access to expert opinion • Agent provocateur • Debate conflicting views • Group work—projects • Encourage conflict of schema	• Coffee shop • Pub • Water cooler • Student initiated • Instructor initiated

have not accessed the setting in a timely manner, it is necessary to employ additional communications media, including the telephone, to ensure that students gain access. In light of the importance of this factor, it is recommended that the responsibility for providing students with course access details be allocated to the K-12 teacher.

Essential in the formation of all communities is the purpose that the community serves in the lives of its members (Hawley, 1950; Sarason, 1974). This purpose may be based in the resolution of a common problem or attainment of a common goal (McMillan, 1996; Tönnies, 1955; Worsley, 1991). Purpose may also rest with perceived benefits received for membership (Lott & Lott, 1965; McMillan, 1996) which may include an increase in both intellectual (Stewart, 1997) and social capital (Putnam, 2000) or unspecified individualized benefits (McMillan, 1996). It has been suggested that a *significant purpose* may be instrumental in the formation of communities in circumstances where antipathy is the norm (Tönnies, 1955). While it is not practical (and perhaps not possible) to identify what constitutes a *significant purpose* for all participants (McMillan, 1996), it is feasible to establish purpose at multiple levels in an attempt to attract members with variable needs.

Purpose may reflect the manner in which student participation is encouraged. Suggestions include mandated participation through the allocation of grades (Hiltz, 1998; Palloff & Pratt, 1999), providing an increase in intellectual resources through guest experts (Hiltz, 1994), presenting a problem or disorientating dilemma (Moore & Brooks, 2001), and linking activities to the *lived-in world* (Palloff & Pratt, 1999). Further support may be attained through setting complex ill-defined problems that reflect authentic activities (Herrington & Oliver, 1995). Providing a disorientating dilemma, a controversial issue (Moore & Brooks, 2001), or an onerous workload (Brook & Oliver, 2003) may also encourage participation. The purpose and context may also be established through encouraging the collaborative construction of knowledge (e.g., Gunawardena et al., 1997) facilitated through group work or projects (Brook & Oliver, 2002) or the instructor acting as an agent provocateur (Hiltz, 1998). It is also possible to stimulate purpose and context through actively promoting social activities (e.g., Hill & Raven, 2000).

Establishing a significant reason and context for participation in group meetings is a valuable process strategy in attracting individuals who may not be inclined to seek community membership under usual conditions. It has been shown that in the event these individuals perceive a significant benefit to themselves as a result of participation, they are likely to seek membership even in the event that feelings of antipathy are the norm (Tönnies, 1955). These factors are likely to resolve many of the troubling barriers to community development in online settings presented by presage factors (see Table 4).

Enabling Communication

An essential requirement for community development in all settings is regular and meaningful meetings (e.g., Tönnies, 1955). In the online setting these meetings may be facilitated through technology tools such as discussion boards, chat facilities, e-mail, or instant messaging (Isenhour, Carroll, Neale, Rosson, & Dunlap, 2000). It is important to remember, however, that this technology does not by necessity prevent the use of other more traditional meeting methods such as face to face and telephone. In essence the nature of these meetings, the meeting schedule, and the manner in which participants take part will reflect the perceived purpose of participation.

Communication may be encouraged through grading participation based on the quality or quantity of communications (e.g., Palloff & Pratt, 1999), requesting responses (Hiltz, 1994), establishing a sense of positive outcome as a result of belonging, and encouraging members to pay their dues (McMillan, 1996). Setting an appropriate pace and schedule for participation that maintains active engagement without dominating the learning experience may provide further support (Collison, Elbaum, Haavind, & Tinker, 2000). Establishing the nature of communicaton—including the tools to be used, roles, and responsibilities—enables communication (Palloff & Pratt, 1999), as does establishing a sense of connectedness. Strategies that promote connectedness include engendering the human elements of community (e.g., Eastmond, 1995) and establishing user profiles (Kim, 2000). Additional strategies include welcoming new members, sharing wisdom, resolving problems, and sharing success (Moore & Brooks, 2001). Allowing for growth and change, two characteristics of community (Sarason, 1974), also promotes a sense of connectedness among members.

Enabling communication is a valuable process factor that may be seen to overcome barriers to community development presented at the presage phase of the Learning Community Development Model. Allowing unrestricted use of communication media can overcome limitations that result from unreliable technical solutions; establishing an appropriate pace maintains the momentum of communication, reinforcing the benefits that result from participation; and establishing roles and responsibilities helps to develop the structure central to the community experience.

These instructional strategies are likely to resolve many of the troubling barriers to community development in online settings presented by presage factors. Table 5 lists the instructional strategies described in preceding paragraphs.

Supporting Communication

Resulting from the more independent nature of the online learning setting, there is a need to support students in managing their own learning experience, including setting goals and prioritizing tasks (Hill & Raven, 2000). Supporting communication includes assisting students in becoming proficient with the technology (Berge & Collins, 1995), developing text-based communication skills (Suler, 2000), and instituting a sequencing of activities (Salmon, 2000).

Table 5. Enable communication to promote community development

Participation	Schedule	Nature	Connectedness
• Required • Recommended • Suggested • Optional • Necessary • As needed • Request responses • Establish a sense of positive outcomes—the result of belonging • Membership pay their dues	• Establish and maintain appropriate pace • Fixed meeting schedule • Meetings as required • Daily meetings • Weekly meetings • Instructor initiated meetings • Student-initiated meetings	• Small group • Whole class • Set topics • Social discussion • Course issues • Report presentation • Student as expert • General discussion • Reflective (application) • Global (instructor to all) • Student initiated • Instructor initiated • Student to student • Student to instructor • Instructor to student • Role play • E-mail (available) • Chat (available) • Instant messaging (available) • Telephone (available) • Face to face (available)	• Ensure all participants are present and active • Establish user profiles • Make member responsibilities explicit • Encourage prompt and timely responses • Weave comments • Normalize and permit disagreement (resolved by participants) • Allow for differing roles • Allow subgroups • Welcome new members • Value all members • Take time to think and reflect • Share wisdom • Resolve problems collectively • Share success • Allow growth and change

Assisting students in coping with the technology includes providing support for the resolution of technical problems and stating the technology requirements (Palloff & Pratt, 1999). Providing multiple means of access (Hill, 2000) also assists students in coping with technology, as does normalizing problems and the appropriate use of humour (Brook & Oliver, 2003). Given the importance of non-verbal factors in communication (Dunn, 1999), which are to a large extent absent in text-based environments (Donath, n.d.), helping students develop text-based communication skills may also support community development (Suler, 2000). There is also a need to prepare students for the possibility of both conflict and tension (Palloff & Pratt, 1999). It is also useful to provide weekly reminders (Brook & Oliver, 2003) and clearly state roles and responsibilities (Palloff & Pratt, 1999).

Contemporary literature suggests that essential requirements for the development of community are the provision of a safe environment where participants can express themselves free from shame (McMillan, 1996). McMillan (1996) emphasizes the need to develop trust through establishing structure. Members must know what they can expect from each other, what power relationships exist, and who holds power and when. Any breakdown in these structures is likely to result in *anomie* (Durkheim, 1964). Trust may be promoted through establishing a code of conduct (McMillan, 1996), avoiding anonymity (Palloff & Pratt, 1999), and providing for the development of an *electronic self* (Kim, 2000; Palloff & Pratt, 1999), as may establishing leadership (e.g., Palloff & Pratt, 1999; Paulsen, 1995).

Providing support for communication is a valuable process factor that is likely to overcome barriers to community development present at the presage phase of the Learning Community Development Model, including technical problems presented by poorly supported technical systems, the uncertainty associated with communication in text-based settings, and assisting students to develop the skills required to regulate their own learning experience.

These factors may resolve many of the troubling barriers to community development in online settings presented by presage factors. Table 6 lists the identified supporting factors.

Moderating Communication

The role of teacher in the development of an online learning community requires activity involvement and a willingness to share leadership.

"Community refers among other things to one's sense of place, its people, their interrelationships, their shared caring for one another and their sense of belonging" (Lorion & Newbrough, 1996, p. 312). The importance of *social presence* in supporting online learning has also been suggested (Frazey & Frazey, 2001; Leh, 2001; McIsaac & Gunawardena, 1996; Richardson & Swan, 2001; Stacey, 2002). It is the sense of *place* and *social presence* that are required in online learning communities. Suggested strategies for developing these include incorporating human elements such as welcoming messages and acknowledging members individually (e.g., Eastmond, 1995). Other suggestions include establishing member profiles, developing a common symbol system (Kim, 2000; Palloff & Pratt, 1999), and including rituals from the *lived-in* world (Kim, 2000; Suler, 2000). The *tone* is also a critical factor, and a range of suggestions have been made, including using a friendly, open, and polite voice, and being curious, analytical, and informal (Collison et al., 2000). Encouraging sharing is also an essential strategy in effective moderation. Sharing takes the form of *trade* in a community economy (McMillan, 1996). Trade is based on self-disclosure and must be perceived as fair (McMillan, 1996) in an environment that provides an abundance of desired resources (Hawley, 1950). It is also suggested that trade progress from safe to risky (McMillan,

Table 6. Supporting communication

Technology Skills	Communication Skills	Management Skills	Behaviour guidelines
• State technology requirements • Provide multiple means of access • Provide links to required downloads • Offer face-to-face help • Provide online help • Encourage peer support • Normalize problems • Use humour	• Modelling • Text-based communication strategies • Express and normalize feelings of uncertainty • Prepare the participants for the possibility of both tension and conflict (normalize this experience) • Establish identity and avoid anonymity	• State expectations • Provide time management tips • Provide tips for prioritizing tasks • State roles and responsibilities • Post weekly reminders	• Establish safety • Outline a code of conduct based on mutual respect • Establish that it is OK to be yourself and tell the truth • Introduce the community economy • Identify expectations of participation in community activities (fair trade)

1996) in order to build trust and progress the group through stages of group development (Salmon, 2000).

Moderating communication is a valuable process factor that may overcome barriers to community development present at the presage phase of the Learning Community Development Model, including an individual disposition not to seek membership and feelings of isolation derived from limited peer support.

These instructional strategies are presented in the process component of the Learning Community Development Model and may resolve many of the troubling barriers to community development in online settings presented by presage factors. Table 7 shows the instructional strategies described in contemporary literature to guide instructors in moderating online interactions and community development.

SUMMARY AND CONCLUSION

Online technologies afford clear educational advantage to teachers, students, and parents through providing the flexibility to meet the needs of non-traditional students, the increased ability to customize learning for special needs groups, and the potential to expand learning opportunities of students living in isolated and remote locations. However, this advantage is unlikely to be achieved through simply transferring the way we have always taught to online settings. Course design requires careful consideration of the needs of the students, parents, and teachers, as well as appropriate pedagogic practice to facilitate learning in online settings.

There is strong support for the supposition that the social phenomenon of community may be put to good use in the support of online learning. This is well supported by theories of learning that highlight the role of social interaction in the construction of knowledge. Some debate continues as to the role that social interaction plays in the construction of knowledge, but it appears that apparently conflicting views may be complementary under certain conditions. The processes and procedures for developing such a community remain largely unknown, with much of current thinking based on the anecdotal records of professional working in the field. Analysis of contemporary literature suggests the possibility of describing the processes of community development as a model describing a chain of events that consists of presage, process, and product factors. Presage factors outline the conditions for community development, process factors outline the strategies employed by the instructor to develop sense of community, and the student

Table 7. Moderating communication

Human Elements	Tone	Sharing
• Welcome members individually • Establish member profiles • Establish identity • Establish guidelines for communication • Allow for a range of roles • Allow a common symbol system • Integrate rituals of community life • Include social elements	• Friendly • Open • Inviting • Polite • Neutral • Humorous • Imaginative • Nurturing • Curious • Analytical • Informal • Whimsical	• Knowledge is the commodity for exchange • Trade is based on self-disclosure • Trade must be fair (defined by members) • Progress trade from safe to risky • Provide an abundance of desired resources

response and product outline the sense of community experience, among other outcomes.

While the Learning Community Development Model represents an integrated system suggesting factors critical to community development, it does not indicate the relative importance of any of the factors, nor those that may be considered essential or simply desirable. Further enquiry to develop an understanding of instructional emphasis and how to design learning settings that promote community development in K-12 settings is required. This enquiry may be assisted through adopting the proposed framework to explore community development and the link between the proposed factors and sense of community.

REFERENCES

Alberta, L. (1999). *Alberta Learning: Best practice for Alberta School jurisdictions.* Retrieved May 3, 2004, from http://www.edc.gov.ab.ca/technology/bestpractices/pdf/onlinelearning.pdf

Allen, C. (2004). *The Dunbar number as a limit to group sizes.* Retrieved April 7, 2004, from http://www.lifewithalacrity.com/2004/03/the_dunbar_numb.html

Barab, S. A., Thomas, T. K., & Merrill, H. (2001). Online learning: From information dissemination to fostering collaboration. *Journal of Interactive Learning Research, 12*(1), 105-143.

Belenky, M. F., Clinchy, B. M., Golberger, N. R., & Tarule, J. M. (1986). *Women's ways of knowing*. New York: Basic Books.

Benbunan-Fich, R. (1997). *Effects of computer-mediated communication systems on learning, performance and satisfaction: A comparison of groups and individuals solving ethical case scenarios.* Unpublished Dissertation, Rutgers University, USA.

Berge, Z., & Collins, M. (1995). *Computer-mediated communications and the online classroom: An introduction.* Cresskill, NJ: Hampton Press.

Berge, Z., Muilenburg, L., & Haneghan, V. (2002). Barriers to distance education and training: Survey results. Retrieved May 12, 2003, from http://www.emoderators.com/barriers/barriers2002.shtml

Biggs, J. B. (1989). Approaches to the enhancement of tertiary teaching. *Higher Education Research and Development, 8*(1), 7-25.

Bonk, C. J., & Cunningham, D. J. (1998). Searching for learner-centred, constructivist, and sociocultural components of collaborative educational learning tools. In C. J. Bonk & K. S. King (Eds.), *Electronic collaborators*. Mahwah, NJ: Lawrence Erlbaum.

Bonk, C. J., & Wisher, R. A. (2000). *Applying collaborative and e-learning tools to military distance learning: A research framework.* Retrieved July 2, 2002, from http://www.publicationshare.com/docs/Dist.Learn(Wisher).pdf

Brook, C., & Oliver, R. (2002). Supporting the development of learning communities in online settings. *Proceedings of the Ed-Media Conference,* Denver, Colorado.

Brook, C., & Oliver, R. (2003). Online learning communities: Investigating a design framework. *Australian Journal of Educational Technology, 19*(2), 139-160.

Bruner, J. (2001). *Constructivist theory.* Retrieved May 10, 2001, from http://tip.psychology.org/bruner.html

Carol, L.A. (1997). *Masters of sociological thought: Ideas in historical and social context* (2nd ed.). Forth Worth, TX: Harcourt Brace Jovanovich.

Cho, S. K., & Berge, Z. (2002). *Overcoming barriers to distance training and education.* Retrieved May 12, 2003, from http://www.usdla.org/html/journal/JAN02_issue/article01.html

Cobb, P. (1994). Where is mind? *Educational Researcher, 23*(7), 13-20.

Collins, M., & Berge, Z. (1996). *Facilitating interaction in computer mediated online courses.* Retrieved May 10, 2001, from http://www.emoderators.com/moderators/flcc.html

Collison, G., Elbaum, B., Haavind, S., & Tinker, R. (2000). *Facilitating online learning.* Madison, WI: Atwood Publishing.

Conklin, E. J. (1996). *Designing organisational memory: Preserving intellectual assets in a knowledge economy.* Retrieved October 20, 2002, from http://cognexus.org/dom.pdf

Constant, D., Sproul, L., & Kieser, S. (1996). The kindness of strangers: On the usefulness of weak ties for technical advice. *Organisational Science, 7,* 119-135.

Cunningham, D. J. (1996). Time after time. In W. Spinks (Ed.), *Semiotics 95* (pp. 263-269). New York: Lang Publishing.

Dewey, J. (1929). *The sources of a science of education.* New York: Liveright.

Donath, J. S. (n.d.). *Body language without the body: Situating verbal cues in the virtual world.* Retrieved July 7, 2003, from http://duplox.wz-berlin.de/docs/panel/judith.html

Dunn, L.J. (1999). *Non-verbal communication: Information conveyed through the use of body language.* Retrieved July 7, 2003, from http://clearinghouse.mwsc.edu/manuscripts/70.asp

Durkheim, E. (1964). *The division of labour in society.* Free Press of Glencoe.

Eastmond, D.V. (1995). *Alone but together: Adult distance education through computer conferencing.* Cresskill, NJ: Hampton Press.

Elkind, D. (1967). *Six psychological studies of Jean Piaget.* London: University of London Press.

Frazey, D. M., & Frazey, J. A. (2001). The potential for autonomy in learning: Perceptions of competence, motivation and locus of control in first year undergraduate students. *Studies in Higher Education, 26*(3), 345-361.

Gilligan, C. (1982). *In a different voice: Psychological theory and women's development.* Cambridge: MA: Harvard University Press.

Glassman, M. (2001). Dewy and Vygotsky; society experience and inquiry in education practice. *Educational Researcher, 30*(4), 3-14.

Goth, D. D. (1992). *Communities: An exploratory study of the existential and transpersonal dimensions of a psychological sense of community as found in the community building workshop.* Unpublished Dissertation, Institute of Transpersonal Psychology, USA.

Gougeon, T. (2002). Participation in computer based curriculum. *International Electronic Journal For Leadership in Learning, 6*(22).

Granovetter, M. (1973). The strength of weak ties. *American Journal of Psychology, 73,* 1361-1380.

Gunawardena, C. N., Lowe, C. A., & Anderson, T. (1997). Analysis of a global online debate and the development of an interaction analysis model for examining social construction of knowledge in computer conferencing. *Journal of Educational Computing Research, 17*(4), 397-431.

Gusfield, J. R. (1975). *The community: A critical response*. New York: Harper Colophon.

Hawley, A. H. (1950). *Human ecology: A theory of community structure*. New York: The Ronald Press Company.

Herrington, J., & Oliver, R. (1995). *Critical characteristics of situated learning: Implications for the instructional design of multimedia.* Paper presented at the Ascelite, University of Melbourne, Australia.

Hildreth, P. J., & Kimbe, C. (2002). The duality of knowledge. *Information Research, 8*(1).

Hill, J. L. (1996). Psychological sense of community: Suggestions for future research. *Journal of Community Psychology, 24*(4), 431-437.

Hill, J. R., & Raven, A. (2000). *Online learning communities: If you build them, will they stay?* Retrieved March 2001 from http://it.coe.uga.edu/itforum/paper46/paper46.htm

Hillery, G. A. (1964). Villages, cities, and total institutions. *American Sociological Review, 28*, 32-42.

Hiltz, S.R. (1994). *Online communities: A case study of the office of the future*. Norwood, NJ: Ablex Publishing.

Hiltz, S. R. (1997). *Impacts of college courses via asynchronous learning networks: Some preliminary results.* Retrieved from http://eies.njit.edu/~hiltz

Hiltz, S. R. (1998). Collaborative learning in asynchronous learning environments: Building learning communities. *Proceedings of the WebNet 98 World Conference of the WWW, Internet and Intranet Proceedings,* Orlando, Florida.

Isenhour, P. L., Carroll, J. M., Neale, D. C., Rosson, M. B., & Dunlap, D. R. (2000). *The virtual school: An integrated collaborative environment for the classroom.* Retrieved April 25, 2002, from http://ifets.ieee.org/periodical/vol_3_2000/a03.html

Johnson, W. D. (1991). Student-student interaction: The neglected variable in education. *Educational Research, 10*(1), 5-10.

Kafai, Y., & Resnick, M. (1996). *Constructionism in practice*. Mahwah, NJ: Lawrence Erlbaum.

Kim, A. J. (2000). *Community building on the Web*. Berkeley, CA: Peachpit Press.

Kruat, R., Patterson, M., Lundmark, V., Kieser, S., Mukophadhyay, T., & Scherlis, W. (1998). Internet paradox: A social technology that reduces social involvement and psychological well being? *American Psychologist, 53*(9), 1017-1031.

Leh, A. S. (2001). Computer mediated communication and social presence in a distance learning environment. *International Journal of Educational Telecommunications, 7*(2), 109-128.

Lorion, R. P., & Newbrough, J. R. (1996). Psychological sense of community: The pursuit of a field's spirit. *Journal of Community Psychology, 24*(4), 311-314.

Lott, A. J., & Lott, B. E. (1965). Group cohesiveness as interpersonal attraction: A review of relationships with antecedents and variables. *Psychological Bulletin, 64,* 259-309.

Lounsbury, J. W., & DeNeui, D. (1996). Collegiate psychological sense of community in relation to size of college/university and extroversion. *Journal of Community Psychology, 24*(4), 381-394.

Maxwell, W. E. (1998). Supplemental instruction, learning communities and students studying together. Retrieved March 19, 2001, from http://www.findarticles.com/cf_0/m0HCZ/2_26/53420232/p1/article.jhtml

McIsaac, M. S., & Gunawardena, C. N. (1996). Distance education. In D. Jonassen (Ed.), *Handbook for research on educational communications and technology* (pp. 403-437). New York: Scholastic Press.

McLellan, H. (1998). The Internet as a virtual learning community. *Journal of Computing in Higher Education, 9*(2), 92-112.

McMillan, D. W. (1996). Sense of community. *The Journal of Community Psychology, 24*(4), 315-325.

McMillan, D. W., & Chavis, D. M. (1986). Sense of community: A definition and theory. *Journal of Community Psychology, 14*, 6-23.

Moore, A. B., & Brooks, R. (2000). Learning communities and community development: Describing the process. *International Journal of Adult and Vocational Learning,* (1), 1-15.

Nonaka, I. (1991). The knowledge creating company. *Harvard Business Review, 69*(November-December), 96-104.

Obst, P., Zinkiewicz, L., & Smith, S. (2002). Sense of community in science fiction fandom part 1: Understanding sense of community in an international community of interest. *Journal of Community Psychology, 30*(1), 87-103.

Palloff, R., & Pratt, K. (1999). *Building learning communities in cyberspace*. San Francisco: Jossey-Bass.

Panitz, T. (1997). *The case for student centered instruction via collaborative learning paradigms*. Retrieved November 29, 2001, from http://home.capecod.net/~tpanitz/tedsarticles/coopbenefits.htm

Papert, S. (1990). *Constructionist learning*. Cambridge: MIT Media Laboratory.

Paulsen, M. F. (1995). Moderating educational computer conferences. In Z. L. Berge & M. P. Collins (Eds.), *Computer mediated communication and the online classroom. Volume three: Distance learning* (pp. 81-104). Cresskill, NJ: Hampton Press.

Polanyi, M. (1967). *The tacit dimension*. London: Routledge and Kegan Paul.

Puddifoot, J. E. (1996). Some initial considerations in the measurement of community identity. *The Journal of Community Psychology, 24*(4), 327-334.

Putnam, R. D. (2000). *Bowling alone*. New York: Simon and Schuster.

Redfield, R. (1960). *The little community and peasant society and culture*. Chicago: The University of Chicago Press.

Richardson, J., & Swan, K. (2001). An examination of social presence in online leaning: Student's perceived learning and satisfaction. *Proceedings of the Conference of the American Educational Research Association,* Seattle, Washington.

Rovai, A. (2002). Development of an instrument to measure classroom community. *The Internet and Higher Education, 5,* 197-211.

Royal, M. A., & Rossi, R. (1996). Individual-level correlates of sense of community: Findings from workplace and school. *Journal of Community Psychology, 24*(5), 395-416.

Salmon, G. (2000). *E-moderating: The key to teaching and learning online*. London: Kogan Page.

Sarason, S. B. (1974). *The psychological sense of community*. San Francisco: Jossey-Bass.

Slavin, R. E. (1990). *Cooperative learning: Theory, research and practice*. Needham Heights, MA: Allyn and Bacon.

Sonn, C., Bishop, B., & Drew, N. (1999). Sense of community: Issues and considerations from a cross-cultural perspective. *Community, Work & Family, 2*(2), 205-218.

Stacey, E. (2002). Quality online participation: Establishing social presence. In T. Evans (Ed.), *Research in distance education* (5th ed., pp. 138-153). Melbourne: Deakin University.

Stewart, T. A. (1997). *Intellectual capital*. New York: Doubleday/Currency.

Stoll, C. (1995). *Silicon snake oil*. New York: Doubleday.

Suler, J. (2000). *Maximising the well being of online groups: The clinical psychology of virtual communities*. Retrieved May 7, 2003, from http://www.rider.edu/~suler/psycyber/clinpsygrp.html

Surratt, C.G. (1998). *Internet citizens and their communities*. New York: Nova Science.

Tannen, D. (1990). *You just don't understand: Women and men in conversation*. New York: William Morrow and Company.

Tannen, D. (1994). *Talking from 9 to 5: How women's and men's conversation styles affect who gets heard, who gets credit, and what gets done at work*. New York: William Morrow and Company.

Tannen, D. (1995). *Gender and discourse*. New York: Oxford University Press.

Tönnies, F. (1955). *Community and association* (C. P. Loomis, Trans.). London: Routland & Kegan Paul.

Triandis, H. C. (1996). The psychological measurement of cultural systems. *American Psychologist, 51*(4), 407-415.

Von Krogh, G. (1998). Care in knowledge creation. *California Management Review, 40*(3), 133-153.

Von Krogh, G., Ichijo, K., & Nonaka, I. (2000). *Enabling knowledge creation*. Oxford, NY: Oxford University Press.

Vygotsky, L. S. (1978). *Mind in society: The development of higher psychological processes* (M. Cole, V. John-Steiner, S. Scribner, & E. Souberman, Trans.). Cambridge, MA: Harvard University Press.

Wegerif, R. (1998). The social dimension of asynchronous learning environments. *JALN, 2*(1), 34-49.

Wellman, B., & Wortley, S. (1990). Different strokes for different folks: Community ties and social support. *American Journal of Sociology, 96,* 558-588.

Wenger, E. (1998). *Communities of practice: Learning meaning and identity*. Cambridge: Cambridge University Press.

Wenger, E., McDermott, R., & Snyder, W. (2002). *Cultivating communities of practice*. Boston: Harvard Business School Press.

Wiesenfeld, E. (1996). The concept of 'we': A community social psychology myth? *The Journal of Community Psychology, 24*(4), 337-346.

Worsley, P. (1991). *The new modern sociology readings* (2nd ed.). New York: Penguin Books.

Chapter XXXII
Knowledge Management, Communities of Practice, and the Role of Technology: Lessons Learned from the Past and Implications for the Future

Lee Tan Wee Hin
National Institute of Education, Singapore

Thiam-Seng Koh
National Institute of Education, Singapore

Wei-Loong David Hung
National Institute of Education, Singapore

ABSTRACT

This chapter reviews the current work in knowledge management (KM) and attempts to draw lessons from research work in situated cognition about the nature of knowledge which can be useful to the field of KM. The role of technologies and the issues of literacy in technology are discussed in the context of communities of practice (CoPs) and the KM framework with some examples described for K-12 settings. Implications are drawn in terms of how teachers and students can be a community of learners-practitioners through technologies which support their work and learning processes.

INTRODUCTION

As countries in the world compete globally in a knowledge and technology-driven environment where national and global business boundaries continue to dissolve at an unprecedented rate, education is seen as one of the key strategies in meeting the challenges ahead. Educating citizens with attributes such as innovation, creativity, and enterprise has become the rallying call of many governments to make their economies competitive. In order for K-12 schools to produce citizens with such attributes to meet the challenges ahead, school leaders and teachers must be able to fully exploit and share critical pedagogical knowledge with one another. Knowledge management (KM) is a key enabler of a successful school today. In this chapter, KM is discussed in the context of teachers sharing knowledge as a community of practitioners, and when such a community can be facilitated through technologies, the technology literacy levels of teachers are developed in the process. This chapter will begin with a discussion of KM and how lessons from situated cognition can be drawn to inform the field of KM and the issues of CoPs, and technology literacy in K-12 schools are drawn as implications to KM as found in the later parts of this chapter. As the field of KM may be new to many readers of this book, a relatively large section of this chapter will be devoted to the discussion of KM.

In essence, KM is an attempt to understand what works in organizations and institutions such as K-12 schools—their best educational practices, expert practitioners' thinking, and other processes that seem obvious to the experienced school leaders and teachers, but would be alien to beginning teachers. In the past, knowledge management practices focused primarily on the management of data and information. But, more recently, KM practices increasingly revolve around facilitating dialogue and forming collaborative groups within the organization that leverage on innovative information technology (IT) tools to create, capture, and use that information to facilitate communication among individuals to meet organizational goals (Duffy, 2000; Petrides & Guiney, 2002). Thus, the appropriate adoption of technologies in KM in schools should increase the technology literacy levels of teachers. Much of the difficulty in KM for schools lies in the fact that these KM processes may be very much hidden as tacit or implicit knowledge. The difficulty with tacit knowledge is that there is only an extent through which that knowledge can be made explicit. Take for example the case of riding a bicycle. If someone were to ask you to describe the process of riding a bicycle (or how to ride it), you would probably begin to tell about how to balance, how to position the steering, how to pedal, and so forth. However, this merely describes the how-to procedure of bicycle riding. You would probably agree that there is more to riding than the procedure of how-to. Even if we could articulate all about our experiences of riding a bicycle, it is still not the same as the actual skill of riding one. There is a fundamental difference between descriptions of experiences and the actual experience. In other words, tacit knowledge may not necessarily be fully described in explicit terms. The irony is that even if one can fully describe the tacit knowledge, you would not know if what is articulated is the fullest description ever possible.

Tacit knowledge about teaching and learning processes is thus the knowledge gained through experience of managing K-12 schools and designing appropriate learning opportunities for students. Expert teachers and school leaders gain a whole wealth of tacit knowledge as they encounter numerous cases and problem-solving experiences during the course of their work. In the past decade, artificial intelligence (AI) attempted to create expert systems such as intelligent tutoring systems by trying to make explicit the tacit knowledge of expert teachers through codifying this knowledge in the form of computer programs. Two decades of research yielded the fact that expert systems (containing the rules of expert thinking) are not isomorphic or equivalent to experts. Polyani (1964) stressed that we know much more than we can say. Expert teachers and school leaders, in others words, know much more than they can tell or articulate what they know. For example, reading a book written by a successful school leader in turning around a poorly performing school is still miles apart from what the person actually knows from the wealth of his personal experiences in turning around such a school.

KM is an attempt to make explicit what is implicit. Of course, not all implicit or tacit knowledge is useful to a school. A school's tacit knowledge is probably better known as the collective wisdom of the school (c.f., Choo, 2001). Taking a corporate example, "...what HP knows which it cannot even describe as best practice and transfer these processes" to another HP plant (Brown & Duguid, 2000). The explicit knowledge of a school would include, among other things, routines, rules, standard operating procedures, strategic planning documents, and curricular-related materials such as schemes of work and syllabuses. In essence, KM's intent should be to make the school a more effective educational organization by attempting to make explicit as much of what "works best" in terms of practices and processes, and to consciously formulate them, either into documentations and/or through imparting the knowledge to others by explicit sharing.

APPROACHES TO KNOWLEDGE MANAGEMENT

This chapter explores the application of the situated cognition concepts and principles (Lave & Wenger, 1991) to KM within a K-12 school context. It highlights fundamental issues on translating the tacit to explicit and institutional knowledge in a school context. It is important to appreciate that not all tacit knowledge about educational practices and processes can be made explicit. In developing KM within a school, there should be a consideration of *context* (the work environment or learning environment in relation to persons, tasks, functions, and others) to understand the limit to which there is meaningful translation of tacit knowledge to explicit knowledge—that is, to be mindful of the sociocultural perspective. While it may not always be possible to fully extract out the tacit knowledge residing in an expert teacher or school leader, it should be possible to engineer processes within a school that could eventually lead to the distillation of key tacit knowledge for the purpose of either "training" (codified knowledge—explicit or established knowledge) or improving educational practices. We suggest that there are two approaches that could be adopted in KM within a school context:

- **Approach 1:** In situations where the tacit knowledge could be understood and clarified, it should be, as far as possible, be made explicit through the process of externalization such as dialogue, reflection, abstraction, and so forth. This extracted knowledge could be improved upon through appropriate re-packaging with the aim of 'imparting' them to practitioners. This 'imparting' process could be through training, documented in descriptions such as books, or reified into the form of artifacts (Choo, 2002, 2003).

- **Approach 2:** In situations where the tacit knowledge is difficult to codify (or made objective), the alternative approach is to design a social setting or environment for apprenticeship (imitation and modeling) where experts and novices could interact and where novices could be mentored. Through such a process, tacit knowledge is "internalized" (both tacit and explicit) through enculturation. Through such an enculturation process, members interact on the basis of problems encountered, stories or narratives of situations and cases experienced, and the co-production of artifacts (Brown & Duguid, 2000).

As not all dimensions of tacit knowledge within a school context could be fully understood objectively, we can never be certain that the first approach is always the better approach. We suggest that KM (in its current state of understanding as a field) move towards the second approach of "designing for knowledge stealing" (as it were), and in the process of communities of practice (CoPs) engage in a systematic process of understanding the known processes (first stance). The two stances can be adopted in tandem. Before we elaborate on the two approaches, we reiterate that Approaches 1 and 2 can be adopted for any generic model of training and learning where technology literacy is a topical issue that teachers and students need to address explicitly or whether CoPs can be fostered around issues of literacy where tacit knowledge is negotiated. If technology literacy is not the subject-context or content, teachers and students could be engaged in issues of concern through the facilitation of technologies, and as a consequence, heighten their technology literacy levels in K-12 contexts.

Approach 1 is commonly known as the objectivist worldview, where the assumption is that knowledge is objective and that there is a one version of truth in reality out there. The entire thrust would be to try to find out scientifically what is the truth (in the form of explicit knowledge) and represent it in some codified form. The worldview assumes that there is a singular perspective to "truth" or reality and thus could be represented into forms of language—albeit through multi-modal means.

From an expert system point of view, AI has, in the past, attempted to distil or make explicit in the form of programmable rules and language the behavior, skills, and thinking of experts. By and large, much of this knowledge is well-structured knowledge, whereas what is considered as ill-structured remains difficult to make explicit. In a sense, one can say that attempts at understanding the well structured or well defined is noteworthy. By doing so, we could improve educational practices by "automating" what is well understood through, for example, off-loading what is well defined from teachers and letting them concentrate on what are more ill-structured demands in the classroom. This approach would make them more efficient.

The situated cognition view taken in Approach 2 above differs from the objectivist worldview. The situated view espouses that knowledge is deeply contextualized, and any attempt to decontextualize knowledge from either the person or context makes that knowledge meaningless. Such a deep interwovenness of context to knowledge makes the argument of abstraction problematic. In addition, the situated cognition view claims that because knowledge is contextualized, all knowledge is deeply influenced by one's interpretation of a reality or phenomena, giving rise to multiple perspectives (of which all perspectives are valid). In other words, even in KM's attempt to codify knowledge, this codified knowledge is via a perspective—the perspective prevalently held by the interpreters of the phenomena. Hence, the situated view is "against" notions or attempts at over-representing knowledge; rather it favors the perspective of designing situated contexts where improvement in educational practices occurs between teachers in schools through interactions such as negotiations of meanings, apprenticeship, and other methods espoused by social-cultural psychology.

Importance of Product and Process in KM

There is much that KM can learn from the lessons of the last two decades—the work of AI arising from the objectivist view, and the subsequent attempts by situated cognition. Our sense is that clearly both views (Approaches 1 and 2) are at the two ends of the knowledge continuum—one espousing that there is one singular objective view, and the other where multiple views exist—and that all these views may be equally valid. On one end, all attempts made are to codify knowledge as *products*, and on the other end, knowledge codification is not emphasized, but rather *process*-interactions are.

Both product and process are important. KM's attempts at putting structures in place within organizations to promote sharing and collaborations, and rewarding process-oriented activities are noteworthy. Such a perspective is very much aligned with the situated cognition view. In addition to facilitating sharing, KM could focus on the design of learning environments or rather the design of workplace environments to facilitate apprenticeship forms of learning and social interactions. The examples of studies done on Xerox engineers by Julian Orr (1996) and others are good examples of such design. John Seely Brown relates the example of how the coffee brewer, once fresh coffee is made, sends a signal to all staff via the Web, and members in the organization flock to the coffee corner (space for informal interactions) where informal knowledge is transacted. However, all these interactions assume that knowledge remains in the organization, as they are held by the individuals and processes within the community. But when these 'experts' leave, the collective knowledge held by these teams within the organizations would most likely be leaked out of the organization as well.

Problem Formulation for KM

Hence, the key issue for schools would be to adopt certain KM practices or processes to ensure that knowledge of best educational practices and tacit knowledge about teaching and learning are retained within the school as far as possible. With an understanding of these practices, knowledge can then be institutionalized (in the form of cultural knowledge) and formalized procedures can then be made explicit. However, when educational contexts change, these institutional norms/practices would have to change accordingly. The challenge for any school is really to know when and how to change.

In summary, the problem of KM can be formulated as an attempt to: (a) understand practice through putting structures in place for knowledge sharing and interactions (still largely explicit knowledge made overt); (b) improve practice by knowing what it knows, retaining as much of what it knows, and by changing what it knows based on changing demands and contexts (establishing and adjusting cultural knowledge); and (c) design community interactions that would facilitate members to appropriate or "steal" knowledge (particularly tacit knowledge).

For this chapter, we would like to concentrate on point (c) that is concerned with the design of community interactions. Based on

KM literature, we recognize that this is one significant area of which the literature on situated cognition and the work in education and learning can inform the field of knowledge management (Choo, 2001, 2002, 2003).

DESIGNING COMMUNITY INTERACTIONS

The concept of communities is not new (Vygotsky, 1978). The earliest communities were tribal in nature, where people form societies underpinned by traditions and beliefs of varied orientations. Communities include all kinds of professional practices, religious communities, networks of people, and so forth. More recently the concept of communities of practice (CoPs) is a prevalent concept that has dominated the field of learning, education, and business management. CoPs arise based on professional practices such as the scientific, mathematics, engineering, law, or accounting practices. CoPs are a community of people who practice a profession oriented towards a code of conduct, ethics, history, and peculiar culture. This community of people shares similar concerns and passions, allowing them to collectively evolve the necessary structures and processes to deepen their expertise and knowledge through engaging one another on an ongoing basis (Barab & Duffy, 2000). So, when we try to model the concepts and strategies of CoPs into schools and learning contexts, we have "CoLs" or communities of learners. Similar to CoPs and CoLs is the concept of schools as learning organizations. Within this concept, Fullan (1999) suggests collaborative organizations to: (a) value diversity, (b) bring conflicts into the open, (c) value the quality of relationships as being central to success, (d) accept emotional responses as a complement to rationality and logic, and (e) recognize the value of quality ideas. These tenets are in the same vein as the kinds of orientations CoPs aim to foster.

Principles for Growing and Sustaining Communities

Four principles are observed to be necessary for growing and sustaining communities. These principles, derived from CoPs, would have to find their counterpart-principle in a community of learners.

First, within CoPs there ought to be mutuality, where members share overlapping histories, values, and beliefs. Because of mutual benefits to one another, interdependence of actions is crucial to sustain activities within the CoP.

Second, the community has to, over time, develop a repertoire of artifacts for mutual enterprise. These artifacts represent the knowledge and skills of the members in that community. Over time, the CoP develops increasingly efficient and innovative mechanisms of production and reproduction, giving rise to common practices.

Third, the CoP has to be organized in ways where there would be plenty of opportunities for interactions, active participation, and meaningful relationships to arise. Within these relationships, respect for diverse perspectives and views would be necessary. In other words, CoPs are connected by intricate, socially constructed webs of beliefs and ways of thinking. The authentic activities arising from CoPs are framed by their culture and demands—usually mooted by society needs. Meanings are socially constructed within CoPs through negotiations among present and past members.

Lastly, within CoPs, there has to be a growth and renewal process of new members with past and present persons within the community. Throughout the history of CoPs, one important phenomena observed is new members joining as legitimate peripheral participants (Lave & Wenger, 1991) where they act as novices elbowing experts (or older members) of the CoP. This process begins with peripheral participants (as novices) appropriating an identity through observing masters at work. After extended opportunities of practicing the trades of the community, these novices begin to behave and think like the experts in the community of practice (Wertsch & Rupert, 1993). Lave and Wenger (1991) espoused that by exposing a newcomer to the practices of a community and providing him or her with the opportunities to engage in those practices, the newcomer would move from *peripheral* participation to a more central participation. In other words, there are levels of participation and contribution through which members in a community make advancements over time. The hypothesis is that, through each advancement, members appropriate a fuller identity similar to the central participants of that particular community of practice. Gradually, over time, these members become significant members in the community and take on central participation within these CoPs.

To summarize the principles underpinning CoPs, Brown and Duguid (2000) characterize learning to be: (a) demand driven, as CoPs create needs of mutual interdependency and outcomes; (b) a social act, as members interact and relate to one another with specific roles and functions in order to achieve the CoP's goals; and (c) an identity formation, where members develop "ways of seeing" meanings, beliefs, and ethics according to the professional practice, for example, mathematicians see patterns in numbers.

The Role of Technology in a CoP Framework

First, within the CoP framework as articulated above, we believe technology can play a crucial role in enhancing learning in the context of relationship-building—that is, *technology as a collaborative tool* (not just for task collaboration, but socializations). The basic question to ask is how technology can bridge the mutual enterprise of members within the CoPs that involves interdependency of engaging in joint tasks and projects for learning among one another. A key question for research in technology is how to develop scaffolds and structures in collaborative applications where it would help members to co-develop mutual interdependency with one another. Today we have examples of online communities which capitalize on the Internet to establish environments where members are always dependent on each other. One such example is the "Tapped In" Community (http://tappedin.org/tappedin/), a community of education professionals established since 1997. This community is an online forum where groups of teachers are brought together to engage one another based on their specific needs and demands. Professionals participate in sites such as "Tapped In" because there is a rather quick way of receiving some (personal) benefit such as having some query answered by someone, somewhere in the Internet world. Members bother to answer questions, and when their answers are unique and "stand out" from others, these "experts" rise to fame and indirectly receive consultancies and celebrity. It all works on the principles of mutual gain and benefit. Within these thriving online communities, members begin to adopt an online identity and sense of belonging to this community of like-minded people with similar interests and passion. Learning occurs as a consequence of participation in these online communities.

Second, a significant use of technology within the community context is what we are all familiar with—*technology as a productivity tool*. We are all familiar with how technology enables us to maximize our outputs and production of resources and artifacts for teaching, to communicate with one another either synchronously or asynchronously, to search for information, and to automate our administrative tasks. There is no need to further elaborate on this point.

Thirdly, technology can be used to support a learning community by assisting learners in seeing meanings and concepts without which it would be difficult. Here, we have *technology as a (collaborative and individual) sense-making tool*. In science, the microscope is an example of a technology which enables learners and scientists to see cells beyond the naked eye. When learners are engaged in meaningful discovery of concepts, technology can open up new vistas of concept visualizations. Computer graphics today can simulate phenomena which traditional media of 2-D charts are unable to do. Chemical bonds and molecular structures can be visualized through appropriate simulations. Simulations are also useful in dangerous or very expensive situations where actual use of equipment may prove unwise.

Thus far, the examples mentioned above described the role of technology in communities. Communities can be formed around teachers who are interested in its use for enhancing their professional practices. In order to leverage on such professional communities to enhance learning, there would be a need to adapt the KM approach to capturing, understanding, and retaining the knowledge and insights gained from the interactions arising from within these communities.

For example, in an ongoing professional development program for Heads of Department for Information Technology (HoDs in IT) in Singapore schools, we had a group of HoDs (IT) who became interested in knowledge-building pedagogies (Hung, Tan, Hedberg, & Koh, 2005; see Scardamalia, 2002 for knowledge-building pedagogies) using computer-supported collaborative learning (CSCL) systems, and who decided to collaborate and support one another in the implementation of knowledge-building activities in their respective schools. One of the HoDs (IT) who was initially observed to be rather passive both during face-to-face sessions and online sessions (when these teachers were attending a heads of department course at the National Institute of Education) developed into an active participant at the end of the program. She mentioned in an interview that she was the least knowledgeable of the three members in her knowledge-building team in terms of the use of CSCL systems for teaching and learning. She described how the other members in the group, more experienced in knowledge-building activities and CSCL systems, supported her during the planning and implementation stages, sharing ideas with her and helping her solve technical problems. Through this enculturation process, she developed from being a peripheral participant (Lave & Wenger, 1991) at the beginning of the program into a more central participant. She has since presented in two sharing sessions to other HoDs (IT), an ICT conference, as well as a school cluster (a group of schools within a particular district) sharing session. She has plans to implement this knowledge-building concept to more levels in her school and tried to tie up with other schools in implementing knowledge-building activities using CSCL systems.

In this example, we could see the development of this HoD (IT) in identity formation, from a peripheral participant at the beginning of the program to a central participant after one year into the program. This process of transformation is facilitated by the use of online discussion tools, used at the beginning of the program for the members to raise issues on teaching and learning using technology, and using e-mail for discussion when the HoDs (IT) were back in school during the implementation period. Such identity formation in a community arose from uncertainty in the use of technology for innovative use of technology in teaching and learning. The KM approach, had it been adapted for use in the above example, would have yielded more useful learning that could be used to grow and sustain other similar communities.

The above roles of technology are not exhaustive, but indicative of technology as an enabler to facilitate learning interactions within communities of learners. We emphasize that educational technology is a means to an end, and not an end in itself. They can be used effectively to enhance social relationships in learning, assist learners in their projects, and help learners explore and deepen their conceptual understanding through simulations and visualizations.

A summary of the role of technology in forming and sustaining communities discussed thus far is provided in Table 1. From Table 1, it is obvious that there is a lack of *technology tools for identity formations* such as the learning or enculturation of beliefs and shared values.

The ability of members such as teachers within a CoP to adopt technology in the above ways enhances their literacy levels with regards to: (1) using language, symbols, and text for knowledge creation and dialogue; (2) interacting between their knowledge-understanding and the information they receive from others and elsewhere; and (3) adopting different technology tools whenever they need to achieve goals at hand (Istance, 2003).

FUTURE DIRECTIONS FOR KM: INVOLVING THE LARGER COMMUNITY

The approach taken by the CoP framework is clearly more *process* than *product* oriented from the KM perspective. So, what's next? Knowledge from Polanyi's perspective is largely tied to context and the person. In other words, knowledge differs from "information" because when information is applied to a context, it becomes contextualized, and the person applying that knowledge gained personal experiences and knowledge. If that person leaves the community (or organization), knowledge leaves with that person(s). We offer one recommendation to KM: invest heavily on person and context-process development. There is a need to form networks of teams across organizations and communities, and expand the global pro-

Table 1. Synthesizing CoP principles with technology tools

CoP-CoL Principles	3 principles of Learning	Technology Tools
Shared beliefs and history	Learning is an identity formation	*Gap area*
Mutual enterprise and production of joint artifacts	Learning is demand driven (based on needs to understand); Learning is a social act	Technology as a productivity tool; Technology as a sense-making tool; Technology as a collaborative tool
Interactions, activities, and relationship building (mutual trust)	Learning is a social act	Technology as a collaborative tool
Cycle of renewal—new members joining the community	Learning is a social act; Learning is an identity formation	Technology as a collaborative tool *Gap area*

cess of knowledge management. Technology today can mediate and connect people and expertise (not just information), and knowledge can still be managed across networks of people. We recommend a community of communities suggestion.

Academia serves as a good example of these community of communities networks. In academia, professors and practitioners belong to different universities and organizations, but generally many are affiliated to societies and professional bodies. Yearly, many of these societies organize activities such as conferences where expertise is being shared. In addition, many of these societies have dissemination means such as Web sites, e-mail communications, journals, and others. The unique part of universities in terms of their rewarding mechanisms to their faculty members is that academics are rewarded when they contribute to new knowledge to the larger community (and not just to the university). And this recognition mechanism is common across most universities. Thus knowledge leaks (as it were, when individuals move out of these respective institutions or organizations) are contained to within the larger communities and mediated by societies and communication means (such as journals). Importantly, in academia, the reward criteria for academics are relatively consistent across universities. These networks of networks represent the non-linearity of interactions that are needed for knowledge creation and management—denoting "complex forms of negotiation and interaction between people, some of whom will offer different kinds of knowledge" (Southworth, 2000, p. 290).

Technologies have in the past supported attempts at codifying knowledge, and these efforts can be seen in databases being created to manage information. More recently, technologies that support processes of knowledge creation and sharing include computer-supported collaborative environments and computer-supported communication tools. These technologies support the process rather than the product-oriented views of knowledge. Similarly, authoring or constructive tools enable individuals to engage in the process of knowledge creation and meaning making rather than receiving knowledge. Increasing, technology must now support the process of story creation or narrative creation as we recognize that storytelling is one effective tool in KM. Moreover, how can technologies support the social construction process of story creation, refining, and reconstruction—the process of which enables individuals to construct a coherent understanding of "what they are doing in practice"? The commonly told accounts of Xerox's technicians constructing stories of the machine problems they encounter, by relating to one another pieces of troubleshooting data, account for their experimentations on certain problems, getting more data to support their conjectures and co-constructing their "story," attempting to being coherent to their "story," and defending that their stories are all part of the social construction process.

CONCLUSION

The fundamental issue of KM is to understand the "ways things work," improve on them, establish and formulate processes, and possibly scale up and sustain these so-called good practices. Within this whole process of attempting to understand practice, we expect dimensions of tacit, explicit, and cultural (or institutional) knowledge to be interwoven and to manifest in various forms. Moving from "if only an organization knows what it knew" to "knowing as much of what it knows and setting processes to sustain its better processes" and "knowing when to change its processes when needed" seem to be what the business of KM is all about. All these attempts presume the assumption that we tease out as much of the tacit (individual and organizational levels) to the explicit (individual and organizational), and formulate cultural knowledge (organizational level) to improve ways of thinking and doing.

We need to recognize that there is a limit to what KM can achieve. Social relations, cultural and contextual underpinnings (or overpinnings), and knowledge as tied to persons are part of the capital of what distinguishes one organization-individual from another. The key lies in striking a balance between codifying knowledge and creating the processes which support knowledge creation, sharing, and transfer. Technology can only facilitate the processes enabled for both and enhance the KM efforts. The CoP frameworks are efforts to create such process-oriented KM processes, but as in any approach, there would be advantages and disadvantages. Certainly, we recognize that both process- and product-oriented KM is necessary, and we see both as dialectically informing each other. We envisage that in the near future, this concept of CoPs will pervade the teaching profession and educational community as its popularity increases (Gee, 2000; Hung et al., 2005). Technology will inevitably be an integral part of CoPs—both as a content-issue to be dialogued upon by members of the CoPs and thus heighten technology literacy, and as a means through which members engage in interactions within CoPs and across disciplines through knowledge brokering. As a result of such adoptions, teachers will become more technology competent both in terms of its awareness and in terms of its use as a tool. Issues in the adoption of technology in learning are dialogued upon as a "product" and adopted as a tool in the "process" of being members in CoPs. CoPs and the concepts of KM can transform our traditional notions of pedagogy and radically shift our mindsets from transmission notions of learning to transformative possibilities where knowledge for the learner (both students and teachers) is not only an entity-product to be absorbed, but as an emerging-process to be constructed and understood both individually and socially.

Schools can no longer afford to be structured in entrenched hierarchical structures, but must be able to respond quickly to the external environment in order to meet the increasingly competitive demands of the global society. Fostering members into CoPs, where every member contributes to others and receives from others within a KM framework, would enable knowledge to flow dynamically within a school (or across schools) (Wenger, McDermott, & Snyder, 2002). In this way, members of CoPs need to assume increased responsibility, ownership, and accountability to the school organization and to knowledge. Inevitably, we believe that schools would have to be "transformed" into dynamic communities and sub-communities. Information technologies would be an integral part of the knowledge flow processes, and as a consequence, literacy levels rise in tandem. Through such a transformative stance, schools become "knowledge-creating schools" (Hargreaves, 1999) where learning is understood as a collaborative effort, and in the process, artifacts are developed within a KM framework which could be knowledge-products, practices, and ideas—both conceptual and material (Paavola, Lipponen, & Hakkarainen, 2004).

Finally, summarizing the entire chapter, we recognize that (as we think through the issues of KM, CoPs, CoLs, and the related underlying theoretical foundations of situated cognition) CoPs and/or CoLs can be seen as ways of organizing teachers (or professionals) for continuous (organizational) learning and for propagating good professional practices. The tools of

CoPs/CoLs are frameworks or strategies available to facilitate interactions among members in some structured manner to give focus and to achieve goals and results. KM is a product and process methodology that captures and retains the knowledge and insights gained for future mining to minimize possible mistakes (which were done in the past) and lower the entry of learning needed to participate in the community (that is, to speed up the process of learning within the community—from novice to expert). Technologies (including AI) are enablers to facilitate KM, CoPs/CoLs, and the implementation of derived framework-strategies to facilitate interactions and processes more conveniently.

REFERENCES

Barab, S., & Duffy, T. (2000). From practice fields to communities of practice. In D. Jonassen & S. Land (Eds.), *Theoretical foundations of learning environments.* Mahwah, NJ: Lawrence Erlbaum.

Brown, J., & Duguid, P. (2000). *The social life of information.* Boston: Harvard Business School Press.

Choo, C. W. (2001). Knowledge management. In J. R. Schement (Ed.), *Encyclopedia of communication and information.* New York: Macmillan Reference.

Choo, C. W. (2002). Sensemaking, knowledge creation, and decision making: Organizational knowing as emergent strategy. *Strategic management of intellectual capital and organizational knowledge.* New York: Oxford University Press.

Choo, C. W. (2003). Perspectives on managing knowledge in organizations. In N.J. Williamson & C. Beghtol (Eds.), *Knowledge organization and classification in international information retrieval.* Binghamton, NY: Haworth Press.

Duffy, J. (2000). Knowledge management: To be or not to be? *Information Management Journal, 34*(1), 64-67.

Fullan, M. (1999). *Change forces: The sequel.* London: Cassell.

Gee, J.P. (2000). Communities of practice in the new capitalism. *Journal of the Learning Sciences, 9*(4), 515-523.

Hargreaves, D. (1999). The knowledge-creating school. *British Journal of Educational Studies, 47*(2), 122-144.

Hung, D., Tan, S. C., Hedberg, J., & Koh, T. S. (2005). A framework for fostering a community of practice: Scaffolding learners through an evolving continuum. *British Journal of Educational Technology, 36*(2), 159-176.

Istance, D. (2003). Schooling and lifelong learning: Insights from OECD analyses. *European Journal of Education, 38*(1), 85-98.

Lave, J., & Wenger, E. (1991). *Situated learning: Legitimate peripheral participation.* Cambridge: Cambridge University Press.

Orr, J. (1996). *Talking about machines: An ethnography of a modern job.* Ithaca, NY: IRL Press.

Paavola, S., Lipponen, L., & Hakkarainen, K. (2004). Models of innovative knowledge communities and three metaphors of learning. *Review of educational research, 74*(4), 557-576.

Petrides, L., & Guiney, S. (2002). Knowledge management for school leaders: An ecological framework for thinking schools. *Teachers College Record, 104*(8), 1702-1717.

Polanyi, M. (1964). *Personal knowledge: Towards a post-critical philosophy.* New York: Harper & Row.

Scardamalia, M. (2002). Collective cognitive responsibility for the advancement of knowledge. In B. Smith (Ed.), *Liberal education in the knowledge society* (pp. 67-98). Chicago: Open Court.

Southworth, G. (2000). How primary schools learn. *Research Papers in Education, 15*(3), 275-291.

Vygotsky, L. S. (1978). *Mind in society: The development of higher psychological processes.* Cambridge: Harvard University Press.

Weber, S. (2003). Boundary-crossing in the context of intercultural learning. In T. Tuomi-Grohn & Y. Engestrom (Eds.), *Between school and work: New perspectives on transfer and boundary-crossing* (pp. 157-177). The Netherlands: Elsevier Science.

Wenger, E., McDermott, R., & Snyder, W. (2002). *Cultivating communities of practice.* Boston: Harvard Business Press.

Wertsch, J. V., & Rupert, L. J. (1993). The authority of cultural tools in a sociocultural approach to mediated agency. *Cognition and Instruction, 11*(3&4), 227-239.

Chapter XXXIII
The Emerging Use of E-Learning Environments in K-12 Education: Implications for School Decision Makers

Christopher O'Mahony
Saint Ignatius' College, Australia

ABSTRACT

Virtual learning environments (VLEs) and managed learning environments (MLEs) are emerging as popular and useful tools in a variety of educational contexts. Since the late 1990s a number of 'off-the-shelf' solutions have been produced. These have generally been targeted at the tertiary education sector. In the early years of the new millennium, we have seen increased interest in VLEs/MLEs in the primary and secondary education sectors. In this chapter, a brief overview of e-learning in the secondary and tertiary education sectors over the period from 1994 to 2004 is provided, leading to the more recent emergence of VLEs and MLEs. Three models of e-learning are explored. Examples of solutions from around the world are considered in light of these definitions. Through the case of one school's journey towards an e-learning strategy, we look at the decisions and dilemmas facing schools and school authorities in developing their own VLE/MLE solutions.

INTRODUCTION

The history of adoption of technological innovations in schools is characterised by a mixture of enthusiasm and apprehension. The adoption of information and communications technology (ICT) in schools is no exception. Governments, educational authorities, individual schools, and educationalists have recognised the tremendous potential of ICT to transform teaching and learning. At the same time, there has been a collective intake of breath as social, financial, industrial, political, pedagogical, and logistical implications have emerged (Cuban, 2000). Increasingly, ICT literacy is a requirement in the K-12 education sector, for both staff and students.

In educational ICT, change is the one constant. For the most part, educational institutions have been on the receiving end of ICT innovation, responding to change rather than driving change. As a result, the journey towards literacy with ICT innovations in schools more often follows ad-hoc diffusion models, rather than as an outcome of specific decision-making strategies. Thus, investments by schools in products/solutions such as school administration systems, e-mail systems, local area networks, laptop programmes, intranets, virtual private networks (VPNs), and the like, can be isolated decisions rather than forming elements of some wider e-learning strategy (Jones, 2003).

Over the past 10 years, many schools have worked hard to begin integrating these disparate solutions and streamline their ICT management. With the increasing ubiquity of the World Wide Web and browser-based educational resources, the integration of various e-learning components became a possibility. In the late 1990s, early versions of integrated learning management systems emerged, predominantly targeted at the tertiary education sector. Now, in the first decade of the new millennium, a variety of solutions are being developed with the primary and secondary sector in mind (BECTA, 2001a, 2001b, 2001c).

This chapter reviews the emergence of e-learning technology components over the period from 1994 to 2004 and their implementation in the K-12 education sector, with particular reference to attempts to integrate these various components into a broader e-learning strategy. By analysing literature concerning models of e-learning in schools, it is shown that many schools, although mapping closely to these models, do so more by coincidence than design. It is suggested that school e-learning strategies evolve to accommodate specific ICT components and capabilities as they emerge. The challenge for schools, as always, is to have the agility to respond appropriately to these innovations, while at the same time exercising wisdom and discernment in their implementation (Dowling, 2003).

LITERATURE

The concepts of computer-based training (CBT) and computer-assisted learning (CAL) have been in circulation since the 1980s, initially in industry. The reality of distance teaching and distance learning has been with us much longer. It has only been since the ubiquity of the World Wide Web in the early 1990s that we have seen a convergence of these domains. A number of factors have assisted this convergence. Increasingly sophisticated Web browsers; increasingly sophisticated Web scripting languages; increasing bandwidth; improved data compression techniques; reducing costs; increased access to powerful personal computing devices; and increased levels of user knowledge and understanding are some of these factors.

A selective chronology of events related to e-learning innovations is shown in Table 1.

Definitions

A variety of definitions of e-learning exist, but most have a common theme. The key distinction between definitions of e-learning and previous definitions of CAL, CBT, and the like are a focus on Web-enabled technologies. For instance:

E-learning is online training that is delivered in a synchronous (real-time, instructor-led) or asynchronous (self-paced) format. (Jones, 2003)

Table 1. A brief chronology of e-learning innovations, 1990-2003

Year	Event
1990	1. Tim Berners-Lee proposes his idea for a World Wide Web
Mid-1994	2. Mosaic communications founded by Mark Andreesen and Jim Clark 3. Schoolsnet Australia launched—early ISP
October 1994	4. Early version of Netscape launched
1995	5. Netscape v 1.1 launched, quickly followed by v 1.2
Late 1995	6. Windows 95 launched, which included the first version of Internet Explorer
1996	7. Early version of WebCT launched—Vancouver 8. Internet Explorer 3 vs. Netscape Navigator 3
1997	9. Blackboard Inc launched—Washington, DC 10. digitalbrain plc incorporated—London 11. Impaq launched as an ISP—Australia 12. Internet Explorer 4 vs. Netscape Communicator
1998	13. AOL buys Netscape
1999	14. Microsoft launches Internet Explorer 5
2000	15. Netscape 6 vs. Internet Explorer 5.5
2001	16. Microsoft releases Internet Explorer 6.0 17. Granada launches its "Learnwise" portal—UK 18. Schoolsnet Australia re-badged as 'myinternet'
2002	19. Netscape 7 released
2003	20. Thousands of schools working with e-learning products

E-learning is:

> .... *an innovative approach for delivering electronically mediated, well-designed, learner-centred and interactive learning environments to anyone, anyplace, anytime by utilising the Internet and digital technologies in concert with instructional design principles.* (Morrison & Khan, 2003)

More functional definitions describe the typical components found in an e-learning system. Barnes and Greer (2002) suggest that Web-based learning environments include "Web browsing and authoring, file transfer, e-mail, chat, discussion groups (communities) and shared whiteboards." A more detailed description is provided by BECTA (2001a) as follows:

> *Although there is some confusion about the definition of Virtual Learning Environments, they are generally a combination of some or all of the following features:*
>
> - *communication tools such as e-mail, bulletin boards and chat rooms*
> - *collaboration tools such as online forums, intranets, electronic diaries and calendars*
> - *tools to create online content and courses*
> - *online assessment and marking*
> - *integration with school management information systems*
> - *controlled access to curriculum resources*
> - *student access to content and communications beyond the school.*

As well as multiple definitions, multiple labels exist for describing e-learning systems. Such terms as *learning management system* (LMS), *course management system* (CMS), *virtual learning environment* (VLE), *managed learning environment* (MLE), and *portal* are often used interchangeably. In the UK, the Joint Information Systems Steering Committee (JISC) makes a clear distinction between a virtual learning environment and a managed learning environment, thus:

> *While recognising that the world at large will continue to use terminology in different and often ambiguous ways, the term* ***Virtual Learning Environment (VLE)*** *is used to refer to the 'online' interactions of various kinds which take place between learners and tutors. The JISC MLE Steering Group has said that VLE refers to the components in which learners and tutors participate in 'online' interactions of various kinds, including online learning. The JISC MLE Steering Group has said that the term* ***Managed Learning Environment (MLE)*** *is used to include the whole range of information systems and processes of a college (including its VLE if it has one) that contribute directly, or indirectly, to learning and the management of that learning.* (JISC, 2001)

A useful model displaying the components encapsulated in the JISC definition is shown in Figure 1.

The JISC model thus separates VLE and MLE components as seen in Example 1.

Whereas such platforms as WebCT and Blackboard have dominated the e-learning market in the tertiary sector (Beshears, 2000), solutions in the primary and secondary education sectors are not so homogeneous. As seen in Table 1, a number of vendors supplying e-learning components and complete solutions have emerged since the late 1990s. Impaq and myinternet (in Australia) and digitalbrain and Learnwise (in the UK) are just some of the vendors in this domain providing off-the-shelf solutions for schools. Other vendors such as Microsoft (Sharepoint and ClassServer) and

Figure 1. The JISC managed learning environment model (JISC, 2001)

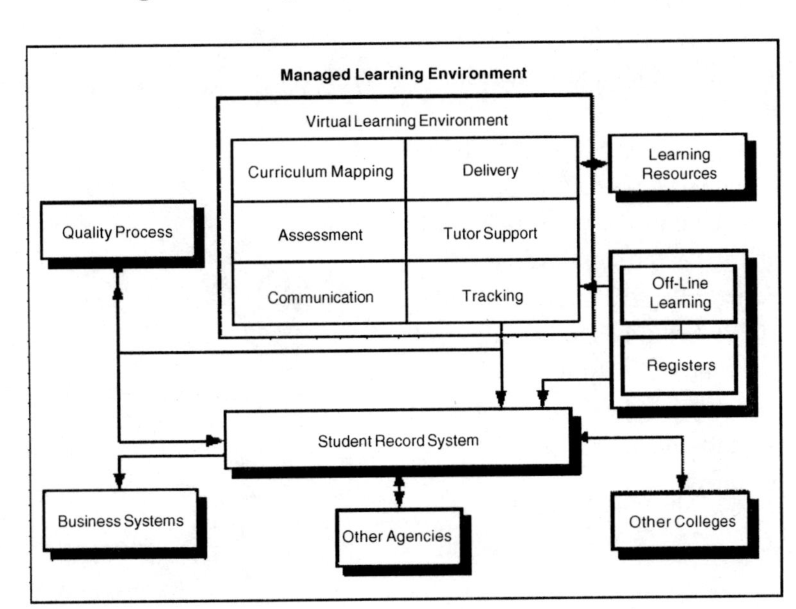

Example 1. Components of VLE and MLE

VLE Components:		MLE Components:
Curriculum Mapping		Learning Resources
Assessment		Student Record System
Communication		Business Systems
Delivery		Off-Line Learning
Tutor Support		Registers
Tracking		Quality Process

Novell (Extend) offer development platforms for highly customisable e-learning solutions. Concerns have been raised by some commentators, however, that some schools are attempting to implement e-learning solutions without sufficient cognizance of issues for learners, administrators, support staff, and the institution itself (Barnes & Greer, 2002; Kilmurray, 2003; Morrison & Khan, 2003).

In the early years of the new millennium, educational authorities in developed countries and third-party vendors simultaneously recognised that if e-learning environments were to succeed, then appropriate content needed to be developed. Furthermore, to ensure interoperability and re-usability of this content, standards needed to be developed. Initiatives such as Curriculum Online (UK) and The Learning Federation (Australia) are indicative of government-sponsored approaches to the development of e-learning content, now commonly known as learning objects. Commercial developers of learning objects include XSIQ, SchoolKit Enactz, and Granada LearnWise. In terms of e-learning standards, we are now seeing the deployment of such standards as SIF (the Student Interoperability Framework) and SCORM (the Scalable Content Object Reference Model). At the time of writing this chapter, the boundary between developers of e-learning platforms and the developers of e-learning objects is blurred.

Components of an E-Learning Strategy

Badrul Khan has been an active commentator in the e-learning domain since 1997 (Khan, 1997). He notes: "A successful e-learning system involved a systematic process of planning, design, development, evaluation and implementation to create an online environment where learning is actively fostered and supported" (Morrison & Khan, 2003).

Khan has proposed a model that identifies eight dimensions organisations need to address in developing an e-learning strategy (see Figure 2). These dimensions are as follows:

- Institutional
- Pedagogical

Figure 2. Khan's eight-point model of e-learning

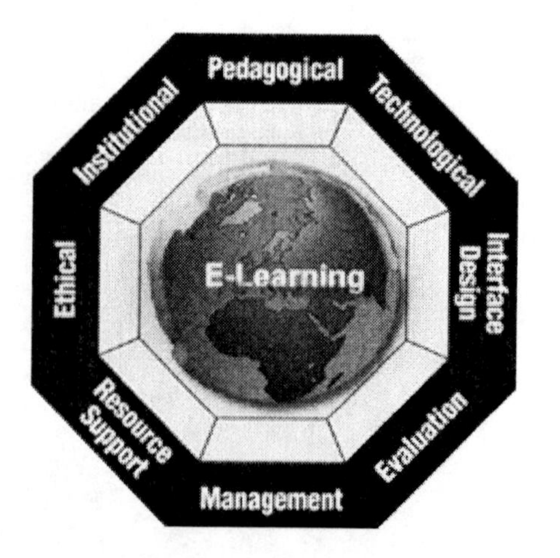

- Technological
- Interface Design
- Management
- Ethical
- Resource Support
- Evaluation

Each dimension is broken down into issues related to specific aspects of a successful e-learning environment. Khan's model is not specific to schools or universities, but rather is applicable to any organisation engaged in e-learning for its constituent community, and thus is put forward as a "global e-learning framework" (Morrison & Khan, 2003).

Requirements for E-Learning Support

Another model proposed by Hitch and MacBrayne (2003) concentrates on the support required for a successful e-learning strategy. Their model has three key components: (a) Faculty Support and (b) Student Support, "glued" together by (c) Technological Infrastructure (see Figure 3). They note:

> *Despite the rising use of information technology in instruction, both in the traditional classroom and at a distance, there remains a substantial gap in providing off-campus students with an array of academic and support services equivalent to the on-campus services...Institutions that provide e-learning, whether they offer totally online or hybrid (i.e., blending face-to-face with online instruction) courses, must provide concurrent e-student support mechanisms.* (Hitch & MacBrayne, 2003)

Potential Benefits of E-Learning Environments

The implementation of an e-learning environment in a school or school authority offers potential benefits to a variety of stakeholders. These can be summarised as follows:

General Benefits

- Can make it easier for staff and students to use ICT within an integrated environment.

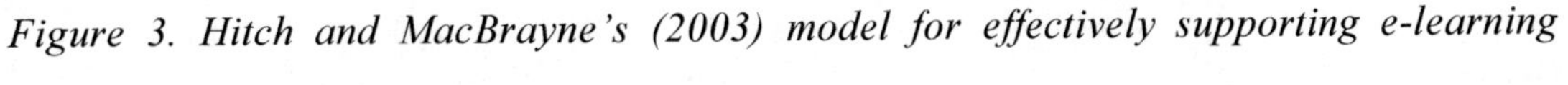

Figure 3. Hitch and MacBrayne's (2003) model for effectively supporting e-learning

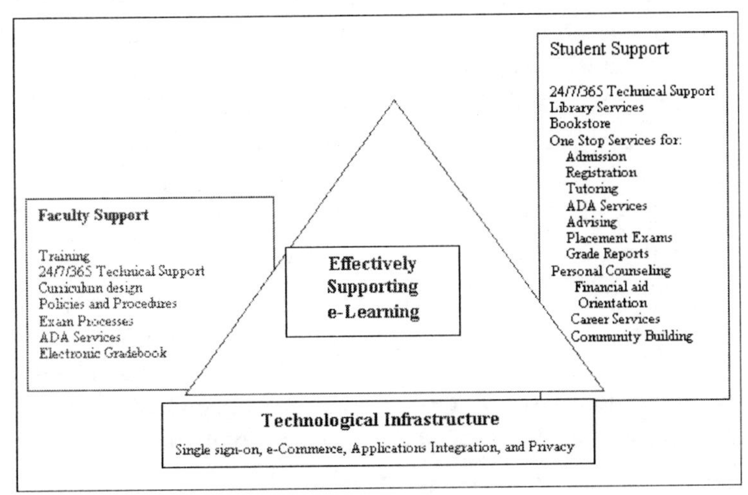

- Brings together a variety of features in one piece of software with a consistent look and feel, which is consequently easier to learn and manage.
- Offers a different communication dimension through e-mail, discussion groups, and chat rooms, in addition to face-to-face classroom interaction.
- Can improve the learning environment and standard of discussion, if the communications are managed effectively by the teacher.

Benefits for Students

- Offers the flexibility of "anytime, anywhere" access.
- Encourages gains in student ICT skills in general, and in journalistic writing, understanding, and presentation skills.
- Encourages development of higher levels of deep and strategic learning styles.
- Encourages the discovery of successful approaches to learning through trial and error in discussion, and through expressing ideas in a written but public way.

Benefits for Teachers

- Supports teacher confidence, and enhances practice and collaboration.
- Fosters self-study by teachers willing to make the commitment to the technology and to sharing personal views and experiences.
- Increases teacher participation in online seminars, which can lead to increased performance in group work.

Benefits for Parents

- Provides a communication gateway between home and school.
- Allows parents to monitor their children's progress.
- Provides access to online content that can help parents to support homework studies out of school hours.

To summarise this section, it can be seen that the literature as of 2004 describes an e-learning milieu that is not fully mature. In this context, attempts by schools to devise e-learning strategies and implement them are necessarily subject to the vagaries of an evolving marketplace. The following section describes the journey of one school on the path towards a full e-learning solution.

ONE SCHOOL'S JOURNEY TOWARDS AN E-LEARNING STRATEGY

Background

The case study school is an independent day and boarding college for boys in Sydney, Australia. Although initially hesitant to embrace ICT innovations in the early 1990s, the school's management realised in 1994 that a number of push and pull factors were at work which required a whole-school strategy for ICT. After engaging external consultants, the school tabled its first ICT Strategic Plan in late 1995. This plan made provision for an extensive rollout of fibre-optic and category5 cabling throughout the school site, an ongoing programme of investment in end-user hardware and software, a review of curriculum outcomes to incorporate ICT elements, and provision for staff training and support.

It is interesting to note the extent to which ICT has become embedded in the school's culture by considering the comparisons between 1994 and 2004 (see Table 2).

Table 2. Evolution of ICT in case study school

Dimension	1994	2004
Students	1,100	1,550
Academic Staff	120	170
Support Staff	50	80
Student Computers	60	450
Student-Computer Ratio	1:19	1:3.5
Staff Computers	15	120
(Academic) Staff-Computer Ratio	1:8	1:1.5
% Computers with Internet Access	5%	100%
% Students with E-Mail Accounts	0%	100%
% Staff with E-Mail Accounts	0%	100%
Servers	2	35
ICT Support Staff	1	10
Annual ICT Spent (as % of total spent)	1%	8%

In addition to a high level of ICT provision within the school, members of the school community (staff, students, and parents) also exhibited high levels of access to ICT outside school. In 2003, 95% of staff reported access to the Internet and e-mail from home. Students and parents reported high levels of access to computers (97%), and high levels of access to the Internet and e-mail in the home (89%). These metrics are consistent with similar statistics from other developed countries around the world (Research Machines, 2000; DfES, 2001; National Statistics, 2002; Ofsted, 2002).

ICT Management Model

There has been an emerging perception throughout the school that ICT is an enabler, providing increased efficiencies and effectiveness in administration, and adding value to teaching and learning (Mumtaz, 2000; Kennewell, Parkinson, & Tanner, 2000; Passey, 2002). Work by O'Mahony (2000) has noted that in educational institutions, ICT can become either a bridge or a chasm. ICT can be a bridge insofar as it has the potential to:

- directly support pedagogical efforts in the teaching and learning context, in terms of delivering educational and applications software to the classroom;
- directly support the back-office functions of the school; and
- indirectly minimise the impact of necessary administrative functions that teachers, students, and parents are required to perform.

ICT can be a chasm in the sense that it:

- so frequently falls short of the expectations and overblown promises of suppliers, developers, and purchasers; and
- is often perceived as a weapon for administrative control, rather than a tool for educational empowerment.

Arising from the school's efforts to embed ICT into its operations is a growing recognition of key factors that enable ICT to flourish, and thus for the school to be more effective in its overall development (Kirkman, 2000). These factors are presented here as a six-point model for achieving confident use with ICT, as seen in

Figure 4. Core components of the school's ICT management

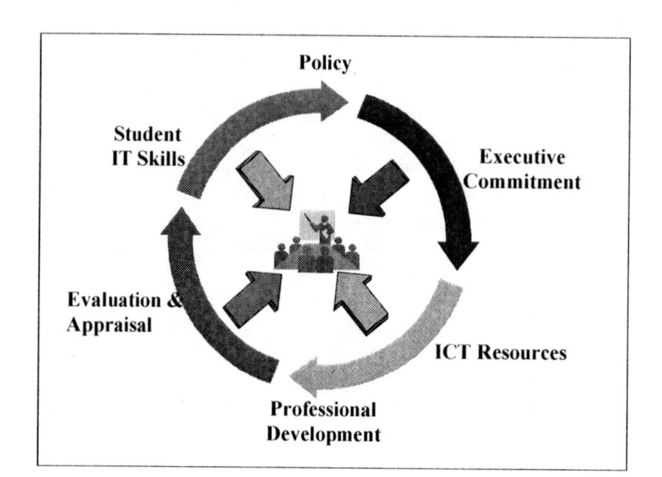

Figure 4. Within the school, this model exists in a specific organisational context, whereby ICT has increasingly become a fulcrum for change in the organisation's culture.

In summary, the six-point-model comprises:

1. Policy
2. Executive Commitment
3. ICT Resources
4. Professional Development
5. Evaluation/Appraisal
6. Student Learning

Details of the 6-point model are outlined in the following sections.

Curriculum ICT Policy (Strategic)

The school, through its ICT Strategic Plan, makes a clear statement of intent and direction concerning the use of information and communication technologies in curriculum areas. This statement is visible in school documentation at the senior management level, and is internalised throughout the curriculum (Kennewell et al., 2000). "In...successful schools, senior management do more than provide support for the IT Coordinator's policy; rather the IT policy is viewed as emanating from the senior management" (ACCAC, 1999). The curriculum ICT policy seeks to articulate well with the school's business and strategic development plans.

Department Commitment (Tactical)

At the department level, ICT policies exist which articulate with the wider ICT strategy, and provide necessary detail and context for the respective curriculum area. These policies express the department's commitment to ICT professional development, and specify expectations of ICT use in the classroom, both in terms of minimum hours and ICT-based tasks (Newton, 2003; Lambert & Nolan, 2003).

ICT Resources

A prerequisite to success with school ICT is the provision of sufficient, reliable, and up-to-date resources (ACCAC, 1999). These resources include network infrastructure, workstation and peripheral hardware, software, and human resources. Table 2 indicates the school's investment in resources over a 10-year period. Strong project management methodologies have been applied to ensure that the school gains good value for money, recognising that inferior products do more damage than good. Rigorous criteria are used in selecting hardware and software applications, such as ease of integration and ease of use (Stevenson, 1997).

Teacher Professional Development

Hiring external trainers has often been the only option for schools, but increasingly, schools are considering the appointment of dedicated training staff within the overall ICT function (Watson, 2001), as is the case with the subject school (O'Mahony, 2002). As well as having ICT resources and policies regarding the use of ICT in teaching, learning, and administration, the school has implemented a robust and measurable professional development programme (Donnelly, 2000; Russell, Finger, & Russell, 2000). This programme has six main components: (1) initial ICT orientation, (2) formal timetabled ICT training (one period per fortnight), (3) ICT surgeries, (4) a regular 'ICT tips & tricks' newsletter, (5) evening master classes (once per term), and (6) provision for external ICT courses.

Staff Appraisal and Review

The appraisal and review process gives crucial feedback for all aspects of the model. To drive home the message concerning the school's commitment to ICT, effective classroom use of ICT is a performance indicator for staff. The reviewer can flag the reviewee's ICT training needs, which is communicated to the training function/coordinator, who organises/delivers the required training. Once completed, confirmation of training is passed back along the chain. As well as providing feedback on staff ability, the appraisal and review process offers the opportunity to flag any issues concerning ICT resourcing or access. These issues, too, are forwarded to the relevant person/function. Collectively, these items will assist in the formation of subsequent ICT strategies.

Student Learning

The ultimate aim of this model, and in particular the Staff ICT professional development programme, is the improvement of student learning. Thus, complementary to a Staff ICT skills programme is a cross-curricular student ICT skills programme. Transcending the use of ICT in specific subjects, this programme provides broad-based exposure to generic ICT skills, including keyboard familiarity, word processing, spreadsheets, presentation graphics, Internet searching, critical analysis of Web-based data, and "appropriate ICT use" (NGfL, 2002a, 2002b; Hruskocy, Cennamo, Ertmer, & Johnson, 2000).

The Evolving E-Learning Strategy

As mentioned in the first section of this chapter, the development of an e-learning strategy at the subject school has been an evolutionary process, which has been modified over time as new e-learning technologies have emerged. Between 1994 and 1996, the school's main focus was on baseline ICT connectivity. That is, fundamental infrastructure decisions were a priority, establishing and maintaining a reliable local area network and associated intra-mural ICT services. Although early Web browsers became available in 1995, the school was not at that stage in a position to leverage these new technologies.

The school's first triennial ICT Strategic Plan (1997-1999) began to focus on value-added ICT functions, including early stages of 'extra-mural' connectivity. The school was fortunate in gaining access to the Internet through a broadband cable modem, and strategies were agreed for implementing e-mail for staff, a rudimentary intranet, and a school Web site. At the same time, the school continued its investment in end-user hardware and software, maintaining a focus on traditional classroom-based teaching and learning.

A second triennial ICT Strategic Plan covered the 2000-2002 period. This plan identified three core challenges:

- responding proactively to the pace of technological change,
- providing opportunities for staff ICT training, and
- providing extra-mural services for school staff and students.

In this plan, there was a clear recognition by management that the school needed to be continually renewing and reviewing its ICT efforts, replacing an earlier belief that one-off capital investment would be sufficient. Parallel to this was an understanding that staff needed further encouragement to build ICT into their teaching and administration. The third core challenge recognised the value to be gained from implementing remote access technologies, and the school actively sought out potential solutions for both staff and students.

In 2001, the school entered into an agreement with an e-learning solution provider (Impaq Australia) for the provision of VPN remote access and portal services. This initiative was the beginning of the school's attempts to integrate previously disparate ICT components into a more cohesive single e-learning solution (Weatherly & McDonald, 2003). Although not without challenges and frustrations (see Boydell & Kane, 2002), by the end of 2002 the school had implemented e-learning components according to Table 3.

The school is now in the middle year of its third triennial ICT Strategic Plan (2003-2005). Key strategies identified in this plan include:

- Identity management and database integration
- Network security
- Continuing commitment to staff training and support

Table 3. Chronology of integration of e-learning components in case study school as of 2002

Components (from BECTA, 2001a)	Installed for Staff	Installed for Students	Integrated
Communication tools such as e-mail, bulletin boards, and chat rooms	1996	2002	2002
Collaboration tools such as online forums, intranets, electronic diaries, and calendars	1997	2001	2001
Tools to create online content and courses	2001	N	N
Online assessment and marking	2001	N	N
Integration with school management information systems (target for 2005)	N	N	N
Controlled access to curriculum resources	1996	1996	2000
Student access to content and communications beyond the school	2001	2001	2001

Table 4. Staff reflections on e-learning environments

Stakeholder	Reflections
Benefits of utilising an e-learning environment for me as a teacher:	• Allows me to be organised, which encourages my students to be organised. • Reduces my photocopying/paper waste. • Students can't 'lose' handouts, thus improving use of my time. • Material is 'always' available for revision purposes. • Easy to update on-the-fly before, during, and after class.
Benefits of utilising an e-learning environment for my faculty:	• Allows the sharing of resources. • Common files can be loaded into one area for all classes to access. • Excellent medium for communicating outcomes, programmes, and so forth between faculty members. • Reduces photocopying/paper waste.
Benefits of utilising an e-learning environment for my students:	• They have an area where they know they can find all the information needed for a specific subject. • They can see important due dates quickly and easily. • Assignment/handout information can never be 'lost'. • Material can be collected/viewed even if they miss a class. • Test content/exam revision/classwork can be accessed anytime, anywhere. • Forgot your homework? Look it up on the portal.
Benefits of utilising an e-learning environment for the wider community:	• Parents can access information about the course, such as programs, assessment schedules, and so forth. • Parents can look at assignment details and homework tasks. • Parents can see what is happening in their child's class. • Wider community can access photos, current events, school newsletter, and so forth.

- Continuing commitment to remote access services

Although the school continued to integrate e-learning components during 2003, a significant setback occurred late in the year when the e-learning solution provider ceased trading. This meant the potential loss of some critical remote services and, by extension, a potential loss of confidence in the school's e-learning efforts. By early 2004, however, the school had engaged alternative providers. It is indicative of the rapid development of e-learning solutions that this 'second time around' implementation was much smoother than the first.

One of the projects being addressed by the school involves integrating a variety of administrative databases (such as student administration, library, business operations, and human resources) with the main network directory service (network authentication). This project has multiple phases and constraints, but has the potential to deliver increased efficiency and effectiveness across the whole school. It will also add value to the school's e-learning efforts by enabling integration of other e-learning components to the school's management information systems through the implementation of SIF standards (as mentioned earlier; see Table 3).

The school has encouraged staff to reflect on changes to their teaching practice as a result of utilising e-learning elements. One example of this reflective process is shown in Table 4.

ISSUES AND IMPLICATIONS

A number of considerations can be perceived for an institution planning the implementation of an e-learning strategy. Articulating with the

Table 5. Considerations for implementing an e-learning strategy

	Pedagogical	Technological	People Factors
Students	How will this help me learn?	How do we provide 1-to-1 access to e-learning resources?	What is the most appropriate 'class size'?
Teachers	How will this help me teach?	How can I ensure the engagement of the students and reliability of the technology?	How do we build staff skill and enthusiasm for these initiatives?
School Managers	How will this improve our outcomes?	How do we measure the cost benefits?	What 'critical mass' of people is required before this will work?
Teacher Educators	What pedagogical models are best suited to this environment?	What skills do we need to include in our training programmes?	How do we ensure our people have the agility to respond positively to change?
Institutions	How does this fit our core business?	Is this sustainable?	How do we retain our unique 'brand' of education?

work of Khan (Morrison & Khan, 2003) as discussed above, Maguire (2003) has identified an analysis grid for online teaching and learning that compares three main categories across five main stakeholders. This grid assists institutions to analyse issues and implications, as in Table 5. For each institution, the grid would be populated differently, according to local needs and contexts. This example table is populated with a typical response drawn from the case study school.

In terms of infrastructure, one of the fundamental questions facing schools that are planning the implementation of an e-learning environment is where to locate host servers and services. Some providers of e-learning solutions offer hosting services as part of their solution. In this instance, client school(s) are only obliged to provide Internet access to the provider's wide area network. This remote hosting model can offer cost-effective advantages to schools, as well as enabling multiple schools to share common resources. An alternative to this is the local hosting model, whereby individual schools buy a licence to operate the e-learning solution, and provide all hardware and network infrastructure themselves. This model gives the client school more control over the development and deployment of the solution, and can deliver cost savings under certain upload/download conditions.

Another major issue facing school decision makers is the degree of customisation required in an e-learning environment. As discussed earlier, suppliers of e-learning solutions fall into two main categories—those that provide an out-of-the-box solution requiring no customisation by the school (the buy option), and those that provide an open platform requiring heavy customisation by the school (the 'build' option). Both the build and the buy options have their positive and negative dimensions, depending on a wide variety of institutional factors. For instance, there has been a trend over the past few years for state-run schools to favour suppliers offering a buy solution, whereas independent schools have tended to favour platforms offering a build solution.

A third option gaining prominence in 2003 and 2004 takes the best of both options. Some institutions find that 'blending' an off-the-shelf framework with D-I-Y content offers a happy compromise. A number of vendors are happy to offer conversion/migration of existing intranet content, however significant issues need to be addressed to enable the "blend" approach to be successful. Other implementation questions facing schools are summarised in Table 6.

Schools considering the introduction of an e-learning solution need to reflect deeply, consult widely, and plan carefully using rigorous project methodologies in order to ensure a successful and effective implementation.

CONCLUSION

This chapter has looked closely at the emergence of e-learning environments in the K-12 education sector. A review of the literature has traced the development of separate e-learning components over the period from 1994 to 2004, and the parallel development of models used to describe, prescribe, and predict these phenomena.

An effective e-learning strategy brings together components that have evolved separately, and integrates them in a way that successfully supports learning. E-learning environments, when implemented well, can greatly assist staff and students to achieve ICT literacy. By integrating components into one environment, with a seamless and attractive interface, students and staff can become engaged, enthusiastic, and competent with a variety of information and communication technologies.

Reflecting on the case described in the third section of this chapter, it is possible to identify elements in common with models proposed by other commentators. Although individual components had been implemented separately as technologies emerged, the process of integration since 2002 demonstrates strong parallels with the e-learning components described by BECTA (2001a). Furthermore, the three-dimension support model suggested by Hitch and MacBrayne (2003) has clear synergies with the case school's emerging strategy (see Figure 3). That is, there is a close mapping between the school's policy and practice, and the model's three dimensions of faculty support, student support, and technological infrastructure.

Table 6. E-learning implementation questions for schools

Can your school sustain an e-learning environment financially, technically, and administratively?	Will introducing an e-learning environment add value to the teaching and learning process, over and above current systems?
Should your school look for intermediate solutions first, such as providing the content only?	Is it worth considering working in a consortium with other schools, to share costs and resources?
Penetration of Internet/e-mail access among target community	Extent of existing intranet/extranet services
IT competence of staff	Available bandwidth for upload and download
Cost/total cost of ownership	Reliability
Training requirements	Robustness
Service level agreements	User interface
Currency of content	Upgrade path
Maintenance	Security

Table 7. E-learning model concordance

Khan's Eight-Point Model	The Case Study School's Six-Point Plan
Institutional	ICT Policy (Strategic)
Ethical	ICT Policy (Strategic)
Management	Department Commitment (Tactical)
Technological	ICT Resources
Interface Design	ICT Resources
Resource Support	ICT Resources/Professional Development
Pedagogical	Student Learning
Evaluation	Appraisal and Review

Similarly, a close relationship can be perceived between the school's six-point Strategic ICT model and Khan's eight-point model of e-learning development. Table 7 demonstrates a concordance between the two, showing some overlap where some items of the school's plan incorporate more than one of Khan's dimensions.

Despite the parallels between theoretical models and the lived experience described in the case study, it is suggested that this is more the result of coincidence than design. School managers devise their e-learning strategies using the best information at hand at the time, but by necessity these strategies must be modified to accommodate a rapidly changing ICT landscape. In addition, the implementation of e-learning strategies requires significant changes to teaching practice (McLoughlin, 2000; Kilmurray, 2003; Weatherly & McDonald, 2003). A recent newspaper article notes:

With issues such as teaching effective new media literacy and critical analysis, achieving depth and breadth in students' e-research and ensuring fair assessment and equity, it is apparent there is not only a gap between the generations in e-learning, there

Figure 8. Relevant Web sites

Organisation	Web Site
The British Educational Communications and Technology Agency—BECTA	www.becta.org.uk
Blackboard (USA)	www.blackboard.com
Curriculum Online (UK)	www.curriculumonline.gov.uk
The Department for Education and Skills (UK)—DfES	www.dfes.gov.uk
Digitalbrain (UK)	www.digitalbrain.com
Granada LearnWise (UK)	www.learnwise.com
The Joint Information Systems Committee (UK)—JISC	www.jisc.ac.uk
Microsoft ClassServer (USA)	www.microsoft.com/education/ClassServer.aspx
Microsoft Sharepoint (USA)	www.microsoft.com/sharepoint
Myinternet (Australia)	www.myinternet.com.au
SchoolKit	www.schoolkit.com
StudentNet (Australia)	http://portals.studentnet.edu.au/studentnet/
The Learning Federation (Australia)	www.thelearningfederation.edu.au
WebCT (Canada)	www.webct.com
XSIQ (Australia)	www.xsiq.com.au

is also a gap between teachers and administrators. (Friedlander, 2004)

As some commentators have noted: "E-learning is by no means as simple to develop as it appears" (Bechervaise & Chomley, 2003). By the same token, the implementation of e-learning strategies can be an epiphany event for some educational institutions (Paoletti, 2003; McKay & Merrill, 2003). The roles of teachers and educational leaders are especially critical to the success of these innovations (Yee, 2000; Schiller, 2002; Webb & Downes, 2003).

Like many other ICT innovations, e-learning can be a mixed blessing. As these innovations mature and the requirements for achieving true ICT literacy evolve, school strategies will continue to be reactive rather than proactive, responding with agility and discernment to this volatile domain. One thing is certain: "The human act of teaching is more than the sum of its parts" (Dowling 2003).

REFERENCES

ACCAC. (1999). *Whole school approaches to developing ICT capability*. Cardiff: ACCAC.

Barnes, A., & Greer, R. (2002, July). Factors affecting successful R-12 learning communities in Web-based environments. *Proceedings of the Australian Computers in Education Conference* (ACEC2002).

Bechervaise, N. E., & Chomley, P. M. M. (2003, November). E-lusive learning: Innovation, forced change and reflexivity. *Proceedings of the E-Learning Conference on Design and Development.* Melbourne: RMIT.

BECTA (British Educational Communications and Technology Agency). (2001a). A review of the research literature on the use of managed learning environments and virtual learning environments in education, and a consideration of the implications for schools in the United Kingdom. Retrieved from http://www.becta.org.uk/page_documents/research/VLE_report.pdf

BECTA. (2001b). *Primary schools of the future—achieving today. A report to the DfEE by BECTA.* Coventry: British Educational Communications and Technology Agency.

BECTA. (2001c). *The secondary school of the future. A preliminary report to the DfEE by BECTA.* Coventry: British Educational Communications and Technology Agency.

Beshears, F. M. (2000). *Web-based learning management systems.* Retrieved from http://ist-socrates.berkeley.edu/~fmb/articles/web_based_lms.html

Boydell, S., & Kane, J. (2002, July). All aboard!—Creating a ubiquitous intranet. *Proceedings of the Australian Computers in Education Conference* (ACEC2002).

Cuban, L. (2000). *Oversold and underused: Computers in the classroom*. Cambridge, MA: Harvard University Press.

DfES. (2001). *ICT access and use: Report on the benchmark survey.* DfES Research Report No 252. London: Department for Education and Skills.

Donnelly, J. (2000). *Information management strategy for schools and local education authorities—Report on training needs.* Retrieved from http://dfes.gov.uk/ims/JDReportfinal.rtf

Dowling, C. (2003). The role of the human teacher in learning environments of the future. *Proceedings of the IFIP Working Groups 3.1 and 3.3 Working Conference: ICT and the Teacher of the Future,* Melbourne.

Friedlander, J. (2004). Cool to be wired for school. *Sydney Morning Herald,* (April 16).

Hitch, L. P., & MacBrayne, P. (2003). A model for effectively supporting e-learning. Retrieved from http://ts.mivu.org/default.asp?show=article&id=1016

Hruskocy, C., Cennamo, K. S., Ertmer, P. A., & Johnson, T. (2000). Creating a community of technology users: Students become technology experts for teachers and peers. *Journal of Technology and Teacher Education, 8,* 69-84.

JISC. (2001). *MLEs and VLEs explained.* Retrieved from http://www.jisc.ac.uk/index.cfm?name=mle_briefings_1

Jones, A. J. (2003). ICT and future teachers: Are we preparing for e-learning? *Proceedings of the IFIP Working Groups 3.1 and 3.3 Working Conference: ICT and the Teacher of the Future,* Melbourne.

Kennewell, S., Parkinson, J., & Tanner, H. (2000). *Developing the ICT-capable school.* London: Routledge Falmer.

Khan, B. H. (1997). *Web-based instruction.* Englewood Cliffs, NJ: Educational Technology Publications.

Kirkman, C. (2000). A model for the effective management of information and communications technology development in schools derived from six contrasting case studies. *Journal of IT for Teacher Education, 9*(1).

Kilmurray, J. (2003). *E-learning: It's more than automation.* Retrieved from http://ts.mivu.org/default.asp?show=article&id=1014

Lambert, M. J., & Nolan, C. J. P. (2003). Managing learning environments in schools: Developing ICT capable teachers. In I. Selwood, A. Fung, & C. O'Mahony (Eds.), *Management of education in the Information Age—The role of ICT.* London: Kluwer for IFIP.

Maguire, M. (2003, October). Questions for Web-based teaching and learning. *Proceedings of the Australian Catholic University School of Education Seminar Series* (unpublished).

McKay, E., & Merrill, M. D. (2003, November). Cognitive skill and Web-based educational systems. *Proceedings of the E-Learning Conference on Design and Development.* Melbourne: RMIT.

McLoughlin, C. (2000). Creating partnerships for generative learning and systematic change: Redefining academic roles and relationships in support of learning. *International Journal for Academic Development, 5*(2).

Morrison, J. L., & Khan, B. H. (2003). *The global e-learning framework: An interview with Badrul Khan.* Retrieved from http://ts.mivu.org/default.asp?show=article&id=1019

Mumtaz, S. (2000). Factors affecting teachers' use of information and communications technology: A review of the literature. *Journal of Information Technology for Teacher Education, 9*(3).

National Statistics. (2002). Retrieved from www.dfes.gov.uk/statistics/db/sbu/b0360/sb07-2002.pdf

Newton, L. (2003). Management and the use of ICT in subject teaching—Integration for learning. In I. Selwood, A. Fung, & C. O'Mahony (Eds.), *Management of education in the Information Age—the role of ICT.* London: Kluwer for IFIP.

NGfL. (2002a). *Impact2: The impact of information and communication technologies on pupil learning and attainment* (ICT in School Research and Evaluation Series—No 7). Annesley: DfES.

NGfL. (2002b). *Impact2: Learning at home and school: Case studies* (ICT in School Research and Evaluation Series—No 8). Annesley: DfES.

Ofsted. (2002). *ICT in schools, effect of government initiatives.* Retrieved from www.ofsted.gov.uk/public/docs01/ictreport.pdf

O'Mahony, C. D. (2000). *The evolution and evaluation of information systems in NSW secondary schools in the 1990s: The impact of values on information systems.* Unpublished PhD Thesis, Macquarie University, Australia.

O'Mahony, C. D. (2002). Managing ICT access and training for educators: A case study. *Proceedings of the Information Technology for Educational Management Conference* (ITEM2002), Helsinki.

Paoletti, J. B. (2003). *Wanted: Course revision without pain.* Retrieved from http://ts.mivu.org/default.asp?show=article&id=1034

Passey, D. (2002). *ICT and school management: A review of selected literature.* Unpublished Research Report, Department of Educational Research, Lancaster University, UK.

Research Machines PLC. (2000). *The RM G7 (8) Report 2000 comparing ICT provision in schools.* Abingdon: RMplc.

Russell, G., Finger, G., & Russell, N. (2000). Information technology skills of Australian teachers: Implications for teacher education. *Journal of IT for Teacher Education, 9*(2).

Schiller, J. (2002). Interventions by school leaders in effective implementation of information and communications technology: Perceptions of Australian principals. *Journal of Information Technology for Teacher Education, 11*(3).

Stevenson, R. (1997). Information and communications technology in UK schools: An independent inquiry. *The Stevenson Report.*

Watson, G. (2001). Models of information technology teacher professional development that engage with teachers' hearts and minds. *Journal of IT for Teacher Education, 10*(1-2).

Weatherly, G., & McDonald, R. (2003). *Where technology and course development meet.* Retrieved from http://ts.mivu.org/default.asp?show=article&id=951

Webb, I., & Downes, T. (2003). Raising the standards: ICT and the teacher of the future. *Proceedings of the IFIP Working Groups 3.1 and 3.3 Working Conference: ICT and the Teacher of the Future,* Melbourne.

Yee, D.L. (2000). Images of school principals' information and communications technology leadership. *Journal of IT for Teacher Education, 9*(3).

Chapter XXXIV
A Socio-Technical Analysis of Factors Affecting the Integration of ICT in Primary and Secondary Education

Charoula Angeli
University of Cyprus, Cyprus

Nicos Valanides
University of Cyprus, Cyprus

ABSTRACT

We live in a world that is constantly impacted by information and communication technology (ICT). ICT is considered an important catalyst and tool for inducing educational reforms and progressively extending and modifying the concept of literacy. With the extensive use of ICT in schools and everyday life, the term computer literate has already been established. Schools are open systems that interact with their environment, and the effective use and integration of technology is directly associated with the role of various socio-technical factors that may impact the integration of ICT in schools. In this chapter, we report on an exploratory study undertaken in Cyprus schools to examine the status of using ICT from the perspective of socio-technical systems. Specifically, teachers' knowledge of ICT, frequency of using ICT for personal purposes, frequency of using ICT for instructional purposes in different subject matters, attitudes toward ICT, self-confidence in using ICT in teaching and learning, and school climate were examined. The findings provide useful guidance to policymakers for planning, implementing, managing, and evaluating the integration of ICT in schools. Implications for the concept of computer literacy are discussed.

INTRODUCTION

Due to rapid technological advancements, we live in a world that is constantly impacted by information and communication technologies (UNESCO, 1999). Some key-markers that characterize differences between 19th-century societies (i.e., industrial-age societies) and 20th-century[1] societies (i.e., information age societies) are: (a) standardization vs. customization, (b) bureaucratic organizations vs. team-based organizations, (c) adversarial relationships vs. cooperative relationships, (d) parts oriented vs. process oriented, (e) compliance vs. initiative, and (f) conformity vs. diversity (Reigeluth & Garfinkle, 1994). By virtue of these differences, we are obliged to evaluate once more the worth of our existing educational systems. Are our current educational systems, with their emphasis on content coverage and teacher-centered classroom practices, conducive to preparing students to survive in a changing world that is steadily shaped by developments in information technologies? How do we prepare our future citizens to become computer or technology literate? Do new computer technologies herald the beginning of an era of broader literacy, and if we are educating children to be active citizens in an information society, what forms of literacy are required? What does it mean to be literate, an active reader, a writer, and a communicator of meaning in the information society?

Countries in North America, South America, Europe, Asia, and Africa have all identified a significant role for information and communication technology (ICT[2]) in improving education and reforming curricula for the purpose of preparing future citizens to be productive and actively involved in an information society (Kozma & Anderson, 2002; Pelgrum, 2001). ICT is considered by many not only to be the "backbone of the Information Society, but also to be an important catalyst and tool for inducing educational reforms that change our students into productive knowledge workers" (Pelgrum, 2001, p. 165). For these reasons, schools have made major investments and continue to invest heavily in increasing the number of computers in schools and the networking of classrooms.

ICT is thus steadily becoming part of classroom life, and it progressively changes the concept of literacy (Brindley, 2000; Watt, 1980). The traditional concept of literacy as the ability to read and write (Crystal, 1987x) is changing, and ICT opens up a further definition of literacy—one that goes beyond the acquisition of basic skills. Brindley (2000) argues that "schooled literacy, which traditionally sees the acquisition of the ability to construct and interpret text as largely an individual activity, bounded by the concept of text as linear and fixed, is no longer adequate" (p. 13). With the enduring introduction of computers in schools and the extensive use of ICT in our everyday life, the term computer literate has been established and flourished.

For many, being computer literate simply means acquiring technical expertise to be able to competently use computer software and hardware. In this chapter, we consider a much more complex and exciting concept of computer literacy—one that is directly associated with the affordances of ICT and the concept of visual literacy. "Visual literacy refers to the use of visuals for the purposes of communication, thinking, learning, constructing meaning, creative expression, [and] aesthetic enjoyment" (Baca, 1990, p. 65). Thus, the extensive use of multimedia in schools and everyday life opens up the way to an extended concept of literacy. For example, ICT reinvents the text and leads us to a new form of literacy, which encompasses a range of media by which students learn and communicate, such as graphics, video, and sound (Papert, 1993). Similarly, McFarlane

(2000) argues that multimedia allow students to record and present their own meaning using multiple media. Thus, the technology of multimedia does not restrict reading and writing to the mere coding and decoding of text. Using a computer, children can represent their creativity with text, graphics, speech, video, animation, and more. Technology offers us new forms of representation and expression that extend the traditional and limited concept of literacy. In the information society one has to be able to "read the text and the image and the moving image and the ability to secure understanding through reading the pictures as well as text in a rich and organic way" (Kempster, 2000, p. 25). For these reasons, Papert (1993) argues for an emerging model of literacy, which encompasses a range of media by which students learn to express themselves and communicate. In the emerging model of literacy, the ability to read and write are enhanced and sometimes replaced by images, graphics, and sound. These new forms of expression and communication can be used by students to construct meaning and represent their understandings after selecting the most appropriate form from multiple alternatives.

McFarlane (1997) also argues that with computers, literacy extends beyond simple manual encoding and decoding of text. "It involves the habit of viewing writing as a way of developing and communicating a child's thoughts. The use of word processors, for example, helps to present text as something to be experimented with, redrafted and developed as ideas develop, or the demands of purpose or audience change. It liberates the writer from the heavy burden of manual editing and presentation" (McFarlane, 1997, p. 119). Word processors are not only great productivity tools to write faster, or to make fewer mistakes, but also tools that fundamentally change the authoring process children are engaged in (Heppell, 2000).

Moreover, a literate person in the 21st century makes judgments about the quality and value of information. The Internet presents new challenges in the way text is presented, and requires a new set of reading skills that go beyond linear book print to screen print. The inclusion of graphics, hyperlinks, and bookmarks also requires an understanding of how text in a Web page is structured in a non-linear way. Thus, a computer-literate person should be able to access this non-linear text with speed and accuracy and construct understanding. Similarly, the National Grid for Learning in the United Kingdom (DfEE, 1998) refers to the new form of literacy as information or network literacy that is defined as the capacity to use electronic networks to access resources, create resources, and communicate with others. The use of e-mail, for example, to electronically communicate with others dramatically extends and alters students' written language in many ways. E-mail can rapidly expand the audience that children write for, and secondly, it can engage children in a different type of writing than the traditional one (Easingwood, 2000). It can also be used to support students' collaborative work with other students in different schools. The notion of children working collaboratively using the new technologies enables them to develop skills that are needed in the workplace and modern society. Thus, it is clear that computers have the potential to make new things possible in new ways, creating new forms of literacy that are critical to our present and future society.

It therefore becomes important to examine whether schools offer the education needed so that students develop the new forms of literacy that are so important for surviving in the information age. Several researchers (Eraut &

Hoyles, 1989; Snyder, 1994) argue that teachers need to be trained in order to utilize ICT appropriately in teaching and learning, and that computers by themselves cannot develop the new form of literacy. Pelgrum (2001) also suggests that ICT in education is an area that is in turmoil and in which many participants play a role. For example, forces operating in schools and in classrooms may be influential in bringing about changes or inhibiting them. Hence, it is important to regularly monitor the status of ICT in education in order to not only account for the financial investments, but also to inform policymakers regarding the content and direction of future policies.

Apple (1986) states that technology in the schools has usually been seen as an autonomous process. "It is set apart and viewed as if it had a life of its own, independent of social intentions, power, and privilege" (p. 105). Along the same line of reasoning, Street (1987) also argues that computer literacy erroneously rests on the assumption that ICT is a neutral tool that can be detached from other specific and social contexts. According to Kling (2000), the integration of ICT in the school system should be examined within a socio-technical framework, where people need to sufficiently interact with the technological tools within the system for the change to be effective. The term "socio-technical systems" was coined in the 1950s (e.g., Trist & Bamforth, 1951; Trist, 1982) to capture the interdependencies between the social and technical aspects of a system. Put simply, a socio-technical system is a mixture of people and technology, hence the concept of "socio-technical system" was established to stress the reciprocal interrelationship between humans and machines. More importantly, the socio-technical systems approach provides us with a comprehensive and systemic methodology for holistically examining factors that may impact the integration of ICT in elementary and secondary education.

In this chapter, we present an exploratory study undertaken in Cyprus schools to examine the current status of using ICT in primary and secondary education, and discuss implications for the development of the new form of literacy given the current status of ICT in Cyprus schools.

THEORETICAL FRAMEWORK

Schools, like other organizations, are open systems that continuously interact with their environment (Getzels & Cuba, 1957; Hanna, 1997; Hoy & Miskel, 2001). The interaction of a system with its external environment is vitally important, because as a result of this interaction, the system receives feedback and appropriately adapts to new demands and circumstances. A system that does not adapt to the needs of its external environment will gradually become extinct (Hoy & Miskel, 2001).

Thus, the introduction of ICT in the school system has created a need for extending the theory of social systems into a theory of socio-technical systems such that the interaction between teachers and technology can be examined and understood. Social informatics is an area of research, which systematically examines the design, uses, and consequences of technology, taking into consideration the context of the organization, the people who work within the organization, and the interactions between people and technology (Denning, 2001; Friedman, 1998; Kling, 2000). One key idea of social informatics is that ICT, in practice, is socially shaped and the uses of technology in an organization are contingent upon several social and technical dependencies. The concept of socio-technical networks or systems is used to describe the interdependencies between technology and people, and to explain that the culture of an organization and people's beliefs, attitudes, and feelings play an important role in shaping the organization's mood and determining the effectiveness of the integration of technology in the organization (Kling & Lamb, 2000; Kling, 2000; Markus & Benjamin, 1987). Thus, the effective use of technology in different organizational settings is directly associated with the intertwining of technical and social elements (Friedman, 1998; Heracleous & Barrett, 2001; Kling, 2000).

Trach and Woodman (1994) support that the socio-technical model constitutes a flexible model for successfully implementing systemic changes in an organization. Thus, technology should be viewed as a catalyst for pursuing systemic as opposed to piecemeal changes across the different subsystems of an educational system (Angeli, 2003; Valanides & Angeli, 2002). Teachers, for example, constitute an important subsystem of every educational system. According to Fullan (1991), every reform effort should take into consideration the knowledge, skills, beliefs, and attitudes of the people who will implement the changes. In general, a precondition for a successful implementation of any change effort is adaptation, which includes all adjustments an organization makes in order to realize the changes (Hoy & Ferguson, 1985). Adaptation is a broad term and may include multiple criteria. In this chapter, six areas of adaptation are examined: (1) teachers' knowledge of ICT, (2) teachers' frequency of using ICT for personal purposes, (3) teachers' frequency of using ICT for instructional purposes in different subject matters, (4) teachers' attitudes toward ICT, (5) teachers' self-confidence in using ICT in teaching and learning, and (6) school climate.

According to Huberman and Miles (1984), many educational reform efforts died out because teachers were not supported in their change efforts and thus never accepted or understood the changes they had to implement. Therefore, the adoption of changes requires educating teachers to understand and accept the nature of the restructuring effort, and develop the knowledge, skills, and attitudes that are required for bringing about the change in their classrooms (Fullan, 1991; Louis & Miles, 1990). As Barth (1990) characteristically stated, nothing influences students more than their teachers' own professional and personal development. The process of integrating ICT in teaching and learning is demanding and requires teachers' continuous professional development (Picciano, 2002). In addition, Picciano (2002) believes that school principals and inspectors also need to participate in ICT training so that they understand how ICT integration affects the classroom (micro level) and the school (macro level). Fullan (1991) explains that educational change efforts often create feelings of uncertainty that are not only related to lack of knowledge and skills, but also confidence, as teachers often feel inadequate and uncertain about their new roles (Fullan & Hargreaves, 1992). For these reasons, teachers' ICT professional development is important not only for the development of ICT knowledge and skills, but also for the development of positive attitudes and confidence in using ICT in teaching and learning. Hoy and Miskel (2001) point out that the style of leadership in an organization is also an important factor. For example, a principal who encourages the use of ICT in teaching and allows teachers to create collaborations within the school and between schools for the exchange of ideas will play an important role in successfully institutionalizing the change effort. Thus, as Hoy and Miskel (2001) state, what is needed is transformational leadership. Transformational leaders are those who foresee the need for change, create new visions for education, and encourage teachers to take responsibility for their professional and personal development in order to successfully fulfill their new obligations and roles.

METHODOLOGY

The Context of the Study

The public educational system in Cyprus consists of the primary and secondary levels, while just recently new attempts have been made to also include pre-primary (3-5½-year-old children) education. Grades 1 to 6 constitute the primary level, and grades 7 to 12 the secondary level. Education is free for all grade levels and mandatory until grade 9 or the age of 16, but an overwhelming 95% of students complete all grade levels. The majority of students attend public schools, and there are only a small number of private schools mainly at the secondary level. Fifty teachers were randomly selected from each one of the 12-grade levels in primary and secondary public education. During the spring semester of 2004, a questionnaire and a pre-stamped self-addressed envelope were delivered to each individual teacher with the help of research assistants. Each teacher was asked to individually complete the questionnaire and return it to the researchers. The majority of the teachers (520) returned their completed questionnaires within a week, and only 22 teachers did not return their questionnaires at all, even after a second reminder by telephone two weeks after the questionnaires were delivered to them. Thus, the data from 578 questionnaires were used in the study.

Data Collection

The questionnaire consisted of seven parts. The first part collected demographic data related to teachers' age, number of computer labs in each school, number of computers in each lab and teachers' classrooms, teachers' ownership of a personal computer, and teachers' participation in an ICT professional development training program. The other six parts collected data related to: (1) teachers' knowledge of computer software, (2) teachers' frequency of software use for personal purposes, (3) teachers' attitudes towards integrating ICT in teaching and learning, (4) teachers' self-confidence in integrating ICT, (5) teachers' frequency of using ICT for instructional purposes in the classroom, and (6) school climate and support.

Specifically, the second part of the questionnaire used a Likert-type scale from 1 to 5 (I do not know how to use it, I somewhat know how to use it, I know how to use it satisfactorily, I know how to use it well, I know how to use it very well) to measure teachers' knowledge of various software, and the third part used a Likert-type scale from 1 to 5 (never, once or twice every three months, once or twice a month, once or twice a week, almost every day) to measure frequency of software use for personal purposes. Similarly, the fourth part measured teachers' attitudes with a Likert-type scale from 1 to 5 (absolutely disagree, disagree, neither disagree nor agree, agree, absolutely agree), and the fifth part used the same Likert-type scale that was used in the fourth part to measure teachers' self-confidence. For the sixth part, which measured teachers' frequency of using various computer programs in classroom practices, teachers had to write how many times a week they were using different software in their teaching. Finally, the seventh part of the questionnaire measured school climate and support. A Likert-type scale from 1 to 5 (absolutely disagree, disagree, neither disagree nor agree, agree, absolutely agree) was also used. Thus, the questionnaire collected demographic data and data related to the six areas of adaptation from the perspective of the socio-technical systems perspective.

RESULTS

Demographic Data

Among the 578 teachers, 446 (77.14%) of them were females and 132 (22.86%) were males. The average age of the participating teachers was 31.98 (SD = 8.107), but the average age was significantly smaller (t = -5.501, p = .000) for female teachers (*Mean* = 30.94 years) than for male teachers (*Mean* = 35.24 years). All teachers owned their own personal home computer (*Mean* = 1.05, SD = 1.05), and in some cases, there were teachers who owned two computers at home. Also, 67.5% of teachers recently participated in an ICT teacher professional training program where they learned how to use several computer programs, such as Word, Excel, PowerPoint, and the Internet. Teachers taught in schools where there was a computer lab with an average of 5.55 computers in each lab and at most one computer in each classroom. The computer-student ratio differed for each school, but in all schools computer access was not prolific.

These data signify the tremendous effort that has been undertaken for successfully integrating ICT in the teaching-learning environment, considering the fact that not too long ago there were no computers in the classrooms and no computer labs in most of the schools in Cyprus. It is however more important to investigate how and to what extent ICT is used in Cyprus schools in relation to teachers' ICT knowledge, frequency of software use for personal purposes, attitudes, self-confidence, frequency of using ICT for instructional purposes in the classroom, and various factors pertaining to the socio-technical character of the school system.

Teachers' Knowledge of Computer Software and Frequency of Use for Personal Purposes

Descriptive statistics related to teachers' knowledge of computer software and teachers' frequency of using software for personal purposes are presented in Tables 1 and 2, respectively.

The results in Table 1 indicate that teachers' knowledge of software varied according to the type of software. Specifically, teachers appeared to be more familiar with Word than with any of the other software such as PowerPoint, the Internet, e-mail, Excel, graphics, authoring software, and databases. More analytically, the results in Table 1 indicate that

Table 1. Descriptive statistics of teachers' knowledge of software

Software	M	SD	n
Word Processing (i.e., Word)	4.17	1.11	577
Databases (i.e., Access)	2.01	1.16	569
Spreadsheets (i.e., Excel)	2.76	1.33	572
Graphics (i.e., Paint)	2.68	1.34	570
Presentation (i.e., PowerPoint)	3.34	1.43	575
Authoring Software (i.e., Hyperstudio)	2.17	1.35	576
Internet	3.28	1.53	570
E-Mail	3.19	1.60	575
Knowledge of Software	**2.95**	**1.03**	**560**

teachers knew how to use well only Word, while their knowledge about PowerPoint, the Internet, and e-mail was rated just above the level of satisfactory use. Teachers' knowledge regarding the rest of the software was rather poor and below the level of satisfactory use. It is also important to mention that these data refer to teachers' self-reported estimates of their knowledge and do not necessarily represent their actual knowledge.

The results in Table 2 indicate that teachers used Word most frequently and rarely used databases. The Internet, e-mail, educational CD-ROMs, and PowerPoint were less frequently used than Word, but more frequently used than graphics, spreadsheets, authoring software, and databases. For example, teachers used Word once or twice a week, while they almost never used databases or authoring software. Thus, the collective results from Tables 1 and 2 clearly indicate that teachers' knowledge of computer software was dependent upon the type of software, and that their frequency of software use followed almost the same order, in terms of magnitude, as their knowledge of computer software. The means of the variables related to knowledge of software and frequency of software use were found to be highly and significantly correlated ($r = .905, p = .01$). Correlation, of course, does not mean a direct causal relation, and it cannot explain whether better knowledge of a computer program causes its frequent use, or whether the need to frequently use a computer program causes better knowledge of it. The only valid conclusion is that the participants reported better knowledge for some kinds of software and higher frequency of use for the same kinds of software.

Teachers' Attitudes

Dealing effectively with ICT relates not only to knowledge of ICT tools, but also to individuals' attitudes and perceptions regarding ICT tools. Attitudes and perceptions act as a filter through which all learning occurs (Marzano, 1992), and are considered as a constituent part of learners' "self-esteem" that oversees all other systems (Markus & Ruvulo, 1992). Thus, learners continually filter their behaviors through their self-belief system to the extent that they even attempt to modify the "outside world" and make it more consistent with the "inside world" (Glaser, 1981). The limited teachers' knowledge or skills about the use of several software and computer applications seem to have an impact on their attitudes and concerns.

For these reasons, teachers' attitudes towards ICT were also examined in this study.

Table 2. Descriptive statistics of teachers' frequency of software use for personal purposes

Software	M	SD	n
Word Processing (i.e., Word)	4.11	1.16	576
Databases (i.e., Access)	1.03	.44	569
Spreadsheets (i.e., Excel)	1.92	1.08	576
Graphics (i.e., Paint)	2.07	1.06	576
Presentation (i.e., PowerPoint)	2.15	1.06	576
Authoring Software (i.e., Hyperstudio)	1.28	.64	576
Internet	3.70	1.38	574
E-Mail	3.20	1.53	573
Educational CD-ROMs	2.55	1.24	572
Frequency of PCU	**2.95**	**1.03**	**560**

Table 3. Descriptive statistics of teachers' attitudes

Item	M	SD	n
I feel comfortable with the computer as a tool in teaching and learning.	3.78	1.07	577
The use of computers makes me stressful.	3.84	1.06	575
If something goes wrong with the computer, I will not know what to do.	2.63	1.12	576
The use of computers in teaching and learning makes me skeptical.	2.70	1.13	576
The use of computers in teaching and learning makes me enthused.	3.78	.83	574
The use of computers in teaching and learning interests me.	3.99	.93	575
The use of computers in teaching and learning scares me.	2.28	1.14	572
I believe the computer is a useful tool for my profession.	4.33	.81	575
The use of computers in teaching and learning will mean more work for me.	3.61	1.00	574
Computers will change the way I teach.	3.77	.83	575
Computers will change the way my students learn.	3.74	.88	571
Whatever the computer can do, I can do it equally well in another way.	2.50	.90	572
Computers make learning harder because they are not easy in their use.	2.16	1.71	575
Computers make learning harder because often times there are technical problems associated with them that students cannot resolve.	2.67	1.07	574
The computer supports and enhances student learning.	4.05	.72	574
The computer makes learning more meaningful.	3.86	.75	576
The computer helps students represent their thinking better.	3.85	.74	577
The computer is a meaningful tool for the teacher because it can help him/her teach a topic more effectively.	3.92	.79	575
The computer hinders teaching because of the technical problems it may cause.	3.62	.92	576
ICT Attitudes	**3.36**	**.26**	**545**

Table 3 shows descriptive statistics related to the 19 items measuring teachers' attitudes towards the use of ICT in education. The results in Table 3 indicate that the majority of teachers expressed rather positive attitudes towards the use of ICT tools in education. Teachers felt rather comfortable in using ICT for instructional purposes, and expressed positive attitudes towards applying ICT in teaching and learning, because ICT could make learning easier, meaningful, and useful. However, there were a lot of teachers who expressed skepticism or even fear, because they felt incompetent to resolve potential technical problems with the computer. In general, teachers expressed a somewhat overall positive attitude towards the use of computers in education, although some of them also expressed concerns pertaining to technical problems that might hinder their work and students' learning. Of course, even though ICT-related attitudes seem to play an important role in how ICT is used in teaching and learning (Levine & Donitsa-Schmidt, 1998), research indicates that positive attitudes alone are not always good indicators of teachers' eventual use of ICT in the classroom (Wild, 1996). This is due to the fact that teachers often times have positive attitudes about ICT integration without realizing how difficult the task is, or how much effort they need to invest to successfully complete the task. Thus, despite teachers' rather positive disposition towards ICT integration, they may still find the task of integrating computers in the classroom difficult, once they realize what it really entails. This also seems to be a reasonable conclusion from the current results, taking into consideration that the participants of the present study had limited knowledge of the full range of affordances of several ICT tools and

Table 4. Descriptive statistics of teachers' confidence

Item	M	SD	n
I feel confident in selecting appropriate software to use in my teaching.	3.51	1.12	572
I feel confident in preparing classroom activities with ICT for my students.	3.30	1.20	569
Confidence	**3.40**	**1.04**	**568**

their applications in the teaching-learning environment.

Teachers' Confidence

Table 4 shows descriptive statistics related to teachers' confidence. The results indicate that teachers felt somewhat confident in selecting appropriate software to be used in their teaching, and felt about the same with designing and implementing classroom activities with ICT tools. Several factors seem to play an important role in affecting how individuals use ICT (Fullan, 1991). These factors include not only ICT knowledge and the amount and nature of prior ICT experience, but also ICT-related attitudes and learners' beliefs in their ability to work successfully with ICT tools (self-confidence or self-efficacy) (Levine & Donitsa-Schmidt, 1998; Liaw, 2002; Murphy, Coover, & Owen, 1989). Attitudes and beliefs are considered as predictors of behaviors and behavioral intentions that are linked to self-confidence. Beliefs about an object usually lead to attitudes towards it, and in turn, attitudes lead to behavioral intentions regarding the object, which affect actual behaviors towards the object. Finally, there is a feedback loop where behavioral experience modifies preexisting beliefs about the object. In terms of ICT use, attitudes toward ICT affect users' intentions or desire to use ICT. Intentions in turn affect actual ICT usage or experience, which modifies beliefs and consequent behaviors or behavioral intentions (future desire), and self-confidence or self-efficacy in employing ICT in learning. Thus, teachers' actual ICT usage in the classroom is directly associated with their knowledge, attitudes, and self-confidence, although attitudes and confidence are directly dependent on knowledge and improve with success and frequent use. It seems that teachers' self-confidence was delimited by teachers' knowledge of software and frequency of software use.

Teachers' Frequency of Using ICT in the Classroom for Instructional Uses

Table 5 shows descriptive statistics of teachers' frequency of using ICT for instructional purposes in the classroom. The results in Table 5 draw a rather pessimistic picture in terms of actual instructional use of ICT in the class-

Table 5. Descriptive statistics of teachers' frequency of using ICT in the classroom

Software	M	SD	n
Internet	1.233	2.46	578
Word Processing (i.e., Word)	1.606	2.78	578
Spreadsheets (i.e., Excel)	.309	.97	578
Databases (i.e., Access)	.031	.21	578
Presentation (i.e., PowerPoint)	.711	1.45	578
Educational CD-ROMs (e.g., drill and practice, tutorials, etc.)	1.524	2.75	578
Frequency of Instructional Use	**5.413**	**8.22**	**576**

room. First, none of the teachers reported any use of electronic communication (i.e., E-mail), authoring software (i.e., Hyperstudio), or graphics (i.e., Paint) in their teaching. Second, computer applications such as spreadsheets, databases, and PowerPoint were minimally used, and teachers' reported mean frequencies of use for these software were .309, .031, and .711, respectively. Third, only the mean frequencies of use for the Internet, Word, and educational CD-ROMs had values higher than one indicating that they were used infrequently and only by very few teachers. Finally, in comparison with the results in Table 2, teachers were using the same software much less frequently for instructional purposes than for personal purposes. This seems to suggest that teachers' knowledge, attitudes, and self-confidence had probably less impact on teachers' instructional use of the software or that there were some other reasons inhibiting the instructional uses of ICT in teaching and learning. For example, the existing socio-technical character of a school could substantially constitute a significant factor in supporting or inhibiting both use and frequency of use of certain software, despite teachers' technical expertise, attitudes, and self-confidence in employing ICT in teaching and learning.

Socio-Technical Environment

Table 6 shows descriptive statistics of various factors related to the socio-technical character of a school. Based on the results, there were participants who felt that their superiors, the computer coordinator, and other colleagues tended to encourage them to use ICT in the classroom, but there were also other participants who did not share the same point of view. Also, teachers in general neither agreed nor disagreed about the availability of technical or instructional support in their school, or whether there was adequate computer equipment or software available. Lastly, teachers expressed mixed views on whether the subject of ICT integration was sufficiently discussed in faculty meetings. The only valid conclusion from the results in Table 6 is that at least teachers did not feel discouraged for their attempts to integrate ICT tools in their classrooms. They perceived a rather neutral socio-technical environment in their schools. It is also possible that teachers did not have a clear understanding of the situation in their school, and thus were unsure about it. Obviously, there was not a strong momentum, nor systematic plan of action for effectively integrating ICT in the participants' schools. Teachers' somewhat positive attitudes

Table 6. Descriptive statistics of socio-technical factors

Item	M	SD	n
There are teachers in my school who help me integrate ICT in my teaching.	3.12	1.120	564
The computer coordinator encourages me to use ICT in my classroom.	3.38	1.171	565
The principal encourages me to use ICT in my classroom.	3.15	1.083	566
The inspector encourages me to use ICT in my classroom.	3.43	1.087	565
We often talk about ICT integration during our faculty meetings.	2.92	1.109	567
There are many software available in my school.	3.04	1.072	566
There is technical support readily available in my school.	3.09	1.086	565
There is ICT instructional support readily available in my school.	2.89	1.080	566
There is adequate computer equipment in my school.	3.04	1.223	566

and perceived self-confidence were rather compatible with the existing socio-technical environment, which, as the results indicated, was not overwhelmingly supportive, but rather neutral.

Personal Computer Use and Instructional Computer Use

The fact that teachers reported rather infrequent instructional use or no instructional use for most of the software is contradictory to teachers' frequency of using computer software for personal purposes, as well as contradictory to their subjective self-confidence, attitudes, reported knowledge of several software, and the socio-technical environment.

In order to better examine the existing discrepancy between the frequency of ICT use for personal purposes and frequency of ICT use for instructional purposes, five composite variables were created, namely, teachers' knowledge (KNOW), frequency of personal computer use (PCU), attitudes (ATT), self-confidence (SCF), and frequency of instructional computer use (ICU). These five variables represented the mean value of the single items in Tables 1, 2, 3, and 4, respectively, with the exception of ICU, which represented the sum of the individual items in Table 5. Two regression analyses were consequently conducted with the frequencies of PCU and ICU as the dependent variables for the first and the second analyses, respectively. The independent variables for both analyses were the other three composite variables (KNOW, ATT, and SCF), teachers' age, participation in an ICT professional development training program, and the nine individual items shown in Table 6 measuring aspects of the socio-technical environment (STE_1 to STE_9). Other variables from the first part of the questionnaire were excluded from the analyses, because they were found not to be discriminating nor redundant, as they were highly and significantly correlated with other variables. For example, *years of teaching experience* was considered to be a redundant variable, because it was highly and significantly correlated with age ($r = .960, p = .01$). *Ownership of a personal computer* was also not a discriminating variable, because all teachers owned a personal computer, with the exception of some of them who reported that they owned two personal computers. Table 7 shows the correlations between all possible variables (dependent and independent) that were used in the two regression analyses and some additional variables from the first part of the questionnaire.

Regarding the items in the first part of the questionnaire (demographic information), it was considered more appropriate to use them as individual variables, since they could not be considered dimensions of the same construct, as it was, for example, the case with teachers' self-confidence or attitudes. Regarding the items in the last part of the questionnaire, although they were measuring aspects of the socio-technical environment, we used them as individual variables, because we were interested in identifying which dimensions of the socio-technical environment seemed to play an important role in ICT integration. In most cases, the guiding principle for including or excluding a variable in the regression analyses was their overlapping meaning and high significant (positive or negative) correlation with other variables. Table 7 shows the correlations between all possible pairs of dependent and independent variables, as well as some additional variables that were considered important to further clarify and interpret the results of the regression analyses.

Table 8 displays the results of the first stepwise multiple regression analysis with the frequency of PCU as the dependent variable

Table 7. Correlations between all possible pairs of criterion, predictor, and selected demographic variables

Variable	1	2	3	4	5	6	7	8
Age (1)	1.00							
ICT Inservice (2)	-.190**	1.00						
KNOW (3)	-.401**	-.075	1.00					
PCU (4)	-.274**	-.077	.772**	1.00				
ICU (5)	-.046	-.087*	.283**	.354**	1.00			
SCF (6)	-.201**	-.134**	.597**	.592**	.303**	1.00		
ATT (7)	-.185**	-.056	.537**	.527**	.309**	.592**	1.00	
STE_1 (8)	-.015	-.046	.085*	.128**	.202**	.220**	.136**	1.00
STE_2 (9)	-.003	-.069	.166*	.195**	.202**	.321**	.232**	.548**
STE_3 (10)	-.019	-.086*	.191*	.242**	.211**	.325**	.273**	.517**
STE_4 (11)	.005	-.046	.160*	.195**	.191**	.285**	.238**	.420**
STE_5 (12)	.031	-.094*	.036	.044	.048	.168**	.036	.294**
STE_6 (13)	-.079	-.104*	.094*	.068	.110**	.185**	.119*	.275**
STE_7 (14)	.003	-061	-.008	-.023	.074	.149**	.052	.273**
STE_8 (15)	.016	-.104*	.042	.050	211**	.188**	.059	.343**
STE_9 (16)	.069	-.111*	.077	.035	.057	.173**	.108*	.236**
Gender (17)	.225**	-.034	.132**	.207**	.118**	.082	.171**	-.008
TE (18)	.960**	-.200**	-.407**	-.290**	-.057	-.217**	-.205**	-.020

Variable	9	10	11	12	13	14	15	16	17
Age (1)									
ICT Inservice (2)									
KNOW (3)									
PCU (4)									
ICU (5)									
SCF (6)									
ATT (7)									
STE_1 (8)									
STE_2 (9)	1.00								
STE_3 (10)	.575**	1.00							
STE_4 (11)	.452**	.656**	1.00						
STE_5 (12)	.282**	.440**	.400**	1.00					
STE_6 (13)	.353**	.361**	.267**	.302**	1.00				
STE_7 (14)	.375**	.267**	.252**	.278**	.438**	1.00			
STE_8 (15)	.436**	.375**	.316**	.354**	.363**	.560**	1.00		
STE_9 (16)	.310**	.321**	.210**	.248**	.437**	.448**	.384**	1.00	
Gender (17)	.108*	.044	.039	-.0119	.003	.003	.044	.102*	1.00
TE (18)	-.024	-.002	.002	.036	-.078	-.001	.022	.047	.146*

Note: SCF = teachers' self-confidence, KNOW = teachers' knowledge, ATT = teachers' attitudes, STE_1 to STE_9 = individual items of the socio-technical environment corresponding to the items in Table 6, TE = years of teaching experience, ICT inservice = participation in ICT professional training, Gender = 1 for females and 2 for males.

and the independent variables that were determined as significant predictors of the dependent variable. The independent variables that contributed significantly to the prediction of frequency of PCU were teachers' knowledge, self-confidence, and gender, and from the socio-technical factors those related to inspector support and computer equipment in the school. Teachers' knowledge was found to be the best predictor of teachers' frequency of PCU and alone explained 59% of the variance.

Teachers' self-confidence was found to be the second best predictor that contributed to a significant increment in R^2, from .590 to .617. There were three other variables, namely gender, inspector support, and computer equipment that also contributed significantly to the prediction of teachers' frequency of PCU, but the amount of variance in frequency of PCU attributable to each one of them was much smaller, namely, .9%, .4%, and .6%, respectively. Although the correlations between teachers' frequency of PCU and some other variables (i.e., age, ICT inservice, and the remaining socio-technical factors) were significant, these variables were not found to be significant predictors of PCU.

For the purpose of further clarifying and interpreting the results of the first regression analysis, a careful examination of the pair-wise relationships between teachers' age, gender, knowledge, and ICT inservice training in Table 7 indicates that these variables were highly and significantly correlated, but only teachers' knowledge and gender proved to be significant predictors of the frequency of PCU. This can be explained by the fact that the younger teachers tended to have more knowledge than the older ones ($r = -.274, p = .01$), female teachers were in general younger than male teachers ($r = .225$, $p = .01$), and younger teachers were more inclined to participate in inservice ICT training ($r = -.190, p = .01$). Similarly, as shown in Table 7, all items corresponding to the socio-technical factors were highly and significantly correlated among each other, but not all of them were found to be significant predictors, as the unique contribution of many of them was not found to be significant. In multiple regression:

> *It is possible for a variable to appear unimportant in the solution when it actually is highly correlated with the dependent variable. If the area of that correlation is whittled away by other independent variables, the unique contribution of the independent variable is often very small despite a substantial correlation with the dependent variable.* (Tabachnick & Fidell, 1989, p. 143)

Table 8. Multiple regression analysis of factors predicting teachers' frequency of personal computer use

Model	Variables	R	R^2	Adjusted R^2	Adjusted ΔR^2	F Change	Significance
1	Knowledge	.768	.590	.590	.590	693.205	.000
2	Confidence	.786	.617	.616	.026	33.594	.000
3	Gender	.792	.628	.625	.009	13.349	.000
4	Inspector Support	.795	.632	.629	.004	5.568	.019
5	Computer Equipment	.799	.635	.635	.006	9.361	.002

Table 9. Multiple regression analysis of factors predicting teachers' frequency of instructional use

Model	Variables	R	R^2	Adjusted R^2	ΔR^2	F Change	Significance
1	Self-Confidence	.301	.091	.089	.089	48.778	.000
2	Instructional Support	.342	.111	.113	.024	14.358	.000
3	Attitudes	.378	.143	.138	.025	14.785	.000
4	Colleagues	.391	.152	.146	.008	5.448	.020
5	Knowledge	.404	.163	.154	.008	6.125	.014
6	**Self-Confidence (removed)**	**.400**	**.160**	**.153**	**-.003**	**1.761**	**.185**
7	Age	.410	.168	.160	.008	4.743	.030

Thus, from the list of the socio-technical factors, only inspector support and availability of computer equipment in the school were found to be significant predictors of the frequency of PCU. Interestingly, computer equipment was found to be an important predictor of PCU, although there was not a significant positive correlation between the two variables. This outcome is really difficult to explain, because computer equipment in the schools does not seem to be directly related to teachers' frequency of PCU outside the classroom. One possible explanation is that teachers began to use computers for personal purposes after the introduction of computers in the schools, which possibly served as the impetus for teachers to learn how to use computers.

Table 9 displays the results of the second stepwise multiple regression analysis between frequency of ICU as the dependent variable and the significant independent variables. The independent variables that contributed significantly to the prediction of the frequency of ICU were teachers' self-confidence, instructional support in the school, teachers' attitudes, support from colleagues, and knowledge. Teachers' self-confidence was the best predictor of the frequency of ICU, but it could explain only 9.1% of the variance. The variables of instructional support, teachers' attitudes, support from other colleagues, and teachers' knowledge also contributed to a significant increment in R^2. These variables could explain 2.4%, 2.5%, 0.8%, and 0.8% of the variance in the frequency of ICU, respectively. The total amount of variance attributable to these variables was only 16.8%.

Interestingly enough, the significant predictor of self-confidence was removed from the regression equation in step six, as its unique contribution was no longer significant. When, in step 7, age was introduced, there was a significant increase of ΔR^2 that made the total amount of variance, after partialling out the contribution of teachers' self-confidence, significantly higher. In stepwise regression, "independent variables are added one at a time if they meet statistical criteria, but they also may be deleted at any step where they no longer contribute significantly to prediction" (Tabachnick & Fidell, 1989, p. 147). Thus, the total amount of variance attributable to instructional support, teachers' attitudes, and support from colleagues, knowledge, and age, was found to be 16.8%. Teachers' self-confidence that was initially found to be the best predictor of ICU was in the end excluded from the list of significant predictors for two reasons. First, teachers' self-confidence, after the inclusion of four other variables (instructional support, attitudes, support from colleagues, and knowledge), did not contribute significantly to the prediction of the frequency of ICU and could be excluded without any significant decrease in R^2. Second, the combination of teachers' knowledge and age was a better predictor of the frequency of ICU than teachers' self-confidence.

DISCUSSION AND IMPLICATIONS

The socio-technical systems model provides us with a framework to systematically identify factors that could possibly affect the integration of ICT in education. From this perspective, a questionnaire was used in this study to collect demographic data and information related to teachers' knowledge of ICT, frequency of using ICT for personal and instructional purposes, attitudes toward ICT, self-confidence in using ICT in teaching and learning, and school climate. The findings tend to support that female teachers were more inclined to participate in ICT inservice training; had better knowledge, attitudes and self-confidence related to ICT; and used ICT tools more frequently both for personal and instructional uses than male teachers. Teachers' knowledge of computer software and frequency of use for personal and instructional purposes were dependent on the type of software. For example, teachers' knowledge and frequency of use for personal purposes was mainly restricted to word processing, and to a much smaller extent to the Internet, e-mail, and educational CD-ROMs, while other software were almost unknown and rarely used. Teachers expressed somewhat positive attitudes towards the use of computers in education, but they also expressed concerns pertaining to technical computer problems that might hinder their work and students' learning. They also felt, to some extent, self-confident in selecting appropriate software to be used in their teaching, and somewhat confident in designing and implementing classroom activities with ICT tools. Teachers' attitudes and self-confidence seem to be delimited by their restricted knowledge of software and frequency of software use for personal and instructional purposes.

Teachers also reported infrequent instructional use or no instructional use even for software that they frequently used for personal purposes outside the classroom. This discrepancy seems to be attributable to the rather neutral socio-technical environment that existed in their schools, but also to other factors. Specifically, a stepwise regression analysis indicated that 63.5% of the variance in teachers' frequency of personal computer use could be predicted by teachers' knowledge, self-confidence, gender, inspector support, and computer equipment. However, according to a second stepwise regression analysis, only 16.8% of teachers' frequency of instructional computer use could be predicted by instructional support from officials, teachers' attitudes, support from the other teachers in the school, knowledge, and age.

The findings indicate that teachers in Cyprus are not illiterate in terms of having ICT skills and in terms of using ICT for personal purposes. What the findings clearly show is that teachers do not feel empowered to actively use ICT in authentic teaching and learning activities. Teachers need to develop confidence in their own professional activities and realize that what they are doing is right and important for their students' education. As Gable and Easingwood (2000) state, it will take time to train teachers to fully appreciate the power of ICT, but it is crucial to invest in such efforts, so that teachers fully appreciate the philosophical aspects of what they are doing, rather than just learning how to use the computer. These results indicate that policymakers in Cyprus have to seriously consider the lack of learning opportunities for the development of new literacy in Cyprus schools, and make coordinated efforts for providing a different and better kind of training to teachers. This training should pay attention not only to teachers' technical expertise, but also their attitudes, self-confidence, and in-depth understanding of ICT's affordances and added value in teaching and learning targeting an extended concept of literacy. Teacher professional development about the instructional uses of ICT in the classroom and about computers as learning tools for providing us with new forms of media that can enrich learner communication and expression is absolutely in great need.

Along the same line of reasoning, teacher education departments must also consider the quality of their curricula and adapt them appropriately, so that they adequately prepare teachers to integrate ICT in teaching and learning. We argue that teacher preparation, inservice or preservice, should focus on new interactive computer-based technologies, such as electronic communication systems, visualization and dynamic systems modeling tools, simulations, and networked multimedia environments, for scaffolding and amplifying students' thinking (Bransford, Brown, & Coccking, 2001). These tools are known as cognitive tools or mindtools (Jonassen, 2000), because they engage learners in meaningful thinking to analyze, critically think about the content they are studying, and organize and represent what they know. Jonassen, Carr, and Yueh (1998) state that "learning with mindtools depends on the mindful engagement of learners in the tasks afforded by these tools and that there is the possibility of qualitatively upgrading the performance of the joint system of learner plus technology" (p. 40). Therefore, mindtools require learners to think harder about the content being studied, and engage them in thinking that would be impossible without the tools. Finally, the tools we use and the way we use them shape our experiences and our thinking (Vygotsky, 1978) and impact our literacy. Thus, "if technology is to be viewed as an add-on in the learning environment that is pursued for the sake of technology alone, then it will not change education" (Valanides, 2003, p. 45), because technology, in and of itself, cannot influence learning, no matter how powerful it might be. On the other hand, if technology is utilized as a cognitive tool that has added value in certain instructional situations, then it will become a driving force for systemic educational change to help teachers and students to experience deep learning and acquire an extended concept of literacy that is compatible with the needs of our society.

The overall findings of the study indicate that ICT is not systematically integrated in Cyprus schools and is not an important part of everyday classroom practices. This seems to be related to several reasons, such as teachers' limited knowledge of a variety of software, limited instructional support provided to teachers by the Ministry of Education, teachers' somewhat positive attitudes, lack of a true community of practice in the schools where teachers help each other to integrate ICT in teaching and learning, and teachers' age. It seems that a supportive school environment could play an important role in effectively and successfully integrating ICT in teaching and learning. Teacher support can be provided in each school by the more experienced teachers in the school or even by more experienced teachers in different schools, by inspectors who visit the school in order to assist teachers in their ICT integration efforts, or by an Instructional Support Service in the Ministry of Education that is responsible for providing instructional guidance to practicing teachers. Moreover, a supportive school environment can eliminate teachers' feelings of isolation in the school, and can encourage effective communication and collaboration among teachers for achieving common goals and literacy in education.

In addition, when ICT is integrated into the classroom environment, the learning environment becomes more learner centered than before, and new assessment strategies are needed in order to capture the essence of learning that takes place in these environments. Traditionally, assessment has been used to sort out students, as well as distinguish the good students from the weak students and, as the end-point of instruction, to assess students' understandings after the instruction ended (Graue, 1993). Hence, the focus of evaluating student learning has been on the products or outcomes of learning, such as facts and information, and not the processes of learning. In ICT-enhanced classrooms, learning objectives vary from achieving deep understanding of concepts to developing critical thinking, decision making, and problem-solving skills, to cultivating positive attitudes towards learning. Therefore, as Shepard argues (2000), the form and content of assessment must change to "capture important learning goals and processes and to more directly connect assessment to ongoing instruction" (p. 5). If the focus of assessment does not change and if new assessment strategies are not developed and accepted as valid methods for assessing student performance, then, as we strongly believe, teachers will hesitate to generously use ICT in their teaching.

Another factor that we consider important, even though it was not found to be a significant predictor in this study, is the lack of adequate computer access in Cyprus schools. For example, given the current situation in Cyprus schools, a teacher who wants to use ICT in a lesson must first make special arrangements to reserve the computer lab in the school in order to be allowed to use it. It seems, however, that because at this point teachers do not use ICT regularly in their classroom practices, they feel that the one computer lab in the school provides them with sufficient computer access. McFarlane (2000) also argues that computer access is a key factor in inhibiting teaching with ICT and states that "until children come to school with a powerful portable computer of their own, access will remain a key brake on the use of digital media in school" (p. 22).

In conclusion, the schools in Cyprus do not seem to be adequately preparing students to develop the new forms of literacy skills that are needed in the information society, and have not been affected to a great extent by new modes of communication, new tools for expression, and new ways of the representation of knowledge. Given the current situation, it is hard to see how new forms of literacy can be satisfactorily developed in Cyprus schools. These findings have implications for Cyprus' international competitiveness. If the educational system in Cyprus will not invest in learning with ICT, then the students in Cyprus will not develop the competencies and the literacy skills that are needed to fairly compete with the students of other countries, which have a better status of ICT in education. Technological illiteracy "could lead to becoming a member of an underclass with a similar status to those who, in previous generations, could not read and write" (Easingwood, 2000, p. 97).

The development of an extended concept of literacy is not an easy matter and many factors seem to affect its development. The implications of this study for the development of an extended concept of literacy are important and need to be seriously taken into consideration by policymakers. It seems that policymakers and government officials have to systemically approach the issue of ICT integration in primary and secondary education, so that ICT is infused in a system that is ready to accept the new educational change. The results imply that a systemic effort for the development of an extended concept of literacy should include a focus on creating a supportive school environment, and a revised focus on teaching and teacher training.

Another implication for the development of an extended concept of literacy is that plans of action have to be developed to identify areas in the curriculum that can be enhanced with the use of ICT. Currently, the curriculum in Cyprus does not include a focus on ICT, and does not appear to have a direction and urgency in systematically integrating ICT. Specifically, the official curriculum in primary and secondary education does not currently include the use of ICT in the teaching of the subject domains despite the fact that ICT integration has been proclaimed as a top priority in the agenda of policymakers. Thus, in the present system, the teacher has to decide how and when to integrate ICT in teaching and learning. Curriculum restructuring efforts need to be undertaken so that teachers receive better guidance about how ICT can be integrated in different subject matters and how ICT can extend the traditional concept of literacy.

CONCLUSION

The purpose of this chapter was to examine factors that may affect teaching with ICT in primary and secondary education, and thus ultimately hinder or delay the development of new literacy skills that are important for citizens to survive in a rapidly changing world. Based on the findings of the study, the development of an extended concept of literacy is not easy and many factors seem to affect its growth, such as teachers' knowledge of ICT, attitudes, self-confidence, age, and instructional support from colleagues and superiors. We argued in this chapter that policymakers need to carefully plan the development of the new forms of literacy in Cyprus schools by systemically integrating ICT in the schools so teachers and students together can develop an extended concept of literacy that is critical for surviving in the information society.

REFERENCES

Angeli, C. (2003). A systemic model of technology integration. Paper presented at the *American Educational Research Association Conference,* Chicago.

Apple, M.W. (1986). *Teachers and texts: A political economy of class and gender relations in Education.* London: Routledge and Kegan Paul.

Baca, J.C. (1990). *Identification by consensus of the critical constructs of visual literacy: A Delphi study.* Unpublished doctoral dissertation, East Texas State University, USA.

Barth, R. (1990). *Improving schools from within: Teachers, parents and principals can make the difference.* San Francisco: Jossey-Bass.

Bransford, J. D., Brown, A. L., & Cocking, R. R. (Eds.). (2001). *How people learn: Brain, mind, experience, and school.* Washington, DC: National Academy Press.

Brindly, S. (2000). ICT and literacy. In N. Gamble & N. Easingwood (Eds.), *ICT and literacy: Information and communications technology, media, reading and writing* (pp. 11-18). London: Continuum.

Crystal, D. (1987). *The Cambridge encyclopedia of language.* Cambridge: Cambridge University Press.

Denning, P. J. (2001). The IT schools movement. *CACM, 44*(8), 19-22.

DfEE. (1998). *The national literacy strategy: Framework for teaching.* London: HMSO.

Easingwood, N. (2000). Electronic communication in the twenty-first-century classroom. In N. Gamble & N. Easingwood (Eds.), *ICT and literacy: Information and communications technology, media, reading and writing* (pp. 45-57). London: Continuum.

Eraut, M., & Hoyles, C. (1989). Group work with computers. *Journal of Computer Assisted Learning, 5,* 12-24.

Friedman, B. (Ed.). (1998). *Human values and the design of computer technology.* Cambridge: Cambridge University Press.

Fullan, M. (1991). *The new meaning of educational change.* New York: Teachers College Press.

Fullan, M., & Hargreaves, A. (1992). *What's worth fighting for in your school?* Buckingham: Open University Press.

Gamble, N., & Easingwood, N. (Eds.). (2000). *ICT and literacy: Information and communications technology, media, reading and writing.* London: Continuum.

Getzels, J. W., & Cuba, E. G. (1957). Social behavior and the administrative process. *Social Review, 65,* 423-441.

Glaser, W. (1981). *Stations of the mind.* New York: Harper & Row.

Graue, M. E. (1993). Integrating theory and practice through instructional assessment. *Educational Assessment, 1,* 293-309.

Hanna, D. (1997). The organization as an open system. In A. Harris, N. Bennett, & M. Preedy (Eds.), *Organizational effectiveness and improvement in education* (pp. 13-20). Philadelphia: Open University Press.

Heppell, S. (2000). Foreword. In N. Gamble & N. Easingwood (Eds.), *ICT and literacy: Information and communications technology, media, reading and writing* (pp. xi-xv). London: Continuum.

Heracleous, L., & Barrett, M. (2001). Organizational change as discourse: Communicative actions and deep structures in the context of informational technology implementation. *Academy of Management Journal, 44*(4), 755-778.

Hoy, W. K., & Ferguson, J. (1985). A theoretical framework and exploration of organizational effectiveness in schools. *Educational Administration Quarterly, 21,* 117-134.

Hoy, W. K., & Miskel, G. C. (2001). *Educational administration: Theory, research, and practice* (6th ed.). New York: McGraw Hill.

Huberman, M., & Miles, M. B. (1984). *Innovation up close.* New York: Plenum.

Kemster, G. (2000). Skills for life: New meanings and values for literacies. In N. Gamble & N. Easingwood (Eds.), *ICT and literacy: Information and communications technology, media, reading and writing* (pp. 25-30). London: Continuum.

Jonassen, D. H. (2000). *Computers as mindtools for schools: Engaging critical thinking* (2nd ed.). Upper Saddle River, NJ: Prentice-Hall.

Jonassen, D. H., Carr, C., & Yueh, H. -P. (1998). Computers as mindtools for engaging learners in critical thinking. *TechTrends, 34*(2), 24-32.

Kling, R. (2000). Learning about information technologies and social change: The contribution of social informatics. *The Information Society, 16*(3), 217-232.

Kling, R., & Lamb, R. (2000). IT and organizational change in digital economies: A socio-technical approach. In B. Kahin & E. Brynjolfsson (Eds.), *Understanding the digital economy: Data, tools, and research.* Boston: MIT Press.

Kozma, R., & Anderson. R. E. (2002). Qualitative case studies of innovative pedagogical practices using ICT. *Journal of Computer Assisted Learning, 18,* 387-394.

Lenine, T., & Donitsa-Schmidt, S. (1998). Computer use, confidence, attitudes, and knowledge: A causal analysis. *Computers in Human Behavior, 14*(1), 125-146.

Liaw, S. -S. (2002). Understanding user perceptions of World Wide Web environments. *Journal of Computer Assisted Learning, 18,* 137-148.

Louis, K. S., & Miles, M. M. (1991). *Improving the urban high school: What works and why.* London: Cassell.

Markus, H., & Ruvulo, A. (1992). "Possible selves." Personalized representation of goals. In L. Pervin (Ed.), *Goal concepts in psychology.* Hillsdale, NJ: Lawrence Erlbaum.

Markus, M. L., & Benjamin, R. I. (1987). The magic bullet theory in IT-enabled transformation. *Sloan Management Review, 38*(2), 55-68.

Marzano, R. J. (1992). *A different kind of classroom: Teaching with dimensions of learning.* Alexandria, VA: ASCD.

McFarlane, A. (2000). Communicating meaning—Reading and writing in a multimedia world. In N. Gamble & N. Easingwood (Eds.), *ICT and literacy: Information and communications technology, media, reading and writing* (pp. 19-24). London: Continuum.

McFarlane, A. (Ed.). (1997). *Information technology and authentic learning: Realizing the potential of computers in the primary classroom.* London: Routledge.

Murphy, C. A., Coover, D., & Owen, S. V. (1989). Development and validity of the computer self-efficacy scale. *Educational and Psychological Measurement, 49,* 893-899.

Papert, S. (1993). *The children's machine: Rethinking school in the age of the computer.* New York: Basic Books.

Pelgrum, W. (2001). Obstacles to the integration of ICT in education: Results from a worldwide educational assessment. *Computers and Education, 37,* 163-178.

Picciano, A. G. (2002). *Educational leadership and planning for technology* (3rd ed.). Upper Saddle River, NJ: Prentice-Hall.

Reigeluth, C. M., & Garfinkle, R. J. (Eds.). (1994). *Systemic change in education.* Englewood Cliffs, NJ: Educational Technology Publications.

Salomon, G., Perkins, D. N., & Globerson, T. (1991). Partners in cognition: Extending human intelligence with intelligent technologies. *Educational Researcher, 20*(3), 2-9.

Shepard, L. (2000). The role of assessment in a learning culture. *Educational Researcher, 29*(7), 1-14.

Snyder, I. A. (1994). Writing with word processors: A research overview. *Journal of Curriculum Studies, 26,* 43-62.

Street, B. V. (1987). Models of computer literacy. In R. Finnegan (Ed.), *Information technology social issues.* London: Sevenoaks, Hodder and Stoughton.

Tabachnick, B. G., & Fidell, L. S. (1989). *Using multivariate statistics* (2nd ed.). New York: Harper & Row.

Trach, L., & Woodman, R. (1994). Organizational change and information technology: Managing on the edge of cyberspace. *Organizational Dynamics, 23,* 30-46.

Trist, E. L. (1982). The development of sociotechnical systems as a conceptual framework and as an action research program. In A. H. Van de Ven & W. F. Joyce (Eds.), *Perspectives on organizational change and behavior* (pp. 19-75). New York: John Wiley & Sons.

Trist, E. L., & Bamforth, K. W. (1951). Some social and psychological consequences of the Longwall method of goal setting. *Human Relations, 4,* 3-38.

UNESCO. (1999). *The science agenda—Framework for action.* Paris: UNESCO.

Valanides, N. (2003). Learning, computers, and science education. *Science Education International, 14*(1), 42-47.

Valanides, N., & Angeli, C. (2002). Challenges in achieving scientific and technological literacy: Research directions for the future. *Science Education International, 13*(1), 2-7.

Vygotsky, L. S. (1978). *Mind in society.* Cambridge, MA: Harvard University Press.

Watt, D. H. (1980). Computer literacy: What should schools be doing about it? *Classroom Computer News, 1*(2), 1-26.

Wild, M. (1996). Technology refusal: Rationalizing the future of student and beginning teachers to use computers. *British Journal of Educational Technology, 27*(2), 134-143.

ENDNOTES

1. We consider the last part of the 20th century to be the beginning of immense developments in information, communication, and network technologies.
2. The term *ICT* is used in this study interchangeably with *computer applications*, and includes the Internet, the World Wide Web, and all types of computer software.

Chapter XXXV
Teaching English as a Second Language with Technology:
Making Appropriate Pedagogical Choices

Kate Mastruserio Reynolds
University of Wisconsin-Eau Claire, USA

Ingrid Schaller
University of Wisconsin-Eau Claire, USA

Dale O. Gable
University of Wisconsin-Eau Claire, USA

ABSTRACT

All U.S. states have standards that require the inclusion of technology into the classroom (Rodriguez & Pelaez, 2002; Abdal-Haqq, 1995; Wright, 1980). Kindergarten-12 teachers face bourgeoning state-mandated curricula that they are required to teach each year. These curricula leave little room for specific computer literacy or technology instruction. Therefore, teachers must achieve both sets of expectations, (i.e., standardized curriculum and technology standards integration), simultaneously and without losing sight of the main content-area focus. This situation is more complex for ESL teachers. ESL instructors' goals are two-fold: (1) the content or subject of instruction, and (2) language acquisition.This chapter will outline the various constraints and challenges that K-12 teachers face when attempting to include technology into their classrooms. Then, a variety of ways to integrate technology while maintaining a content and language acquisition focus by providing practical, accessible, and user-friendly resources, activities, and tips for inclusion will be proposed. Finally, learner encountered resulting from a productive mixture of content and technology in English as a second language (ESL) classes will be shared.

INTRODUCTION

K-12 levels of all schools in the United States are required to integrate technology into the curriculum. Although all states have standards that require the inclusion of technology into the classroom (Rodriguez & Pelaez, 2002; Abdal-Haqq, 1995; Wright, 1980), the types of technology and the level of sophistication are not delineated.

Teacher education standards also mandate a degree of computer literacy awareness (Wright, 1980) in the hope that the teachers will thus be equipped to integrate technology into their K-12 classrooms on a regular, if not daily, basis.

The pressure teachers feel is compounded because K-12 teachers in the U.S. must incorporate changes to the extensive state-mandated curricula that they are required to teach each year. Some curricula have hundreds of standards or detailed subject matter topics that need to be taught within a given year so that learners can perform successfully on state-mandated, high-stakes standardized tests. These curricula leave little room for specific computer literacy or technology instruction. Therefore, teachers must achieve both sets of expectations, (i.e., standardized curriculum and technology standards integration) simultaneously and without losing sight of the main content-area focus.

This chapter will outline the various constraints that K-12 teachers face when attempting to incorporate technology into their classrooms. The chapter will then outline a variety of ways of integrating technology while maintaining a content focus by providing practical, accessible, and user-friendly resources, activities, and tips for inclusion. Finally, it will highlight the learner outcomes that were encountered as a result of a productive mixture of content and technology in English as a second language (ESL) classes.

THE CHALLENGING CONSTRAINTS TEACHERS FACE

The mandate to include technology in K-12 has been a challenge and source of constraint for teachers. All K-12 teachers, be they new teachers a year or two out of colleges or the 20-year veterans, face issues regarding their personal knowledge and computer skills, environmental or context-related challenges, technological and financial issues, as well as time constraints (Hughes, 2003).

Personal Knowledge and Skills

Although computer literacy courses have been accessible to learners in the U.S. K-12 since the mid-1980s, the daily use of computer technology for teachers became a real expectation in the mid-1990s (Barker & Howley, 1997).

One of the main challenges for all teachers during this period was to become computer literate and savvy. Personal knowledge and computer skills had to be developed (Adkins-Bowling, Brown, & Mitchell, 2001; Johns, & Torrez, 2001), because the teachers' technology knowledge directly impacts their willingness to employ it in the classroom (Hughes, 2003). In order to facilitate teacher training, since the mid-1990s there have been numerous local, state, and federal grants available to provide computer instruction for teachers in the commonly used software packages, such as Microsoft Word, Excel, and PowerPoint, and others. Teachers have primarily employed this training to facilitate e-mail for internal communication purposes with other teachers, administrators, and teacher aides. They have also used it to assist in their lesson planning, activity creation (e.g., using clip art and creating graphic organizers), and grading by maintaining their grade books, scores, or norming test results.

As a consequence, the computer training was implemented into teachers' daily operations more for internal school communication, lesson planning, and development purposes than for direct instructional purposes for school-age learners. Some tech-savvy and interested teachers did attempt to integrate more technology into their lessons; however, the remaining teachers were not sufficiently well-trained in the technology to feel competently prepared to instruct with the authoring technology, hypermedia and Moo's, for example.

Since the mid-1990s teachers' use of technology and knowledge of the Internet's resources have increased. Teachers are using the Internet quite frequently for lesson ideas, activities, and inexpensive resources (Raya, 2003). They see the very practical and valuable worth of the Internet to further classroom discussions (Little, 1996; Stevens, 1992), to widen the learners' experiences, and to connect to the real world outside of the classroom (Sutherland-Smith, 2002; Rong-Chang & Hart, 1996).

Environmental Considerations

Another challenge encountered by the K-12 teachers in the U.S. is environmental or context related. In many districts, schools have one or two computers per classroom, or at best each school has a computer lab onsite. This scenario presents a couple of interesting choices. First, if there are only one or two computers in the classroom, who can make use of the computers, when, and for what purposes? Should the use of the computers be a reward for the faster learners? It may be a convenient choice, but is it a fair instructional decision? All learners should be allowed to utilize the computers. As a result, teachers are uncertain as to who ought to have priority to operate the computers, and when. In addition, teachers are unsure of how to manage a classroom in which there are a couple of learners using the computers while the rest of the students lack access. As there are simply not enough computers for all the learners, a management and logistics challenge for the K-12 teachers is inevitable.

Second, in the event that the school has a computer lab, the teachers may bring their classes to the lab when they can schedule time in it. Scheduling may or may not be a problem depending on how many classes need access to this resource and the degree to which the resource is limited. On the other hand, this arrangement solves one quandary of how to instruct all learners on the computers simultaneously.

A final consideration in terms of environment is whether the learners have access to computer technology outside of the school. In some districts, learners have access to computers at home (Barker & Howley, 1997), whereas some learners do not, but they may have to access one at their local library. In some cases, the learners do not have any access to computers outside the academic environment. Because of this variability, it is unfair for teachers to assign computer homework for completion outside of school hours. Therefore, these teachers must plan opportunities for learners to do the computer-related work on school grounds or within school hours.

Hopefully, school districts, administrators, and teachers have been attempting to solve these issues of access and to mediate access to technology for all learners, in spite of the financial status of the district or the socioeconomic level of the learners.

Technological Factors

For a plethora of reasons, it is likely that there are technology-related issues which K-12 teachers predictably encounter. First, according to Boswood (1997), 99% of language teachers do not have access to cutting-edge technology. Second, the individual teacher may not have access to a technology expert at the school who is responsible for managing, maintaining, and training on the hardware or software. In order for a teacher to utilize new software or other technology (e.g., scanners, digital camera, photo software, etc.), the teacher may need to become the technology expert by installing the software or technology, maintaining it, and training others on it. This is a challenging situation for most individuals who do not have the specialized HTML or network training or who have many other demands on their time and resources.

Ideally, the K-12 schools have a technology resource person in-house who can facilitate installation, maintenance, and training on new hardware and software for the teachers.

Financial Issues

As alluded to in the environmental factors section above, there are certain financial constraints that may to some degree impact the use of technology in the K-12 public schools. The financial capabilities of the district will influence the number of computers available, whether they are located in individual classrooms, a computer lab, or the library. The fiscal wherewithal will also be a factor in whether there is a computer technology support person or staff member available. Finally, and most importantly for teachers, the financial flexibility may limit the type of technology (hardware and software) available or able to be purchased.

Time Constraints

Ultimately, all teachers encounter issues of time management due to their never-ending workload of preparing and teaching classes, grading and assessing learners, modifying instruction for all learners, tackling learners' performance and/or behavioral issues, engaging in staff and parental meetings, and participating in professional development obligations.

Adding the need to become computer savvy, identify technology resources, prepare lessons with technology included, interact with the learners and the technology, and remain abreast of all the technological changes, can be an unwelcome or frustrating addition.

All these factors mingle to significantly impact who can apply technology on a daily basis; how the technology can be used; and when, where, and for what purposes technology can be used.

WAYS OF INTEGRATING TECHNOLOGY WHILE MAINTAINING A CONTENT FOCUS

Faced with the constraints, many instructors have been working for several years to include appropriate technology into their English as a Second Language classes in an English language program: (1) within the personal, environmental, technological, financial, and time constraints typical of all classroom contexts; and (2) without losing the principal focus of the learning of ESL.

This chapter will share activities and strategies, which are financially, technologically, and contextually accessible to any teacher in any environment, that were successfully employed in English as a Second Language classes with learners ranging in age from 15 to 25 years old. Learners in this context were at the ACTFL intermediate low and the advanced low levels of language proficiency. The instructors were content-area and ESL-trained instructors with an average of five years of instructional experience. All instructors held master's degrees. None of the instructors considered themselves techies.

Importance of Technology

These instructors recognized the importance of technology in teaching, and view the inclusion of all types of technology into ESL classes as a vital component of instruction and program philosophy. It would be absurd to claim that learners and teachers do not need to know about technology or access it within their classrooms or schools.

First, from the learners' perspective, it is vital that technology be integrated into lessons, not only so that they become familiar with the technology and resources, but also so they are motivated and feel that the information presented in lessons is useful and meaningful to their real-world experiences (Busch, 2003). The exploitation of technology also extends the classroom boundaries and brings world events into classroom learning. Ultimately, this creates a much more authentic classroom experience that is more contextualized or more integrated with the outside world (Busch, 2003).

From the teachers' perspective, the use of technology is highly important, because it provides teachers with ample resources (particularly via the Internet), without the responsibility of authoring or networking knowledge. It also allows teachers to integrate useful up-to-date knowledge and information for little or no cost to the school or the learners. It stimulates and motivates the learners while facilitating communication (e.g., listening/speaking, reading/writing, etc.) and language learning (Busch, 2003).

Even more interestingly, technology allows for certain interactive types of activities including peer revising, listserv discussions, and e-mail pen pal exchanges (Johns & Torrez, 2001; Leu, Leu & Len, 1997; Parsell & Wilhelm, 1997; Warschauer, 1995), to this point not easily conducted. The interactive nature of technology also allows for immediate reactions, feedback, and information exchange that facilitate all kinds of classroom practices, particularly learner-to-learner spoken and written communication.

Most importantly, though, it is invaluable to include technology into classrooms and programs because it allows for the establishment and maintenance of forward-thinking, cutting-edge, competitive instruction, programs, and institutions.

Importance of Content

Above all, the most important concern for K-12 teachers, naturally, is that when technology is included in a classroom and the learners need to be trained, the focus of the classroom shifts from the content or topic of instruction to the technology itself. Since learners and sometimes teachers can lose sight of the primary task at hand because the technology is stimulating or difficult; it is possible that the pedagogical usefulness of the technology is questioned. This is particularly true when learners are taught authoring of Web sites (Reynard, 2003) or more advanced technological activities and skills, for instance.

Since all K-12 teachers have closely controlled curricula in the United States, it is of vital importance to them to employ technology without sacrificing the instructional content. It is critical for the learners as well, since high-stakes tests are influencing choices on knowledge, abilities, and promotion (Abrams & Madaus, 2003, pp. 31-35).

In the case of ESL classrooms, there is typically a dual emphasis on: (1) a core content area, such as language arts, math, social studies, or science, as well as (2) English language acquisition (i.e., the language learning is an additional area of content instruction) (Echevarria & Graves, 2003; Kasper, 2002; National Center for ESL Literacy Education, BBB 36588, 2002; Snow & Brinton, 1997; Chamot & O'Malley, 1994).

So, the addition of technology to this mélange increases the pressure on the teacher with regard to the subject matter and the language instruction. Consequently, the best teacher choice in this situation is to ensure that the technology portion enhances the presentation of the subject matter and language instruction, without replacing either.

Theoretical Underpinnings

ESL instructors are trained to create a myriad of spoken and written interactional opportunities in order to facilitate language acquisition (Brown, 2003; Gass & Selinker, 2001; Brown & Attardo, 2000; Lightbown & Spada, 2000; Nunan, 1999; Gass, 1997) and learners' construction and negotiation of meaning of linguistic knowledge and performance (Healey & Klinghammer, 2002; Sotillo, 2002). According to Vygotsky (1978), learners construct knowledge through social interaction called "instructional conversation." Instructional conversation, or IC, is the verbal or written interaction between a novice or a learner with less information on a given topic, and an expert, the teacher, or another learner with more information on the given topic (Lantolf, 2000). It is through this spoken and/or written communication during an instructional conversation that the expert provides scaffolded information (i.e., information with links to the novice's prior knowledge), so that the novice learner may move from his or her actual level of development (i.e., current state of knowledge) to his or her proximal level of development (or potential knowledge level after instruction) (Ohta, 2000; Lantolf, 2000; Wertsch, 1985, 1991).

These instructors' pedagogic goals for using technology are, primarily, to increase learners' communication (i.e., interaction) opportunities, create an information gap between the expert and novice learners, extend learners' engagement in the ESL classroom (Sutherland-Smith, 2002), and provide abundant exposure for the English language learners to the native-speaker use of English (Sutherland-Smith, 2002; National Center for ESL Literacy Education, BBB 36588, 2002). Reynard (2003), Johns and Torrez (2001), and Terrill (2000) all found that the use of technology, and specifically the Internet, can serve effectively to enhance and promote language learning. Furthermore, Da (2003), Hayward and Tuzi (2003), Reynard (2003), Sotillo, (2002), Huang (1999), and Rong-Chang and Hart (1996) all demonstrated that computer technology facilitated interaction and enhanced the writing instruction and development of linguistic skills of English language learners (ELLs).

Technology-related activities create countless learner-to-learner interaction opportunities (Freiermuth, 2002; Healey & Klinghammer, 2002; Sotillo, 2002), which result in instructional conversation between novice and expert learners. The instructional conversation, in turn, focuses on the subject matter knowledge construction *and* provides scaffolded interaction while allowing learners to learn by using the language (Hatch, 1979) to move toward automaticity (Bialystok, 1978). Furthermore, this social interaction allows English language learners to develop their language competence or knowledge of the grammar, phonology, lexicon, syntax, and pragmatics of English while improving their performance of the language in terms of their speaking, listening, reading, and writing (Brown, 2000).

TYPES OF ACTIVITIES AND STRATEGIES TO INCLUDE TECHNOLOGY WITHOUT SACRIFICING CONTENT

CNN Online Streaming Video in ESL Current Global Events Class

The Global Current Events ESL class in this Midwestern language program had been taught without an assigned text (see Appendix A for course description and objectives), because instructors chose to employ CNN Student News Broadcasts as the primary source of course materials and content. Broadcasts, via online streaming video each night, were taped daily or weekly in order to have ample selection of timely topics to cover in class. The variety of topics served as a springboard for pursuing semester themes, and inspired further investigation for student projects and presentations. Instructors chose to link the topic of the ESL program's supplemental texts, and other materials to recent events and developments worldwide, so as to explore all program themes and topics in depth from a contemporary perspective. Other objectives included familiarizing learners with news formats, as well as the rate of speech and vocabulary used in news broadcasting.

The instructors and students were able to locate CNN Student News broadcasts via CNN's Web site (www.cnn.com). An instructor could request that the show be sent to an e-mail address daily. Archived transcripts and discussion questions for three weeks' worth of broadcasts were available, which was useful in making a unit out of an issue or topic, in the event that a similar topic had previously aired. Web site links to additional Web sites were provided, both during the broadcast and online, with the transcripts, as well as discussion questions and ideas for further investigation—all of which proved useful when teaching the class, and allowed the flexibility to pursue student interests further.

As in any classroom, teachers implemented techniques to introduce, present, assess, and further the students' understanding and linguistic performance which were employed while exploiting the technology of a video or Internet activity. When presenting the individual stories in CNN Student News, the students often brainstormed about a particular topic, compared their observations with that of the narrator, and used the vocabulary that was introduced. In addition, they provided more ideas and highlighted traditions that occur in their cultures prior to, during, and after they successively listened to the broadcast segment. The students practiced listening for global meaning, and then for attention to details, discussed vocabulary, postulated questions, and fielded comments from their peers. On several occasions, they followed a particular story throughout the semester, and then capitalized on its importance in assigned group and individual presentations. All of these activities supported the verbal and written interaction among the learners, and are widely recognized as furthering second-language acquisition. Jun (2002) found that "presentations involving reporting, discussing, writing, and commenting provide students with opportunities to enhance their oral and written communication skills in the target language" when employed in conjunction with Web resources.

In past classes, students have conversed about news stories related to the European Union status, debated Haitian politics versus the United States' politics and the political process in learners' home countries, and investigated prejudice in the news by doing research in various online newspapers. Learners interacted linguistically and worked collaboratively on in-class presentations about Haitian politics, having been initially introduced to this topic while viewing the CNN broadcasts. Each group of students employed the news stories as a starting point for discovering more about differing aspects of the Haitian political situation, by stating what they specifically wanted to know, researching those questions, and then reporting their findings to the class.

The Global Current Events course was an intermediate-level class which allowed instructors to challenge learners with new, more advanced vocabulary, and listening and speaking skills targeted to their linguistic input needs.

As with most contexts, the students' learning curve was steepest at the commencement of the course while they adjusted to teaching styles, materials, their classmates and the expectations of their teacher. At the beginning of the semester, the students generally needed more background information and vocabulary activities than toward the end of the semester, particularly when a topic was introduced or was unusual or highly abstract. It was sometimes useful to provide students with a transcript so they could read along while listening to the broadcast for details. Since students were immersed in English, both in and outside the classroom, it should not be surprising that their listening gradually showed vast improvement.

As the semester continued, the instructors and students quickly became aware of the agenda of the news station. At times, it seemed that the broadcasts were not as international as they could have been and seemed to be focused on the United States' agenda, or a particular issue plaguing it. Consequently, it was difficult to exclusively use the broadcasts as the only source for content. Given the possibilities of the Internet and the Web sites mentioned, however, it is easy to branch off and study a topic of interest to the students.

The technology-related activities employed, in conjunction with the CNN news broadcasts, established a fluid, motivating, and learner-centered language learning environment, in which the learners had an intrinsic need to communicate in spoken and written form with peers. This environment facilitated the development of their top-down (i.e., global listening) and bottom-up (i.e., listening for detail) listening skills, as well as their speaking, reading, writing, grammar, note-taking, vocabulary, and pronunciation skills. Instructors would typically assign increasingly challenging task-based, extension activities to recycle the knowledge, vocabulary, and related grammar, thus allowing learners to practice the linguistic knowledge in an appropriate and meaningful context, without recourse to rote repetition. This aspect of the instruction solidified cognitive links and provided opportunities for learners to monitor their production for improvement purposes. Finally, instructors assigned a considerably more challenging written task for homework, in order to extend learners' discourse production on the topic, and to assess learners' cumulative content knowledge and acquired linguistic skills/improvement. The employment of strategies continued to reinforce the knowledge, and required learners to stretch their knowledge and linguistic performance in the formal written discourse on the topic.

Setting up a taping system that did not inconvenience the instructor proved crucial. Preparation time varied from a couple of days prior to class, to a couple of hours, depending on whether the instructor availed herself of the most current broadcast, just before the class was scheduled to meet. To keep frustration to a minimum, the instructor needed to fetch the video from its taping source or depend on someone else to have taped it correctly and bring it to school on that day. Video recorders for automatic recording could also be set for the appropriate time and day. The technology itself is simple and easily accessible, only requiring an Internet link for recording, and a VCR and television for viewing.

The application of technology presented certain constraints. In the Global Current Events course, the instructors needed to be able to record the news segment or have them recorded. Sometimes this was more convenient for the individual teacher to do rather than having a university tech person record. Regardless of how the program was taped, the instructors of each class must request the right to copy the program from CNN, which is possible through the Web site. For this particular course, the only financial implications included having VHS tapes available for copying, and the necessary audio-visual equipment. Reusing videotapes also saved money. In this case, it was necessary to look for more fluid options. An avenue one might pursue for more accessible technology might be to have each program taped on a digital system and available to a classroom.

As to the use of CNN Student News broadcasts, one ought to be mindful of time constraints, in that, whereas the section one used in class might only be two or three minutes long, it is shown numerous times at the outset of the course. The activities surrounding the particular story frequently overlap into the next class, thus making for more in-depth study of the issue, but decreasing one's ability to cover a diversity of topics during the semester. Additionally challenging for teachers who would like to prepare all their material in advance is the transient nature of a course with the words "current events" in the title. Choosing topics to be covered in the given time is vital for smooth integration, but not impossible for instructors to manage.

Internet at the National Public Radio (NPR) Web Site

Teachers also employed archived news radio reports on the Internet from the National Public Radio (NPR) Web site (www.npr.org) for extending the learning in an ESL Listening and Note-Taking class and others. The intent in using this technology is to provide plentiful opportunities for extensive and intensive listening and note-taking (see course description and objectives, Appendix B) so as to further the English language learners' aural proficiency, help them understand diverse English accents, assist their comprehension of various rates of oral language production, and enhance their mainstream learning experiences.

Using the technology was quite simple. One can visit the NPR Web site (www.npr.org), click on the "Archives" link, and access all previously recorded radio broadcasts. One of the benefits of using the NPR archives was that they have radio stories in their coffers that span numerous topics and are recent in nature. Another advantage was that their archives are stable; they retained broadcasts in their online collection for 10 years or more. Therefore, teachers can rely on the fact that the news broadcasts will be available to the learners without worrying that the story in question will be deleted.

Teachers need only visit the Web site while planning their course syllabus (see Appendix B for a portion of a sample syllabus) in order to search the archives for topics that enhance and extend the textbook chapters' topics. One challenge with this is that teachers *must* note the title of the piece, the radio show that it was broadcast on, and most importantly, the date it aired, as some of the reports share identical titles (e.g., globalization).

Learners, on the other hand, need to know how to input a URL in the address bar in order to access the Web site. They may need to be shown how to find a specific story. At the onset of the semester, teachers may choose to demonstrate how to access the NPR reports to the whole group, by either walking learners through the process using screen prints, or by using a computer and projector to simultaneously display the access path. Learners can access the reports by either using the archives (and completing the three search questions about the piece), or by using the search function (inputting the report title) at the top of the screen. They would then click on the title link; verify the date, show, and story title; and click on the audio speaker symbol to hear the report. They should double check the sound quality and volume as well. In order to do this in-class or as a homework task, the learners would need access to computers with Internet access, Real Player or other audio-type software, and headphones that work with the computer.

A specific activity performed by students in listening class when they covered the unit entitled, "Actions Speak Louder than Words," focused on non-verbal behavior in society. The students were shown how to find www.npr.org and search for the radio broadcast on *All Things Considered* entitled, "Seeing Anger," which aired on June 18, 2002. Learners listened to the broadcast for homework and made a detailed outline in their notes. They were permitted to listen to the four-minute piece as many times as they wanted or needed. This helped the learners to listen globally to the information once or twice so that they could outline the segment, then listen additional times to glean supporting points and examples, and important details. They were able to listen at their own rate and choose the number of repetitions. They were able to stop, start, and move forward or backward in the report at any point. (Depending on the computer literacy level of the learners, this may need to be demonstrated).

Prior to class, the teacher was able to listen to the report and make detailed notes in outline form, or purchase the transcript, if funds were available.

In class, the teacher and learners shared information orally or in writing about the report. They discussed viewpoints, themes, details, and connections to the chapter and the world at large. Learners were trained in an assortment of focused listening and note-taking strategies. There were also a multitude of extension activities that could be based on the broadcasts. For instance, the group discussed the overall conclusions of the report, how it impacted the discussion of non-verbal behavior from the chapter, and any specifics that the teacher desired to emphasize. The teacher could arrange a debate or values clarification activity to further the discussion as well. Students could write an opinion article in response or further investigate the topic in the library. Teachers could employ the learners' outlines of the piece to assess learners' comprehension or to verify aural language proficiency progress.

Later in that unit, the Listening and Note-Taking class furthered its conversation of non-verbal behavior by exploring how Botox numbs portions of one's face (*Weekend Edition* radio talk entitled, "Botox," which aired on September 19, 1998, on NPR). They were instructed to listen as many times as needed and to take organized, detailed, academic-style notes. Through readings, essay writing and reflective journaling, and discussion, the concepts were connected and enhanced so that learners were learning speaking and listening in a detailed, focused manner. They developed their learning strategies for getting the gist and listening for specifics, and they synthesized all the information into a fully rounded and developed schema with all of their background knowledge and second-language information intertwined.

These technology-related activities met course objectives, and assisted the teacher to create dynamic, motivating, and learner-centered tasks in which the learners had an intrinsic need to communicate in spoken and written form with peers. Information they learned through discussion and brainstorming prior to exploiting the Web site served as an introduction to the topic and the vocabulary specific to the discourse domain, and developed cognitive connections for learners to their prior knowledge on the topic. This introduction prepared learners for whole-group or individual investigation of the Web site, and for the in-depth reading required of these intermediate and advanced proficiency ESL learners. Once they had accessed the Web site and furthered their knowledge on the topic through repeated listening to the audio segment, they were then able to take notes. This process facilitated development of their top-down and bottom-up listening skills. When learners returned to class, they would then discuss what they had heard and construct connections to the subject matter in the textbook. Instructors would then typically assign increasingly challenging task-based, extension activities to recycle the knowledge, vocabulary, and related grammar, thus causing learners to practice the linguistic knowledge in an appropriate context, and without rote repetition. This portion of the instruction solidified cognitive links and provided opportunities for learners to monitor their production for improvement purposes. Finally, instructors would assign a slightly more challenging written task related to the subject for homework to extend learners' discourse production on the topic. This strategy would continue the knowledge solidification and require learners to stretch their knowledge and performance in the formal written discussion of the topic.

The only challenges encountered in using the NPR Web site were related to the fact that learners needed to be shown how to access the Web site and search for the reports. Once that was demonstrated by screen prints or visually with a computer and projector, learners were very well equipped to find the other broadcasts employed in the class. Another issue was when there were two stories with the same title. The best remedy for this is being specific about the title, show, and the date the story aired. Finally, there may be a problem with sound quality and headphones. Learners should be instructed to use a private computer and listen from the computer's speakers, or by using a public computer and headphones. Luckily, most commonly used headphones work with most computers. Learners should also be aware that they may need to test that the sound quality is clear and the volume is appropriate. Since the learners can listen to the reports many times, this is not necessarily an important challenge. Typical time, cost, and training constraints did not apply when using this Web site (or other Web sites mentioned), because teachers can access and prepare appropriate Web sites and topics during pre-term course planning, as well as use printouts or a computer lab for access. The training for the instructor and the student is minimal, since most educators and students in the United States have working knowledge and experience with the Internet.

New York Times Web Site

Multiple readings from the New York Times Web site (www.nytimes.com) supported the latest financial information which enriched economic discussions and investigations in the ESL Business English course (see Appendix C for course description and objectives). In the Business English course, the advanced proficiency learners were required to access weekly readings from this site, and relate them to the chapter of the week. Prior to accessing the

Web site, the teachers built vocabulary and background knowledge by using the textbook and activities. Then, learners were given directions to the URL, the title of the article, and the date of publication. Their directions were to read for the gist of the article, to preview it for the next class meeting, or to look up new vocabulary words and record them in their vocabulary notebooks.

In class, the learners would discuss overview questions and vocabulary, and then silently read the article. They would discuss opinions, perspectives, or predict the next events. They would also further develop their vocabulary by analyzing collocations, new vocabulary words, or colloquial expressions used in the article. At times, the teacher would assign topically related articles so that the learners could participate in an information-gap activity. Alternatively, the teacher would bring in articles related to a specific grammatical, pronunciation, language skill, language function, or competency to work on with the class. They would analyze the piece for the language construct's application and meanings. They would then employ the same language construct in other extension activities and tasks to further their productive abilities, such as memo writing, telephone, e-mail, sales pitch, advertisement writing, or Internet inquiries, reports, and presentations.

In one learning activity, the business English learners used an article to discuss the current status of the hotel industry, in the U.S. and worldwide, and the possible reasons for this situation. Learners shared vocabulary that challenged them, and the class discussed possible meanings and connections to other business and financial ventures. They analyzed the colloquial expressions and the manner in which the author described the hotel executive's efforts. They strategized about potential marketing schemes and gimmicks. As an extension activity for homework, learners wrote a marketing strategy and advertisement that would represent one of the hotel chains listed in the article, to target the hotel's new market focus and ideas.

Again, instructors used the Web site in relation to language activities, to cause learners to want to share their perspectives and communicate through speaking and writing about the topic with peers. Prior to accessing the Web sites, instructors scaffolded instruction so that learners' knowledge of the subject was peaked and sufficiently developed. Instructors consider this "best practice" and do this so the learners experience success without struggle when accessing the information from the Web site. Instructors would choose the means by which to present the Web site article. For example, the instructor might offer two thematically connected articles (via hard copy or on the computer depending on resources) for learners to read in class and summarize for the other group. Teachers could then exploit both readings for new thematically related vocabulary with a cloze or multiple-choice activities and one grammar construction heavily utilized in the articles, for a five-minute focus-on-grammar activity. They would further their language learning and acquisition through text- and topic-related, progressively more demanding, extension activities which could employ reading, writing, speaking, and listening in task-based activities like values clarification activities, information-gap exercises, and decision-making tasks.

These instructional strategies and techniques are currently considered 'best practice' in ESL/EFL instruction, because these activities allow learners to use the language (e.g., lexical and grammatical constructions) related to the subject matter, and facilitate interaction between novices and experts in a variety of modes.

The technology use was not difficult to teach. Again, learners only needed to know how to access a specific URL, and search within that Web site. One problem, though, was in the presentation of the *New York Times* homepage. If you are a business person, approaching the *New York Times* Web site is probably very natural and commonplace; however, for the typical educator, there is a great deal of financial information presented on the *New York Times* homepage that can be a bit overwhelming. It was important not to overwhelm the English language learners by the language format, style, and information presented on the page, so that they could actually access the particular information necessary.

Public Broadcasting System (PBS) Web Site

Textbooks are only as useful as teachers can make them; they have their advantages and disadvantages. One way to overcome the issue of an outdated or an overly simplified text is to provide supplementary materials. The Public Broadcasting System Web site (www.pbs.org) allowed instructors to expand the texts in an Integrated Language Skills course (for theme, course description, and objectives, please see Appendix D) at the intermediate and advanced proficiency levels. A variety of multimedia from the PBS Web site (www.pbs.org) helped to stimulate discussion, further connections, deepen understanding, and build learners' English speaking, listening, reading, writing, grammar, pronunciation, and vocabulary by relating the information and resources to the topics used in the Integrated Language Skill course.

The manipulation of the PBS Web site is quite trouble-free. One only needed to access the URL and to search for the appropriate story. This can be conducted in the exact same manner as outlined in the NPR section above. The PBS Web site also keeps long-term archives of resources, so that teachers can rely on their availability.

There are two examples of how the instructors utilized the wealth of information that the PBS Web site provided. First, while doing a chapter on Montezuma, Cortez, and the end of the Aztec Empire in Mexico, they discovered that the native Spanish-speaking learners knew far more about the topic than the textbook discussed. Not all the learners in this ethnically diverse ESL course were as familiar with this material, but they needed to intellectually challenge *all* the learners in the classroom (Sotillo, 2002). Using the PBS Web site (http://www.pbs.org/opb/conquistadors/espanol/home.htm) enhanced the discussions of the material by learning in more depth; viewing in-depth reports about the timeline of the conquest; the social, political, and health-related issues that influenced the conquest; as well as identifying supplemental readings and photos to enhance the reading/writing, speaking/listening activities. They were then able to exploit this supplemental, and more timely and detailed, information to concentrate on the specific use of past tense in the reports for language analysis, and by expanding the learners' lexicons. The Public Broadcasting System explored the topic in depth and in a culturally unbiased manner, which is of immeasurable benefit to ESL classes. Finally, the site provides middle and high school teachers with lesson plans and activities pertinent to the topic.

A second example of how they used the PBS Web site was the documentary about Japan from 1500-1800 entitled, "Japan: Memoirs of a Secret Empire" (http://www.pbs.org/empires/japan/). In Integrated Language Skills courses that emphasize content-areas and multicultural subject matter, the PBS Web site helped to investigate this historical topic with the learners. They viewed the PBS television broadcast in the class, initially, to build learners' background knowledge. Then, the heterogeneous class of learners from all over the world shared their prior beliefs about this timeframe and context. For instance, one learner was surprised that wives and daughters of samurai were also trained samurai as well. Others expressed similarities between the culture's class system and that of their home countries during this timeframe. Next, the class utilized the Web site to investigate the interesting and provocative character composites ranging from courtesan, merchant, samurai, farmer, geisha, and daimyo to shogun. They wrote reflections on one of these character types and responded to a query concerning what their impressions of this character's life would have been like. They then read excerpts from *Memoirs of a Geisha*, which furthered the learners' emerging understanding of the realities of life in Edo, Japan, from the 1500s to the 1800s. They had discussions of daily living conditions, status, and politics within this society, and the life options available to individuals living within this system.

In the previous section on the use of the CNN and NPR Web sites, one can read in detail how the learners would acquire language and how the teachers would scaffold interaction and instruction with the use of a Web site that is primarily written text and graphics. The PBS Web site is similarly designed and employed. One note, however: learners' outcomes resulting from this unit were outstanding; they wrote exemplary character description papers that demonstrated insight and understanding, while demonstrating their improved written communication skills (i.e., speaking, pronunciation, listening, reading, and writing), grammar, and subject-related vocabulary.

The only challenges that they encountered with the use of this technology occurred when using online streaming video. There were technological constraints in terms of gaining access without error messages, and being able to watch the video without the computers locking up or freezing. This situation did not require sophisticated technology knowledge or the assistance of a technician, only some patience. Also, for some students there was a learning curve in use of the technology. The curve varied from the simple and easily rectifiable problem of mistyping the URL, or making adjustment to the volume level on the computers, to the more difficult matter of time constraints. In certain cases, the students' low language level made the timely completion of the task challenging. Aside from the processor speed, the technology challenges were easily rectified.

LEARNER OUTCOMES

Although it is difficult to isolate variables in educational research, there were some indirect learner outcomes that were attributable to the use of these technologically related activities in these content-based ESL courses. First, learners' performance on post-tests (using the ACT COMPASS ESL computerized test) indicated a 30-point average increase in the ESL learners' listening comprehension for the Listening and Note-Taking course mentioned previously. Learners' abilities at the beginning of the 16-week term were listed in the 60-point range (intermediate level proficiency), whereas at the end of the term, they all performed within the 90-point range (advanced-level proficiency). Second, learners' performance on final authentic and performance assessments, in all classes listed, improved by an average of 15%. Their overall language proficiency development demonstrated abilities which shifted from the intermediate to advanced levels of proficiency, while the advanced learners advanced to the mainstream courses. These anecdotal findings are supported by the research finding of Rodriguez and Pelaez (2002), who demonstrated that "All courses share the common theme of students' achievement with language acquisition through the integration of technologies into the teaching and learning process."

Other important learner outcomes were that students displayed more in-depth schema development, an enhanced ease of learning the language, and a greater understanding for the culture, material, and academic expectations. Additionally, the students anecdotally reported that the electronic materials made the courses' content up to date, meaningful, and more authentic (Rong-Chang & Hart, 1996) than the use of texts alone. In their exit interviews, they frequently related a high level of contentment with the program and their learning of English.

TEACHERS' REACTIONS TO INTERACTION THROUGH TECHNOLOGY

The ESL instructors in these content-based ESL courses reported that through the use of these technological resources and the supportive, interactional, and constructivist instructional methodology: (1) learners' content depth of knowledge increased; (2) learners' language skills within multiple discourse domains improved, in terms of their contextual, lexical, and conceptual learning; and (3) in-class spoken and written interaction increased because learners had an area of expertise to share. This allowed for a more learner-centered classroom where learners benefited from a tremendous amount of interaction. Instructors also noted: (4) an increase in language competence and performance in all learners; (5) an improvement of learners' abilities to perform the novice and expert roles, while they built self-esteem and lowered their anxiety about communicating in public in their non-native language; (6) an augmentation of the learners' motivation because the technology was interesting, contemporary, inspiring, and stimulating; and finally, (7) an extension of classroom discourse, texts, and materials through the technology that made the classroom interaction more authentic, detailed, relevant, and learner driven. All these objectives were met without instructors feeling that time was being wasted or content was not being covered.

TIPS FOR TEACHERS TO OVERCOME TECHNOLOGICALLY RELATED CONSTRAINTS

The Web sites and interconnected interactional activities and tasks allowed the teachers in these ESL classes to overcome the fears and constraints pertaining to their personal knowledge and computer skills, environmental or context-related challenges, technological and financial issues, as well as time constraints. One reason for this is the Internet did not require training for teachers or sophisticated technology knowledge, but incorporated technology into instruction. Other choices instructors made to accomplish this feat were:

1. *Choosing easily accessed, understood, and stable Internet resources.* In doing so, they could rely on the site's stability and ease of use so that learners can work independently within a short period of time without re-teaching how to use it, where to find it, or what to do with it.
2. *Providing written access instructions for ESL/EFL learners.*
3. *Including the same type of technology into classroom routines so one only needed to train the learners once on how to access materials, such as URLs, via the technology.* Teachers could then employ Web sites with a plethora of re-

sources (e.g., PBS or NPR) or different URLs for different topics and objectives without retraining the learners or taking too much class time. The learners' familiarity with the routine increases quickly; therefore, within a short time, technology training required few precious class minutes.

4. *Omitting authoring of Web sites and HTML code unless they really had the time and knowledge.* Teaching authoring of Web sites and HTML code and dealing with the technology-related quirks and bugs for each individual student's project begins to devour class time. These instructors chose wisely and practically to employ technology that would not require more than a short classroom introduction. This tactic saved them time, not just on the first day using the technology, but throughout the term.
5. *Integrating technology and activities as a means to provide input, output, and negotiation of meaning/interaction opportunities, so the learners can learn the subject matter content and the language simultaneously.* This strategy saved time and money, while meeting all expectations, goals, and standards. According to the National Center for ESL Literacy Education (2002), teachers should choose "technology that supports and complements the approaches, needs and goals of instruction." If the technology is congruent with the instruction, and does not push aside the content and the language instruction, then the learners will benefit while teachers meet their goals. McDonell (2001) shared a discussion with Mark Warschauer in which he supported this recommendation "...project-based learning...allows learners to pursue their own interests and concerns while learning new language and technology skills" (p. 38).
6. *Using public computer hardware and technology resources, such as the school's or public library's computer lab.*
7. *Assigning rotating teams to collaborate on conducting research on specific Web sites during class time.*

CONCLUSION

Although teachers encounter many challenges and constraints to including technology into the ever-burgeoning standardized curriculum, there are accessible, useful, and beneficial ways to enhance learners' knowledge with a couple of mainstay Web sites and resources. Learners are then in a position to use the technology, develop their linguistic knowledge and performance, and learn the content in more depth. If teachers have a little preparation time at the beginning of a term to plan for including relatively simple computer technology, choosing simple and routine technology, and identifying easy ways of facilitating learners' access to computer hardware, the learning can remain the chief area of concentration in the classroom, while meeting the divergent pedagogical expectations.

In sum, we hope that the challenges facing educational professionals will be minimized through the ongoing discussion of appropriate incorporation of technology in classrooms, permitting them, at the same time, to realize fundamental pedagogic goals, in our case, the learning of English as a Second Language.

REFERENCES

Abdal-Haqq, I. (1995). *Infusing technology into preservice teacher education.* East Lansing, MI: National Center for Research on Teacher Learning. (ERIC Document Reproduction Service No. ED 389 699).

Abrams, L. M., & Madaus, G. F. (2003, November). The lessons of high-stakes testing. *Educational Leadership, 61*(3), 31-35.

Adkins-Bowling, T., Brown, S., & Mitchell, T. L. (2001, November 8-10). The utilization of instructional technology and cooperative learning to effectively enhance the academic success of students with English-as-a-Second-Language. *Proceedings of the 43rd Biennial Meeting of Kappa Delta Pi,* Orlando, Florida (ERIC Document Reproduction Service No. ED 472699).

Barker, B., & Howley, C. (1997). *The national information infrastructure: Keeping rural values and purposes in mind.* East Lansing, MI: National Center for Research on Teacher Learning (ERIC Document Reproduction Service No. ED 458208).

Beglar, D., Murray, N., & Rost, M. (2002). *Contemporary topics 3: Advanced listening and note-taking skills* (2nd ed.). White Plains, NY: Longman.

Bialystok, E. (1978). A theoretical model of second language learning. *Language Learning, 28,* 69-83.

Boswood, T. (1997). *New ways of using computers in language teaching.* New Ways in TESOL Series II, Innovative Classroom Techniques. Alexandria, VA: Teaching English to Speakers of Other Languages (TESOL).

Brown, H. D. (2003). *Principles of language learning and teaching* (4th ed.). Englewood Cliffs, NJ: Longman.

Brown, S., & Attardo, S. (2000). *Understanding language structure, interaction, and variation.* Ann Arbor: University of Michigan Press.

Busch, H. J. (2003). Computer based readers for intermediate foreign language students. *Educational Media International, 40*(3-4), 277.

Cable News Network. (n.d.). Retrieved May 25, 2004, from www.cnn.com

Cameron, C. (1989). *Computer assisted language learning.* Worcester, UK: Billing & Sons.

Chamot, A. U., & O'Malley, J. M. (1994). *The CALLA handbook: Implementing the cognitive academic language learning approach.* Boston: Longman.

Da, J. (2003, March 30-April 1). The use of online courseware in foreign country instruction and its implication for classroom pedagogy. *Teaching, Learning & Technology: The Challenge Continues. Proceedings of the 8th Annual Mid-South Instructional Technology Conference,* Murfeesboro, Tennessee (ERIC Document Reproduction Service No. ED 479245).

Dunkel, P. A., Pialorsi, F., & Kozyrev, J. (1996). *Advanced listening comprehension: Developing aural and note taking skills* (2nd ed.). Boston: Heinle & Heinle.

Echevarria, J., & Graves, A. (2003). *Sheltered content instruction: Teaching English-language learners with diverse abilities* (2nd ed.). Boston: Longman.

Freiermuth, M. R. (2002). Internet chat: Collaborating and learning via e-conversations. *TESOL Journal, 11*(3), 36-40.

Gass, S.M. (1997). *Input, interaction, and the second language learner.* Mahwah, NJ: Lawrence Erlbaum.

Gass, S. M., & Selinker, L. (2001). *Second language acquisition: An introductory course* (2nd ed.). Mahwah, NJ: Lawrence Erlbaum.

Harris, J. (1992-1995). Mining the Internet for educational resources [Column]. *The Computing Teacher/Learning and Leading with Technology,* 20-23.

Harris, J. (1994). *Way of the ferret: Finding educational resources on the Internet* [Column]. Eugene, OR: International Society for Technology in Education.

Hayward, N. M., & Tuzi, F. (2003). Confessions of a technophobe and a technophile: The changing perspective of technology in ESL. *TESOL Journal, 12*(1), 3-8.

Healey, D., & Klinghammer, S. J. (2002). Constructing meaning with computers. *TESOL Journal, 11*(3), 3.

Huang, S. C. (1999). *Internet assisting EFL writing learning: From learners' perspectives.* Washington, DC: Educational Research International Clearinghouse (ERIC Document Reproduction Service No. ED 429460).

Hughes, J. E. (2003). Toward a model of teachers' technology-learning. *Action in Teacher Education, 24*(4), 10-17.

Johns, K. M., & Torrez, N. (2001). *Helping ESL learners succeed.* Washington, DC: Educational Research International Clearinghouse (ERIC Document Reproduction Service No. ED 466101).

Jones, C. (1987). *Using computers in the language classroom.* Essex, UK: Longman.

Jun, W. (2002). A trip to Tahiti. *TESOL Journal, 11*(3), 45-46.

Kasper, L. F. (2002). Technology as a tool for literacy in the age of information: Implications for the ESL classroom. *Teaching English in the Two-Year College, 30*(2), 129-144.

Lantolf, J. P. (2000). Introducing sociocultural theory. In J. P. Lantolf (Ed.), *Sociocultural theory and second language acquisition.* Hong Kong: Oxford University Press.

Leu, D. J. Jr., Leu, D. D., & Len, K. R. (1997). *Teaching with the Internet: Lessons from the classroom.* Washington, DC: National Clearinghouse for ESL Literacy Education (BBB31500) (ERIC Document Reproduction Service No. ED412922).

Lightbown, P. M., & Spada, N. (2000). *How languages are learned* (2nd ed.). New York: Oxford University Press.

Little, D. (1996). Freedom to learn and compulsion to interact: Promoting learner autonomy through the use of information systems and information technologies. In R. Pemberton, E. S. L. Li, W. W. F. Orr, & H. Pierson (Eds.), *Control: Autonomy in language learning* (pp. 203-218). Hong Kong: Hong Kong University Press.

McDonell, T. B. (2001). Technology and literacy: A discussion with Mark Warschauer. *TESOL Journal, 10*(4), 38-39.

National Center for ESL Literacy Education, BBB 36588. (2002). *Report.* Washington, DC: Educational Research International Clearinghouse (ERIC Document Reproduction Service No. ED 461306).

National Public Radio. Retrieved May 25, 2004, from www.npr.org

New York Times. Retrieved May 25, 2004, from www.nytimes.com

Nunan, D. H. (1999). *Second language teaching and learning.* Boston: Heinle & Heinle.

Ohta, A. S. (2000). Rethinking interaction in SLA: Developmentally appropriate assistance in the zone of proximal development and the acquisition of L2 grammar. In J.P. Lantolf

(Ed.), *Sociocultural theory and second language acquisition*. Hong Kong: Oxford University Press.

Parsell, J., & Wilhelm, K. H. (1997). Whole class to individualized Internet use: An overview for ESL/EFL instructors. *College ESL, 7*(2), 81-100.

Public Broadcasting System. Retrieved May 25, 2004, from *www.pbs.org*

Raya, M. J. (2003). Multimedia in modern language teacher education. *Educational Media International, 40*(3-4), 305-317.

Reynard, R. (2003, March 30-April 1). Using the Internet as an instructional tool: ESL distance learning. *Teaching, Learning & Technology: The Challenge Continues. Proceedings of the 8th Annual Mid-South Instructional Technology Conference,* Murfeesboro, Tennessee (ERIC Document Reproduction Service No. ED 479246).

Rodriguez, D., & Pelaez, G. (2002, November 22-23). Reaching TESOL teachers through technology. *Proceedings of the 29th Annual Meeting of TESOL,* Puerto Rico (ERIC Document Reproduction Service No. ED 472699).

Rong-Chang, L., & Hart, R. S. (1996). What can the World Wide Web offer ESL teachers? *TESOL Journal, 6*(2), 5-10.

Snow, M. A., & Brinton, D. M. (1997). *The content-based classroom: Perspectives on integrating language and content*. Boston: Longman.

Sotillo, S. M. (2002). Constructivist and collaborative learning in a wireless environment. *TESOL Journal, 11*(3), 16-20.

Stevens, V. (1992). Humanism and CALL: A coming of age? In M. C. Pennington & V. Stevens (Eds.), *Computers in applied linguistics* (pp. 11-38). Clevedon, UK: Multilingual Matters.

Sutherland-Smith, W. (2002). Integrating online discussion in an Australian intensive English language class. *TESOL Journal, 11*(3), 31-35.

Terrill, L. (2000). *Benefits and challenges in using computers and the Internet with adult English learners.* Washington, DC: National Clearinghouse for ESL Literacy Education (BBB31500) (ERIC Document Reproduction Service No. EDD000013).

Vygotsky, L. S. (1978). *Mind in society: The development of higher psychological processes*. Cambridge, MA: Harvard University Press.

Warschauer, M. (1995). *E-mail for English teaching: Bringing the Internet and computer learning networks into the language classroom.* Alexandria, VA: Teaching English to Speakers of Other Languages (TESOL).

Wertsch, J. V. (1985). *Vygotsky and the social foundation of mind.* Cambridge, MA: Harvard University Press.

Wertsch, J. V. (1991). *Voices of the mind: A sociocultural approach to mediated action.* Cambridge, MA: Harvard University Press.

Wright, A. (1980). Developing standards and norms for computer literacy. East Lansing, MI: National Center for Research on Teacher Learning (ERIC Document Reproduction Service No. ED 208 847).

APPENDIX A: CURRENT GLOBAL EVENTS COURSE DESCRIPTION AND OBJECTIVES

Course Description

In this course, you will gather information from lectures and videos, especially "CNN Student News," which is produced weekdays and contains the latest news from around the world. You will also strengthen listening and concentration skills by listening and taking notes from these broadcasts and mini-lectures. You will engage in a variety of activities including out-of-class research, in-class listening exercises and quizzes, and small- and large-group discussions of the global news reports. Active attendance, participation, and timely completion of assignments are essential for you to satisfactorily complete this course.

Course Objectives

Through this course, you, the student, will:

- Improve your ability to gather information from lecture and videos using authentic speech.
- Develop your ability to listen to longer segments of speech with improved comprehension.
- Build your confidence in gathering information from lecture situations.
- Improve your note-taking skills.
- Increase cultural awareness of American and global cultures.
- Develop research techniques (library and Internet)
- Build your confidence in participating in and contributing to class discussions
- Practice speaking alone or with other group members in front of the class.
- Develop your ability to evaluate classmates' presentations and provide feedback for future presentations.

APPENDIX B: LISTENING AND NOTE-TAKING COURSE DESCRIPTION AND OBJECTIVES

Course Description

Emphasizes academic functions of aural/oral language. In listening: understanding academic lectures and natural speech; applying note-taking and critical-thinking strategies. In speaking: communication strategies; pronunciation; discussion; presentation and information transfer skills.

Course Objectives

Through this course, students will:

- Consistently demonstrate ability to discriminate sounds in connected speech.
- Identify main points, supporting ideas, and details in talks.
- Demonstrate ability to take concise lecture notes and restate lectures from notes.
- Keep weekly Listening Logs in which students will record a summary of listening, and will provide cultural insights regarding what they learned from TV, radio, film, conversations, meetings, etc.
- Present formally in front of a group and give a short oral report of 2-5 minutes.
- Ask and respond to questions on personal topics; express and comment on a range of emotions and subjective reactions.
- Initiate and contribute to general conversations on a limited number of topics in academic and non-academic situations.
- Respond and contribute to classroom discussions.

Homework and Readings—25%

You will be assigned homework tasks for which you will need to visit a Web site http://www.npr.org and listen to radio or video broadcasts. You will need to listen and take notes. You must listen to the broadcast only once. You must bring your notes to class the following day for discussion.

Portion of the Course Schedule

M 9-30 — Conduct activities from Unit 4 on "Actions Speak Louder than Words."
HW: Visit www.npr.org. Find the radio broadcast on *All Things Considered* entitled, "Seeing Anger," which aired on 6/18/2002.

T 10-1 — Conduct activities from Unit 4 on "Actions Speak Louder than Words."

W 10-2 — Conduct activities from Unit 4 on "Actions Speak Louder than Words."
HW: *Visit www.npr.org.* Find the radio broadcast on *Weekend Edition* entitled, "Botox," which aired on 9/19/1998.

R 10-3 — Conduct activities from Unit 4 on "Actions Speak Louder than Words."
HW: Preview Chapters 3 & 4 in Dunkel.

APPENDIX C: BUSINESS ENGLISH COURSE DESCRIPTION AND OBJECTIVES

Course Description

This course is designed for non-native English speaking students. It is intended to equip English language learners with the English writing, speaking, and listening skills necessary to communicate effectively on the job while building learners' English vocabulary for the business discourse domain. Communication and culture are tightly interwoven; thus, it will also introduce learners to the mainstream values of the American culture and business milieu.

Course Objectives

In this course, you will:

- Strengthen your general writing abilities by focusing on sentence and paragraph structure.
- Understand the functions of and create various types of business correspondence: memos, letters, reports, cover letters, and résumés.
- Master the rhetorical requirements for delivering successful business presentations.
- Employ culturally appropriate verbal and non-verbal behavior patterns for meetings and other job-related interactions.
- Acquire up-to-date career strategies for job hunting, writing effective résumés, and preparing for employment interviews.
- Become aware of mainstream American values and buzzwords associated with the corporate world.

APPENDIX D: INTEGRATED LANGUAGE SKILLS COURSE DESCRIPTION AND OBJECTIVES

Fall Semester 2004 ESL ELA Theme: Justice

Course: ESL 201—Integrated Skills for ESL Students 8 credits

Course Description

This intermediate-level, multi-skill ESL course is designed to improve listening, speaking, reading, and writing skills as well as grammar, pronunciation, and vocabulary while preparing students for academic learning.

Course Objectives/Competencies

In this course, students will:

- Recognize the structure of simple sentences using a variety of tenses.
- Recognize correct word order.
- Know conventions of capitalization and punctuation.
- Recognize correct usage of the basic auxiliary system.
- Read brief prose involving simple sentences related to everyday needs.
- Have a basic understanding of common idioms and expressions.
- Draw simple conclusions from reading passages.
- Understand brief questions and answers related to everyday needs.
- Understand short conversations supported by context.
- Begin to comprehend main ideas and details.
- Begin to understand tense shifts and more complex sentence structures.
- Be able to engage in extended discourse (i.e., give a report).
- Be able to express an opinion and support it.
- Learn basic note-taking skills.
- Be able to summarize what is heard and read.
- Employ mnemonic devices.
- Be aware of differences in English accents.
- Understand reductions.
- Use a variety of brainstorming techniques.
- Improve comprehension of numbers.

About the Authors

Leo Tan Wee Hin has a PhD in marine biology. He holds the concurrent appointments of director of the National Institute of Education, professor of biological sciences in Nanyang Technological University (Singapore), and president of the Singapore National Academy of Science. Prior to this, he was director of the Singapore Science Centre. His research interests are in the fields of marine biology, science education, museum science, telecommunications, and transportation. He has published numerous research papers in international refereed journals.

R. Subramaniam has a PhD in physical chemistry. He is an associate professor at the National Institute of Education in Nanyang Technological University (Singapore) and honorary secretary of the Singapore National Academy of Science. Prior to this, he was acting head of physical sciences at the Singapore Science Center. His research interests are in the fields of physical chemistry, science education, theoretical cosmophysics, museum science, telecommunications, and transportation. He has published several research papers in international refereed journals.

* * *

Paul Adams is a senior lecturer of education and professional studies at Newman College of Higher Education, Birmingham, UK. As a University of Warwick graduate, he became a middle school teacher and subsequently developed a passion for learning beyond "the mere acquisition of previously articulated givens." Following master's study at King's College, London, he worked as an advisory teacher for Northumberland LEA. In 2001 he moved to Newman, where he works with students so that they might construct personal understandings of learning theory, health, and citizenship education, and mediate this into the social space. He is married with three small children.

John Anderson is the education technology strategy coordinator, Education Technology Strategy Management Group (Northern Ireland), responsible for the overall strategy for information and communications technology for the Northern Ireland school service.

Charoula Angeli earned BS and MS degrees in computer science, and a PhD in instructional systems technology, from Indiana University-Bloomington, USA. She has worked as a post-doctoral fellow at the Learning Research and Development Center at the University of Pittsburgh, USA (1998-1999). She is now at the University of Cyprus as a lecturer in instructional technology in the Department of Education. Her research interests include the utilization of educational technologies in K-12, the design of computer-enhanced curricula, educational software design, teacher training, teaching methodology, and the design of learning environments for the development of critical and scientific thinking skills.

Stephen Atkinson is a researcher at the Centre for Studies in Literacy, Policy, and Learning Cultures at the University of South Australia and a sessional lecturer in the Department of International Studies and Politics at the Flinders University of South Australia. In addition to work on digital media and ICTs in education, his current research interests include the cross-cultural study of reconciliation processes, the political uses of horror, local television histories, and the function of memory and amnesia in film.

Roger Austin is the head of the School of Education at the University of Ulster in Northern Ireland. He has published widely on the role of information communication technology in inter-cultural education and is currently co-directing the Dissolving Boundaries Programme.

Ann E. Barron is a professor in the instructional technology program at the University of South Florida (USA), where she teaches graduate-level courses in multimedia, instructional design, Web design, and telecommunications. She is the author of numerous books and articles related to instructional technology. Her professional presentations focus on the design and development of technology-based instruction in academic, industrial, and military environments.

As a teacher of online teaching at James Cook University (Australia), **Colin Baskin** describes himself as the "necessary lubricant and sometimes coolant" between teaching staff and the pressures generated by institutional moves towards new learning technologies. As a senior lecturer in the School of Education and (past) head of a multi-campus teaching development unit, he has published widely in the area of e-learning and e-learning uptake, focusing more recently on textually mediated communities of practice and the roles of information and communication technologies on transforming learning. He lives in tropical North Queensland with his family, and is based at the Cairns Campus of James Cook University.

Catherine Beavis is an associate professor of education at Deakin University (Australia). Her research interests focus on literacy, popular culture and ICT, the changing nature of literacy and text as reflected in online popular culture, and the implications of young people's engagement with digital culture for curriculum and pedagogy in schools. Recent and current research projects include studies of the use of computer games as text in the secondary English classroom; young people's literacy and communication practices as they engage with online multiplayer computer games; intersections between literacy, community, and identity in online popular culture; and the gendered dimensions of computer gaming .

Jared V. Berrett has enjoyed a career in the information technology industry working directly for or consulting for three major IT corporations in the United States. Eventually he and his wife decided to follow his reoccurring impressions to become a teacher, and after teaching for three different high schools in the state of Utah, he decided to pursue a PhD in teaching and learning from the University of Illinois Urbana-Champaign. At Illinois, he coordinated an online master's degree program, was an NSF-sponsored technology trainee, and consulted with engineering faculty to improve their teaching and use of technology. He is currently a faculty member at Brigham Young University (USA) in the technology teacher education program. His research interests are in technological literacy in the K-12 setting, teaching pedagogy that promotes higher-order thinking skills, creativity in the classroom, and technology that improves student learning.

Marcie J. Bober is an associate professor of educational technology at San Diego State University (USA). There, she teaches field-based courses where students work directly with community-based clients (SeaWorld, San Diego City Schools, Kyocera Wireless, International Rescue Center) to conduct both voluntary and mandated evaluation studies. Since 1995, Dr. Bober has served as lead evaluator on several federal grants targeting our public schools—for the most part, initiatives to stimulate effective use of advanced technologies in the classroom and positively affect student learning and workforce readiness. She currently co-directs a Fund for the Improvement of Post-secondary Education (FIPSE) grant in which visualizing software has been fully integrated into the undergraduate engineering curriculum to foster engaged learning and build critical teaming skills. Dr. Bober is a frequent presenter at technology conferences focused on the design, development, and evaluation of technologies aimed at improved human performance. She consults extensively with private and public organizations interested in assessing how their interventions impact both work processes and employees (skills, knowledge, attitudes). She is also active in the school community, having taught traditionally underserved students in two local-area school districts and managed a technology-based program for students at serious academic risk.

Chris Brook is a post-doctoral research scholar in the School of Education at Edith Cowan University (Australia). He has teaching background as a classroom practitioner in the K-12 setting and has worked as a curriculum consultant in 52 schools. In addition, he has worked as an instructional designer in tertiary settings and has been actively engaged in the design and development of online learning materials. Current research includes the exploration of online learning communities and the use of ICT in school settings.

Elizabeth A. Buchanan is an assistant professor at the University of Wisconsin-Milwaukee School of Information Studies (USA), where she is also co-director of the school's Center for Information Policy Research. Her research focuses on information ethics, ethics and technology, Internet research ethics, and distance education pedagogy, specifically improving the student experience in virtual environments, including the role of the library to support distance education. She is the editor of the 2004 *Readings in Virtual Research Ethics*, among many scholarly articles and book chapters.

Karen Cadiero-Kaplan is an assistant professor at San Diego State University (USA) in the Department of Policy Studies in Language and Cross-Cultural Education. Her research interests include literacy ideologies of the arts and technology, democratic practices for teacher development, and policy that impacts programming for biliteracy and English language development. Recent publications include the book, *The Literacy Curriculum and Bilingual Education: A Critical Examination.*

Yin Cheong Cheng is professor and director of the Centre for Research and International Collaboration of the Hong Kong Institute of Education. He is also head of the Asia-Pacific Cente for Education Leadership and School Quality, and president of the Asia-Pacific Educational Research Association (APERA). Professor Cheng has published 14 academic books and nearly 180 book chapters and academic journal articles. Some of his publications have been translated into Chinese, Hebrew, Korean, Spanish, and Thai languages. He currently serves on the advisory boards of eight international journals. Professor Cheng's research has won him a number of international awards and recognition, including the Awards for Excellence from the Literati Club in the UK in 1994, 1996-1998, and 2001. In recent years, he has been invited to give over 30 keynote/plenary presentations by national and international organizations.

Niki Christodoulou received her bachelor's degree in international relations and French (1985) and her master's degree in education and TESOL (1987) from the American University in Washington, DC. From 1987-1992 she taught English as a second language to immigrants in New Jersey. From 1993-2003 she was a senior lecturer at Intercollege, Nicosia, Cyprus, and as of January 2004 she is an assistant professor teaching English at the same institution. Her research interests lie in the areas of bilingualism/bidialectalism, English language acquisition, reflective teaching, communication strategies, learning styles, learning strategies, and the teaching of Greek as a foreign language.

Christopher Essex is coordinator of instructional design and development for the Indiana University School of Education Office of Instructional Consulting (USA). He has published articles in *Educational Technology*, *TechTrends*, *The Quarterly Review of Distance Education,* and *Distance Education in China.* He has presented numerous times at educational technology conferences. He has taught at the elementary school level in Boulder, Colorado, and Bloomington, Indiana. He received his BA in English and diploma in education from the University of Colorado-Boulder, an MS in instructional systems technology from Indiana University, and in 2005 was expected to complete his PhD in instructional systems technology from Indiana University.

Moti Frank is head of the Center for the Advancement of Teaching at the Technion-Israel Institute of Technology and a senior lecturer in the Department of Education in Technology and Science. He is in charge of Technion's e-learning project. His research interests are educational technology, e-learning, computers in education, systems thinking, and engineering education.

Dale O. Gable currently manages a Title III Professional Development Grant, which enables mainstream teachers to earn their add-on license in TESOL. She teaches ESL and Spanish at the University of Wisconsin-Eau Claire (USA), where she enjoys being a mentor to international students and faculty, and moonlighting as an interpreter in medical and legal settings. She received her master's degree in Spanish from Middlebury College and TEFL certification from Hamline University.

Virginia E. Garland is coordinator of the graduate programs in educational administration and supervision at the University of New Hampshire (USA). She has also been a visiting professor in the Cross-Cultural Studies Department of Kobe University, Japan, and the Shanghai, Beijing, and Tianjin Institutes of Education, China. Her teaching and research interests include technology integration and international education. Dr. Garland has authored or co-authored more than 20 published articles and book chapters. She recently created an interactive Web site, "Space Exploration: Case Based Pedagogy" (http://www.ali.apple.com/ali_sites/nhli/exhibits/1000577/), selected by the Apple Learning Interchange as an exhibit of best teaching practice.

Susan E. Gibson is an associate professor in the Department of Elementary Education at the University of Alberta, Canada. Her subject area specialization is social studies education. Prior to her work in the university setting, she was an elementary and junior high school social studies teacher. Dr. Gibson's research interests include learning to teach social studies, problem-based learning and learning to teach, preservice teachers' historical knowledge and its influence on their thinking about pedagogy, the integration of computer technologies into teacher education, and using computers in ways that support constructivist learning principles.

J. Christine Harmes has a PhD with dual majors in instructional technology and educational measurement and research. She does extensive consulting for the Florida Department of Education, Office of Educational Technology, including supervising the development and deployment of a statewide technology survey. In the past year, she has led an instrument development project to create a Web-based technology literacy assessment to be administered to all teachers in Florida. Her research reports have twice been nominated for distinguished paper awards by the Florida Educational Research Association, and her doctoral dissertation was a finalist in the AERA-Division D Mary Outstanding Dissertation competition.

Kendall Hartley is an assistant professor of educational computing and technology at the College of Education at the University of Nevada, Las Vegas (USA). Dr. Hartley conducts research in the area of effective instructional uses of technology, and recently published articles related to hypermedia instruction in *Educational Researcher,* the *Journal of Educational Multimedia and Hypermedia,* and the *Journal of Educational Computing Research.* Before coming to the University of Nevada, Dr. Hartley taught high school science in Bellevue, Nebraska.

Stephenie M. Hewett is an assistant professor in the School of Education at The Citadel, The Military College of South Carolina located in Charleston, South Carolina (USA). She is a learner-centered educator dedicated to the mastery approach to teaching. She is a national and international speaker, and has published numerous articles on teacher preparation and technology.

Lyn C. Howell has taught in middle and high schools in states from Alaska to Florida. After earning her PhD at the University of New Mexico, she moved to Tennessee where she is an assistant professor of education at Milligan College (USA). She teaches educational technology among other classes and works with teachers and pre-service teachers helping them to integrate technology into their curriculum.

David A. Huffaker is a PhD student in media, technology, and society at Northwestern University, USA. He holds a master's degree in communication, culture, and technology from Georgetown University. His research includes children and technology, online communities, computer-mediated communication, and educational technology. For more information, visit http://www.soc.northwestern.edu/gradstudents/huffaker.

Wei-Loong David Hung is an associate professor at the National Institute of Education (NIE) and the current head of the Learning Sciences and Technologies Academic Group at NIE (Singapore). He is associate editor of the *International Journal of Learning Technologies,* contributing editor of *Educational Technology,* and has recently been invited as an advisory board member of a newly formed *Knowledge Management* journal. Dr. Hung is one key initiator and member of the Learning Sciences Lab at NIE.

Tamara L. Jetton currently teaches in the Secondary Program at James Madison University (USA). She has held positions at the University of Utah and Texas A&M University. Her research interests include understanding how adolescents learn with text in a variety of learning environments and how they engage in discussions both in the classroom and in online environments. She has published in several journals and edited volumes that include *Reading Research Quarterly, Journal of Educational Psychology, Review of Educational Research, Educational Psychology, Educational Psychology Review, Handbook of Reading Research,* and *Handbook of Discourse Processes.*

Cushla Kapitzke is a senior lecturer in the School of Education at the University of Queensland (Australia). Her publications include *Literacy and Religion* (John Benjamins) and an edited volume, *Difference and Dispersion: Educational Research in a Postmodern Context* (PostPressed). She has published in *Educational Theory, Teachers College Record, Journal of Adolescent & Adult Literacy,* and *Educational Technology & Society.* Current research interests focus on the social and educational implications for Australia of intellectual property rights law as framed by the bi-lateral Australia/U.S. Free Trade Agreement.

Katherine J. Kemker is completing her PhD in curriculum and instruction with an emphasis on instructional technology. Currently, she provides technical assistance with districts throughout the state of Florida on technology planning, laptop computer initiatives, and training on the integration of technology in the curriculum. In addition, she has spoken at national, state, and local conferences, such as the Annual Conference for the Association for Supervision and Curriculum Development, Florida Educational Technology Conference, and the Florida Title I Conference on "Closing the Achievement Gap with Technology."

Thiam-Seng Koh is currently director of the Educational Technology Division at the Ministry of Education (MOE). He oversees all IT in education projects in research and development, ensuring that all schools in Singapore embrace IT for learning and instruction. Dr. Koh's work is also central to the Learning Sciences Lab initiative, where he is a key collaborator from MOE to ensure that the research efforts are translated to the schools.

Elena Landone graduated with a degree in Spanish language and literature from the University of Milan, Italy, and attended a post-graduate master's of online learning and communication course at the University of Florence, Italy. After a research grant for specialization in distance learning of Spanish as a second language at the Utrecht University, The Netherlands, she became a researcher at the University of Milan in 2000. Since then, she is a professor of didactics of Spanish, educational technology, Spanish culture, and Spanish language at the University of Milan. Her research activities and publications are in the fields of electronic cooperative groups and cooperative learning, collaborative electronic writing, input enhancement in language teaching, alternative evaluation, and portfolios. She is Coordinator of *the Project of Study of the Electronic European Language Portfolio (ELP),* co-funded by the EU Socrates/Minerva Action group (2003-2005).

Wei-Ying Lim is a research associate at the Learning Sciences Lab (LSL) in the National Institute of Education, Nanyang Technological University, Singapore. She has keen research interest in sustainability of educational reform, teacher epistemology, learning communities, and inquiry learning.

Tomas A. Lipinski completed his Juris Doctor (JD) from Marquette University Law School, before receiving his Master of Laws degree (LLM) from The John Marshall Law School and his PhD from the University of Illinois at Urbana-Champaign. Professor Lipinski has worked in a variety of legal settings including the private, public, and non-profit sectors. In summers he is a visiting associate professor at the Graduate School of Library and Information Science, University of Illinois at Urbana-Champaign. Professor Lipinski currently teaches, researches, and speaks frequently on various topics within the areas of information law and policy, especially copyright issues in schools and libraries, and serves as co-director of the Center for Information Policy Research at the University of Wisconsin-Milwaukee and an associate professor at its School of Information Studies. He has authored many articles and publications of interest to educators and librarians on the design and use of Web-based materials in education, as well as copyright and other legal issues pertinent to educators.

Chee-Kit Looi is head of the Learning Sciences Lab (LSL) in the National Institute of Education, Nanyang Technological University, Singapore. He is an editor of the *Journal of Computer-Assisted Learning* as well as the *International Journal of CSCL.* His research interests include scalability and sustainability issues of educational reform, computer-supported collaborative environments, inquiry learning, and knowledge organization. He has published widely in conferences, journals, and chapters of books.

Kristina Love lectures in the Department of Language, Literacy, and Arts Education at the University of Melbourne, Australia, having worked as a secondary English teacher prior to that. Her research interests include analysis of the spoken, written, and electronic discourses of secondary English and applications of computer technologies in literacy contexts. Her CD-ROM, "Building Understandings in Literacy and Teaching (BUILT)," uses QuickTime video of best teaching practice across the K-12 context to illustrate key principles of language and learning in pre-service and teacher professional development.

Kate Mastruserio Reynolds has served as director of TESOL and campus coordinator of ESL programming at the University of Wisconsin-Eau Claire (USA) since 2001. She holds degrees in French language and literature, and teaching English to speakers of other languages (TESOL) and literacy from the University of Cincinnati. She has taught language courses at a variety of levels in ESL and teacher preparation courses in Southern Connecticut State University, Harvard University's Intensive English Language Program, and the University of Cincinnati. She has presented at TESOL, AAAL, ConnTESOL, OhioTESOL, Wisconsin TESOL, and written articles for the *TESOL Quarterly* and the *Journal of Educational Technology*.

Steven C. Mills is the director and chief executive officer of the Ardmore Higher Education Center. Formerly, he was a research professor at the University of Kansas, conducting research focusing on integrating technology and online learning in K-12 and college classrooms. Dr. Mills has made numerous national presentations on the subjects of technology integration and Web site design, and has authored several grants, journal articles, and two college textbooks on using technology to support teaching and learning in the classroom. He holds a PhD in instructional psychology and technology from the University of Oklahoma.

Aidan Mulkeen is a senior lecturer in education at the National University of Ireland, Maynooth. He has a specialist interest in the use of information and communications technology in education. He has done extensive research on the use and impact of ICT in schools, in Ireland and in OECD countries. Also, he is co-founder of the Dissolving Boundaries project, which uses ICT to bring learners in different communities together to develop mutual understanding and engage in educational projects.

Wan Ng is the science education lecturer at the School of Educational Studies, La Trobe University, Australia. Her experiences include working as a research fellow in Biochemistry, and serving in various capacities for the Victorian Department of Education & Training and the Victorian Curriculum & Assessment Authority. Her research interests lie in the use of ICT in education, creative and motivating science curriculum (including the Sun & Science project¾http://www.latrobe.edu.au/solar), gifted education, and language/cultural influences in the learning of the sciences.

Helen Nixon is an associate professor in the Center for Studies in Literacy, Policy, and Learning Cultures in the School of Education at the University of South Australia. Her research interests include literacy and the new media, and the pedagogies of global media culture. She is particularly interested in how children's out-of-school media popular culture interests might be used within a critical literacy/English curriculum. She is editor with Brenton Doecke and David Homer of the book *English Teachers at Work* (AATE/Wakefield Press, 2003) and has published widely for English/literacy teachers and researchers.

Christopher O'Mahony is currently head of the Center-Information Technology at Saint Ignatius' College Riverview (Australia), an independent day and boarding school for boys in Sydney, Australia. Dr. O'Mahony has been a practitioner and researcher in the field of school information systems since 1992. His doctoral thesis, through Macquarie University, concerned the evolution and evaluation of information systems in Australian schools throughout the 1990s. He is an active member of the Australian Computer Society and also of the IFIP Working Group 3.7–Information Technology for Educational Management (ITEM). Dr. O'Mahony has presented the results of his work at both national and international conferences, as well as in a number of refereed academic journals.

Ron Oliver is professor of interactive multimedia in the School of Communications and multimedia at Edith Cowan University (Australia). He has a research background in multimedia and e-learning, and currently leads a major research team at ECU in this field. He has experience in the design, development, and evaluation of multimedia and computer-based learning materials. Current research projects involve explorations and investigations of effective online teaching and learning in higher education, and exploring scalability and reusability of e-learning materials.

Ingrid Schaller is a lecturer of Spanish and English as a second language at the University of Wisconsin-Eau Claire. Her training includes a master's degree in Spanish language and literature, a K-12 teaching license, and a number of pedagogy and methodology classes for both foreign languages and English as a second language. As a lecturer, she finds that the inherently flexible content of the global current events class described in this text is enhanced by the natural diversity of student perspectives. Outside the classroom, Ms. Schaller travels, teaches children Spanish, and is raising her own bilingual child.

Cathy Soenksen has taught middle school, high school, and college English for the past 20 years in Virginia and West Virginia, as well as teaching graduate courses for West Virginia University. She is currently employed in Harrisonburg City Public Schools (VA), where she teaches English in Project Achieve, a program designed for high school freshmen who are reading three to five years below grade level. Her most recent challenges as an educator have been in learning to address the specific reading needs of English language learners.

Agni Stylianou-Georgiou earned her BA in elementary education from the University of Cyprus in 1998 and her PhD in educational psychology from the University of Connecticut in 2003. She has worked as a teaching and research assistant at the University of Connecticut (1999-2003) and at Cyprus College, Cyprus. Currently she is an assistant professor of educational psychology at Intercollege, Nicosia. Her research interests are in the areas of hypertext, metacognition, reading comprehension, and science education. Her current research is focused on designing navigation support for hypertext environments in science, and studying issues (i.e., role of metacognition) in navigation and reading comprehension of nonlinear text. She has published in scholarly journals and conference proceedings in her areas of study.

Nicos Valanides is an associate professor of science education at the University of Cyprus. He earned a BA in physics and a BA in law from the Aristotelian University of Thessaloniki, along with a Teaching Diploma and MA in education, teaching sciences from the American University of Beirut. He also earned an MSc in instructional supervision and a PhD in curriculum and instruction and educational research from the University of Albany, State University of New York. His research interests include teacher training, methodology of teaching and curricula for science education, development of logical and scientific thinking, science-and-technology literacy, the utilization of ICT in science education, and the design of educational interventions and learning environments.

Mark van 't Hooft conducts research for the Research Center for Educational Technology at Kent State University (USA) and provides technical support in RCET's SBC Ameritech Classroom. His current research focus is on the use of handheld computers in K-12 education. Prior to his work at RCET, he taught middle school and high school social studies and language arts. He holds a BA in American studies from the Catholic University of Nijmegen, The Netherlands, and an MA in history from Southwest Texas State. He is currently finishing his doctoral degree with a dual major in curriculum and instruction, and evaluation and measurement at Kent State University.

Charalambos Vrasidas is executive director of CARDET-Center for the Advancement of Research and Development in Educational Technology (Cyprus) (http://www.cardet.org) and an associate professor of learning technologies at Intercollege. His areas of interest include evaluation of educational technologies, teacher professional development, e-learning, technology integration, and media literacy. He is the author of three books and of more than 100 research papers published in scholarly journals, books, and presented at international conferences around the world. He is also a member of the executive committee of the International Council of Educational Media (http://www.icem-cime.com), a UNESCO-affiliated organization.

Leonard J. Waks is professor of educational leadership and policy studies at Temple University (USA). He earned a PhD in philosophy at the University of Wisconsin in 1968, and an EdD in social and organizational psychology from Temple in 1984. He taught philosophy at Purdue and Stanford before joining the education faculty at Temple in 1971. He is the author of *Technology's School* (JAI, 1995) and more than 75 peer-reviewed scholarly articles and book chapters on educational theory, technology education, philosophy of technology, and American pragmatism. He serves on the editorial boards of the *Journal of Curriculum Studies,* the *International Journal of Technology and Design Education, and Education and Culture: The Journal of the John Dewey Society.*

Patrick Wong is the head teacher of computer studies, The Mission Covenant Church Holm Glad College, Hong Kong, as well as the teacher in charge of the IT in Education Committee in his school since 1999. He is a part-time lecturer at the CITE, teaching refresher training courses for school teachers. He has been the system administrator of a QEF project, Kwun Tong Inter-School Education Net (KTIEN). His main duty was to maintain an electronic platform for sharing educational resources among five different schools in Kwun Tong.

Allan Yuen is an associate professor and head of the Division of Information and Technology Studies, Faculty of Education, The University of Hong Kong. He is also deputy director of the Center for Information Technology in

Education (CITE) since its establishment in 1998, and vice president of the Hong Kong Association for Educational Communications and Technology (HKAECT). Dr. Yuen has engaged in a number of research and development projects on information technology in education. His research interests include computer-supported collaborative learning, information technology leadership and management in education, computer studies education, and teacher education.

Michalinos Zembylas is an associate professor of Education at Intercollege, Cyprus, and an adjunct professor of teacher education at Michigan State University. He is also a member of the board of CARDET, a non-profit research and development center based in Cyprus and Singapore. His research interests are in the areas of science and technology education, curriculum theory, and philosophy of education. He has conducted extensive research in the area of emotions in teaching and learning, and participated in several research programs funded by the National Science Foundation, the European Union, and private corporations in the United States. His recent articles have appeared in *Educational Theory, Journal of Research in Science Teaching, Science Education and Teaching,* and *Teacher Education.*

Index of Key Terms

F

G

H

I

J

K

L

M

N

O

P

Q

R

S

T

Z